T0292379

Mathematical and Computational Approaches in Advancing Modern Science and Engineering

Jacques Bélair • Ian A. Frigaard • Herb Kunze
Roman Makarov • Roderick Melnik
Raymond J. Spiteri
Editors

Mathematical and Computational Approaches in Advancing Modern Science and Engineering

Springer

Editors
Jacques Bélair
Department of Mathematics and Statistics
University of Montreal
Montreal, QC
Canada

Ian A. Frigaard
Department of Mathematics
University of British Columbia
Vancouver, BC
Canada

Herb Kunze
Department of Mathematics and Statistics
University of Guelph
Guelph, ON
Canada

Roman Makarov
Department of Mathematics
Wilfrid Laurier University
Waterloo, ON
Canada

Roderick Melnik
MS2Discovery Institute
Wilfrid Laurier University
Waterloo, ON
Canada

Raymond J. Spiteri
Department of Computer Science
University of Saskatchewan
Saskatoon, SK
Canada

ISBN 978-3-319-30377-2 ISBN 978-3-319-30379-6 (eBook)
DOI 10.1007/978-3-319-30379-6

Library of Congress Control Number: 2016943639

Mathematics Subject Classification (2010): 00A69, 00A71, 00A79, 92-XX, 35Qxx, 81T80, 97M10, 47N60, 49-xx, 91Axx, 62Pxx, 97Pxx, 70-xx

© Springer International Publishing Switzerland 2016
This work is subject to copyright. All rights are reserved by the Publisher, whether the whole or part of the material is concerned, specifically the rights of translation, reprinting, reuse of illustrations, recitation, broadcasting, reproduction on microfilms or in any other physical way, and transmission or information storage and retrieval, electronic adaptation, computer software, or by similar or dissimilar methodology now known or hereafter developed.
The use of general descriptive names, registered names, trademarks, service marks, etc. in this publication does not imply, even in the absence of a specific statement, that such names are exempt from the relevant protective laws and regulations and therefore free for general use.
The publisher, the authors and the editors are safe to assume that the advice and information in this book are believed to be true and accurate at the date of publication. Neither the publisher nor the authors or the editors give a warranty, express or implied, with respect to the material contained herein or for any errors or omissions that may have been made.

Printed on acid-free paper

This Springer imprint is published by Springer Nature
The registered company is Springer International Publishing AG Switzerland

Preface

This book consists of five parts covering a wide range of topics in applied mathematics, modeling, and computational science (AMMCS). It resulted from two highly successful meetings held jointly in Waterloo (Canada) on the main campus of Wilfrid Laurier University. It is the oldest university in the Cambridge-Kitchener-Waterloo-Guelph area, a beautiful part of Canada, just west of the city of Toronto. The main campus of the university is located in a comfortable driving distance from some of North America's most spectacular tourist destinations, including the Niagara Escarpment, a UNESCO World Biosphere Reserve. Over the years, this university has become a traditional venue for the International Conference on Applied Mathematics, Modeling and Computational Science, and in 2015 it was held jointly with the annual meeting of the Canadian Applied and Industrial Mathematics (CAIMS) from June 7–12, 2015. The AMMCS interdisciplinary conference series runs biannually. Focusing on recent advances in applied mathematics, modeling, and computational science, the 2015 AMMCS-CAIMS Congress drew some of the top scientists, mathematicians, engineers, and industrialists from all over the world and was a true celebration of interdisciplinary research and collaboration involving mathematical, statistical, and computational sciences within a larger international community.

The book clearly demonstrates the importance of interdisciplinary interactions between mathematicians, scientists, engineers, and representatives from other disciplines. It is a valuable source of the methods, ideas, and tools of mathematical modeling, computational science, and applied mathematics developed for a variety of disciplines, including natural and social sciences, medicine, engineering, and technology. Original results are presented here on both fundamental and applied levels, with an ample number of examples emphasizing the interdisciplinary nature and universality of mathematical modeling.

The book contains 70 articles, arranged according to the following topics represented by five parts:

- Theory and Applications of Mathematical Models in Physical and Chemical Sciences

Fig. 1 Participants of the 2015 International AMMCS-CAIMS Congress, Canada (Photo taken by Tomasz Adamski on the Waterloo Campus at Wilfrid Laurier University)

- Mathematical and Computational Methods in Life Sciences and Medicine
- Computational Engineering and Mathematical Foundation, Numerical Methods, and Algorithms
- Mathematics and Computation in Finance, Economics, and Social Sciences
- New Challenges in Mathematical Modeling for Scientific and Engineering Applications

These chapters are based on selected refereed contributions made by the participants of both meetings. The AMMCS-CAIMS Congress featured over 30 special and contributed sessions with mini-symposia ranging from mathematical models in nanoscience and nanotechnology to statistical equilibrium in economics and to mathematical neuroscience, the embedded Conference of the Computational Fluid Dynamics Society of Canada, and the 2nd Canadian Symposium on Scientific Computing and Numerical Analysis, as well as larger sessions around such scientific themes as applied analysis and dynamical systems, industrial mathematics, mathematical biology, financial mathematics, and much more. Over 600 participants from all continents attended the Congress and shared the latest achievements, ideas, insights, and theories about modern problems in science, engineering, and society that can be approached with new advances in mathematical modeling and mathematical, computational, and statistical methods.

This book presents a selected sample of the above topics and can serve as a reference to some of the state-of-the-art original works on a range of such topics. It

Fig. 2 Members of the local organizing committee and student volunteers (Photo taken by Dr. Shyam Badu on the Waterloo Campus at Wilfrid Laurier University)

has a strong multidisciplinary focus, supported by fundamental theories, rigorous procedures, and examples from applications. Furthermore, the book provides a multitude of examples accessible to graduate students and can serve as a source for graduate student projects.

Taking this opportunity, we would like to thank our colleagues on the AMMCS-CAIMS Congress organizing team, as well as our sponsors and partners, in particular the Fields Institute and PIMS, and the Centre de Recherches Mathématiques, as well as Wilfrid Laurier University, NSERC, and the Government of Ontario. Among others, traditional supporters of the AMMCS Interdisciplinary Conference series were Maplesoft and SHARCNET, as well as Springer, De Gruyter, and CRC Press. The Congress was held under the auspices of the MS2Discovery Interdisciplinary Research Institute based at Wilfrid Laurier University and in cooperation with the Society of Industrial and Applied Mathematics and the American Institute of Mathematical Sciences.

The Congress scientific committee included 15 internationally known researchers. We would like to thank them, as well as the Congress referees whose help in the refereeing process was invaluable. Among them we had some of the leading researchers from all parts of the world, and their assistance was decisive in completing this project. Our technical support committee and students' team were exemplary, and we are truly grateful for their efforts. Last but not least, we are also grateful to the editorial team at Springer, in particular Martin Peters and Ruth Allewelt, whose continuous support during the entire process was at the highest professional level.

We believe that the book will be a valuable addition to the libraries, as well as to private collections of university researchers and industrialists, scientists and engineers, graduate students, and all of those who are interested in the recent progress in mathematical modeling and mathematical, computational, and statistical methods applied in interdisciplinary settings.

Montreal, Canada	Jacques Bélair
Vancouver, Canada	Ian Frigaard
Guelph, Canada	Herb Kunze
Waterloo, Canada	Roman Makarov
Waterloo, Canada	Roderick Melnik
Saskatoon, Canada	Raymond Spiteri

Contents

Part III Computational Engineering and Mathematical Foundation, Numerical Methods and Algorithms

Part I
Theory and Applications of Mathematical Models in Physical and Chemical Sciences

Compressibility Coefficients in Nonlinear Transport Models in Unconventional Gas Reservoirs

Iftikhar Ali, Bilal Chanane, and Nadeem A. Malik

Abstract Transport models for gas flow in unconventional hydrocarbon reservoirs possess several model parameters such as the density (ρ), the permeability (K), the Knudsen number (K_n), that are strongly dependent upon the pressure p. Each physical parameter, say γ, in the system has an associated compressibility factor $\zeta_\gamma = \zeta_\gamma(p)$ (which is the relative rate of change of the parameter with respect to changes in the pressure, Ali I et al. (2014, Time-fractional nonlinear gas transport equation in tight porous media: an application in unconventional gas reservoirs. In: 2014 international conference on fractional differentiation and its applications (ICFDA), Catania, pp 1–6, IEEE)). Previous models have often assumed that $\zeta_\gamma = Const$, such as Cui (Geofluids 9(3):208–223, 2009), and Civan (Transp Porous Media 86(3):925–944, 2011). Here, we investigate the effect of selected compressibility factors (real gas deviation factor (ζ_Z), gas density (ζ_ρ), gas viscosity (ζ_μ), permeability (ζ_K), and the porosity (ζ_ϕ) of the source rock) as functions of the pressure upon rock properties such as K and ϕ. We also carry out a sensitivity analysis to estimate the importance of each model parameter. The results are compared to available data.

1 Introduction

Unconventional gas reservoirs include tight gas, coalbed methane, and shale gas. Shale gas is distributed over large areas and is found in discrete largely unconnected gas pockets. Different methods are applied to induce fractures inside the rocks to release the gas, such as hydraulic fracturing, but this is very expensive. Hence, an initial guess is required before drilling. Reservoir simulations can be crucial in

I. Ali (✉) • B. Chanane • N.A. Malik
Department of Mathematics & Statistics, King Fahd University of Petroleum and Minerals, P. O. Box 5046, Dhahran 31261, Saudi Arabia
e-mail: namalik@kfupm.edu.sa; nadeem_malik@cantab.net; chanane@kfupm.edu.sa; iali@kfupm.edu.sa

© Springer International Publishing Switzerland 2016

J. Bélair et al. (eds.), *Mathematical and Computational Approaches in Advancing Modern Science and Engineering*, DOI 10.1007/978-3-319-30379-6_1

assisting this process for economical recovery. This requires accurate determination of fluid and rock properties, and a realistic transport model, [2, 5, 11, 15].

Unconventional gas reservoirs are characterized by extremely low permeability, in the nano- to micro-Darcy range, and low porosity, in the 4 %–15 % range. The gas extraction process is very complex and involves new technologies, and takes a lot of time, money and human resources, [18]. The science and technology of tight gas transport and extraction is still in its infancy, and field data urgently required especially from shale gas reservoirs in order to test the newly emerging theories.

Reservoir simulations typically solve model transport equations in the form of advection-diffusion partial differential equations (PDE). Some of the latest models are highly non-linear, where the apparent diffusivity $D(p)$ and the apparent velocity $U(p, p_x)$ are strongly non-linear functions of the pressure and its derivative, [7]. D and U involve compressibility factors ζ_γ of various physical parameters,

$$\zeta_\gamma = \frac{\partial \ln \gamma}{\partial p} = \frac{1}{\gamma}\frac{\partial \gamma}{\partial p}. \tag{1}$$

and these must be known as functions of p and p_x. However, most applications to date have been simplified by assuming constant compressibility factors. The impact of this important assumption has not been assessed to date.

The aim here is to assess the importance of using fully pressure dependent model parameters. This is done through numerical simulations of the transport equation and matching the results against the data from Pong et al. [17]. A sensitivity analysis is also carried out to assess the importance of each physical parameter in the system.

2 Physical Properties of Shale Gas Reservoirs

Various flow regimes occur in the gas transport process through tight shale rock formations [10]. They are classified by a Knudsen number, see Table 1 and [17, 19], which is the ratio of mean free path of gas molecules (λ) to the radius (R) of the flow channels, $K_n = \lambda/R$. λ is given by [13], $\lambda = \frac{\mu}{\rho}\sqrt{\frac{\pi}{2R_gT}}$, where ρ is gas density, T is temperature, R_g is universal gas constant, and μ is gas viscosity. R is given by, [4, 6], $R = 2\sqrt{2\tau}\sqrt{\frac{K}{\phi}}$, where τ is the tortuosity and ϕ is the porosity of porous media and K is intrinsic permeability. Several recent works have focused transport on the so-called four flow regimes, Table 1.

Table 1 Classification of flow regimes based on Knudsen number, [19]

Knudsen number	Flow regimes
$K_n < 0.01$	Continuous flow
$0.01 < K_n < 0.1$	Surface diffusion or slip flow
$0.1 < K_n < 10$	Transition flow
$K_n > 10$	Knudsen diffusion or free molecular flow

The correlation between porosity and intrinsic permeability is given by the Kozeny-Carman equation [8]

$$\sqrt{\frac{K}{\phi}} = \Gamma_{KC}\left(\frac{\phi}{\alpha_{KC} - \phi}\right)^{\beta_{KC}}, \tag{2}$$

where $\phi < \alpha_{KC} \leq 1, 0 \leq \beta_{KC} < \infty$ and $\Gamma_{KC} \geq 0$. α_{KC}, β_{KC}, and Γ_{KC} are empirical constants which must be determined, or estimated, before hand.

For the simulation purposes, we use the following porosity-pressure correlation,

$$\phi = a_\phi \exp(-b_\phi p^{c_\phi}), \tag{3}$$

where a_ϕ, b_ϕ and c_ϕ are model constants. Tortuosity is related to porosity by,

$$\tau = 1 + a_\tau(1 - \phi), \tag{4}$$

where a_τ is also a model constant.

There is a difference between the intrinsic permeability, K, and the apparent permeability, K_a. K is the measured permeability from rock samples, but due to various physical effects such as slip flow, the quantity appearing in transport equations is K_a. Beskok [3] has derived an formula that relates the two quantities,

$$K_a = Kf(K_n) \tag{5}$$

where $f(K_n)$ is the flow condition function given by

$$f(K_n) = (1 + \sigma K_n)\,(1 + (4 - b_{SF})K_n)\,(1 - b_{SF}K_n)^{-1}, \tag{6}$$

where σ is called the Rarefaction Coefficient Correlation [6] given by

$$\sigma = \sigma_o\left(1 + A_\sigma K_n^{-B_\sigma}\right)^{-1}, \tag{7}$$

where A_σ and B_σ are empirical constants and b_{SF} in Eq. 6 is the slip factor.

Some of the gas adheres (clings) to pore surfaces due to the diffusion of gas molecules. Cui [9] and Civan [7] developed a formula for estimating the amount of adsorbed gas based on Langmuir isotherms and is given by

$$q = \frac{\rho_s M_g}{V_{std}} q_a = \frac{\rho_s M_g}{V_{std}} \frac{q_L p}{p_L + p}, \tag{8}$$

where ρ_s (kg/m^3) denotes the material density of the porous sample, q (kg/m^3) is the mass of gas adsorbed per solid volume, q_a (std m^3/kg) is the standard volume of gas adsorbed per solid mass, q_L (std m^3/kg) is the Langmuir gas volume, V_{std} (std m^3/kmol) is the molar volume of gas at standard temperature (273.15 K) and pressure (101,325 Pa), p (Pa) is the gas pressure, p_L (Pa) is the Langmuir gas pressure, and M_g (kg/kmol) is the molecular weight of gas.

Gas density ρ (kg/m^3) is given by the real-gas equation of state,

$$\rho = \frac{M_g p}{Z R_g T} \tag{9}$$

where Z (dimensionless) is the real gas deviation factor [12] and it can be found by using the correlation developed by Mahmoud [14] and it is given by

$$Z = a p_r^2 + b p_r + c \tag{10}$$

$$a = 0.702 \exp(-2.5 t_r) \tag{11}$$

$$b = -5.524 \exp(-2.5 t_r) \tag{12}$$

$$c = 0.044 T_r^2 - 0.164 t_r + 1.15 \tag{13}$$

where p_c is the critical pressure and t_c is the critical temperature, and $p_r = p/p_c$ and $t_r = t/t_c$ are the reduced pressure and temperature respectively.

Mahmoud [14] also gave correlations for determining the gas viscosity,

$$\mu = \mu_{S_c} \exp(A \rho^B) \tag{14}$$

$$A = 3.47 + 1588 T^{-1} + 0.0009 M_g$$

$$B = 1.66378 - 0.04679 A$$

$$\mu_{S_c} = \frac{1}{(10.5)^4} \left[\frac{M^3 p_c^4}{T_c} \right]^{1/6} \times$$
$$\left[0.807 T_r^{0.618} - 0.357 \exp(0.449 T_r) + 0.34 \exp(-4.058 T_r) + 0.018 \right]$$

3 Mathematical Formulation

The ultra low permeability and the occurrence of various flow regimes are key features of unconventional gas reservoirs (UGR). The PDE's that are used to describe transport process in conventional gas reservoirs (CGR) are based on Darcy's law $u = (-K/\mu)dp/dx$ and continuity equation $-(\rho u)_x = 0$, where K, μ, and ρ are constants, but such models do not produce satisfactory results in UGRs. Civan [7] has proposed a transport model for gas flow through tight porous media which incorporates all flow regimes that occur in the reservoirs. Civan's model is a non-linear advection-diffusion PDE for the pressure field $p(x, t)$, which is given by,

$$\frac{\partial p}{\partial t} + U(p, p_x)\frac{\partial p}{\partial x} = D(p)\frac{\partial^2 p}{\partial x^2}. \tag{15}$$

The apparent diffusivity D (m^2/s) is given by,

$$D = \frac{\rho K_a}{\mu} \{\rho\phi\zeta_1(p) + (1-\phi)q\zeta_2(p)\}^{-1}, \tag{16}$$

and the apparent convective flux (velocity) U (m/s) is given by,

$$U = -\zeta_3(p)D\frac{\partial p}{\partial x}. \tag{17}$$

where the ζ_1, ζ_2 and ζ_3 appearing in D and U are given by

$$\zeta_1(p) = \zeta_\rho(p) + \zeta_\phi(p), \tag{18}$$

$$\zeta_2(p) = \zeta_q(p) - \left(\frac{\phi}{1-\phi}\right)\zeta_\phi(p), \tag{19}$$

$$\zeta_3(p) = [\zeta_\rho(p) + \zeta_{K_a}(p) - \zeta_\mu(p)]. \tag{20}$$

where $\zeta_{K_a} = \zeta_K + \zeta_f$ which is obtained from Eq. (5).

A numerical solver for the system equations (15), (16), (17), (18), (19), and (20) has been developed. We use a finite volume implicit method with constant grid size and constant time step. The system is linearised and iterated to convergence before advancing to the next time step. The implicit nature of the solver gives stability to the solver which is essential for such a highly non-linear system. The solver can also be applied to the steady state system, see below.

4 Model Validation Under Steady State Conditions

The steady state solution for the pressure field is obtained by solving, (see [1, 7]),

$$L_a\left(\frac{\partial p}{\partial x}\right) = \frac{\partial^2 p}{\partial x^2}, \qquad 0 \le x \le L, \tag{21}$$

where

$$L_a = -\left[\zeta_\rho(p) + \zeta_K(p) + \zeta_f(p) - \zeta_\mu(p)\right]\frac{\partial p}{\partial x}, \tag{22}$$

with boundary conditions, $p(0) = p_L$ and $p(L) = p_R$; p_L and p_R assumed known.

Sixteen different models were considered, Table 2. An entry of '0' means that the compressibility factor is zero, $\zeta_\gamma = 0$; an entry of 'p' means that $\zeta_\gamma \neq 0$ and the associated physical parameter is a function of pressure, $\gamma = \gamma(p)$. The final column shows the relative error between the simulated values and the experimental values of Pong et al. [16], given by,

$$\text{Relative Error} = \sum_{i=1}^{N}\left[\frac{p_i^{cal} - p_i^{meas}}{p_i^{cal}}\right]^2. \tag{23}$$

where the summation is over the $N = 30$ data-points in [16]. Case 1 in Table 2 corresponds to the Darcy law where all the physical parameters are constant and

Table 2 List of models considered. In columns 2–5, an entry of 0 means that the compressibility factor is zero; an entry of p means that it is nonzero and the associated physical parameter is function of pressure p. The final column shows the relative error from simulations using Eq. (23)

Cases	ζ_ρ	ζ_K	ζ_f	ζ_μ	Error
1	0	0	0	0	2.69e−02
2	p	0	0	0	2.68e−02
3	0	p	0	0	4.05e−03
4	0	0	p	0	3.16e−01
5	0	0	0	p	2.69e−02
6	p	p	0	0	1.17e−01
7	p	0	p	0	3.19e−02
8	p	0	0	p	2.68e−02
9	0	p	p	0	1.84e+00
10	0	p	0	p	4.05e−03
11	0	0	p	p	3.17e−01
12	p	p	p	0	1.37e−04
13	p	p	0	p	1.17e−01
14	0	p	p	p	3.19e−02
15	p	0	p	p	1.84e+00
16	p	p	p	p	1.36e−04

$\zeta_\gamma = 0$. Case 16 is the fully pressure-dependent case. An additional case, from Civan [8] with constant factors for ζ_K, ζ_ϕ, ζ_μ, and ζ_τ, was also carried out.

Figure 1 shows the comparisons of the simulated results (solid lines) for the pressure against the distance for selected models (see captions) against the data of Pong et al. [16] (symbols). The inlet pressures for the different simulations are, respectively from bottom to top, 135, 170, 205, 240, and 275 kPa. Figure 1a compares with Darcy's Law Case 1, and it shows significant errors. Figure 1b compares with Civan's case, and Fig. 1c compares with Case 16, which is the best fit to the data. Figure 1d shows the relative error on log-scale for the 16 cases in Table 2. We refer to Case 16 as the Base Case henceforth (Table 3).

It is important to note that although the Civan case Fig. 1b and the Base Case Fig. 1c appear to yield similar results, the rock properties obtained in the two cases are quite different. Civan used $\phi = 0.2$ independent of pressure and he predicted

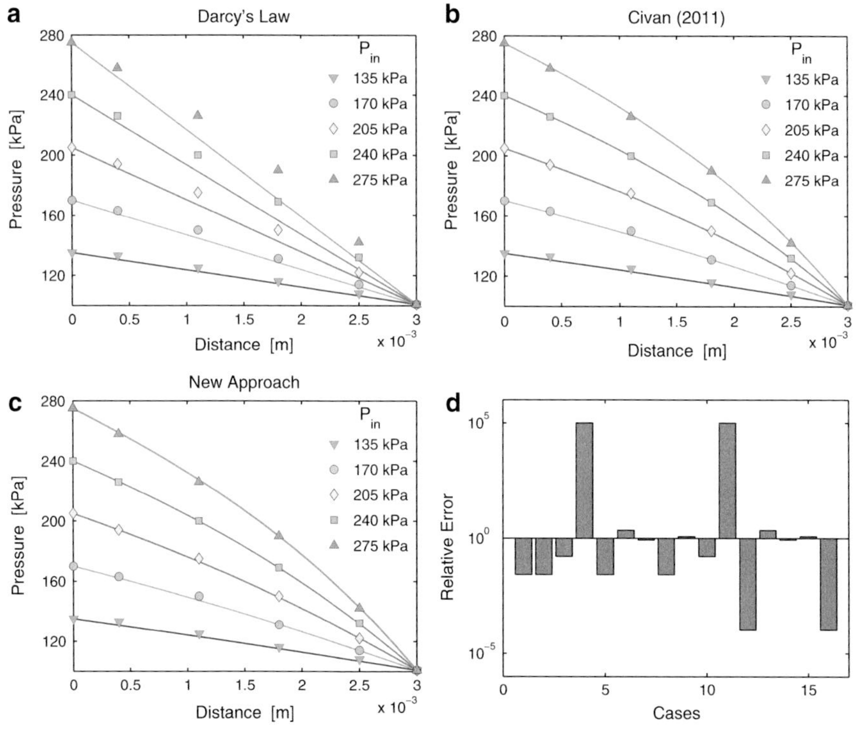

Fig. 1 Pressure, p against the distance along the core sample, x, from numerical solutions of the Steady State Model, Eqs. (21), (22), for different inlet pressures, P_{in}, as indicated by color. *Solids lines* are from the simulations, and symbols are data from Pong et al. [16]. (**a**) Darcy's law, Case 1 in Table 2, with compressibility factors, $\zeta_\gamma = 0$. (**b**) Civan's model with constant compressibility factors, $\zeta_\gamma = Const$, for some parameters, see [7]. (**c**) Case 16 in Table 2 (new model) with pressure dependent parameters and non-constant compressibility factors, $\zeta_\gamma(p)$. (**d**) Relative errors for the 16 cases in Table 2

Table 3 Reservoir model parameters used in the Base Case, Case 16 in Table 2

Parameter		Parameter	
L (m)	0.003	α_{KC}	1
N_x	100	β_{KC}	1
R_g (J kmol^{-1} K^{-1})	8314.4	Γ_{KC}	1
M_g (kg kmol^{-1} K^{-1})	16	a_ϕ	0.2
T (K)	350	b_ϕ	-1×10^{-6}
p_c (kPa)	3.1×10^3	c_ϕ	1.96
t_c (K)	125	σ_0	10
b_{SF}	-1	A_σ	0.2
a_τ	1.5	B_σ	0.4

Table 4 The range of parameters that are used to determine the values of permeability, and porosity in Fig. 3, from Eqs. (2) and (3)

Cases	α_{KC}	β_{KC}	Γ_{KC}	a_ϕ	b_ϕ	c_ϕ
1	1.0	1.0	1.0	0.20	−1e−6	1.96
2	1.0	0.65	1e−7	0.08	−1e−6	2.09
3	0.75	0.66	1e−7	0.15	−1e−6	2.09
4	0.25	0.4	1e−8	0.15	−1e−6	1.96
5	0.5	0.5	0.1	0.10	−1e−6	2.1112
6	0.5	1.5	1.0	0.05	−1e−8	2.90
7	0.45	0.65	1e−6	0.01	−1e−8	2.88

$K = 1 \times 10^{-15}\,\text{m}^2$. From the present calculations the porosity is pressure dependent and in the range $0.01 \le \phi \le 0.2$, and the permeability is also pressure dependent and in the range $10^{-20} \le K \le 10^{-3}\,\text{m}^2$, which are more realistic (Table 4).

5 Sensitivity Analysis and Estimation of Model Parameters

It is important to determine how much the results and predicted rock properties change due to small changes in model parameters. A sensitivity analysis was carried out by adjusting one model parameter at a time by factors of 2 and $1/2$, starting with Case 16 as the base case – One-at-a-Time (OAT) methodology. Sensitivity is measured by monitoring the changes in the model output.

Figure 2 shows sensitivity to selected parameters: (a) p_c (critical pressure), (b) T (temperature), (c) a_τ (constant in the tortuosity model), (d) a_ϕ (constant in the porosity model). Except for the temperature, Fig. 2b, all results show significant sensitivity to changes in the selected parameter especially at higher inlet pressures.

Figure 3 illustrates the sensitivity of the calculated permeability, and porosity against the pressure, for different combinations of α_{KC}, β_{KC}, and Γ_{KC}, Eq. (2).

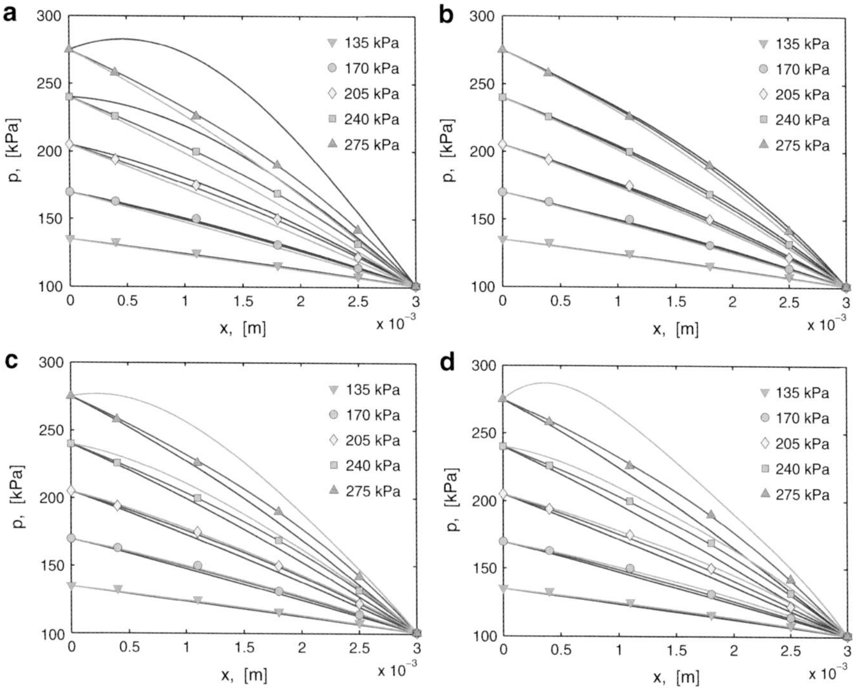

Fig. 2 OAT sensitivity analysis of the new model. Symbols are the data from Pong et al. [16] (see Fig. 2 for details). Sensitivity to the following parameters: (**a**) Critical pressure p_c, (**b**) Temperature T, (**c**) Tortuosity parameter a_τ in Eq. (3), (**d**) porosity parameter a_ϕ in Eq. (4). *Red lines* are the Base Case parameter values in Table 3. *Blue lines*: the specific parameter is divided by 2. *Green lines*: the specific parameter is multiplied by 2

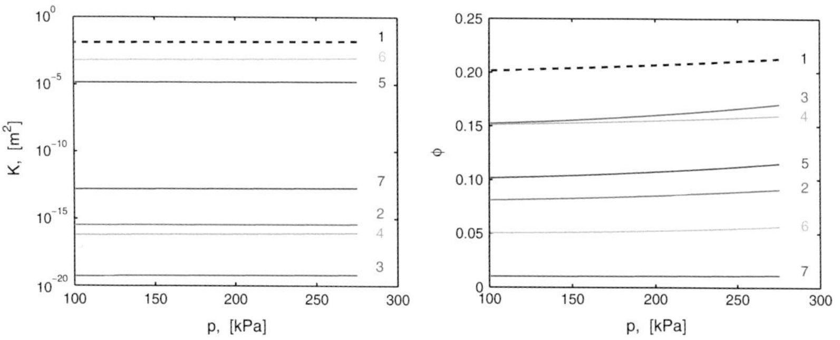

(a) Permeability (b) Porosity

Fig. 3 Permeability (K), and porosity (ϕ), against the pressure p, based upon the parameter values in Table 3. Seven cases for each parameter, shown in Table 4, are considered and are indicated on the plots. (Case 1 is the Base Case, shown as *black dashed line*.) (**a**) permeability curves are obtained using data from columns 2 to 4 of Table 4. (**b**) Porosity curves are obtained using data from columns 5 to 7

6 Summary

The Base Case 16 in Table 2, which is the fully pressure-dependent non-linear model, performs better than other models giving the smallest error against available data, Fig. 1c. Darcy's Law performs the worst illustrating its limitations for gas transport in tight porous media. A OAT sensitivity analysis shows that rock properties such as the porosity, and permeability, are very sensitive to most of the model parameters, Figs. 2 and 3. In the future, the sensitivity analysis for all of the model parameters will be completed.

Acknowledgements The authors would like to acknowledge the support provided by King Abdulaziz City for Science and Technology (KACST) through the Science Technology Unit at King Fahd University of Petroleum and Minerals (KFUPM) for funding this work through project No. 14-OIL280-04.

References

1. Ali, I., Malik, N.A., Chanane, B.: Time-fractional nonlinear gas transport equation in tight porous media: an application in unconventional gas reservoirs. In: 2014 International Conference on Fractional Differentiation and Its Applications (ICFDA), Catania, pp. 1–6. IEEE (2014)
2. Aziz, K., Settari, A.: Petroleum Reservoir Simulation, vol. 476. Applied Science Publishers, London (1979)
3. Beskok, A., Karniadakis, G.E.: Report: a model for flows in channels, pipes, and ducts at micro and nano scales. Microsc. Thermophys. Eng. **3**(1), 43–77 (1999)
4. Carman, P.C., Carman, P.C.: Flow of Gases Through Porous Media. Butterworths Scientific Publications, London (1956)
5. Chen, Z.: Reservoir Simulation: Mathematical Techniques in Oil Recovery. CBMS-NSF Regional Conference Series in Applied Mathematics, vol. 77. SIAM, Philadelphia (2007)
6. Civan, F.: Effective correlation of apparent gas permeability in tight porous media. Transp. Porous Media **82**(2), 375–384 (2010)
7. Civan, F., Rai, C.S., Sondergeld, C.H.: Shale-gas permeability and diffusivity inferred by improved formulation of relevant retention and transport mechanisms. Transp. Porous Media **86**(3), 925–944 (2011)
8. Civan, F., et al.: Improved permeability equation from the bundle-of-leaky-capillary-tubes model. In: SPE Production Operations Symposium, Oklahoma City. Society of Petroleum Engineers (2005)
9. Cui, X., Bustin, A., Bustin, R.M.: Measurements of gas permeability and diffusivity of tight reservoir rocks: different approaches and their applications. Geofluids **9**(3), 208–223 (2009)
10. Cussler, E.L.: Diffusion: Mass Transfer in Fluid Systems. Cambridge University Press, Cambridge/New York (2009)
11. Darishchev, A., Rouvroy, P., Lemouzy, P.: On simulation of flow in tight and shale gas reservoirs. In: 2013 SPE Middle East Unconventional Gas Conference & Exhibition, Muscat (2013)
12. Kumar, N.: Compressibility factors for natural and sour reservoir gases by correlations and cubic equations of state. M.Sc. Thesis, Texas Tech University (2004)
13. Loeb, L.B.: The Kinetic Theory of Gases. Courier Dover Publications (2004)

14. Mahmoud, M.: Development of a new correlation of gas compressibility factor (z-factor) for high pressure gas reservoirs. J. Energy Resour. Technol. **136**(1), 012903 (2014)
15. Peaceman, D.W.: Fundamentals of Numerical Reservoir Simulation. Elsevier, New York (1977)
16. Pong, K.C., Ho, C.M., Liu, J., Tai, Y.C.: Non-linear pressure distribution in uniform microchannels. ASME public. FED **197**, 51–51 (1994)
17. Rathakrishnan, E.: Gas Dynamics. PHI Learning, New Delhi (2013)
18. Wang, Z., Krupnick, A.: A retrospective review of shale gas development in the United States. What led to the boom? Pub. Resources, Washington (2013)
19. Ziarani, A.S., Aguilera, R.: Knudsen's permeability correction for tight porous media. Transp. Porous Media **91**(1), 239–260 (2012)

Solutions of Time-Fractional Diffusion Equation with Reflecting and Absorbing Boundary Conditions Using Matlab

Iftikhar Ali, Nadeem A. Malik, and Bilal Chanane

Abstract The main objective of this work is to develop Matlab programs for solving the time-fractional diffusion equation (TFDE) with reflecting and absorbing boundary conditions on finite and infinite domains. Essentially, there are three major codes, one for finding the exact solution of the TFDE and other two are for finding the numerical solution of the TFDE. The code for finding the exact solutions is based on the fundamental solution of the TFDE, whereas the codes for finding the numerical solutions are based on the explicit and the implicit finite difference schemes, respectively. Finally, we illustrate the effectiveness of the codes by applying them to TFDEs with sharp initial data and for various reflecting and absorbing boundary conditions both on finite and infinite domains. The results show the difference of solutions between the standard diffusion equation and the time-fractional diffusion equation.

1 Introduction

Many physical processes evolve in spaces that are heterogeneous in nature, such as, crowded system, protein diffusion within cells, anomalous diffusion through porous media, see [3, 4, 6, 17]. Mathematical models, based on standard calculus, have failed to describe such intricate processes whereas mathematical models, based on fractional calculus techniques, have proven their effectiveness in explaining such complex processes, [1, 2, 5, 10, 15].

Time-fractional diffusion equation have been derived in the framework of Continuous Time Random Walk (CTRW) model. It is based on the idea of considering the transport processes as the flow of particles in the form of packets and then assigning a probability of locating a packet at position x at time t. Law of Total Probability is used to determine probability $P(x, t)$. Luchko has derived the time-fractional diffusion equation by using these concept, see the details in [8, 9]. The

I. Ali (✉) • N.A. Malik • B. Chanane
King Fahd University of Petroleum and Minerals, Dhahran, Saudi Arabia
e-mail: iali@kfupm.edu.sa

© Springer International Publishing Switzerland 2016

J. Bélair et al. (eds.), *Mathematical and Computational Approaches in Advancing Modern Science and Engineering*, DOI 10.1007/978-3-319-30379-6_2

equation is given by,

$$\tau \partial_t P(x,t) = \partial_t^{1-\alpha}[-v\partial_x P(x,t) + k^2 \partial_x^2 P(x,t)] \tag{1}$$

as $t \to \infty$ and $|x| \to \infty$. Equation (1) is called time-fractional advection-diffusion equation and in the case $v = 0$ it reduces to time-fractional diffusion equation. For more detailed derivation, see [9, 11].

In this work, we develop Matlab programs for finding exact and numerical solutions of the time fractional diffusion equation (TFDE) on finite and infinite domains, and also with various boundary conditions. The manuscript is organized as follows; in Sect. 2, procedure for finding the fundamental solution of the TFDE is explained; in Sect. 3, the numerical schemes are discussed; in Sect. 4, Matlab codes are provided; in Sect. 5, several examples are given to illustrate the effectiveness of Matlab programs; finally, in Sect. 6, conclusions are given.

2 Fundamental Solution of Time Fractional Diffusion Equation

Consider the time fractional diffusion equation, in Caputo form, over the whole real line with given initial data,

$$\frac{\partial^\alpha}{\partial t^\alpha} u(x,t) = \frac{\partial^2}{\partial x^2} u(x,t), \qquad 0 < \alpha \leq 1 \tag{2}$$

$$u(x,0) = f(x). \tag{3}$$

Equation (2) can be written in the integral form as follows,

$$u(x,t) = f(x) + \frac{1}{\Gamma(\alpha)} \int_0^t (t-\tau)^{\alpha-1} u_{xx}(x,\tau)d\tau. \tag{4}$$

Application of Laplace transform yields a second order linear differential equation

$$\tilde{u}_{xx}(x,p) - p^\alpha \tilde{u}(x,p) = -f(x)p^{\alpha-1}. \tag{5}$$

The solution of Eq. (5) is given by

$$\tilde{u}(x,p) = \int_{-\infty}^{\infty} \tilde{k}(|x-y|, p^{\alpha/2}) p^{\alpha-1} f(y)dy, \tag{6}$$

where $k(|x|,\lambda) = \frac{1}{\sqrt{2\pi\lambda}}|x|^{1/2}k_{1/2}(\lambda|x|)$ is modified Bessel function of second kind [16]. Furthermore, Eq. (6) can be expressed as

$$\tilde{u}(x,p) = \int_{-\infty}^{\infty} \tilde{G}^{\alpha}(|x-y|,p)f(y)dy, \tag{7}$$

where $\tilde{G}^{\alpha}(|x|,p) = \tilde{k}(|x|,p^{\alpha/2})p^{\alpha-1}$.

Note that directly taking the inverse Laplace transform is not feasible, so we use the relationship between the Laplace and Mellin transforms to obtain

$$\begin{aligned}\tilde{G}^{\alpha}(|x|,s) &= \frac{1}{\Gamma(1-s)}\int_0^{\infty} p^{-s}\tilde{G}^{\alpha}(|x|,p)dp \\ &= \frac{|x|^{1/2}}{\sqrt{2\pi}\Gamma(1-s)}\int_0^{\infty} p^{3\alpha/4-s-1}\tilde{k}_{1/2}(|x|p^{\alpha/2})dp.\end{aligned} \tag{8}$$

Using the results,

$$\mathscr{M}[x^{\lambda}f(ax^{b})] = \frac{1}{b}a^{-\frac{s+\lambda}{b}}\tilde{f}\left(\frac{s+\lambda}{b}\right),$$

$$\tilde{k}_{\sigma}(s) = 2^{s-2}\Gamma\left[\frac{s-\sigma}{2}\right]\Gamma\left[\frac{s+\sigma}{2}\right],$$

Equation (8) becomes

$$\tilde{G}^{\alpha}(|x|,s) = \frac{1}{\alpha\sqrt{\pi}}2^{-2s/\alpha}|x|^{2s/\alpha-1}\frac{\Gamma[1-s/\alpha]\Gamma[1/2-s/\alpha]}{\Gamma[1-s]}. \tag{9}$$

Taking the inverse Mellin transform and using Fox function, we obtain

$$G^{\alpha}(|x|,t) = \frac{1}{\alpha\sqrt{\pi}}|x|^{-1}H_{12}^{20}\left[\frac{|x|^{2/\alpha}}{2^{2/\alpha}t}\middle|\begin{matrix}(1,1)\\(1/2,1/\alpha),(1,1/\alpha)\end{matrix}\right]. \tag{10}$$

The general solution of the time fractional diffusion equation is given by

$$u(x,t) = \int_{-\infty}^{\infty} G^{\alpha}(|x-y|,t)f(y)dy. \tag{11}$$

If the initial data is given as delta potential, that is, $u(x,0) = \delta(x)$, then the solution (11) becomes

$$u(x,t) = \frac{1}{\alpha\sqrt{\pi}}|x|^{-1}H_{12}^{20}\left[\frac{|x|^{2/\alpha}}{2^{2/\alpha}t}\middle|\begin{matrix}(1,1)\\(1/2,1/\alpha),(1,1/\alpha)\end{matrix}\right]. \tag{12}$$

For more details, readers are referred to Wyss [18] and Schneider & Wyss [14].

3 Numerical Solutions

In this section, we provide two numerical schemes based on finite difference methods, one is explicit and other one is implicit. First, we give an explicit finite difference scheme which was derived by Yuste [19]. A uniform grid is placed on the space-time domain, that is, $x_j = j\Delta x$ and $t_m = m\Delta t$. The numerical approximation of the unknown function $u(x,t)$ at the point (x_j, t_m) is denoted by U_j^m and it is obtained by $U_j^m \approx u_j^m \equiv u(x_j, t_m)$. The time-fractional Riemann-Liouville derivative is discretized by Grunwald-Letnikov derivative [13]. The explicit finite difference scheme is given by the following formula

$$u_j^{m+1} = u_j^m + S_\gamma \sum_{k=0}^{m} \omega_k^{(1-\gamma)} \left(u_{j-1}^{m-k} - 2u_j^{m-k} + u_{j+1}^{m-k}\right), \tag{13}$$

where

$$\omega_0^{(1-\gamma)} = 1, \qquad \omega_k^{(1-\gamma)} = \left(\frac{k-2+\gamma}{k}\right) \omega_{k-1}^{(1-\gamma)}. \tag{14}$$

The numerical scheme (13) is stable on the interval $0 \le S_\gamma \le 1/2^{(2-\gamma)}$, where $S_\gamma = k_\gamma \Delta t^\gamma / \Delta x^2$, for more details about the stability condition, see [19]. Note that in this work, we only develop Matlab codes for uniform grid; for the case of non-uniform grid, see for example [20].

We use the implicit finite difference scheme derived by Langlands and Henry [7]. The numerical scheme is given by

$$\begin{aligned} -S_\gamma u_{j-1}^m + (1+2S_\gamma)u_j^m - S_\gamma u_{j+1}^m &= u_j^{m-1} \\ + S_\gamma \sum_{k=1}^{m} \omega_k^{(1-\gamma)} \left(u_{j-1}^{m-k} - 2u_j^{m-k} + u_{j+1}^{m-k}\right). \end{aligned} \tag{15}$$

The above scheme (15) is unconditionally stable, see Theorem 2.1 in [12]. For more technical details, readers are referred to [7, 12].

3.1 Discretization of Boundary Conditions

Dirichlet boundary conditions, $u(x,t) = 0$, are discretized in the standard way and implemented naturally, where x represents left or right boundary. However, Neumann boundary conditions, $u_x(x,t) = 0$, are discretized by the second order difference formula, that is, $(u_{j+1}^m - u_{j-1}^m)/2\Delta x$, where x represents left or right boundary.

In the case of left Neumann boundary condition, we have $u_1^m = u_{-1}^m$ and letting $j = 0$ in the explicit scheme (13), we obtain

$$u_0^{m+1} = u_0^m + 2S_\gamma \sum_{k=0}^{m} \omega_k^{(1-\gamma)} \left(u_1^{m-k} - u_0^{m-k}\right). \tag{16}$$

On the other hand, the implicit scheme (15) yields

$$(1 + 2S_\gamma)u_0^m - 2S_\gamma u_1^m = u_0^{m-1} + 2S_\gamma \sum_{k=1}^{m} \omega_k^{(1-\gamma)} \left(u_1^{m-k} - u_0^{m-k}\right). \tag{17}$$

Similarly, we can obtain the schemes at the right boundary.

4 Matlab Codes

The following Matlab code is used for the computation of the exact solution (12).

Listing 1 Matlab code for exact solution

```
clc, clear all
format long e
syms x xx c_n
n  = 55;  L   =  10;       T = [0.1 1 5];
dx = 0.1; x1  = (0:dx:L);
x0 = 1;   gma =  0.5;      K_gma = 1; v = 1;
ExactSol = zeros(length(x1), length(T));
for k  = 1:length(T)
     a1 = 1/sqrt(4*K_gma*T(k)^gma);
     gama = (-1).^((1:n)-1) ...
              ./(factorial((1:n)-1).*gamma(1-0.5*gma*(1:n)));
     xx = ((x-x0).^2/(K_gma*T(k)^gma)).^(0.5*((1:n)-1)) ...
          - ((x+x0).^2/(K_gma*T(k)^gma)).^(0.5*((1:n)-1));
     c_n = a1*sum(gama.*xx);
     ExactSol(:,k) = subs(c_n, x1);
end
plot(x1,ExactSol),
axis([0 5 0 0.72])
```

The following Matlab code is used for the computation of the numerical solutions which is based on the explicit finite difference scheme (13). The code can be easily modified for various initial and boundary conditions.

Listing 2 Matlab code for explicit finite difference scheme

```
clc, clear all
gma = 0.5; S_gma = 0.33;
K_gma = 1; L = 10;
dt = 0.001; T = 5;
t = (0:dt:T);    M = length(t);
dx = sqrt(K_gma*dt.^gma./S_gma);
x1 = (0:dx:1); x2 = (1+dx:dx:L);
x = [x1 1 x2]; J = length(x);
for a = 1:J
    if x(a)==1; aa = a; break, end
end
w = ones(1,M);
for k = 2:M
    w(k) = (1-(2-gma)/(k-1))*w(k-1);
end
V = zeros(J,M); V(aa,1) = 1/dx;
for m = 2:M
   for j = 2:J-1
      V(j,m) = V(j,m-1) + ...
          S_gma*sum(flipud(w(1:m-1)')'.*(V(j-1, 1:m-1)...
               - 2*V(j,1:m-1) + V(j+1,1:m-1)));
   end
end
plot(x,V(:,t==0.1),'o', x, V(:,t==1),'o', x, V(:,t==5),'o')
axis([0 5 0 0.72])
```

The following Matlab code is used for the computation of the numerical solutions which is based on the implicit finite difference scheme (15). The code can be easily modified for various initial and boundary conditions.

Listing 3 Matlab code for implicit finite difference scheme

```
function ImplicitSolution
clc, clear all
alfa = 0.5; K_alfa = 1; L = 10;
dt = 0.01; T = 5;
t = (0:dt:T);    M = length(t);
dx = 0.1; x = (0:dx:L); J = length(x);
S_alfa = K_alfa*dt^alfa/dx^2;
for a = 1:J
    if x(a)==1; aa = a; break, end
end
w = ones(1,M);
for k = 2:M
    w(k) = (1-(2-alfa)/(k-1))*w(k-1);
end
V = zeros(J,M); V(aa,1) = 1/dx;
d = S_alfa*ones(1,J-2);
E = diag(1+2*d) + diag(-d(1:J-3),1) + diag(-d(2:J-2),-1);
for m = 2:M
```

```
    V(2:J-1, m) = BTCS(E, S_alfa, V(:, 1:m), w(2:m));
end
plot(x,V(:,t==0.1), x, V(:,t==1), x, V(:,t==5))
axis([0 5 0 1])
end
%##################################################
function vnew = BTCS(E, s, U, v)
v = flipud(v'); [J, m] = size(U);
uold = U(2:J-1,m-1);
V1 = U(1:J-2, 1:m-1);
V2 = U(2:J-1, 1:m-1);
V3 = U(3:J, 1:m-1);
V4 = V1 - 2*V2 + V3;
lhs = uold + s*(V4*v);
vnew = E\lhs;
end
```

5 Numerical Experiments

In this section, we provide several examples which are solved by using above Matlab codes on an Intel Core-i7 machine. The computation time is given for each problem.

Example 1 Consider the TFDE

$$\frac{\partial u}{\partial t} = K_\gamma \, {}_0D_t^{1-\gamma}\left(\frac{\partial^2 u}{\partial x^2}\right), \tag{18}$$

on the domain $x > 0$ and $t > 0$ with the initial data is taken as a delta function at $x = 1$. Absorbing and reflecting boundary conditions are taken (one by one) at left boundary $x = 0$, where as $u(x, t) \to 0$ as $x \to \infty$. Exact solution is given by

$$u(x, t) = W(x - x_0, t) \pm W(x + x_0, t), \tag{19}$$

where $W(x, t)$ is the solution of TFDE (18) over the whole real line with decaying boundary conditions when $|x|$ becomes large. In series form, $u(x, t)$ is expressed as

$$u(x,t) = \frac{1}{4K_\gamma t^\gamma}\sum_{n=0}^{\infty}\frac{(-1)^n}{n!\Gamma(1-\gamma(1+n)/2)}\left(\frac{(x-x_0)}{K_\gamma t^\gamma}\right)^{n/2} \pm \left(\frac{(x+x_0)}{K_\gamma t^\gamma}\right)^{n/2} \tag{20}$$

Note the minus sign is taken in the case of absorbing BC and plus sign is taken in the case of reflecting BC. Figure 1 shows the numerical solutions at times $t = 0.1, 1, 5$. The data used for numerical computation is $\Delta t = 0.001$, $T = 5$, $\gamma = 0.5$, $S_\gamma = 0.33$; and the computational time is 9.91 s.

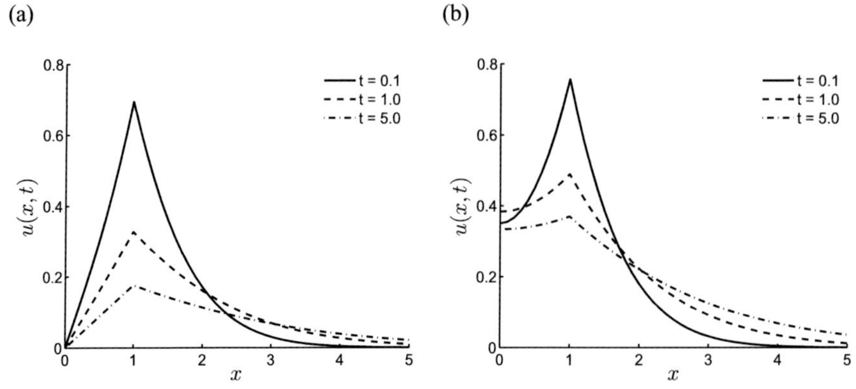

Fig. 1 Solutions of time fractional diffusion equation 18 with absorbing and reflecting boundary conditions at the left boundary. Initial condition is taken as delta function at $x = 1$, where $\gamma = 0.5$. (**a**) Absorbing boundary condition. (**b**) Reflecting boundary condition

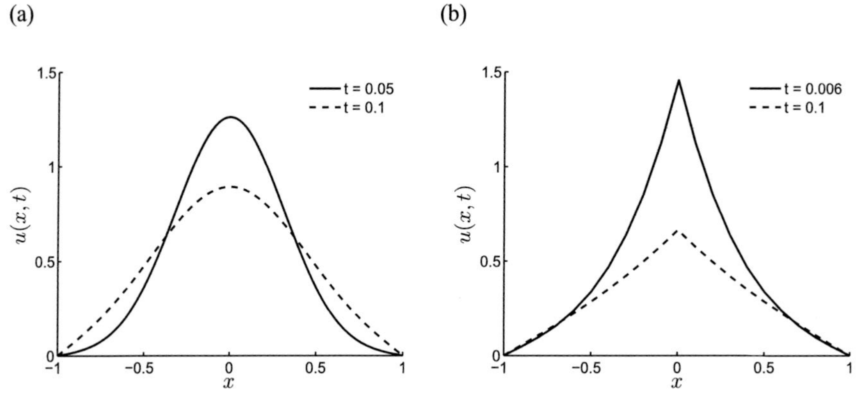

Fig. 2 Solutions of time fractional diffusion equation 18 with absorbing boundary conditions at left and right boundaries. Initial condition is taken as delta function at $x = 0$. *Cusp shape* is the distinct feature of the curve in the case $\gamma = 0.5$. (**a**) Standard diffusion $\gamma = 1.0$. (**b**) Anomalous diffusion $\gamma = 0.5$

Example 2 Consider the TFDE (18) on a box $-1 \leq x \leq 1$ and $t > 0$ with absorbing boundaries at $x = -1$ and 1. Initial data is given by delta function at $x = 0$. Exact solution of the problem is given by

$$u(x,t) = \sum_{n=-\infty}^{\infty} \left[W(x + 4n, t) - W(4n - x + 2, t) \right]. \tag{21}$$

Figure 2 shows the numerical solutions at times $t = 0.05, 0.1$ for $\gamma = 1$ and at times $t = 0.006, 0.1$ for $\gamma = 0.5$. The data used for numerical computation is

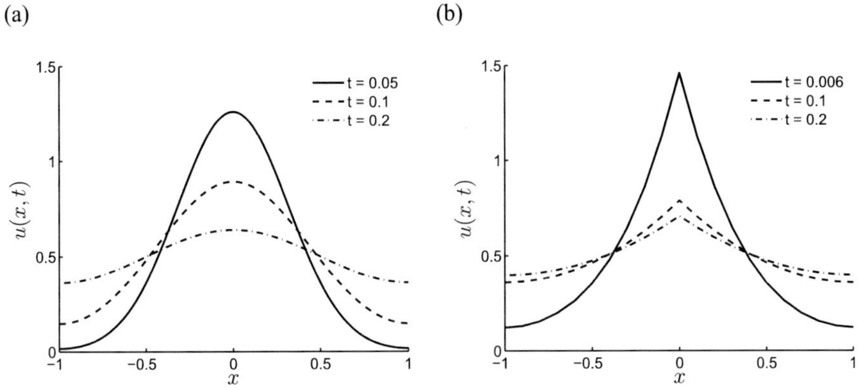

Fig. 3 Solutions of time fractional diffusion equation 18 with reflecting boundary conditions at left and right boundaries. Initial condition is taken as delta function at $x = 0$. *Cusp shape* is the distinct feature of the curve in the case $\gamma = 0.5$. (**a**) Standard diffusion $\gamma = 1.0$. (**b**) Anomalous diffusion $\gamma = 0.5$

$\Delta x = 0.1$, $T = 0.1$, $\gamma = 1.0$, $S_\gamma = 0.49$; and the computational time is 9.5 s. In the case of $\gamma = 0.5$, we choose $S_\gamma = 0.33$, and the computational time is 14.71 s.

Example 3 We solve TFDE (18) on a box $-1 < x < 1$ and $t > 0$ with reflecting boundaries at $x = -1$ and 1. Initial data is given by delta function at $x = 0$. Exact solution of the problem is given by

$$u(x,t) = \sum_{n=-\infty}^{\infty} \left[W(x+4n,t) + W(4n-x+2,t) \right]. \tag{22}$$

Figure 3 shows the numerical solutions at times $t = 0.05, 0.1, 0.2$ for $\gamma = 1$ and at times $t = 0.006, 0.1, 0.2$ for $\gamma = 0.5$. The data used for numerical computation is $\Delta x = 0.01$, $T = 0.2$, $\gamma = 1.0$, $S_\gamma = 0.49$; and the computational time is 50.12 s. In the case of $\gamma = 0.5$, we choose $S_\gamma = 0.33$, $\Delta x = 0.1$, and the computational time is 99.97 s.

6 Conclusions

Time fractional advection-diffusion equation is derived under the CTRW framework [8, 9]. The fundamental solution of the time-fractional diffusion equation is obtained by applying Fourier and Laplace transforms and which is finally expressed in terms of Fox's function by using Mellin transform [14, 18].

Matlab codes are developed for obtaining the exact and the numerical solutions of time-fractional diffusion equation. The code for the exact solution is based on the work of Wyss [18] and Schneider & Wyss [14], whereas the code for finding the numerical solution is based on the work of Yuste & Acedo (explicit case) [19] and Langlands & Henry (implicit case) [7].

We have given several examples that illustrate the effectiveness of the codes. Essentially, numerical solution are found both on finite and infinite domains with absorbing and reflecting boundary conditions. Initial conditions are taken as nonsmooth functions (delta functions) and it is observed that in the case of anomalous diffusion the cusp shape remains in the solutions compared to standard diffusion where the solution smooths up as time increases.

Acknowledgements The author would like to acknowledge the support provided by King Abdulaziz City for Science and Technology (KACST) through the Science Technology Unit at King Fahd University of Petroleum and Minerals (KFUPM) for funding this work through project No. 14-OIL280-04.

References

1. Ali, I., Malik, N.A.: Hilfer fractional advection–diffusion equations with power-law initial condition; a numerical study using variational iteration method. Comput. Math. Appl. **68**(10), 1161–1179 (2014)
2. Ali, I., Malik, N.A., Chanane, B.: Time-fractional nonlinear gas transport equation in tight porous media: an application in unconventional gas reservoirs. In: 2014 International Conference on Fractional Differentiation and Its Applications (ICFDA), Catania, pp. 1–6. IEEE (2014)
3. Chen, W., Sun, H., Zhang, X., Korošak, D.: Anomalous diffusion modeling by fractal and fractional derivatives. Comput. Math. Appl. **59**(5), 1754–1758 (2010)
4. Fedotov, S., Iomin, A.: Migration and proliferation dichotomy in tumor-cell invasion. Phys. Rev. Lett. **98**(11), 118,101 (2007)
5. Jeon, J., Tejedor, V., Burov, S., Barkai, E., Selhuber-Unkel, C., Berg-Sørensen, K., Oddershede, L., Metzler, R.: In vivo anomalous diffusion and weak ergodicity breaking of lipid granules. Phys. Rev. Lett. **106**(4), 048,103 (2011)
6. Koch, D., Brady, J.: Anomalous diffusion in heterogeneous porous media. Phys. Fluids (1958–1988) **31**(5), 965–973 (1988)
7. Langlands, T., Henry, B.: The accuracy and stability of an implicit solution method for the fractional diffusion equation. J. Comput. Phys. **205**(2), 719–736 (2005)
8. Luchko, Y.: Anomalous diffusion: models, their analysis, and interpretation. Adv. Appl. Anal. 115–145 (2012)
9. Luchko, Y., Punzi, A.: Modeling anomalous heat transport in geothermal reservoirs via fractional diffusion equations. GEM Int. J. Geomath. **1**(2), 257–276 (2011)
10. Malik, N., Ali, I., Chanane, B.: Numerical solutions of non-linear fractional transport models in unconventional hydrocarbon reservoirs using variational iteration method. In: 5th International Conference on Porous Media and Their Applications in Science (Engineering and Industry, Eds, ECI Symposium Series, Volume) (Hawaii, 2014). http://dc.engconfintl.org/porous_media_V/43
11. Metzler, R., Klafter, J.: The random walk's guide to anomalous diffusion: a fractional dynamics approach. Phys. Rep. **339**(1), 1–77 (2000)
12. Murio, D.A.: Implicit finite difference approximation for time fractional diffusion equations. Comput. Math. Appl. **56**(4), 1138–1145 (2008)

13. Podlubny, I.: Fractional Differential Equations: An Introduction to Fractional Derivatives, Fractional Differential Equations, to Methods of Their Solution and Some of Their Applications, vol. 198. Academic, New York (1998)
14. Schneider, W., Wyss, W.: Fractional diffusion and wave equations. J. Math. Phys. **30**(1), 134–144 (1989)
15. Tabei, S., Burov, S., Kim, H., Kuznetsov, A., Huynh, T., Jureller, J., Philipson, L., Dinner, A., Scherer, N.: Intracellular transport of insulin granules is a subordinated random walk. Proc. Natl. Acad. Sci. **110**(13), 4911–4916 (2013)
16. Watson, G.N.: A Treatise on the Theory of Bessel Functions. Cambridge University Press, Cambridge/New York (1995)
17. Weiss, M., Elsner, M., Kartberg, F., Nilsson, T.: Anomalous subdiffusion is a measure for cytoplasmic crowding in living cells. Biophys. J. **87**(5), 3518–3524 (2004)
18. Wyss, W.: The fractional diffusion equation. J. Math. Phys. **27**(11), 2782–2785 (1986)
19. Yuste, S., Acedo, L.: An explicit finite difference method and a new von Neumann-type stability analysis for fractional diffusion equations. SIAM J. Numer. Anal. **42**(5), 1862–1874 (2005)
20. Yuste, S.B., Quintana-Murillo, J.: A finite difference method with non-uniform timesteps for fractional diffusion equations. Comput. Phys. Commun. **183**(12), 2594–2600 (2012)

Homoclinic Structure for a Generalized Davey-Stewartson System

Ceni Babaoglu and Irma Hacinliyan

Abstract In this study, we analyze the homoclinic structure for the generalized Davey-Stewartson system with periodic boundary conditions. This system involves three coupled nonlinear equations and describes $(2 + 1)$ dimensional wave propagation in a bulk medium composed of an elastic material with coupled stresses. We first provide linearized stability analysis of the plane wave solutions of the generalized Davey-Stewartson system. Then, give an analytic description of the characteristics of homoclinic orbits near the fixed point by finding soliton type solutions. These solutions are derived via Hirota's bilinear method. We also show that two of these solutions form a pair of symmetric homoclinic orbits and all these symmetric homoclinic orbit pairs construct the homoclinic tubes.

1 Introduction

Homoclinic orbits are important for the study of chaos in deterministic nonlinear dynamics. In a neighborhood of such an orbit, an extended knowledge of geometric structures lead one to a better understanding of chaotic dynamics. The homoclinic structure of nonlinear Schrodinger (NLS) equation is considered by Ablowitz and Herbst in [1]. They have shown how this structure associated with the cubic NLS equation may be obtained from the N-soliton solutions of the defocusing NLS equation. We follow a similar approach and first observe that the fixed point in the GDS system is hyperbolic. Then, we analyze the homoclinic structure for the generalized Davey-Stewartson (GDS) equations.

C. Babaoglu (✉) • I. Hacinliyan
Faculty of Science and Letters, Department of Mathematics, Istanbul Technical University, 34469 Maslak-Istanbul, Turkey
e-mail: ceni@itu.edu.tr; hacinliy@itu.edu.tr

© Springer International Publishing Switzerland 2016

J. Bélair et al. (eds.), *Mathematical and Computational Approaches in Advancing Modern Science and Engineering*, DOI 10.1007/978-3-319-30379-6_3

The generalized Davey-Stewartson system has been introduced in [2] to study $(2 + 1)$ dimensional wave propagation in a bulk medium composed of an elastic material with coupled stresses. They are given by

$$
\begin{aligned}
&i\tilde{A}_\tau + p\tilde{A}_{\xi\xi} + r\tilde{A}_{\eta\eta} = q|\tilde{A}|^2 A + \frac{k^2}{2\omega}(\gamma_3\tilde{\phi}_{1,\xi} + \gamma_1\tilde{\phi}_{2,\eta})\tilde{A}, \\
&(c_g^2 - c_1^2)\tilde{\phi}_{1,\xi\xi} - c_2^2\tilde{\phi}_{1,\eta\eta} - (c_1^2 - c_2^2)\tilde{\phi}_{2,\xi\eta} = \gamma_3 k^2(|\tilde{A}|^2)_\xi, \\
&(c_g^2 - c_2^2)\tilde{\phi}_{2,\xi\xi} - c_1^2\tilde{\phi}_{2,\eta\eta} - (c_1^2 - c_2^2)\tilde{\phi}_{1,\xi\eta} = \gamma_1 k^2(|\tilde{A}|^2)_\eta,
\end{aligned}
\tag{1}
$$

where ξ and η are spatial coordinates and τ is time; $\tilde{A}$ is the complex amplitude of the short transverse wave mode, and $\tilde{\phi}_1$ and $\tilde{\phi}_2$ are long longitudinal and long transverse wave modes, respectively. The coefficients that appear in (1) can be given as follows

$$
\begin{aligned}
&c_1^2 = \frac{\lambda + 2\mu}{\rho_0}, \quad c_2^2 = \frac{\mu}{\rho_0}, \quad c_g = c_2^2(k + 8m^2k^3)/\omega, \\
&\gamma_1 = c_1^2 - 2c_2^2 + \frac{\mathscr{B}}{\rho_0}, \quad \gamma_3 = c_1^2 + \frac{\mathscr{A} + 2\mathscr{B}}{2\rho_0}, \\
&p = -\frac{1}{2\omega}(c_g^2 - c_2^2 - 24m^2c_2^2k^2), \quad r = \frac{c_2^2}{2\omega}(1 + 8m^2k^2), \quad q = \frac{k^6\gamma_3^2}{\omega D_1(2k, 2\omega)},
\end{aligned}
$$

where k is the wave number, ω is the frequency, m, λ, μ, $\mathscr{A}$ and $\mathscr{B}$ are material constants, c_g is the group speed of transverse waves whereas c_1 and c_2 are phase speeds of longitudinal and transverse waves, respectively. Also the frequency and the dispersion relation is as follows

$$
\omega = c_2k(1 + 4m^2k^2)^{\frac{1}{2}}, \quad D_1(k, \omega) = \omega^2 - c_1^2k^2 - 4(1 + \nu)c_2^2m^2k^4. \tag{2}
$$

A simple algebra shows that the coefficients p, q and r are all positive. In terms of dimensionless variables, $\tilde{A} = u$, $\tilde{\phi}_1 = \varphi_1$, $\tilde{\phi}_2 = \varphi_2$, and

$$
\tau = \frac{\gamma_1^2(c_g^2 - c_1^2)^2}{\gamma_3^4k^4r}t, \quad \xi = \frac{c_g^2 - c_1^2}{\gamma_3k^2}x, \quad \eta = \frac{\gamma_1(c_g^2 - c_1^2)}{\gamma_3^2k^2}y, \tag{3}
$$

the three coupled evolution equations take the form

$$
\begin{aligned}
&iu_t + \delta u_{xx} + u_{yy} = \chi|u|^2u + b(\varphi_{1,x} + \varphi_{2,y})u, \\
&\varphi_{1,xx} + m_2\varphi_{1,yy} + n\varphi_{2,xy} = (|u|^2)_x, \\
&\lambda\varphi_{2,xx} + m_1\varphi_{2,yy} + n\varphi_{1,xy} = (|u|^2)_y,
\end{aligned}
\tag{4}
$$

where the non-dimensional coefficients are

$$\delta = \frac{p}{r}(\frac{\gamma_1}{\gamma_3})^2, \quad \chi = \frac{q\gamma_1^2(c_g^2 - c_1^2)^2}{r\gamma_3^4 k^4}, \quad b = \frac{c_g^2 - c_1^2}{2\omega r}(\frac{\gamma_1}{\gamma_3})^2,$$

$$m_1 = \frac{c_1^2}{c_1^2 - c_g^2}(\frac{\gamma_3}{\gamma_1})^2, \quad m_2 = \frac{c_2^2}{c_1^2 - c_g^2}(\frac{\gamma_3}{\gamma_1})^2, \quad \lambda = \frac{c_2^2 - c_g^2}{c_1^2 - c_g^2}, \quad n = \frac{c_1^2 - c_2^2}{c_1^2 - c_g^2}(\frac{\gamma_3}{\gamma_1}),$$

so $(1-\lambda)(m_1 - m_2) = n^2$, $m_1 > m_2$ and $\lambda < 1$. It should be also noted that

$$c_g^2 - c_2^2 = 4m^2c_2^4k^4(3 + 16m^2k^2)/\omega^2 \tag{5}$$

is always positive. There exists a critical wave number

$$k_c^2 = [c_1^2 - 4c_2^2 + c_1(c_1^2 + 8c_2^2)^{\frac{1}{2}}]/(32c_2^2m^2) \tag{6}$$

such that $c_1^2 - c_g^2 < 0$ if $k > k_c$ and $c_1^2 - c_g^2 > 0$ if $k < k_c$ because $c_1^2 > c_2^2$. (The case where $k = k_c$ corresponds to long-wave short-wave resonance since the phase speed of longitudinal wave, c_1, is equal to the group speed of the transverse wave, c_g.) Thus, depending on the wave number k chosen, the coefficients of the second and third equations of the GDS system may change their sign. For example, the respective sign of (m_1, m_2, λ) is $(-,-,+)$ if $k > k_c$ and is $(+,+,-)$ if $k < k_c$.

Now the GDS system will be classified according to the values of parameters. In fact, since $\delta > 0$ the first equation is always elliptic. The classification of the last two coupled equations of (4) is based on eigenvalues of the coefficient matrix of a first order linear system with four equations equivalent to the second-order linear system, $(4)_2$ and $(4)_3$. Therefore, system (4) can be classified as elliptic-elliptic-elliptic, elliptic-elliptic-hyperbolic, and elliptic-hyperbolic -hyperbolic according to the respective sign of (m_1, m_2, λ): $(+,+,+)$, $(+,+,-)$, and $(-,-,+)$ [3]. The descriptions given above lead that m_1, m_2 and λ cannot be positive at the same time. Thus, the last two cases correspond to physical cases.

2 Linearized Stability Analysis

In this section, the GDS system (4) is considered with the following boundary

$$\begin{aligned} u(x,y,t) &= u(x + l_1, y + l_2, t), \\ \varphi_i(x,y,t) &= \varphi_i(x + l_1, y + l_2, t), \quad i = 1, 2, \end{aligned} \tag{7}$$

and initial conditions

$$u(x,y,0)=u_0(x,y),\quad u_0(-x,-y)=u_0(x,y),$$
$$\varphi_i(x,y,0)=\varphi_{i0}(x,y),\quad \varphi_{i0}(-x,-y)=\varphi_{i0}(x,y),\quad i=1,2,\quad x,y\in\Omega. \tag{8}$$

The starting point for developing homoclinic-type solutions is to find a suitable fixed point. In this case, for (u,φ_1,φ_2), the fixed point will be chosen as $(a\,\mathrm{e}^{-i\chi|a|^2t},0,0)$, where a is any complex number.

Next, we investigate the stability of fixed point by considering small perturbations of the form,

$$u=a\,\mathrm{e}^{-i\chi|a|^2t}(1-\varepsilon(x,y,t)),\quad \varphi_{1,x}=\varphi_1^{\varepsilon}(x,y,t),\quad \varphi_{2,y}=\varphi_2^{\varepsilon}(x,y,t). \tag{9}$$

Substituting (9) into (4) and keeping linear terms leads one to

$$i\varepsilon_t+\varepsilon_{xx}+\varepsilon_{yy}=\chi|a|^2(\varepsilon+\varepsilon^*)-b(\varphi_1^{\varepsilon}+\varphi_2^{\varepsilon}),$$
$$\varphi^{\varepsilon}_{1,xx}+m_2\varphi^{\varepsilon}_{1,yy}+n\varphi^{\varepsilon}_{2,xx}=-|a|^2(\varepsilon+\varepsilon^*)_{xx},$$
$$\lambda\varphi^{\varepsilon}_{2,xx}+m_1\varphi^{\varepsilon}_{2,yy}+n\varphi^{\varepsilon}_{1,yy}=-|a|^2(\varepsilon+\varepsilon^*)_{yy}. \tag{10}$$

Assuming the following form of solutions for the linearized system we obtain

$$\varepsilon(x,y,t)=A[\mathrm{e}^{i(\mu_nx+\bar{\mu}_ny)+\sigma_nt}]+B[\mathrm{e}^{-i(\mu_nx+\bar{\mu}_ny)+\sigma_nt}],$$
$$\varphi_1^{\varepsilon}(x,y,t)=C[\mathrm{e}^{i(\mu_nx+\bar{\mu}_ny)+\sigma_nt}+\mathrm{e}^{-i(\mu_nx+\bar{\mu}_ny)+\sigma_nt}], \tag{11}$$
$$\varphi_1^{\varepsilon}(x,y,t)=D[\mathrm{e}^{i(\mu_nx+\bar{\mu}_ny)+\sigma_nt}+\mathrm{e}^{-i(\mu_nx+\bar{\mu}_ny)+\sigma_nt}],$$

where the growth rate σ_n of the nth mode is

$$\sigma_n^{\pm}=\pm|\delta\mu_n^2+\bar{\mu}_n{}^2|[\frac{2|a|^2(m_1-1)\chi}{(\delta-1)(\mu_n^2+m_1\bar{\mu}_n{}^2)}-1]^{1/2}. \tag{12}$$

For the proper choice of the parameters we see that the fixed point is hyperbolic.

3 Homoclinic Structure of the GDS System

Homoclinic structure of the GDS system will be studied in this section. One of the two physical cases, i.e. the respective sign of (m_1,m_2,λ) is $(+,+,-)$ with $m_1>m_2$, will be given in detail, since they have similar approach and result.

In order to write the GDS system in Hirota bilinear form we need to make the following restrictions on the parameter values

$$m_2 = m_1 - n, \quad \lambda = 1 - n, \quad b = \frac{\chi(1-\delta m_1)}{\delta - 1}. \tag{13}$$

We have to note that under these conditions imposed on the parameters generalized Davey-Stewartson equations are isomorphic to that of the standard integrable Davey-Stewartson equations.

Since we are interested in a homoclinic orbit of the fixed point, the following substitutions will be made for (4)

$$u = a\,\frac{G}{F}\,\mathrm{e}^{-i\chi a^2 t}, \quad \varphi_1 = \frac{2(1-\delta)}{\chi(1-m_1)}(\log F)_x, \quad \varphi_2 = \frac{2(1-\delta)}{\chi(1-m_1)}(\log F)_y, \tag{14}$$

where G is complex, a and F are real. After straightforward calculations we write the system in Hirota bilinear form

$$\begin{aligned}&(i\mathscr{D}_t + \delta\mathscr{D}_x^2 + \mathscr{D}_y^2 + a^2\chi)G\cdot F = \beta G\cdot F,\\ &(\mathscr{D}_x^2 + m_1\mathscr{D}_y^2)F\cdot F - 2a^2|G|^2 = \frac{-2\beta}{\chi}F\cdot F.\end{aligned} \tag{15}$$

Now, Eqs. (15) are assumed to possess the following solution functions:

$$\begin{aligned}G &= 1 + [b_1\mathrm{e}^{i(p_1x+p_2y)} + b_2\mathrm{e}^{-i(p_1x+p_2y)}]\mathrm{e}^{\Omega t+\gamma} + b_3\mathrm{e}^{2(\Omega t+\gamma)},\\ F &= 1 + b_4[\mathrm{e}^{i(p_1x+p_2y)} + \mathrm{e}^{-i(p_1x+p_2y)}]\mathrm{e}^{\Omega t+\gamma} + b_5\mathrm{e}^{2(\Omega t+\gamma)},\end{aligned} \tag{16}$$

where $p_1, p_2, \Omega, \gamma, b_4, b_5$ are real and b_1, b_2, b_3 are complex. Collecting coefficients leads to the following relations among the constants

$$\begin{aligned}b_1 &= b_2 = -\frac{\delta p_1^2 + p_2^2 + i\Omega}{\delta p_1^2 + p_2^2 - i\Omega}\,b_4, \beta = a^2\chi\\ b_3 &= \frac{(\delta p_1^2 + p_2^2 + i\Omega)^3}{\Omega^2(\delta p_1^2 + p_2^2 - i\Omega)}\,b_4^2, \quad b_5 = [1 + (\frac{\delta p_1^2 + p_2^2}{\Omega})^2]\,b_4^2,\\ \Omega_\pm &= \pm|\delta p_1^2 + p_2^2|[\frac{2a^2(m_1-1)\chi}{(\delta-1)(p_1^2 + m_1p_2^2)} - 1]^{1/2}.\end{aligned} \tag{17}$$

Here the result obtained for Ω coincides with the result in the linearized stability analysis. At this step, p_1 and p_2 will be chosen as

$$p_1 = a\sqrt{\frac{(m_1-1)\chi}{(\delta-1)}}\,\sin\theta, \quad p_2 = a\sqrt{\frac{(m_1-1)\chi}{(\delta-1)m_1}}\,\sin\theta, \tag{18}$$

where ψ_0 is a constant. Then, we get the following solutions for the GDS system (4) as

$$
\begin{aligned}
u_{\pm} &= a\mathrm{e}^{-i\chi a^2 t} \frac{1 + 2b_{\pm} \cos(p_1 x + p_2 y)\mathrm{e}^{\Omega_{\pm} t + \gamma} + b_{\pm}^2 \mathrm{e}^{2(\Omega_{\pm} t + \gamma)} \sec^2 \theta}{1 + 2\cos(p_1 x + p_2 y)\mathrm{e}^{\Omega_{\pm} t + \gamma} + \mathrm{e}^{2(\Omega_{\pm} t + \gamma)} \sec^2 \theta}, \\
\varphi_{1,x}^{\pm} &= \mathrm{e}^{\Omega_{\pm} t + \gamma} \frac{4(\delta - 1)p_1^2}{\chi(1 - m_1)} \frac{2\mathrm{e}^{\Omega_{\pm} t + \gamma} + \cos(p_1 x + p_2 y)(1 + \mathrm{e}^{2(\Omega_{\pm} t + \gamma)} \sec^2 \theta)}{(1 + 2\cos(p_1 x + p_2 y)\mathrm{e}^{\Omega_{\pm} t + \gamma} + \mathrm{e}^{2(\Omega_{\pm} t + \gamma)} \sec^2 \theta)^2}, \\
\varphi_{2,y}^{\pm} &= \mathrm{e}^{\Omega_{\pm} t + \gamma} \frac{4(\delta - 1)p_2^2}{\chi(1 - m_1)} \frac{2\mathrm{e}^{\Omega_{\pm} t + \gamma} + \cos(p_1 x + p_2 y)(1 + \mathrm{e}^{2(\Omega_{\pm} t + \gamma)} \sec^2 \theta)}{(1 + 2\cos(p_1 x + p_2 y)\mathrm{e}^{\Omega_{\pm} t + \gamma} + \mathrm{e}^{2(\Omega_{\pm} t + \gamma)} \sec^2 \theta)^2},
\end{aligned}
\tag{19}
$$

where $b_{\pm} = \cos(2\theta) \mp \sin(2\theta)$.

4 Conclusions

We will conclude by discussing the homoclinic structure of the GDS system (4). The solutions (19) have the characteristics of a homoclinic orbit having spatially periodic fixed points with periods $p_1 = 2\pi m/L_1$ and $p_2 = 2\pi m/L_2$. The solution $(u_+, \varphi_{1,x}^+, \varphi_{2,y}^+)$ leaves the ring of fixed points $(a\ \mathrm{e}^{-i\chi a^2 t}, 0, 0)$ as $t \to -\infty$ and returns to the ring $(a\ \mathrm{e}^{-i\chi a^2 t} b_-^2, 0, 0)$ as $t \to \infty$. On the other hand, the solution $(u_-, \varphi_{1,x}^-, \varphi_{2,y}^-)$ leaves the ring of fixed points $(a\ \mathrm{e}^{-i\chi a^2 t}, 0, 0)$ as $t \to \infty$ and returns to the ring $(a\ \mathrm{e}^{-i\chi a^2 t} b_+^2, 0, 0)$ as $t \to -\infty$. Moreover, phase shift is seen between the solutions $(u_+, \varphi_{1,x}^+, \varphi_{2,y}^+)$ and $(u_-, \varphi_{1,x}^-, \varphi_{2,y}^-)$. Besides, while $(u_+, \varphi_{1,x}^+, \varphi_{2,y}^+)(x_0, y_0, t)$ forms homoclinic orbit $(u_+, \varphi_{1,x}^+, \varphi_{2,y}^+)(x_0 + 2\pi m/p_1, y_0 + 2\pi m/p_2, t)$ forms homoclinic orbit as well and this result also holds for $(u_-, \varphi_{1,x}^-, \varphi_{2,y}^-)$. Finally, we can say that the solutions $(u_+, \varphi_{1,x}^+, \varphi_{2,y}^+)$ and $(u_-, \varphi_{1,x}^-, \varphi_{2,y}^-)$ form a pair of symmetric homoclinic orbits and all of these orbit pairs construct homoclinic tubes. As an illustration of the dynamical behavior separated by the homoclinic orbits, the amplitude $\rho_+ = |u_+|$ for $\gamma = -10$ and $\rho_- = |u_-|$ for $\gamma = 10$ are shown in Fig. 1.

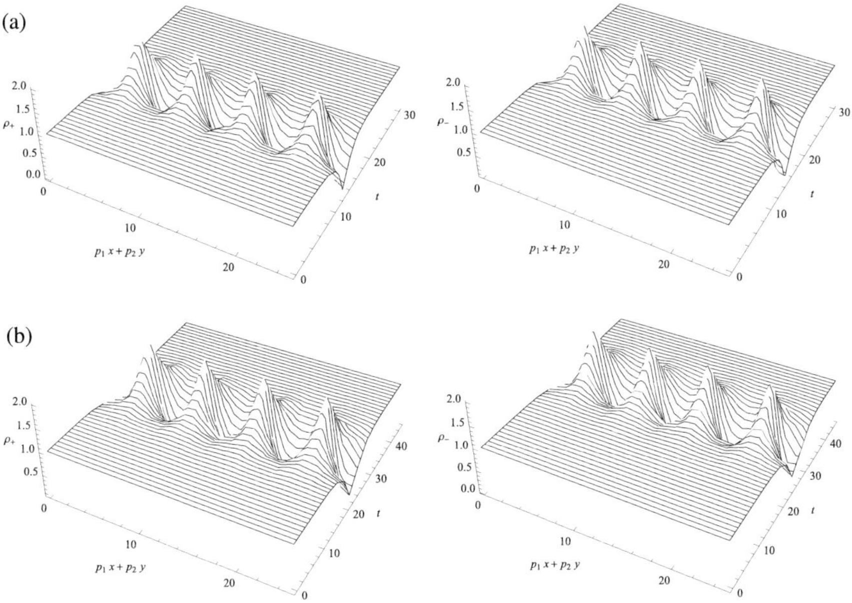

Fig. 1 The solution of (4) for (**a**) $\delta = 0.5$, $\theta = \pi/3$, $\lambda = -0.5$, $\chi = -1$, $a = 1$, $m_1 = 1.75$, $m_2 = 0.25$ and $n = 1.5$ for (**b**) $\delta = 1.5$, $\theta = \pi/3$, $\lambda = -0.1$, $\chi = 1$, $a = 1$, $m_1 = 1.2$, $m_2 = 0.1$ and $n = 1.1$

References

1. Ablowitz, M.J., Herbst, B.M.: On homoclinic structure and numerically induced chaos for the nonlinear Schrodinger equation. SIAM J. Appl. Math. **50**, 339–351 (1990)
2. Babaoglu, C., Erbay, S.: Two-dimensional wave packets in an elastic solid with couple stresses. Int. J. Non-Linear Mech. **39**, 941–949 (2004)
3. Babaoglu, C., Eden, A., Erbay, S.: Global existence and nonexistence results for a generalized Davey-Stewartson system. J. Phys. A Math. Gen. **37**, 11531–11546 (2004)

Numerical Simulations of the Dynamics of Vortex Rossby Waves on a Beta-Plane

L.J. Campbell

Abstract Observational analyses of hurricanes in the tropical atmosphere indicate the existence of spiral rainbands which propagate outwards from the eye and affect the structure and intensity of the hurricane. These disturbances may be described as vortex Rossby waves. It has been suggested that vortex Rossby waves may play a role in the eyewall replacement cycle observed in tropical cyclones in which concentric rings of high-intensity wind develop and propagate in towards the centre of the cyclone. In previous work with Nikitina, we investigated the dynamics of vortex Rossby waves in a cyclonic vortex in a two-dimensional configuration on a beta-plane and derived analytical solutions. In the current work, some results of numerical simulations are presented.

1 Introduction

A tropical cyclone is an axisymmetric vortex of rotating fluid flow that originates over the tropical ocean and is driven by heat transfer from the ocean. An intense tropical cyclone with high sustained wind speed tends to develop an eye, an area of relative calm at the centre of the circulation surrounded by a ring of intense rainbands with maximum tangential wind speed called the eyewall. A large percentage of tropical cyclones with wind speeds of at least $60\,\mathrm{m s^{-1}}$ undergo an eyewall replacement cycle [8] in which an outer ring of rainbands called a secondary eyewall develops outside the original eyewall [25], strengthens and contracts inwards, while the original inner eyewall weakens and is replaced by the secondary eyewall. This process repeats over a time frame of up to 2 days [20].

Vortex Rossby waves are oscillations that occur within cyclonic vortices as a result of the radial gradient of cyclonic vorticity. In the context of hurricanes or tropical cyclones, vortex Rossby waves have been observed as spiral rainbands propagating outwards from the eye. It has been suggested that vortex Rossby wave mean-flow interactions may play a role in the secondary eyewall formation [3, 13, 14], although this role appears to be less significant [1, 10] to that played

L.J. Campbell (✉)
Carleton University, Ottawa, ON, Canada
e-mail: campbell@math.carleton.ca

© Springer International Publishing Switzerland 2016

J. Bélair et al. (eds.), *Mathematical and Computational Approaches in Advancing Modern Science and Engineering*, DOI 10.1007/978-3-319-30379-6_4

by the dominant system-scale processes such as surface friction and boundary layer effects [15, 21].

Vortex Rossby wave mean-flow interactions take place primarily in the vicinity of the critical radius where the mean angular velocity of the vortex is equal to the phase speed of the waves. This suggests that vortex Rossby wave critical layer theory could be helpful in advancing our understanding of the mechanisms by which these waves could contribute to the development of secondary eyewalls in tropical cyclones. The theory for the analogous problem of barotropic planetary Rossby waves in a rectangular domain on a beta-plane is well-developed and that problem has been studied extensively using analytical and numerical methods, e.g. [2, 4–6, 22–24]. Recent analytical investigations with Nikitina [16–18] of the barotropic vortex Rossby wave configuration indicate several qualitative similarities between the two problems.

Nikitina and Campbell [17] presented analytical solutions for a configuration involving a cyclonic vortex with angular velocity $\bar{\Omega}(r)$ on a horizontal plane defined in terms of polar coordinates r and λ. The f-plane approximation was made in this preliminary investigation, i.e., the Coriolis parameter was approximated by a constant. A wave of the form $\cos(k\lambda - \omega t)$ was forced at a fixed radius $r = r_1$ representing the location of the primary eyewall and the linearized barotropic vorticity equation was solved to determine the amplitude of the forced wave as a function of the radial variable r and time t. For a special quadratic profile of $\bar{\Omega}(r)$, exact analytical solutions were obtained in terms of hypergeometric functions for waves with steady amplitude and these solutions were then used to find late-time asymptotic solutions for waves with time-dependent amplitude, first in the outer region away from the critical radius and then in the inner region in the vicinity of the critical radius. The solutions obtained show that the wave amplitude is greatly attenuated at the critical radius as the wave propagates outwards from the eyewall. This can be interpreted as wave absorption by the mean flow. In the limit of infinite time, the time-dependent solution in the outer region approaches the corresponding steady solution, but the inner solution grows with time. These conclusions are consistent with the situation that is attained in the case of forced planetary Rossby waves in a rectangular domain on a β-plane where the Coriolis parameter is considered to be a linear function of latitude.

Nikitina and Campbell [18] extended the investigation of [17] to include the nonlinear terms in the governing equations, as well as the terms arising from the latitudinal gradient of the Coriolis parameter which give the so-called β-effect. The wave amplitude was then considered as an expansion in powers of two small parameters representing nonlinearity and the β-effect with the leading-order terms in the expansion being given by the outer and inner solutions derived in [17]. It was found that nonlinearity gives rise to higher wavenumbers in multiples of the forced wavenumber k, a zero wavenumber component which represents a divergence of momentum flux into the mean flow, and an inward displacement of the instantaneous critical radius. The variation of the Coriolis force gives rise to wave modes with wavenumbers $(k \pm 1)$. In the case where the forced wavenumber $k = 1$, the variation of the Coriolis force thus introduces an additional zero wavenumber component and hence contributes to the evolution of the mean flow.

The model employed by Nikitina and Campbell [17, 18] is highly idealized considering that it is two-dimensional and does not take into account effects such as diabatic heating and boundary layer friction which form the basis of the conventional theory of tropical cyclone secondary eyewall generation. But on the other hand, the simplicity of the model allowed us to examine Rossby wave mean-flow interaction mechanisms in isolation of other effects and the analytical solutions obtained gave us some insight into the temporal evolution of the solution. The main features of the solutions, namely the critical layer absorption of the waves, the development of concentric rings of high wave activity and the changes in the location of these rings with time, are consistent with the hypothesis that vortex Rossby waves contribute to the secondary eyewall replacement cycle. However, the weakly-nonlinear analysis of [18] is valid only for finite time since the higher-order terms in the perturbation expansion for the solution grow with time. Multiple-time-scale asymptotic analyses would be needed to continue the analytical investigation to later time and to higher orders in the expansion parameters. In analogy with the classical rectangular configuration [22, 24], it can be anticipated that there would be wave reflection at the critical radius at late time, possibly leading to eventual wave breaking and instabilities.

The purpose of the current investigation is to use numerical methods to further elucidate the critical layer behaviour of forced vortex Rossby waves propagating outwards in a cyclonic vortex and investigate the effects of the critical layer interaction on the evolution of the mean vortex. The model used is described in Sect. 2 and some preliminary numerical results are presented in Sect. 3.

2 Configuration and Equations

The numerical simulations presented here make use of a two-dimensional model for barotropic vortex Rossby waves in a cyclonic vortex on a β-plane, a horizontal plane on which the Coriolis parameter is approximated by a linear function of latitude. The flow is described in terms of a streamfunction Ψ and is represented by the nondimensional barotropic vorticity equation (see, e.g., [9]). This can be written in polar coordinates r and λ as [17]

$$\nabla^2\Psi_t - \frac{1}{r}\Psi_\lambda\nabla^2\Psi_r + \frac{1}{r}\Psi_r\nabla^2\Psi_\lambda - \frac{\beta}{r}\Psi_\lambda \sin\lambda + \beta\Psi_r\cos\lambda = 0, \tag{1}$$

where the subscripts denote partial differentiation with respect to r and λ. The non-dimensional parameter β is the latitudinal gradient of the Coriolis parameter f. The corresponding dimensional quantity is [9]

$$\beta^* = \frac{2\Omega_{Earth}}{R_{Earth}}\cos\theta_0, \tag{2}$$

where θ_0 is the latitude of the centre of the vortex, $\Omega_{Earth} \approx 7 \times 10^{-5}\,\mathrm{s}^{-1}$ is the angular velocity of the Earth's rotation, and $R_{Earth} \approx 6.3\times10^6$ km is the radius of the Earth. Near the equator where θ_0 is close to zero, β^* is close to $2.2 \times10^{-11}\,\mathrm{m}^{-1}\mathrm{s}^{-1}$. For a typical tropical cyclonic vortex radius L of about $2 - 3 \times 10^5$ m, and a typical tangential wind speed U of about 30-100 ms^{-1}, $\beta = L^2\beta^*/U \sim 10^{-2}$ and can thus be considered as a small parameter in the nondimensional problem. The limit of $\beta \to 0$ gives the f-plane approximation, in which the Coriolis parameter is approximated by a constant.

The vortex wave is represented as a small-amplitude perturbation to the basic flow of the cyclonic vortex. The streamfunction $\bar{\psi}(r)$, the angular velocity $\bar{\Omega}(r)$ and the azimuthal component $\bar{v}(r)$ of the velocity of the basic flow are related by

$$\bar{v}(r) = \bar{\psi}'(r), \quad \bar{v}(r) = r\bar{\Omega}(r), \tag{3}$$

where the prime denotes differentiation with respect to r. The total streamfunction is written as

$$\Psi(r, \lambda, t) = \bar{\psi}(r) + \varepsilon\psi(r, \lambda, t), \tag{4}$$

where the parameter ε is the ratio of the dimensional magnitude of the perturbation to that of the basic flow. Observations of the asymmetric spiral bands that are described as vortex Rossby waves in hurricanes indicate small deviations from circular symmetry. For example, analyses [7, 19] of Doppler wind and Omega dropwindsonde data from Hurricane Gloria (1985) show that within 500 km of the hurricane centre, the asymmetric components of the tangential wind corresponding to azimuthal wavenumbers 1, 2, 3 and 4 are much smaller than the symmetric tangential wind. This would suggest that ε can be considered as a small parameter in the nondimensional problem and thus justifies the use of linear [14, 17] and weakly-nonlinear analyses [18] as a means to provide insight into a fully nonlinear configuration that can only be examined numerically.

Substituting (4) into (1) gives a nonlinear equation for the perturbation

$$\begin{aligned}\left(\frac{\partial}{\partial t} + \frac{\bar{v}}{r}\frac{\partial}{\partial \lambda}\right)\nabla^2\psi - \frac{1}{r}\psi_\lambda \frac{d}{dr}\left(\bar{v}' + \frac{\bar{v}}{r}\right)\\ -\frac{\beta}{r}\psi_\lambda \sin\lambda + \beta\psi_r\cos\lambda + \frac{\beta}{\varepsilon}\bar{v}\cos\lambda = -\frac{\varepsilon}{r}(\psi_r\nabla^2\psi_\lambda - \psi_\lambda\nabla^2\psi_r).\end{aligned} \tag{5}$$

This equation is examined in the annular region $r_1 \le r < \infty,\ 0 \le \lambda < 2\pi$ shown in Fig. 1 for $t \ge 0$. The waves are forced at $r = r_1$ by a boundary condition of the form $\psi(r_1, \lambda, t) = \cos(k\lambda - \omega t)$, where k is the azimuthal wave number and ω is the circular frequency of the wave.

The mean angular velocity profile is taken to be either $\bar{\Omega}(r) = \Omega_0 e^{-\alpha r^2}$, as suggested by Martinez et al. [11, 12] as a reasonable representation of typical flows

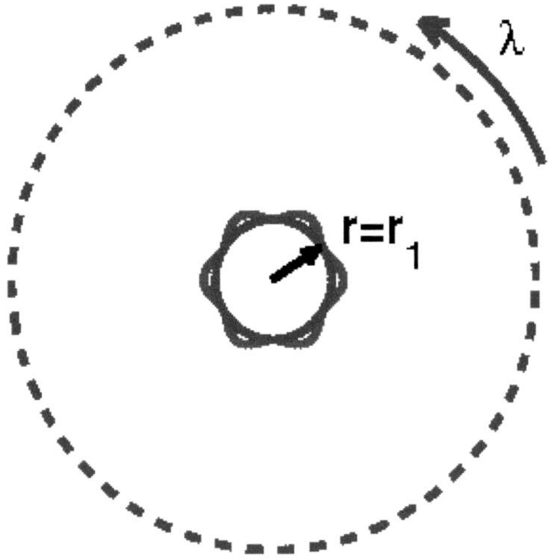

Fig. 1 Configuration: the waves are forced by a sinusoidal boundary condition at $r = r_1$ and propagate outwards [17]

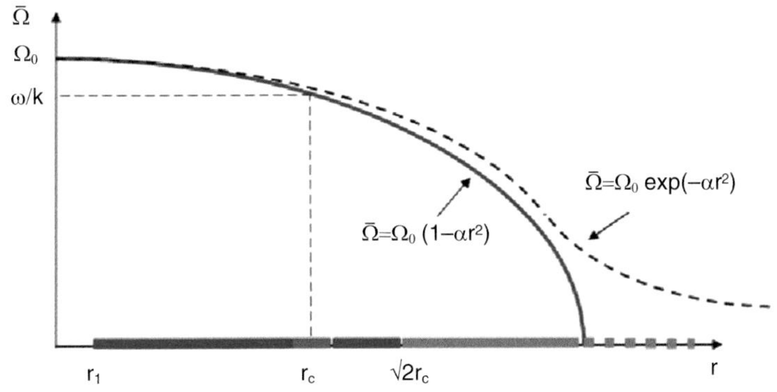

Fig. 2 The mean flow angular velocity profiles used in the numerical simulations and the intervals in which the analytical solutions obtained in [17, 18] are valid

observed in tropical cyclones, or $\bar{\Omega}(r) = \Omega_0(1 - \alpha r^2)$, as suggested by Brunet and Montgomery [3] as an approximation for the exponential profile of [11, 12]. These functions are shown in Fig. 2. The positive constants α and Ω_0 are chosen so that there is a critical radius within the domain at some value $r = r_c$ where $\bar{\Omega}(r_c) = \omega/k$, as shown in Fig. 2. The numerical method used for the solution of (5) involves a pseudo-spectral approximation and is based on that described in [4].

3 Results

The results presented here are for the quadratic profile shown in Fig. 2. The waves are forced at $r = r_1 = 2$ with a frequency of $\omega = 1$ and a wavenumber of $k = 2$ or $k = 1$. The nonlinear parameter is set to $\varepsilon = 0.05$ and the gradient of planetary vorticity is $\beta = 0$ or $\beta = 0.05$. In each case, the nondimensional parameters α and Ω_0 are chosen so that the critical radius is at $r = r_c = 12$.

Three configurations are examined here: $\beta = 0$, $k = 2$ (f-plane); $\beta = 0.05$, $k = 2$ (β-plane); $\beta = 0.05$, $k = 1$ (β-plane). In each case, contour plots of the perturbation streamfunction and the perturbation vorticity and a graph of the Fourier or wavenumber spectrum of the perturbation streamfunction are presented at nondimensional time $t = 100$. The perturbation streamfunction is written as

$$\psi(r, \lambda, t) = \sum_{\kappa=-\infty}^{\infty} \phi(r, \kappa, t) e^{i\kappa\lambda} \tag{6}$$

and $|\phi(r, \kappa, t)|$ is shown as a function of wavenumber κ at $r = r_c = 12$ and $t = 100$. The zero wavenumber component in the sum (6) gives the change in the total mean flow. The total mean angular velocity at location r and time t is $\Omega(r, t) = \bar{\Omega}(r) + \Delta\Omega(r, t)$, where the change in angular velocity is $\Delta\Omega(r, t) = \varepsilon\phi_r(r, 0, t)/r$. A plot of $\Delta\Omega(r, t)$ at $r = r_c = 12$ and $t = 100$ is presented for each configuration. From the asymptotic analysis of [18], it is expected that, at least within the early time frame in which the analysis is valid, the total mean angular velocity decreases as a result of nonlinearity and, in the case where $k = 1$, the β-effect also affects the mean angular velocity.

Figures 3, 4, 5, and 6 show results obtained with $\beta = 0$ and $k = 2$. In each of the contour plots shown, there are 20 equally-spaced levels. The analytical results of [17, 18] tell us that the steady term in the solution varies like $r^{-k} = r^{-2}$ between the inner boundary and the critical radius. This attenuation of the waves and their absorption by the mean flow at the critical radius is seen in the contour plot of the wave streamfunction in Fig. 3a. A ring of high vorticity is seen in the vicinity of the critical radius in Fig. 3b. Figure 4 shows the Fourier spectrum of the perturbation streamfunction at the critical radius and time $t = 100$. The black circles indicate the forced wave modes and the unfilled circles indicate the modes generated by the nonlinear effects. There is a zero wavenumber component of the solution ($\kappa = 0$) as well as higher harmonics ($\kappa = \pm 4, \pm 6, \ldots$), which are nonzero but of small amplitude. According to Fig. 5, the mean flow decreases initially and then starts to increase until the change becomes positive indicating a decrease in the extent of wave absorption at late time and an evolution towards a reflecting state. This is an analogous result to that obtained in the classical case of forced planetary Rossby waves in a rectangular domain on a β-plane [2, 4].

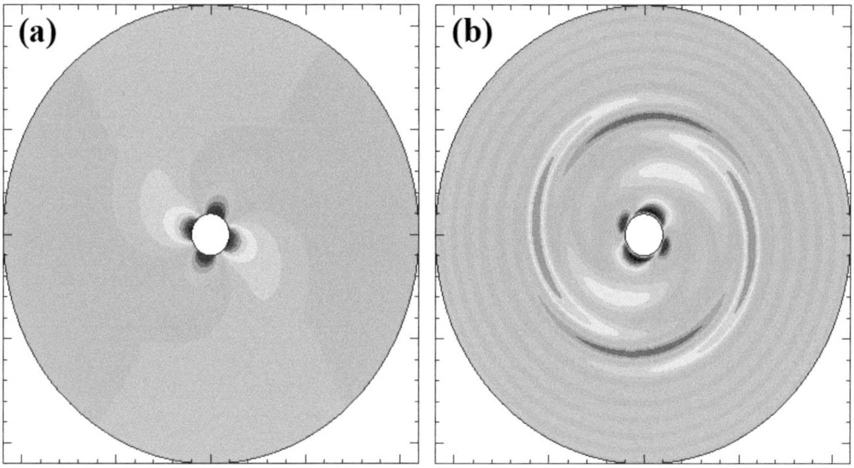

Fig. 3 Nonlinear numerical simulations on an f-plane with $k = 2$: (**a**) Wave streamfunction; (**b**) Wave vorticity

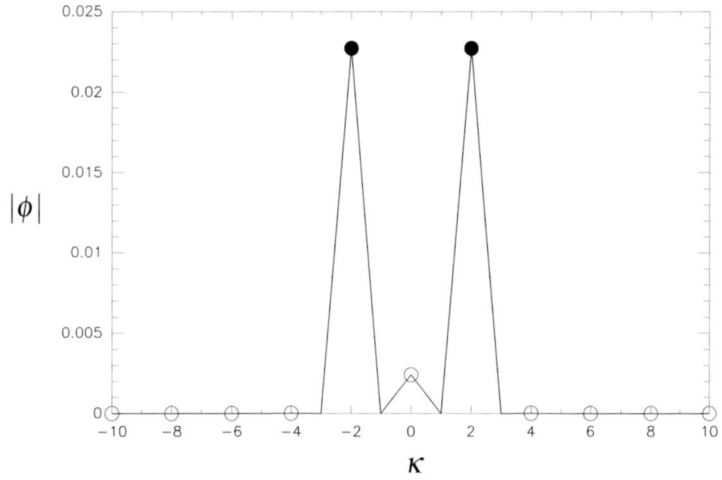

Fig. 4 Nonlinear numerical simulations on an f-plane with $k = 2$: Fourier spectrum of the streamfunction amplitude at $r = r_c = 12$

Figures 6, 7, and 8 show results obtained with $\beta = 0.05$ and $k = 2$. With the addition of the β-effect, Fig. 6a, b indicate that there is a greater degree of transmission beyond the critical radius than is seen in the corresponding plots for the f-plane configuration. The graph of the Fourier spectrum given in Fig. 7 shows that there are modes corresponding to $\kappa = \pm 1, \pm 3, \ldots$ (indicated by

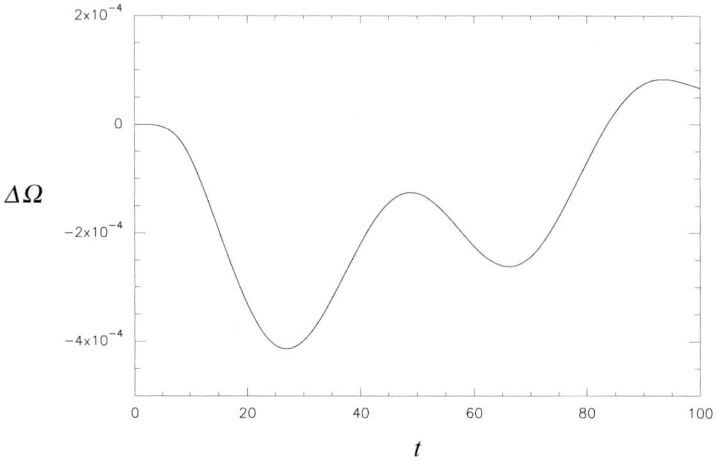

Fig. 5 Nonlinear numerical simulations on an f-plane with $k = 2$: change in the mean flow as a function of time at $r = r_c = 12$

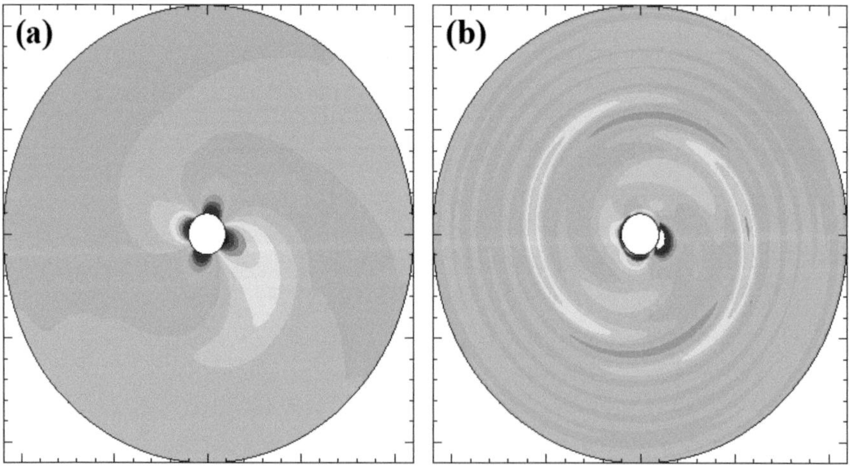

Fig. 6 Nonlinear numerical simulations on a β-plane with $k = 2$: (**a**) Wave streamfunction; (**b**) Wave vorticity

asterisks), which are generated by the β-effect, in addition to those corresponding to $\kappa = 0, \pm 2, \pm 4, \ldots$. The presence of these additional modes results in a larger wave amplitude at the critical radius than that in the f-plane case shown in Fig. 4. Figure 8 indicates again a decrease in the extent of wave absorption at late time.

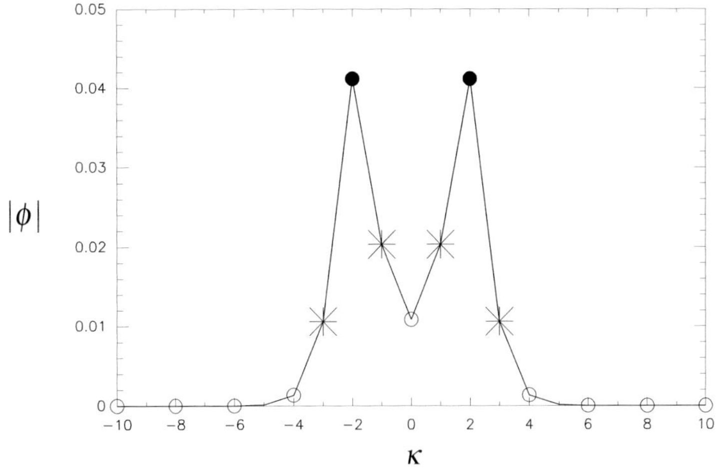

Fig. 7 Nonlinear numerical simulations on an β-plane with $k = 2$: Fourier spectrum of the streamfunction amplitude at $r = r_c = 12$

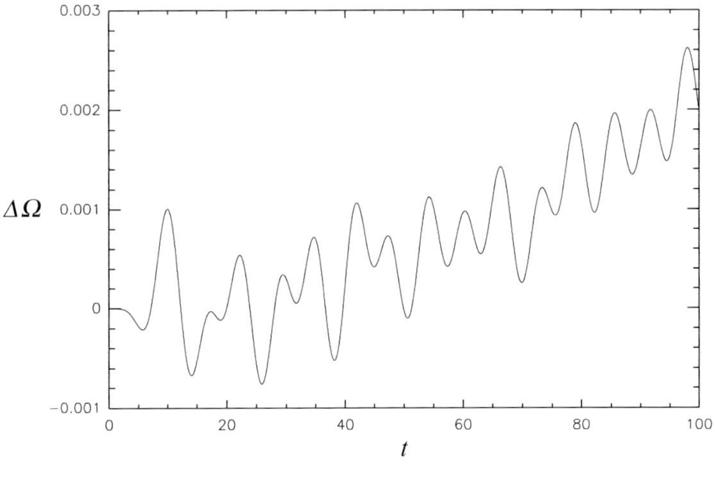

Fig. 8 Nonlinear numerical simulations on a β-plane with $k = 2$: change inthe mean flow as a function of time at $r = r_c = 12$

Figures 9, 10, and 11 show results obtained with $\beta = 0.05$ and $k = 1$. With this smaller value of $k = 1$, the wave amplitude decreases less rapidly with radial distance from the centre followed by a more rapid attenuation at the critical radius. This is seen in Fig. 9a, b. From Fig. 10 it is also seen that the β-effect generates modes corresponding to $\kappa = \pm(k-1), \pm(k+1) = 0, \pm 2, \ldots$, which coincide

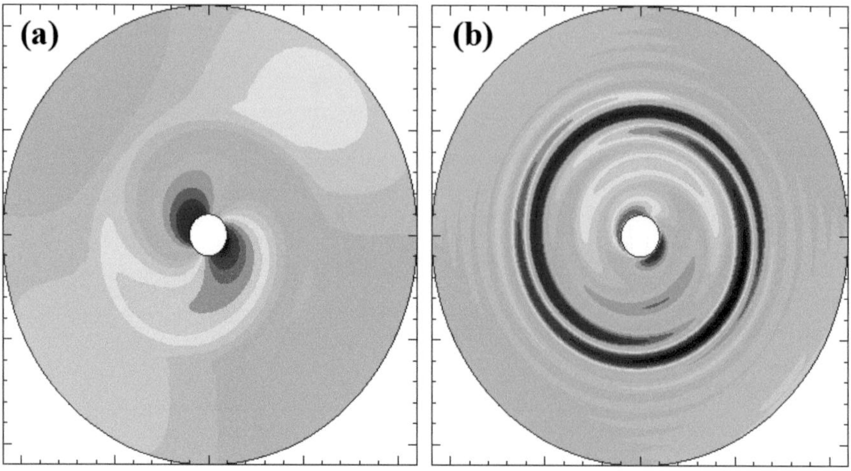

Fig. 9 Nonlinear numerical simulations on a β-plane with $k = 1$: (**a**) Wave streamfunction; (**b**) Wave vorticity

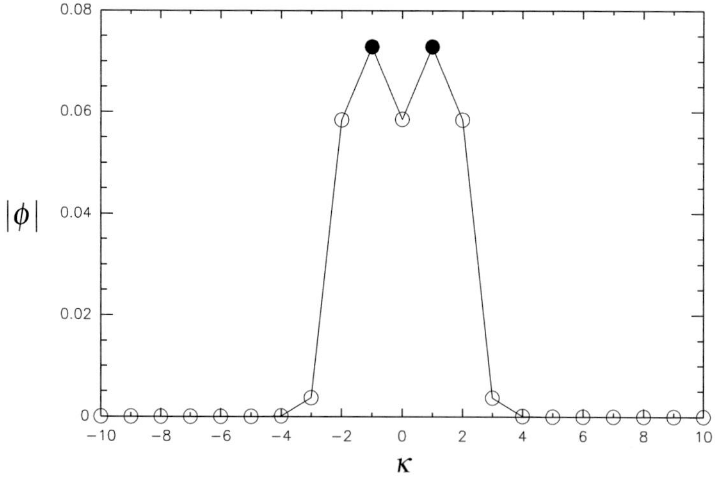

Fig. 10 Nonlinear numerical simulations on an β-plane with $k = 1$: Fourier spectrum of the streamfunction amplitude at $r = r_c = 12$

with the modes generated by the effect of nonlinearity to produce terms with greater magnitude than those shown in Fig. 7. Figure 11 shows that the mean flow evolution takes the form of oscillations with an amplitude that increases slowly with time.

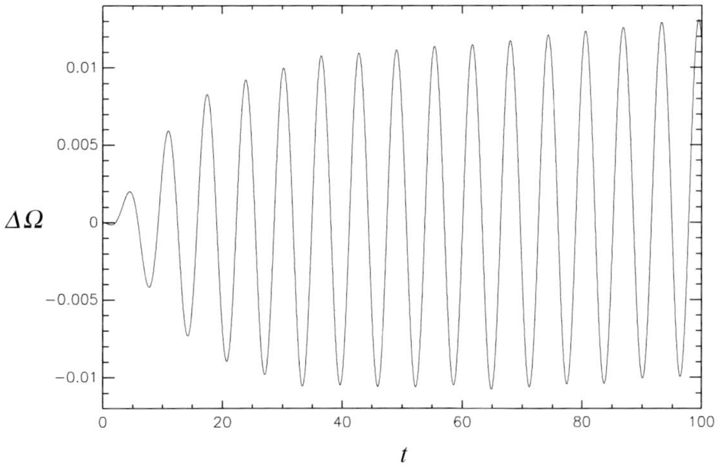

Fig. 11 Nonlinear numerical simulations on a β-plane with $k = 1$: change in the mean flow as a function of time at $r = r_c = 12$

4 Concluding Remarks

Numerical methods were used to investigate vortex Rossby wave mean-flow interactions in a barotropic model of a tropical cyclone. Consistent with the analytical solutions presented in [17, 18], it was found that nonlinearity gives rise to higher wavenumbers, momentum flux divergence into the mean flow and an (initial) inward displacement of the critical radius. The ring of high vorticity around the critical radius resembles a developing secondary wall. The variation of the Coriolis force gives rise to wave modes with wavenumbers $\pm(k \pm 1)$ and affects the mean flow if $k = 1$. In each case, the dominant contributions to the vortex perturbation come from the zero wavenumber and low wavenumber components, at least within the time frame shown. This is consistent with observations of Hurricane Gloria (1998) [7] analyzed and discussed by Shapiro and Montgomery [19].

These numerical results are preliminary; further numerical experimentation and analyses are needed. Ultimately, vertical variation, diabatic heating and other effects need to be included for a more realistic representation, but the observations from the results obtained with this simplified barotropic configuration can be used as a starting point for further studies.

Acknowledgements The author would like to thank Dr. L. Nikitina for drawing the schematic diagrams shown in Figs. 1 and 2 and the anonymous reviewer for helpful comments.

References

1. Abarca, S.F., Montgomery, M.T.: Essential dynamics of secondary eyewall formation. J. Atmos. Sci. **70**, 3216–3230 (2013)
2. Béland, M.: Numerical study of the nonlinear Rossby wave critical level development in a barotropic zonal flow. J. Atmos. Sci. **33**, 2066–2078 (1976)
3. Brunet, G., Montgomery, M.T.: Vortex Rossby waves on smooth circular vortices: part 1. Theory. Dyn. Atmos. Oceans **35**, 135–177 (2002)
4. Campbell, L.J.: Wave mean-flow interactions in a forced Rossby wave packet. Stud. Appl. Math. **112**, 39–85 (2004)
5. Campbell, L.J., Maslowe, S.A.: Forced Rossby wave packets in barotropic shear flows with critical layers. Dyn. Atmos. Oceans **28**, 9–37 (1998)
6. Dickinson, R.E.: Development of a Rossby wave critical level. J. Atmos. Sci. **27**, 627–633 (1970)
7. Franklin, J.L., Lord, S.J., Feuer, S.E., Marks, Jr., F.D.: The kinematic structure of Hurricane Gloria (1985) determined from nested analyses of dropwindsonde and Doppler radar data. Mon. Weather Rev. **121**, 2433–2451 (1993)
8. Hawkins, J.D., Helveston, M.: Tropical cyclone multiple eyewall characteristics. In: 28th Conference of Hurricanes and Tropical Meteorology, Orlando, 28 April – 2 May. American Meteorological Society (2008)
9. Holton, J.R.: An Introduction to Dynamic Meteorology, 3rd edn. Academic, San Diego (1992)
10. Huang, Y.-H., Montgomery, M.T., Wu, C.-C.: Concentric eyewall formation in Typhoon Sinlaku (2008). Part II: axisymmetric dynamical processes. J. Atmos. Sci. **69**, 662–674, (2012)
11. Martinez, Y., Brunet, G., Yau, M.K.: On the dynamics of two-dimensional hurricane-like vortex symmetrization. J. Atmos. Sci. **67**, 3559–3580 (2010)
12. Martinez, Y., Brunet, G., Yau, M.K.: On the dynamics of two-dimensional hurricane-like concentric rings vortex formation. J. Atmos. Sci. **67**, 3253–3268 (2010)
13. Martinez, Y., Brunet, G., Yau, M.K., Wang, X.: On the dynamics of concentric eyewall genesis: space-time empirical normal modes diagnosis. J. Atmos. Sci. **68**, 457–476 (2011)
14. Montgomery, M.T., Kallenbach, R.J.: A theory of vortex Rossby-waves and its application to spiral bands and intensity changes in hurricanes. Q. J. R. Meteorol. Soc. **123**, 435–465 (1997)
15. Montgomery, M.T., Smith, R.K.: Paradigms for tropical cyclone intensification. Aust. Meteorol. Oceanograph. J. **64**, 37–66 (2014)
16. Nikitina, L.: Dynamics of vortex Rossby waves in tropical cyclones. Ph.D. thesis, Carleton University (2013)
17. Nikitina, L.V., Campbell, L.J.: Dynamics of vortex Rossby waves in tropical cyclones, part 1: linear steady-state exact solutions on an f-plane. Stud. Appl. Math. **135**, 377–421 (2015)
18. Nikitina, L.V., Campbell, L.J.: Dynamics of vortex Rossby waves in tropical cyclones, part 2: nonlinear time-dependent asymptotic analysis on a β-plane. Stud. Appl. Math. **135**, 422–446 (2015)
19. Shapiro, L.J., Montgomery, M.T.: A three-dimensional balance theory for rapidly rotating vortices. J. Atmos. Sci. **50**, 3322–3335 (1993)
20. Sitkowski, M., Kossin, J.P., Rozoff, C.M.: Intensity and structure changes during hurricane eyewall replacement cycles. Mon. Weather Rev. **139**, 3829–3847 (2011)
21. Smith, R.K., Montgomery, M.T., Nguyen, V.S.: Tropical cyclone spin up revisited. Q. J. R. Meteorol. Soc. **135**, 1321–1335 (2009)
22. Stewartson, K.: The evolution of the critical layer of a Rossby wave. Geophys. Astro. Fluid Dyn. **9**, 185–200 (1978)
23. Warn, T., Warn, H.: On the development of a Rossby wave critical level. J. Atmos. Sci. **33**, 2021–2024 (1976)
24. Warn, T., Warn, H.: The evolution of a nonlinear critical level. Stud. Appl. Math. **59**, 37–71 (1978)
25. Willoughby, H.E., Clos, J.A., Shoreibah, M.G.: Concentric eyewalls, secondary wind maxima, and the evolution of the hurricane vortex. J. Atmos. Sci. **39**, 395–411 (1982)

On the Problem of Similar Motions of a Chain of Coupled Heavy Rigid Bodies

Dmitriy Chebanov

Abstract This chapter contributes to the study of dynamic properties of a chain of n heavy Hess tops coupled by ideal spherical joints by constructing a new class of particular solutions of the equations of chain's motion. The new class describes the chain's motions under which all bodies move similar to each other. We establish conditions for existence of the new solutions and show how the equations of motion can be reduced to quadratures in the case when these conditions are fulfilled.

1 Introduction

One of the classic directions of investigation in the theory of motion of a system of several coupled rigid bodies is concerned with particular cases of integrability of the equations of system's motion [2, 9, 16]. In comparison with the Euler and Poisson equations describing the dynamics of a single rigid body about a fixed point, the analytical study of mathematical models for a system of hinge-connected rigid bodies is a much more complicated problem; as a consequence, there are very few known cases of integrability in the dynamics of many-body systems [1, 15].

In this chapter we study dynamic properties of a mechanical system consisting of an arbitrary number of heavy rigid bodies coupled by ideal spherical joints so that the system constitutes a chain of rigid bodies. The chain rotates about a fixed point. Under assumption that, for every chain's link, the axis connecting its attachment points passes through its center of mass, we construct a class of particular solutions of the equations of chain's motion that describes a rotational motion of the chain with the following property: while the chain is in motion, its skeleton composed of the segments of the abovementioned axes bounded by the corresponding attachment points belongs to a vertical plane rotating about the vertical line while the skeleton's segments change their position with respect to the plane identically in time, i.e., all the bodies move similarly. The similar motions were first discovered for a chain of n heavy Lagrange tops [14, 18]. Some properties of these motions and their generalizations can be found in the works [4, 6, 7, 14]. Recently, paper [8] suggested

D. Chebanov (✉)
City University of New York – LaGCC, Long Island City, NY, USA
e-mail: dchebanov@lagcc.cuny.edu

© Springer International Publishing Switzerland 2016

J. Bélair et al. (eds.), *Mathematical and Computational Approaches in Advancing Modern Science and Engineering*, DOI 10.1007/978-3-319-30379-6_5

a generic approach for establishing the existence of such motions under no initial assumption regarding the mass distribution in the bodies.

We present a new particular solution of the problem described above that is analogous to the solution of the Euler and Poisson equations found by Hess [10]. We describe the structure of the new solution, establish conditions for its existence, prove compatibility of these conditions, and, in the case when the conditions are fulfilled, reduce the equations of motion to quadratures. Then, we explore some aspects of the geometry of the chain's motion described by the new solution.

2 Description of the Model

In this section, we recall the terminology and notations of [8] complementing them with the ones needed for the purpose of this contribution.

We consider a mechanical system S consisting of n heavy rigid bodies $B_1, B_2, \ldots, B_n$. The bodies B_i and $B_{i+1} (i = 1, 2, \ldots, n-1)$ are coupled by an ideal spherical joint at a common point O_{i+1} so that the system S constitutes a chain of rigid bodies. One of the chain's end links, B_1, is absolutely fixed at one of its points $O_1 (\neq O_2)$. It is assumed that the line $l_i (i = 1, 2, \ldots, n-1)$ connecting the attachment points O_i and O_{i+1} of the body B_i passes through its center of mass C_i. For the body B_n, l_n denotes the line passing through the body's attachment point O_n and its center of mass C_n. If the position of the points C_i and O_{i+1} relative to O_i are determined by the vectors $\mathbf{c}_i$ and $\mathbf{s}_i$, respectively, then, due to the above assumptions, $\mathbf{c}_i = c_i \mathbf{e}_i$ and $\mathbf{s}_i = s_i \mathbf{e}_i$, where $\mathbf{e}_i$ is a unit vector directed along l_i.

While studying the motion of system S, it is convenient to deal with mass characteristics of so-called augmented body B_i^* instead of mass characteristics of B_i. Let us denote the mass of B_i by m_i. By definition, the body B_i^* consists of the body B_i and the point mass $m_i^* = \sum\limits_{j=i+1}^{n} m_j$ which is rigidly attached to B_i at point O_{i+1}. Since it has been assumed that the points O_i, C_i, and O_{i+1} lie on $l_i (i = 1, 2, \ldots, n-1)$, this definition implies that the mass center C_i^* of body B_i^* lies on l_i as well. It also follows from the definition of an augmented body that the absolute angular velocities of bodies B_i and B_i^* are equal, while the first-order mass momentum $\mathbf{a}_i$ and the tensor of inertia $\mathbf{I}_i^*$ of B_i^* at O_i can be expressed as

$$\mathbf{a}_i = (m_i + m_i^*)\mathbf{c}_i^* = m_i \mathbf{c}_i + m_i^* \mathbf{s}_i = a_i \mathbf{e}_i, \tag{1}$$

$$\mathbf{I}_i^* = \mathbf{I}_i + m_i^* \left(\mathbf{s}_i^2 \boldsymbol{\delta} - \mathbf{s}_i \otimes \mathbf{s}_i \right) = \mathbf{I}_i + m_i^* s_i^2 \left(\boldsymbol{\delta} - \mathbf{e}_i \otimes \mathbf{e}_i \right), \tag{2}$$

where $\mathbf{c}_i^* = \mathbf{O}_i \mathbf{C}_i^*$, $a_i = m_i c_i + m_i^* s_i$, $\mathbf{I}_i$ is the tensor of inertia of B_i at O_i, $\boldsymbol{\delta}$ is the 3×3 identity matrix, and $\otimes$ denotes a dyadic product of two tensors. Since the body B_n has no descendant chain links connected to it, we have $m_n^* = 0$ implying that the bodies B_n and B_n^* coincide. Therefore, $a_n = m_n c_n$ and $\mathbf{I}_n^* = \mathbf{I}_n$.

The vector equations of motion for the system S under consideration can be written as follows [5, 13]:

$$(\mathbf{I}_i^* \boldsymbol{\omega}_i)^{\cdot} + a_i \sum_{j=1}^{i-1} s_j \left[\mathbf{e}_i \times \left(\boldsymbol{\omega}_j \times \mathbf{e}_j\right)^{\cdot}\right] + s_i \sum_{j=i+1}^{n} a_j \left[\mathbf{e}_i \times \left(\boldsymbol{\omega}_j \times \mathbf{e}_j\right)^{\cdot}\right] + a_i g \mathbf{e}_i \times \boldsymbol{\nu} = \mathbf{0}, \tag{3}$$

where $\boldsymbol{\omega}_i$ is the absolute angular velocity of body B_i, $\boldsymbol{\nu}$ is the upward vertical unit vector, and the dot denotes absolute derivative.

Let $\Sigma = \{O_1, \boldsymbol{\nu}_1 \boldsymbol{\nu}_2 \boldsymbol{\nu}_3\}$ be a Cartesian reference frame whose vectors are fixed in inertial space so that $\boldsymbol{\nu}_3 = \boldsymbol{\nu}$. Let also $\widetilde{\Sigma}_i = \{O_i, \tilde{\mathbf{e}}_1^{(i)} \tilde{\mathbf{e}}_2^{(i)} \tilde{\mathbf{e}}_3^{(i)}\}$ be an orthonormal moving frame whose axes are the principal axes of B_i at O_i. Then, the inertia matrix $\widetilde{\mathbf{I}}_i$ of B_i in this frame takes the form $\widetilde{\mathbf{I}}_i = \operatorname{diag}\left(\widetilde{I}_i^x, \widetilde{I}_i^y, \widetilde{I}_i^z\right)$, where $\widetilde{I}_i^x, \widetilde{I}_i^y$, and $\widetilde{I}_i^z$ are the principal moments of inertia of B_i at O_i.

In the rest of the chapter, we study a case when, for every i, the center of mass of B_i belongs to one of its principal planes. In this case, without loss of generality, we assume that C_i is located in the plane formed by the vectors $\tilde{\mathbf{e}}_1^{(i)}$ and $\tilde{\mathbf{e}}_3^{(i)}$, implying that $\mathbf{e}_i = (e_i^x, 0, e_i^z)$ in the principal axes frame, and choose $\Sigma_i = \{O_i, \mathbf{e}_1^{(i)} \mathbf{e}_2^{(i)} \mathbf{e}_3^{(i)}\}$ to be a so-called "special" frame [11, 12] whose base vectors are given by

$$\mathbf{e}_3^{(i)} = \mathbf{e}_i, \qquad \mathbf{e}_2^{(i)} = \frac{\tilde{\mathbf{e}}_3^{(i)} \times \mathbf{e}_i}{\left|\mathbf{e}_i \times \tilde{\mathbf{e}}_3^{(i)}\right|} = \tilde{\mathbf{e}}_2^{(i)}, \qquad \mathbf{e}_1^{(i)} = \mathbf{e}_2^{(i)} \times \mathbf{e}_3^{(i)} = \frac{\mathbf{e}_i \times \left(\mathbf{e}_i \times \tilde{\mathbf{e}}_3^{(i)}\right)}{\left|\mathbf{e}_i \times \tilde{\mathbf{e}}_3^{(i)}\right|},$$

i.e., the special frame Σ_i can be obtained from the principal axes frame $\widetilde{\Sigma}_i$ by rotating the latter about its second axis. Hence, $\mathbf{e}_2^{(i)}$ is a unit vector of a principal axis of B_i at O_i and the rotation matrix $\mathbf{R}_i$ describing the above frame transformation is given by

$$\mathbf{R}_i = \begin{pmatrix} e_i^z & 0 & -e_i^x \\ 0 & 1 & 0 \\ e_i^x & 0 & e_i^z \end{pmatrix}.$$

Then, in the special frame, $\mathbf{e}_i = (0, 0, 1)$ and the inertia tensor $\mathbf{I}_i$ can be represented as

$$\mathbf{I}_i = \mathbf{R}_i \widetilde{\mathbf{I}}_i \mathbf{R}_i^T = \begin{pmatrix} \widetilde{I}_i^x \left(e_i^z\right)^2 + \widetilde{I}_i^z \left(e_i^x\right)^2 & 0 & \left(\widetilde{I}_i^x - \widetilde{I}_i^z\right) e_i^x e_i^z \\ 0 & \widetilde{I}_i^y & 0 \\ \left(\widetilde{I}_i^x - \widetilde{I}_i^z\right) e_i^x e_i^z & 0 & \widetilde{I}_i^x \left(e_i^x\right)^2 + \widetilde{I}_i^z \left(e_i^z\right)^2 \end{pmatrix} = \begin{pmatrix} I_i^x & 0 & I_i^{xz} \\ 0 & I_i^y & 0 \\ I_i^{xz} & 0 & I_i^z \end{pmatrix}.$$

In order to determine the position of body B_i with respect to the reference frame, we use Euler angles θ_i, ψ_i, and φ_i, where $\theta_i (0 \leq \theta_i < \pi)$ is the angle of nutation,

$\psi_i (0 \le \psi_i < 2\pi)$ is the angle of precession, and $\varphi_i (0 \le \varphi_i < 2\pi)$ is the angle of proper rotation. The motion of system S is a superposition of the motion of its skeleton $O_1 O_2 \ldots O_n C_n$, that is composed of the segments of axes l_i bounded by the corresponding attachment points, and the pure rotation of each body about l_i. The former motion is completely determined by all angles θ_i, ψ_i, while the rotation of B_i about l_i is described by the angle φ_i.

3 Formulation of the Problem

We will say that system S performs similar motions if it moves so that its skeleton belongs to a vertical plane Π rotating about the vertical axis defined by $\boldsymbol{\nu}$ in accordance with a non-stationary law $\psi(t)$ while the skeleton's segments change their position with respect to Π identically in time, i.e., all the bodies move similarly. For such motions, it is fulfilled that

$$\theta_i = \theta(t), \qquad \psi_i = \psi(t) + \delta_i \pi, \tag{4}$$

where $\theta(t)$ and $\psi(t)$ are functions of time to be determined, $\delta_i \in \{-1, 0, 1\}$, and $i = 1, 2, \ldots, n$.

The similar motions (4) were first discovered in [14, 18] for a chain of n heavy Lagrange tops. Below we follow the strategy for analyzing the problem on similar motions suggested in [8].

Projecting equations (3) onto the axes of the corresponding body-fixed frames and substituting (4) into the equations so obtained yields the following overdetermined system of $3n$ second-order differential equations with respect to $n + 2$ unknowns $\theta(t)$, $\psi(t)$, and $\varphi_i(t)$:

$$\begin{gathered} J_i^x \dot{p}_i + I_i^{xz} \dot{r}_i + \left(I_i^z - J_i^y\right) q_i r_i + I_i^{xz} p_i q_i - \left[a_i g + H_i (\cos\theta)^{\cdot\cdot}\right] \sin\theta \cos\varphi_i = 0, \\ J_i^y \dot{q}_i + \left(J_i^x - I_i^z\right) p_i r_i + I_i^{xz} \left(r_i^2 - p_i^2\right) + \left[a_i g + H_i (\cos\theta)^{\cdot\cdot}\right] \sin\theta \sin\varphi_i = 0, \\ I_i^{xz} \dot{p}_i + I_i^z \dot{r}_i + \left(I_i^y - I_i^x\right) p_i q_i - I_i^{xz} q_i r_i = 0, \end{gathered} \tag{5}$$

where $p_i = \dot{\theta} \cos\varphi_i + \dot{\psi} \sin\theta \sin\varphi_i$, $q_i = -\dot{\theta} \sin\varphi_i + \dot{\psi} \sin\theta \cos\varphi_i$, $r_i = \dot{\varphi}_i + \dot{\psi} \cos\theta$,

$$J_i^\alpha = I_i^\alpha + m_i^* s_i^2 + L_i \quad (\alpha = x, y), \tag{6}$$

$$L_i = a_i \sum_{j=1}^{i-1} s_j \varepsilon_{ij} + s_i \sum_{j=i+1}^{n} a_j \varepsilon_{ij}, \qquad H_i = a_i \sum_{j=1}^{i-1} s_j \kappa_{ij} + s_i \sum_{j=i+1}^{n} a_j \kappa_{ij}, \tag{7}$$

ε_{ij} is equal to 1, if $\psi_i = \psi_j$, and -1, otherwise, and $\kappa_{ij} = 1 - \varepsilon_{ij}$.

In what follows, we construct a new class of particular solutions of the system of equations (5).

4 Structure of the Solution

Let us assume that each body B_i is a Hess top [10], i.e. a rigid body whose center of mass lies on the perpendicular to the circular cross-section of the gyration ellipsoid. In the principal axes frame $\widetilde{\Sigma}_i$, the condition on the mass parameters of a Hess top B_i is given by

$$\left(e_i^x\right)^2 \widetilde{I}_i^x \left(\widetilde{I}_i^y - \widetilde{I}_i^z\right) = \left(e_i^z\right)^2 \widetilde{I}_i^z \left(\widetilde{I}_i^x - \widetilde{I}_i^y\right). \tag{8}$$

When expressed in Σ_i, it can be written as

$$\left(I_i^{xz}\right)^2 = I_i^z \left(I_i^x - I_i^y\right). \tag{9}$$

Below we construct a solution of (5) that, as the Hess solution [10] of the Euler and Poisson equations, is characterized by the invariant relation

$$I_i^{xz} p_i + I_i^z r_i = 0. \tag{10}$$

By virtue of (9), (10), equations (5) reduce to the system of $2n$ equations

$$J_i^y \left[\dot{p}_i + I_i^{xz} \left(I_i^z\right)^{-1} p_i q_i\right] - \left[a_i g + H_i \left(\cos\theta\right)^{\cdot\cdot}\right] \sin\theta \cos\varphi_i = 0, \tag{11}$$

$$J_i^y \left[\dot{q}_i - I_i^{xz} \left(I_i^z\right)^{-1} p_i^2\right] + \left[a_i g + H_i \left(\cos\theta\right)^{\cdot\cdot}\right] \sin\theta \sin\varphi_i = 0, \tag{12}$$

which possesses the following first integrals

$$J_i^y \left(\dot{\theta}^2 + \dot{\psi}^2 \sin^2\theta\right) + H_i \dot{\theta}^2 \sin^2\theta + 2a_i g \cos\theta = h_i, \quad J_i^y \dot{\psi} \sin^2\theta = k_i, \tag{13}$$

where h_i and k_i are integration constants.

Solving (13) for $\dot{\theta}^2$ and $\dot{\psi}$ yields

$$\dot{\theta}^2 = \Theta_i(\theta), \qquad \dot{\psi} = \Psi_i(\theta), \tag{14}$$

where $\Theta_i(\theta) = \left[J_i^y \left(h_i - 2a_i g \cos\theta\right) \sin^2\theta - k_i^2\right] / \left[J_i^y \left(J_i^y + H_i \sin^2\theta\right) \sin^2\theta\right]$, $\Psi_i(\theta) = k_i / \left(J_i^y \sin^2\theta\right)$.

To ensure that, for every $i = 1, 2, \ldots, n$, equations (11), (12) reduce to the set of integrals (14) with the same right-hand sides, we require

$$\Theta_1(\theta) \equiv \Theta_2(\theta) \equiv \ldots \equiv \Theta_n(\theta), \qquad \Psi_1(\theta) \equiv \Psi_2(\theta) \equiv \ldots \equiv \Psi_n(\theta). \tag{15}$$

Each of the relations $\Theta_i(\theta) \equiv \Theta_1(\theta)$ and $\Psi_i(\theta) \equiv \Psi_1(\theta)$ $(i = 2, 3, \ldots, n)$ can be transformed to the form $P(\cos\theta) \equiv 0$, where $P(\cos\theta)$ is a polynomial in $\cos\theta$ which needs to be satisfied identically in $\cos\theta$. Equating the coefficients of all powers of $\cos\theta$ to zero in the identities obtained from (15) leads to the conditions

$$\frac{J_i^y}{J_1^y} = \frac{H_i}{H_1} = \frac{a_i}{a_1} = \frac{h_i}{h_1} = \frac{k_i}{k_1}. \tag{16}$$

Proposition 1 *If the conditions (9) and (16) are fulfilled, the system of equations (3) has a class of exact solutions with properties (4).*

Indeed, we infer from the previous discussion that, under the assumptions of this proposition, the system (5) is compatible. To find the dependence of the variables θ, ψ, and φ_i on time, one can proceed as follows. From the first equation in (14), we find

$$t = t_0 + \int_{\cos\theta_0}^{\cos\theta} \sqrt{\frac{J_i^y + H_i - H_i\xi^2}{(h_i - 2a_i g\xi)(1 - \xi^2) - k_i^2/J_i^y}}\, d\xi. \tag{17}$$

Thus, $\cos\theta$ can be obtained as the inverse of the hyperelliptic integral (17), in the form of an hyperelliptic function. Let $\cos\theta = F(t - t_0)$. We can get $\theta(t)$ by solving the last equation for θ. Then, we find $\psi(t)$ from (14):

$$\psi(t) = \psi_0 + \frac{k_i}{J_i^y} \int_{t_0}^{t} \frac{d\xi}{1 - F^2(\xi - t_0)}.$$

Finally, we observe that, for each i, the relation (10) can be transformed, by virtue of (14), (17), into a Ricatti equation with respect to the variable $u_i(t) = \tan(\varphi_i(t)/2)$:

$$\dot{u}_i = \left(\mu_i^{(1)}(t) - \mu_i^{(2)}(t)\right) u_i^2 - \mu_i^{(3)}(t) u_i - \mu_i^{(1)}(t) - \mu_i^{(2)}(t), \tag{18}$$

where

$$\mu_i^{(1)}(t) = \frac{I_i^{xz}}{2I_i^z} \sqrt{\frac{(h_i - 2a_i g F(t - t_0))(1 - F^2(t - t_0)) - k_i^2/J_i^y}{(J_i^y + H_i - H_i F^2(t - t_0))(1 - F^2(t - t_0))}},$$

$$\mu_i^{(2)}(t) = \frac{k_i F(t - t_0)}{2J_i^y (1 - F^2(t - t_0))}, \qquad \mu_i^{(3)}(t) = \frac{I_i^{xz} k_i}{I_i^z J_i^y \sqrt{1 - F^2(t - t_0)}}.$$

Using the change of variables $u_i(t) = \dot{w}_i(t) / \left[\left(\mu_i^{(2)}(t) - \mu_i^{(1)}(t)\right) w_i(t)\right]$, equation (18) can be reduced to the second order linear differential equation

$$\ddot{w}_i - P_i^{(1)}(t)\dot{w}_i + P_i^{(2)}(t) w_i = 0, \tag{19}$$

where

$$P_i^{(1)}(t) = \frac{\dot{\mu}_i^{(1)} - \dot{\mu}_i^{(2)}}{\mu_i^{(1)} - \mu_i^{(2)}} - \mu_i^{(3)}, \qquad P_i^{(2)}(t) = \left(\mu_i^{(2)}\right)^2 - \left(\mu_i^{(1)}\right)^2.$$

Given a solution $w_i(t)$ of (19), one can find $\varphi_i(t)$ from

$$\varphi_i(t) = 2\tan^{-1}\left\{\dot{w}_i(t)/\left[\left(\mu_i^{(2)}(t) - \mu_i^{(1)}(t)\right)w_i(t)\right]\right\}.$$

Note, however, that, in a general case, equation (19) cannot be integrated in quadratures.

5 On Compatibility of the Conditions (9) and (16)

In this section we show that there exist physically meaningful values of the multibody chain parameters making the conditions (9), (16) compatible in the case when $s_i \neq 0$ and $a_i \neq 0$. For the sake of brevity, we consider a simplest case of a two-body system assuming that $\psi_2 = \psi_1 + \pi$ in (4). Then, due to (2), (6), (7), $\varepsilon_{12} = -1, \kappa_{12} = 2, J_1^y = I_1^y + m_2 s_1^2 + L_1, J_2^y = I_2^y + L_2, L_1 = L_2 = -a_2 s_1, H_1 = H_2 = 2a_2 s_1$, the conditions (9) are equivalent to (8), and the conditions (16) become

$$J_2^y = J_1^y, \qquad a_2 = a_1, \qquad h_2 = h_1, \qquad k_2 = k_1. \tag{20}$$

Taking into account (1) and (6), we observe that the relations (8), (20) form an algebraic system of 6 equations with respect to 19 parameters $\widetilde{I}_i^j(>0), e_i^j((e_i^x)^2 + (e_i^z)^2 = 1), I_i^{yc}(>0), m_i(>0), c_i, h_i, k_i\ (i = 1, 2; j = x, z), s_1$. Here I_i^{yc} is the central moment of inertia of body B_i with respect to the axis defined by vector $\tilde{\mathbf{e}}_2^{(i)}$, and hence

$$I_i^y = \widetilde{I}_i^y = I_i^{yc} + m_i c_i^2. \tag{21}$$

In the rest of this section we solve the following problem: if the parameters defining the mass distribution in the bodies are known, find possible ways for coupling the bodies as well as the initial conditions of their motion.

From the second equation in (20), we obtain that

$$c_2 = \frac{m_1}{m_2}c_1 + s_1. \tag{22}$$

Substituting the expression for c_2 into the first equation in (20) and solving the equation so obtained for s_1 yields

$$s_1 = \frac{I_1^{yc} - I_2^{yc}}{2m_1 c_1} + \frac{m_2 - m_1}{2m_2} c_1. \tag{23}$$

Using (23), the formula (22) for determining c_2 becomes

$$c_2 = \frac{I_1^{yc} - I_2^{yc}}{2m_1 c_1} + \frac{m_2 + m_1}{2m_2} c_1. \tag{24}$$

In order to satisfy the system of relations (9), (16), one can now select the parameters of system S as follows. Assuming that the masses m_i and parameter c_1 have been chosen arbitrarily, the inertia moments $\widetilde{I}_i^x, \widetilde{I}_i^z, I_i^{yc}$ and parameters e_i^x, e_i^z can be selected to comply with (8). Then, the values for s_1 and c_2 can be found from (23) and (24), respectively. Finally, the values of h_i and k_i are to be determined to make the last two relations in (20) true. Knowing the values of the integration constants, one can obtain the initial conditions of motion, using (4) and (14).

As follows from the above analysis, in the case under consideration, there is a unique way of coupling the tops B_1 and B_2 that guarantees the existence of the motion of interest. If $c_1 > 0, m_2 > m_1$, and $I_1^{yc} \geq I_2^{yc}$, then $s_1 > 0$ and $c_2 > 0$, i.e. the conditions (9), (16) can be satisfied by positive values of c_1, c_2, and s_1. Moreover, $s_1 > c_1$, when $0 < c_1 < \sqrt{m_1^{-1} m_2 (m_1 + m_2)^{-1} \left(I_1^{yc} - I_2^{yc}\right)}$.

Thus, we have proven that it is possible to select physically meaningful values of parameters characterizing the chain of rigid bodies under consideration so that the conditions (9), (16) are fulfilled. This completes our proof of the existence of the similar motions for the chain S of Hess tops.

6 Geometry of the Motion of the Chain's Skeleton

The class of particular solutions of equations (3) that is constructed in the previous sections of this chapter describes a relatively complex motion of system S; therefore, a complete analysis of geometry of the system's motion in this case is a quite complicated problem. Since the motion of S is a superposition of the motion of its skeleton $O_1 O_2 \ldots O_n C_n$ and the pure rotation of each body about l_i, then, in order to understand how the system moves, one can study the rotational motion of each of the above components of its motion and then put them together to get a complete picture. In this section, we give some properties of the skeleton's motion.

As was noted before, the skeleton's motion is completely determined by the angles θ_i, ψ_i. For the similar motions (4), however, it is sufficient to analyze the rotation of any of the skeleton's segments $O_i O_{i+1}$ about O_i in order to get an understanding of the nature of the skeleton's motion. Once such an analysis is complete, one can get a clear idea of how the plane Π rotates about the vertical line through O_1 as well as how the barycentric axes l_i move relative to Π. A general description of the motion of any segment can be obtained from the properties of differential equations (14) without integration, employing the methods that are usually used for studying the rotation of the symmetry axis of a symmetric top about a fixed point [17]. In what follows, we analyze the rotation of O_1O_2 about O_1.

Equations (14) can be somewhat simplified by taking $\gamma = \cos\theta$, $a = a_i g/J_i^y$, $h = h_i/J_i^y$, $H = H_i/J_i^y$, and $k = k_i/J_i^y$. Then, the equations become

$$\dot{\gamma}^2 = \frac{(h - 2a\gamma)\left(1 - \gamma^2\right) - k^2}{1 + H - H\gamma^2} = \frac{2af_1(\gamma)}{f_2(\gamma)}, \qquad \dot{\psi} = \frac{k}{1 - \gamma^2}. \tag{25}$$

We leave it to the reader to verify that, due to (6), (7), (16), (21), $f_2(\gamma) \neq 0$ at any moment of time, i.e. there are no singularities in the right-hand side of the first equation in (25) and thus $\dot{\theta}(t)$ is a bounded function of time.

The polynomial $f_1(\gamma)$ in (25) is negative for $\gamma = -\infty, -1$, and $+1$, and positive for $\gamma = +\infty$. Since $\gamma = \cos\theta$ and for real motion θ is real, there should be two real roots, γ_2 and γ_3, of $f_1(\gamma)$ between -1 and $+1$, and a third root, γ_1, is greater than $+1$. The former can be achieved, for example, by requiring $h > k^2$. We therefore conclude that γ oscillates between the values γ_2 and γ_3.

From the second equation in (25), we observe that $\dot{\psi}$ has same sign as k at any moment of time. Hence, when $k \neq 0$, the angle of precession monotonically increases/decreases over time implying that, while the system S performs the similar motions, the plane Π always rotates in the same direction.

Let us consider a unit sphere drawn at O_1 as a center. The l_1-axis intersects the unit sphere in a point P. As the body B_1 moves, this point describes a curve C on the sphere. Let θ_2 and θ_3 be the angles corresponding to γ_2 and γ_3, respectively. If two cones are constructed so that the apex of each cone is at O_1 and the cones' generating angles are θ_2 and θ_3, respectively, these two cones intersect the sphere in two circles, C_2 and C_3. Since $\gamma_3 < \gamma_2$, then $\theta_3 > \theta_2$ and therefore the circle C_2 lies above the circle C_3. The curve C lies on the sphere between these two circles, touching the first one and then the other.

Let O_1X, O_1Y, and O_1Z be the coordinate axes of Σ and let O_1x, O_1y, and O_1z be the coordinate axes of Σ_1. Following [17], we represent the position of the point P on the unit sphere by the arc vector $Z\mathsf{P} = \theta$ and the angle $XZ\mathsf{P}$, which is the longitude, λ, of the pole of the xy-plane. Then, $\dot{\lambda} = \dot{\psi}$ and

$$\frac{d\lambda}{d\gamma} = \frac{k\sqrt{f_2(\gamma)}}{(1 - \gamma^2)\sqrt{2af_1(\gamma)}}, \tag{26}$$

which is the differential equation of the curve C.

Fig. 1 Trajectory of the point P of the barycentric axis l_1 on the unit sphere

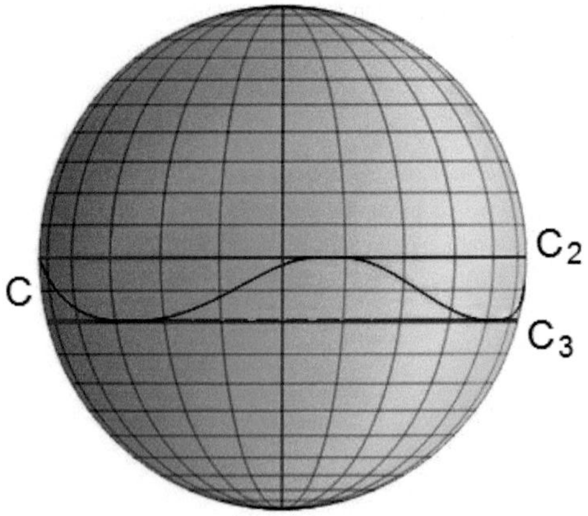

Let α be the angle which the curve C makes with the arc vector ZP. Using the results of [17] and (26), we derive that

$$\tan\alpha = \frac{d\lambda}{d\theta}\sin\theta = -\left(1-\gamma^2\right)\frac{d\lambda}{d\gamma} = -\frac{k\sqrt{f_2(\gamma)}}{\sqrt{2af_1(\gamma)}}. \qquad (27)$$

As follows from (26) and (27), $\tan\alpha$ is infinite only when γ is equal to γ_2 or γ_3 and therefore the curve C is tangent to the circle C_2 or C_3 whenever γ takes on one of these values. A typical curve C is shown in Fig. 1. By analogy with [3], one can show that the time it takes for P to reach C_2 starting from C_3 is equal to the time it takes to get back from C_3 to C_2.

Thus, while chain S of Hess tops rotates about O_1 so that its bodies move similar to each other, the plane Π rotates about the vertical line through O_1 in the same direction and the barycentric axis of each body moves along a spherical curve C oscillating between C_2 and C_3.

Acknowledgements Support for this project was provided by a PSC-CUNY Award, jointly funded by The Professional Staff Congress and The City University of New York (Award # 67306-00 45).

References

1. Bolgrabskaya, I., Lesina, M., Chebanov, D.: Dynamics of Systems of Coupled Rigid Bodies. Naukova Dumka, Kyiv (2012)
2. Borisov, A., Mamaev, I.: Dynamics of a Rigid Body. Hamiltonian Methods, Integrability, Chaos. Institute of Computer Science, Moscow/Izhevsk (2005)
3. Bukhgolts, N.: An Elementary Course in Theoretical Mechanics, vol. 2. Nauka, Moscow (1969)

4. Chebanov, D.: On a generalization of the problem of similar motions of a system of Lagrange gyroscopes. Mekh. Tverd. Tela **27**, 57–63 (1995)
5. Chebanov, D.: New dynamical properties of a system of Lagrange gyroscopes. Proc. Inst. Appl. Math. Mech. **5**, 172–182 (2000)
6. Chebanov, D.: Exact solutions for motion equations of symmetric gyros system. Multibody Syst. Dyn. **6**(1), 30–57 (2001)
7. Chebanov, D.: A new class of nonstationary motions of a system of heavy Lagrange tops with a non-planar configuration of the system's skeleton. Mekh. Tverd. Tela **41**, 244–254 (2011)
8. Chebanov, D.: New class of exact solutions for the equations of motion of a chain of n rigid bodies. Discret. Cont. Dyn. Syst. **Supplement**, 105–113 (2013)
9. Gashenenko, I., Gorr, G., Kovalev, A.: Classical Problems of the Rigid Body Dynamics. Naukova Dumka, Kyiv (2012)
10. Hess, W.: Über die Eulerschen Bewegungsgleichungen und über eine neue particulare Lösung des Problems der Bewegung eines starren Körpers un einen festen Punkt. Math. Ann. **37**(2), 178–180 (1890)
11. Kharlamov, P.: On the Equations of Motion for a Heavy Rigid Body with a Fixed Point. J. Appl. Math. Mech. **27**(4), 1070–1078 (1963)
12. Kharlamov, P.: Lectures on Rigid Body Dynamics. Novosibirsk University, Novosibirsk (1965)
13. Kharlamov, P.: The equations of motion of a system of rigid bodies. Mekh. Tverd. Tela **4**, 52–73 (1972)
14. Kharlamov, P.: Some classes of exact solutions of the problem of the motion of a system of Lagrange gyroscopes. Mat. Fiz. **32**, 63–76 (1982)
15. Kharlamova, E.: A survey of exact solutions of problems of the motion of systems of coupled rigid bodies. Mekh. Tverd. Tela **26**(II), 125–138 (1998)
16. Leimanis, E.: The General Problem of the Motion of Coupled Rigid Bodies About Fixed Point. Springer, Berlin (1965)
17. MacMillan, W.: Dynamics of Rigid Bodies. Dover Publications, New York (1960)
18. Savchenko, A., Lesina, M.: A particular solution for motion equations of Lagrange gyroscopes system. Mekh. Tverd. Tela **5**, 27–30 (1973)

On Stabilization of an Unbalanced Lagrange Gyrostat

Dmitriy Chebanov, Natalia Mosina, and Jose Salas

Abstract In this contribution we consider a chain of two coupled Lagrange gyrostats moving about a fixed point in a non-homogeneous gravitational field and show the existence of the following stabilization effect: for a gyrostat that is in its unstable equilibrium position, there is another gyrostat such that, when the two of them are coupled to form this chain, the rotation of the latter gyrostat can be used to stabilize the equilibrium of the former one. We establish and analyze stabilization conditions in the space of mechanical parameters characterizing the chain.

1 Introduction

In this contribution we investigate dynamic properties of a chain of two coupled gyrostats rotating about a fixed point in a gravitational field. By definition, a gyrostat is a mechanical system consisting of a rigid carrier and other bodies connected to it such that their motion relative to the carrier does not alter the distribution of masses of the mechanical system. Examples of such systems include a rigid body to which axes of several symmetric rotors are connected, or a rigid body with cavities completely filled with a homogeneous fluid [12, 18]. Mechanical models involving gyrostats have a number of applications; for instance, they are used for controlling the attitude dynamics of spacecraft and for stabilizing its rotations [1, 8], play an important role in studying fluid dynamical systems [4, 5], can be employed in the analysis of the mechanical models of DNA molecules [16], open-loop dynamic characteristics of smooth air gap brushless dc motors [7], instabilities associated with the operation of lasers [6], microscopic dynamics of deterministic chemical chaos [10], etc.

Assuming that, for each gyrostat, its mass distribution is analogous to the one of a Lagrange top, its attachment point(s) lie on its dynamic symmetry axis, and its gyrostatic moment is directed along this axis, we study a problem of stability of the state of chain's motion with the following property: one of the gyrostats permanently rotates about a vertical axis, whereas the other one is at rest. While

D. Chebanov (✉) • N. Mosina • J. Salas
City University of New York – LaGCC, Long Island City, NY, USA
e-mail: dchebanov@lagcc.cuny.edu; nmosina@lagcc.cuny.edu; jasn_en14@hotmail.com

© Springer International Publishing Switzerland 2016

J. Bélair et al. (eds.), *Mathematical and Computational Approaches in Advancing Modern Science and Engineering*, DOI 10.1007/978-3-319-30379-6_6

studying this problem, we show the existence of an interesting stabilization effect: for any Lagrange gyrostat that is in its unstable equilibrium position, there is another Lagrange gyrostat such that, when the two gyrostats are coupled to form a chain, the rotation of the latter gyrostat can be used to stabilize the former one. A similar stabilization effect was discovered for a system of two heavy Lagrange tops moving about a fixed point [9, 15, 17]. This contribution extends the results of [9, 15, 17] to a case of a system of two gyrostats moving in a non-homogeneous gravitational field.

2 Description of the Model

We consider a mechanical system S consisting of two gyrostats G_1 and G_2. Each gyrostat G_k is composed by a rigid body B_k and other (variable or rigid) bodies $\widetilde{B}_k$ which are connected to it. The motion of these bodies relative to B_k does not change the distribution of masses of the gyrostat G_k. We assume that the body B_1 is attached to an immovable base at one of its points O_1, while the bodies B_1 and B_2 are coupled at a common point $O_2(\neq O_1)$ by an ideal spherical joint.

Let m_k denote the mass of G_k. Let also B_1^* denote an augmented body [2] consisting of B_1 and the mass point m_2 located at O_2. Similarly, the augmented gyrostat G_1^* consists of G_1 and the mass point m_2 located at O_2. For our convenience, we use the notation $B_2^* = B_2$ and $G_2^* = G_2$. Note that the absolute angular velocities of bodies B_k and B_k^* are equal, while the first-order mass momentum $\mathbf{a}_k$ and the tensor of inertia $\mathbf{I}_k$ of G_k^* at O_k can be expressed as

$$\mathbf{a}_1 = (m_1 + m_2)\mathbf{b} = m_1\mathbf{c}_1 + m_2\mathbf{s}, \qquad \mathbf{a}_2 = m_2\mathbf{c}_2, \tag{1}$$

$$\mathbf{I}_1 = \mathbf{J}_1 + m_2\left(\mathbf{s}^2\boldsymbol{\delta} - \mathbf{s}\otimes\mathbf{s}\right), \qquad \mathbf{I}_2 = \mathbf{J}_2, \tag{2}$$

where $\mathbf{b} = \mathbf{O}_1\mathbf{C}_1^*$, C_1^* is the barycenter of G_1^*, $\mathbf{c}_k = \mathbf{O}_k\mathbf{C}_k$, C_k is the center of mass of G_k, $\mathbf{s} = \mathbf{O}_1\mathbf{O}_2$, $\mathbf{J}_k$ is the tensor of inertia of G_k at point O_k, $\boldsymbol{\delta}$ is the 3×3 identity matrix, and $\otimes$ denotes a dyadic product of two tensors.

We further assume that each gyrostat G_k has the mass distribution analogous to the one of a Lagrange top, the attachment point(s) of B_k lie on the axis of dynamic symmetry of G_k, and the gyrostatic moment $\boldsymbol{\lambda}_k$ of G_k is constant relative to B_k and directed along the G_k's symmetry axis, i.e., $O_1, O_2, C_1, C_1^* \in l_1, O_2, C_2 \in l_2$, and $\boldsymbol{\lambda}_k || l_k$, where l_k denotes the dynamic symmetry axis of G_k.

Suppose that system S moves under the gravitational attraction of a point mass μ located at a fixed point P whose position with respect to O_1 is defined by the vector $R\boldsymbol{\nu}$ where $\boldsymbol{\nu}$ is the unit vector in the direction of $\mathbf{O}_1\mathbf{P}$ and R is the distance between O_1 and P.

We introduce a Cartesian reference frame $\{O_1, \boldsymbol{\nu}_1\boldsymbol{\nu}_2\boldsymbol{\nu}_3\}$ whose axes are fixed in inertial space such that $\boldsymbol{\nu}_3 = \boldsymbol{\nu}$ and a Cartesian frame $\{O_k, \mathbf{e}_1^{(k)}\mathbf{e}_2^{(k)}\mathbf{e}_3^{(k)}\}$ that is rigidly embedded in body B_k such that its coordinate axes are the principal axes of inertia of G_k and $\mathbf{e}_3^{(k)} || l_k$. We also assume that, in the corresponding moving frame, $\mathbf{I}_k = diag\{I_k, I_k, I_k^z\}, \mathbf{J}_k = diag\{J_k, J_k, J_k^z\}, \mathbf{s} = s\mathbf{e}_3^{(1)}, \mathbf{a}_k = a_k\mathbf{e}_3^{(k)}$, and $\boldsymbol{\lambda}_k = \lambda_k \mathbf{e}_3^{(k)}$.

We determine the position of gyrostat G_k with respect to the reference frame by Bryan-Krylov angles α_k, β_k, and γ_k [11, 18]. The equations of motion of system S in terms of the Bryan-Krylov angles can be written in the form [3]

$$F_k^{(m)} + a_2 s G_{kj}^{(m)} + \varepsilon a_2 s H_{kj}^{(m)} = 0, \qquad \ddot{\gamma}_k - \ddot{\alpha}_k \sin\beta_k - \dot{\alpha}_k \dot{\beta}_k \cos\beta_k = 0, \tag{3}$$

where $\varepsilon = g/R, m = 1, 2; j, k = 1, 2$ with $j \neq k$,

$$\begin{aligned}
F_k^{(1)} &= I_k \left(\ddot{\beta}_k + \dot{\alpha}_k^2 \sin\beta_k \cos\beta_k\right) + I_k^z \left(\dot{\gamma}_k - \dot{\alpha}_k \sin\beta_k\right) \dot{\alpha}_k \cos\beta_k + \lambda_k \dot{\alpha}_k \cos\beta_k \\
&\quad + a_k g \cos\alpha_k \sin\beta_k + 3\varepsilon \left(I_k - I_k^z\right) \cos^2\alpha_k \sin\beta_k \cos\beta_k, \\
G_{kj}^{(1)} &= \left(\ddot{\beta}_j \cos\beta_j - \dot{\beta}_j^2 \sin\beta_j\right) \cos\beta_k \\
&\quad + \left(2\dot{\alpha}_j \dot{\beta}_j \sin\beta_j - \ddot{\alpha}_j \cos\beta_j\right) \sin\beta_k \sin(\alpha_k - \alpha_j) \\
&\quad + \left(\ddot{\beta}_j \sin\beta_j + \left(\dot{\alpha}_j^2 + \dot{\beta}_j^2\right) \cos\beta_j\right) \sin\beta_k \cos(\alpha_k - \alpha_j), \\
H_{kj}^{(1)} &= \left(2\cos\alpha_k \cos\alpha_j - \sin\alpha_k \sin\alpha_j\right) \sin\beta_k \cos\beta_j + \sin\beta_j \cos\beta_k, \\
F_k^{(2)} &= I_k \left(\ddot{\alpha}_k \cos\beta_k - 2\dot{\alpha}_k \dot{\beta}_k \sin\beta_k\right) - I_k^z \left(\dot{\gamma}_k - \dot{\alpha}_k \sin\beta_k\right) \dot{\beta}_k \\
&\quad - \lambda_k \dot{\beta}_k + a_k g \sin\alpha_k + 3\varepsilon (I_k - I_k^z) \sin\alpha_k \cos\alpha_k \cos\beta_k, \\
G_{kj}^{(2)} &= \left(\ddot{\alpha}_j \cos\beta_j + 2\dot{\alpha}_j \dot{\beta}_j \sin\beta_j\right) \cos(\alpha_k - \alpha_j) \\
&\quad + \left(\ddot{\beta}_j \sin\beta_j + \left(\dot{\alpha}_j^2 + \dot{\beta}_j^2\right) \cos\beta_j\right) \sin(\alpha_k - \alpha_j), \\
H_{kj}^{(2)} &= \cos\beta_j \left(2\sin\alpha_k \cos\alpha_j + \sin\alpha_j \cos\alpha_k\right).
\end{aligned}$$

3 The Problem of Interest

The motion of a gyrostat is a permanent rotation if, while the gyrostat is in motion, its angular velocity vector is constant. This vector defines an axis of permanent rotation which is fixed in both an inertial space and the gyrostat's carrier. When the permanent axis is a vertical line, the gyrostat permanently rotates about a vertical axis.

In this contribution we investigate a problem of the stability of the state of motion of system S with the following property: G_1 permanently rotates about its dynamic symmetry axis coinciding with the vertical axis passing through O_1, whereas G_2 is at rest so that $\mathbf{e}_3^{(2)}||\boldsymbol{\nu}$. One can check that equations (3) have a particular solution

$$\alpha_k \equiv 0, \qquad \beta_k \equiv 0, \qquad \gamma_1 = \omega t, \qquad \gamma_2 \equiv 0, \qquad \omega = const \tag{4}$$

that describes the motion of interest; the angular velocity ω can be chosen in (4) arbitrarily. Below we discuss a case when the parameters c_1, s, and $c_2 \left(= \mathbf{c}_2 \cdot \mathbf{e}_3^{(2)}\right)$ are negative. In this case, $a_1 < 0$ and $a_2 < 0$.

The solution (4) is a special case of a more general solution ($\alpha_k \equiv 0, \beta_k \equiv 0, \gamma_k = \omega_k t, \omega_k = const$) of equations (3) that describes permanent rotations of system S about a vertical axis. The necessary conditions for stability of such rotations have been recently established in [3]. For the motion of interest, these conditions assume the form

$$\mu_2^2 - \mu_1\mu_3 > 0, \qquad 12\left(\mu_2^2 - \mu_1\mu_3\right)^2 - \mu_1^2\left(\mu_1\mu_5 - 4\mu_2\mu_4 + 3\mu_3^2\right) > 0, \tag{5}$$

$$\left(\mu_1\mu_5 - 4\mu_2\mu_4 + 3\mu_3^2\right)^3 - 27\left(\mu_1\mu_3\mu_5 + 2\mu_2\mu_3\mu_4 - \mu_1\mu_4^2 - \mu_2^2\mu_5 - \mu_3^3\right)^2 > 0,$$

where $\mu_1 = I_1 I_2 - a_2^2 s^2$, $\mu_2 = (I_1\lambda_2 + I_2 w)/4$, $\mu_3 = \left(w\lambda_2 - I_1 d_2 - I_2 d_1 + 2\varepsilon a_2^2 s^2\right)/6$, $\mu_4 = -(wd_2 + \lambda_2 d_1)/4$, $\mu_5 = d_1 d_2 - \varepsilon^2 a_2^2 s^2$, $w = I_1^z\omega + \lambda_1$, $d_k = 2\varepsilon a_2 s + a_k g + 3\varepsilon\left(I_k - I_k^z\right)$.

While analyzing the conditions (5) in the rest of the contribution, we aim to establish the existence of the following stabilization effect. Suppose that the parameters of G_2 are selected so that, if G_2 is decoupled from the chain S and then coupled to an immovable base at point O_2, its equilibrium position would be unstable. (Using Lyapunov's indirect method, it is possible to show that this is the case when $\lambda_2^2 + 4J_2\left[m_2 c_2 g + 3\varepsilon\left(J_2 - J_2^z\right)\right] < 0$ or, by virtue of (1),

$$\lambda_2^2 + 4I_2\left[a_2 g + 3\varepsilon\left(I_2 - I_2^z\right)\right] < 0. \tag{6}$$

Special cases of (6) are given in [1, 13, 14].) We seek to justify that the parameters of G_1 can be chosen so that, when both gyrostats form chain S, the motion of S described by (4) is stable, i.e., the permanent rotation of G_1 stabilizes G_2.

4 Analysis of Necessary Stability Conditions

The conditions (5) depend on 12 parameters: angular velocity ω, mass characteristics I_k, I_k^z, a_k, the magnitudes of the gyrostatic moments λ_k, parameters ε and g, and the distance s between the gyrostats' attachment points. Recall that $a_k < 0, s < 0$ and assume additionally that $I_k < I_k^z$. Then, for sufficiently small ε, we have $d_k < 0$. In order to reduce the number of parameters in the problem of study, we introduce dimensionless parameters $x, \kappa_1, \kappa_2, \kappa_3, \kappa_4$ by the formulas:

$$w = I_1 x\sqrt{\frac{|d_2|}{I_2}},\ I_2 d_1 = \kappa_1 I_1 d_2,\ a_2^2 s^2 = \kappa_2 I_1 I_2,\ \varepsilon I_2 = \kappa_3 d_2,\ \lambda_2 = \kappa_4\sqrt{I_2|d_2|}. \tag{7}$$

The conditions (5) now deduce to the form

$$\Lambda_j > 0, \qquad j = 1, 2, 3, \tag{8}$$

where

$$\begin{aligned}
\Lambda_1 &= 3x^2 + 2(4\kappa_2 - 1)\kappa_4 x + 3\kappa_4^2 - 8(1-\kappa_2)(\kappa_1 - 2\kappa_2\kappa_3 + 1),\\
\Lambda_2 &= 3x^4 + 4(4\kappa_2 - 1)\kappa_4 x^3 + \left[2\left(8\kappa_2^2 + 1\right)\kappa_4^2 - 16(1-\kappa_2)(\kappa_1 - 2\kappa_2\kappa_3 + \kappa_2)\right]x^2\\
&\quad + 4\left[(4\kappa_2 - 1)\kappa_4^2 - 4(1-\kappa_2)\left((\kappa_1 + 1)(3\kappa_2 - 1) - 4\kappa_2^2\kappa_3\right)\right]\kappa_4 x\\
&\quad + 3\kappa_4^4 - 16(1-\kappa_2)(\kappa_1\kappa_2 - 2\kappa_2\kappa_3 + 1)\kappa_4^2\\
&\quad + 16(1-\kappa_2)^2\left((1-\kappa_1)^2 + 4\kappa_2(\kappa_1 - \kappa_3)(1-\kappa_3)\right),\\
\Lambda_3 &= \mu_6 x^6 + \mu_5 x^5 + \mu_4 x^4 + \mu_3 x^3 + \mu_2 x^2 + \mu_1 x + \mu_0.
\end{aligned}$$

The expressions for $\mu_6, \mu_5, \ldots, \mu_0$ in terms of the parameters $\kappa_1, \kappa_2, \kappa_3, \kappa_4$ are given in Appendix.

The conditions (8) form a system of three algebraic inequalities (with respect to x) whose left-hand sides, $\Lambda_1(x)$, $\Lambda_2(x)$, and $\Lambda_3(x)$, are polynomials of the second, fourth, and six order in x, respectively. The coefficients of these polynomials depend on four parameters $\kappa_1, \kappa_2, \kappa_3$, and κ_4. It is quite difficult to conduct a thorough analytical study of the conditions (8) due to their high nonlinearity and tediousness. Therefore, we have studied them numerically, seeking to find the values of the parameters under which the system (8) is compatible.

Note that due to the assumptions stated at the beginning of this section and (7), the parameters κ_1, κ_2 are positive, while κ_3 is negative and $|\kappa_3|$ is relatively small. Moreover, $\kappa_2 < 1$. (Indeed, by (7), this inequality is equivalent to $I_1 I_2 - a_2^2 s^2 > 0$. In order to prove the latter, we observe that, due to (1), (2), $a_2 = m_2 c_2, I_1 = J_1 + ms^2, I_2 = J_2$. Furthermore, $J_2 = J_2^c + m_2 c_2^2$, where J_2^c is the central equatorial moment of inertia of G_2. Then, $I_1 I_2 - a_2^2 s^2 = \left(J_1 + m_2 s^2\right)\left(J_2^c + m_2 c_2^2\right) - m_2^2 c_2^2 s^2 = J_1 J_2^c + m_2\left(J_1 c_2^2 + J_2^c s^2\right) > 0$.) Since each of the polynomials Λ_j is invariant with respect to the simultaneous replacement of x and κ_4 with $-x$ and $-\kappa_4$, respectively, it is enough to study only a case when $\kappa_4 > 0$.

We also note that the inequality (6) can be written in terms of κ_2, κ_3, and κ_4 as follows

$$\frac{\kappa_4}{\sqrt{1 - 2\kappa_3\sqrt{\kappa_2}\sqrt{I_1/I_2}}} < 2. \tag{9}$$

Since κ_3 is negative and relatively small and $0 < \kappa_2 < 1$, the expression $1 - 2\kappa_3\sqrt{\kappa_2}\sqrt{I_1/I_2}$ is positive for all possible values of the parameters it contains and its second term is negligible. Moreover, κ_4 provides an upper bound for the left-hand side in (9). Hence, requiring

$$\kappa_4 < 2 \tag{10}$$

guarantees the fulfillment of (9).

Thus, based on the above discussion, the suitable values of κ_1, κ_2, and κ_4 belong to the set $D = \{(\kappa_1, \kappa_2, \kappa_4) | \kappa_1 > 0, 0 < \kappa_2 < 1, 0 < \kappa_4 < 2\}$. Setting $\kappa_3 = -0.01$, we have conducted a numerical analysis of the conditions (8) for the members of D. Our analysis reveals that there exists a simply connected region $D_* \subset D$ in the parameter space where these conditions are fulfilled.

The existence of the interval(s) of x for which the conditions (8) are fulfilled depends on the number of real zeros of $\Lambda_j(x)$ and their relative position along the x-axis. Our study indicates that, when parameters κ_1, κ_2, and κ_4 are selected from D_*, the number of zeros of $\Lambda_1(x)$ can vary from 0 to 2, while $\Lambda_2(x)$ and $\Lambda_3(x)$ have ether two zeros or no zeros at all. (The number of zeros of $\Lambda_1(x)$ is completely determined by the sign of the expression $L = (2\kappa_2 + 1)\kappa_4^2 - 3(\kappa_1 - 2\kappa_2\kappa_3 + 1)$. In more details, $\Lambda_1(x)$ has two zeros, when $L < 0$, no zeros, when $L > 0$, and one zero, otherwise). We also observe that the leading coefficients of $\Lambda_1(x)$, $\Lambda_2(x)$, and $\Lambda_3(x)$ take on positive, positive, and negative values, respectively, for any $(\kappa_1, \kappa_2, \kappa_4) \in D$. Figures 1 and 2 demonstrate various cases of the relative position of the intervals of positiveness of $\Lambda_i(x)$ that have been encountered in our analysis. It is interesting to note that for any set of parameters from D_*, we have been able to find only one stabilization interval $[x_*, x^*]$ whose endpoints are always the zeros of $\Lambda_3(x)$.

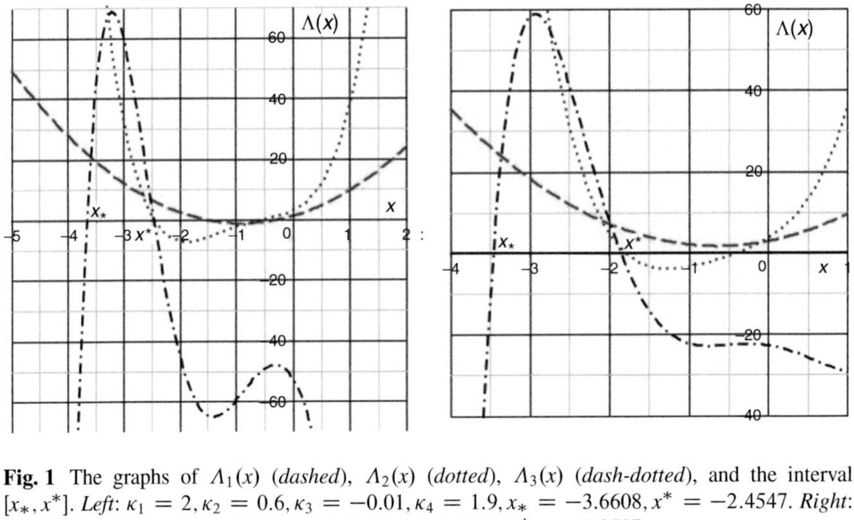

Fig. 1 The graphs of $\Lambda_1(x)$ (*dashed*), $\Lambda_2(x)$ (*dotted*), $\Lambda_3(x)$ (*dash-dotted*), and the interval $[x_*, x^*]$. *Left*: $\kappa_1 = 2, \kappa_2 = 0.6, \kappa_3 = -0.01, \kappa_4 = 1.9, x_* = -3.6608, x^* = -2.4547$. *Right*: $\kappa_1 = 1, \kappa_2 = 0.5, \kappa_3 = -0.01, \kappa_4 = 1.9, x_* = -3.4564, x^* = -1.8527$

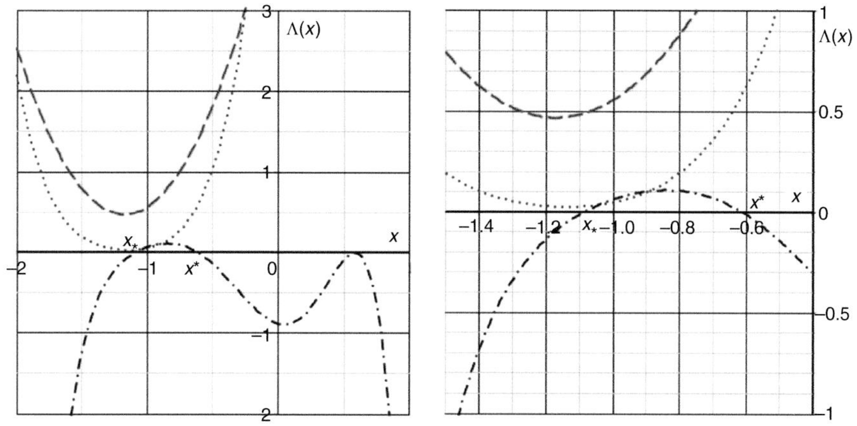

Fig. 2 The graphs of $\Lambda_1(x)$ (*dashed*), $\Lambda_2(x)$ (*dotted*), $\Lambda_3(x)$ (*dash-dotted*), and the interval $[x_*, x^*] = [-1.0816, -0.6153]$ for $\kappa_1 = 0.1, \kappa_2 = 0.9, \kappa_3 = -0.01$, and $\kappa_4 = 1.35$. *Left*: zoom on $-2 \leq x \leq 1$. *Right*: zoom on $-1.5 \leq x \leq -0.4$

Table 1 Numerical estimates for $\kappa_1^*, \kappa_2^*, x_*$, and x^* for some values of κ_4

κ_4	κ_1^*	κ_2^*	$[x_*, x^*]$
1.40	0.11	0.75	[−0.6668, −0.6473]
1.40	0.17	0.80	[−0.7914, −0.8154]
1.40	0.25	0.85	[−0.9535, −0.9378]
1.65	0.20	0.60	[−0.9651, −0.8665]
1.65	0.40	0.70	[−1.2547, −1.1815]
1.65	0.70	0.80	[−1.4966, −1.4746]
1.90	2.00	0.50	[−2.5674, −2.5324]
1.90	2.50	0.60	[−3.2193, −2.7011]
1.90	3.80	0.70	[−3.1720, −3.0646]

We have also observed that, for a fixed κ_4, there are some values κ_k^* such that only the values κ_k satisfying $0 < \kappa_1 < \kappa_1^*$ and $\kappa_2^* < \kappa_2 < 1$ belong to D_*; moreover, κ_k^* depends on both κ_4 and $\kappa_i (i, k = 1, 2, i \neq k)$. Numerical estimates for κ_k^* as well as for the interval $[x_*, x^*]$ of the fulfillment of (8) are given in Table 1. When κ_4 is close to its upper bound (10), there is a relatively wide range of values for κ_1 and κ_2 in D_*. As κ_4 starts getting smaller, this range narrows down significantly at a high rate. In particular, we have observed that in this case κ_1^* tends to drop its value significantly for the same κ_2.

Thus, it is possible to select the dimensionless parameters x and κ_1 through κ_4 to achieve the desired stabilization effect. Once the mass characteristics and the gyrostatic moment of G_2 are given, and the dimensionless parameters are chosen to comply with (8), the distance s, the angular velocity, mass parameters, and gyrostatic moment of G_1 can be determined from (7). A similar analysis can be conducted to justify that it is possible to stabilize an unstable equilibrium position of G_1 by a permanent rotation of G_2.

5 Conclusion

In this contribution, we have conducted a dynamic analysis of a multibody chain consisting of two Lagrange gyrostats and moving about a fixed point in a non-homogeneous gravitational field. We have established and analyzed the conditions for stability of the state of chain's motion when the gyrostat connected to the fixed point permanently rotates about a vertical axis while the other one is at rest. It follows from our analysis that that permanent rotation of the former gyrostat can be used to stabilize an unstable equilibrium position of the latter one.

Acknowledgements Support for this project was provided by a PSC-CUNY Award, jointly funded by The Professional Staff Congress and The City University of New York (Award # 68091-00 46). The work of the second and third authors was supported by the NASA New York Space Grant CCPP Program.

Appendix

The coefficients $\mu_6, \mu_5, \ldots, \mu_0$ of Λ_3 in (8) admit the following representation in terms of the dimensionless parameters $\kappa_1, \kappa_2, \kappa_3, \kappa_4$.

$$\mu_6 = \kappa_4^2 - 4,$$

$$\mu_5 = -2\kappa_4\left[\left(\kappa_1 - 2\kappa_2\left(1+\kappa_3^2\right)+1\right)\kappa_4^2 + 9\kappa_2\left(1+\kappa_3^2\right) - 2\left(2\kappa_1 - \kappa_2\kappa_3 + 2\right)\right],$$

$$\begin{aligned}\mu_4 = {} & \left[8\kappa_2\kappa_3^2\left(2\kappa_2-1\right) + \kappa_1\left(\kappa_1 - 8\kappa_2 + 4\right) + 1\right]\kappa_4^4 - 2\big[12\kappa_2^2\kappa_3^3 + 3\kappa_1\kappa_2\kappa_3^2 \\ & +40\kappa_2^2\kappa_3^2 - 8\kappa_1\kappa_2\kappa_3 + 12\kappa_2^2\kappa_3 - 19\kappa_2\kappa_3^2 + \kappa_1^2 - 19\kappa_1\kappa_2 - 8\kappa_2\kappa_3 \\ & +12\kappa_1 + 3\kappa_2 + 1\big]\kappa_4^2 - 27\kappa_2^2\kappa_3^4 + 36\kappa_2^2\kappa_3^3 + 36\kappa_1\kappa_2\kappa_3^2 - 2\kappa_2^2\kappa_3^2 \\ & -40\kappa_1\kappa_2\kappa_3 + 36\kappa_2^2\kappa_3 - 12\kappa_2\kappa_3^2 - 8\kappa_1^2 - 12\kappa_1\kappa_2 - 27\kappa_2^2 - 40\kappa_2\kappa_3 \\ & +32\kappa_1 + 36\kappa_2 - 8,\end{aligned}$$

$$\begin{aligned}\mu_3 = {} & -2\kappa_4\big[\left(\kappa_1^2 + \kappa_1 - 2\kappa_1^2\kappa_2 - 2\kappa_2\kappa_3^2\right)\kappa_4^4 + \big(64\kappa_2^3\kappa_3^3 + 8\kappa_1\kappa_2^2\kappa_3^2 - 40\kappa_2^2\kappa_3^3 \\ & +2\kappa_1^2\kappa_2\kappa_3 - 40\kappa_1\kappa_2^2\kappa_3 + 7\kappa_1\kappa_2\kappa_3^2 + 8\kappa_2^2\kappa_3^2 + \kappa_1^3 + 7\kappa_1^2\kappa_2 + 24\kappa_1\kappa_2\kappa_3 \\ & +7\kappa_2\kappa_3^2 - 9\kappa_1^2 + 7\kappa_1\kappa_2 + 2\kappa_2\kappa_3 - 9\kappa_1 + 1\big)\kappa_4^2 + 48\kappa_2^3\kappa_3^4 + 6\kappa_1\kappa_2^2\kappa_3^3 \\ & -160\kappa_2^3\kappa_3^3 - 18\kappa_2^2\kappa_3^4 + 3\kappa_1^2\kappa_2\kappa_3^2 - 38\kappa_1\kappa_2^2\kappa_3^2 + 48\kappa_2^3\kappa_3^2 + 130\kappa_2^2\kappa_3^3 \\ & -2\kappa_1^2\kappa_2\kappa_3 + 130\kappa_1\kappa_2^2\kappa_3 - 26\kappa_1\kappa_2\kappa_3^2 - 38\kappa_2^2\kappa_3^2 - 4\kappa_1^3 + 7\kappa_1^2\kappa_2 \\ & -18\kappa_1\kappa_2^2 - 92\kappa_1\kappa_2\kappa_3 + 6\kappa_2^2\kappa_3 + 7\kappa_2\kappa_3^2 + 20\kappa_1^2 - 26\kappa_1\kappa_2 - 2\kappa_2\kappa_3 + 20\kappa_1 \\ & +3\kappa_2 - 4\big],\end{aligned}$$

$$\mu_2 = \kappa_1^2\kappa_4^6 - 2\big(12\kappa_1^2\kappa_2^2\kappa_3 + 40\kappa_1\kappa_2^2\kappa_3^2 + 12\kappa_2^2\kappa_3^3 + 3\kappa_1^3\kappa_2 - 8\kappa_1^2\kappa_2\kappa_3$$

$$
\begin{aligned}
&-19\kappa_1\kappa_2\kappa_3^2+\kappa_1^3+38\kappa_1^2\kappa_2-8\kappa_1\kappa_2\kappa_3+3\kappa_2\kappa_3^2+12\kappa_1^2+\kappa_1\big)\kappa_4^4\\
&+\big(256\kappa_2^4\kappa_3^4-64\kappa_1\kappa_2^3\kappa_3^3-320\kappa_2^3\kappa_3^4-66\kappa_1^2\kappa_2^2\kappa_3^2-320\kappa_1\kappa_2^3\kappa_3^2\\
&+76\kappa_1\kappa_2^2\kappa_3^3-64\kappa_2^3\kappa_3^3-2\kappa_2^2\kappa_3^4-4\kappa_1^3\kappa_2\kappa_3+76\kappa_1^2\kappa_2^2\kappa_3+40\kappa_1^2\kappa_2\kappa_3^2\\
&+696\kappa_1\kappa_2^2\kappa_3^2+76\kappa_2^2\kappa_3^3+\kappa_1^4+40\kappa_1^3\kappa_2-2\kappa_1^2\kappa_2^2-108\kappa_1^2\kappa_2\kappa_3+76\kappa_1\kappa_2^2\kappa_3\\
&-192\kappa_1\kappa_2\kappa_3^2-66\kappa_2^2\kappa_3^2-4\kappa_1^3-192\kappa_1^2\kappa_2-108\kappa_1\kappa_2\kappa_3+40\kappa_2\kappa_3^2+102\kappa_1^2\\
&+40\kappa_1\kappa_2-4\kappa_2\kappa_3-4\kappa_1+1\big)\kappa_4^2+4\big(64\kappa_2^4\kappa_3^5-24\kappa_1\kappa_2^3\kappa_3^4-128\kappa_2^4\kappa_3^4\\
&-72\kappa_2^3\kappa_3^5-6\kappa_1^2\kappa_2^2\kappa_3^3-56\kappa_1\kappa_2^3\kappa_3^3+36\kappa_1\kappa_2^2\kappa_3^4+64\kappa_2^4\kappa_3^3+152\kappa_2^3\kappa_3^4\\
&+\kappa_1^3\kappa_2\kappa_3^2+40\kappa_1^2\kappa_2^2\kappa_3^2+152\kappa_1\kappa_2^3\kappa_3^2+52\kappa_1\kappa_2^2\kappa_3^3-56\kappa_2^3\kappa_3^3-12\kappa_2^2\kappa_3^4\\
&+6\kappa_1^3\kappa_2\kappa_3-14\kappa_1^2\kappa_2^2\kappa_3-49\kappa_1^2\kappa_2\kappa_3^2-72\kappa_1\kappa_2^3\kappa_3-192\kappa_1\kappa_2^2\kappa_3^2-24\kappa_2^3\kappa_3^2\\
&-14\kappa_2^2\kappa_3^3-\kappa_1^4-15\kappa_1^3\kappa_2-12\kappa_1^2\kappa_2^2+26\kappa_1^2\kappa_2\kappa_3+52\kappa_1\kappa_2^2\kappa_3+31\kappa_1\kappa_2\kappa_3^2\\
&+40\kappa_2^2\kappa_3^2+12\kappa_1^3+31\kappa_1^2\kappa_2+36\kappa_1\kappa_2^2+26\kappa_1\kappa_2\kappa_3-6\kappa_2^2\kappa_3-15\kappa_2\kappa_3^2\\
&-22\kappa_1^2-49\kappa_1\kappa_2+6\kappa_2\kappa_3+12\kappa_1+\kappa_2-1\big),
\end{aligned}
$$

$$
\begin{aligned}
\mu_1 = &-2\kappa_4\big[\big(9\kappa_1^3\kappa_2+2\kappa_1^2\kappa_2\kappa_3+9\kappa_1\kappa_2\kappa_3^2-4\kappa_1^3-4\kappa_1^2\big)\,\kappa_4^4+\big(48\kappa_1^2\kappa_2^3\kappa_3^2\\
&-160\kappa_1\kappa_2^3\kappa_3^3+48\kappa_2^3\kappa_3^4+6\kappa_1^3\kappa_2^2\kappa_3-38\kappa_1^2\kappa_2^2\kappa_3^2+130\kappa_1\kappa_2^2\kappa_3^3-18\kappa_2^2\kappa_3^4\\
&+3\kappa_1^4\kappa_2-18\kappa_1^3\kappa_2^2-2\kappa_1^3\kappa_2\kappa_3+130\kappa_1^2\kappa_2^2\kappa_3+7\kappa_1^2\kappa_2\kappa_3^2-38\kappa_1\kappa_2^2\kappa_3^2\\
&+6\kappa_2^2\kappa_3^3-4\kappa_1^4-26\kappa_1^3\kappa_2-92\kappa_1^2\kappa_2\kappa_3-26\kappa_1\kappa_2\kappa_3^2+20\kappa_1^3+7\kappa_1^2\kappa_2+3\kappa_2\kappa_3^2\\
&-2\kappa_1\kappa_2\kappa_3+20\kappa_1^2-4\kappa_1\big)\kappa_4^2+4\big(8\kappa_1^2\kappa_2^3\kappa_3^3-32\kappa_1\kappa_2^3\kappa_3^4+8\kappa_2^3\kappa_3^5+2\kappa_1^3\kappa_2^2\kappa_3^2\\
&-32\kappa_1^2\kappa_2^3\kappa_3^2-2\kappa_1^2\kappa_2^2\kappa_3^3+96\kappa_1\kappa_2^3\kappa_3^3+20\kappa_1\kappa_2^2\kappa_3^4-32\kappa_2^3\kappa_3^4-13\kappa_1\kappa_2\kappa_3^2\\
&-2\kappa_1^3\kappa_2^2\kappa_3-3\kappa_1^3\kappa_2\kappa_3^2+8\kappa_1^2\kappa_2^3\kappa_3+54\kappa_1^2\kappa_2^2\kappa_3^2-32\kappa_1\kappa_2^3\kappa_3^2-92\kappa_1\kappa_2^2\kappa_3^3\\
&+8\kappa_2^3\kappa_3^3+20\kappa_2^2\kappa_3^4-3\kappa_1^4\kappa_2+20\kappa_1^3\kappa_2^2-4\kappa_1^3\kappa_2\kappa_3-92\kappa_1^2\kappa_2^2\kappa_3-13\kappa_1^2\kappa_2\kappa_3^2\\
&+54\kappa_1\kappa_2^2\kappa_3^2-2\kappa_2^2\kappa_3^3+4\kappa_1^4-13\kappa_1^3\kappa_2+20\kappa_1^2\kappa_2^2+72\kappa_1^2\kappa_2\kappa_3-2\kappa_1\kappa_2^2\kappa_3\\
&+2\kappa_2^2\kappa_3^2-4\kappa_1^3-13\kappa_1^2\kappa_2-4\kappa_1\kappa_2\kappa_3-3\kappa_2\kappa_3^2-4\kappa_1^2-3\kappa_1\kappa_2+4\kappa_1\big)\big],
\end{aligned}
$$

$$
\begin{aligned}
\mu_0 = &-4\kappa_1^3\kappa_4^6+\big(-27\kappa_1^4\kappa_2^2+36\kappa_1^3\kappa_2^2\kappa_3-2\kappa_1^2\kappa_2^2\kappa_3^2+36\kappa_1\kappa_2^2\kappa_3^3-27\kappa_2^2\kappa_3^4\\
&+36\kappa_1^4\kappa_2-40\kappa_1^3\kappa_2\kappa_3-12\kappa_1^2\kappa_2\kappa_3^2-8\kappa_1^4-12\kappa_1^3\kappa_2-40\kappa_1^2\kappa_2\kappa_3\\
&+36\kappa_1\kappa_2\kappa_3^2+32\kappa_1^3-8\kappa_1^2\big)\kappa_4^4+\big(256\kappa_1^2\kappa_2^4\kappa_3^3-512\kappa_1\kappa_2^4\kappa_3^4+256\kappa_2^4\kappa_3^5\\
&-96\kappa_1^3\kappa_2^3\kappa_3^2-224\kappa_1^2\kappa_2^3\kappa_3^3+608\kappa_1\kappa_2^3\kappa_3^4-288\kappa_2^3\kappa_3^5-24\kappa_1^4\kappa_2^2\kappa_3\\
&-288\kappa_1^3\kappa_2^3\kappa_3+160\kappa_1^3\kappa_2^2\kappa_3^2+608\kappa_1^2\kappa_2^3\kappa_3^2-56\kappa_1^2\kappa_2^2\kappa_3^3-224\kappa_1\kappa_2^3\kappa_3^3\\
&-48\kappa_1\kappa_2^2\kappa_3^4-96\kappa_2^3\kappa_3^4+4\kappa_1^5\kappa_2+144\kappa_1^4\kappa_2^2+24\kappa_1^4\kappa_2\kappa_3+208\kappa_1^3\kappa_2^2\kappa_3\\
&-60\kappa_1^3\kappa_2\kappa_3^2-768\kappa_1^2\kappa_2^2\kappa_3^2+208\kappa_1\kappa_2^2\kappa_3^3+144\kappa_2^2\kappa_3^4-4\kappa_1^5-196\kappa_1^4\kappa_2
\end{aligned}
$$

$$
\begin{aligned}
&-48\kappa_1^3\kappa_2^2 + 104\kappa_1^3\kappa_2\kappa_3 - 56\kappa_1^2\kappa_2^2\kappa_3 + 124\kappa_1^2\kappa_2\kappa_3^2 + 160\kappa_1\kappa_2^2\kappa_3^2 - 24\kappa_2^2\kappa_3^3 \\
&+48\kappa_1^4 + 124\kappa_1^3\kappa_2 + 104\kappa_1^2\kappa_2\kappa_3 - 196\kappa_1\kappa_2\kappa_3^2 - 88\kappa_1^3 - 60\kappa_1^2\kappa_2 \\
&+24\kappa_1\kappa_2\kappa_3 + 4\kappa_2\kappa_3^2 + 48\kappa_1^2 - 4\kappa_1\big)\kappa_4^2 + 256\kappa_1^2\kappa_2^4\kappa_3^4 - 512\kappa_1\kappa_2^4\kappa_3^5 \\
&+256\kappa_2^4\kappa_3^6 - 128\kappa_1^3\kappa_2^3\kappa_3^3 - 512\kappa_1^2\kappa_2^4\kappa_3^3 - 128\kappa_1^2\kappa_2^3\kappa_3^4 + 1024\kappa_1\kappa_2^4\kappa_3^4 \\
&+512\kappa_1\kappa_2^3\kappa_3^5 - 512\kappa_2^4\kappa_3^5 - 256\kappa_2^3\kappa_3^6 + 16\kappa_1^4\kappa_2^2\kappa_3^2 - 128\kappa_1^3\kappa_2^3\kappa_3^2 \\
&+128\kappa_1^3\kappa_2^2\kappa_3^3 + 256\kappa_1^2\kappa_2^4\kappa_3^2 + 1152\kappa_1^2\kappa_2^3\kappa_3^3 - 128\kappa_1^2\kappa_2^2\kappa_3^4 - 512\kappa_1\kappa_2^4\kappa_3^3 \\
&-1536\kappa_1\kappa_2^3\kappa_3^4 + 256\kappa_2^4\kappa_3^4 + 512\kappa_2^3\kappa_3^5 + 128\kappa_1^4\kappa_2^2\kappa_3 - 16\kappa_1^4\kappa_2\kappa_3^2 \\
&+512\kappa_1^3\kappa_2^3\kappa_3 - 64\kappa_1^3\kappa_2^2\kappa_3^2 - 1536\kappa_1^2\kappa_2^3\kappa_3^2 - 640\kappa_1^2\kappa_2^2\kappa_3^3 + 1152\kappa_1\kappa_2^3\kappa_3^3 \\
&+512\kappa_1\kappa_2^2\kappa_3^4 - 128\kappa_2^3\kappa_3^4 - 16\kappa_1^5\kappa_2 - 128\kappa_1^4\kappa_2^2 - 128\kappa_1^4\kappa_2\kappa_3 - 256\kappa_1^3\kappa_2^3 \\
&-640\kappa_1^3\kappa_2^2\kappa_3 + 192\kappa_1^3\kappa_2\kappa_3^2 + 512\kappa_1^2\kappa_2^3\kappa_3 + 1632\kappa_1^2\kappa_2^2\kappa_3^2 - 128\kappa_1\kappa_2^3\kappa_3^2 \\
&-640\kappa_1\kappa_2^2\kappa_3^3 - 128\kappa_2^3\kappa_3^3 - 128\kappa_2^2\kappa_3^4 + 16\kappa_1^5 + 192\kappa_1^4\kappa_2 + 512\kappa_1^3\kappa_2^2 \\
&+128\kappa_1^3\kappa_2\kappa_3 - 640\kappa_1^2\kappa_2^2\kappa_3 - 352\kappa_1^2\kappa_2\kappa_3^2 - 64\kappa_1\kappa_2^2\kappa_3^2 + 128\kappa_2^2\kappa_3^3 - 64\kappa_1^4 \\
&-352\kappa_1^3\kappa_2 - 128\kappa_1^2\kappa_2^2 + 128\kappa_1^2\kappa_2\kappa_3 + 128\kappa_1\kappa_2^2\kappa_3 + 192\kappa_1\kappa_2\kappa_3^2 + 16\kappa_2^2\kappa_3^2 \\
&+96\kappa_1^3 + 192\kappa_1^2\kappa_2 - 128\kappa_1\kappa_2\kappa_3 - 16\kappa_2\kappa_3^2 - 64\kappa_1^2 - 16\kappa_1\kappa_2 + 16\kappa_1.
\end{aligned}
$$

References

1. Beletsky, V.: The Motion of a Satellite About Its Centre of Mass in a Gravitational Field. Moscow University Press, Moscow (1975)
2. Chebanov, D.: Precessional motions of a chain of coupled gyrostats in a central Newtonian field. In: Proceedings of the ASME 2014 International Design and Engineering Technical Conferences (DETC2014), Buffalo (2014). Paper DETC2014-35590
3. Chebanov, D., Salas, J.: On permanent rotations of a system of two coupled gyrostats in a central Newtonian force field. In: Proceedings of the ASME 2015 International Design and Engineering Technical Conferences (DETC2015), Boston (2015). Paper DETC2015-47434
4. Gluhovsky, A.: Modeling turbulence by systems of coupled gyrostats. In: Fitzmaurice, N., Gurarie, D., McCaughan, F., Woyczynski, W. (eds.) Nonlinear Waves and Weak Turbulence. Progress in Nonlinear Differential Equations and Their Applications, vol. 11, pp. 179–197. Birkhäuser, Boston (1993)
5. Gluhovsky, A.: Energy-conserving and Hamiltonian low-order models in geophysical fluid dynamics. Nonlinear Proc. Geoph. **13**(2), 125–133 (2006)
6. Haken, H.: Analogy between higher instabilities in fluids and lasers. Phys. Lett. A **53**(1), 77–78 (1975)
7. Hemati, N.: Strange attractors in brushless DC motors. IEEE Trans. Circuits Syst. I Fundam. Theory Appl. **41**(1), 40–45 (1994)
8. Hughes, P.: Spacecraft Attitude Dynamics. Wiley, New York (1986)
9. Lesina, M.: Stabilization of an unbalanced Lagrange gyroscope at rest. Mekh. Tverd. Tela **11**, 88–92 (1979)
10. Li, Q., Wang, H.: Has Chaos implied by macrovariable equations been justified? Phys. Rev. E **58**, 1191–1194 (1998)

11. Lurie, A.: Analytical Mechanics. Springer, Berlin (2002)
12. Moiseyev, N., Rumyantsev, V.: Dynamic Stability of Bodies Containing Fluid. Springer, New York (1968)
13. Rubanovskii, V.: On bifurcation and stability of stationary motions in certain problems of dynamics of a solid body. J. Appl. Math. Mech. **38**(4), 573–584 (1974)
14. Rumyantsev, V.: On the stability of motion of gyrostats. J. Appl. Math. Mech. **25**(1), 9–19 (1961)
15. Savchenko, A., Bolgrabskaya, I., Kononyhin, G.: Stability of Motion of Systems of Coupled Rigid Bodies. Naukova Dumka, Kyiv (1991)
16. Starostin, E.: Three-dimensional shapes of looped DNA. Meccanica **31**(3), 235–271 (1996)
17. Varkhalev, I., Savchenko, A., Svetlichnaya, N.: On the stabilization of an unbalanced Lagrange gyroscope at rest. Mekh. Tverd. Tela **14**, 105–109 (1982)
18. Wittenburg, J.: Dynamics of Multibody Systems. Springer, Berlin (2008)

Approximate Solution of Some Boundary Value Problems of Coupled Thermo-Elasticity

Manana Chumburidze

Abstract We consider a non-classical model of a pseudo oscillation system of partial differential equations of coupled thermo-elasticity in the Green-Lindsay formulation. The matrices of fundamental and singular solutions for isotropic homogeneous elastic materials have been obtained. We propose and justify a technique of approximate method for the solution of boundary value problems with mixed boundary conditions. The tools applied in this development are based on singular integral equations, the potential method and the generalized Fourier series analysis.

Mathematics Subject Classifications (2010) 26A33 60G22 35R60 34K37

1 Introduction

As we know, non-classical theories of thermo-elasticity have been developed in order to remove the paradox of physically impossible phenomenon of infinite velocity of thermal signals in the conventional coupled thermo-elasticity. Lord-Shulman [17, 23] theory and Green-Lindsay [14] theory are important generalized theories of thermo-elasticity that become centre of interest of recent research in this area. In Lord-Shulman theory, a flux rate term into the Fourier's law [2] of heat conduction is incorporated (with one relaxation time) and formulated a generalized theory admitting finite speed for thermal signals. Green-Lindsay theory called as temperature rate-dependent is included among the constitutive variables with two constants that act as two relaxation times, which does not violate the classical Fourier law of heat conduction when the body under consideration has a center of symmetry. There are special classes of thermo-elasticity problems such as coupled thermo-elasticity, which require entirely different mathematical approaches and means of analysis [11, 20–22].

M. Chumburidze (✉)
Ak.Tsereteli State University, Kutaisi, Georgia
e-mail: maminachumb02@gmail.com

© Springer International Publishing Switzerland 2016

J. Bélair et al. (eds.), *Mathematical and Computational Approaches in Advancing Modern Science and Engineering*, DOI 10.1007/978-3-319-30379-6_7

The dynamical problems for a conjugated system of two-dimensional linear theory coupled thermo-elasticity (CPTE) [4, 12, 16] in the Green-Lindsay formulation for isotropic homogeneous elastic materials with a center of symmetry are investigated in work [4]. Green's tensors for basic boundary value problems are derived. The boundary integral method in combination with the harmonic potentials theory and Laplace transform [3, 7, 16] are applied to solve the problems.

This paper is devoted to the development of approximate methods for the construction of solutions in the form that admit efficient numerical evaluation [3, 6, 8, 16]. In particular, we consider two-dimensional stationary problems of CPTE in Green-Lindsay formulation. The formulation is given for isotropic homogeneous elastic materials with a center of symmetry. Boundary value problems for a finite domain have been derived, when the couple-stresses components, displacement components, rotation, heat flux and temperature are given on the surface of Holder class [19].

Introduce the notations: R^2 is two-dimensional Euclidean space, $x = \left(x_j\right)$ $(j = 1, 2)$ is point of this space. $\Omega \subset R^2$ is finite domain, bounded by the closed surface $\partial\Omega$ of Holder class.

We have [4]:

$$\begin{gathered}(\mu+\alpha)\,\Delta u\,(x,\tau)+(\lambda+\mu-\alpha)\,graddivu+2\alpha rotu_3-\gamma_\tau gradu_4-\\ -\rho\tau^2 u=h^{(1)}\,(x)\\ (\nu+\beta)\,\Delta u_3\,(x,\tau)+2\alpha rotu-4\alpha u_3-I\tau^2 u_3=h_3(x)\\ \Delta u_4\,(x,\tau)-\tfrac{\tau}{\Re_\tau}u_4-\eta\tau divu=h_4(x)\end{gathered} \tag{1}$$

where $u = (u_1, u_2)$ is the displacement vector, u_3 is a characteristic of the rotation, u_4 is the temperature variation, $\gamma_\tau = \gamma(1 + \tau_2\tau)$, $\dfrac{1}{\Re_\tau} = \frac{1}{\Re}(1 + \tau_1\tau), \tau_2 > \tau_1 > 0$ are constants of relaxation [3],$\rho > 0, \alpha > 0, \mu > 0, 3\lambda + 2\mu > 0, \nu > 0, \beta > 0, \gamma > 0, \Re > 0, I > 0$ are constants of elasticity [4, 16, 20], Δ is two-dimensional Laplacian operator [13], $\tau = \sigma + iq, \sigma > 0$ corresponds to the general dynamic problems [4], $H = \left(h^{(1)}, h_3, h_4\right) = (h_1, h_2, h_3, h_4) \in C^{0,\alpha}(\overline{\Omega}^+) rotu_3 = \left(\frac{\partial u_3}{\partial x_2}, -\frac{\partial u_3}{\partial x_1}\right)^T$, $rotu = \frac{\partial u_2}{\partial x_1} - \frac{\partial u_1}{\partial x_2}$.

Let us construct the matrix of differential operators:

$$P_{(q)}(\partial x, n(x)) = \left|\begin{matrix}\|T(\partial x, n(x))\|_{3x3}\| - N\gamma_\tau\|_{3x1}\\ \|\delta_{4k}[-(\delta_{1q}+\delta_{3q})+(\delta_{0q}+\delta_{3q})\frac{\partial}{\partial n}]\|_{1x4}\end{matrix}\right|, q = 0, 3$$

$$Q_{(q)}(\partial x, n(x)) = \left|\begin{matrix}\|\delta_{jk}\|_{3x4}\\ \|\delta_{4k}[(\delta_{0q}+\delta_{3q})+(\delta_{1q}+\delta_{3q})\frac{\partial}{\partial n}]\|_{1x4}\end{matrix}\right|, q = 0, 3$$

where $N(x) = (n, 0), n = (n_1, n_2), T(\partial x, n(x)) = \|T_{jk}(\partial x, n(x))\|_{3x3}$ is the matrix of stress operator on the plain of couple-stress elasticity:

$$T_{jk}(\partial x, n(x)) = \lambda n_j(x)\frac{\partial}{\partial x_k} + (\mu - \alpha)n_k(x)\frac{\partial}{\partial x_j} - (\mu + \alpha)\delta_{kj}\frac{\partial}{\partial nx}, j, k = 1, 2$$

$$T_{jk}(\partial x, n(x)) = -2\alpha \sum_{p=1}^{2} \xi_{jkp} n_p(x), j = 1, 2, k = 3$$

$$T_{jk}(\partial x, n(x)) = (\nu + \beta)\delta_{kj}\frac{\partial}{\partial nx}, j = 3, k = 1, 3$$

Differential equations seem to be well suited as models for systems. Thus an understanding of differential equations are at least as important as an understanding of matrix equations.

Allow us introduce the matrix $L(\partial x, \tau) = |L_{jk}(\partial x, \tau)|_{4x4}$ of differential operator of pseudo oscillation:

$$L(\partial x, \tau) = \begin{vmatrix} \|L^{(1)}\|_{2x2} & \|L^{(2)}\|_{2x1} & \| - \gamma_\tau G^T(\partial x)\|_{2x1} \\ \|L^{(3)}\|_{1x2} & L^{(4)} & 0 \\ \| - \eta\tau G(\partial x)\|_{1x2} & 0 & \Delta - \frac{\tau}{\Re_\tau} \end{vmatrix}$$

where

$$L_{ij}^{(1)}(\partial x) = \delta_{ij}\left[(\mu + \alpha)\Delta - \rho\tau^2\right] + (\lambda + \mu - \alpha)\frac{\partial^2}{\partial x_i \partial x_j}, i, j = 1, 2$$

$$L_{ij}^{(2)}(\partial x) = L_{ij}^{(3)}(\partial x) = -2\alpha \sum_{p=1}^{2} \xi_{ijp}\frac{\partial}{\partial x_p}(x), i = 1, 2 j = 3; j = 1, 2, i = 3$$

$$L^{(4)}(\partial x) = (\nu + \beta)\Delta - 4\alpha - I\tau^2, G(\partial x) = (\partial x_1, \partial x_2)$$

where δ_{ij} is Kronecker's symbol, ξ_{ijp} is Levi-Chivita' s symbol.

Now the system (1) can be written in the form:

$$L(\partial x, \tau)U(x, \tau) = H(x) \tag{2}$$

In view of this the conjugated with $L(\partial x, \tau)$ operator will be considered:

$$\widehat{L}(\partial x, \tau) = \begin{vmatrix} \|L^{(1)}\|_{2x2} & \|L^{(2)}\|_{2x1} & \| - \eta\tau G^T(\partial x)\|_{2x1} \\ \|L^{(3)}\|_{1x2} & L^{(4)} & 0 \\ \| - \gamma_\tau G(\partial x)\|_{1x2} & 0 & \Delta - \frac{\tau}{\Re_\tau} \end{vmatrix}$$

The matrices $L(\partial x, \tau)$ and $\widehat{L}(\partial x, \tau)$ we use in technical point of view efficiently solving the system of partial differential equations of CPTE in Green-Lindsay formulation. In next section we will solve the matrix equations (2).

2 Fundamental and Singular Solutions

In our investigations the fundamental [16] and singular solutions together with basic potentials becomes an useful tool for the development of approximate methods solution of stationary problems.

Let us construct the matrix of fundamental solutions [4, 16] of the operator (2):

$$\Phi(x,\tau) = |\Phi_{ij}(x,\tau)|_{4x4} = \overline{L}(\partial x,\tau)\varphi(x,\tau) \tag{3}$$

where $\overline{L}$ is the associated with $L(\partial x, \tau)$ matrix:

$$L(\partial x,\tau)\overline{L}(\partial x,\tau)\varphi(x,\tau) = \overline{L}(\partial x,\tau)\varphi(x,\tau)L(\partial x,\tau) = IdetL(\partial x,\tau) \tag{4}$$

where I is the 4×4 dimensional unit matrix. According to (2) and (3) we have:

$$detL(\partial x,\tau) = (\lambda + 2\mu)(\mu + \alpha)(\nu + \beta)\prod_{k=1}^{4}(\Delta + \sigma_k^2)\varphi(x,\tau) = 0 \tag{5}$$

From (5)we shall get:

$$\varphi(x,\tau) = \sum_{k=1}^{4} a_k H_0^{(1)}(\sigma_k|x|) \tag{6}$$

where $H_0^{(1)}(\sigma_k|x|)$ is Hankel Function of the first kind (the zero order) [18], $|x| = \sqrt{x_1^2 + x_2^2}, \sigma_k(k = 1,4)$ are parameters of thermo-elasticity [4, 16], $\varphi(x,\tau)$ is unknown scalar, $a_k(k = 1,4)$ are constants, they are sought in such manner that partial derivatives of the eight order of function $\varphi(x,\tau)$ has an isolated singularity of the kind $ln|x|$. If the matrix of fundamental solutions of the operator $\overline{L}(\partial x,\tau)$ we denote as $\widehat{\Phi}(x,\tau)$ by a direct check we can make sure that: $\widehat{\Phi}(x,\tau) = \Phi^T(-x,\tau)$.

Let us construct the basic potentials:

$$V(x,\varphi) = \int_l \Phi(y - x,\tau)\varphi(y)d_y l$$

$$M^{(1)}(x,\varphi) = \int_l [\widehat{P}_{(q)}(\partial y, n)\widehat{\Phi}(y - x,\tau)]^T \varphi(y)d_y l$$

$$M^{(2)}(x,\varphi) = \int_l [\widehat{Q}_{(q)}(\partial y, n)\widehat{\Phi}(y-x,\tau)]^T \varphi(y) d_y l$$

$$U(x,\varphi) = \int_\Omega \Phi(y-x,\tau)\varphi(y)dy$$

where $l \in L_2(\alpha)$, $\alpha > 0$, $V(x,\varphi)$ is the single-layer potential, $M^{(1)}(x,\varphi), M^{(2)}(x,\varphi)$ are the mixed-type potentials, $U(x,\varphi)$ is the volume potential. Conjugated operators $\widehat{P}_{(q)}, \widehat{Q}_{(q)}$ can be obtained from corresponding operators $P_{(q)}, Q_{(q)}$ replacing $-\gamma_\tau$ and $-\eta\tau$.

3 Approximate Solutions

The aim of this section is to obtain approximate solutions of stationary problems of CPTE in Green-Lindsay formulation. Similar methods of these techniques which are extended to certain classes of problems are developed in the recent papers: [3, 10, 16]. In particular, we formulate basic boundary value problems for two-dimensional isotropic homogeneous elastic materials with a center of symmetry. It is assumed that surfaces are sufficiently smooth.

Problem $\mathbf{P}_{(q)}(\mathbf{1\tau})$. It is required to find regular solution $U = (u, u_3, u_4)$ – with the following conditions:

$$\forall x \in \Omega : \quad L(\partial x, \tau)U(x,\tau) = H(x)$$
$$\forall z \in \partial\Omega : P_{(q)}(\partial z, n(z))U(z) = F^{(1)}(z)$$

Problem $\mathbf{Q}_{(q)}(\mathbf{1\tau})$. It is required to find regular solution $U = (u, u_3, u_4)$ – with the following conditions:

$$\forall x \in \Omega : \quad L(\partial x, \tau)U(x,\tau) = H(x)$$
$$\forall z \in \partial\Omega : Q_{(q)}(\partial z, n(z))U(z) = F^{(2)}(z)$$

The existence and uniqueness of this solution has been proved in [4].

Solution of the **Problem $\mathbf{P}_{(q)}(\mathbf{1\tau})$** will be found by the formula:

$$U(x) = -\frac{1}{2}\int_\Omega \Phi(y-x,\tau)H(y)dy + \int_l [\widehat{Q}_{(q)}(\partial y, n)\Phi^T(y-x,\tau)]^T \varphi(y) d_y l \qquad (7)$$

where $\varphi(y)$ is solution of the singular integral equations of normal type with zero (total) index:

$$\varphi(z) + \int_l P_{(q)}(\partial z, n)[\widehat{Q}_{(q)}(\partial y, n)\Phi^T(y-z,\tau)]^T \varphi(y) d_y l = F^{(1)}(z) -$$
$$-\frac{1}{2}\int_\Omega P_{(q)}(\partial z, n)\Phi(y-z,\tau)H(y)dy \qquad (8)$$

Equation (8) is the singular integral equations with a Cauchy kernel [10] of normal type, which have an index equal to zero (total) and in this case the Fredholm theorems hold [4, 5, 15].

Let us represent solutions in the following form:

$$U(x,\tau) = -\frac{1}{2}\int_{\Omega} G^{T}_{P_{(q)}}(x,y;\tau,\Omega)H(y)dy + \\ + \int_{l} [\widehat{Q}_{(q)}(\partial y,n)G^{T}_{P_{(q)}}(x,y;\tau,\Omega)]^{T}F^{(1)}(y)d_{y}l \tag{9}$$

where $G_{P_{(q)}}(x,y;\tau,\Omega)$ is the tensor of Green of **Problem $\mathbf{P}_{(q)}(\imath\boldsymbol{\tau})$** [4, 16].

Analogically we will get:

$$U(x,\tau) = -\frac{1}{2}\int_{\Omega} G^{T}_{Q_{(q)}}(x,y;\tau,\Omega)H(y)dy + \\ + \int_{l} [\widehat{P}_{(q)}(\partial y,n)G^{T}_{Q_{(q)}}(x,y;\tau,\Omega)]^{T}F^{(2)}(y)d_{y}l \tag{10}$$

where $G_{Q_{(q)}}(x,y;\tau,\Omega)$ is the tensor of Green of **Problem $\mathbf{Q}_{(q)}(\imath\boldsymbol{\tau})$**

Formulas (9),(10) have an essential meaning to verify the generalized Fourier series method [8, 16] for **Problem $\mathbf{P}_{(q)}(\imath\boldsymbol{\tau})$** and **Problem $\mathbf{Q}_{(q)}(\imath\boldsymbol{\tau})$**.

Let us construct auxiliary domains and surfaces. Let us consider $\widehat{\Omega} \subset \Omega$ finite domain,bounded by the closed surfaces of Holder class :$\partial\widehat{\Omega}$ be sufficiently smooth surface – boundary of $\widehat{\Omega}$, $\{x_k\}_{k=1}^{\infty} \subset \partial\widehat{\Omega}$ be everywhere accounted set of points.

The next theorems are proved there [3, 4, 6, 16]:

Theorem 1 *Accounted set of the vectors* $\{P_{(q)}(\partial y,n)\Phi^{(j)}(y-x^{k})\}_{k=1}^{\infty}, j = 1,2,3,4, y \in \partial\Omega$, *is linearly independent and full in the Hilbert space* $L_2(\partial\Omega)$.

Theorem 2 *Accounted set of the vectors* $\{Q_{(q)}(\partial y,n)\Phi^{(j)}(y-x^{k})\}_{k=1}^{\infty}, j = 1,2,3,4, y \in \partial\Omega$, *is linearly independent and full in the Hilbert space [1]* $L_2(\partial\Omega)$.

Let us prove the following theorem:

Theorem 3 *For any* $\varepsilon > 0$ *can be found the natural number* N_0 *such, that when* $N > N_0$ *in any domain* $\overline{\Omega} \subset \Omega$, *uniformly holds the inequality*

$$|U(x) - U^{N}(x)| < \varepsilon, x \in \overline{\Omega},$$

where

$$U^{N}(x) = \sum_{k=1}^{N}\sum_{j=1}^{k} X_{k}^{(1)} d_{k}^{j}\Phi^{e_j}(y - x^{[\frac{j+3}{4}]}, i\tau) - \frac{1}{2}\int_{\Omega}\Phi(y-x,\tau)H(y)dy$$

Proof Introduce the following notations:

$$\psi^k(y) = P_{(q)}(\partial y, n)\Phi^{e_k}(y - x^{[\frac{k+3}{4}]}), k = \overline{1, \infty}$$

where

$$e_k = k - 4\left[\frac{k-1}{4}\right]$$

Let us consider that $\varphi^k(y)$ is orthonormal system of vectors $\psi^j(y)$ on $\partial\Omega$, then

$$\varphi^k(y) = \sum_{j=1}^{k} a_k^j \psi^j(y), k = \overline{1, \infty}, y \in \partial\Omega \tag{11}$$

where a_k^j are coefficients of the orthonormalization.

According to (3), we have:

$$\varphi^k(y) = \sum_{j=1}^{k} a_k^j P_{(q)}(\partial y, n)\Phi^{e_j}(y - x^{[\frac{j+3}{4}]}), k = \overline{1, \infty}, y \in \partial\Omega \tag{12}$$

Let us consider that $U(x)$ be the direct solution of the **Problem** $\mathbf{P}_{(q)}(\imath\boldsymbol{\tau})$ (existences this solutions are proved [4]). Let us consider the vector:

$$V(x) = U(x) - \frac{1}{2}\int_{\Omega} \Phi(y - x, \tau)H(y)dy$$

Obviously $V(x)$ is the regular solution of the following boundary value problem:

$$\begin{aligned} &\forall x \in \Omega : L(\partial x, \tau)V(x, \tau) = 0, x \in \Omega \\ &\forall z \in \partial\Omega : P_{(q)}(\partial z, n(z))V(z) = X^{(1)}(z) \end{aligned} \tag{13}$$

where

$$X^{(1)}(z) = F^{(1)}(z) - \frac{1}{2}\int_{\Omega} P_{(q)}(\partial z, n)\Phi(y - z, \tau)H(y)dy$$

According to the above $X^{(1)}(z) \in C^{(0,\alpha)}(\partial\Omega), \alpha > 0$ we have:

$$X^{(1)}(y) \approx \sum_{k=1}^{n} X_k^{(1)}\varphi^{(k)}(y), y \in \partial\Omega$$

where

$$X_k^{(1)} = \int_{\partial\Omega} X^{(1)}(y)\varphi^{(k)}(y)dl$$

Allow us introduce the vectors:

$$V^N(x) = \sum_{k=1}^{N} X_k^{(1)}\varphi^{(k)}(y) = \sum_{k=1}^{N} X_k^{(1)} \sum_{j=1}^{k} d_k^j \psi^j(y) = \sum_{k=1}^{N}\sum_{j=1}^{k} X_k^{(1)} d_k^j \Phi^{e_j}(y - x^{[\frac{j+3}{4}]}, i\tau)$$

Let us show that it is approximation solution of **Problem** $\mathbf{P}_{(q)}(\imath\boldsymbol{\tau})$:
Obviously $V^N(x)$ is the regular solution of the following boundary value problem:

$$\begin{aligned} \forall x \in \Omega : \qquad & L(\partial x, \tau)V^N(x,\tau) = 0, x \in \Omega \\ \forall z \in \partial\Omega : & P_{(q)}(\partial z, n(z))V^N(z) = \textstyle\sum_{k=1}^{N} X_k^{(1)}\varphi^{(k)}(z) \end{aligned} \tag{14}$$

But, according to (9) we have:

$$V^N(x,\tau) = -\frac{1}{2}\int_l [\widehat{Q}_{(q)}(\partial y, n) G^T_{P_{(q)}}(x, y; \tau, \Omega)]^T \sum_{k=1}^{N} X_k^{(1)}\varphi^{(k)}(y) d_y l \tag{15}$$

and also we have:

$$V(x,\tau) = -\frac{1}{2}\int_l [\widehat{Q}_{(q)}(\partial y, n) G^T_{P_{(q)}}(x, y; \tau, \Omega)]^T X^{(1)}(y) d_y l \tag{16}$$

Hence, from the formulas (15) and (16) and according to the assumptions of Green tensor [4] and applying the Cauchy-Bunyakovski inequality [9], we have:

$$V(x) = \lim_{N \longrightarrow \infty} V^N(x)$$

Analogically we can show that approximate solution of **Problem** $\mathbf{Q}_{(q)}(\imath\boldsymbol{\tau})$ has following form:

$$U^N(x) = \sum_{k=1}^{N}\sum_{j=1}^{k} X_k^{(2)} b_k^j \Phi^{e_j}(y - x^{[\frac{j+3}{4}]}, i\tau) - \frac{1}{2}\int_{\Omega} \Phi(y - x, \tau) H(y) dy$$

where

$$X_k^{(2)} = \int_{\partial\Omega} X^{(2)}(y)\omega^{(k)}(y)dl,$$

$$X^{(2)}(z) = F^{(2)}(z) - \frac{1}{2}\int_{\Omega} Q_{(q)}(\partial z, n)\Phi(y - z, \tau)H(y)dy$$

$$\omega^{(k)}(y) = \sum_{j=1}^{k} b_k^j \mu^j(y),$$

$$\mu^j(y) = Q_{(q)}(\partial y, n)\Phi^{e_j}(y - x^{[\frac{j+3}{4}]})$$

4 Conclusion

Thus, two-dimensional stationary problems of CPTE in the Green-Lindsay formulation have been investigated by applying the boundary integral method in combination with the harmonic potentials theory and generalized Fourier series analysis. Approximate solutions of boundary value problems for a finite domain bounded by closed surface of Holder class have been constructed.

As a result, it is shown that the approximate method is reliable for obtaining effective solutions (explicitly) of boundary value problems for isotropic homogeneous elastic materials with a center of symmetry.

Acknowledgements I would like to express my special gratitude to memories of my professors Tengis Burchuladze and Davit Gelashvili.

References

1. Bachman, G., Narici, L.: Fourier and Wavelet Analysis. Universitext. Springer, Berlin/New York (2000)
2. Bonetto, F., Lebowitz, J.L., Rey-Bellet, L.: Fourier's Law: A Challenge to Theorists. Mathematical Physics. Imperial College Press, London (2000)
3. Burchuladze, T., Gegelia, T.: Development of the Potential Method in Elasticity Theory. Mecniereba, Tbilisi (1985)
4. Chumburidze, M.: Non-classical Models of the Some Theory of Boundary Value Problems. LAP LAMBERT Academic Publishing, Saarbrucken (2014)
5. Chumburidze, M., Lekveishvili, D.: Effective Solution of Boundary Value Problems of the Theory of Thermopiezoelasticity for a Half-Plane. American Institute of Physics (2013). doi:10.1063/1.485476
6. Chumburidze, M. Lekveishvili, D.: Approximate solution of some mixed boundary value problems of the generalized theory of couple-stress thermo-elasticity. Int. J. Math. Comput. Nat. Phys. Eng. (2014). WASET. http://waset.org/Publication/9998377

7. Chumburidze, M., Lekveishvili, D., Khurcia, Z.: Solutions Of boundary-value problems of the generalized theory of couple-stress thermo-diffusion. J. Math. Syst. Sci. **3**(7), 365–370 (2013). David Publishing, New York
8. Constanda, C.: Generalized Fourier Series. Mathematical Methods for Elastic Plates. Springer, London (2014)
9. Dragomir, S.: A survey on Cauchy–Bunyakovsky–Schwarz type discrete inequalities. J. Inequal. Pure Appl. Math. (JIPAM) **4**(3), 222–226 (2003)
10. Eshkuvatov, Z., Long, N.: Approximate solution of singular integral equations of the first kind with Cauchy kernel. Appl. Math. Lett. (2009). doi:10.1016/j.aml.2008.08.001
11. Ezzat, M., Zakaria, M.: Generalized thermoelasticity with temperature dependent modulus of elasticity under three theories. J. Appl. Math. Comput. **14**, 193–212 (2004)
12. Gelashvili, D.: To Ward the Theory of Dynamic Problems of Couple-Stress Thermodiffusion of Deformable Solid Micropolar Elastic Bodies. North-Holland, Amsterdam (1979)
13. Gilbarg, D., Trudinger, N.: Elliptic Partial Differential Equations of Second Order. Springer, Berlin (2001)
14. Green, E., Lindsay, K.A.: Thermoelasticity. J. Elast. **2**, 1–7 (1972)
15. Hakl, R., Zamora, M.: Fredholm-type theorem for boundary value problems for systems of nonlinear functional differential equations. Bound. Value Probl. **2014**(1), 113 (2014). Springer
16. Kupradze, V. et al.: Three-Dimensional Problems of the Mathematical Theory of Elasticity and Thermo Elasticity. North-Holland, Amsterdam-New York (1983)
17. Lord, H., Shulman, Y.: A generalized dynamical theory of thermoelasticity. J. Mech. Phys. Solids **15**, 299–309 (1967)
18. Orlando, F.: Hankel Functions. Mathematical Methods for Physicists, 3rd edn. Academic, Orlando (1985)
19. Qing, H., Fanghua, L.: Elliptic Partial Differential Equations. Courant Institute of Mathematical Science, New York (1997)
20. Rezazadeh, G., Vahdat, A.: Thermoelastic damping in a micro-beam resonator using modified couple stress theory. Acta Mechanica **223**(6), 1137–1152 (2012). Springer
21. Sherief, H., Hamza, F., Saleh, H.: The theory of generalized thermoelastic diffusion. Int. J. Eng **5**, 591–608 (2004). Elsevier
22. Tripathi, J., Kedar, G., Deshmukh, K.: Dynamic problem of generalized thermoelasticity for a semi-infinite cylinder with heat sources. J. Thermoelast. **2**(1), 01–08 (2014)
23. Youssef, HM.: Theory of two-temperature-generalized thermoelasticity. J. Appl. Math. IMA **71**(3), 383–390 (2006)

Symmetry-Breaking Bifurcations in Laser Systems with All-to-All Coupling

Juancho A. Collera

Abstract We consider a system of n semiconductor lasers with all-to-all coupling that is described using the Lang-Kobayashi rate equations. The lasers are coupled through their optical fields with delay arising from the finite propagation time of the light from one laser to another. As a consequence of the coupling structure, the resulting system of delay differential equations is equivariant under the symmetry group $\mathbf{S}_n \times \mathbf{S}^1$. Since symmetry gives rise to eigenvalues of higher multiplicity, implementing a numerical bifurcation analysis to our laser system is not straightforward. Our results include the use of the equivariance property of the laser system to find symmetric solutions, and to correctly locate steady-state and Hopf bifurcations. Additionally, this method identifies symmetry-breaking bifurcations where new branches of solutions emerge.

1 Introduction

Semiconductor lasers are highly sensitive to optical feedback that even a small amount of optical feedback is enough to produce chaotic instabilities [7, 8]. In 1980, Lang and Kobayashi [9] examined the influences of external optical feedback on semiconductor laser properties. A single mode laser was examined where a portion of the laser output is reflected back to the laser cavity from an external mirror. The dimensionless form of the Lang-Kobayashi (LK) rate equations derived in [2] are given by the following system of differential equations

$$\begin{aligned} \dot{E}(t) &= (1 + i\alpha)N(t)E(t) + \kappa e^{-i\Omega\tau}E(t-\tau), \\ T\dot{N}(t) &= P - N(t) - (1 + 2N(t))|E(t)|^2, \end{aligned} \tag{1}$$

J.A. Collera (✉)
Department of Mathematics and Computer Science, University of the Philippines Baguio, Gov. Pack Road, Baguio City 2600, Philippines
e-mail: jacollera@up.edu.ph

© Springer International Publishing Switzerland 2016

J. Bélair et al. (eds.), *Mathematical and Computational Approaches in Advancing Modern Science and Engineering*, DOI 10.1007/978-3-319-30379-6_8

where $E(t)$ is the complex electric field, $N(t)$ is the excess carrier number, and the fixed delay time τ represents the external cavity roundtrip time of the feedback light. The parameters α, κ, Ω, T and P correspond to the linewidth enhancement factor, feedback strength, angular frequency of the solitary laser, electron decay rate, and pump parameter, respectively. System (1) was shown to correctly describe the dominant effects observed experimentally, see for example [10–12].

In 2006, Erzgraber et al. [4] studied a model of two mutually delay-coupled semiconductor lasers in a face-to-face configuration. The two lasers are coupled through their optical field, and the finite propagation time of the light from one laser to the other constitute the time delay. A special case with zero detuning given by the following LK-type rate equations

$$\begin{aligned} \dot{E}_1(t) &= (1 + i\alpha)N_1(t)E_1(t) + \kappa e^{-iC_p}E_2(t-\tau), \\ \dot{E}_2(t) &= (1 + i\alpha)N_2(t)E_2(t) + \kappa e^{-iC_p}E_1(t-\tau), \\ T\dot{N}_1(t) &= P - N_1(t) - (1 + 2N_1(t))|E_1(t)|^2, \\ T\dot{N}_2(t) &= P - N_2(t) - (1 + 2N_2(t))|E_2(t)|^2, \end{aligned} \tag{2}$$

where C_p is called the coupling phase parameter, was shown to have great importance as this case organizes the dynamics for small non-zero detuning. Notice that (2) is in fact an extension of the LK equations in (1) into the case of two mutually delay-coupled semiconductor lasers.

In this paper, we consider a system of n semiconductor lasers with all-to-all coupling. The lasers are coupled through their optical fields with time delay arising from the finite propagation time of the light from one laser to another. This generalizes the one-laser case in (1) and the zero-detuning two delay-coupled lasers case in (2). As a consequence of the coupling structure, the resulting system of delay differential equations (DDEs) is equivariant under a symmetry group. Symmetric systems are known to give rise to eigenvalues with higher multiplicity [6] and this makes numerical bifurcation analysis of such systems harder to implement. Our results include employing a group-theoretic approach from [6] to overcome this problem. This method not only locates steady-state and Hopf bifurcations correctly but also identifies symmetry-breaking (SB) bifurcations where branches of new solutions emerge.

The paper is organized as follows. In Sect. 2, we introduce our model, its symmetry properties and basic solutions. Then, in Sect. 3 we show how to find symmetric solutions and use numerical continuation to find branches of solutions and their stability. In Sect. 4, we give our main result which is a method of finding and classifying steady-state and Hopf bifurcations. We also look at the symmetry group of bifurcating branches of solutions from SB bifurcations. Then, lastly we give our conclusions.

2 Laser Systems with All-to-All Coupling

We now consider the n-laser system with all-to-all coupling as described by the following LK rate equations, for $j = 1, \ldots, n$,

$$\begin{aligned} \dot{E}_j(t) &= (1 + i\alpha)N_j(t)E_j(t) + \kappa e^{-iC_p} \sum_{k=1,\ k\neq j}^{n} E_k(t-\tau), \\ T\dot{N}_j(t) &= P - N_j(t) - (1 + 2N_j(t))|E_j(t)|^2. \end{aligned} \tag{3}$$

To determine the symmetry group of system (3), we first define the action of the permutation group $\mathbf{S}_n$ and the circle group $\mathbf{S}^1$ to the state variables $(E_j(t), N_j(t))$, $j = 1, \ldots, n$, as follows:

$$\rho \cdot (E_j(t), N_j(t)) = (E_{\rho^{-1}(j)}, N_{\rho^{-1}(j)}), \qquad \text{and} \qquad \vartheta \cdot (E_j(t), N_j(t)) = (E_j e^{i\vartheta}, N_j),$$

where $\rho \in \mathbf{S}_n$ and $\vartheta \in \mathbf{S}^1$. Notice that the action of $\mathbf{S}_n$ permutes the position of the lasers, while $\mathbf{S}^1$ acts only on the optical fields $E_j(t)$ for all j. Moreover, observe that if $(E_j(t), N_j(t))$ is a solution to (3) for all j, then $\rho \cdot (E_j(t), N_j(t))$ and $\vartheta \cdot (E_j(t), N_j(t))$ are also solutions to (3) for all j. Hence, system (3) is equivariant under the group $\mathbf{S}_n \times \mathbf{S}^1$.

Basic solutions to (3), called *compound laser modes* (CLMs), are of the form $E_j(t) = R_j e^{i\omega t + i\sigma_j}$ and $N_j(t) = N_j$ for $j = 1, \ldots, n$, where ω, $R_j > 0$, σ_j, and N_j are all real-valued for all j, and with $\sigma_1 = 0$. We refer to CLMs with $R_j = R$, $\sigma_j = 0$, and $N_j = N$ for all j, as the *fully symmetric CLMs* and are given by

$$E_j(t) = Re^{i\omega t}, \qquad \text{and} \qquad N_j(t) = N, \tag{4}$$

for $j = 1, \ldots, n$. These fully symmetric CLMs are fixed by elements $(\rho, 0)$ of $\mathbf{S}_n \times \mathbf{S}^1$ for all $\rho \in \mathbf{S}_n$. We use the symbol $\mathbf{S}_n^0$ to denote this symmetry group of CLMs in (4). Note that this subgroup $\mathbf{S}_n^0$ of $\mathbf{S}_n \times \mathbf{S}^1$ is isomorphic to $\mathbf{S}_n$ and is generated by all elements of the form $(\rho, 0)$ where $\rho \in \mathbf{S}_n$. We call such symmetry group of CLMs as *isotropy subgroup* [5].

3 Symmetric CLMs

In this section, we provide a method of finding symmetric CLMs, that is, CLMs fixed by an isotropy subgroup of $\mathbf{S}_n \times \mathbf{S}^1$. In particular, we use the case when the isotropy subgroup is $\mathbf{S}_n^0$ to find the fully symmetric CLMs in (4).

Substituting the ansatz in (4) to system (3) and then following a similar computation in [4], we obtain the following transcendental equation in ω

$$\omega + (n-1)\kappa\sqrt{1+\alpha^2}\sin(C_p + \omega\tau + \tan^{-1}\alpha) = 0. \tag{5}$$

Once a value for ω is found by solving (5), we can then compute for corresponding values of $N = \omega/(\alpha + \tan(C_p + \omega\tau))$ and $R = \sqrt{(P-N)/(1+2N)}$. Fully symmetric CLMs in (4) are obtained using these values of R, ω, and N. It is worth noting that the isotropy subgroup of the CLMs that we seek provides relations amongst R_j, N_j, and σ_j, and this to some extent simplifies the form of the transcendental equation, such as in (5), which is key in finding symmetric CLMs.

A branch of CLMs can be obtained through numerical continuation by varying a single parameter. We use DDE-Biftool [3] to obtain such branch of solutions. We employ the same technique as in [4] to follow a CLM as an equilibrium in DDE-Biftool. We choose to vary the coupling phase parameter C_p for two reasons. First, because system (3) has a 2π-translational symmetry in C_p, and secondly, because the coupling phase can be changed accurately in experiments [1].

Example 1 Consider system (3) with $n = 4$, and parameters $\alpha = 2.5$, $T = 392$, $P = 0.3$, $\tau = 20$, $\kappa = 0.1$, and $C_p = 10$. Solving for ω in (5) gives 11 CLMs which are shown in the left panel of Fig. 1 as dots. By following any of these CLMs in DDE-Biftool, a branch of fully symmetric CLMs is obtained which is the ellipse in the left panel of Fig. 1. The stability of this branch is also determined using DDE-Biftool. The right panel of Fig. 1 shows the stable and unstable parts of the branch in dashed line and solid line, respectively.

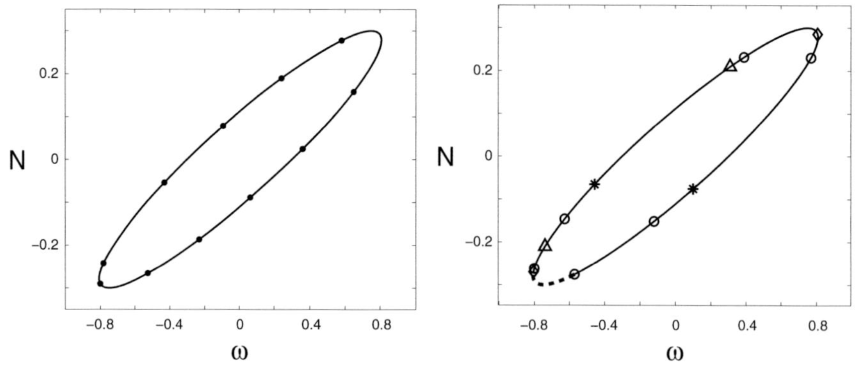

Fig. 1 (*Left*) A branch of fully symmetric CLMs obtained in DDE-Biftool by varying the coupling phase parameter C_p. (*Right*) Stable part of the branch is shown in *dashed line* while unstable part of the branch is shown in *solid line*. Regular steady-state and Hopf bifurcations are marked with ($\diamond$) and ($\circ$), respectively. For SB bifurcations, we use (Δ) and ($*$) to marked pitchfork and SB Hopf bifurcations, respectively

Symmetric systems, such as (3), are known to have eigenvalues of higher multiplicity. Hence, implementing a numerical bifurcation analysis on such system is not straight-forward. In the section that follows, we develop a method to locate steady-state and Hopf bifurcations along a branch of solutions and classify these bifurcations into regular and symmetry-breaking.

4 Symmetry-Breaking Bifurcations

We now give our main result which is a method of finding and classifying steady-state and Hopf bifurcations.

We first determine the linearized system corresponding to (3) around the fully symmetric CLM (4). We follow a similar computation done in [13] for the one-laser model in (1). In polar form, the LK rate equations in (3) are given by

$$\begin{aligned} \dot{R}_j(t) &= N_j(t)R_j(t) + \kappa \sum_{k=1,\, k\neq j}^{n} R_k(t-\tau)\, \cos\big(-C_p + \varphi_k(t-\tau) - \varphi_j(t)\big), \\ \dot{\varphi}_j(t) &= \alpha N_j(t) + \kappa \sum_{k=1,\, k\neq j}^{n} \frac{R_k(t-\tau)}{R_j(t)}\, \sin\big(-C_p + \varphi_k(t-\tau) - \varphi_j(t)\big), \\ \dot{N}_j(t) &= \frac{1}{T}\Big[P - N_j(t) - (1 + 2N_j(t))\, \big|R_j(t)\big|^2\Big], \end{aligned} \tag{6}$$

for $j = 1, \ldots, n$. If we let $X_j(t) = \big[R_j(t), \varphi_j(t), N_j(t)\big]^T$ and $Y_j(t) = X_j(t-\tau)$, then (6) can be written in the form $\dot{X}_j(t) = f(X_j(t), Y_1(t), \ldots, Y_{j-1}(t), Y_{j+1}(t), \ldots, Y_n(t))$, for $j = 1, \ldots, n$. Now, let $X(t) = [X_1(t), \ldots, X_n(t)]^T$ and $Y(t) = [Y_1(t), \ldots, Y_n(t)]^T$ so that the full system can be written as $\dot{X}(t) = F(X(t), Y(t))$. The fully symmetric CLM in (4) written in polar form is given by $X^* = [X_1^*(t), \ldots, X_n^*(t)]^T$ where $X_j^*(t) = [R, \omega t, N]^T$ for all j. To obtain the linear variational equation around X^*, we first compute for $\overline{A} := d_{X_j(t)} f(X_j^*)$ and $\overline{B} := d_{Y_j(t)} f(X_j^*)$. Now, let M_1 be the block-diagonal matrix with the block $\overline{A}$ on the main diagonal and the block 0 elsewhere, and let M_2 be the block matrix with the block 0 on the main diagonal and the block $\overline{B}$ on all off main diagonal entries. Then, the Jacobian matrix evaluated at X^* is $dF(X^*) = [M_1 \mid M_2]$. The linear variational equation around the fully symmetric CLM is given by $\dot{X}(t) = M_1 X(t) + M_2 X(t-\tau)$, and its corresponding characteristic equation is $\det \Delta(\lambda) = 0$ where $\Delta(\lambda) = \lambda I_n - M_1 - e^{-\lambda\tau} M_2$. Notice that if we let $A := \lambda I_3 - \overline{A}$ and $B := -e^{-\lambda\tau}\overline{B}$, then

$$L := \Delta(\lambda) = \begin{bmatrix} A & B & \cdots & B \\ B & A & \cdots & B \\ \vdots & & \ddots & \vdots \\ B & B & \cdots & A \end{bmatrix}$$

where blocks A and B are 3×3 matrices.

We now use a method from [6] to examine the eigenvalues of L. Let $\zeta = e^{\frac{2\pi}{n}i}$ and define $V_k = \left\{[v, \zeta^k v, \zeta^{2k} v, \ldots, \zeta^{(n-1)k} v]^T \mid v \in \mathbf{R}^3\right\}$, for $k = 0, 1, 2, \ldots, n-1$. Observe that $V_0 = \{[v, v, v, \ldots, v]^T \mid v \in \mathbf{R}^3\}$ and the action of L on V_0 is given by $L[v, \ldots, v]^T = [(A + (n-1)B)v, \ldots, (A + (n-1)B)v]^T$. This means that the eigenvalues of $L|_{V_0}$ are those of $A + (n-1)B$. In general, the eigenvalues of $L|_{V_k}$, for $k = 0, 1, 2, \ldots, n-1$, are those of $A + \left(\sum_{j=1}^{n-1} \zeta^{jk}\right) B$. When n is odd and $k \neq 0$, the complex numbers $\zeta^k, \zeta^{2k}, \ldots, \zeta^{(n-1)k}$ can be grouped into pairs of conjugates. Observe that $\zeta^{jk} + \zeta^{(n-j)k} = 2\cos(2\pi kj/n)$ since ζ^{jk} and $\zeta^{(n-j)k}$ are conjugates. Hence, $\sum_{j=1}^{n-1} \zeta^{jk} = \sum_{j=1}^{(n-1)/2} 2\cos(2\pi kj/n) = -1$ using the Dirichlet kernel identity $1 + \sum_{j=1}^{N} 2\cos(\theta j) = \sin(N\theta + \frac{1}{2}\theta)/\sin(\frac{1}{2}\theta)$. When n is even and $k \neq 0$, we have $\zeta^{n/2} = -1$. Consequently,

$$\sum_{j=1}^{n-1} \zeta^{jk} = (-1)^k + \sum_{j=1}^{\frac{n}{2}-1} 2\cos(2\pi kj/n) = (-1)^k + \frac{\sin(\pi k - \pi k/n)}{\sin(\pi k/n)} - 1$$

by pairing up conjugates, and then using the Dirichlet kernel identity. Now, since $\sin(\pi k - \pi k/n) = (-1)^{k+1}\sin(\pi k/n)$ and $\sin(\pi k/n) \neq 0$ for $1 \leq k \leq n-1$, we get $\sum_{j=1}^{n-1} \zeta^{jk} = -1$. This is essentially the same for the case when n is odd. Therefore, the eigenvalues of $L|_{V_k}$, for $k = 1, 2, \ldots, n-1$, are those of $A - B$. This means that the problem of solving the characteristic equation $\det \Delta(\lambda) = 0$, reduces to solving the equations $\det(A + (n-1)B) = 0$ and $\det(A - B) = 0$. Furthermore, the eigenvalues of L from $A + (n-1)B$ are simple while those from $A - B$ are of multiplicity $n - 1$. Also, notice that the symmetry group $\mathbf{S}_n^0$ acts trivially on V_0 while its action on V_k, for $k = 1, 2, \ldots, n-1$, is non-trivial. This means that SB bifurcations are obtained from block $A - B$ while bifurcations from block $A + (n-1)B$ are regular bifurcations.

We now illustrate the above technique in finding ordinary and SB bifurcations.

Example 2 Continuing from Example 1, we first look for SB bifurcations. Pitchfork bifurcations along the branch of fully symmetric CLMs are found by looking at the intersections of the curves $\det(A - B)|_{\lambda=0} = 0$ and (5), while SB Hopf bifurcations are found by finding the intersections of the curves $\det(A - B)|_{\lambda=i\beta} = 0$ with $\beta > 0$, and (5). Two pitchfork bifurcations and two SB Hopf bifurcations were found and are shown in the right panel of Fig. 1 with markers ($\triangle$) and ($*$), respectively. Similarly, we can get the regular bifurcations from the block $A + 3B$. Saddle-node bifurcations are obtained from the intersections of the curves $\det(A + 3B)|_{\lambda=0} = 0$ and (5), while regular Hopf bifurcations were found by intersecting the curves $\det(A + 3B)|_{\lambda=i\beta} = 0$ with $\beta > 0$, and (5). Two saddle-node bifurcations and six regular Hopf bifurcations were obtained and are shown in the right panel of Fig. 1 using markers ($\diamond$) and ($\circ$), respectively.

Identification of SB bifurcations is important because they give rise to new branches of solutions. We now examine the symmetry group of branches of solutions that emerge from the SB bifurcations obtained in Example 2. To do this,

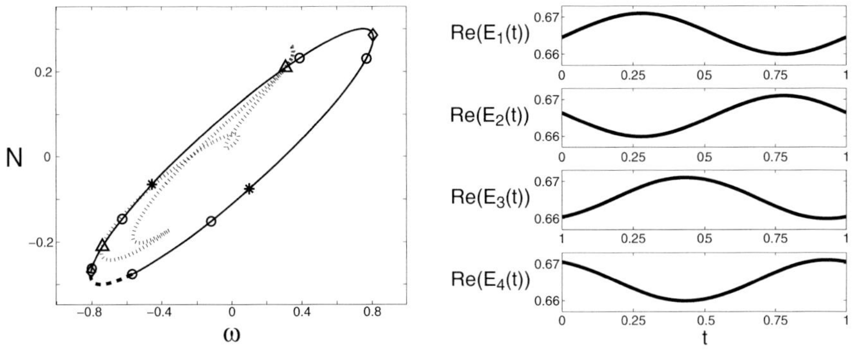

Fig. 2 (*Left*) A new branch of CLMs shown in *dotted curve* emanates from the pitchfork bifurcations (△) and has symmetry group $\mathbf{S}_3^0$. (*Right*) Time-series plots showing that a branch of periodic solutions emerging from the SB Hopf bifurcations (∘) has spatio-temporal symmetry $\mathbf{Z}_2(\mu, \pi)$

we use DDE-Biftool to perform numerical continuation, again varying the coupling phase parameter C_p. A new branch of symmetric CLMs shown as the dotted curve in the left panel of Fig. 2 is obtained. This new branch of CLMs emanates from the pitchfork bifurcations (△) and has a smaller symmetry group $\mathbf{S}_3^0$ compared to the symmetry group $\mathbf{S}_4^0$ of the fully symmetric CLMs.

Similarly, a branch of periodic solutions is obtained by following a SB Hopf bifurcation ($*$) in DDE-Biftool. This branch of periodic solutions has isotropy subgroup $\mathbf{Z}_2(\mu, \pi)$, which is a subgroup of $\mathbf{S}_4 \times \mathbf{S}^1$ that is isomorphic to $\mathbf{Z}_2$ and is generated by the order-two element (μ, π). Here, μ is an element of $\mathbf{S}_4$ whose action interchanges lasers 1 and 2, and lasers 3 and 4. Hence, the element (μ, π) interchanges lasers 1 and 2 (respectively, lasers 3 and 4) and then shifts the phase of each laser by half a period. The right panel of Fig. 2 shows a time series of the real part of the $E_j(t)$ for $j = 1, 2, 3, 4$ where the interval for t is one period. Notice that interchanging the curves for Re $E_1(t)$ and Re $E_2(t)$ (respectively, Re $E_3(t)$ and Re $E_4(t)$) and then shifting them by half a period preserves the plots. Indeed, the bifurcating branch of periodic solutions emanating from the SB Hopf bifurcations has isotropy subgroup $\mathbf{Z}_2(\mu, \pi)$.

5 Conclusion

The main contribution of this research is the identification of symmetry-breaking bifurcations in symmetric systems. The n-laser system (3) with all-to-all coupling is an example of a symmetric system. Symmetry gives rise to eigenvalues of higher multiplicity and this makes numerical bifurcation analysis of such system harder to implement. However, we showed that symmetry actually aids us in finding symmetric solutions in Example 1, and in locating steady-state and Hopf bifurcations along the branch of fully symmetric CLMs as illustrated in Example 2. The action of the symmetry group decomposes the physical space into invariant subspaces. This, in turn, gives us a means to identify symmetry-breaking bifurcations where new branches of solutions emerge.

Acknowledgements This work was funded by the UP System Emerging Interdisciplinary Research Program (OVPAA-EIDR-C03-011). The author also acknowledged the support of the UP Baguio through RLCs during the A.Y. 2014–2015.

References

1. Agrawal, G.P., Dutta, N.K.: Long-Wavelength Semiconductor Lasers. Van Nostrand Reinhold, New York (1986)
2. Alsing, P.M., Kovanis, V., Gavrielides, A., Erneux, T.: Lang and Kobayashi phase equation. Phys. Rev. A **53**, 4429–4434 (1996)
3. Engelborghs, K., Luzyanina, T., Samaey, G.: DDE-BIFTOOL v. 2.00: A MATLAB package for bifurcation analysis of delay differential equations. Technical report TW-330. Department of Computer Science, K.U. Leuven, Leuven (2001)
4. Erzgraber, H., Krauskopf, B., Lenstra, D.: Compound laser modes of mutually delay-coupled lasers. SIAM J. Appl. Dyn. Syst. **5**, 30–65 (2006)
5. Golubitsky, M., Stewart, I.: The Symmetry Perspective: From Equilibrium to Chaos in Phase Space and Physical Space. Birkhauser Verlag, Basel (2002)
6. Golubitsky, M., Stewart, I., Schaeffer, D.G.: Singularities and Groups in Bifurcation Theory, vol. II. Springer, New York (1988)
7. Gray, G.R., Ryan, A.T., Agrawal, G.P., Gage, E.C.: Control of optical-feedback-induced laser intensity noise in optical data recording. Opt. Eng. **32**, 739–745 (1993)
8. Gray, G.R., Ryan, A.T., Agrawal, G.P., Gage, E.C.: Optical-feedback-induced chaos and its control in semiconductor lasers. In: SPIE's 1993 International Symposium on Optics, Imaging, and Instrumentation, San Diego, pp. 45–57. International Society for Optics and Photonics (1993)
9. Lang, R., Kobayashi, K.: External optical feedback effects on semiconductor injection laser properties. IEEE J. Quantum Electron **16**, 347–355 (1980)
10. Mork, J., Tromborg, B., Mark, J.: Chaos in semiconductor lasers with optical feedback: theory and experiment. IEEE J. Quantum Electron **28**, 93–108 (1992)
11. Sano, T.: Antimode dynamics and chaotic itinerancy in the coherence collapse of semiconductor lasers with optical feedback. Phys. Rev. A **50**, 2719–2726 (1994)
12. van Tartwijk, G.H.M., Levine, A.M., Lenstra, D.: Sisyphus effect in semiconductor lasers with optical feedback. IEEE J. Select. Top. Quantum Electron. **1**, 466–472 (1995)
13. Verduyn Lunel, S.M., Krauskopf, B.: The mathematics of delay equations with an application to the Lang-Kobayashi equation. In: Krauskopf, B., Lenstra, D. (eds.) Fundamental Issues of Nonlinear Laser Dynamics. AIP Conference Proceedings, vol. 548. American Institute of Physics, New York (2000)

Effect of Jet Impingement on Nano-aerosol Soot Formation in a Paraffin-Oil Flame

Masoud Darbandi, Majid Ghafourizadeh, and Mahmud Ashrafizaadeh

Abstract In this paper, the effects of mico-jet impingement on the formation of soot nano-particles, CO, CO_2, and C_6H_6 species in a turbulent paraffin-oil flame are investigated numerically. In this regard, we use a two-equation κ-ϵ turbulence model, a PAH-inception two-equation soot model, a detailed chemical kinetic consisting of 121 species and 2613 elementary reactions, and steady flamelet combustion model. We take into account the turbulence-chemistry interaction by using presumed-shape probability density functions PDFs. We also take into account the radiation heat transfer of soot and gases assuming optically-thin flame. In the first place, we solve a documented experimental test case and compare the flame structure to evaluate our numerical results. Then, we embed a micro-scale injector at the burner wall, split the incoming air-flow between primary and secondary streams, inject the secondary air into the burner via the embeded injector, and compare the results for different values of impinging-jet mass flow rate. Our results show that the mico-jet impingement affects the reactive flow behavior within the burner and results in a compact-flame near the fuel-injector nozzle. Our calculations also show that the emission of CO_2 is slightly affected by the mico-jet impingement. They show that increasing the mass flow rate of the micro-impinging-jet effectively reduces the formation and emission of the soot nano-aerosol and CO, C_6H_6 pollutants. It is also found that mico-jet impingement leads to a more uniform profiles of temperature, soot volume fraction, and mass fractions of CO, CO_2, and C_6H_6 at the burner outlet.

M. Darbandi (✉)
Department of Aerospace Engineering, Center of Excellence in Aerospace Systems, Institute for Nanoscience and Nanotechnology, Sharif University of Technology, Tehran, P. O. Box 11365-8639, Iran
e-mail: darbandi@sharif.edu

M. Ghafourizadeh
Department of Aerospace Engineering, Center of Excellence in Aerospace Systems, Sharif University of Technology, Tehran, P. O. Box 11365-8639, Iran

M. Ashrafizaadeh
Department of Mechanical Engineering, Isfahan University of Technology, Isfahan, P. O. Box 84156-83111, Iran
e-mail: mahmud@cc.iut.ac.ir

© Springer International Publishing Switzerland 2016
J. Bélair et al. (eds.), *Mathematical and Computational Approaches in Advancing Modern Science and Engineering*, DOI 10.1007/978-3-319-30379-6_9

1 Introduction

There are several types of flame holders to maintain the combustion process inside a burner. Combustion can be stabilized using (1) an external energy supply such as heated plate, torch, highly reactive chemicals, lasers, electric arcs, chemical additives, (2) bluff bodies such as V-gutters, cylinders, spheres, flat plates, (3) steps such as rearward facing steps, forward facing steps, side dumps, cavities, (4) flow instabilities such as swirls, cyclones, (5) jets such as reverse flow jets, opposed jets, and fuel jet blockages [1]. The choice of flame holding method depends on the burner size, weight, pressure loss, flammability limits, material temperature, and structural limitations. Most of these methods stabilize flames through establishing recirculation zones which carry high-temperature products upstream to separate flow region for igniting the incoming flows. Among the above mentioned methods, the opposed jet, i.e. an impinging jet, is an interesting flame-holding method regarding the limitations. So, a deep understanding on the behavior of impinging jets would help to keep a stable combustion in combustors burning reactants and forming products.

Many researchers have studied the formation of CO and CO_2 pollutants in combustion processes while nano-particulate soot, incorporated with hazardous aromatics, i.e. C_6H_6, needs special attention and considerations. Even a low emission of these pollutants can cause serious health problems, i.e. different organs cancer. On the other hand, C_6H_6 has been widely used as an important component of gasoline to enhance the octane number of fuels. So, this aromatic compound requires to be understood very well due to vast applications in engines as well as its carcinogenicity.

Literature shows aerosol modeling of soot formation in laminar flames has been studied in last decades [2, 3]. Few researchers also studied soot formation in turbulent flames fueled by simple hydrocarbons [4, 5]. There is a lack of resources studying the formation of nano-particulate soot aerosol in turbulent flames burning common fuels such as paraffin-oil, gasoline, jet fuel, or diesel. Brooks and Moss [4] proposed soot acetylene-inception model and showed that soot formation in simple fuels is limited by inception rate of particles from single-ring aromatics formed from acetylene. However, soot formation in practical fuels, which have higher amounts of carbon atoms as well as aromatics compounds, is limited by growth rate of aromatics. In this regard, Hall et al. [6] proposed a soot model, i.e. soot PAH-inception model, with inception rate based on two and three-ringed aromatics formed from single-ring aromatic species, i.e. C_6H_6 and C_6H_5 radical.

Back to our past publications, we have already simulated the nano-aerosol soot formation in turbulent non-premixed flames fed with methane [7, 8], ethylene [9, 10], and propane [11, 12]. In this paper, we use and extend our previous studies and simulate the nano-aerosol soot formation in a turbulent non-premixed flame burning paraffin-oil as its fuel. In this regard, we employ a two-equation soot model to solve for the soot mass fraction and the number density considering the soot formation and its oxidation based on PAHs and OH agents, respectively. We utilize a steady flamelet model as our combustion model considering a large-detailed kinetic reaction mechanism with 2613 reversible chemical reactions and 121 chemical species. We use a two-equation κ-ϵ turbulence model with round-jet corrections and take into account the turbulence-chemistry interaction using some presumed-shape PDFs. We also take into account the radiation heat transfer of the soot and gases assuming an optically-thin flame and calculate their radiations locally only by emissions. To evaluate our numerical solution, we simulate a benchmark turbulent paraffin-oil non-premixed flame inside a burner and compare the obtained results with those of experiment as well as another numerical study. The obtained results indicate our numerical simulation can predict soot volume fraction, mixture fraction, and temperature distributions of the flame. Then, we embed a micro-scale injector at burner wall, split incoming air-flow between primary- and secondary-air streams and inject the secondary-air into burner via the micro-scale injector and compare the results. We also study mass-flow-rate effect of the micro-scale injector on nano-aerosol soot formation, the emissions of CO, CO_2, and C_6H_6 from turbulent non-premixed paraffin-oil flame inside burner.

2 The Governing Equations

In the cylindrical coordinates, i.e. r, z, the fluid flow conservation laws consisting of continuity, r-momentum, and z-momentum are given by

$$\nabla \cdot (\rho \mathbf{V}) + \rho \frac{u}{r} = 0 \tag{1}$$

$$\nabla \cdot (\rho \mathbf{V} u) = -\frac{\partial p}{\partial r} + \nabla \cdot (\mu_e \nabla u) - \mu_e \frac{u}{r^2} + \frac{\mu_e}{r}\frac{\partial u}{\partial r} \tag{2}$$

$$\nabla \cdot (\rho \mathbf{V} v) = -\frac{\partial p}{\partial z} + \nabla \cdot (\mu_e \nabla v) + \frac{\mu_e}{r}\frac{\partial v}{\partial r} - \rho g \tag{3}$$

The radial and axial components of the velocity vector are u and v, respectively. The mixture density, velocity vector, pressure, and effective viscosity are represented by ρ, $\mathbf{V}$, p, and μ_e, respectively. The transport equations for turbulence quantities, i.e. turbulence kinetic energy κ and its dissipation rate ϵ, are given by

$$\nabla \cdot (\rho \mathbf{V} \kappa) = \nabla \cdot \left(\frac{\mu_e}{\sigma_\kappa} \nabla \kappa \right) + \frac{\mu_e}{\sigma_\kappa r}\frac{\partial \kappa}{\partial r} + G_\kappa - \rho \epsilon \tag{4}$$

$$\nabla \cdot (\rho \mathbf{V} \epsilon) = \nabla \cdot \left(\frac{\mu_e}{\sigma_\epsilon} \nabla \epsilon \right) + \frac{\mu_e}{\sigma_\epsilon r}\frac{\partial \epsilon}{\partial r} + \frac{\epsilon}{\kappa}(c_1 G_\kappa - c_2 \rho \epsilon) \tag{5}$$

In a confined jet, the turbulence model constants for Eqs. (4) and (5) are taken from Ref. [13]. To model combustion in a turbulent diffusion flame, we use the steady flamelet model. In this study, we choose a detailed kinetic scheme, i.e., 121 chemical species and 2613 chemical reactions, to perform our simulations. The transport equations for the first two moments of mixture fraction, i.e., f and f''^2, are given by

$$\nabla \cdot (\rho \mathbf{V} f) = \nabla \cdot \left(\frac{\mu_e}{\sigma_f} \nabla f \right) + \frac{\mu_e}{\sigma_f r} \frac{\partial f}{\partial r} \tag{6}$$

$$\nabla \cdot \left(\rho \mathbf{V} f''^2\right) = \nabla \cdot \left(\frac{\mu_e}{\sigma_f} \nabla f''^2 \right) + \frac{\mu_e}{\sigma_f r} \frac{\partial f''^2}{\partial r} + c_g \mu_e (\nabla f)^2 - \rho c_\chi \frac{\epsilon}{\kappa} f''^2 \tag{7}$$

As is known, the results from pre-computed laminar flamelets and turbulent statistics can be tabulated as a 3D lookup table in terms of mixture fraction, mixture fraction variance, and the scalar dissipation rate. If so, all thermo-chemical quantities in the solution domain can be obtained from this 3D lookup table [14].

To simulate the aerosol dynamics, we need to solve two transport equations for the soot mass fraction m^* and the soot number density n^* [7]. These equations are given by

$$\nabla \cdot (\rho \mathbf{V} m^*) = \nabla \cdot \left(\frac{\mu_e}{\sigma_{soot}} \nabla m^* \right) + \frac{\mu_e}{\sigma_{soot} r} \frac{\partial m^*}{\partial r} + S_{m^*} \tag{8}$$

$$\nabla \cdot (\rho \mathbf{V} n^*) = \nabla \cdot \left(\frac{\mu_e}{\sigma_{nuc}} \nabla n^* \right) + \frac{\mu_e}{\sigma_{nuc} r} \frac{\partial n^*}{\partial r} + S_{n^*} \tag{9}$$

Assuming a unit Lewis number, the total enthalpy equation h is given by

$$\nabla \cdot (\rho \mathbf{V} h) = \nabla \cdot \left(\frac{\mu_e}{\sigma_h} \nabla h \right) + \frac{\mu_e}{\sigma_h r} \frac{\partial h}{\partial r} + q_{rad} \tag{10}$$

The radiation source term in the energy conservation law, can be determined locally only by emission assuming an optically thin flame. Finally, the density is obtained from the equation of state as $p = \rho RT \sum_{m=1}^{n} Y_m / W_m$ where m counts the number of chemical species in the mixture from 1 to n total number of species. The temperature, gas constant, mass fraction, and molecular weight are represented by T, R, Y, and W, respectively.

3 Computational Method

Back to our past experiences [7–12], we choose a hybrid finite-volume-element FVE discipline and break the solution domain into a large number of quadrilateral elements. We use our past experiences in treating non-staggered (collocated) grid arrangement and implement finite element shape functions and physical influence upwind scheme PIS for the diffusion and convection terms, respectively. Using PIS scheme we also handle the pressure-velocity coupling effectively.

4 The Benchmark Test Case and Validation

A paraffin-oil turbulent non-premixed flame is chosen to verify our numerical solutions. We employ the experimental conditions of Young et al. [15] to perform our simulations. Figure 1 shows the configuration of the burner consuming paraffin-oil. Because of the symmetry of problem, we consider a rectangular solution domain applying the symmetry boundary conditions at the center line. The computational domain has 0.0775×0.6 m dimensions, i.e. $R_0 = 0.0775$ m and $L = 0.6$ m, see Fig. 1. The fuel nozzle diameter is 1.5 mm, i.e. $R_1 = 0.75$ mm. This fuel nozzle injects the gaseous paraffin-oil, which consists of 80 % *n*-decane, i.e. $C_{10}H_{22}$, and 20 % toluene, i.e. C_7H_8, as fuel at a speed of 22.28 m/s into the burner. The oxidizer, i.e. co-flow air stream, which consists of 23.3 % oxygen and 76.7 % nitrogen, enters the burner at a speed of 0.234 m/s. The initial temperatures of paraffin-oil and air are 598 K and 288 K, respectively. As understood, the paraffin-oil is evaporated and delivered through a heated line and injected into the burner. The turbulence intensity and eddy length scale for both the paraffin-oil and air streams are 3 % and 0.02 m at their nozzle exits, respectively.

To evaluate the accuracy of our numerical solution in simulating turbulent reacting flow and aerosol modeling of soot nano-particles, we solve the test case given in Young et al. [15] and compare the predicted flame structure, i.e. the distributions of mixture fraction, temperature, and species concentrations, with the data collected by this reference as well as another available numerical solution [16]. Wen et al. [16] numerically solved this test case considering a detailed chemical kinetic mechanism with 141 species and 1015 elemental reactions to construct the flamelet library. Figure 2 presents the axial distributions of mixture fraction, temperature, and soot volume fraction at centerline, i.e. $r = 0$. It also shows the radial distributions of mixture fraction, temperature, and soot volume fraction at $z = 0.3$ m above the burner. The figure shows that there are good agreements with the data reported by Young et al. [15]. Considering the experimental data [15], our results are more satisfactory than those of obtained by Wen et al. [16] in the context of same assumptions and models for turbulence, combustion, radiation, and soot. Our numerical results predict the flame length of 0.3 m, i.e. the stoichiometric value of mixture fraction for *n*-decane, i.e. $f_{st} = 0.0615$, which also corresponds to the peak of axial temperature distribution. As seen in Fig. 2, in far downstream of burner exit, the mixture fraction at centerline is lower than its stiochiometric value, i.e. the fuel-lean region, which indicates the fuel is fully depleted and that the burner is long enough to result in a full burn of paraffin-oil injected into the burner. As seen, the peak of axial temperature distribution is lower than the adiabatic flame temperature of paraffin-oil/air flame, i.e. 2366 K. There is a discrepancy between our results with

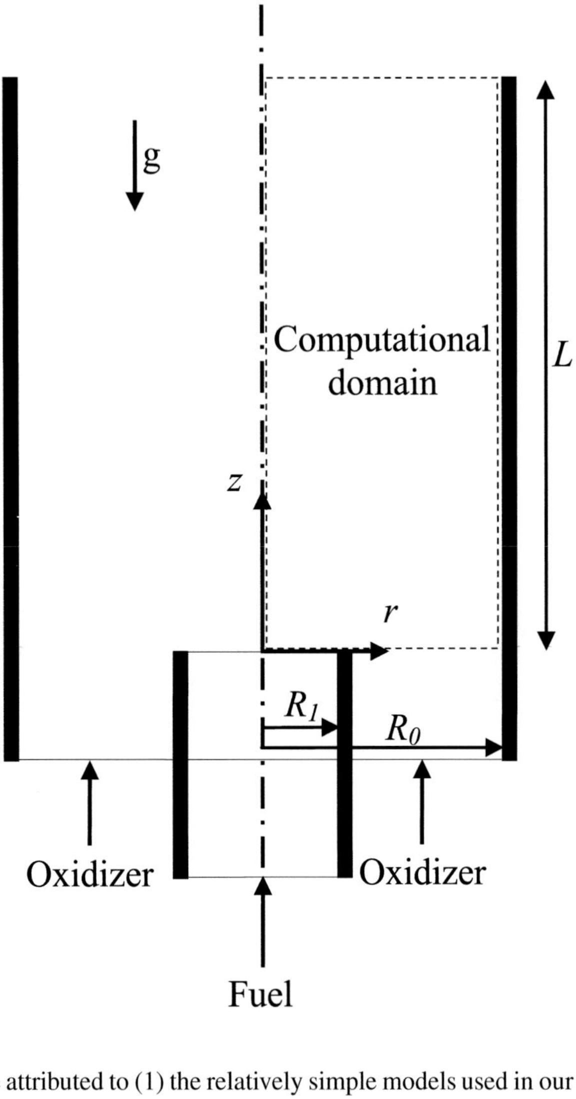

Fig. 1 The configuration of the burner consuming paraffin-oil [15]

the measured data. It can be attributed to (1) the relatively simple models used in our calculations, e.g., the turbulence model, the radiation model, and the soot model, (2) the assumptions made to simplify the problem, e.g., the gas-phase nucleation and the free-molecular-regime coagulation assumptions used in soot modeling, and the optically-thin flame assumption used in calculating the radiative heat transfer rate, which took into account only the most radiating species. Evidently, the use of more sophisticated radiation and soot models would help to overcome such shortcomings and predict the soot characteristics and flame structure more accurately, which is beyond the scope of this paper.

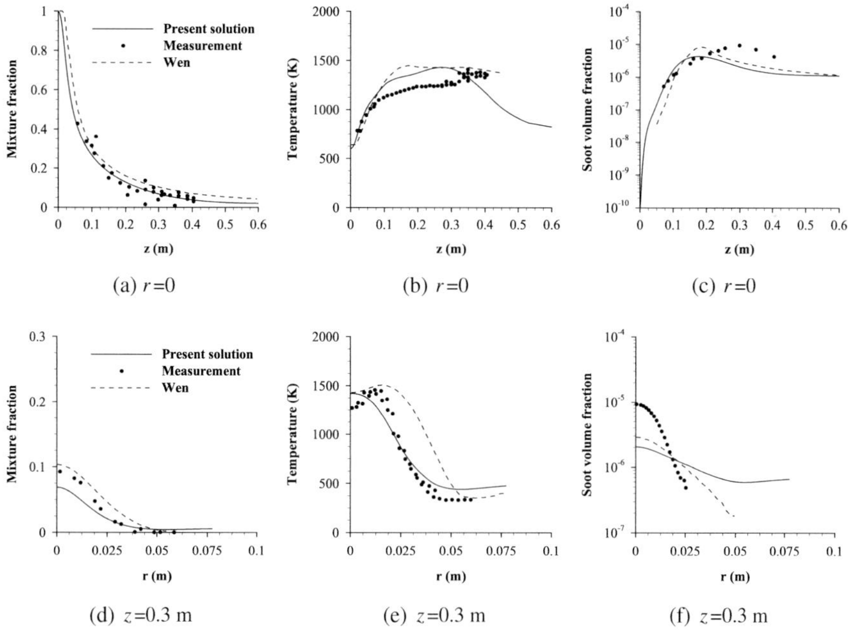

Fig. 2 The current axial and radial distributions of mixture fraction, temperature, and soot volume fraction in the flame and comparison with the experimental data [15] and numerical solution [16]; (**a**) $r = 0$, (**b**) $r = 0$, (**c**) $r = 0$, (**d**) $z = 0.3$ m, (**e**) $z = 0.3$ m, and (**f**) $z = 0.3$ m

5 The Results and Discussion

In this section, we study the effect of mico-jet impingement on the reactive flow behavior and the resulting emissions of CO, CO_2, C_6H_6 species and soot nano-aerosol using the current developed numerical method. We first study the effect of mico-jet impingement on the above parameters comparing with the benchmark test case. So, we embed a micro-scale injector at the burner wall, split the incoming air-flow between the primary- and secondary-air streams and inject the primary air into the burner via the embedded micro-scale injector. The dimension of micro-scale injector is 100 μm embedded at a height of $z = 0.1$ m above the fuel nozzle exit. We inject a mass flow rate of 2.7 g/s of the total incoming-air via the micro-scale injector while the rest of incoming air would enter the burner as primary-air. The other parameters, i.e. the geometry and boundary conditions (BCs), are similar to the benchmark test case.

Figure 3 shows the effect of mico-jet impingement on the distributions of stream function, temperature, OH and C_6H_6 mass fractions in the turbulent paraffin-oil flame. In each subfigure, the left part depicts the distributions for benchmark test case, i.e., without any mico-jet impingement, while the right part depicts the distribution for the burner incorporated with the mico-jet impingement, which

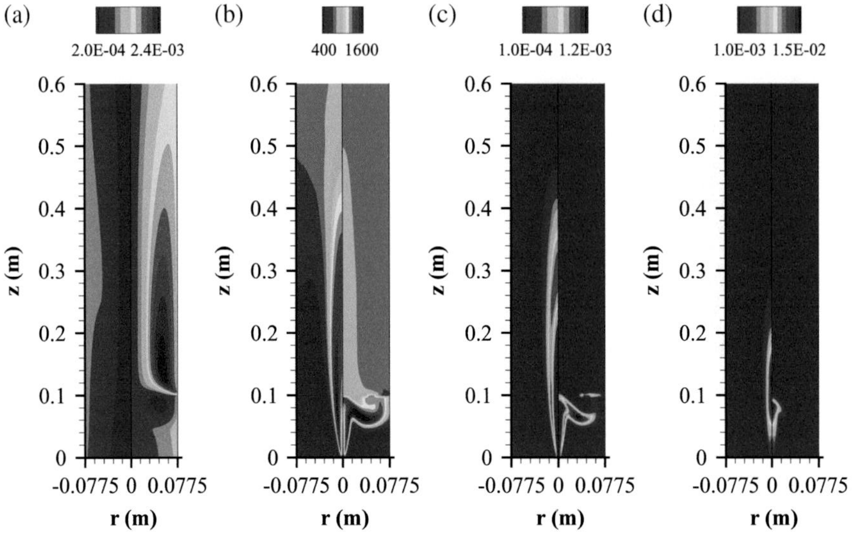

Fig. 3 The effect of mico-jet impingement on the distributions of (**a**) stream function (kg/s), (**b**) temperature (K), (**c**) OH mass fraction, and (**d**) C_6H_6 mass fraction in the turbulent paraffin-oil flame

injects a mass flow rate of 2.7 g/s via the embedded injector inserted at the burner wall. As seen, the mico-jet impingement causes two recirculation zones upstream and downstream of the micro-injector, which affect the flame envelop. The upstream recirculation, which is due to the injected air micro-jet, would cause the entrance of fresh air to the flame and this results in a well mixing of that with the gaseous high-temperature products. The downstream recirculation, which is due to the injected air micro-jet, would cause the return of high-temperature products at regions downstream of the burner, which is eventually mixed with the low-temperature gaseous reactants, i.e. the exhaust gas recirculation EGR. Such enhancement in mixing performance due to the appeared recirculation zones, which are formed upstream and downstream of the micro-impinging jet, would result in high-temperature combustion products at the burner outlet and a boost up in the combustion efficiency. The mixing enhancement also leads to a uniform temperature of exhaust gases at the burner outlet. So, the mico-jet impingement improves the combustion efficiency via enhancing the existing mixing. As seen, the mico-jet impingement also results in a compact distribution of OH mass fraction near the fuel injector exit. There are several definitions for the flame length. In the numerical simulations, this parameter is defined as the location of maximum OH concentration, the stoichiometric line, or the location of maximum temperature gradient. Considering the aforementioned definition, it is observed that the mico-jet impingement reduces the flame length and results in a compact flame. The compact flame together with the resulting EGR phenomenon would cause the by-products to be fully-burnt at the regions near the fuel-injector exit. So, C_6H_6 species would be

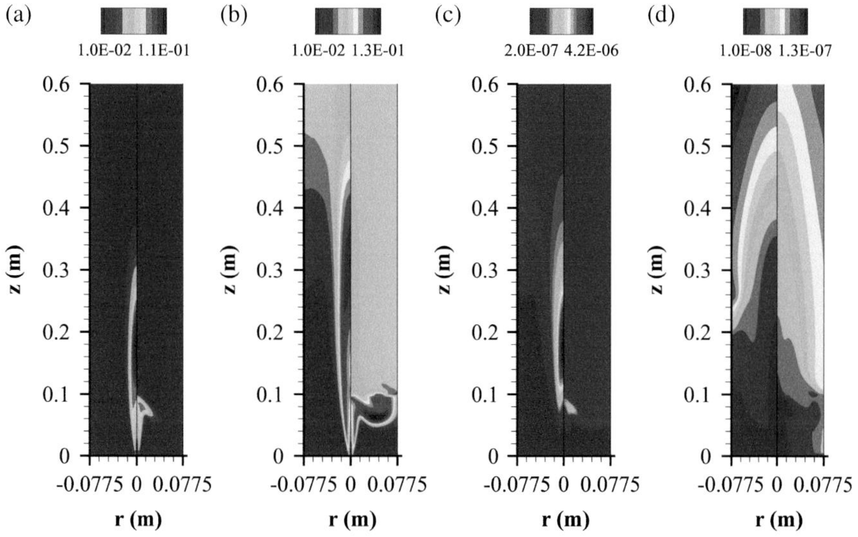

Fig. 4 The effect of mico-jet impingement on the distributions of (**a**) CO mass fraction, (**b**) CO_2 mass fraction, (**c**) soot volume fraction, and (**d**) soot particles diameter (m) in the turbulent paraffin-oil flame

formed within the burner at regions near fuel-injector exit and so their emissions would be reduced at the outlet, i.e., a very low concentrations release into the surrounding ambient.

As seen in Fig. 4, the EGR, due to the appeared two recirculation zones of micro-impinging-jet, would cause the rebrurn of combustion by-products, i.e. the CO, and the combustion products, i.e. CO_2, at the regions near the fuel-injector exit. So, the mico-jet impingement would reduce the emission of CO at the outlet. As seen, the appeared recirculation zones of micro-impinging-jet would result in a well mixing of combustion products with the unburnt reactants, which in turn results in a uniform distribution of combustion products, i.e. the CO_2, within the burner and at the outlet. As seen also, the nano-aerosol soot formation within the burner is reduced as a result of EGR, i.e. due to the appearance of two recirculation zones. The soot depletion within the burner would result in low concentrations of the soot in the exhaust gases released into the surrounding atmosphere.

Now, we summarize, compare, and study the effect of different values of micro-impinging-jet mass flow rate on the exhaust-gases temperature and the emissions of CO, CO_2, C_6H_6, and the soot nano-aerosol released from the current turbulent paraffin-oil flame. In this regard, we consider different values of micro-impinging-jet mass flow rate, i.e. 1.8, 2.7, 3.2, and 3.6 g/s. The other parameters, i.e. the geometry and BCs, are similar to the benchmark test case. As observed in Table 1, the emissions of CO, C_6H_6, and the soot nano-aerosol would decrease as the micro-impinging-jet mass flow rate increases. It is also observed that the exhaust-gases temperature would increase as the micro-impinging-jet mass flow rate increases.

Table 1 The mass-flow-rate effect of mico-jet impingement on temperature and emissions of pollutants at the burner outlet

Jet mass flow rate (g/s)	Temperature (K)	CO mass fraction	CO_2 mass fraction	Soot volume fraction	Soot particles diameter (m)	C_6H_6 mass fraction
–	767	3.60E−06	0.0576	1.13E−06	1.20E−07	5.84E−11
1.8	774	3.73E−06	0.0573	9.33E−07	1.10E−07	9.04E−11
2.7	747	0	0.0579	5.46E−07	9.37E−08	0
3.2	787	0	0.0579	3.38E−07	7.63E−08	0
3.6	807	0	0.0578	1.90E−07	5.97E−08	0

This indicates the role of mico-jet impingement in the amount of combustion efficiency enhancement.

6 Conclusion

Using a numerical study, we have investigated the effect of mico-jet impingement on the formation of soot nano-aerosol and the emissions of CO, CO_2, C_6H_6 from a turbulent paraffin-oil flame. We have first evaluated our numerical simulation by simulating a benchmark turbulent paraffin-oil non-premixed flame in a burner and comparing the flame structure with those of measured and another available numerical solution. We have compared the distributions of mixture fraction, temperature, and soot volume fraction in the flame. The comparison showed a good agreement with those of experiment. Then, we embedded a micro-scale injector in the burner wall, split the incoming air-flow between primary and secondary streams, injected the secondary-air into the burner via the micro-scale injector, and compared the results obtained for different values of micro-impinging-jet mass flow rates. Our numerical study have shown that mico-jet impingement changes the reactive flow pattern within the burner and reduces the flame length. They have also shown that as the mass flow rate of micro-impinging-jet increases, the emissions of soot nano-aerosol, CO, and C_6H_6 reduce. The profiles of exhaust-gas temperature, soot volume fraction, mass fractions of CO, CO_2, and C_6H_6 in exhaust gases also become more uniform in the case of a mico-jet impingement.

Acknowledgements The authors would like to thank the financial support received from the Deputy of Research and Technology in Sharif University of Technology. Their financial support and help are greatly acknowledged.

References

1. Civilian, D.L.D.: Numerical analysis of two and three dimensional recessed flame holders for scramjet applications. PhD thesis, Air University, Air Force Institute of Technology (1996)
2. Leung, K.M., Lindstedt, R.P., Jones, W.P.: A simplified reaction mechanism for soot formation in nonpremixed flames. Combust. Flame **87**, 289–305 (1991)
3. Kennedy, I.M., Yam, C., Rapp, D.C.: Modeling and measurements of soot and species in a laminar diffusion flame. Combust. Flame **107**, 368–382 (1996)
4. Brookes, S.J., Moss, J.B.: Predictions of soot and thermal radiation properties in confined turbulent jet diffusion flames. Combust. Flame **116**, 486–503 (1999)
5. Kronenburg, A., Bilger, R.W., Kent, J.H.: Modeling soot formation in turbulent methane-air jet diffusion flames. Combust. Flame **121**, 24–40 (2000)
6. Hall, R.J., Smooke, M.D., Colket, M.B.: Predictions of soot dynamics in opposed jet diffusion flames, chapter 8. In: Sawyer, R.F., Dryer, F.L. (eds.) Physical and Chemical Aspects of Combustion: A Tribute to Irvin Glassman. Combustion Science and Technology Book Series, vol. 189, pp. 189–230. Gordon and Breach, Amsterdam (1997)
7. Darbandi, M., Ghafourizadeh, M., Schneider, G.E.: Aerosol modeling of soot nanoparticles in a turbulent diffusion flame using an extended detailed kinetic scheme. In: 52nd Aerospace Sciences Meeting, National Harbor. AIAA 2014-0580 (2014)
8. Darbandi, M., Ghafourizadeh, M., Jafari, S.: Simulation of soot nanoparticles formation and oxidation in a turbulent nonpremixed methane-air flame at elevated pressure. In: 13th IEEE International Conference on Nanotechnology, Beijing, pp. 608–613. IEEE-NANO 2013-0382 (2013)
9. Darbandi, M., Ghafourizadeh, M., Schneider, G.E.: Extending a numerical procedure to simulate the micro/nanoscale soot formation in ethylene-air turbulent flame using acetylene-route nucleation. In: 11th AIAA/ASME Joint Thermophysics and Heat Transfer Conference, Atlanta. AIAA 2014-2385 (2014)
10. Darbandi, M., Ghafourizadeh, M., Schneider, G.E.: Numerical study on surface oxidation of carbonaceous nano- and micro-particles in a heavily sooting ethylene turbulent jet flame. In: 53rd AIAA Aerospace Sciences Meeting, Kissimmee. AIAA 2015-2084 (2015)
11. Darbandi, M., Ghafourizadeh, M., Schneider, G.E.: Prediction of soot nano/micro particles emission (smoke pollution) in a turbulent propane nonpremixed flame. In: The CSME International Congress, Toronto (2014)
12. Darbandi, M., Ghafourizadeh, M., Schneider, G.E.: Two-phase flow simulation of nano- and micro-particulate soot in an air-preheated nonpremixed turbulent propane flame. In: 22nd Annual Conference of the CFD Society of Canada, Toronto (2014)
13. Kent, J.H., Honnery, D.: Soot and mixture fraction in turbulent diffusion flames. Combust. Sci. Technol. **54**, 383–397 (1987)
14. Sozer, E., Hassan, E.A., Yun, S., Thakur, S., Wright, J., Ihme, M., Shyy, W.: Turbulence-chemistry interaction and heat transfer modeling of H_2/O_2 gaseous injector flows. In: 48th AIAA Aerospace Sciences Meeting, Orlando. AIAA 2010-1525 (2010)
15. Young, K.J., Stewart, C.D., Moss, J.B.: Soot formation in turbulent nonpremixed kerosine-air flames burning at elevated pressure: experimental measurement. Proc. Combust. Inst. **25**(1), 609–617 (1994)
16. Wen, Z., Yun, S., Thomson, M.J., Lightstone, M.F.: Modeling soot formation in turbulent kerosene/air jet diffusion flames. Combust. Flame **135**, 323–340 (2003)

Normalization of Eigenvectors and Certain Properties of Parameter Matrices Associated with The Inverse Problem for Vibrating Systems

Mohamed El-Gebeily and Yehia Khulief

Abstract Solutions of the equation of motion of an n-dimensional vibrating system $M\ddot{q} + D\dot{q} + Kq = 0$ can be found by solving the quadratic eigenvalue problem $L(\lambda)x := \lambda^2 Mx + \lambda Dx + Kx = 0$. The inverse problem is to identify real definite matrices $M > 0, K > 0$ and $D \leq 0$ from a specified pair (Λ, X_c) of n-eigenvalues and their corresponding eigenvectors of the eigenvalue problem. We assume here that $\Lambda = U + iW$, where $U \leq 0$ and $W > 0$ are diagonal matrices. The well posedness of the inverse problem requires that the matrix X_c be specially normalized. It is known that for such specially normalized X_c, there exist a nonsingular matrix X_R and an orthogonal matrix Θ, both real, such that $X_c = X_R(I - i\Theta)$. The identified matrices depend on a matrix polynomial $P_r(\Theta) = U_r + W_r\Theta^T + \Theta W_r - \Theta W_r \Theta^T, r = -1, 0, 1$, where $U_r = \Re(\Lambda^r)$ and $W_r = \Im(\Lambda^r)$. In this work we give an explicit characterization of normalizers of X_c, introduce some new results on the class of admissible orthogonal matrices Θ and characterize the invertibility of the polynomials $P_r(\Theta)$ in terms of the invertibility of Λ^r. For $r = -1, 1$ this is equivalent to identifying $M > 0, K > 0$. For $r = 2$ but U_r not strictly negative, we give an example to show that $P_r(\Theta)$ is indefinite for all Θ.

1 Introduction

In this article we are interested in the identification problem associated with a free vibrating beam or pipe. The discretized equation of motion that represents the transverse vibration of an Euler-Bernoulli beam with no axial deformations has the general form [11].

$$M\ddot{q} + D\dot{q} + Kq = 0, \tag{1}$$

where M, D, K are called the mass matrix, damping matrix and stiffness matrix, respectively. For a physical stably vibrating system, M and K are strictly positive

M. El-Gebeily (✉) • Y. Khulief
King Fahd University of Petroleum and Minerals, Dhahran 31261, Saudi Arabia
e-mail: mgebeily@kfupm.edu.sa; khulief@kfupm.edu.sa

© Springer International Publishing Switzerland 2016

J. Bélair et al. (eds.), *Mathematical and Computational Approaches in Advancing Modern Science and Engineering*, DOI 10.1007/978-3-319-30379-6_10

definite ($M > 0, K > 0$) while D is negative definite ($D \leq 0$). All these matrices are assumed to be of dimension n. It is well known [9] that solving (1) is equivalent to solving the eigenvalue problem

$$L(\lambda)x = 0, \tag{2}$$

where $L(\lambda)$ is the second order matrix polynomial

$$L(\lambda) = \lambda^2 M + \lambda D + K. \tag{3}$$

Solutions (λ, x) of (2) appear in conjugate pairs and for vibrating beams or pipes, λ has a non-positive real part: $\Re(\lambda) \leq 0$ and a nonzero imaginary part: $\Im(\lambda) \neq 0$. Hence, it suffices to consider the case $\Im(\lambda) > 0$. Let $\lambda_1, \lambda_2, \cdots, \lambda_n$ be eigenvalues and $x_1, x_2, \cdots, x_n$ be their corresponding eigenvectors. Following the notation in [8], we let

$$\begin{array}{llll} U = & \operatorname{diag}(\Re\,(\lambda_k))_{k=1}^n, & W = & \operatorname{diag}(\Im\,(\lambda_k))_{k=1}^n, \\ \Lambda = & U + iW, & X_c = & \begin{bmatrix} x_1 \; x_2 \; \ldots \; x_n \end{bmatrix} \end{array}, \tag{4}$$

The inverse problem for (1) is one of identifying the real matrices $M > 0, K > 0$ and $D \leq 0$ from a specified pair (Λ, X_c). There is a great deal of interest and vast literature devoted to this inverse problem for vibrating beams and other related vibrating systems including system monitoring and fault detection, inverse Sturm-Liouville problems, applied physics, and signal processing. The reader is referred to [1–3, 9, 10, 12] and the references therein.

The well posedness of the inverse problem requires that the matrix X_c be specially normalized [8] in the sense that there is a complex diagonal matrix γ such that the normalized matrix

$$\tilde{X}_c := X_c \gamma \tag{5}$$

satisfies

$$\tilde{X}_R \tilde{X}_R^T = \tilde{X}_I \tilde{X}_I^T, \tag{6}$$

where $\tilde{X}_R = \Re(\tilde{X}_c)$ and $\tilde{X}_I = \Im(\tilde{X}_c)$. To the best of our knowledge, no formula has been given for finding a normalization matrix γ except through solving case by case a quadratic system of equations involving the elements of γ, see, e.g. [4–6, 8]. In this work we give an explicit formula for γ, which turns out to be also a characterization of such normalization matrices. This is done in Sect. 3 in which we also give an illustrative example of an actual vibrating beam problem.

Furthermore, it follows from the polar decomposition of $\tilde{X}_c$ that its real and imaginary parts are related by

$$\tilde{X}_I = -\Theta \tilde{X}_R,$$

where Θ is a real orthogonal matrix. Therefore,

$$\tilde{X}_c = \tilde{X}_R(I - i\Theta). \tag{7}$$

We also show that the spectrum of Θ satisfies

$$\sigma(\Theta) \subset \{z \in \mathbf{C} : \arg(z) < \pi/2\}. \tag{8}$$

This result is then used to show that a certain matrix Q associated with the so called Jordan triples is invertible, which eliminates the need to check this condition from Algorithm 1 in [8]. Other properties of Θ and a certain matrix polynomial defined in terms of it are discussed in Sect. 4, where we characterize the invertibility of this polynomial in terms of the given parameter Λ. This has direct consequence on the identification of positive definite M, K. We also provide an example to show that the identification of a $D \leq 0$, may not be possible to in general.

In Sect. 2, we collect some facts about discretized vibrating systems which we will need in this paper.

2 Preliminaries

Let U, W, Λ be as defined in (4) and, for $r \in \mathbf{Z}$, let

$$\Lambda_r = \Lambda^r, U_r = \Re(\Lambda_r), \text{ and } W_r = \Im(\Lambda_r). \tag{9}$$

Here, $\Lambda_0 = I$. We will always assume that $U \leq 0, W > 0$. It is well known that $\sigma(L(\lambda)) = \Lambda \cup \bar{\Lambda} = \sigma(\lambda A - B)$, where

$$A = \begin{bmatrix} D & M \\ M & 0 \end{bmatrix}, \quad B = \begin{bmatrix} -K & 0 \\ 0 & M \end{bmatrix}.$$

With X_c, Λ, M defined as above, given matrices $X \in \mathbf{C}^{n\times 2n}, J \in \mathbf{C}^{2n\times 2n}, Y \in \mathbf{C}^{2n\times n}$, the triple (X, J, Y) is called a Jordan triple for $L(\lambda)$ if

$$X = \begin{bmatrix} X_c \ \bar{X}_c \end{bmatrix}, \quad J = \begin{bmatrix} \Lambda & 0 \\ 0 & \bar{\Lambda} \end{bmatrix}, \quad Y = Q^{-1}\begin{bmatrix} 0 \\ M^{-1} \end{bmatrix}, \tag{10}$$

where

$$Q = \begin{bmatrix} X \\ XJ \end{bmatrix} \tag{11}$$

and $\bar{\ }$ denotes complex conjugate. To determine real symmetric matrices K and D from Λ and X_c, it suffices that (X, J, X^T) be a Jordan triple for $L(\lambda)$. In this case, we must have

$$\widetilde{X}_R\widetilde{X}_R^T = \widetilde{X}_I\widetilde{X}_I^T, \tag{12}$$

for some normalization $\widetilde{X}_c = \widetilde{X}_R + i\widetilde{X}_I$ of X_c. When (12) is satisfied, we have the representations

$$\widetilde{X}_I = -\widetilde{X}_R\widetilde{\Theta}, \qquad \widetilde{X}_c = \widetilde{X}_R\left(I - i\widetilde{\Theta}\right) \tag{13}$$

for some real orthogonal matrix $\widetilde{\Theta}$.

3 Eigenvector Normalization

In this section we address the question: which normalizations (if any) of X_c allow (12) to hold?

Define the complex matrix Θ by

$$\Theta = X_c^{-1}\bar{X}_c \tag{14}$$

Proposition 1 *Assume $X_c = X_R + iX_I$ is a given $n \times n$ matrix whose columns are eigenvectors of a symmetric system and let Θ be defined by (14). There exists a complex diagonal $n \times n$ normalization matrix γ such that $\widetilde{X}_c = X_c\gamma$ satisfies (12) if and only if there exists an $n \times n$ invertible matrix A such that*

1. *$-\Theta A$ is diagonal,*
2. *$-A^T = \overline{A}$ and*
3. *$\gamma^2 = -\Theta A$.*

Proof Suppose γ exists. Then $\widetilde{X}_c = X_c\gamma$ satisfies (12). The latter can be written in matrix form as

$$\left[X_R\ X_I\right]\begin{bmatrix}\gamma_R^2 - \gamma_I^2 & -2\gamma_R\gamma_I \\ -2\gamma_R\gamma_I & \gamma_I^2 - \gamma_R^2\end{bmatrix}\begin{bmatrix}X_R^T \\ X_I^T\end{bmatrix} = 0.$$

Observing that $\begin{bmatrix}I & I \\ iI & -iI\end{bmatrix}\begin{bmatrix}I & -iI \\ I & iI\end{bmatrix} = 2I_{2n}$, we can rewrite the last equation as

$$\left[X_c\ \overline{X}_c\right]\begin{bmatrix}0 & \gamma^2 \\ \overline{\gamma}^2 & 0\end{bmatrix}\begin{bmatrix}\overline{X}_c^T \\ X_c^T\end{bmatrix} = 0$$

Multiplying on the left and right by X_c^{-1} and X_c^{-T}, respectively, and putting $\Theta := X_c^{-1}\bar{X}_c$ we get

$$\begin{bmatrix} I \; \Theta \end{bmatrix} \begin{bmatrix} 0 & \gamma^2 \\ \bar{\gamma}^2 & 0 \end{bmatrix} \begin{bmatrix} \Theta^T \\ I \end{bmatrix} = 0 \tag{15}$$

Clearly, the kernel of the operator $\begin{bmatrix} 0 & \gamma^2 \\ \bar{\gamma}^2 & 0 \end{bmatrix} \begin{bmatrix} \Theta^T \\ I \end{bmatrix} : \mathbf{C}^n \to \mathbf{C}^{2n}$ is $\{0\}$. Furthermore, the operator $\begin{bmatrix} I \; \Theta \end{bmatrix} : \mathbf{C}^{2n} \to \mathbf{C}^n$ has rank n, and therefore, its kernel also has rank n. For (15) to hold, the operator $\begin{bmatrix} 0 & \gamma^2 \\ \bar{\gamma}^2 & 0 \end{bmatrix} \begin{bmatrix} \Theta^T \\ I \end{bmatrix}$ must map onto the kernel of the operator $\begin{bmatrix} I \; \Theta \end{bmatrix}$. The latter is spanned by the columns of $\begin{bmatrix} -\Theta \\ I \end{bmatrix}$. Therefore, there exists an $n \times n$ change of base matrix A such that

$$\begin{bmatrix} 0 & \gamma^2 \\ \bar{\gamma}^2 & 0 \end{bmatrix} \begin{bmatrix} \Theta^T \\ I \end{bmatrix} = \begin{bmatrix} -\Theta \\ I \end{bmatrix} A.$$

Hence, we must have

$$\begin{aligned} \gamma^2 &= -\Theta A, \\ \bar{\gamma}^2 &= A\Theta^{-T}. \end{aligned}$$

Since $\Theta^{-1} = \bar{\Theta}$, we may rewrite the above equations as

$$\gamma^2 = -\Theta A = \bar{A}\Theta^T. \tag{16}$$

Furthermore, since the matrices in the last two equalities of (16) must be diagonal matrices and Θ is invertible, we conclude that $\bar{A} = -A^T$.

The only if part of the proof can be done by reversing the above steps. □

Proposition 1 suggests the following algorithm to find a normalization matrix γ. The steps parallel the three parts of the proposition.

Algorithm 1 *To compute a normalization diagonal matrix γ:*

1. *Perform column reduction operations on Θ to reduce it to a lower triangular form. (This fixes the upper triangular part of A and, by Proposition 1-2, the lower triangular part. At this point, the matrix $-\Theta A$ is actually diagonal.)*
2. *Ensure that the diagonal part of A is purely imaginary so that Proposition 1-2 is satisfied.*
3. *Calculate γ from Proposition 1-3.*

Since column reduction operations are not unique, we expect that the normalization matrix γ is not unique. The following example shows that this is the case. It also explores an alternative approach for finding γ that works in special situations.

Example 1 The coefficients of a finite element discretization of a vibrating beam (see [7], Chapter 8) are given by

$$M = \begin{bmatrix} .0929 & 0 & 0161 & -.0967 \\ 0 & 1.4881 & .0967 & -.558 \\ .0161 & .0967 & .0464 & -.1637 \\ -.0967 & -.558 & -.1637 & .7448 \end{bmatrix},$$

$$K = 10^4 \times \begin{bmatrix} .0052 & 0 & -.0026 & .0326 \\ 0 & 1.088 & -.0326 & .272 \\ -.0026 & -.0326 & .0026 & -.0326 \\ .0326 & .272 & -.0326 & .544 \end{bmatrix}$$

and $D = 0.2M + .005K$. The eigenvalues and 'un-normalized' eigenvectors of this system are

$$\Lambda = \begin{bmatrix} -128.86 & 0 & 0 & 0 \\ 0 & -15.46 & 0 & 0 \\ 0 & 0 & -1.43 & 0 \\ 0 & 0 & 0 & -0.13 \end{bmatrix} + i \begin{bmatrix} 186.81 & 0 & 0 & 0 \\ 0 & 76.84 & 0 & 0 \\ 0 & 0 & 23.05 & 0 \\ 0 & 0 & 0 & 3.60 \end{bmatrix},$$

$$10^3 X_c = \begin{bmatrix} -0.6 & -0.3 & -2.3 & 20.1 \\ -0.3 & 0.4 & 0 & 1.4 \\ -2.5 & -2.5 & 3.2 & 59.5 \\ -1.0 & -0.5 & 0.3 & 1.6 \end{bmatrix} + i \begin{bmatrix} 0.9 & 1.3 & 30.8 & 71.8 \\ 0.4 & -1.9 & -0.4 & 4.9 \\ 3.6 & 12.5 & -42.6 & 212.4 \\ 1.4 & 2.4 & -4.1 & 5.9 \end{bmatrix}$$

The real part of this matrix is invertible, therefore, we will put

$$\Theta = X_R^{-1} X_I = \begin{bmatrix} -1.4497 & 0 & 0 & 0 \\ 0 & -4.9605 & 0 & 0 \\ 0 & 0 & -13.187 & 0 \\ 0 & 0 & 0 & 3.5725 \end{bmatrix},$$

which happens to be a diagonal matrix. In this rather simple case, (12) simplifies to

$$\gamma_R = \gamma_I (I + \Theta)(I - \Theta)^{-1}.$$

This equation shows the arbitrariness in assigning, say γ_I. If we choose $\gamma_I = I$, then

$$\gamma_R = \begin{bmatrix} -5.4473 & 0 & 0 & 0 \\ 0 & -1.5050 & 0 & 0 \\ 0 & 0 & -1.1641 & 0 \\ 0 & 0 & 0 & -0.5626 \end{bmatrix}.$$

The normalized eigenvectors are

$$10^3\widetilde{X}_c = \begin{bmatrix} 2.5 & -0.9 & -2.8 & -83.1 \\ 1.0 & 1.3 & 3.0 & -5.7 \\ 10.0 & -8.7 & 38.8 & -245.9 \\ 3.9 & -1.7 & 3.7 & -6.8 \end{bmatrix} + i \begin{bmatrix} -5.6 & -2.2 & -38.2 & -20.3 \\ -2.3 & 3.3 & 0.5 & -1.4 \\ -22.3 & -21.3 & 52.8 & -60.0 \\ -8.6 & -4.1 & 5.1 & 1.7 \end{bmatrix}. \quad \square$$

4 Consequences of Normalization

From now on, we will assume that X_c is properly normalized so that (12) is satisfied. It was shown in [8] that, to determine a mass matrix $M > 0$, it is necessary that

$$\Theta + \Theta^T > 0, \tag{17}$$

where Θ is a real orthogonal matrix satisfying (13). In this section, we will discuss some consequences of this condition. Let $\eta_1, \eta_2, \cdots, \eta_n$ be an orthonormal system of eigenvectors of Θ^T corresponding to its eigenvalues $e^{i\varphi_1}, e^{i\varphi_2}, \cdots, e^{i\varphi_n}$ with $-\pi \le \varphi_1, \varphi_2, \cdots, \varphi_n < \pi$. Define the matrices

$$\Pi = \operatorname{diag}\left(e^{i\varphi_1}, e^{i\varphi_2}, \cdots, e^{i\varphi_n}\right) \quad \text{and} \quad \Phi = [\eta_1, \eta_2, \cdots, \eta_n]. \tag{18}$$

Notice that $\Theta^T\Phi = \Phi\Pi$ and $\Phi^*\Theta = \Pi^*\Phi^*$. If matrix Q of (11) is nonsingular, then (X, J, X^T) is a Jordan triple for $L(\lambda)$ and we have

$$M = (XJX^T)^{-1}, \tag{19}$$

$$D = -M(XJ^2X^T)^{-1}M, \tag{20}$$

$$K = (XJ^{-1}X^T)^{-1}. \tag{21}$$

Equations (19), (20) and (21) determine a symmetric system whose eigenvalues and eigenvectors are Λ and X_c, respectively. We shall see later (see Lemma 2 below) that (17) ensures the invertibility of Q. This will require a characterization of the eigenvalues of the matrix Θ^T, which will be given in Lemma 1.

The set of orthogonal matrices Θ can be parameterized using real skew-symmetric matrices C as

$$\Theta = (I - C)(I + C)^{-1} \tag{22}$$

Condition (17) restricts the class of parameters as in the following lemma. We should observe here that skew-symmetric matrices have pure imaginary eigenvalues.

Lemma 1 *Let Θ be a given orthogonal matrix and define C by (22). The following are equivalent:*

1. *Equation (17) holds*
2. $\sigma(C) \subset i(-1/\sqrt{2}, 1/\sqrt{2})$, *that is, every eigenvalue $i\delta$ of the parameterization matrix C satisfies*

$$|\delta| < 1/\sqrt{2} \tag{23}$$

3. $\sigma\left(\Theta^T\right) \subset \mathbf{C}^+ := \{z \in \mathbf{C} : \Re(z) > 0\}$.

Proof 1 $\Longleftrightarrow$ 2. Let Θ be an orthogonal matrix and let C be its corresponding parameter matrix defined by (22). Then

$$\Theta + \Theta^T = (I - C)(I + C)^{-1} + (I - C)^{-1}(I + C).$$

If $i\delta$ is an eigenvalue of C, then by the spectral mapping theorem,

$$\frac{1 - i\delta}{1 + i\delta} + \frac{1 + i\delta}{1 - i\delta}$$

is a corresponding eigenvalue of $\Theta + \Theta^T$. Then, by (17),

$$\frac{1 - 2\delta^2}{1 + \delta^2} > 0,$$

which is equivalent to (23).

1 $\Longleftrightarrow$ 3. If (17) holds, then

$$0 < \Phi^*\left(\Theta + \Theta^T\right)\Phi = \Pi\Phi^*\Phi + \Phi^*\Phi\Pi^* = 2\Re(\Pi),$$

which implies 3. If 3. holds, then for any $y \in \mathbf{C}^n$, we may write $y = \Phi\alpha$ for some $\alpha \in \mathbf{C}^n$. Then

$$y^*\left(\Theta + \Theta^T\right)y = \alpha^*\Phi^*\left(\Theta + \Theta^T\right)\Phi\alpha = 2\alpha^*\Re(\Pi)\alpha > 0,$$

which implies 1.□

Lemma 2 *The matrix Q defined by (11) is invertible. Consequently, (X, J, X^T) is a Jordan triple for $L(\lambda)$.*

Proof It follows from the invertibility of X_c and (13) that both X_R and $I - i\Theta$ are invertible. Furthermore, $I + i\Theta$ is also invertible since its eigenvalues are conjugates of those of $I - i\Theta$. Now,

$$Q = \begin{bmatrix} X \\ XJ \end{bmatrix} = \begin{bmatrix} X_R (I - i\Theta) & X_R (I + i\Theta) \\ X_R (I - i\Theta)\, \Lambda & X_R (I + i\Theta)\, \overline{\Lambda} \end{bmatrix}$$

$$= \begin{bmatrix} X_c & 0 \\ 0 & X_c \end{bmatrix} \begin{bmatrix} I & (I - i\Theta)^{-1} (I + i\Theta) \\ \Lambda & (I - i\Theta)^{-1} (I + i\Theta)\, \overline{\Lambda} \end{bmatrix}.$$

Put

$$\Gamma = -i\, (I - i\Theta)^{-1} (I + i\Theta)\,. \tag{24}$$

A straightforward calculation shows that $\sigma(\Gamma) \subset \mathbf{R}^+$ and hence, Γ has a well defined square root. Then

$$\begin{bmatrix} -iI & 0 \\ i\Lambda & -iI \end{bmatrix} \begin{bmatrix} I & (I - i\Theta)^{-1} (I + i\Theta) \\ \Lambda & (I - i\Theta)^{-1} (I + i\Theta)\, \overline{\Lambda} \end{bmatrix} = \begin{bmatrix} iI & \Gamma \\ 0 & -\Lambda\Gamma + \Gamma\overline{\Lambda} \end{bmatrix}.$$

Therefore, the invertibility of Q is equivalent to the invertibility of $-\Lambda\Gamma + \Gamma\overline{\Lambda}$, which, in turn, is equivalent to the invertibility of $\Gamma - \Lambda^{-1}\Gamma\overline{\Lambda}$. To show the invertibility of the latter, we write

$$\begin{aligned} \Gamma - \Lambda^{-1}\Gamma\overline{\Lambda} &= \Gamma^{1/2} \left(I - \left(\Gamma^{-1/2}\Lambda^{-1}\Gamma^{1/2}\right)\left(\Gamma^{1/2}\overline{\Lambda}\Gamma^{-1/2}\right)\right) \Gamma^{1/2} \\ &= \Gamma^{1/2} \left(I - \left(\Gamma^{1/2}\Lambda\Gamma^{-1/2}\right)^{-1} \overline{\left(\Gamma^{1/2}\Lambda\Gamma^{-1/2}\right)}\right) \Gamma^{1/2} \\ &= \Gamma^{1/2} \left(I - B^{-1}\overline{B}\right) \Gamma^{1/2}, \end{aligned}$$

where $B = \Gamma^{-1/2}\Lambda\Gamma^{1/2}$ and where we made use of the fact that $\overline{\Gamma} = \Gamma^{-1}$. Therefore, we have to show that $\left(I - B^{-1}\overline{B}\right)$ is invertible. If not, then, invoking the spectral mapping theorem, we have, for some $\delta \in \sigma(B) = \sigma(\Lambda)$,

$$0 = 1 - \frac{\overline{\delta}}{\delta} = 2i\frac{\Im(\delta)}{\delta},$$

which contradicts the assumption that $W > 0$.□

Next, we investigate the invertibility of the matrices M, D, K as identified from (21), (22) and (23). For $r \in \mathbf{Z}$, let Λ_r, U_r, W_r be as defined in (9) and let

Θ be a real orthogonal matrix. Define the matrix function

$$\begin{aligned} P_r(\Theta) &= \Re\left((I - i\Theta)\Lambda_r\left(I - i\Theta^T\right)\right) \\ &= U_r + W_r\Theta^T + \Theta W_r - \Theta U_r\Theta^T. \end{aligned}$$

Then

$$P_r(\Theta) = \begin{bmatrix} I + i\Theta & I - i\Theta \end{bmatrix} \begin{bmatrix} 0 & \overline{\Lambda}_r \\ \Lambda_r & 0 \end{bmatrix} \begin{bmatrix} I - i\Theta^T \\ I + i\Theta^T \end{bmatrix}$$

and we have the following proposition.

Proposition 2 *$P_r(\Theta)$ is invertible if and only if Λ_r is invertible.*

Proof $P_r(\Theta)$ is invertible if and only if $\ker P_r(\Theta) = \{0\}$. Note that $P_r(\Theta) = (I - i\Theta)\begin{bmatrix} i\Gamma & I \end{bmatrix}\begin{bmatrix} 0 & \overline{\Lambda}_r \\ \Lambda_r & 0 \end{bmatrix}\begin{bmatrix} I \\ i\Gamma^T \end{bmatrix}\left(I - i\Theta^T\right)$, where Γ is given by (24). Therefore, $\ker P_r(\Theta) = \ker\begin{bmatrix} i\Gamma & I \end{bmatrix}\begin{bmatrix} 0 & \overline{\Lambda}_r \\ \Lambda_r & 0 \end{bmatrix}\begin{bmatrix} I \\ i\Gamma^T \end{bmatrix}$. Arguing as in Proposition 1, the operator $\begin{bmatrix} 0 & \overline{\Lambda}_r \\ \Lambda_r & 0 \end{bmatrix}\begin{bmatrix} I \\ i\Gamma^T \end{bmatrix} : \mathbf{C}^n \to \mathbf{C}^{2n}$ has rank n, and the operator $\begin{bmatrix} i\Gamma & I \end{bmatrix}$: $\mathbf{C}^{2n} \to \mathbf{C}^n$ has rank n and kernel space of dimension n. It follows that

$$\ker\begin{bmatrix} i\Gamma & I \end{bmatrix}\begin{bmatrix} 0 & \overline{\Lambda}_r \\ \Lambda_r & 0 \end{bmatrix}\begin{bmatrix} I \\ i\Gamma^T \end{bmatrix} = \{0\} \tag{25}$$

if and only if $\begin{bmatrix} 0 & \overline{\Lambda}_r \\ \Lambda_r & 0 \end{bmatrix}\begin{bmatrix} I \\ i\Gamma^T \end{bmatrix}$ maps into the orthogonal complement of the kernel of $\begin{bmatrix} i\Gamma & I \end{bmatrix}$. The kenel of $\begin{bmatrix} i\Gamma & I \end{bmatrix}$ is spanned by the columns of $\begin{bmatrix} I \\ -i\Gamma \end{bmatrix}$ and it is straightforward to see that its orthogonal complement is spanned by the columns of $\begin{bmatrix} i\Gamma^T \\ I \end{bmatrix}$. Therefore, (25) holds if and only if there exists an invertible matrix B such that

$$\begin{bmatrix} 0 & \overline{\Lambda}_r \\ \Lambda_r & 0 \end{bmatrix}\begin{bmatrix} I \\ i\Gamma^T \end{bmatrix} = \begin{bmatrix} i\Gamma^T \\ I \end{bmatrix} B.$$

The second of the above equations gives

$$\Lambda_r = B. \square$$

Under our assumptions on U and W, $P_{-1}(\Theta), P_1(\Theta)$ are always invertible. This, in turn is equivalent [8] to the positive definiteness of M, K. However,the following example shows that $P_2(\Theta)$ may be indefinite if U is not strictly negative definite.

Example 2 Let

$$U = \begin{bmatrix} 2 & 0 \\ 0 & 0 \end{bmatrix}, W = \begin{bmatrix} 1 & 0 \\ 0 & 2 \end{bmatrix}.$$

Then

$$\Lambda^2 = \begin{bmatrix} 3 & 0 \\ 0 & -4 \end{bmatrix} + i \begin{bmatrix} -4 & 0 \\ 0 & 0 \end{bmatrix}.$$

All orthogonal 2×2 matrices Θ have the form

$$\begin{bmatrix} \cos\theta & -\sin\theta \\ \sin\theta & \cos\theta \end{bmatrix} or \begin{bmatrix} \cos\theta & \sin\theta \\ \sin\theta & -\cos\theta \end{bmatrix}.$$

The eigenvalues of the first type are $e^{\pm i\theta}$ and by Lemma 1, Θ is admissible only if $\theta \in (0, \pi/2)$. The eigenvalues of the second kind are $-1, 1$ and thus, they do not correspond to an admissible Θ. For the first kind, we calculate

$$P_2(\Theta) = \begin{bmatrix} -8\cos\theta - 7/2\cos 2\theta + 7/2 & -4\sin\theta - 7/2\sin 2\theta \\ -4\sin\theta - 7/2\sin 2\theta & 7/2\cos 2\theta - 7/2 \end{bmatrix}$$

which has eigenvalues $\lambda_{1,2} = -4\cos\theta \mp 1/2\sqrt{162 - 98\cos 2\theta}$. It can be shown that $\lambda_1 < 0, \lambda_2 > 0$ for all $\theta \in (0, \pi/2)$. Therefore, $P_2(\Theta)$ is non-definite for any admissible Θ.□

Acknowledgements This work is funded by KACST-NSTIP Project No. 12-ADV3005-04. The authors acknowledge the support provided by King Abdulaziz City for Science & Technology and King Fahd University of Petroleum & Minerals.

References

1. Bai, Z.J.: Constructing the physical parameters of a damped vibrating system from eigendata. Linear Algebra Appl. **428**(2–3), 625–656 (2008)
2. Chu, M.T.: Inverse eigenvalue problems. SIAM Rev. **40**, 1–39 (1998)
3. Chu, M.T., Golub, G.H.: Inverse Eigenvalue Problems: Theory, Algorithms, and Applications. Oxford University Press, Oxford (2005)
4. De Angelis, M., Imbimbo, M.: A procedure to identify the modal and physical parameters of a classically damped system under seismic motions. Advances in Acoustics and Vibration, Hindawi Publishing Corporation, Article ID 975125, 11 pages (2012). doi:10.1155/2012/975125

5. Friswell, M.I., Prells, U.: A measure of non-proportional damping. Mech. Syst. Sig. Process **14**(2), 125–137 (2000)
6. Garvey, S.D., Penny, J.E.T.: The relationship between the real and imaginary parts of complex modes. J. Sound Vib. **212**(1), 64–72 (1998)
7. Kwon, Y.W., Bang, H.: The Finite Element Method Using Matlab. CRC Press, Boca Raton (1997)
8. Lancaster, P., Prells, U.: Inverse problems for damped vibrating systems. J. Sound Vib. **283**, 891–914 (2005)
9. Lancaster, P., Tismenetsky, M.: The Theory of Matrices. Academic, Orlando, Academic Press, New York (1985)
10. Lancaster, P., Zaballa, I.: On the inverse symmetric quadratic eigenvalue problem. SIAM J. Matrix Anal. Appl. **35**(1), 254–278 (2014)
11. Reddy, J.N., Wang, C.M.: Dynamics of fluid-conveying beams, CORE Report No. 2004-03, Centre for Offshore Research and Engineering, National University of Singapore (2004)
12. Xu, S.F.: An Introduction to Inverse Algebraic Eigenvalue Problems. Peking University Press/Friedr. Vieweg & Sohn, Beijing/Braunschweig/Wiesbaden (1998)

Computational Aspects of Solving Inverse Problems for Elliptic PDEs on Perforated Domains Using the Collage Method

H. Kunze and D. La Torre

Abstract The treatment of an inverse problem on a perforated domain is complicated heavily by the presence of the perforations or holes. We present several theoretical results that provide relationships between the problem on the perforated domain and the same problem on the corresponding unperforated/solid domain. The results establish that we can approximate the solution of the inverse problem on the perforated domain by instead solving the inverse problem on the associated solid domain. Examples are provided.

1 Introduction

In this paper, we are interested in the parameter estimation inverse problem for elliptic PDEs. One physical setting for such problem is the estimation of the thermal diffusivity in a lamina based on (perhaps noisy) observational data of its equilibrium temperature. Indeed, the examples in the final section of the paper include images that can be thought of as isotherm plots in a lamina. In Sect. 2, we present the Collage Method approach for solving such inverse problems. In the past [5], we have established that the Collage Method compares favourably with established methods, such a Tikhonov regularization [9]. The complication in the current work is that we wish to consider such an inverse problem on a *perforated* domain.

A perforated domain (or porous medium) is a material characterized by a partitioning of the total volume into a solid portion often called the "matrix" and a pore space usually referred to as "holes." Mathematically speaking, these holes can be either materials different from that of the matrix or real physical holes. When

H. Kunze (✉)
Department of Mathematics and Statistics, University of Guelph, Guelph, ON, Canada
e-mail: hkunze@uoguelph.ca

D. La Torre
Department of Economics, Management, and Quantitative Methods, University of Milan, Milan, Italy

Department of Applied Mathematics and Sciences, Khalifa University, Abu Dhabi, UAE
e-mail: davide.latorre@unimi.it

© Springer International Publishing Switzerland 2016
J. Bélair et al. (eds.), *Mathematical and Computational Approaches in Advancing Modern Science and Engineering*, DOI 10.1007/978-3-319-30379-6_11

formulating differential equations over porous media, the term "porous" implies that the state equation is written in the matrix only while boundary conditions should be imposed on the whole boundary of the matrix, including the boundary of the holes. Solving differential equations over a perforated domain is typically a complicated task because the size and distribution of the holes within the material play an important role in its characterization. Simulations conducted over a perforated domain that includes a large number of matrix-hole interfaces present numerical challenges since a very fine discretization mesh and a large computation time are required. The direct problem is of great interest in various areas of science, engineering, and industry (see, for example, [2] and [3] for discussion of real-world problems). One way to treat these (perhaps idealized) problems with rigorous mathematics is called homogenization (see [3] and [8]), which takes advantage of the multiscale nature of the perforated domain.

The inverse problem over a perforated domain inherits all of these challenges. In this paper, we seek to avoid the complications of working with the inverse problem solution machinery on the perforated domain. In Sect. 3, we present several theoretical results that connect the problem on the perforated domain with the same problem on the corresponding unperforated (solid) domain. To frame the situation, we set up the two types of problems here.

Let $\Omega \in \mathbb{R}^2$ be compact and convex and Ω_B be collection of circular m holes $\cup_{i=1}^m B(x_j, \varepsilon_j)$, where $x_j \in \Omega$, the radii $\varepsilon_j > 0$, and the holes $B(x_j, \varepsilon_j)$ are non-overlapping and strictly inside Ω. We let $\varepsilon = \max_j \varepsilon_j$, the maximum hole radius. Let Ω_ε denote the closure of $\Omega \setminus \Omega_B$. Then, we consider the problem (P_ε) on the perforated domain

$$\begin{cases} \nabla \cdot (K^\lambda(x,y)\nabla u(x,y)) = f^\lambda(x,y) & \text{in } \Omega_\varepsilon \\ u(x,y) = 0 & \text{on } \partial\Omega_\varepsilon \end{cases} \qquad (P_\varepsilon)$$

and the associated problem (P) on the corresponding unperforated domain

$$\begin{cases} \nabla \cdot (K^\lambda(x,y)\nabla u(x,y)) = f^\lambda(x,y) & \text{in } \Omega \\ u(x,y) = 0 & \text{on } \partial\Omega \end{cases}, \qquad (P)$$

where $\lambda \in \Lambda \subset \mathbb{R}^n$ is a parameter belonging to the compact set Λ. The Dirichlet boundary conditions above can be replaced by Neumann boundary conditions, in particular on the boundaries of the holes ($\frac{\partial u}{\partial n} = 0$ on $\partial\Omega_B$).

In each case, the inverse problem is to estimate the parameter λ (perhaps coefficients defining the diffusivity K and/or the source/sink function f) given observational data values of the solution $u(x,y)$.

Given observational data for the solution of problem (P_ε), we solve the inverse problem for (P); the results we present in Sect. 3 show that the obtained parameter values solve (approximately) the inverse problem for (P_ε).

2 The Collage Theorem for Elliptic PDEs

The variational equation associated with an elliptic PDE can be written as

$$a(u, v) = \phi(v), \qquad v \in H, \tag{1}$$

where $\phi(v)$ and $a(u, v)$ are linear and bilinear maps, respectively, both defined on a Hilbert space H. We denote by $\langle \cdot, \cdot \rangle$ the inner product in H, $\|u\|^2 = \langle u, u \rangle$ and $d(u, v) = \|u - v\|$, for all $u, v \in H$.

The inverse problem of interest may now be viewed as follows: Suppose that we have an observed solution u and a given (restricted) family of bounded, coercive bilinear functionals $a^\lambda(u, v)$, $\lambda \in \Lambda$, and a family of bounded linear functionals ϕ^λ. Then, by the Lax-Milgram theorem, for each $\lambda \in \Lambda$ there exists a unique $u^\lambda \in H$ such that $\phi^\lambda(v) = a^\lambda(u^\lambda, v)$ for all $v \in H$. We would like to determine if there exists a value of the parameter λ such that $u^\lambda = u$ or, more realistically, such that $\|u^\lambda - u\|$ is small enough. The following theorem is useful for the solution of this problem.

Theorem 1 (Generalized Collage Theorem [6]) *For all $\lambda \in \Lambda$, suppose that $a^\lambda(u, v) : \Lambda \times H \times H \to \mathbb{R}$ is a family of bilinear forms and $\phi^\lambda : \Lambda \times H \to \mathbb{R}$ is a family of bounded linear functionals. Let u^λ denote the solution of the equation $a^\lambda(u, v) = \phi^\lambda(v)$ for all $v \in H$, as guaranteed by the Lax-Milgram theorem. Then, given a target element $u \in H$,*

$$\|u - u^\lambda\| \leq \frac{1}{m^\lambda} F^\lambda(u), \tag{2}$$

where

$$F^\lambda(u) = \sup_{v \in H,\ \|v\|=1} \left| a^\lambda(u, v) - \phi^\lambda(v) \right| \tag{3}$$

and $m^\lambda > 0$ is the coercivity constant of a^λ.

In order to ensure that the approximation u^λ is close to a target element $u \in H$, we can, by the Generalized Collage Theorem, try to make the term $F^\lambda(u)/m_\lambda$ as close to zero as possible. If $\inf_{\lambda \in \Lambda} m^\lambda \geq m > 0$ then the inverse problem can be reduced to the minimization of the function $F^\lambda(u)$ on the space Λ, that is,

$$\min_{\lambda \in \Lambda} F^\lambda(u). \tag{4}$$

We refer to the minimization of the functional $F^\lambda(u)$ as a "generalized collage method" because of philosophical connection with our earlier approach to inverse problems for ODEs [7]; in that work, the word "collage" was used due to connections to ideas in fractal imaging [1]. Such an optimization problem has a solution that can be approximated with a suitable discrete and quadratic program, derived from the application of the Generalized Collage Theorem and the use of an orthonormal basis in the Hilbert space H [6].

3 Theoretical Results

We introduce the Sobolev spaces $H = H_0^1(\Omega)$ and $H_\varepsilon = H_0^1(\Omega_\varepsilon)$. Since any function in H_ε can be extended to be zero over the holes, it is trivial to prove that H_ε can be embedded in H. We let $\Pi_\varepsilon u$ be the projection of $u \in H$ onto H_ε; it follows that

$$\|u - \Pi_\varepsilon u\|_H \to 0 \text{ whenever } \varepsilon \to 0.$$

When Neumann boundary conditions are considered, it is still possible to extend a function in H_ε to a function of H: these extension conditions are well studied (see [8]) and they typically hold when the domain Ω has a particular structure.

The variational formulations of (P_ε) and (P) are

$$\text{find } u \in H_\varepsilon \text{ such that } a_\varepsilon^\lambda(u, v) = \phi_\varepsilon^\lambda(v), \ \forall v \in H_\varepsilon \tag{P_ε}$$

and

$$\text{find } u \in H \text{ such that } a^\lambda(u, v) = \phi^\lambda(v), \ \forall v \in H. \tag{P}$$

The generalized collage distance associated to (P) is stated in (3) in Theorem 1, while the generalized collage distance associated to (P_ε) is

$$F_\varepsilon^\lambda(u) = \sup_{v \in H_\varepsilon, \ \|v\|_{H_\varepsilon}=1} \left| a_\varepsilon^\lambda(u, v) - \phi_\varepsilon^\lambda(v) \right|. \tag{5}$$

In the results that follow, we assume that the continuous and bilinear forms a_ε^λ and a^λ are uniformly coercive and bounded with respect to λ and ε, namely there exists two positive constants m and M such that for all $\lambda \in \Lambda$

$$\begin{cases} a_\varepsilon^\lambda(u, u) \geq m\|u\|^2 & \forall u \in H_\varepsilon \\ a_\varepsilon^\lambda(u, v) \leq M\|u\|\|v\| & \forall u, v \in H_\varepsilon \\ a^\lambda(u, u) \geq m\|u\|^2 & \forall u \in H \\ a^\lambda(u, v) \leq M\|u\|\|v\| & \forall u, v \in H \end{cases} \tag{H1}$$

We also assume that the linear functionals ϕ_ε^λ and ϕ^λ are uniformly bounded with respect to λ and ε, namely there exists a positive constant μ such that

$$\begin{cases} \phi_\varepsilon^\lambda(u) \leq \mu\|u\| & \forall u \in H_\varepsilon \\ \phi^\lambda(u) \leq \mu\|u\| & \forall u \in H \end{cases} \tag{H2}$$

Under the hypotheses (H1) and (H2), (P_ε) and (P) have unique solutions u_ε^λ and u^λ, respectively, for each $\lambda \in \Lambda$ and for each fixed choices of $\varepsilon_j, j = 1, \ldots, m$.

We now state three results relating (P_ε) and (P), first presented with proofs in [4].

Proposition 1 *The following estimate holds:*

$$\|\Pi_\varepsilon u - u_\varepsilon^\lambda\|_{H_\varepsilon} \leq \frac{F^\lambda(u)}{m} + \frac{M}{m}\|u - \Pi_\varepsilon u\|_H \tag{6}$$

Proposition 2 *There exists a constant C, that does not depend on ε, such that the following estimate holds:*

$$F^\lambda(\Pi_\varepsilon u) \leq F_\varepsilon^\lambda(\Pi_\varepsilon u) + C\varepsilon \tag{7}$$

for all $\lambda \in \Lambda$, $\varepsilon > 0$.

Proposition 3 *Suppose that $F^\lambda(u), F_\varepsilon^\lambda(v) : \Lambda \to \mathbb{R}_+$ are continuous for all $u \in H$, $v \in H_\varepsilon$, and $\varepsilon > 0$. Let λ_ε be a sequence of minimizers of $F_\varepsilon^\lambda(u)$ over Λ. Then there exists $\varepsilon_n \to 0$ and $\lambda^* \in \Lambda$ such that $\lambda_{\varepsilon_n} \to \lambda^*$, with λ^* a minimizer of $F^\lambda(u)$ over Λ.*

The practical fallout of these results is that we have some justification in minimizing $F^\lambda(u)$ (the generalized collage distance for the problem on the domain with no holes) to obtain estimates of the parameter values for (P_ε) (the problem on the domain with holes). The quality of the estimate may be affected by the size of the largest hole; certainly, when all holes are sufficiently small, we expect the approximation to be good.

4 Examples

Example 1 We place nine holes of assorted sizes inside $\Omega = [0, 1]^2$, as in Fig. 1 We choose $K(x, y) = K_{true}(x, y) = 8 + 3x^2 + y^2$ and consider the problem

$$\begin{cases} \nabla \cdot (K(x, y)\nabla u(x, y)) = 2x^2 + y^2, & \text{in } \Omega_\varepsilon, \\ u(x, y) = 0, & \text{on } \partial\Omega, \\ \frac{\partial u}{\partial n}(x, y) = 0, & \text{on } \partial\Omega_B, \end{cases} \tag{8}$$

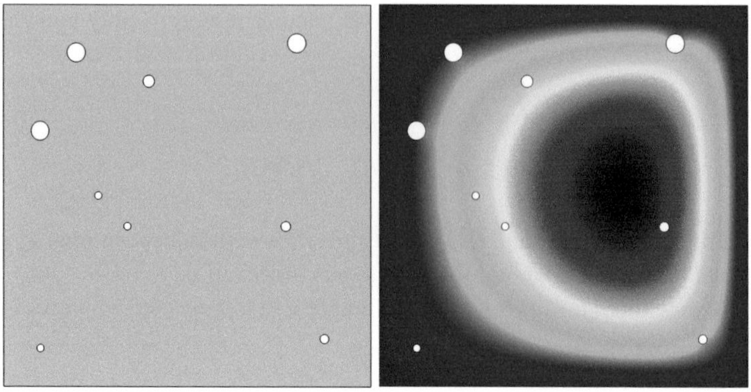

Fig. 1 The domain and level curves of solutions for Example 1

Table 1 Results for the inverse problem in Example 1. Due to the normalization of λ_0, the true values are $(\lambda_1, \lambda_2, \lambda_3) = (8, 3, 1)$

δ (%)	λ_1	λ_2	λ_3
0	7.1935	2.9783	1.2431
1	7.1586	3.0466	1.2573
3	7.0737	3.1945	1.2934

where Ω_B is the union of the nine holes. We solve the diffusion problem numerically and sample the solution u_ε at 49 uniformly-distributed points strictly inside Ω. The level curves of the solution are illustrated in Fig. 1. If a sample point lies inside a hole, we obtain no information at the point.

Now, beginning with the observational data points, we consider the inverse problem of estimating K and f. To this end, we define $K^\lambda(x, y) = \lambda_0 + \lambda_1 x^2 + \lambda_2 y^2$. Using the data values and $f(x, y) = 4x^2 + y^2$, we seek to estimate the values of λ_i in $K^\lambda(x, y)$ by applying the generalized collage theorem to solve the related inverse problem on Ω with no holes.

The results for various cases, with relative noise of δ% added, are presented in Table 1. The results worsen as noise is added, but remain reasonably good.

Example 2 For $\varepsilon \in \{0.1, 0.025, 0.01\}$, define $N_\varepsilon = \frac{1}{10\varepsilon}$ and

$$\Omega_B = \bigcup_{i,j=1}^{N_\varepsilon} B_\varepsilon\left(\left(i - \frac{1}{2}\right)\varepsilon, \left(j - \frac{1}{2}\right)\varepsilon\right),$$

a domain with N_ε^2 uniformly-distributed holes all of radius ε. Choosing $K(x,y) = K_{true}(x,y) = 9 + 3x + 2y$, we consider the steady-state diffusion problem

$$\begin{cases} \nabla \cdot (K(x,y)\nabla u(x,y)) = 2x^2 + y^2, & \text{in } \Omega_\varepsilon, \\ u(x,y) = 0, & \text{on } \partial\Omega, \\ \frac{\partial u}{\partial n}(x,y) = 0, & \text{on } \partial\Omega_B. \end{cases} \tag{9}$$

For each fixed value of ε, we solve the diffusion problem numerically and sample the solution at 49 uniformly-distributed points strictly inside Ω, obtaining no information at the point if it lies in a hole. Using the data values, with relative noise of $\delta\%$ added, we use the generalized collage theorem to solve the related inverse problem, knowing $f(x,y) = 2x^2 + y^2$ and seeking a diffusivity function of the form $K(x,y) = \lambda_0 + \lambda_1 x + \lambda_2 y$.

The level curves of each solution are illustrated in Fig. 2. When the hole is too large, as in the $N = 1$ case, the estimates are very poor. In this case, the hole needs to be incorporated into the macroscopic-scale model, as it can't be considered part of the smaller-scale model. In the other cases of the table, the estimates are good.

We see that as the size of the holes decreases (even while the number increases), the solution to the inverse problem produces better estimates of the parameters. As the noise increases, the results worsen but remain good (Table 2).

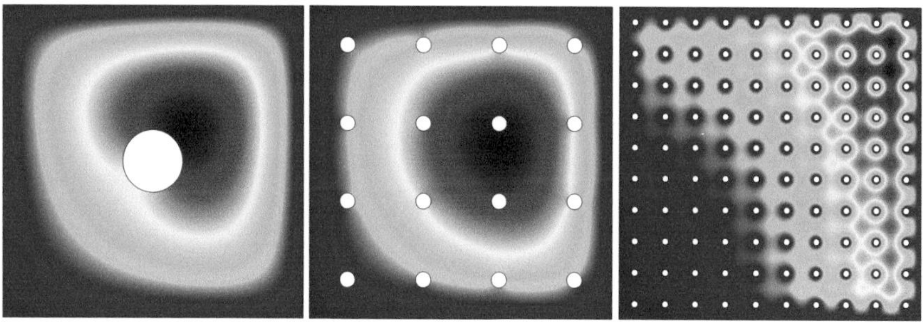

Fig. 2 Level curves of solutions in Example 2, with $\varepsilon = 0.1, 0.025$, and 0.01

Table 2 Results for the inverse problem in Example 2. True values are $(\lambda_0, \lambda_1, \lambda_2) = (10, 3, 2)$

			Recovered parameters		
ε	N_ε	δ (%)	λ_0	λ_1	λ_2
0.1	1	0	14.4169	10.1594	−1.8807
		1	14.2991	10.2965	−1.8130
		3	14.0300	10.5812	−1.6543
0.025	4	0	8.7683	4.1071	2.4944
		1	8.7188	4.1853	2.5131
		3	8.6024	4.3496	2.5587
0.01	10	0	8.8155	3.7757	2.3408
		1	8.7645	3.8531	2.3661
		3	8.6453	4.0158	2.4252

Acknowledgements H. Kunze thanks the Natural Sciences and Engineering Research Council for supporting this research.

References

1. Barnsley, M.F.: Fractals Everywhere. Academic, New York (1989)
2. Civan, F.: Porous Media Transport Phenomena. Wiley, Hoboken (2011)
3. Espedal, M.S., Fasano, A., Mikelić, A.: Filtration in Porous Media and Industrial Application. Lecture Notes in Mathematics, vol. 1734. Springer, New York (1998)
4. Kunze, H., La Torre, D., Collage-type approach to inverse problems for elliptic PDEs on perforated domains. Electron. J. Differ. Equ. **48**, 1–11 (2015)
5. Kunze, H., La Torre, D., Levere, K., Ruiz Galán, M.: Inverse problems via the "generalized collage theorem" for vector-valued Lax-Milgram-based variational problems. Math. Probl. Eng. (2015). http://dx.doi.org/10.1155/2015/764643
6. Kunze, H., La Torre, D., Vrscay, E.R.: A generalized collage method based upon the Lax–Milgram functional for solving boundary value inverse problems. Nonlinear Anal. **71**(12), 1337–1343 (2009)
7. Kunze, H., Vrscay, E.R.: Solving inverse problems for ordinary differential equations using the Picard contraction mapping. Inverse Probl. **15**, 745–770 (1999)
8. Marchenko, V.A., Khruslov, E.Y.: Homogenization of Partial Differential Equations. Birkhauser, Boston (2006)
9. Tychonoff, A.N.: Solution of incorrectly formulated problems and the regularization method. Dokl. Akad. Nauk SSSR **151**, 501–504 (1963)

Dynamic Boundary Stabilization of a Schrödinger Equation Through a Kelvin-Voigt Damped Wave Equation

Lu Lu and Jun-Min Wang

Abstract In this paper, we study an interconnected system of a Schrödinger and a wave equation with Kelvin-Voigt (K-V) damping, where the K-V damped wave equation performs as a dynamic feedback controller. We show that the system operator generates a C_0-semigroup of contractions in the energy state space, and the system is well-posed. By detailed spectral analysis, we know that the spectral of the system operator composes of two parts, point spectrum and continuous spectrum. Moreover, the points in the spectra all have negative real parts. It follows that the C_0-semigroup generated by the system operator achieves asymptotic stability.

1 Introduction

There has been much interests in the problem of dynamic feedback controller, which is modeled by coupled partial differential equations (PDEs) or interconnected PDEs. This kind of problems are quite challenging, yet have became attractive to researchers these days. Control design and stability analysis for such systems have become active over the past 5 years, see [8, 9] and the references therein. In [9], an interconnected system of Schrödinger equation and heat equation is carefully studied, which replaces the static feedback by dynamic feedback governed by a heat equation. It shows the exponential stability of the system and the Gevery class property of the semigroup. The boundary and internal stabilizations for Schrödinger equation are considered in [6], where the stability of the systems is achieved by using multiplier techniques. Two kinds of boundary controllers for Schrödinger equation are concerned in [2], which shows that a simple proportional collocated boundary controller can exponentially stabilize the system but the decay rate cannot be prescribed, while the backstepping method can ensure it to have arbitrary decay rate.

It is also known that viscoelastic materials have been widely used in engineering and lots of researchers have put much efforts to analyze the dynamic behavior of

L. Lu (✉) • J.-M. Wang
School of Mathematics and Statistics, Beijing Institute of Technology, Beijing 100081, P.R. China
e-mail: lulu0408100@126.com; jmwang@bit.edu.cn

© Springer International Publishing Switzerland 2016
J. Bélair et al. (eds.), *Mathematical and Computational Approaches in Advancing Modern Science and Engineering*, DOI 10.1007/978-3-319-30379-6_12

vibration for elastic structures with viscoelasticity over the past several decades. Kelvin-Voigt (K-V) damping is one of those most commonly used viscoelastic model, due to its easy and huge applications in modern technology. And there has been an abundance of literature on the study of elastic system with viscoelastic damping. In [3, 4], it is shown that local K-V damping can ensure the exponential stability of both a string and an Euler-Bernoulli beam system when the material parameter is smooth enough at the interface. However, if the smooth condition cannot be satisfied, the string system is non-exponentially stable even if the material parameter is a constant. Passive control of a wave equation with internal K-V damping is studied in [1]. Results there reveal that the spectrum of the system operator consists of point spectrum and continuous spectrum, due to the fact that the resolvent of the system operator for a viscoelastic system is not compact anymore.

In this paper, we present a dynamic input/output feedback controller, which feeds the K-V damped wave equation and Schrödinger equation into each other through the boundary. The coupled Schrödinger-wave system (as shown in Fig. 1) is written as follows:

$$\begin{cases} y_t(x,t) + iy_{xx}(x,t) = 0, & 0 < x < 1, t > 0, \\ z_{tt}(x,t) - z_{xx}(x,t) - \alpha z_{xxt}(x,t) = 0, & 1 < x < 2, t > 0, \\ y(0,t) = z(2,t) = 0, & t \geq 0, \\ y(1,t) = kz_t(1,t), & t \geq 0, \\ \alpha z_{xt}(1,t) + z_x(1,t) = -iky_x(1,t), & t \geq 0, \\ y(x,0) = y_0(x), & 0 < x < 1, \\ z(x,0) = z_0(x), \quad z_t(x,0) = z_1(x), & 1 < x < 2, \end{cases} \tag{1}$$

where $\alpha > 0$, $k \neq 0$. The two equations are coupled at $x = 1$ with interconnected conditions and fixed at each end.

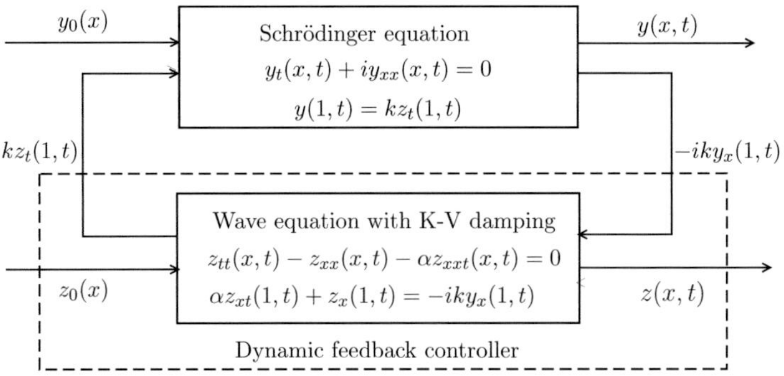

Fig. 1 Block diagram for the dynamic boundary feedback of the coupled system

By introducing the following transformation

$$\begin{cases} w(x,t) = y(1-x,t),\ 0 < x < 1, t > 0, \\ u(x,t) = z(x+1,t),\ 0 < x < 1, t > 0, \end{cases} \tag{2}$$

then (1) becomes

$$\begin{cases} w_t(x,t) + iw_{xx}(x,t) = 0, & 0 < x < 1, t > 0, \\ u_{tt}(x,t) - u_{xx}(x,t) - \alpha u_{xxt}(x,t) = 0, & 0 < x < 1, t > 0, \\ w(1,t) = u(1,t) = 0, & t \geq 0, \\ w(0,t) = ku_t(0,t), & t \geq 0, \\ \alpha u_{xt}(0,t) + u_x(0,t) = ikw_x(0,t), & t \geq 0, \end{cases} \tag{3}$$

Accordingly, the initial conditions for system (3) are $w(x,0) = w_0(x), u(x,0) = u_0(x), u_t(x,0) = u_1(x), 0 < x < 1$.

The energy function for (3) is given by

$$E(t) = \frac{1}{2}\int_0^1 \left[|w(x,t)|^2 + |u_x(x,t)|^2 + |u_t(x,t)|^2\right] dx. \tag{4}$$

In this paper, we analyze the spectrum of (3) in which the system operator has no compact resolvent. We first set up the system operator and show it generates a C_0-semigroup of contractions, and the system is well-posed. By detailed spectral analysis, we obtain that the residual spectrum is empty and the continuous spectrum contains only one negative point. Moreover, all the eigenvalues of the system lie in the open left half plane. Therefore, this controller design moves the eigenvalues of the Schrödinger and wave equations into the second quadrant. It follows that the C_0-semigroup generated by the system operator achieves asymptotic stability.

2 Well-Posedness of System (3)

We consider system (3) in the energy space $\mathscr{H} = L^2(0,1) \times H_E^1(0,1) \times L^2(0,1)$, where $H_E^1(0,1) = \{g \in H^1(0,1) | g(1) = 0\}$. The norm in $\mathscr{H}$ is induced by the inner product

$$\langle X_1, X_2 \rangle = \int_0^1 \left[f_1(x)\overline{f_2(x)} + g_1'(x)\overline{g_2'(x)} + h_1(x)\overline{h_2(x)}\right] dx, \tag{5}$$

where $X_s = (f_s, g_s, h_s) \in \mathscr{H}$, $s = 1, 2$. Define the system operator of (3) by

$$\begin{cases} \mathscr{A}(f,g,h) = (-if'',\ h,\ (g'+\alpha h')'\ ),\ \forall\ (f,g,h) \in D(\mathscr{A}), \\ D(\mathscr{A}) = \left\{ (f,g,h) \in \mathscr{H}, \mathscr{A}(f,g,h) \in \mathscr{H}, \left| \begin{array}{l} g'+\alpha h' \in H^1(0,1), \\ f(1)=0,\ f(0)=kh(0), \\ g'(0)+\alpha h'(0) = ikf'(0) \end{array} \right. \right\}. \end{cases} \tag{6}$$

Then (3) can be written as an evolution equation in $\mathscr{H}$:

$$\begin{cases} \dfrac{dX(t)}{dt} = \mathscr{A}X(t),\quad t > 0, \\ X(0) = X_0, \end{cases} \tag{7}$$

where $X(t) = (w(\cdot,t), u(\cdot,t), u_t(\cdot,t))$.

Theorem 1 *Let $\mathscr{A}$ be given by (6). Then $\mathscr{A}^{-1}$ exists, and hence $0 \in \rho(\mathscr{A})$, the resolvent set of $\mathscr{A}$. Moreover $\mathscr{A}$ is dissipative in $\mathscr{H}$ and $\mathscr{A}$ generates a C_0-semigroup $e^{\mathscr{A}t}$ of contractions in $\mathscr{H}$.*

Proof For any given $(f_1, g_1, h_1) \in \mathscr{H}$ by,

$$\mathscr{A}(f,g,h) = (-if'', h, (g'+\alpha h')') = (f_1, g_1, h_1), \tag{8}$$

we have

$$\begin{cases} f''(x) = if_1(x), \quad h(x) = g_1(x), \quad (g'(x)+\alpha h'(x))' = h_1(x), \\ f(1) = g(1) = 0, \quad f(0) = kh(0), \quad g'(0)+\alpha h'(0) = ikf'(0), \quad g_1(1) = 0, \end{cases} \tag{9}$$

and the solution of (9) is given by

$$\begin{cases} f(x) = f'(0)(x-1) - i\displaystyle\int_0^x (1-x)f_1(\xi)d\xi - i\int_x^1 (1-\xi)f_1(\xi)d\xi, \\ g(x) = ik(x-1)f'(0) - \alpha g_1(x) - \displaystyle\int_x^1 (1-\xi)h_1(\xi)d\xi\ + (x-1)\int_0^x h_1(\xi)d\xi, \\ h(x) = g_1(x) \\ f'(0) = -i\displaystyle\int_0^1 (1-\xi)f_1(\xi)d\xi - kg_1(0). \end{cases} \tag{10}$$

Hence, we get the unique solution $(f, g, h) \in D(\mathscr{A})$ to equation (8), thus $\mathscr{A}^{-1}$ exists. Now we show that $\mathscr{A}$ is dissipative in $\mathscr{H}$. Let $X = (f, g, h) \in D(\mathscr{A})$. Then we have

$$\begin{aligned}
\langle \mathscr{A}X, X\rangle &= \big\langle (-if'', h, (g' + \alpha h')'), (f, g, h)\big\rangle \\
&= \int_0^1 (-if'')\overline{f}dx + \int_0^1 h'\overline{g'}dx + \int_0^1 (g' + \alpha h')'\overline{h}dx \\
&= -if'\overline{f}\Big|_0^1 + i\int_0^1 |f'|^2dx + \int_0^1 h'\overline{g'}dx + (g' + \alpha h')\overline{h}\Big|_0^1 - \int_0^1 (g' + \alpha h')\overline{h'}dx \\
&= -if'(1)\overline{f}(1) + if'(0)\overline{f}(0) + i\int_0^1 |f'|^2dx + \int_0^1 h'\overline{g'}dx + (g'(1) + \alpha h'(1))\overline{h}(1) \\
&\quad -(g'(0) + \alpha h'(0))\overline{h}(0) - \int_0^1 \overline{h'}g'dx - \alpha\int_0^1 |h'|^2dx \\
&= if'(0)\overline{f}(0) + i\int_0^1 |f'|^2dx + \int_0^1 h'\overline{g'}dx - ikf'(0)\overline{h}(0) - \alpha\int_0^1 |h'|^2dx \\
&\quad - \int_0^1 \overline{h'}g'dx \\
&= -\alpha\int_0^1 |h'|^2dx + \left(i\int_0^1 |f'|^2dx + \int_0^1 h'\overline{g'}dx - \int_0^1 \overline{h'}g'dx\right)
\end{aligned} \tag{11}$$

and

$$\mathrm{Re}\langle \mathscr{A}X, X\rangle = -\alpha\int_0^1 |h'|^2dx \le 0. \tag{12}$$

Hence $\mathscr{A}$ is dissipative and $\mathscr{A}$ generates a C_0-semigroup $e^{\mathscr{A}t}$ of contractions in $\mathscr{H}$ by the Lumer-Phillips theorem [7]. The proof is complete. □

3 Spectral Analysis

In this section, we consider the eigenvalue problem of (3). Let $\mathscr{A}X = \lambda X$, where $0 \neq X = (f, g, h) \in D(\mathscr{A})$, then f, g, h satisfy:

$$\begin{cases}
f''(x) - i\lambda f(x) = 0, \\
h(x) = \lambda g(x), \\
(1 + \alpha\lambda)g''(x) - \lambda^2 g(x) = 0, \\
f(1) = g(1) = 0, \;\; f(0) = kh(0), \\
\alpha h'(0) + g'(0) = ikf'(0).
\end{cases} \tag{13}$$

Let $p(\lambda) = 1 + \alpha\lambda$, when $p(\lambda) \neq 0$ i.e. $\lambda \neq -\dfrac{1}{\alpha}$, (13) changes to

$$\begin{cases} f''(x) = i\lambda f(x), \\ g''(x) = \dfrac{\lambda^2}{1+\alpha\lambda} g(x) = \dfrac{\lambda^2}{p(\lambda)} g(x), \\ f(1) = g(1) = 0,\ f(0) = \lambda k g(0), \\ (1+\alpha\lambda) g'(0) = p(\lambda) g'(0) = ikf'(0). \end{cases} \tag{14}$$

We can get

$$f(x) = a_1 e^{\sqrt{i\lambda}x} + b_1 e^{-\sqrt{i\lambda}x}, \quad g(x) = c_1 e^{\sqrt{\frac{\lambda^2}{p(\lambda)}}x} + d_1 e^{-\sqrt{\frac{\lambda^2}{p(\lambda)}}x}, \tag{15}$$

where a_1, b_1, c_1 and d_1 are constants. Substituting these into the boundary conditions of (14), we have

$$\begin{cases} a_1 e^{\sqrt{i\lambda}} + b_1 e^{-\sqrt{i\lambda}} = 0, \\ c_1 e^{\sqrt{\frac{\lambda^2}{p(\lambda)}}} + d_1 e^{-\sqrt{\frac{\lambda^2}{p(\lambda)}}} = 0, \\ a_1 + b_1 = \lambda k (c_1 + d_1), \\ p(\lambda)\sqrt{\frac{\lambda^2}{p(\lambda)}}(c_1 - d_1) = ik\sqrt{i\lambda}(a_1 - b_1). \end{cases} \tag{16}$$

Then (14) has the nontrivial solution if and only if the characteristic equation $\det \Delta(\lambda) = 0$, where

$$\Delta(\lambda) = \begin{bmatrix} e^{\sqrt{i\lambda}} & e^{-\sqrt{i\lambda}} & 0 & 0 \\ 0 & 0 & e^{\sqrt{\frac{\lambda^2}{p(\lambda)}}} & e^{-\sqrt{\frac{\lambda^2}{p(\lambda)}}} \\ 1 & 1 & -k\lambda & -k\lambda \\ ik\sqrt{i\lambda} & -ik\sqrt{i\lambda} & -p(\lambda)\sqrt{\frac{\lambda^2}{p(\lambda)}} & p(\lambda)\sqrt{\frac{\lambda^2}{p(\lambda)}} \end{bmatrix}. \tag{17}$$

Lemma 1 *Let $\mathscr{A}$ be defined by (6). Then for each $\lambda \in \sigma_p(\mathscr{A})$, we have* $\mathrm{Re}\lambda < 0$.

Proof By Theorem 1, since $\mathscr{A}$ is dissipative, we have for each $\lambda \in \sigma(\mathscr{A})$, $\mathrm{Re}\lambda \le 0$. So we only need to show there is not any eigenvalue on the imaginary axis. Let $\lambda = \pm i\mu^2 \in \sigma_p(\mathscr{A})$ with $\mu \in \mathbb{R}^+$ and $X = (f, g, h) \in D(\mathscr{A})$ be its associated eigenfunction of $\mathscr{A}$. Then by (12), we have

$$\mathrm{Re}\langle \mathscr{A}X, X\rangle = -\alpha \int_0^1 |h'|^2 dx = 0.$$

Hence $h'(x) = 0$. By the second and third equations of (13), we have $h(x) = g(x) = 0$. Then by the first equation of (13) and its boundary conditions we have:

$$\begin{cases} f''(x) = i\lambda f(x), \\ f(0) = f'(0) = f(1) = 0. \end{cases}$$

A direct computation yields $f(x) = 0$. Hence, $X = (f, g, h) = 0$. Therefore, there is no eigenvalue on the imaginary axis. This completes the proof. □

Proposition 1 *Let $\mathscr{A}$ be defined by (6). Then $\lambda = -\frac{1}{\alpha} \notin \sigma_p(\mathscr{A})$.*

Proof When $\lambda = -\frac{1}{\alpha}$, then $p(\lambda) = 0$. From the third and second equation of (13), we can easily get $g(x) = h(x) = 0$. Following the same manner as the proof of Lemma 1, we have $f(x) = 0$. Hence, $X = (f, g, h) = 0$. So that $\lambda = -\frac{1}{\alpha} \notin \sigma_p(\mathscr{A})$. □

Theorem 2 *Let $\mathscr{A}$ be defined by (6). The eigenvalues of $\mathscr{A}$ have the following asymptotic expressions:*

(i) When $\lambda \to -\frac{1}{\alpha}$, the asymptotic eigenvalue is given by:

$$\lambda_{2n} = -\frac{1}{\alpha} - \frac{1}{n^2\pi^2\alpha^3} + \frac{\sqrt{\alpha}A}{n^4\pi^4\alpha^5} + \mathscr{O}(n^{-5}), \tag{18}$$

where

$$A = \frac{2}{B(1 + e^{\frac{2i\sqrt{i}}{\alpha}})}\left[\frac{1-B}{\sqrt{\alpha}} - \frac{1+B}{\sqrt{\alpha}}e^{\frac{2i\sqrt{i}}{\alpha}}\right], \quad B = \frac{k^2\sqrt{i}}{\alpha^2}. \tag{19}$$

(ii) When $|\lambda| \to \infty$, there are two families of eigenvalues given by:

$$\lambda_{1n} = -(n\pi + \frac{\theta}{2})|\ln r| + \left[(n\pi + \frac{\theta}{2})^2 - \frac{\ln^2 r}{4}\right]i + \mathscr{O}(n^{-1}), \tag{20}$$

$$\lambda_{3n} = -\alpha(n\pi + \frac{\varphi}{2})^2 + \frac{\alpha}{4}\ln^2 r + \frac{1}{\alpha} - \left[\alpha(n\pi + \frac{\varphi}{2})\ln r\right]i + \mathscr{O}(n^{-1}). \tag{21}$$

Here r, θ and φ are three constants given by

$$0 < r = \frac{\sqrt{\alpha^2 + k^8}}{\alpha + \sqrt{2\alpha}k^2 + k^4} < 1, \quad \ln r < 0, \quad \ln r^{-1} = -\ln r > 0, \tag{22}$$

$$\theta = \begin{cases} \arctan \frac{\sqrt{2\alpha}k^2}{\alpha - k^4}, & \alpha - k^4 > 0, \\ \frac{\pi}{2}, & \alpha = k^4, \\ \pi - \arctan \frac{\sqrt{2\alpha}k^2}{k^4 - \alpha}, & \alpha - k^4 < 0, \end{cases} \tag{23}$$

and

$$\varphi = \begin{cases} \arctan \frac{\sqrt{2\alpha}k^2}{k^4 - \alpha}, & k^4 - \alpha > 0, \\ \frac{\pi}{2}, & \alpha = k^4, \\ \pi - \arctan \frac{\sqrt{2\alpha}k^2}{\alpha - k^4}, & k^4 - \alpha < 0. \end{cases} \tag{24}$$

Moreover,

$$\mathrm{Re}\lambda_{1n}, \ \mathrm{Re}\lambda_{3n} \to -\infty, \quad \text{when } n \to \infty. \tag{25}$$

Proof Due to space limitation, we give the outline of the proof here. From (17), and let $s(\lambda) := \sqrt{\frac{\lambda^2}{p(\lambda)}} \in \mathbb{C}$, a direct computation gives,

$$\det \Delta(\lambda) = \left[e^{-\sqrt{i\lambda}} e^{s(\lambda)} - e^{\sqrt{i\lambda}} e^{-s(\lambda)}\right]\left[p(\lambda)s(\lambda) + ik^2\lambda\sqrt{i\lambda}\right]$$
$$+ \left[e^{-\sqrt{i\lambda}} e^{-s(\lambda)} - e^{\sqrt{i\lambda}} e^{s(\lambda)}\right]\left[p(\lambda)s(\lambda) - ik^2\lambda\sqrt{i\lambda}\right]. \tag{26}$$

Let $\det \Delta(\lambda) = 0$, asymptotic expressions of the eigenvalues can be achieved.

(I) When $\lambda \to -\frac{1}{\alpha}$, let $\varepsilon = \lambda + \frac{1}{\alpha}$, then $\varepsilon \to 0$. We get λ_{2n} as shown in (18) and (19).
(II) When $\lambda \to \infty$, we consider $\lambda := i\rho^2$, $0 \le \arg \rho \le \frac{\pi}{2}$. We then get λ_{1n} and λ_{3n} (given by (20) and (21)), when ρ belongs to $0 \le \arg \rho \le \frac{\pi}{8}$ and $\frac{\pi}{8} < \arg \rho \le \frac{3\pi}{8}$, respectively. There is no high frequency solution in $\frac{3\pi}{8} < \arg \rho \le \frac{\pi}{2}$. □

Proposition 2 *Let $\mathscr{A}$ be defined by (6). Then its adjoint operator $\mathscr{A}^*$ has the following form:*

$$\begin{cases} \mathscr{A}^*(f,g,h) = (if'', \ -h, \ -(g'-\alpha h')'\ ), \ \forall\ (f,g,h) \in D(\mathscr{A}^*), \\ D(\mathscr{A}^*) = \left\{ (f,g,h) \in \mathscr{H}, \mathscr{A}^*(f,g,h) \in \mathscr{H} \left| \begin{array}{l} g' - \alpha h' \in H^1(0,1), \\ f(1) = 0, \ f(0) = kh(0), \\ g'(0) - \alpha h'(0) = ikf'(0) \end{array} \right. \right\}. \end{cases} \tag{27}$$

Proposition 3 *Let $\mathscr{A}$ is defined by (6). Then $\sigma(\mathscr{A}) = \{-\frac{1}{\alpha}\} \bigcup \sigma_p(\mathscr{A})$.*

Proposition 4 *Let $\mathscr{A}$ be defined by (6). Then $\sigma_r(\mathscr{A}) = \emptyset$, and $\sigma_c(\mathscr{A}) = \{-\frac{1}{\alpha}\}$.*

Proof From Propositions 1 and 3, we have $-\frac{1}{\alpha} \notin \sigma_p(\mathscr{A})$ and $\{-\frac{1}{\alpha}\} \bigcup \sigma_p(\mathscr{A}) = \sigma(\mathscr{A})$. The desired results will be got if $-\frac{1}{\alpha} \notin \sigma_r(\mathscr{A})$. Now we suppose $-\frac{1}{\alpha} \in \sigma_r(\mathscr{A})$, then $-\frac{1}{\alpha} \in \sigma_p(\mathscr{A}^*)$. By $\mathscr{A}^*X = -\frac{1}{\alpha}X$, where $X = (f,g,h) \in D(\mathscr{A}^*)$, we get

$$\begin{cases} f''(x) - \dfrac{i}{\alpha} f(x) = 0 \\ h(x) = \dfrac{1}{\alpha} g(x) \\ -(g'(x) - \alpha h'(x))' = -\dfrac{1}{\alpha} h(x) \\ f(1) = g(1) = 0, \ f(0) = kh(0), \\ \alpha h'(0) - g'(0) = -ikf'(0). \end{cases} \tag{28}$$

From the second and third equation of (28), we get $g(x) = h(x) = 0$. Then $f(x)$ satisfy

$$\begin{cases} f''(x) - \dfrac{i}{\alpha} f(x) = 0, \\ f(0) = f'(0) = f(1) = 0. \end{cases} \tag{29}$$

Simple computation shows that $f(x) = 0$. This implies that $X = (f,g,h) = 0$, which is a contradiction. So $-\frac{1}{\alpha} \notin \sigma_r(\mathscr{A})$. The proof is complete. □

4 Asymptotic Stability of System (3)

Definition 1 A C_0-semigroup $T(t)$ is called asymptotically (strongly) stable, if

$$\lim_{t\to\infty} ||T(t)x|| = 0.$$

Theorem 3 *([5]) Let $T(t)$ be a uniformly bounded C_0-semigroup on a Banach space X and let A be its generator. If*

$$\sigma(A)\bigcap i\mathbb{R} \subset \sigma_c(A),$$

and $\sigma_c(A)$ is countable, then $T(t)$ is asymptotically stable.

Theorem 4 *Let $\mathscr{A}$ be defined by (6). Then the system (3) achieves asymptotic stability.*

Proof Since from Lemma 1 and Proposition 3, we know that when $\lambda \in \sigma(\mathscr{A})$, $\mathrm{Re}\lambda < 0$. So we have

$$\sigma(\mathscr{A})\bigcap i\mathbb{R} = \emptyset \subset \sigma_c(\mathscr{A}) = \left\{-\frac{1}{\alpha}\right\},$$

and $\sigma_c(\mathscr{A})$ is obviously countable. Then system (3) achieves asymptotic stability by Theorem 3. □

5 Conclusions

In this paper, we use a wave equation with K-V damping to be a dynamic feedback controller for a Schrödinger equation. We give the asymptotic expression of the eigenvalues, and also the exact composition of the spectrum of the system operator. At last, the asymptotic stability of the system was achieved.

Acknowledgements This work was supported by the National Natural Science Foundation of China.

References

1. Guo, B.Z., Wang, J.M., Zhang, G.D.: Spectral analysis of a wave equation with Kelvin-Voigt damping. Z. Angew. Math. Mech. **90**(4), 323–342 (2010)
2. Krstic, M., Guo, B.Z., Smyshlyaev, A.: Boundary controllers and observers for the linearized Schrödinger equation. SIAM J. Control Optim. **49**(4), 1479–1497 (2011)

3. Liu, K., Liu, Z.: Exponential decay of energy of the Euler-Bernoulli beam with locally distributed Kelvin-Voigt damping. SIAM J. Control Optim. **36**, 1086–1098 (1998)
4. Liu, K., Liu, Z.: Exponential decay of energy of vibrating strings with local viscoelasticity. Zeitschrift für angewandte Mathematik und Physik ZAMP **53**, 265–280 (2002)
5. Luo, Z.H., Guo, B.Z., Morgul, O.: Stability and Stabilization of Infinite Dimensional Systems with Applications. Communications and Control Engineering Series. Springer, London (1999)
6. Machtyngier, E.: Exact controllability for the Schrödinger equation. SIAM J. Control Optim. **32**, 24–34 (1994)
7. Pazy, A.: Semigroups of Linear Operators and Applications to Partial Differential Equations. Springer, New York (1983)
8. Wang, J.M., Krstic, M.: Stability of an interconnected system of Euler-Bernoulli beam and heat equation with boundary coupling. ESAIM: Control, Optim. Calc. Var. **21**(4), 1029–1052 (2015)
9. Wang, J.M., Ren, B., Krstic, M.: Stabilization and Gevrey regularity of a Schrödinger equation in boundary feedback with a heat equation. IEEE Trans. Autom. Control **57**, 179–185 (2012)

Molecular-Dynamics Simulations Using Spatial Decomposition and Task-Based Parallelism

Chris M. Mangiardi and R. Meyer

Abstract This article discusses the implementation of a hybrid algorithm for the parallelization of molecular-dynamics simulations. The hybrid algorithm combines the spatial decomposition method using message passing with a task-based, threaded approach for the parallelization of the workload. Benchmark simulations on a multi-core system and an Intel Xeon Phi co-processor show that the hybrid algorithm provides better performances than the message-passing or threaded approaches alone.

1 Introduction

Molecular-dynamics (MD) is a computer simulation method that is widely used in computational physics, chemistry and materials science. The method is described in detail in Refs. [1, 3]. Since MD is frequently used to perform large-scale simulations on high-performance computers, it is important to develop MD algorithms that make the best use of modern computing architectures.

This article discusses a hybrid approach that uses a two-level approach for the parallelization of MD simulations. The first level is based on the spatial decomposition method [10] and is implemented with the Message-Passing Interface (MPI) [6]. The second parallelization level employs the cell-task method for the parallelization of the work-load within the spatial domains and is implemented with Intel's Threading Building Blocks Library [11].

If a MD simulations uses a short-range force model and the simulated system is sufficiently homogeneous, the well-known spatial decomposition method [10] provides an effective means for the parallelization of the simulation. In this method, the system being simulated is divided into equally sized and shaped sub-volumes, wherein each of these sub-volumes are processed by a processor, in order to improve the performance and reduce the time required to run the simulation. These sub-volumes of the system, however, require utilizing message passing in order to communicate their particles' data to neighbouring sub-volumes. The spatial

C.M. Mangiardi (✉) • R. Meyer
Laurentian University, Sudbury, Canada
e-mail: cmangiardi@laurentian.ca; rmeyer@cs.laurentian.ca

© Springer International Publishing Switzerland 2016

J. Bélair et al. (eds.), *Mathematical and Computational Approaches in Advancing Modern Science and Engineering*, DOI 10.1007/978-3-319-30379-6_13

decomposition method works best when the particles are evenly distributed between sub-volumes, as in homogeneous systems. If the system is inhomogeneous, load balancing problems occur and the parallel efficiency is reduced.

The cell-task method [7–9] is made to work well for inhomogeneous systems, wherein the work load would not be evenly balanced with the spatial decomposition method. This method works in a similar manner to the spatial decomposition method, in that it splits the system into sub-volumes; although for this method these sub-volumes are the width of the interaction range (Verlet-radius), and there is typically thousands of these sub-volumes. Each of these sub-volumes are scheduled to run on a processor core in an order wherein no two of the sub-volumes currently being simulated interact with the same particles, in order to remove the requirement of cache blocking. This method, unlike spatial decomposition, does not require the use of message passing, since the design is for a shared-memory system.

The hybrid method discussed in this article utilizes both of these methods, by first dividing the simulation into large sub-volumes for distribution to separate processors, as in the spatial decomposition method. After this point, the cell-task method is used to create many smaller sub-volumes of the system for dynamic scheduling across the processor cores. Communication between the larger sub-volumes is still required for particles near the border with the other subsections.

The primary rationale for the implementation of the hybrid method is that it extends the range of the cell-task method to more than one compute node. However, even on a single node, the hybrid approach can be advantageous. While the cell-task method is more efficient for inhomogeneous systems, the situations is less clear for homogeneous systems where spatial decomposition works well. In this case, the situation depends on system details since the overhead of the task management in the cell-task method competes with the communication overhead of the spatial decomposition method. Further, the spatial decomposition approach may have a slight advantage through a more localized memory access pattern. In this situation, a hybrid approach can lead to performance enhancements as it allows to interpolate between the cell-task method and spatial decomposition.

This work focuses on the performance of the hybrid model on a single compute node. In Sect. 3, results from a series of benchmark simulations on two test machines are presented. The results show that the hybrid method enables performance gains compared to the pure spatial decomposition or cell-task approaches.

2 Benchmark Procedures

The parallelization methods are tested using a system with multi-core processors and a separate system with an Intel Xeon Phi co-processor. The multi-core compute nodes contain two Intel Xeon E5-2680 processors, each with eight cores using the AVX instruction set architecture. The Intel Xeon Phi co-processors utilized are the 5110P model, with 60 physical cores, using hardware based threading for a total of 240 threads (details on the Xeon Phi many-core architecture can be found in Ref. [4]).

For the purposes of this work, the measurement being utilized is the parallel speedup factor. This represents the number of times faster than the baseline run the current run is, and is measure as the baseline time divided by the current run time.

On both systems, a variety of combinations of threads and MPI ranks are used, in order to gauge and compare performances. Speedups are measured against baseline runs using a single MPI rank and a single thread. In addition to the simulations using the hybrid method, all MPI runs were carried out using a single thread per MPI rank, and all threaded runs on the multi-core system using a single MPI rank. On the Xeon Phi, all threaded runs were based upon a recompiled version of the code without the MPI overhead, including its baseline run, as the speedups on this system are affected by the MPI overhead; conversely, this overhead does not effect the speedups on the multi-core system. On the multi-core systems, tests were then done using one, two, four, eight and sixteen MPI ranks, and varying numbers of threads to the total number of cores available per system. On the Xeon Phi co-processor, tests employed one, two, four, eight, sixteen, thirty, sixty, one-hundred twenty, and two-hundred forty MPI ranks, with varying numbers of threads.

All simulation runs, including the baseline, used vectorized implementations of the potentials [9] for better performance. All tests were run for 1000 time steps (2 femtoseconds per step), with the Verlet-lists (neighbour-lists) regenerated every 10 time steps. Each set of tests were run five times, taking the average time of each, in order to more accurately measure the time required to perform the simulation on a given system.

A number of simulation systems are used to ensure results extend beyond a single system. A bulk copper system with approximately 4.5-million particles, Cu (bulk), is used for its homogeneity, which works well with spatial decomposition. A porous copper system with two-million particles, Cu (porous), is used as it highlights the advantages of the cell-task method. Both these systems use the tight-binding potential [2], which has moderate force calculations. Additionally, an iron system with four-million particles, Fe (bulk), using the Mendelev potential [5] is included, as it has more complex force calculations and is less sensitive to memory access. Lastly, a liquid silver system with four-million particles, Ag (liquid), using the Lennard-Jones potential [1] is used due to its fast force calculations which makes it highly sensitive to memory access speed.

3 Results

As an example for all model systems, Fig. 1 shows the Cu (bulk) system's speedups using different numbers of processor cores on the multi-core system. For the hybrid method, only the data obtained with two MPI ranks is shown. The speedups with four and eight ranks are nearly identical to those obtained with two ranks and have been omitted to avoid crowding of the figure.

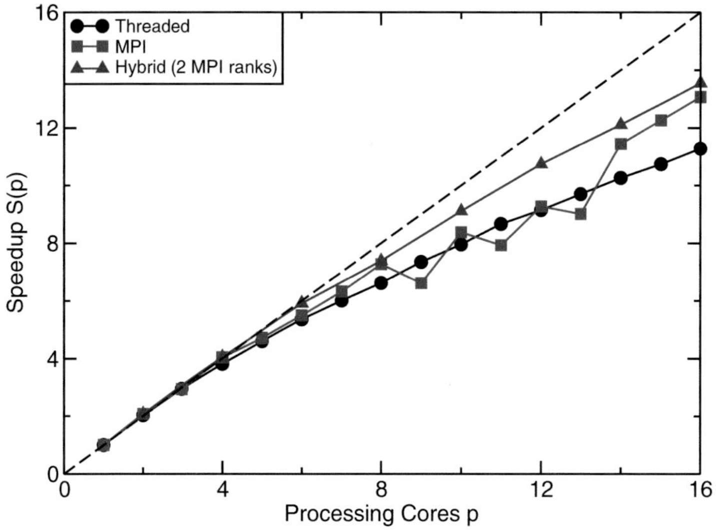

Fig. 1 Parallel speedups S in simulations of the Cu (bulk) system on the multi-core system as a function of the number of processing cores used p. The *dashed line* indicates the ideal speedup $S(p) = p$

Table 1 Best parallel speedup factors in simulations involving different systems on the multi-core processor

System	MPI speedup (ranks)	Threaded speedup (threads)	Hybrid speedup (ranks × threads)
Cu (bulk)	13.1 (16)	11.2 (16)	13.5 (2 × 8)
Cu (porous)	7.9 (16)	12.6 (16)	13.6 (2 × 8)
Fe (bulk)	13.1 (16)	10.4 (16)	13.2 (2 × 8)
Ag (liquid)	7.6 (16)	7.4 (16)	4.7 (8 × 2)

The best speedup factors, together with the number of ranks or threads where they were obtained, are shown in Table 1 for all test systems on the multi-core machine. Data is shown for spatial decomposition method (MPI), the cell-task approach (threaded) and the hybrid method utilizing the optimal combination of ranks and threads.

Figure 2 shows the Cu (bulk) system's speedups on the Intel Xeon Phi system. To avoid cluttering, only data for the cases of eight and thirty MPI ranks are shown for the hybrid method. The best results for the hybrid method were attained using thirty ranks, however, as seen with the eight ranks, the behaviour of the speedups are not monotonous, but instead are dependent upon the combination of ranks and threads. The best speedup factors with the corresponding numbers of ranks and threads are summarized for all test systems in Table 2.

From Fig. 1 and Table 1 it can be seen that when utilizing the tight-binding potentials on the multi-core system, the hybrid method produces better speedups than either the spatial decomposition or cell-task methods alone. For the homo-

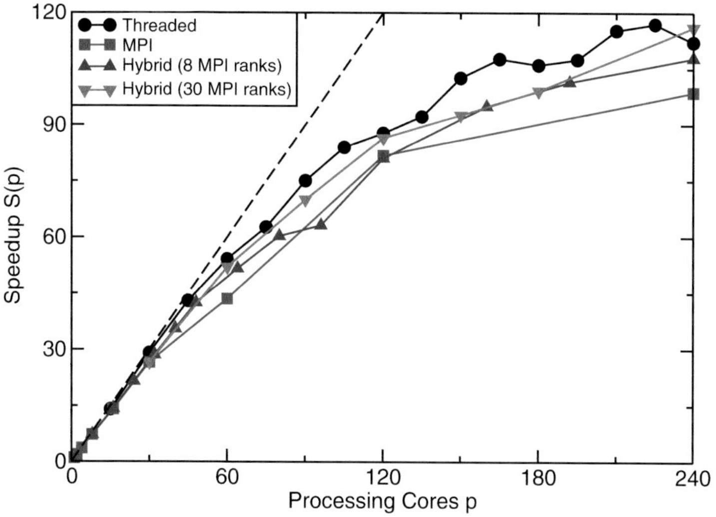

Fig. 2 Parallel speedups S in simulations of the Cu (bulk) system on the Intel Xeon Phi system as a function of the number of processing cores used p

Table 2 Best parallel speedup factors in simulations involving different systems on the Intel Xeon Phi co-processor

System	MPI speedup (ranks)	Threaded speedup (threads)	Hybrid speedup (ranks × threads)
Cu (bulk)	98.8 (240)	117.7 (230)	116.1 (30 × 8)
Cu (porous)	22.8 (240)	86.6 (185)	71.6 (4 × 40)
Fe (bulk)	100.6 (240)	124.1 (240)	118.5 (16 × 15)
Ag (liquid)	81.3 (240)	79.2 (240)	88.4 (120 × 2)

geneous system, the spatial decomposition method works well, allowing particles to be evenly divided amongst the different cores. The task-based method similarly works well, but shows a slower performance then the other two methods for this test system. This is in contrast to previous results [7, 8] where the performance of spatial decomposition and the cell-task method was very similar for homogeneous systems. A likely reason for this discrepancy is the larger size of the test systems in this work. Since the generation of the task schedule for the cell-task method does not parallelize well, and the number of tasks grows with the system size, larger systems might shift the balance towards spatial decomposition.

As shown by Fig. 2 and Table 2, the behaviour is different on the Xeon Phi co-processor where the cell-task method and the hybrid approach outperform spatial decomposition for the Cu (bulk) system. This is in agreement with the design of this architecture which favours threaded programs and is not optimal for large numbers of MPI ranks [4].

When utilizing the Cu (porous) system, the speedups of the hybrid method are more noticeable as compared to the spatial decomposition method. Due to the system being inhomogeneous, the spatial decomposition method does not evenly divide the system's particles amongst the processor cores. This allows for both the threaded and hybrid approaches to significantly outperform this method. When utilizing the Intel Xeon Phi, the system is divided into 240 areas with the spatial decomposition method, resulting in a large portion of the co-processor cores being idle, waiting on other cores to finish their work. This is significantly alleviated by the hybrid method, which divides the system into less areas, and uses the task-based approach to reduce the number of idle cores. The cell-task method, however, outperforms the hybrid method on the Xeon Phi, as it does not split the system into any areas, which allows it to better allocate cores to reduce the effects of the inhomogeneity.

For the Fe (bulk) system both the hybrid and cell-task methods outperform the spatial decomposition method. Despite the spatial decomposition method having a smaller portion the system to work on for each area, as compared to the hybrid method, the overhead associated with MPI and the inter-process communication counteracts the speed improvements attained calculating the forces. This effect is reduced significantly in the hybrid method, as it is uses less MPI ranks therefore reducing the overhead of the spatial decomposition method. Further, the cell-task method has larger speedups compared to the other methods for this system. This is most likely due to the higher complexity of the force calculations for the Mendelev potential. With more time spent per force calculation, the impact of limiting factors like memory access time is reduced.

The Ag (liquid) system performs significantly slower using the hybrid method compared to the other methods on the multi-core system. Due to this being a liquid system, particles are able to more freely move about the system, and as a result, there is a high amount of inter-process communication required. Further, the overhead of creating the schedule utilizing less threads further degrades the performance of the hybrid method. The Lennard-Jones potential is also simpler than other potentials, reducing the amount of time spent in the calculations of forces, and instead emphasizes other sections of the program, such as the Verlet-list generation, scheduler, and inter-process communication. On the Xeon Phi system, however, the hybrid method performs significantly better than the other two methods on the Ag (liquid) system. The different behaviour of this test system on the two machines is not fully understood and requires more studying.

4 Conclusions

The results of the benchmark simulations shown in the previous section demonstrate that the hybrid method utilizing both spatial decomposition and task-based parallelism, is able to provide better performance on molecular-dynamics simulations than either method alone. The hybrid method takes advantage of each method's benefits, whilst minimizing their drawbacks.

For homogeneous systems spatial decomposition works very well and often delivers, in particular on multi-core systems, a better performance than the cell-task method. The Intel Xeon Phi hardware and system software, however, favours threaded programming and is not designed to work optimally with $\approx$240 MPI ranks. For simulations of homogeneous systems, the hybrid model provides the means to optimize the performance by reducing the number of threads and MPI communications.

By its design, the cell-task method outperforms spatial decomposition for inhomogeneous systems since it avoids the load balancing problems that occur with the spatial decomposition method when the number of particles varies between domains. In principle, the hybrid method suffers from the same issue, however since the hybrid methods operates with larger domains, inhomogeneities may average out. By reducing the task management overhead of the cell-task method, the hybrid approach can therefore in certain cases still improve the total performance.

An advantage of the hybrid parallelization method described in this work is that it makes the simulation program adaptable. By choosing the combination of MPI ranks and threads, the speed of the simulation can be optimized for the specific type of the simulated system and the computer system on which the simulation is run.

Overall, the performance gains of the hybrid method are much more pronounced on the Xeon Phi co-processor than they are on the multi-core system. This emphasizes that novel many-core architectures require the development of new algorithms to maximize the performance.

Acknowledgements This work has been made possible by generous allocation of computer time on computer systems managed *by Calcul Québec*, the *Shared Hierarchical Academic Research Network* (SHARCNET) and *Compute/Calcul Canada*. Financial support by Laurentian University and the Natural Sciences and Engineering Research Council of Canada (NSERC) is gratefully acknowledged.

References

1. Allen, M.P., Tildesley, D.J.: Computer Simulations of Liquids. Clarendon, Oxford (1987)
2. Cleri, F., Rosato, V.: Tight-binding potentials for transition metals and alloys. Phys. Rev. B **48**(1), 22–33 (1993)
3. Frenkel, D., Smit, B.: Understanding Molecular Simulation. Academic, San Diego (2002)
4. Jeffers, J., Reinders, J.: Intel Xeon Phi Coprocessor High Performance Programming. Morgan Kaufman, New York (2013)
5. Mendelev, M.I., Han, S., Srolovitz, D.J., Ackland, G.J., Sun, D.Y., Asta, M.: Development of new interatomic potentials appropriate for crystalline and liquid iron. Philos. Mag. **83**(35), 3977–3994 (2003)
6. Message Passing Interface Forum: http://www.mpi-forum.org/ (2016)
7. Meyer, R.: Efficient parallelization of short-range molecular dynamics simulations on many-core systems. Phys. Rev. E **88**(5), 053,309 (2013)
8. Meyer, R.: Efficient parallelization of molecular dynamics simulations with short-ranged forces. J. Phys.: Conf. Ser. **540**(1), 012,006 (2014)

9. Meyer, R., Mangiardi, C.M.: Parallelization of molecular-dynamics simulations using tasks. MRS Proc. **1753**, mrsf14–1753–nn10–09 (2015)
10. Plimpton, S.: Fast parallel algorithms for short-range molecular dynamics. J. Comput. Phys. **117**(1), 1–19 (1995)
11. TBB official website: http://threadingbuildingblocks.org/ (2016)

Modelling of Local Length-Scale Dynamics and Isotropizing Deformations: Formulation in Natural Coordinate System

O. Pannekoucke, E. Emili, and O. Thual

Abstract We propose an algorithm to model anisotropic correlation functions using an approach based on the deformation of locally isotropic ones. In a previous work, a set of equations that allow to calculate the desired deformation was derived for a flat coordinates system. However this strategy is not adapted for curved geometry as the sphere (regional and global atmospheric models), where it is suitable to state the local isotropy in terms of the local Riemannian metric.

This paper introduces the theoretical background to deal explicitly with natural coordinate systems leading to a formulation adapted with the Riemannian metric. It results that the isotropizing deformation is obtained from the resolution of coupled non-linear equations depending on the geometry. This procedure is illustrated within a 2D setting.

1 Introduction

In variational data assimilation, the analysis state minimizes the cost function

$$J(\mathscr{X}) = (\mathscr{X} - \mathscr{X}^b)^T\mathbf{B}^{-1}(\mathscr{X} - \mathscr{X}^b) + (\mathscr{Y}^o - \mathbf{H}\mathscr{X})\mathbf{R}^{-1}(\mathscr{Y}^o - \mathbf{H}\mathscr{X}), \qquad (1)$$

where $\mathscr{X} \in \mathbb{R}^n$ is the state vector, $\mathscr{Y}^o \in \mathbb{R}^p$ the observational vector, $\mathbf{B} = \mathbb{E}[\boldsymbol{\varepsilon}^b \boldsymbol{\varepsilon}^{b^T}]$ denotes the covariance matrix of the background error $\boldsymbol{\varepsilon}^b = \mathscr{X}^b - \mathscr{X}^t$, $\mathscr{X}^b$ ($\mathscr{X}^t$)

O. Pannekoucke (✉)
CERFACS/CNRS URA 1875, Toulouse, France

CNRM/GAME, Météo-France/CNRS UMR 3589, Toulouse, France

INPT-ENM, Toulouse, France
e-mail: olivier.pannekoucke@cerfacs.fr

E. Emili
CERFACS/CNRS URA 1875, Toulouse, France

O. Thual
CERFACS/CNRS URA 1875, Toulouse, France

Université de Toulouse; INPT, CNRS; IMFT; F-31400 Toulouse, France

© Springer International Publishing Switzerland 2016

J. Bélair et al. (eds.), *Mathematical and Computational Approaches in Advancing Modern Science and Engineering*, DOI 10.1007/978-3-319-30379-6_14

is the background state (the true state), $\mathbf{R} = \mathbb{E}[\boldsymbol{\varepsilon}^o \boldsymbol{\varepsilon}^{oT}]$ denotes the covariance matrix of the observational error $\boldsymbol{\varepsilon}^o = \mathscr{Y}^o - \mathbf{H}\mathscr{X}^t$, and $\mathbf{H}$ is the observation operator (assumed linear here) that maps the model state to observation locations. For large dimensional problems, like meteorological applications, the covariance matrix $\mathbf{B}$ cannot be explicitly represented and is often modelled.

Preliminary works have shown that it is possible to estimate and to use the correlation length-scale, to produce a coordinate change that facilitates the modelling of the anisotropic covariance matrix $\mathbf{B}$. The approach, developed in [1], relies on the local metric tensor $\mathbf{g_x}$ defined from the Taylor expansion of a correlation function $\rho(\mathbf{x}, \mathbf{x} + \delta\mathbf{x}) = 1 - \frac{1}{2}||\delta\mathbf{x}||^2_{\mathbf{g_x}} + \mathscr{O}(|\delta\mathbf{x}|^3)$, with $||\mathbf{x}||^2_{\mathbf{E}} = \mathbf{x}^T\mathbf{E}\mathbf{x}$ (this formalism can be extended in the particular case where the gradient of ρ is not zero at $\mathbf{x} = \mathbf{y}$). This local definition for $\mathbf{g_x}$ serves to define a metric field $\mathbf{g}$ over a domain (thereafter the two notations $G_\mathbf{x}$ and $G(\mathbf{x})$ are used to denote the value of a quantity G at point $\mathbf{x}$). In particular, isotropic correlation functions in $\mathbb{R}^n$ takes the form $\rho_{iso}(\boldsymbol{u}_1, \boldsymbol{u}_2) = \rho_{iso}(||\boldsymbol{u}_1 - \boldsymbol{u}_2||_{\mathbb{R}^n})$, where $||\cdot||_{\mathbb{R}^n}$ denotes the Euclidian norm and where $(\boldsymbol{u}_1, \boldsymbol{u}_2)$ are two points in $\mathbb{R}^n$. Considering a Riemannian manifold M embedded in $\mathbb{R}^n$ and denoting by $\boldsymbol{\omega}$ the metric induced by the Euclidian metric of $\mathbb{R}^n$, then the restriction of ρ_{iso} to the manifold defines correlation functions ρ on M [2]. The local behavior of these functions is given by $\rho(\mathbf{x}, \mathbf{x} + \delta\mathbf{x}) = \rho_{iso}\left(||\delta\mathbf{x}||_{\omega_x}\right)$. Then, the local metric tensor $\mathbf{g_x}$ of ρ at point $\mathbf{x}$ is often expressed in terms of the Riemannian metric $\boldsymbol{\omega}_\mathbf{x}$ as $\mathbf{g_x} = \frac{1}{L_h^2}\boldsymbol{\omega}_\mathbf{x}$, where L_h is the constant length-scale, as encountered for, e.g. $\rho_{iso}(r) = e^{-\frac{r^2}{2L_h^2}}$. By analogy with correlation modelling on the sphere, we say that the correlation functions ρ on M are isotropic.

As an example of Riemann manifold, we can consider the sphere of radius a, parametrized thanks to the longitude/co-latitude coordinate system $\mathbf{x} = (\theta, \phi)$ and equipped with the natural metric $\boldsymbol{\omega}$ defined from the square arc-length $ds^2 = \omega_{ij}dx^i dx^j$ (in Einstein's summation convention) with $ds^2 = a^2(\sin\phi)^2 d\theta^2 + a^2 d\phi^2$.

The local metric tensor is a very attractive quantity. It can be diagnosed from ensemble estimation [3], associated with a filtering step to damp spurious sampling noise [4]. This local metric is often associated with the local diffusion tensor $\boldsymbol{\nu}_\mathbf{x}$ [5], defined by $\nu^{ij} = \frac{1}{2}g^{ij}$ where g^{ij} denotes the inverse tensor of $\mathbf{g}$, i.e. $g^{ij}g_{jl} = \delta_{il}$ where δ_{ij} denotes the Kronecker symbol.

In data assimilation it is usual to find anisotropic correlation functions which require sophisticated covariance models, e.g. the wavelets formulation [6], the recursive filter formulation [7] the diffusion equation formulation [8, 9] or the coordinate change of isotropic correlations [10]. This coordinate change is a tricky way to reduce the numerical cost or facilitate the parallelization of the algorithm [1]: ones the computation of the isotropizing coordinate change has been done, anisotropic correlations simply result from low cost interpolations between the two coordinate systems where efficient algorithms exist. For example a locally isotropic correlation field can be efficiently obtained by alternate applications of diffusion equation method, parallelized in the space dimension. Thereafter the application of the coordinate change consists in introducing a coordinate change, able to transform isotropic correlation into the desired anisotropic correlation. The coordinate change can be estimated from ensemble method, e.g. [11] has proposed a procedure that relies on the wavelet estimation of the gradient of the deformation [12]. Legrand and Michel [13] illustrated the feasibility of the isotropization procedure for real data, and the potential for data assimilation. Until now, the general framework that leads to a local isotropy for the curved spaces has not been addressed.

In [1] the coordinate change is obtained from the local metric tensor $\mathbf{g}(\mathbf{x})$: it consists in finding a differential map $\mathbf{x}(\tilde{\mathbf{x}})$ that transforms a coordinate system $\tilde{\mathbf{x}}$, where the local metric tensor $\tilde{\mathbf{g}}(\tilde{\mathbf{x}})$ is isotropic, into a coordinate system, where the local metric $\mathbf{g}(\mathbf{x})$ is the anisotropic diagnosed one, and respecting the chain rule

$$\tilde{g}_{\alpha\beta} = \partial_{\tilde{x}^\alpha} x^i \partial_{\tilde{x}^\beta} x^j g_{ij}. \tag{2}$$

In particular, it has been shown possible to construct a coordinate change leading to a local isotropic tensor $\tilde{g}_{\alpha\beta}(\tilde{\mathbf{x}}) = \frac{1}{L^2(\tilde{\mathbf{x}})}\delta_{\alpha\beta}(\tilde{\mathbf{x}})$. This coordinate change is found in two steps:

(a) first one has to find $\mathbf{x}(\tilde{\mathbf{x}})$, solution of the coupled non-linear Poisson's like equations

$$\Delta x^i + \Gamma^i_{jk} \partial_{\tilde{x}^\alpha} x^j \partial_{\tilde{x}^\alpha} x^k = 0, \tag{3}$$

where Δ is the Laplacian operator, and $\Gamma^i_{jk} = g^{il}\Gamma_{jkl}$ denotes the Christoffel symbols of the second kind associated with the metric $\mathbf{g}$ defined from the Christoffel symbols of the first kind $\Gamma_{ijk} = (\partial_i g_{kj} + \partial_j g_{kj} - \partial_k g_{ij})/2$;

(b) in a second step, the appropriate coordinate change $\tilde{\mathbf{x}}(\mathbf{x})$ is obtained as the inverse of the differential map $\mathbf{x}(\tilde{\mathbf{x}})$.

For general manifold, the metric δ_{ij} is not the one found in natural coordinates. As mentioned above, on a part of the sphere, the metric δ_{ij} must be replaced by the natural metric ω_{ij} and a global isotropic correlation function is locally expressed in terms of ω_{ij} so that $\tilde{g}_{ij}(\tilde{\mathbf{x}}) = \frac{1}{L_h^2}\omega_{ij}(\tilde{\mathbf{x}})$ where L_h is a constant length-scale. The local isotropic version is thus $\tilde{g}_{ij}(\tilde{\mathbf{x}}) = \frac{1}{L^2(\tilde{\mathbf{x}})}\omega_{ij}(\tilde{\mathbf{x}})$, where $L(\tilde{\mathbf{x}})$ is a length-scale field. It follows that the metric ω_{ij} has to be taken into account in the isotropizing process.

The aim of the present contribution is to extend the results from [1], by taking into account natural coordinate systems. We first introduce a geometrical framework for the isotropization issue in Sect. 2, where the isotropizing deformation appears as an harmonic map, solution of a system of non-linear partial differential equations. The isotropization procedure proposed is then tested into a simplified 2D setting in Sect. 3. The conclusions are reported in Sect. 4.

2 General Formalism for the Isotropization Procedure

In this section we describe the theoretical background for constructing the isotropizing coordinate transform. This relies on Riemannian geometry [14, 15] and in particular on the properties of harmonic maps [16].

2.1 Coordinate Change Equation Set

The geographical domain considered here is assumed to be represented by a first Riemannian manifold M of dimension d and equipped with the metric $\boldsymbol{\omega}$, it is denoted by $(M, \boldsymbol{\omega})$. The diagnostic of the local metric tensor field $\mathbf{g}$ endows M with a second Riemannian structure, $(M, \mathbf{g})$. A third Riemannian structure $(M, \tilde{\mathbf{g}})$ is considered, where the metric $\tilde{\mathbf{g}}$ takes the form $\tilde{\mathbf{g}} = \frac{1}{L^2}\boldsymbol{\omega}$ with L is an unknown length-scale field. The isotropizing coordinate change takes the form of a differential map $\mathbf{x}(\tilde{\mathbf{x}})$ from $(M, \tilde{\mathbf{g}})$ to $(M, \mathbf{g})$ so that $\tilde{\mathbf{g}}$ and $\mathbf{g}$ are related according to Eq. (2). From a theoretical point of view, $\mathbf{x}(\tilde{\mathbf{x}})$ can be considered as an isometric diffeomorphism from $(\tilde{M}, \tilde{\mathbf{g}})$ to $(M, \mathbf{g})$, where $\tilde{M}$ is isomorphic to M. Note that when considering a differential map $\boldsymbol{f}$ from $\tilde{M}$ to $(M, \mathbf{g})$, it is possible to define a third metric $\boldsymbol{f}^*\mathbf{g}$ on $\tilde{M}$ by $(\boldsymbol{f}^*\mathbf{g})_{\alpha\beta} = \partial_{\tilde{x}^\alpha} f^i \partial_{\tilde{x}^\beta} f^j g_{ij}$. Then, Eq. (2) also reads $\tilde{\mathbf{g}} = \mathbf{x}^*\mathbf{g}$.

It can be shown from transformation rules of the Christoffel symbols of the second kind [15] $\tilde{\Gamma}^\gamma_{\alpha\beta} = \partial_{\tilde{x}^\alpha} x^j \partial_{\tilde{x}^\beta} x^k \partial_{x^i} \tilde{x}^\gamma \Gamma^i_{jk} + \partial^2_{\tilde{x}^\alpha \tilde{x}^\beta} x^i \partial_{x^i} \tilde{x}^\gamma$, that the coordinate functions x^i are solutions the system of coupled non-linear equations

$$\Delta_{\tilde{\mathbf{g}}} x^i + \Gamma^i_{jk} \partial_{\tilde{x}^\alpha} x^j \partial_{\tilde{x}^\beta} x^k \tilde{g}^{\alpha\beta} = 0, \tag{4}$$

where $\Delta_{\tilde{\mathbf{g}}}$ denotes the Beltrami's Laplacian with Γ^i_{jk} (respectively $\tilde{\Gamma}^i_{jk}$) are the Christoffel symbols of the second kind associated with the metric field $\mathbf{g}$ (respectively $\tilde{\mathbf{g}}$).

Hence, Eqs. (4) provides a necessary condition verified by the isotropizing coordinate change, and can be used to find the unknown coordinate change. However multiple solutions can exist, which means that several length-scale fields could exist for a given metric $\mathbf{g}$. A major difficulty is that these equations directly make use of the unknown metric $\tilde{\mathbf{g}}$.

To better clarify these two statements we introduce a functional which is related to the energy of the deformation $\mathbf{x}(\tilde{\mathbf{x}})$.

A differential map, solution of Eq. (4), is an harmonic map [14, 16], that is a stationary point (critical point), for variations of the differential map $\mathbf{x}(\tilde{\mathbf{x}})$, of the energy functional [17], $E[\mathbf{x}(\tilde{\mathbf{x}}), \tilde{\mathbf{g}}, \mathbf{g}] = \int_{\tilde{M}} \frac{1}{2}||d\mathbf{x}||^2 d\tilde{M}$, where $d\tilde{M} = \sqrt{\tilde{g}} d\tilde{x}^1 \cdots d\tilde{x}^d$ is the invariant volume element of $\tilde{M}$ and $|| \cdot ||$ denotes the Hilbert-Schmidt norm. This implies that the coordinate notation for the energy density $e_{\tilde{\mathbf{x}}}[\mathbf{x}(\tilde{\mathbf{x}}), \tilde{\mathbf{g}}, \mathbf{g}] = \frac{1}{2}||d\mathbf{x}||^2$ reads $e_{\tilde{\mathbf{x}}}[\mathbf{x}(\tilde{\mathbf{x}}), \tilde{\mathbf{g}}, \mathbf{g}] = \frac{1}{2}\tilde{g}^{\alpha\beta} \partial_{\tilde{x}^\alpha} x^i \partial_{\tilde{x}^\beta} x^j g_{ij}$. In particular, Eq. (4) are the Euler-

Lagrange of the energy E for variation of the differential map $\mathbf{x}(\tilde{\mathbf{x}})$ [14]. Note that $E[\mathbf{x}(\tilde{\mathbf{x}}), \tilde{\mathbf{g}}, \mathbf{g}]$ is a quadratic functional but since it is not convex in $\mathbf{x}$ it may have multiple critical points.

2.2 *Conformal Equivalence Classes*

For covariance modelling in data assimilation, a simplified decomposition of linear operators can be introduced [6, 8, 18, 19]. For instance the background error covariance matrix $\mathbf{B}$ as specified in the non-separable spectral decomposition takes the form $\mathbf{B} = \boldsymbol{\Sigma}\mathbf{S}^{-1}\mathbf{D}_h^{1/2}\mathbf{E}_{hv}\mathbf{D}_h^{T/2}\mathbf{S}^{-T}\boldsymbol{\Sigma}^T$ where linear operators are as follows: $\mathbf{E}_{hv}$ encodes the horizontal-vertical correlations, $\mathbf{D}_h$ encodes the horizontal correlations, $\mathbf{S}$ is the grid point to spectral transform operator and $\boldsymbol{\Sigma}$ encodes the grid point standard-deviation [18]. The separable formulation based on the diffusion equation [8, 9, 20] takes the form $\mathbf{B} = \boldsymbol{\Sigma}\mathbf{S}^{-1}\mathbf{L}_h\mathbf{L}_v\mathbf{W}^{-1}\mathbf{L}_v^T\mathbf{L}_h^T\boldsymbol{\Sigma}^T$ where the linear operator $\mathbf{W}$ is the metric tensor, $\mathbf{L}_h$ and $\mathbf{L}_v$ encode the horizontal and the vertical diffusion time integration with the diffusion tensor deduced from objective estimation [3, 9, 20]. Note that, [21] has proposed a non-separable formulation of the diffusion formulation that relies on the wavelets. It results, from the linear operator decomposition described above, that the horizontal correlation can be treated with a linear operator that is 2D for a given level.

A particular property of the 2D case is that the energy functional is conformaly invariant in $\tilde{\mathbf{g}}$, i.e. it remains constant when replacing $\tilde{\mathbf{g}}$ by $\frac{1}{L(\tilde{x})^2}\boldsymbol{\omega}$, i.e. $E[\mathbf{x}, \tilde{\mathbf{g}}, \mathbf{g}] = E[\mathbf{x}, \boldsymbol{\omega}, \mathbf{g}]$ (Weyl transformation and invariance, see [15]). Thanks to this invariance, the unknown length-scale field L can be eliminated by solving the problem within the conformal equivalent class of $\boldsymbol{\omega}$. It results that, for two dimensional problems, one can replace Eq. (4) by equations

$$\Delta_\omega x^i + \Gamma^i_{jk}\partial_{\tilde{x}^\alpha}x^j\partial_{\tilde{x}^\beta}x^k\omega^{\alpha\beta} = 0, \tag{5}$$

this can also be directly obtained from Eq. (4) when replacing $\tilde{\mathbf{g}}$ by $L^{-2}\boldsymbol{\omega}$ in 2D. In the particular case where $\omega_{ij} = \delta_{ij}$, Eq. (5) leads to Eq. (3) as previously found in [1].

The unknown metric $\tilde{\mathbf{g}}$ can be explicitly deduced from the solution $\mathbf{x}(\tilde{\mathbf{x}})$ of Eq. (5) thanks to the metric change Eq. (2) that reads $\tilde{\mathbf{g}} = \mathbf{x}^*\mathbf{g}$, and then provides the length-scale field $L(\tilde{\mathbf{x}}) = \left(\frac{1}{2}Trace\left(\mathbf{x}^*\mathbf{g}_{\tilde{\mathbf{x}}}\ \boldsymbol{\omega}_{\tilde{\mathbf{x}}}^{-1}\right)\right)^{-1/2}$.

We are now able to describe the isotropizing procedure.

2.3 Isotropizing Procedure

The two step isotropizing procedure, detailed in Algorithm 1, is similar to the one described in [1], using a finite difference scheme for spatial derivative, a forward Euler time scheme, and a spline bi-cubic interpolation. First it consists in computing the inverse isotropization transformation, obtained by solving Eq. (5). For that purpose the pseudo-time diffusion scheme (the heat flow) [16]

$$\partial_\tau x^i = \Delta_\omega x^i + \Gamma^i_{jk} \partial_{\tilde{x}^\alpha} x^j \partial_{\tilde{x}^\beta} x^k \omega^{\alpha\beta}, \tag{6}$$

can be employed to find the stationary state, solution of Eq. (5), where the initial condition is the identity map $id(\tilde{\mathbf{x}}) = \tilde{\mathbf{x}}$. This defines a family of differential map $\mathbf{x}_\tau$ indexed by $\tau \in [0, \infty)$, continuously dependent of τ. For diffeomorphic compact manifold without boundary, if the algorithm converges toward a stationary differential solution, then this solution is an harmonic map, and this harmonic map is continuously obtained from the initial condition as the limit $\lim_{\tau\to\infty} \mathbf{x}_\tau$ [22]. Note that the dynamics Eq. (6) is the gradient flow associated with the energy. As a consequence, the tendency must be non-positive, $\frac{dE(\mathbf{x}_\tau)}{d\tau} \leq 0$, so that the energy $E[\mathbf{x}_\tau(\tilde{\mathbf{x}}), \boldsymbol{\omega}, \mathbf{g}]$ is minimized along the path $\mathbf{x}_\tau$.

Algorithm 1 Two-steps isotropizing procedure in curved space

Require: The metric field $\mathbf{g}$ is assumed to be known (e.g. estimated from an ensemble)

Compute $\Gamma_{ijk}(\mathbf{x}) = \frac{1}{2}\left(\partial_j g_{ik}(\mathbf{x}) + \partial_k g_{ji}(\mathbf{x}) - \partial_i g_{jk}(\mathbf{x})\right)$
Compute $\Gamma^i_{jk}(\mathbf{x}) = g^{il}(\mathbf{x})\Gamma_{ljk}(\mathbf{x})$
Set $q_\infty = 500$ and $k_\infty = 2$
Compute $\Gamma^i_{jk}(\mathbf{x}) = g^{il}(\mathbf{x})\Gamma_{ljk}(\mathbf{x})$

\# Step 1: Pseudo-diffusion
$x^i_1(\tilde{\mathbf{x}}) = \tilde{x}^i$
for q from 1 to q_∞ **do**
    for i from 1 to d **do**
        $\gamma^i_{jk}(\tilde{\mathbf{x}}) = \Gamma^i_{jk}[\mathbf{x}_1(\tilde{\mathbf{x}})]$ (Spline interpolation)
        $x^i_2(\tilde{\mathbf{x}}) = x^i_1(\tilde{\mathbf{x}}) + d\tau\left(\Delta_\omega x^i_1 + \gamma^i_{jk}\partial_{\tilde{x}^\alpha} x^j_1 \partial_{\tilde{x}^\beta} x^k_1 \omega^{\alpha\beta}\right)(\tilde{\mathbf{x}})$
    end for
    $\mathbf{x}^i_1(\tilde{\mathbf{x}}) = \mathbf{x}^i_2(\tilde{\mathbf{x}})$
end for
$\mathbf{x}(\tilde{\mathbf{x}}) = \mathbf{x}_1(\tilde{\mathbf{x}})$

\# Step 2: Inverse transform (fixed-point's iterations)
$\delta_1(\tilde{\mathbf{x}}) = 0$
for k from 1 to k_∞ **do**
    $\delta_2(\tilde{\mathbf{x}}) = \frac{1}{2}\{\mathbf{x}[\tilde{\mathbf{x}} - \delta_1(\tilde{\mathbf{x}})] - \tilde{\mathbf{x}}\}$
    $\delta_1(\tilde{\mathbf{x}}) = \delta_2(\tilde{\mathbf{x}})$
end for
$\tilde{\mathbf{x}}(\mathbf{x}) = \mathbf{x} - 2\delta_1(\tilde{\mathbf{x}})$

The isotropizing coordinate change $\tilde{\mathbf{x}}(\mathbf{x})$ is obtained as the inverse of the differential map $\mathbf{x}(\tilde{\mathbf{x}})$. The differential map $\tilde{\mathbf{x}}(\mathbf{x})$ is denoted $\overset{\circ}{D}$.

The isotropization procedure is illustrated within a 2D setting in the next section.

3 Numerical Experiments

Anisotropic correlations are numerically constructed in 2D from globally isotropic ones with a deformation derived from an arbitrary stream function. This construction will be used to test the skill of local isotropization methods. The 2D setting is specified in order to mimic a portion of the sphere, which is often the case of horizontal coordinates employed in regional models.

3.1 Experimental Set-Up

To validate the isotropizing process, the following route is considered: a Riemannian manifold $(M, \boldsymbol{\omega})$ is first introduced, that mimics the situation encountered in atmospheric modelling where the local metric $\boldsymbol{\omega}$ is not flat ; a local isotropic metric $\tilde{\mathbf{g}}$, with respect to the metric $\boldsymbol{\omega}$, is then produced from a given length-scale field L ; a deformation D is constructed and used to generate an anisotropic metric $\mathbf{g} = D_* \tilde{\mathbf{g}}$ (the pushforward metric of $\tilde{\mathbf{g}}$ by D). Using the isotropizing process, the challenge is to find from $\mathbf{g}$ and $\boldsymbol{\omega}$ – the only information available in real applications – a coordinate change that transforms $\mathbf{g}$ into a local isotropic metric with respect to $\boldsymbol{\omega}$.

As a first step, a bi-periodic domain M is considered, parametrized with a single bi-periodic chart $\Big(U = [0, 1]\times[0, 1], \mathbf{x} = (x, y)\Big)$ discretized in $n = 81$ points along the two directions x and y. This coordinate system is similar to an angular chart on the sphere, and is equipped with the metric ω_{ij} defined from the arc-length square

$$ds^2 = R^2\Big[\alpha + (1-\alpha)(\sin \pi y)^2\Big]dx^2 + R^2 dy^2,$$

with $\alpha = 0.2$ and $R = 1000$ km, such that the resolution at $\mathbf{x} = (0.5, 0.5)$ is isotropic with $ds \approx 12$ km, the length of the domain along x for $y = 0.5$ is equal to $R = 1000$ km, while the length for $y = 0$ or for $y = 1$ is equal to $R\sqrt{\alpha} \approx 447$ km. The length along y does not depends on the position x and is equal to R.

A deformation $D(\mathbf{x}) = \mathbf{x} + d(\mathbf{x})$ is constructed as follows. First, a wind field $\mathbf{u}_0(\mathbf{x}) = \mathbf{k} \times \nabla\psi$ is introduced, where ψ is a stream function, and it is normalized so that $\mathbf{u} = (u, v) = (u_0/\max(|u_0|), v_0/\max(|v_0|))$. Then, for each position $\mathbf{x}$, the geodesic curve $\gamma_{\mathbf{x}}(t)$ is computed, starting at $\mathbf{x}$ with the velocity $\dot{\gamma}_{\mathbf{x}}(0) = \mathbf{u}(\mathbf{x})dt$ where $dt = 0.05$ is a magnitude factor. Denoting by $\mathbf{V}(t)$ the velocity $\dot{\gamma}_{\mathbf{x}}(t)$,

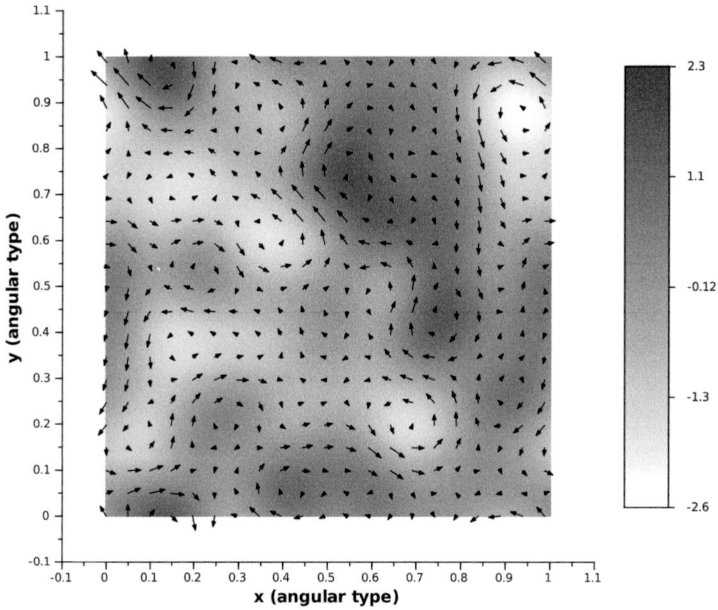

Fig. 1 Stream function (*shading*) and displacement field resulting from the geodesic time integration with initial velocity deduced from the stream current. The *arrows* indicate the direction and the intensity of the associated displacement

the geodesic curve is obtained as the time integration of the geodesic system $\left\{\frac{d\gamma_{\mathbf{x}}^i}{dt} = V^i, \frac{dV^i}{dt} = -\Omega^i_{jk}V^jV^k\right\}$, with initial condition $\left(\gamma_{\mathbf{x}}(0) = \mathbf{x}, \mathbf{V}(0) = \mathbf{u}(\mathbf{x})dt\right)$ and where Ω^i_{jk} are the Christoffel symbols of the second kind associated with the metric field $\boldsymbol{\omega}$. The displacement field is then defined as $d(\mathbf{x}) = \gamma_{\mathbf{x}}(1) - \mathbf{x}$.

The stream function used for numerical experiment is shown in Fig. 1, with the displacement field deduced from the velocity field.

3.2 Anisotropic Correlations and Isotropizing Procedure

In a data assimilation framework, the metric $\mathbf{g}$ is computed from ensemble estimation, and it can be used within the isotropizing procedure we propose. In this work, a simulated experiment is designed to validate the isotropizing procedure and a synthetic anisotropic metric field is generated from the deformation of a locally isotropic metric field.

For this purpose, locally isotropic correlation functions are considered, specified by their local metrics $\tilde{\mathbf{g}}(\tilde{\mathbf{x}}) = \frac{1}{L(\tilde{\mathbf{x}})^2}\boldsymbol{\omega}(\tilde{\mathbf{x}})$ where the varying length-scale field L has been arbitrarily fixed, with average value $L_h = 50\,\text{km}$. The diffusion tensor

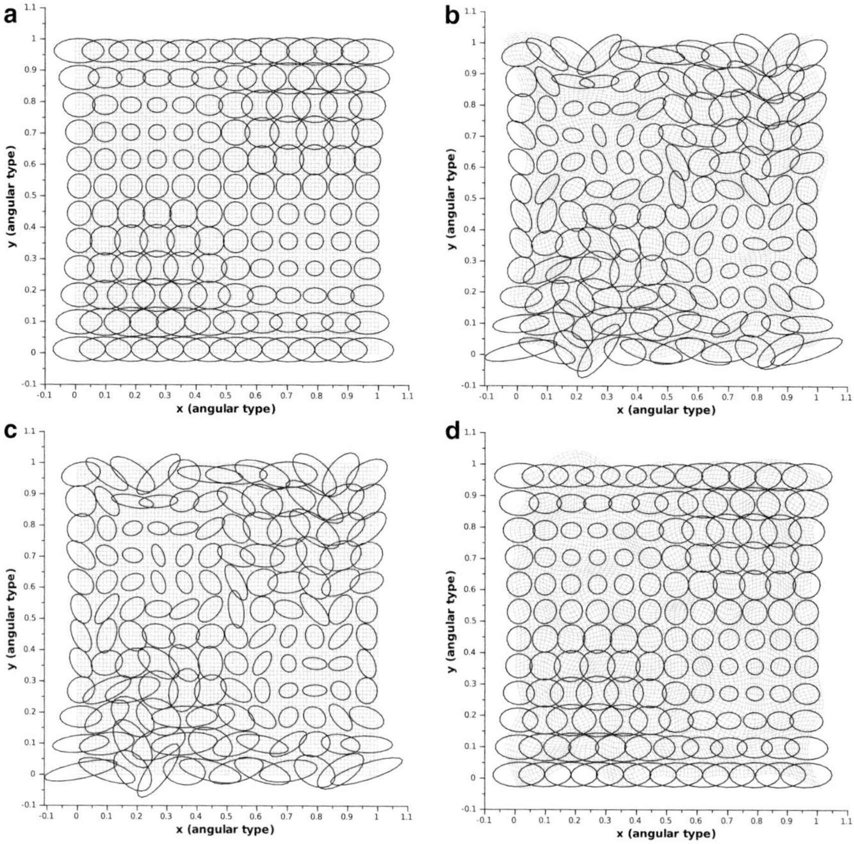

Fig. 2 The initial locally isotropic metric, represented on a regular coordinate system (**a**), is deformed under the action of the deformation D leading to anisotropic metric field (**b**). Then, the anisotropic metric field represented within the natural coordinate system (**c**) is diagnosed and employed in the isotropizing process which provides the isotropizing deformation $\overset{\circ}{D}$ (corresponding theoretically to D^{-1}). The action of $\overset{\circ}{D}$ on the anisotropic metric field is the nearly locally isotropic metric field (**d**). The coordinate systems are reproduced in *gray*

associated with $\tilde{\mathbf{g}}$ is illustrated in Fig. 2a The coordinate change D is then applied on these functions to obtain anisotropic correlations functions diagnosed by their diffusion tensor, Fig. 2b.

The isotropization procedure is considered on the metric $\mathbf{g}$ (diagnosed within a natural coordinate system) as illustrated in Fig. 2c. In a first step, the Christoffel symbols of second kind Γ^i_{jk} are computed from $\mathbf{g}$. Then time integration of Eq. (6), with initial condition $\mathbf{x} = \tilde{\mathbf{x}}$ and time step $\delta\tau = \delta x^2/4$ (see [1] for the details), is achieved to obtain a solution of Eq. (5). In a second step, the inverse transformation is constructed following the fixed-point procedure $\delta = \tilde{d}[x - \delta]$ (valid in curved geometry), computed similarily to a second-order semi-Lagrangian scheme: the fixed-point problem $\delta^* = \frac{1}{2}\tilde{d}[x-\delta^*]$ is iteratively first solved, then the displacement field is deduced as as $\delta = 2\delta^*$ (see Algorithm 1).

The diffusion tensor of the isotropization of the metric $\mathbf{g}$ by $\overset{\circ}{D}$ is given in Fig. 2d, and appears close to the theoretical one in Fig. 2a. Hence, in this experiment, we have shown possible to find a coordinate system in which the isotropizing metric is conformal to the natural metric $\boldsymbol{\omega}$.

4 Conclusion

We have extended the preliminary work of [1] within a geometrical background that offers a new insight on the isotropizing issue. Here, we have shown how the formalism of the isotropization should be stated to incorporate the natural metric on a geographical domain, as those encountered in atmospheric data assimilation.

In particular, we have shown that, in 2D, the diagnosis of the length-scale (i.e. the local metric tensor) can be used to find a coordinate change, solution of a pseudo-diffusion scheme, that provides isotropic local metric. The pseudo-diffusion scheme used to solve the non-linear system of coupled Poisson's equations, can be considered for real application since it relies on classical numerical components of Numerical Weather Prediction Systems or Chemical Transport Model (semi-Lagrangian scheme, first-order derivative and Beltrami's Laplacian).

Further work is to propose and to test more efficient algorithms than the pseudo-diffusion scheme.

Acknowledgements OP would like thanks Joseph Tapia, Jean-Pierre Otal and Marina Ville, John Harlim, Tyrus Berry and Dimitrios Giannakis for interesting discussions ; This work was supported by the French LEFE INSU program and the MACC2 project within the FP7 E.U. reasearch program.

References

1. Pannekoucke, O., Emili, E., Thual, O.: Modeling of local length-scale dynamics and isotropizing deformations. Q. J. R. Meteorol. Soc. **140**, 1387 (2014)
2. Gaspari, G., Cohn, S.: Construction of correlation functions in two and three dimensions. Q. J. R. Meteorol. Soc. **125**, 723 (1999)
3. Pannekoucke, O., Berre, L., Desroziers, G.: Background error correlation length-scale estimates and their sampling statistics. Q. J. R. Meteorol. Soc. **134**, 497 (2008)
4. Raynaud, L., Pannekoucke, O.: Sampling properties and spatial filtering of ensemble background-error lengthscales. Q. J. R. Meteorol. Soc. **139**, 784 (2013)
5. Pannekoucke, O., Massart, S.: Estimation of the local diffusion tensor and normalization for heterogeneous correlation modelling using a diffusion equation. Q. J. R. Meteorol. Soc. **134**, 1425 (2008)
6. Fisher, M.: Background error covariance modeling. In: ECMWF (ed.) Proceedings. ECMWF Seminar on Recent Developments in Data Assimilation for Atmosphere and Ocean, pp. 45–63 (2003)

7. Purser, R., Wu, W.S., Parrish, D., Roberts, N.: Numerical aspects of the application of recursive filters to variational statistical analysis. Part I: Spatially homogeneous and isotropic Gaussian covariances. Mon. Weather Rev. **131**, 1524 (2003)
8. Weaver, A., Courtier, P.: Correlation modelling on the sphere using a generalized diffusion equation (Tech. Memo. ECMWF, num. 306). Q. J. R. Meteorol. Soc. **127**, 1815 (2001)
9. Weaver, A.T., Mirouze, I.: On the diffusion equation and its application to isotropic and anisotropic correlation modelling in variational assimilation. Q. J. R. Meteorol. Soc. **139**(670), 242 (2013)
10. Desroziers, G.: A coordinate change for data assimilation in spherical geometry of frontal structures. Mon. Weather Rev. **125**, 3030 (1997)
11. Michel, Y.: Estimating deformations of random processes for correlation modelling in a limited area model. Q. J. R. Meteorol. Soc. **139**, 534 (2013)
12. Clerc, M., Mallat, S.: The texture gradient equation for recovering shape from texture. IEEE Trans. Pattern Anal. Mach. Intell. **24**, 536 (2002)
13. Legrand, R., Michel, Y.: Modelling background error correlations with spatial deformations: a case study. Tellus **66**, 23984 (2014)
14. Jost, J.: Riemannian Geometry and Geometric Analysis. Springer, Berlin (2005)
15. Nakahara, M.: Geometry, Topology and Physics, 2nd edn. Taylor & Francis, New York (2003)
16. Eells, J., Sampson, J.H.: Harmonic mappings of Riemannian manifolds. Am. J. Math. **86**, 109 (1964)
17. Eells, J., Lemaire, L.: A report on harmonic maps. Bull. Lond. Math. Soc. **10**, 1 (1978)
18. Courtier, P., Andersson, E., Heckley, W., Pailleux, J., Vasiljević, D., Hamrud, M., Hollingsworth, A., Rabier, F., Fisher, M.: The ECMWF implementation of three-dimensional variational assimilation (3D-Var). I: formulation. Q. J. R. Meteorol. Soc. **124**, 1783 (1998)
19. Derber, J., Bouttier, F.: A reformulation of the background error covariance in the ECMWF global data assimilation system. Tellus A **51**, 195 (1999)
20. Massart, S., Piacentini, A., Pannekoucke, O.: Importance of using ensemble estimated background error covariances for the quality of atmospheric ozone analyses. Q. J. R. Meteorol. Soc. **138**, 889 (2012)
21. Pannekoucke, O.: Heterogeneous correlation modeling based on the wavelet diagonal assumption and on the diffusion operator. Mon. Weather Rev. **137**, 2995 (2009)
22. Jost, J.: Harmonic mapping between Riemannian manifolds. In: Proceedings of the Centre for Mathematical Analysis, Australian National University (1983)

Post-Newtonian Gravitation

Erik I. Verriest

Abstract Einstein's field equations relate space-time geometry to matter and energy distribution. These tensorial equations are so unwieldy that solutions are only known in some very specific cases. A semi-relativistic approximation is desirable: One where space-time may still be considered as flat, but where Newton's equations (where gravity acts *instantaneously*) are replaced by a post-Newtonian theory, involving *propagation* of gravity at the speed of light. As this retardation depends on the geometry of the point masses, a dynamical system with state dependent delay results, where delay and state are implicitly related. We investigate several cases with Lagrange's inversion technique and perturbation expansions. Interesting phenomena (entrainment, dynamic friction, fission and orbital speeds) emerge.

1 Introduction

A post-Newtonian (pN) gravitation (see [7]) is developed, with the main assumption that gravity cannot act instantaneously at a distance, but has an interaction speed that is limited by the speed of light, otherwise one would violate relativity [4]. Section 2 establishes the formulas for the retarded potentials in analogy to the Liénard-Wiechert potentials in electromagnetic theory. From it, the retarded field, which determines the motion, is obtained by the gradient of the potential in Sect. 3. This results in a functional differential equation (FDE) with state dependent delay (See [3]). In particular, interesting new phenomena (entrainment, dynamic friction, orbital speeds) emerge from this pN theory. In Sect. 4 we discuss the dynamics for two masses flying apart in a rectilinear way, a (1-D) toy model for a supernova. The main problem is the implicit relation between state and delay occurring in the FDE describing the motion. We discuss several cases where the Lagrange's inversion technique [5] can be applied to render such a relation explicit [6].

E.I. Verriest (✉)
Georgia Institute of Technology, Atlanta, GA 30306, USA
e-mail: erik.verriest@ece.gatech.edu

© Springer International Publishing Switzerland 2016

J. Bélair et al. (eds.), *Mathematical and Computational Approaches in Advancing Modern Science and Engineering*, DOI 10.1007/978-3-319-30379-6_15

2 Retarded Gravitational Potential

Let at time t' the mass distribution be represented by the scalar function $\rho(\mathbf{r}', t')$, the density field. This assumption requires already a notion of simultaneity and therefore cannot be consistent with the theory of relativity. We postulate that this density field generates a gravitational field, which assumes at a fixed position, $\mathbf{r}$, and time t the value due to the superposition of the fields of all infinitesimal contributions. However the contribution from $\mathbf{r}'$ must have traveled a (straight line) distance $|\mathbf{r} - \mathbf{r}'|$, which takes a propagation time $\tau(\mathbf{r}') = |\mathbf{r} - \mathbf{r}'|/c$. Hence only the past value, $\rho(\mathbf{r}', t - \tau(\mathbf{r}'))$, contributes to the field at $\mathbf{r}$ at time t. Again, general relativity implies the warping of space and time, and hence geodesics (along which signals propagate) are not straight lines in general. The following will be approximately valid for weak fields and small speeds relative to c, the speed of light.

Assumption Given the time varying mass density $\rho(\mathbf{r}', t')$, the gravitational potential at a fixed point $\mathbf{r}$ and time t is given by the linear superposition

$$\phi(\mathbf{r}, t) = G \int_{\mathbb{R}^3} \frac{\rho(\mathbf{r}', t - \tau(\mathbf{r}'))}{|\mathbf{r} - \mathbf{r}'|} \, \mathrm{d}V(\mathbf{r}'), \tag{1}$$

where G is the gravitational constant, and $V(\mathbf{r}')$ denotes the volume element at position $\mathbf{r}'$. This equation is coupled with an implicit delay-state relation

$$c\tau(\mathbf{r}') = |\mathbf{r} - \mathbf{r}'|. \tag{2}$$

If ρ is continuous in its arguments, the field at a point $\mathbf{r}$ where $\rho(\mathbf{r}, \cdot) \neq 0$ can be obtained by taking the integral over an ϵ-small ball around $\mathbf{r}$ away, since that part will not contribute to the field by virtue of the asymptotic isotropy.

Let us now assume that the mass distribution is due to a *unit point mass* in motion, prescribed by its trajectory, $\mathbf{r}_p(t)$. This means that the mass density is given by a three-dimensional Dirac delta $\rho(\mathbf{r}', t') = \delta(\mathbf{r}' - \mathbf{r}_p(t'))$. The gravitational potential at the fixed point and time $(\mathbf{r}, t)$ is then

$$\phi(\mathbf{r}, t) = G \int_{\mathbb{R}^3} \frac{\delta(\mathbf{r}' - \mathbf{r}_p(t_b))}{|\mathbf{r} - \mathbf{r}'|} \, \mathrm{d}V(\mathbf{r}'), \tag{3}$$

where $t_b = t - \tau(\mathbf{r}')$ is the retarded time satisfying (2). We evaluate this integral by first principles. First, a Dirac delta peaks where its argument equals zero. However, this information is not sufficient to describe this behavior in its entirety. The rate at which the zero is approached is also required. To figure this out, let first $\mathbf{r}'_0$ be the solution to $\mathbf{r}_0' = \mathbf{r}_p(t - \tau(\mathbf{r}_0'))$ and set $\mathbf{r}' = \mathbf{r}_0' + \tilde{\mathbf{r}}'$. Substituting these expressions in the Dirac delta, we get (I stands for the identity matrix)

$$\delta(\mathbf{r}_0' + \mathbf{r}' - \mathbf{r}_p(t - \tau(\mathbf{r}_0' + \tilde{\mathbf{r}}'))) = \frac{1}{|\det[I + \dot{\mathbf{r}}_p(t - \tau(\mathbf{r}_0'))\nabla_{r'}\tau(\mathbf{r}_0')]|}\,\delta(\tilde{\mathbf{r}}'). \quad (4)$$

Using the identity $\det[I + pq] = \det[I + qp]$, and substituting (4) in (3), yields:

$$\phi(\mathbf{r}, t) = G\int_{\mathbb{R}^3} \frac{\delta(\tilde{\mathbf{r}}')}{|\mathbf{r} - \mathbf{r}'|\,|\det(1 + \nabla_r\tau(\mathbf{r}_0')\dot{\mathbf{r}}_p(t - \tau(\mathbf{r}_0')))|}\,\mathrm{d}V(\mathbf{r}').$$

Taking the gradient w.r.t. $\mathbf{r}'$ of (2) and evaluating at $\mathbf{r}' = \mathbf{r}_p$ gives

$$c\nabla_{r'}\tau = \frac{(\mathbf{r} - \mathbf{r}_\mathbf{p}(t_b))^\top}{|\mathbf{r} - \mathbf{r}_\mathbf{p}(t_b)|}\,[-I],$$

so that finally

$$\phi(\mathbf{r}, t) = \frac{G}{|\mathbf{r} - \mathbf{r}_p(t_b)|\left|1 - \frac{(\mathbf{r} - \mathbf{r}_\mathbf{p}(t_b))^\top \dot{\mathbf{r}}_p(t_b)}{c|\mathbf{r} - \mathbf{r}_\mathbf{p}(t_b)|}\right|}. \quad (5)$$

This result could also have been obtained using the standard trick to derive the Liénard-Wiechert potentials in some physics texts, by reducing the three dimensional Dirac to a one-dimensional time Dirac (see [5]).

3 Gravitational Field Due to a Moving Point Mass

The gravitational field is the gradient of the gravitational potential, $F(\mathbf{r}, t) = \nabla_r\phi(\mathbf{r}, t)$, obtained in Sect. 2. Thus

$$F(\mathbf{r}, t) = -\frac{G}{\left|1 - \frac{(\mathbf{r} - \mathbf{r}_p(t_b))^\top \dot{\mathbf{r}}_p(t_b)}{c|\mathbf{r} - \mathbf{r}_p(t_b)|}\right|}\,\frac{1}{|\mathbf{r} - \mathbf{r}_p(t_b)|^2}\nabla_r|\mathbf{r} - \mathbf{r}_p(t_b)| +$$

$$-\frac{G}{|\mathbf{r} - \mathbf{r}_p(t_b)|}\,\frac{1}{\left|1 - \frac{(\mathbf{r} - \mathbf{r}_p(t_b))^\top \dot{\mathbf{r}}_p(t_b)}{c|\mathbf{r} - \mathbf{r}_p(t_b)|}\right|^2}\,\nabla_r\left|1 - \frac{(\mathbf{r} - \mathbf{r}_p(t_b))^\top \dot{\mathbf{r}}_p(t_b)}{c|\mathbf{r} - \mathbf{r}_p(t_b)|}\right|. \quad (6)$$

From $\nabla_r|\mathbf{r} - \mathbf{r}_0|^2 = 2|\mathbf{r} - \mathbf{r}_0|\,\nabla_r|\mathbf{r} - \mathbf{r}_0|$ and noting that the left hand side is equal to

$$\nabla_r(\mathbf{r} - \mathbf{r}_0)^\top(\mathbf{r} - \mathbf{r}_0) = 2(\mathbf{r} - \mathbf{r}_0)^\top \underbrace{\nabla_r(\mathbf{r} - \mathbf{r}_0)}_{=I} = 2(\mathbf{r} - \mathbf{r}_0)^\top,$$

it follows that

$$\nabla_r |\mathbf{r}-\mathbf{r}_0| = \frac{(\mathbf{r}-\mathbf{r}_0)^\top}{|\mathbf{r}-\mathbf{r}_0|}.$$

Note that the gradient with respect to a column vector is represented by a row vector. Similarly, if $|\mathbf{v}_0| < c$

$$\nabla_r \left|1 - \frac{(\mathbf{r}-\mathbf{r}_0)^\top \mathbf{v}_0}{c|\mathbf{r}-\mathbf{r}_0|}\right| = \nabla_r \left(1 - \frac{(\mathbf{r}-\mathbf{r}_0)^\top \mathbf{v}_0}{c|\mathbf{r}-\mathbf{r}_0|}\right) = -\frac{\mathbf{v}_0^\top}{c}\left[\nabla_r \frac{\mathbf{r}-\mathbf{r}_0}{|\mathbf{r}-\mathbf{r}_0|}\right].$$

Also,

$$\nabla_r \frac{\mathbf{r}-\mathbf{r}_0}{|\mathbf{r}-\mathbf{r}_0|} = \frac{1}{|\mathbf{r}-\mathbf{r}_0|}\left[I - \frac{(\mathbf{r}-\mathbf{r}_0)(\mathbf{r}-\mathbf{r}_0)^\top}{|\mathbf{r}-\mathbf{r}_0|^2}\right].$$

The matrix between the square brackets is a projection operator (projection onto the plane perpendicular to the vector $\mathbf{r}-\mathbf{r}_0$). Putting this all together, the gravitational field due to a moving mass is

$$F(\mathbf{r},t) = -\frac{(\mathbf{r}-\mathbf{r}_p(t_b))^\top}{|\mathbf{r}-\mathbf{r}_p(t_b)|^3} \frac{G}{\left|1 - \frac{(\mathbf{r}-\mathbf{r}_p(t_b))^\top \dot{\mathbf{r}}_p(t_b)}{c|\mathbf{r}-\mathbf{r}_p(t_b)|}\right|} + \tag{7}$$

$$+\frac{G}{|\mathbf{r}-\mathbf{r}_p(t_b)|^2} \frac{1}{\left|1 - \frac{(\mathbf{r}-\mathbf{r}_p(t_b))^\top \dot{\mathbf{r}}_p(t_b)}{c|\mathbf{r}-\mathbf{r}_p(t_b)|}\right|^2} \frac{\dot{\mathbf{r}}_p(t_b)^\top}{c}\left[I - \frac{(\mathbf{r}-\mathbf{r}_p(t_b))(\mathbf{r}-\mathbf{r}_p(t_b))^\top}{|\mathbf{r}-\mathbf{r}_p(t_b)|^2}\right].$$

If the gravitating mass moves in such a way that $\dot{\mathbf{r}}_p$ is parallel to $(\mathbf{r}-\mathbf{r}_p(t_b))$, then the second term in (7) is zero and the field is directed along $\mathbf{r}-\mathbf{r}_p(t_b)$. Aligning the x-axis with this direction, force, position and velocities can be expressed in the x-coordinate. In particular, the x-component of the force acting on a unit mass particle at x is

$$F_x = -\frac{\operatorname{sgn}(x - x_p(t_b))G}{|x - x_p(t_b)|^2 \left|1 - \operatorname{sgn}(x - x_p(t_b))\frac{\dot{x}_p(t_b)}{c}\right|}, \tag{8}$$

where as always, t_b denotes the retarded time.

This form displays two deviations from Newton's law: First the delayed position of the gravitating particle appears, and second, there is an aberration which depends on the (delayed) velocity.

It should also be observed that a naive generalization to Newton's law in the form

$$F_x = -\frac{\operatorname{sgn}(x - x_p(t_b))G}{|x - x_p(t_b)|^2},$$

instead of (8) cannot be correct.

3.1 Expression the Field in Terms of the Predicted Position

First, reorganize (7) in the form

$$\begin{aligned} F(\mathbf{r}, t) = & \frac{G}{|\mathbf{r} - \mathbf{r}_p(t_b)|^2} \frac{1}{\left|1 - \frac{(\mathbf{r}-\mathbf{r_p}(t_b))^\top}{|\mathbf{r}-\mathbf{r}_p(t_b)|} \frac{\dot{\mathbf{r}}_p(t_b)}{c}\right|^2} \frac{\dot{\mathbf{r}}_\mathbf{p}(t_b)^\top}{c} + \\ & - \frac{G(\mathbf{r} - \mathbf{r}_p(t_b))^\top}{|\mathbf{r} - \mathbf{r_p}(t_b)|^3 \left|1 - \frac{(\mathbf{r}-\mathbf{r_p}(t_b))^\top}{|\mathbf{r}-\mathbf{r_p}(t_b)|} \frac{\dot{\mathbf{r}}_\mathbf{p}(t_b)}{c}\right|^2} \\ & \times \left[\left|1 - \frac{(\mathbf{r} - \mathbf{r_p}(t_b))^\top}{|\mathbf{r} - \mathbf{r_p}(t_b)|} \frac{\dot{\mathbf{r}}_\mathbf{p}(t_b)}{c}\right| + \frac{(\mathbf{r} - \mathbf{r_p}(t_b))^\top}{|\mathbf{r} - \mathbf{r_p}(t_b)|} \frac{\dot{\mathbf{r}}_\mathbf{p}(t_b)}{c}\right]. \end{aligned} \tag{9}$$

Introducing $\widehat{\mathbf{r}_p} = \mathbf{r}_p(t_b) + \tau(t_b)\dot{\mathbf{r}}_p(t_b)$, which is the *expected position* of the particle in motion at time t, if the velocity of the mass were fixed at the constant $\mathbf{v}_p(s) = \dot{\mathbf{r}}_p(t_b)$ for $t_b(t) = t - \tau(t_b) \leq s \leq t$, it follows that

$$\left|1 - \frac{(\mathbf{r} - \mathbf{r}_p(t_b))^\top}{|\mathbf{r} - \mathbf{r}_p(t_b)|} \frac{\dot{\mathbf{r}}_p(t_b)}{c}\right| = \frac{1}{|\mathbf{r} - \mathbf{r}_p(t_b)|} |\mathbf{1}_{\mathbf{r}-\mathbf{r}_p}^\top (\mathbf{r} - \widehat{\mathbf{r}_p})|,$$

where $\mathbf{1}_{\mathbf{r}-\mathbf{r}_p} = \frac{(\mathbf{r}-\mathbf{r}_p(t_b))^\top}{|\mathbf{r}-\mathbf{r}_p(t_b)|}$ denotes the unit vector in the direction of $\mathbf{r}-\mathbf{r}_p$. Relativity imposes $|\dot{\mathbf{r}}_p| < c$, which with the Cauchy-Schwarz inequality reduces (9) to

$$\nabla_r \phi(\mathbf{r}, t) = -\frac{G}{|\mathbf{r} - \mathbf{r}_p(t_b)| \left|\mathbf{1}_{\mathbf{r}-\mathbf{r}_p}^\top (\mathbf{r} - \widehat{\mathbf{r}_p})\right|^2} (\mathbf{r} - \widehat{\mathbf{r}_p})^\top. \tag{10}$$

Thus the gravitational force exerted by a particle in motion is directed towards the *predicted position based on a uniform motion* given the delayed information (i.e. delayed position and velocity).

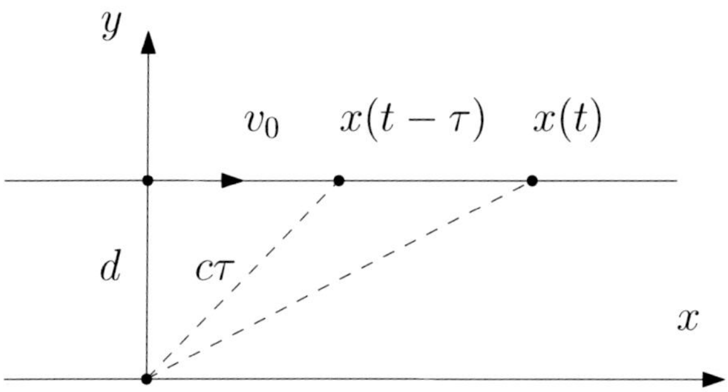

Fig. 1 Uniform motion

3.2 Field Due to Particle with Uniform Velocity

For a particle moving with *uniform* velocity, $\widehat{\mathbf{r}_p}(t)$ is the position of the particle at time t, $\widehat{\mathbf{r}_p}(t) = \mathbf{r}_p(t)$. Without loss of generality, we fix $\mathbf{r}$, the observation point, at the origin, and let the particle move along a line parallel to the x-axis, at a distance d with uniform velocity v_0. For notational simplicity, the x-coordinate at $t = 0$ is taken to be zero. From the geometry of the problem (See Fig. 1): a quadratic equation for τ results: $(c^2 - v_0^2)\tau^2 + 2v_0^2 t\tau - (v_0^2 t^2 + d^2) = 0$. The field magnitude follows

$$F = \frac{G((d^2 + v_0^2 t(t-\tau))^2 + d^2 v_0^2 \tau^2)}{(d^2 + v_0^2 t(t-\tau))^2 \sqrt{(d^2 + v_0^2 (t-\tau)^2)(d^2 + v_0^2 t^2)}}. \tag{11}$$

In Fig. 2 the magnitude is shown as function of time for $d = 1$, $v_0 = 1$ and $c = 1.1$, 2 and the Newtonian case ($c = \infty$). As the mass closes in on the origin for $t < 0$, the delayed effect is quite pronounced as the speed c is increased. When $d = 0$, the delayed gravitational force at time t exceeds Newton's law by a factor $1 + \frac{v_0}{c}$ if the particle is moving towards the origin, and a factor $1 - \frac{v_0}{c}$ when it is moving away from the origin. The same holds for all d in the limits respectively for $t \to -\infty$ and $t \to +\infty$.

3.3 Gravitational Current and Its Entrainment

Let the origin coincide with the observation point, $\mathbf{r}$, and consider an infinitely long straight line mass moving along its axis, parallel to the x-axis, at a distance d with uniform velocity v_0. Thus the field at the origin due to a *gravitational current* is studied. Consider first a mass-element at position (x_0, d) at time $t = 0$. From the geometry of the problem (Fig. 1): $c\tau(t) = \sqrt{(x_0 + v_0(t - \tau(t)))^2 + d^2}$. Letting $\Delta = \sqrt{(x_0 + v_0 t_b)^2 + d^2}$, the x-component of the field at the origin at time 0 due this

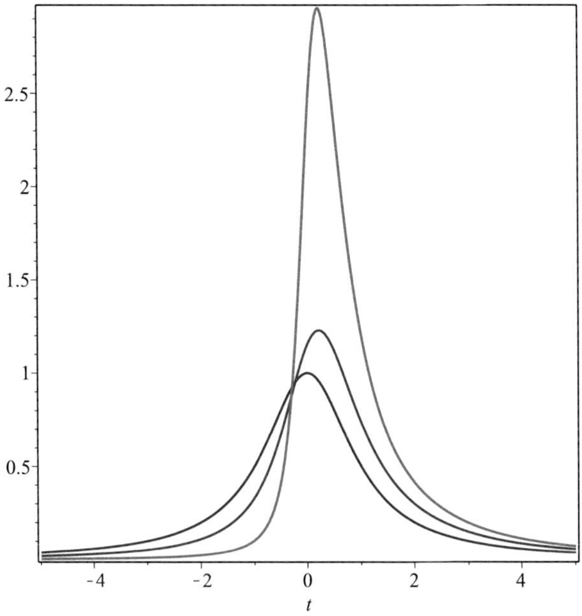

Fig. 2 Field at the origin due to particle with uniform motion. The symmetric curve corresponds to the Newtonian case ($c = \infty$)

elementary current element is

$$F_x(x_0, v_0) = \frac{x_0 + v_0 t_b}{\Delta^3 \left(1 + \frac{v_0(x_0+v_0 t_b)}{c\Delta}\right)} - \frac{v_0}{c} \frac{\left(1 - \frac{(x_0+v_0 t_b)^2}{\Delta^2}\right)}{\Delta^2 \left(1 + \frac{v_0(x_0+v_0 t_b)}{c\Delta}\right)^2}, \tag{12}$$

where t_b is the *backward time* for $t = 0$, i.e., $t_b = -\tau(0)$. Likewise, the y-component

$$F_y(x_0, v_0) = \frac{d}{\Delta^3 \left(1 + \frac{v_0(x_0+v_0 t_b)}{c\Delta}\right)} - \frac{v_0 d}{c} \frac{1}{\Delta^4 \left(1 + \frac{v_0(x_0+v_0 t_b)}{c\Delta}\right)^2}. \tag{13}$$

Integrating over x_0 from $-\infty$ to ∞ gives the total gravitational force of the mass current. The Figs. 3 and 4 show for $d = 1$ respectively the x and y components of the field at the origin as function of $\beta = v/c$. In the static case ($v_0 = 0$) this is

$$F_x(0) = 0, \quad F_y(0) = \int_{-\infty}^{\infty} \frac{Gd}{(x^2 + d^2)^{3/2}} \,\mathrm{d}x = \frac{2G}{d^2}.$$

This static field is directed perpendicular to the current direction, and points towards the linear mass. With $v_0 \neq 0$ it is augmented by a velocity dependent term. For the

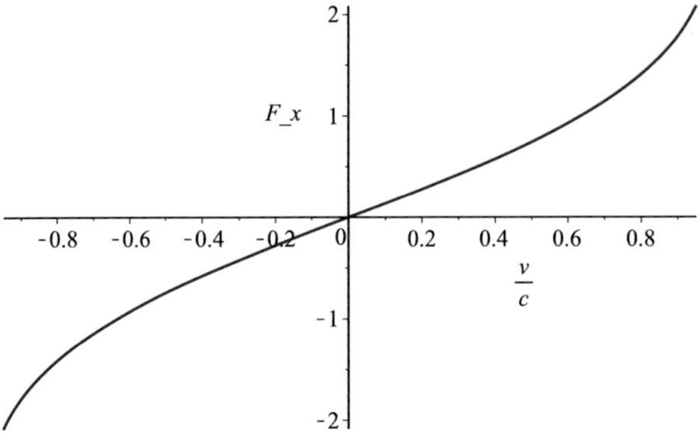

Fig. 3 Parallel field component as function of v/c for $G = d = 1$

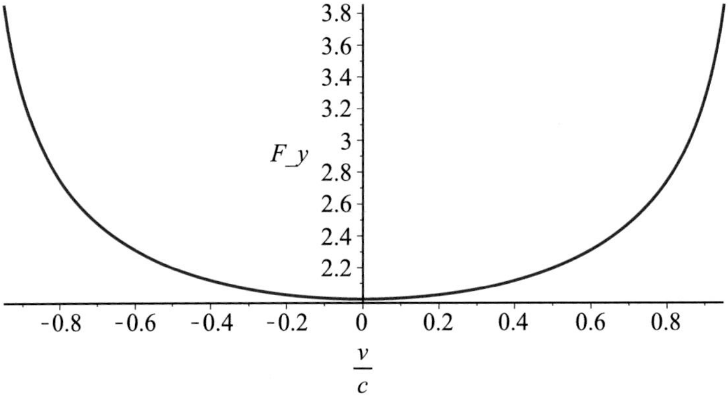

Fig. 4 Orthogonal field component as function of v/c for $G = d = 1$

parallel (to the current) component, this additive term behaves *linearly* near $v = 0$, and *quadratically* for the component directed towards the current line. The effect of the current is to *entrain* the surrounding mass. The large scale structure of the universe is made up of large filaments of galaxy superclusters that may be modelled as gravitational currents.

Remark One may be tempted to approximate $r(t - \tau)$ by its truncated Taylor expansion in terms of the instantaneous delay. It was found that such a naive use of Taylor expansions can not yield solutions that are consistent in the Newtonian limit.

3.4 Dynamic Friction

The results of the previous section imply a phenomenon analogous to *dynamic friction* which was analyzed by Chandrasekhar in a classical setting [1]. A lone star colliding with a galaxy (moving through a uniformly distributed star field) will slow down. As only relative motion is significant, we may consider the problem as that of a stationary mass point (say located at the origin) embedded in a steady flow of mass in the direction of the x-axis. Locally, at O, the problem is isotropic in the (y,z) plane orthogonal to the motion. Hence F_y and F_z experienced by the star are zero. There is however a resulting force in the x-direction. As seen in Sect. 3.3, the star will be entrained by the moving field of stars and follows a law of the form $\ddot{x} = k(v - \dot{x})$, for some k which is computable by a more detailed analysis and must depend on the density of the star field. As before, v is the initial relative speed of the star field with respect to the star. But this means that the galactic medium acts as a *viscous* medium on the star. At equilibrium, the star moves in unison with the medium.

4 Escape from Gravity

Consider the motion of two equal masses, m, moving symmetrically in one dimension, hence having the same speed with opposite directions (gravitational fission, supernova). Let $r(t)$ and $l(t)$ respectively be the position of the masses moving towards the right and the left. The symmetry imposes $l(t) = -r(t)$. See Fig. 5.

Assumption Newton's law $F_r(t) = m\ddot{r}(t)$ holds, where $F_r(t)$ is force acting on the mass moving towards the right at time t. The gravitational force felt by the particle

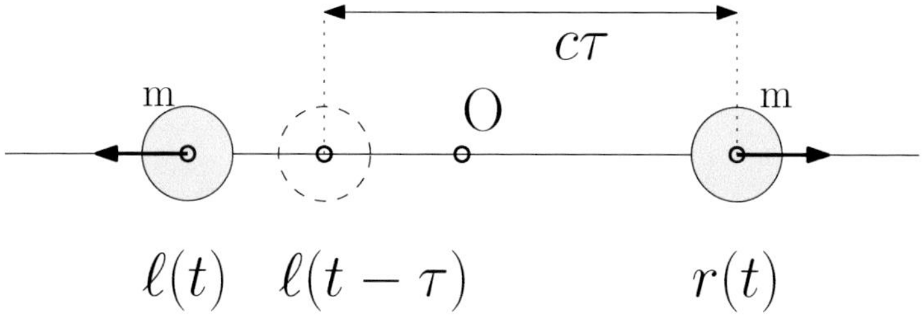

Fig. 5 Gravitational Fission

at $r(t)$ due to the particle moving towards the left leads to the the coupled set

$$\ddot{r}(t) = -\frac{Gm}{\left(1 + \frac{\dot{r}(t-\tau(t))}{c}\right) c^2\tau^2(t)} \tag{14}$$

$$c\tau(t) = r(t) + r(t - \tau(t)). \tag{15}$$

We present two solution methods: Lagrange inversion and perturbation expansion.

4.1 Lagrange Inversion

The explicit form of $\tau(t)$ can be obtained from $c\tau(t) = r(t) + r(t - \tau(t))$ about a point $r(t_0) = r_0$, $\tau(t_0) = \tau_0$, by Lagrange inversion [2] provided that r is analytic, and $|\dot{r}| < c$ at time $t = t_0$, where it holds that $c\tau_0 = r_0 + r(t_0 - \tau_0)$,

$$\tau(t) = (t - t_0 + \tau_0) - \sum_{i\geq 1}^{\infty} \frac{(c(t-t_0) + r_0 - r(t))^i}{i!} \xi_i. \tag{16}$$

With $\mathbf{D} = \mathrm{d}/\mathrm{d}s$, the coefficients of the series expansion are expressed as

$$\xi_i = \left\{ \mathbf{D}^{i-1} \frac{1}{\left(c + \frac{r(t_0-\tau_0+s)-c\tau_0+r_0}{s}\right)^i} \right\}_{s=0}. \tag{17}$$

4.2 Solution via Perturbation Expansion

Postulate a solution to equations (14) and (15) as a series in $\epsilon = 1/c$.

$$r(t) = r_0(t) + \epsilon r_1(t) + \epsilon^2 r_2(t) + \ldots \tag{18}$$

$$\tau(t) = \epsilon\tau_1(t) + \epsilon^2\tau_2(t) + \ldots. \tag{19}$$

The velocity factor $\left(1 + \frac{r(t_b)}{c}\right)$ expands to $1 + \epsilon\dot{r}_0 + \epsilon^2[\dot{r}_1 - \ddot{r}_0\tau_1] + \epsilon^3[\dot{r}_2 - \ddot{r}_0\tau_2 + \frac{1}{2}r_0^{(3)}\tau_1^2 - \ddot{r}_1\tau_1] + \cdots$ Matching the expansions in ϵ for the state dependent delay differential equation and eliminating τ_k, $k = 1, 2, \ldots$ gives

$$\ddot{r}_0 = -\frac{\gamma}{4r_0^2} \tag{20}$$

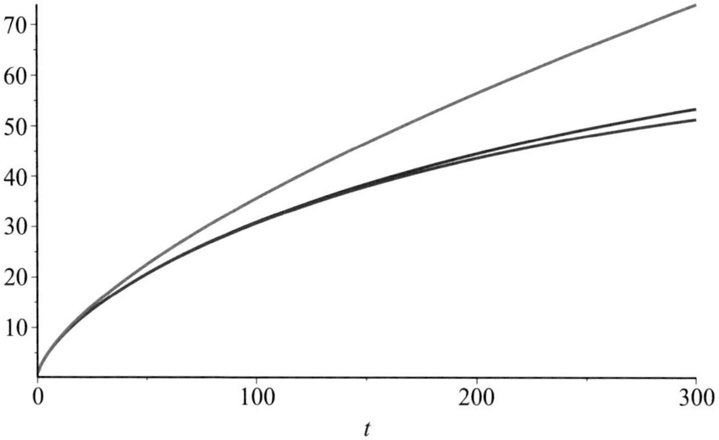

Fig. 6 Position $r(t)$ as function of t for sub-escape with $\gamma = 1$ ($r_0 = 1, v_0 = 1$). The *curves* correspond to the Newtonian case (marginal escape), and first and second order perturbations (which return eventually to r_0)

$$\ddot{r}_1 = \frac{\gamma}{2r_0^3} r_1 - \frac{\gamma}{4r_0^2} \dot{r}_0 \tag{21}$$

$$\ddot{r}_2 = -3\frac{\gamma}{4r_0^4} r_1^2 + 2\frac{g}{r_0^3}(r_1 \dot{r}_0 + r_2) - \frac{\gamma}{4r_0^2} \dot{r}_1 \tag{22}$$

$$\vdots$$

The solutions of order 0 (Newtonian case), and up to order 1 and 2 are shown in Fig. 6 for $\gamma = 4$, $r_0 = 1$ and $v_0 = v_{esc}|_{\text{Newton}} = \sqrt{2\gamma/r_0}$, the escape velocity in the Newtonian case. In pN theory $v_0 = v_{esc}|_{\text{Newton}}$ is *insufficient* for escape. For $\gamma = 1$, $r_0 = 1$ and $v_0 = 1$, the post-Newtonian escape velocity was computed (up to second order for several values of c and it was found that for small $1/c$ the adjusted quantity $\frac{v_{\text{esc}}(c)}{\sqrt{1+\frac{v_{\text{esc}}^2(\infty)}{c^2}}}$ is close to $v_{esc}(\infty) = \sqrt{2}$. This led to the *conjecture*: $v_{\text{esc}}(c) = \frac{v_{\text{esc}}(\infty)}{\sqrt{1-\frac{v_{\text{esc}}^2(\infty)}{c^2}}}$.

5 Conclusions and Beyond

The physical aspect of the paper established post-Newtonian gravitation as a system with state dependent delay. The mathematics centers around an implicit relation of delay and state, which is resolved by Lagrange inversion and perturbation expansions. The field of mass in uniform motion was described in detail. Its integrated form gives rise to the consideration of gravitational currents and their ensuing entrainment. This led to the emergence of dynamic friction. Finally we discussed the escape velocity for splitting masses (fission). Lack of space did not allow the discussion of the Kepler problem in pN theory. As expected, stable orbits exist, which is not the case had one considered the naive generalization (delayed Newton's law) of the gravitational field mentioned in Sect. 3. It should also be pointed out that naive Taylor expansions lead to results that are inconsistent in the Newtonian limit.

References

1. Chandrasekhar, S.: Dynamical friction. I. General considerations: the coefficient of dynamical friction. Astrophys. J. **97**, 255–262 (1943)
2. Good, I.J.: Generalization to several variables of Lagrange's expansion, with applications to stochastic processes. Proc. Camb. Philos. Soc. **56**, 367–380 (1960)
3. Hartung, F., Krisztin, T., Walther, H.O., Wu, J.: Functional differential equations with state-dependent delays: theory and applications. In: Canada, A., Drábek, P., Fonda, A. (eds.) Handbook of Differential Equations, vol. 3, pp. 435–545. Elsevier, Amsterdam (2006)
4. Misner, C.W., Thorne, K.S., Wheeler, J.A.: Gravitation. Freeman, San Francisco (1973)
5. Smith, G.: Classical Electromagnetic Radiation. Cambridge University Press, Cambridge/ New York/Melbourne (1997)
6. Verriest, E.I.: Inversion of state-dependent delay. In: Karafyllis, I., Malisoff, M., Mazenc, F., Pepe, P. (eds.) Recent Results on Nonlinear Delay Control Systems. Springer International Publishing Switzerland, 327–346 (2016)
7. Will, C.: The renaissance of general relativity. In: Davies, P. (ed.) The New Physics, pp. 7–33. Cambridge University Press, Cambridge (1989)

Part II
Mathematical and Computational Methods in Life Sciences and Medicine

A Quantitative Model of Cutaneous Melanoma Diagnosis Using Thermography

Ephraim Agyingi, Tamas Wiandt, and Sophia Maggelakis

Abstract Cutaneous melanoma is the most commonly diagnosed cancer and its incidence is on the rise worldwide. Early detection and differentiation of a malignant melanoma from benign cutaneous lesions provides an excellent chance for treating the disease. Thermography is a non-invasive tool that can be used to detect and monitor skin lesions. We model heat transfer in a skin region containing a lesion. The model which is governed by the Pennes equation uses the steady state temperature at the skin surface to determine whether there is an underlying lesion. Numerical simulations from the model ascertain whether the lesion is malignant or benign.

1 Introduction

The skin is the largest organ in the body and has the most exposure to the external environment. Thus, it is prone to lesions attributed to both internal and external factors. Skin lesions begin when alterations in cellular metabolism allow cells to grow without restriction. Skin lesions do constitute the majority of all cancers and their incidence is on the rise [1]. The majority of skin lesions are benign and harmless. Melanoma is a malignant skin cancer that can easily undergo metastasis and consequently lead to death if not detected early. Melanoma accounts for the most cancer deaths in the United States compared to other cancers [2].

One of the many tools for diagnosing skin cancers is thermography [3]. The procedure is based on established observations that the temperature of the skin directly above a tumor is significantly higher than the one in the absence of a tumor. Thermography uses an infrared camera to map the temperature distribution over the desired skin surface. The FDA in the United States approved thermography as an adjunct tool for diagnosing breast cancers in 1982 [4]. Technological advances in the past decades have improved thermographic imaging so that temperature differences of about 0.025 °C can be detected [5].

E. Agyingi (✉) • T. Wiandt • S. Maggelakis
Rochester Institute of Technology, Rochester, NY 14623, USA
e-mail: eoasma@rit.edu; tiwsma@rit.edu; sxmsma@rit.edu

© Springer International Publishing Switzerland 2016
J. Bélair et al. (eds.), *Mathematical and Computational Approaches in Advancing Modern Science and Engineering*, DOI 10.1007/978-3-319-30379-6_16

The published literature contains mathematical models and simulations illustrating thermography. The majority of the studies are for breast cancer [6–12] and only a few are dedicated to skin cancer [13–16]. Most of these papers model tumors that are vascularized and governed by some version of the Pennes bio-heat equation. The metabolic heat generating rate of tumors are known to range from 0.205–1.645 W/cm^3, which is about 20–200 times the metabolic heat generation rates of normal tissue [6]. Lin et al. [8] considered a tumor metabolic rate of 24.156 W/cm^3, which is about 2500–5000 times that of normal tissue. In a more recent paper [17], the authors studied a prevascular breast tumor, and related the temperature of the skin surface to the tumor size, tumor depth and the tumor metabolic heat generating rate.

The most effective treatment for melanoma is surgical incision before metastasis. Almost all tumors start up as prevascular tumors. The tumor has a single location, and has no vasculature connecting it to the host capillary network. The aim of this paper is to investigate whether melanoma can be distinguished from a benign skin lesion at the prevascular level. To achieve this, we model the tumor as a spheroid, consisting of a necrotic core and surrounded by a viable region of proliferating cells. The work reported here builds on the one-dimensional model reported by [18], which was recently extended in [17]. In [17], it was assumed that the heat diffusion was the dominant effect and the net effect of perfusion was zero in the healthy region. This paper differs from [17] in that the boundary conditions take into account the effect of blood perfusion in the healthy region.

2 Mathematical Model

We present a model of heat transfer in prevascular skin tumors. We assume that the tumors are spheroids and in the interest of simplicity, we shall present the model as a spatially two-dimensional model. We consider a two-dimensional cross section of the skin tissue, containing a circular shaped tumor as depicted in Fig. 1. The temperature of the cross-sectional domain is given by $T(x, y)$, where the x coordinate is the horizontal direction and y is the depth. The two dimensional cut is placed in the (x, y)-plane so that the $y = 0$ level corresponds to the bottom layer of the skin, and $y = d$ corresponds to the surface. The origin is located so that $-a \leq x \leq a$, with the center of the tumor at $(0, d/2)$. The radius of the tumor is R.

The equations governing heat flow in each portion of the entire region are derived from the Pennes equation [19]:

$$\rho \bar{c} \frac{\partial T}{\partial t} = \nabla \cdot (K \nabla T) + m_b c_b (T_A - T) + S, \tag{1}$$

where ρ is the tissue's density, $\bar{c}$ is the tissue's specific heat, K is the tissue's thermal conductivity, m_b is the mass flow rate of blood, c_b is the blood's specific heat, T_A is the arterial blood temperature, and S is the metabolic heat generation rate.

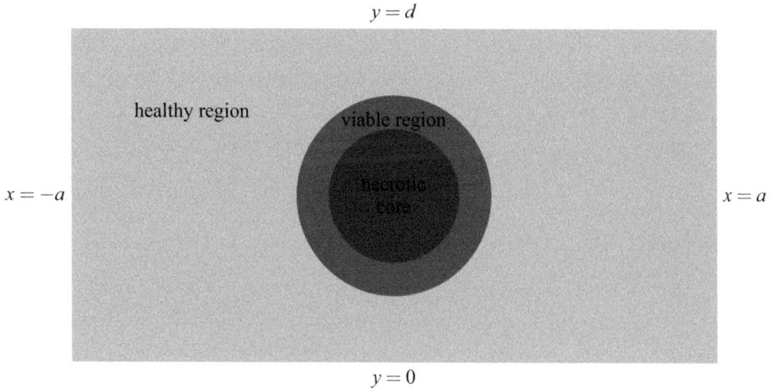

Fig. 1 Cross-sectional sketch of the skin containing a prevascular tumor

We note that at steady state, the time derivative is zero. Therefore the equation describing the temperature $T(x, y)$ in the healthy region is given by

$$\Delta T_h = -\frac{S_h}{K_h} - \frac{m_b c_b (T_A - T_h)}{K_h},$$

where S_h is the metabolic heat generation rate of the healthy tissue and K_h is its thermal conductivity.

For the viable tumor region, we remark that the tumor has no vasculature and therefore the perfusion term in equation (1) vanishes. Consequently, at steady state, the equation describing the temperature $T_t(x, y)$ of the viable region of the tumor is given by

$$\Delta T_t = -\frac{S_t}{K_t},$$

where S_t is the metabolic heat generation rate of the tumor and K_t is its thermal conductivity.

Finally, we note that the necrotic core is comprised of dead cells and generates no heat. This portion is simply described by the Laplace equation $\Delta T_c = 0$.

The boundary conditions of the various parts of the entire domain are listed below. The outer boundary conditions of the healthy region are provided by items (i)–(iii). Items (iv)–(v) provide inner boundary conditions for the healthy region and outer boundary conditions for the viable tumor region, and item (vi) provides the boundary conditions between the viable region and the necrotic core.

(i) At the bottom layer, $y = 0$, $T_h(x, 0) = T_b$ is the temperature of the body.
(ii) At the skin surface, $y = d$. Hence, $-K \left.\frac{\partial T_h}{\partial y}\right|_{y=d} = \lambda(T_h - T_a)$, where T_a is the ambient temperature and λ is the surface heat transfer coefficient.

(iii) At the boundaries $x = \pm a$, $T_h(-a, y) = T_h(a, y)$ is the temperature distribution of healthy tissue. To compute this value, we assume homogeneity of the tissue and therefore reduce the problem into a one-dimensional equation of the form

$$\Delta T_h = T_h''(y) = -\frac{S_h}{K_h} - \frac{m_b c_b (T_A - T_h)}{K_h}$$

and therefore,

$$T_h''(y) - \frac{m_b c_b T_h}{K_h} = -\frac{S_h}{K_h} - \frac{m_b c_b T_A}{K_h},$$

where y is the tissue depth, $T_h(0) = T_b$, and $T_h'(d) = -\lambda(T_h - T_a)/K_h$. The solution of this equation is given by

$$T_h(y) = \eta - T_b + \eta \cosh(\omega y) + \eta \left(\frac{\lambda(T_a - T_b) - K\omega \sinh(\omega d) - \lambda \cosh(\omega d)}{\lambda \sinh(\omega d) + K_h \omega \cosh(\omega d)} \right) \sinh(\omega y)$$

where $\eta = T_b - \frac{S}{m_b c_b} - T_A$ and $\omega = \sqrt{\frac{m_b c_b}{K_h}}$.

(iv) We assume that the temperature is continuous across the interface of the healthy tissue and the viable region of the tumor, i.e. $T_h(x, y) = T_t(x, y)$ when $x^2 + (y - d/2)^2 = R^2$.

(v) The heat flux is continuous across the interface of the healthy tissue and the viable region of the tumor, i.e. $K_t \nabla T_t = K_h \nabla T_h$ when $x^2 + (y - d/2)^2 = R^2$.

(vi) On the interface between the viable region of the tumor and the necrotic core, we assume continuous temperature and flux as well.

3 Results and Discussion

In this section, we present numerical simulations of the model presented above. The results are for a single tumor and multiple tumors for a given skin cross section. The calculations were performed using a MATLAB finite element solver. All thermo-physical parameter values used were chosen within the range of published data. We set $S_h = 0.009\,\text{W/cm}^3$, $K_h = K_t = 0.0042\,\text{W/((cm)°C)}$, $\lambda = 0.0005\,\text{W/((cm}^2\text{)°C)}$, $m_b = 0.0005\,\text{g/(ml s)}$ and $c_b = 4.2\,\text{J/g°C}$. We also set the arterial blood temperature $T_A = 37\,°\text{C}$ and the body temperature, at the level $y = 0$, to be $T_b = 37\,°\text{C}$. Other parameters used were chosen to investigate the behavior of the model.

The first results presented in Fig. 2 are for a single tumor of radius 1 mm and center located at $(0, 0.25)$. Figure 2a provides a contour map and temperature distribution over the entire domain. As expected, we observe that heat diffuses away from the tumor towards the cooler surrounding region. In Fig. 2b we examine the

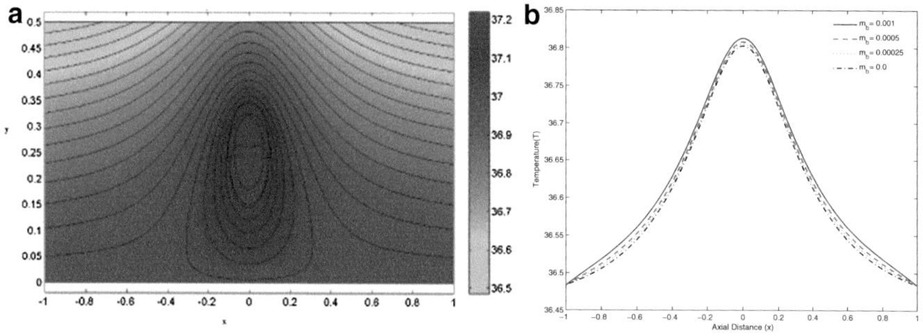

Fig. 2 Numerical simulations of one tumor with radius $R = 1$ mm and $S_t = 20S_h$. The plot in (**a**) is the temperature distribution over the cross section and (**b**) is the skin surface temperature for different perfusion rates (m_b)

steady state temperature at the skin surface for different perfusion rates. We note that the results in Fig. 2b are almost identical even when the perfusion rate was double ($m_b = 0.001$) or neglected ($m_b = 0.0$). This suggests that the effect of perfusion is negligible for prevascular tumors, that is, very small tumors that are close to the skin surface. This may be attributed to the fact that there are no large blood vessels in the skin region. The average temperature increase at the skin surface, caused by the presence of the tumor, was observed to be about 0.33 °C, which is a very significant number that can be easily detected using an infrared camera.

Next, we consider two tumors of the same radii 1 mm, with centers located at $(-0.4, 0.25)$ and $(0.0, 0.25)$. We investigate two cases; firstly, the tumors have the same S_t values, and secondly, one of the values is altered. The results are given in Fig. 3. Figure 3a represents a 3D steady state temperature profile for tumors with the same metabolic heat generating rates $S_t = 20S_h$. In Fig. 3b the metabolic heat generating rate of the right tumor was reduced to $S_t = 10S_h$. Figure 3c gives the steady state temperature at the skin surface for the two cases. The results affirm that a tumor with a higher metabolic heat generating rate will produce more heat and consequently a better temperature profile at the skin surface.

The next results as presented in Fig. 4 are for two tumors with the same metabolic heat generating rates ($S_t = 20S_h$) and different radii. The tumor on the left has a radius of 1 mm with center located at $(-0.4, 0.25)$, while the tumor on the right has a radius of 0.75 mm with center located at $(0.4, 0.25)$. A 3D steady state temperature profile for tumors is given in Fig. 4a and the steady state temperature at the skin surface is given in Fig. 4b. Here we see that a bigger tumor will produce more heat compared to a smaller tumor having the same metabolic heat generating rate.

Finally, Fig. 5 illustrates two tumors with different metabolic heat generating rates and different radii. The bigger tumor (i.e. left tumor) with radius 1 mm and center located at $(-0.4, 0.25)$ was given a smaller metabolic heat generating rate $S_t = 5S_h$. The smaller tumor (i.e. right tumor) with radius 0.75 mm and center located at $(0.4, 0.25)$ was given a higher metabolic heat generating rate $S_t = 20S_h$. Figure 5a shows a 3D steady state temperature profile for tumors over the entire

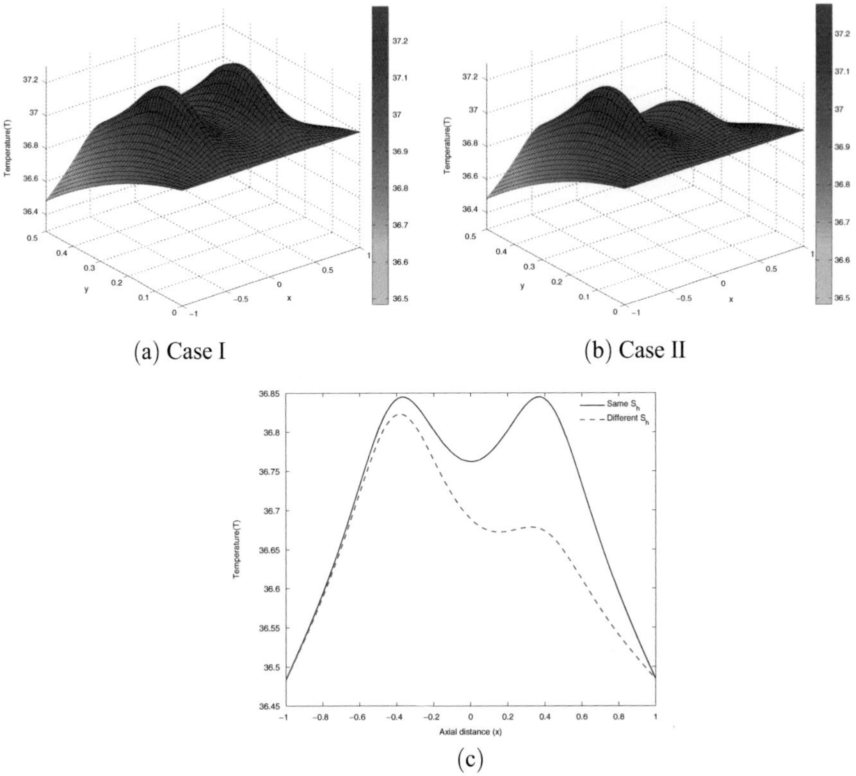

Fig. 3 Numerical simulation of two tumors of the same radii. The results in (**a**) for the same S_t, (**b**) for different S_t and (**c**) the temperature profile at the skin surface for cases I and II

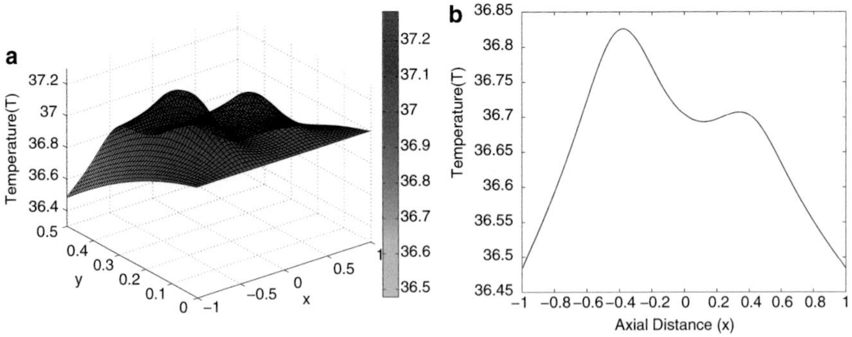

Fig. 4 Numerical simulations of two tumors with same $S_t = 20S_h$ and different radii. The plot in (**a**) is a 3D temperature distribution over the cross section and (**b**) is the skin surface temperature

domain, while Fig. 5b provides the steady state temperature at the skin surface. The results show that a smaller tumor with a high metabolic heat generating rate will

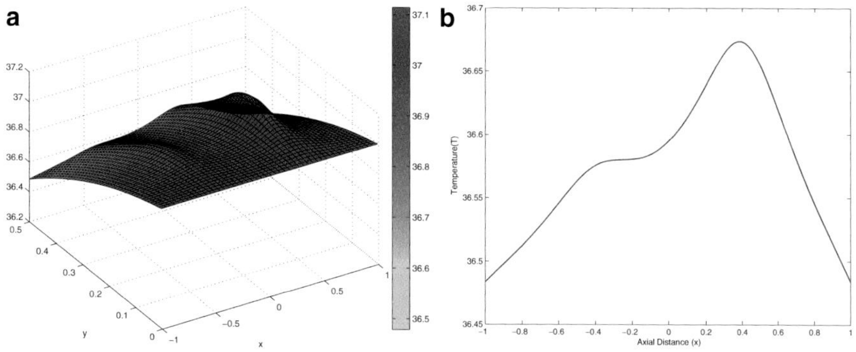

Fig. 5 Numerical simulations of two tumors with different S_t and different radii. The plot in (**a**) is a 3D temperature distribution over the cross section and (**b**) is the skin surface temperature

produce more heat compared to a bigger tumor with a low metabolic heat generating rate.

An analysis of how the results of a variant of the model for breast tumors depend on the underlying parameters was carried out in [17]. Among others, different values of the depth of the tumor, the ambient temperature, and the internal body temperature were investigated. A similar analysis yields analogous results for the current model for melanomas.

3.1 Future Work

There are several directions in which the model presented in this paper can be extended. The case of multiple tumors can be investigated by considering different size tumors at multiple, randomly chosen locations. This would give a better understanding of possible heat signatures. The case of non-spherical tumors can be explored as well. However, this requires a slight modification of the current numerical methods. Another aspect of interest is how the sizes of the blood vessels in the neighborhood of the tumor modify the temperature profile of the tumor at the skin surface.

The extension of the model from the 2-dimensional domain to 3 dimensions is under way. The numerical methods have to be adjusted to deal with the added dimension, but the extended model might discover some new phenomena not present in the 2-dimensional investigations.

The inverse problem requires a different approach to the model. One possible extension of our investigation is to set up an optimization problem to identify the location and size of the tumors for given heat signatures.

4 Conclusion

We have presented a mathematical model for cutaneous melanoma at the prevascular level based on the Pennes bio-heat transfer equation. We studied the effect of blood perfusion rate, tumor size and tumor metabolic heat generating rate. The model suggests that the perfusion rate does not influence the steady state temperature at the skin surface for prevascular tumors. The size and tumor metabolic heat generating rate are shown to be very important factors in determining the steady state temperature distribution of the region under consideration. These two factors are therefore useful bio-markers that can be used to diagnose whether a skin lesion is dormant and harmless or malignant. The results of the model associate melanoma with a tumor that has a high metabolic heat generating rate and to one that is increasing in size.

References

1. Stern, R.S.: Prevalence of a history of skin cancer in 2007: results of an incidence-based model. Arch. Dermatol. **146**, 279–282 (2010)
2. American Cancer Society: Cancer Facts & Figures. http://www.cancer.org/acs/groups/content/@editorial/documents/document/acspc-044552.pdf (2015)
3. Estee, L., Psaty, B.A., Allan, C., Halpern, M.D.: Current and emerging technologies in melanoma diagnosis: the state of the art. Clin. Dermatol. **27**, 35–45 (2009)
4. Arora, N., Martins, D., Ruggerio, D., Tousimis, E., Swistel, A.J., Osborne, M.P., Simmons, R.M.: Effectiveness of a noninvasive digital infrared thermal imaging system in the detection of breast cancer. Am. J. Surg. **196**, 523–526 (2008)
5. Bronzino, J.D.: Medical Devices and Systems. CRC/Taylor & Francis, Boca Raton (2006)
6. González, F.J.: Thermal simulation of breast tumors. Revista Mexicana de Fisica, **53**, 323–326 (2007)
7. González, F.J.: Non-invasive estimation of the metabolic heat production of breast tumors using digital infrared imaging. QIRT J. **8**, 139–148 (2011)
8. Lin, Q.Y., Yang, H.Q., Xie, S.S., Wang, Y.H., Ye, Z., Chen, S.Q.: Detecting early breast tumour by finite element thermal analysis. J. Med. Eng. Technol. **33**, 274–280 (2009)
9. Sudharsan, N.M., Ng, E.Y.K., Teh, S.L.: Surface temperature distribution of a breast with and without tumour. Comput. Methods Biomech. Biomed. Eng. **2**, 187–199 (1999)
10. Agnelli, J.P., Barrea, A.A., Turner, C.V.: Tumor location and parameter estimation by thermography. Math. Comput. Model.: Int. J. **53**, 1527–1534 (2011)
11. Mital, M., Scott, E.P.: Thermal detection of embedded tumors using infrared imaging. J. Biomech. Eng. **129**, 33–39 (2007)
12. Paruch, M., Majchrzak, E.: Identification of tumor region parameters using evolutionary algorithm and multiple reciprocity boundary element method. Eng. Appl. Artif. Int. **20**, 647–655 (2007)
13. Deng, Z., Liu, J.: Mathematical modeling of temperature mapping over skin surface and its implementation in thermal disease diagnostics. Comput. Biol. Med. **34**, 495–521 (2004)
14. Pirtini Cetingül, M., Herman, C.: A heat transfer model of skin tissue for the detection of lesions: sensitivity analysis. Phys. Med. Biol. **55**, 5933–5951 (2010)
15. Pirtini Cetingül, M., Herman, C.: Quantification of the thermal signature of a melanoma lesion. Int. J. Therm. Sci. **50**, 421–431 (2011)

16. Bhowmik, A., Repaka, R., Mishra, S.C.: Thermographic evaluation of early melanoma within the vascularized skin using combined non-Newtonian blood flow and bioheat models. Comput. Biol. Med. **53**, 206–219 (2014)
17. Agyingi, E., Wiandt, T., Maggelakis, S.: Thermal detection of a prevascular tumor embedded in breast tissue. Math. Biosci. Eng. **12**, 907–915 (2015)
18. Maggelakis, S.A., Savakis, A.E.: Heat transfer in tissue containing a prevascular tumor. Appl. Math. Lett. **8**, 7–10 (1995)
19. Pennes, H.H.: Analysis of tissue and arterial blood temperatures in the resting forearm. J. Appl. Physiol. **1**, 93–122 (1948)

Time-Dependent Casual Encounters Games and HIV Spread

Safia Athar and Monica Gabriela Cojocaru

Abstract In Tully et al. (Math Biosci Eng AIMS, 2015, to submitted) the authors model and investigate casual sexual encounters between two members of a population with two possible HIV states: positive and negative, using a Nash game framework in which players try to maximize their expected payoff resulting out of a possible encounter. Each player knows their own HIV status, but do not know the HIV status of a potential partner. They do however have a personal assessment of the risk that the potential partner may be HIV positive. Last but not least, each player has a ranked list of preferences of potential types of sexual outcomes: unprotected, protected, or no sexual outcome. In Tully et al. (Math Biosci Eng AIMS, 2015, to submitted), the game model is studied via 1- and 2-dimensional sensitivity analyses on parameters such as the utility values of unprotected sex of an HIV negative individual with an HIV positive, and values of personal risk (of encountering an HIV positive partner) perception.

In this work, we introduce time as a variable which affects players' risk perceptions, and thus their strategies. Given that HIV transmission happens when an HIV positive player has a non-zero probability (strategy) of having unprotected sex with a HIV negative player, we are also able to keep track of the time evolution of the overall fraction of HIV positive individuals in the population, as reflected as an outcome of repeated casual encounters. We model a continuous time dynamic game (as in Cojocaru et al. (J Optim Theory Appl 127(3):549–563, 2005)) where we compute the stable strategies of each player based on a dynamical system defined on a set of functions. We observe that with change in choices the HIV prevalence in the population increases.

1 Introduction

HIV/AIDS (human immunodeficiency virus/acquired immunodeficiency syndrome) was first observed in California, 1980 [6]. Patients, at that time, were treated by the local physicians for intense fever, diarrhoea, weight loss and swollen lymph nodes.

S. Athar (✉) • M.G. Cojocaru
University of Guelph, 50 Stone Rd E, Guelph, ON N1G 2W1, Canada
e-mail: sathar@uoguelph.ca; mcojocar@uoguelph.ca

© Springer International Publishing Switzerland 2016

J. Bélair et al. (eds.), *Mathematical and Computational Approaches in Advancing Modern Science and Engineering*, DOI 10.1007/978-3-319-30379-6_17

In 1982, 600 cases were reported by CDC, out of which 75 % were identified as homosexual or bisexual males. Therefore, the early name given to these symptoms was 'GRID' (gay related immune deficiency). In late 1982, it was changed to AIDS by CDC after determining the fact that this disease is not exclusive only to the gay population. About 1500 lives were claimed by this disease in 1984 [5]. A poll in 1985 indicated that nearly half of Canadians were concerned about getting infected by this disease [9].

HIV is a unique virus among others as it incorporates its own DNA into the host's cellular DNA. The virus also contains a protein that takes over the host cell's reproduction ability; is then using it as an aid for self replication. This virus impacts the immune system resulting in life threatening infections. It can be transferred through blood, semen,vaginal fluid or breast milk [12].

Nearly 15,000 cases of HIV were discovered in United States in 1985, whereas in Africa the estimated number was approximately half a million. The latest research proved that HIV first appeared in humans in central Africa. First evidence of the evolution of HIV emerged in 1985, when scientists found a virus in Macaque monkeys that was closely related to HIV virus [11, 12].

HIV was classified as a pandemic by the World Health Organization (WHO). According to estimates by WHO and UNAIDS, 35 million people were living with HIV globally at the end of 2013. In the same year, 2.1 million people became newly infected, and 1.5 million died of AIDS-related causes. Although medical treatments have reduced the annual rate of HIV, the drop in new HIV infections is still not significant. The primary mode of transmission, for this disease, is sexual encounters in many countries. For instance, 80 % of the cases in the United States are due to unprotected sexual encounters [13].

According to [11], the spread of HIV infection is largely influenced by people's way of thinking about their sexual encounters. Transmission of HIV in a population may increase if the individuals have unprotected sex. To better understand the spread of HIV, population models have been used. These models are helpful in observing the change in infected populations caused by different parameters. Usually, the concept of probability is used to understand the process of decision-making in relation to unprotected sex. Game theory is an important mathematical tool that has been extensively used to describe the decision-making of individuals in certain non cooperative situations. Different classes for the games are being used to model individual's decision depending upon for example: linear payoff with zero or non-zero sum game as in a 2-player game; or non-linear payoff as in a multi-player game [11].

In this paper, we are interested in studying a model of two players engaged in finding a casual sexual partner. The game here is dynamic, i.e., we consider time-dependency of equilibrium(Nash) strategies under time evolution of utilities and player's risk assessment. The main purpose is to investigate the influence of different parameters involved in the game upon the infected population over a certain period of time. This paper has following structure: Sect. 2 describes briefly the one-shot 2-player game; Sect. 3 presents the dynamic game using the frame work of evolutionary variational inequalities; Sect. 4 discusses the results of the computational work along with the analysis of parameters. We close with conclusions and a few ideas for future work.

2 2-Player Game: A Brief Introduction

Tully et al. [11] described a casual sexual encounter between two individuals as a game. The status of the two players are known only to themselves, while players are aged 15 years and above. Players are denoted by P_1 and P_2 whose HIV status is positive and negative respectively. We denote by ϵ_+ the proportion of HIV positives (HIV_+) in total population, and by ϵ_- the proportion of HIV negatives (HIV_-) in the total population with the condition that $\epsilon_+ + \epsilon_- = 1$.

The authors [11] used game models to find Nash equilibrium probabilities of having unprotected sex (US) in casual encounters. The probabilities of unprotected sex US for the two players are denoted by:

$$x^i \in [0,1]^2; x^i = (x^i_-, x^i_+), i \in \{1,2\}$$

where x^i_- represents the probability of P_i having US with an HIV_- individual and x^i_+ represents the probability of P_i having US with an HIV_+ individual. The expected utility for $P_i, i \in \{1,2\}$, when interacting with $HIV-$ and $HIV+$ individuals, is given by:

$$E^i_- = \rho[x^i_- U(US,+,-) + (1 - x^i_-)U(notUS,+,-)]$$

$$E^i_+ = \rho[x^i_+ U(US,+,+) + (1 - x^i_+)U(notUS,+,+)]$$

where $i \in \{1,2\}$

By the term $notUS$ means either the players have protected sex or no sex at all. Therefore, the overall expected utility for $P_i, i \in \{1,2\}$ is given by:

$$E^i(x^i_-, x^i_+) = (1 - b_+)E^i_- + b_+ E^i_+$$

where ρ represents the activity parameter for player $P_i, i \in \{1,2\}$ and b_+, b_- represent the risk assessments of P_1 and P_2 respectively, regarding the individual they are engaging in a casual encounter with. The risk parameters b_+ and b_- are defined as follows:

$$b_- = \beta_- b^{self}_- + (1 - \beta_-)\epsilon_+$$

$$b_+ = \beta_+ b^{self}_+ + (1 - \beta_+)\epsilon_+. \tag{1}$$

The parameters b_+^{self} and b_-^{self} represent the personal belief of player P_1, respectively P_2 about the level of *HIV* infection in the population. Their values are between 0 and 1, with 1 representing the belief that everyone else in the population is infected.

The parameters β_- and β_+ represent the weight a player places upon personal assumptions of *HIV* prevalences. Their values range again between 0 and 1. With these in mind, ϵ_+ is defined as:

$$\epsilon_+ = \epsilon_+(0) + [x_-^1 \epsilon_+(0) + x_+^2 \epsilon_-(0)]\tau,$$

where $\epsilon_+(0)$ is the initial fraction of HIV_+ in the population. By initial we mean here: before the encounter, where as $\epsilon_-(0)$ represents the initial fraction of HIV_- in the population. We let $\tau = 0.02$ be the transmission rate of HIV in the population [2].

Both players want to optimize their expected payoffs subject to constraints defined as follows: for each $i \in \{1, 2\}$,P_i solves the optimization problem,

$$\begin{cases} \max E^i := E^i(x^1, x^2) \\ \text{s.t } x^i \in K_i := [0, 1]^2 \cap \{x_-^i + x_+^i = 1\} \end{cases}$$

Definition 1 Assume each player is rational and wants to maximize their payoff. Then the Nash equilibrium is a vector

$x^* \in K := K_1 \times K_2$ which satisfies the inequalities: For all i

$f_i(x_i^*, x_{-i}^*) \geq f_i(x_i, x_{-i}^*)$, $\forall x_i \in K_i$ where $x_{-i}^* = (x_1, \ldots, x_{i-1}, x_{i+1}, \ldots, x_2)$.

The authors used variational analysis to find the Nash equilibria for the above game as in Definition 1.

Definition 2 Given a set $K \subset \mathbb{R}^n$, closed, convex, non-empty and given $F : K \longrightarrow \mathbb{R}^n$ is a function, the variational inequality(VI) problem is to find a vector $x^* \in K$ such that

$$\langle F(x^*), y - x^* \rangle \geq 0, \quad \forall y \in K \tag{2}$$

Specifically, the Nash game is reformulated into a variational inequality problem as follows [3, 10]:

Theorem 1 *Provided the utility functions u_i are of class $\mathbf{C}^1$ and concave (meaning $-u_i$ is convex) with respect to the variables x_i, then $x^* \in K$ is a Nash equilibrium if and only if it satisfies the VI:*

$$\langle F(x^*), x - x^* \rangle \geq 0 \quad \forall x \in K \tag{3}$$

where $F(x) = (-\nabla_{x_1} u_1(x), \ldots.. - \nabla_{x_n} u_n(x))$ *and* $\nabla_{x_i} u_i(x) = (-\frac{\partial u_1(x)}{\partial x_1}, -\frac{\partial u_2(x)}{\partial x_2} \ldots. -\frac{\partial u_n(x)}{\partial x_N})$

The variational inequality problem has at least one solution if the constraint set K is closed, bounded and convex and $F : K \longrightarrow \mathbb{R}^n$ is continuous. In other words, K is a compact and convex set then solutions for variational inequality problems exist. Further, solution may be unique if the function F is strictly monotone on K [8].

In our case we have: $F(x^1, x^2) = (- \nabla E^i(x^1, x^2)) = (- \nabla_{x^1} E^1, - \nabla_{x^2} E^2)$. Here, K is closed, bounded and convex and $F(x^1, x^2)$, being linear, is continuous; therefore, the existence of solution for the problem is proved. However, $F(x^1, x^2)$ is not strictly monotone, therefore the strategy applied here, by the authors, is to look at the Nash points $(x^1, x^2) \in K$ as critical points of a set of differential equations driven by the vector field F and constraint set K:

$$\frac{dx}{d\tau} = P_{T_K(x(\tau))}(-F(x(\tau))); \quad x(0) = (x_-^1(0), x_+^1(0), x_-^2(0), x_+^2(0)) \in K = K_1 \times K_2 \tag{4}$$

2.1 *One-Shot Game: Example*

We develop a base case for our further investigations. The parameters of the model are set in such a way that we can compute the Nash equilibrium vector for an ideal situation. Ideal situation is developed to restrict the players to their own groups, i.e., P_1 interacts with HIV_- and P_2 interacts with HIV_+ player. We solve VI problem attached to the game, using the following values in the Table 1 for the parameters involved. The utilities for *US* and *notUS* are also given in the following Table 1.

We start with uniformly distributed initial conditions and find the unique Nash equilibrium for the 2-player game.The value for this equilibrium is (0,1,1,0). This shows that the two players are careful regarding the choice of their partner for casual encounters.

The strategies of the two players are plotted to see the evolution towards the equilibrium point when different initial values have been considered. The plot in Fig. 1 shows the strategy of P_1 and P_2, when selecting their partners for a casual sex encounter.

Table 1 Parameters used in base case

β_-	β_+	b_-^{self}	b_+^{self}
0.5	0.5	0.6	0.3
$\epsilon_-(0)$	$\epsilon_+(0)$	τ	ρ
0.95	0.05	0.02	1
$US(-,-)$	$US(-,+)$	$US(+,-)$	$US(+,+)$
1	0.5	0.5	1
$notUS(-,-)$	$notUS(-,+)$	$notUS(+,-)$	$notUS(+,+)$
0.5	0.5	0.5	0.5

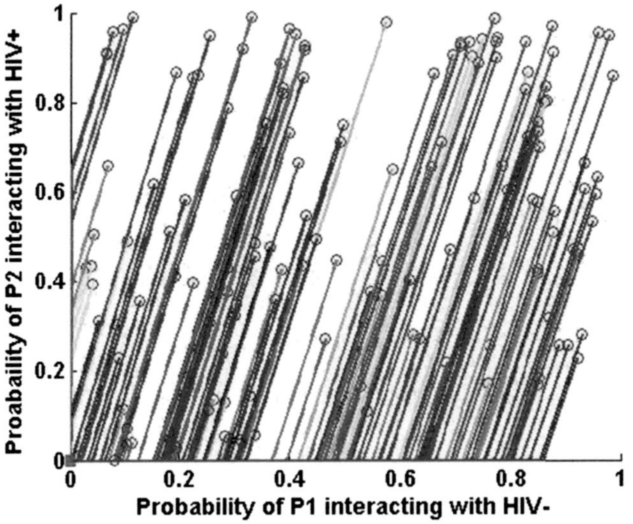

Fig. 1 Phase portrait for the strategy selections of P_1 and P_2; when P_1 being HIV_+ interact with HIV_-, whereas P_2 interact with HIV_+. We started with 150 uniformly distributed initial conditions to solve (4) using Matlab

3 Time-Dependent 2-Player Encounter Game

We convert the one-shot game, as described in Sect. 2, into a time-dependent game. The purpose behind this extension is to see the effect of time on the decision-making process of the two players. Moreover, an additional purpose is to see how this decision-making can affect the change in HIV prevalence (ϵ_+), if the 2-player game is played repeatedly. We use the combined theories of projected dynamical system (PDS) and evolutionary variational inequality (EVI), in order to investigate the dynamics of the 2-player game [1, 4]. These two theories have been applied to a number of other problems in different disciplines, such as operational research, economics and finance etc. [1, 4]. For this purpose, we reformulate the variational inequality problem of our 2-player game into a time-dependent variational inequality, (EVI). We then measure the effects of time dependency on the HIV prevalence and on the strategies used by the players.

First, we start by defining constraint set $\mathbb{K}$, as in Sect. 2, and rewrite it for the EVI.

Our feasible vector $u(t) = (x^1(t), x^2(t)) = (x^1_-(t), x^1_+(t), x^2_-(t), x^2_+(t))$ has to satisfy the time-dependent constraints on $x^i, i \in \{1, 2\}$, and it belongs to the set of functions given by:

$\mathbb{K} = \{x^i(t) \in L^2([0, T], \mathbb{R}^4) : 0 \leq x^i_+(t), x^i_-(t) \leq 1, x^i_+(t) + x^i_-(t) = 1, i \in \{1, 2\} \quad a.e \in [0, T]\}$

Since $F(x^1, x^2) = (-\nabla E(x^1, x^2))$, we write $F(u(t)) = (-\nabla E(u(t)))$ where $F : \mathbb{K} \longrightarrow L^2([0,T], \mathbb{R}^4)$. Then EVI is defined as:

Definition 3 Find $u \in \mathbb{K}$ such that

$$\langle F(u(t)), v(t) - u(t)\rangle \geq 0 \quad \forall \quad v(t) \in \mathbb{K}(t) \text{ for a.a t}$$

where

$$\mathbb{K} = \{x^i(t) \in L^2([0,T], \mathbb{R}^4) : 0 \leq x^i_+(t), x^i_-(t) \leq 1,$$
$$x^i_+(t) + x^i_-(t) = 1, i \in \{1,2\} \text{ a.e in } [0,T]\}$$

This is called point-wise form for EVI.

Theorem 2 *Let H be Hilbert space and let $\mathbb{K} \subset H$ be non-empty, closed and convex subset. Let $F : \mathbb{K} \longrightarrow H$ be the Lipschitz continuous vector field with Lipschitz constant b. Then the solutions of the time dependent variational inequality Definition 3 are the same as the critical points of the projected differential equation:*

$$\frac{du(t,\tau)}{d\tau} = P_{T_{\mathbb{K}(x)}}(u(t,\tau), -F(u(t,\tau)))$$

That is the point $x \in \mathbb{K}$ such that $P_{T_{\mathbb{K}(x)}}(u(t), -F(u(t))) \equiv 0$, and the converse also holds.

If we choose $H = L^2([0,T], \mathbb{R}^4)$ then the solution of EVI in Definition 3 and the critical points of the equation in (2) are the same.

The important point in (2) is the difference between the two times, t, τ. The time $[0,T]$ represents the time interval the game is considered to be played over. As t varies over the interval [0,T], we can obtain one or more curves representing how Nash strategies of both players may change. The time τ is the simulated time of evolving from an initial point of the differential equation (2) towards one of the Nash equilibria on this curve(s). For the computational purpose, a sequence of partition $\{t_n^0, t_n^1, \ldots.., t_n^N\} \in [0,T]$ is defined such that

$$0 = t_n^0 < t_n^1 <, \ldots. < t_n^N = T.$$

Then for each t_n^j, we solve the time dependent variational inequality:

$$\langle F(u(t_n^j)), v - u(t_n^j)\rangle \geq 0 \text{ for all } v \in \mathbb{K}(t_n^j),$$

where

$\mathbb{K}(t_n^j) = \{x^i(t_n^j) \in L^2([0,T], \mathbb{R}^4) : 0 \leq x^i_+(t_n^j), x^i_-(t_n^j) \leq 1, x^i_+(t_n^j) + x^i_-(t_n^j) = 1, i \in \{1,2\} \text{ a.e in } [0,T]\}$.

We compute all solutions of this finite dimensional variational inequality, by finding the critical point of the projected dynamical system [4]:

$$P_{T_{\mathbb{K}(x)}}(u(t_n^j, \tau), -F(u(t_n^j, \tau))) = 0$$

4 Computation of Time-Dependent Nash Strategies and Their Effects on the HIV Prevalence

4.1 Base Case

We use the same values for the parameters in the base case as in the one-shot game. The only difference in this base case is that both the strategies and ϵ_+ are time dependent. The parameters used for the base case are described in Table 1. We run this case for $T = 10$ periods, and reach the same equilibrium point as in the one-shot game. No change in the value of $\epsilon_+(t)$ is observed (Fig. 2).

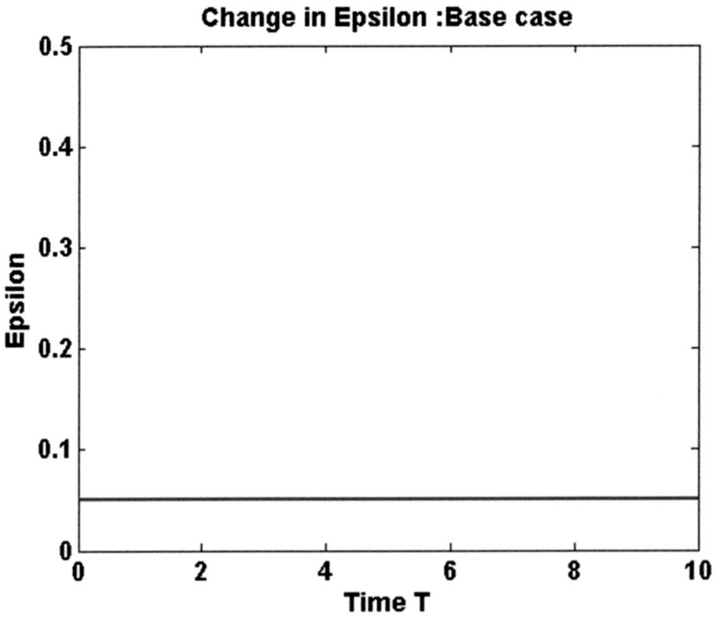

Fig. 2 Increase in HIV prevalence over a period of 10 periods

4.1.1 Influence of $US(-,+)$ and $US(+,-)$ on HIV Prevalence

We vary $US(-,+)$ & $US(+,-)$ as a function of time, whereas the other parameters are kept constant. The change in choices about having *US* over the given period of time can be considered the result of emotional decision-making and/or the level of information (Fig. 3). Gutnik et al. in [7] discussed the role of emotions in decision-making. The role was considered as negative hindrance in the rational decision-making. According to new research, people use analytic and experimental systems to understand and assess the factors of risk. However, the emotions are the most common factor that works in the experimental system. These emotions depend on past experiences or the perceived risks [7]. The players can change their preferences over the given period under the influence of emotion. These decisions under emotional circumstances perhaps occur due to lack of adequate knowledge or due to the "heat of moment" (Table 2).

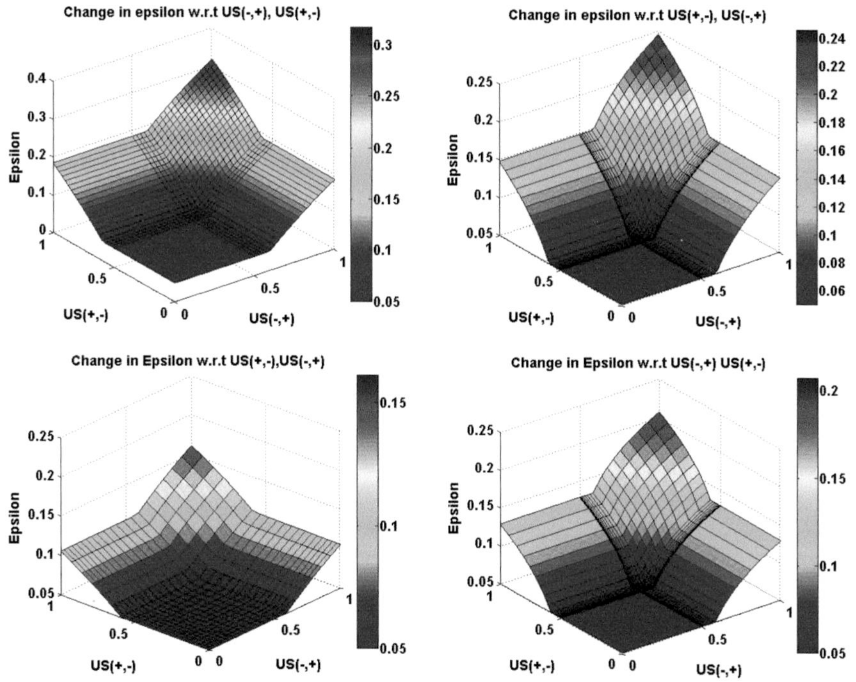

Fig. 3 3-D plots shows the change in HIV prevalence, when the $US(-,+)$ and $US(+,-)$ are made time dependent. *Upper left* shows the change in $US(-,+)$ and $US(+,-)$ are linear function of time; *upper right* shows $US(-,+)$ and $US(+,-)$ are quadratic function of time. *Lower left* shows $US(-,+)$ and $US(+,-)$ are quadratic function as well but in different formation and *lower right* shows $US(-,+)$ and $US(+,-)$ are cubic function of time. The increase is much higher in linear case, $US(-,+)$ and $US(+,-)$ are linear function of time, as compared to the case when these choices are quadratic or cubic function of time

Table 2 Effect of $US(-,+)$ and $US(+,-)$ on HIV prevalence $\epsilon_+(t)$

Cases	$US(-,+)$	$US(+,-)$	$\epsilon(t)$
Case 1	$0.5 + 0.5 \times (t/10)$	$0.5 + 0.5 \times (t/10)$	5 % to 32 %
Case 2	$0.5 + 0.5 \times (t/10)^2$	$0.5(t) + 0.5 \times (t/10)^2$	5 % to 24 %
Case 3	$0.5 + 0.5 \times (t/10)^3$	$0.5 + 0.5 \times (t/10)^3$	5 % to 21 %

The time is taken as 10 periods and Matlab is used for simulations. We observe an increase in HIV prevalence from 5 % to 32 % when $US(+,-)$ is linear function of time; whereas the increase is 24 % when $US(+,-)$ in quadratic in time. The impact of $US(+,-)$ on increase in HIV prevalence is perhaps due to change in preferences of player $1(HIV_+)$ over the time.

5 Conclusion and Future Work

In this work we converted a discrete time 2-player game into a time-dependent game. We investigated the effects of changes in utilities for having unprotected sex between HIV_+ and HIV_- players on the HIV prevalence rate over a given period of time. Also, we examined the influence of individual's personal risk perception about their partners in casual encounters.

In the near future we are interested in formulating a multi-player time-dependent game, when two different types of *HIV* viruses, i.e., $HIV - 1$ and $HIV - 2$ are involved. The players with these viruses interact with HIV_- players from same age groups or different age groups. Further, we want to investigate the impact of prophylactic vaccination against *HIV* in younger population groups in multi-players time-dependent games.

References

1. Barbagallo, A., Cojocaru, M.-G.: Dynamic vaccination games and variational inequalities on time-dependent sets. J. Biol. Dyn. **4**(6), 539–558 (2010)
2. Basar, T., Olsder, G.J.: Dynamic Noncooperative Game Theory, vol. 200. SIAM, Philadelphia, (1995)
3. Cojocaru, M.-G., Greenhalgh, S.: Dynamic games and hybrid dynamical systems. Optim. Eng. Appl. Var. Inequal. Issue **13**(3), 505–517 (2011)
4. Cojocaru, M., Daniele, P., Nagurney, A.: Projected dynamical systems, evolutionary variational inequalities, applications, and a computational procedure. In: Pareto Optimality, Game Theory and Equilibria, pp. 387–406. Springer, New York (2008)
5. Engel, J.: The Epidemic: A History of AIDS. HarperCollins, New York (2009)
6. Grmek, M.D.: History of AIDS: Emergence and Origin of a Modern Pandemic. Princeton University Press, Princeton, NJ (1993)

7. Gutnik, L.A., et al.: The role of emotion in decision-making: a cognitive neuroeconomic approach towards understanding sexual risk behavior. J. Biomed. Inform. **39**(6), 720–736 (2006)
8. Kinderlehrer, D., Stampacchia, G.: An Introduction to Variational Inequalities and Their Applications. Academic, New York (1980)
9. Levon, B.: Facing a fatal disease. Maclean's **1**(6), 48 (1986)
10. Moulin, H.: On the Uniqueness and Stability of Nash Equilibrium in Non-cooperative Games, vol. 130. North-Holland Publishing Company, Amsterdam (1980)
11. Tully, S., Cojocaru, M.-G., Bauch, C.: Multiplayer games and HIV transmission via casual encounters. Math. Biosci. Eng. AIMS (2015, to submitted)
12. World Health Organization: Health topics: HIV/AIDS. http://www.who.int/hiv/pub/guidelines/en. Accessed 6 Mar 2013
13. World Health Organization: Health topics: HIV/AIDS. http://www.who.int/features/qa/71/en/ http://www.who.int/bulletin/volumes/85/11/06-033779/en/index.html. Accessed 6 Mar 2013

Modelling an Aquaponic Ecosystem Using Ordinary Differential Equations

C. Bobak and H. Kunze

Abstract Aquaponic agriculture is a sustainable system which uses interdependent processes and has been growing in popularity. However, relatively little mathematical and other academic research has been conducted in the practice. In this paper, we develop a system of ordinary differential equations to model the population and concentration dynamics of the environment. Our model has an asymptotically stable non-trivial equilibrium, representing the inherent symbiotic relationship of the variables. Values of the nine parameters in the system are estimated from the research literature. We provide simulated results simulated results to illustrate the nature of solutions to the system, and we present and discuss a sensitivity analysis.

1 Introduction

Aquaponics is a closed-loop agricultural system which uses a symbiotic relationship between aquatic organisms and aquatic macrophytes. The system recirculates water through an aquaculture environment (fish in a designated body of water) and a hydroponic structure (aquatic plants in soilless water) to create a sustainable environment which fully conserves water and nutrients. The key motivation behind aquaponics is using waste produced by fish in the system as a nutrient source for the plants. This process not only allows the fish waste to act as a natural fertilizer for the plants but also inherently cleans the water to be returned to the fish [6]. Thus the environment is in a natural state of stability, which we seek to model using a system of differential equations. This paper discusses background theory in aquaponics, develops a compatible aquaponic model in the form of a system of differential equations, discusses equilibria of the model and their stability, presents specific solution simulations for parameter values derived from the research literature, performs a sensitivity analysis, and suggests future research direction.

C. Bobak (✉) • H. Kunze
Department of Mathematics and Statistics, University of Guelph, Guelph, ON, Canada
e-mail: cbobak@uoguelph.ca; hkunze@uoguelph.ca

© Springer International Publishing Switzerland 2016
J. Bélair et al. (eds.), *Mathematical and Computational Approaches in Advancing Modern Science and Engineering*, DOI 10.1007/978-3-319-30379-6_18

2 Background: Research of Aquaponic Environments

Aquaponic agriculture systems have been growing in popularity due to their robust economic and ecological benefits. Both hydroponics and aquaculture have proven to have a detrimental effect on the environment. Some studies have suggested that aquaponics has a 75 % smaller carbon footprint when compared to traditional farming methods [7] and is able to satisfy human demand long-term [3]. Importantly, aquaponics achieves all of these benefits while maintaining economic viability for farmers. The lower cost of resources and lower spatial requirements has led to suggestions that aquaponic agriculture may be a viable solution in densely populated urban regions such as Pakistan and India [3].

The basic aquaponic system consists of two main components: one for fish and one for plants. The system relies heavily on food input as many studies have demonstrated that factors like protein content of food and feeding frequency have the largest effect on the efficiency of the system. These factors highly contribute to fish growth and are also directly related to the amount of fish waste in the environment [6].

Organic nitrogen in fish waste naturally converts to ammonia through biological degradation [4]. Ammonia is highly toxic to fish, and is an inefficient nutrient source for plants [6]. In order for ammonia to be used as fertilizer for the plants, it must go through a natural microbial process called the Nitrogen Cycle, which causes it to convert into nitrate. Nitrate is a nutrient rich food source for plants [4] and studies have shown that plants' uptake efficiency of nitrates ranges from 86 % to 98 %. Whatever concentration of nitrate is left in the water is not harmful to the fish, so the water can be recirculated for fish use [6].

3 Development of the Model

The overarching goal of the model is to capture the symbiotic relationship between the fish and the macrophytes. However, as evidenced by background research on aquaponic agriculture, the aerobic microbial process which converts ammonia to nitrate needs to be considered. The assumpions made are as follows:

i. The aquaponic ecosystem is a closed environment.
ii. The fish population increases at some natural survival rate ($\frac{\text{Births}}{\text{Deaths}}$), hindered by a carrying capacity due to the limited tank space.
iii. There is additional fish decay due to increased ammonia presence in the water until it reaches a critical ammonia level where no fish survive. This can be reasonably modelled using a linear constant of ($\frac{\text{Ammonia}}{\text{Toxic Ammonia Level}}$).
iv. Ammonia is present in the system exclusively due to fish waste and hence grows at a rate proportional to the fish population. It decays due to its conversion to nitrate.

v. Nitrate grows at a rate proporational to the level of ammonia, and decays due to plant uptake.
vi. Plants grow at a constant rate hindered by a carrying capacity indicative of the limited surface area of the system.
vii. Modelling the concentrations of ammonia and nitrate in the system will capture any other relationships between other variables in the nitrogen cycle.
viii. The system is well mixed so the nitrogen cycle occurs naturally and plants have even access to nitrate.

The proposed model is below, with variables F, A, N, and P representing the population of fish, ammonia (in mg), nitrate (in mg) and population of plants respectively:

$$\dot{F} = a_1\left(1 - \frac{F}{K_F}\right)FP - \frac{A}{K_A}F \tag{1}$$

$$\dot{A} = a_2F - a_3A \tag{2}$$

$$\dot{N} = a_4A - a_5NP \tag{3}$$

$$\dot{P} = a_6\left(1 - \frac{P}{K_P}\right)PN \tag{4}$$

where $a_i \geq 0\ \forall i$ are growth and decay rates, and $K_F, K_A, K_P > 0$ are the carrying capacities of fish, ammonia, and plants respectively.

Equation (1) models the evolution of the fish population, using assumptions ii and iii. Equation (2) models the evolution of the ammonia concentration in the system using assumption iv. Equation (3) captures the growth rate of nitrate concentration as it relates to the conversion from ammonia and the decay rate due to plant uptake using assumption v. The final equation (4) captures the growth rate of the plants using assumption vi.

4 Analyzing the Equilibria

Our aquaponic environment model (1), (2), (3) and (4) consists of four equations with nine unknown parameters. The Jacobian matrix of this model is as follows:

$$Df(F,A,N,P) = \begin{bmatrix} -\frac{a_1FP}{K_F} + a_1\left(1-\frac{F}{K_F}\right)P - \frac{A}{K_A} & -\frac{F}{K_A} & 0 & a_1\left(1-\frac{F}{K_F}\right)F \\ a_2 & -a_3 & 0 & 0 \\ 0 & a_4 & -a_5P & -a_5N \\ 0 & 0 & a_6P\left(1-\frac{P}{K_P}\right) & a_6N\left(1-\frac{P}{K_P}\right) - \frac{a_6NP}{K_P} \end{bmatrix} \tag{5}$$

The system (1) (2), (3) and (4) has three equilibria:

$$\{F = 0, A = 0, N = 0, P = P\} \tag{6}$$

$$\{F = 0, A = 0, N = N, P = 0\} \tag{7}$$

$$\left\{F = \frac{a_1 a_3 K_F K_A K_P}{a_1 a_3 K_P K_A + a_2 K_F}, A = \frac{a_1 a_2 K_F K_A K_P}{a_1 a_3 K_A K_P + a_2 K_F}, \right.$$

$$\left. N = \frac{a_1 a_2 a_4 K_F K_A}{a_5 \left(a_1 a_3 K_A K_P + a_2 K_F\right)}, P = K_P \right\} \tag{8}$$

of which the third, in equation (8), since it represents all variables surviving in the environment. We focus on it in the next section.

4.1 *Equilibrium (8): Coexistence*

We analyzed the stability of the nontrivial equilibrium in equation (8) using the research literature to provide estimates for the nine parameters: $a_1 = 0.0124$, $a_2 = 0.1$, $a_3 = 0.94$, $a_4 = 3.6$, $a_5 = 0.92$, $a_6 = 0.056$, $K_F = 250$, $K_P = 300$ and $K_A = 20$ [1–4, 6]. The values selected for our carrying capacities were based on an arbitrary initial tank size of 10L, however, this can easily be scaled up or down to accomodate systems of various sizes. Other initial values were $\{F(0) = 10, A(0) = 0, N(0) = 0, P(0) = 0.5\}$, where $P(0) = 0.5$ was used to represent plants which had not yet reached maturity in the system. Substituting the estimated parameter values in the Jacobian gives:

$$A_* = \begin{bmatrix} -2.77 & -9.31 & 0 & 0.59 \\ 0.1 & -0.98 & 0 & 0 \\ 0 & 3.6 & -276 & -0.228 \\ 0 & 0 & 0 & -0.014 \end{bmatrix} \tag{9}$$

The eigenvalues of A_* in this particular simulation are:

$$\lambda_{31*} = -1.84 + 0.332i \tag{10}$$

$$\lambda_{32*} = -0.014 \tag{11}$$

$$\lambda_{33*} = -276 \tag{12}$$

$$\lambda_{34*} = -1.84 - 0.332i \tag{13}$$

Notably, the real parts of (10) and (13) are negative, and values (11) and (12) are negative, thus an asymptotically stable equilibrium is achieved in this case.

Because (10) and (13) are complex, some spiralling behaviour is present in the system.

Note that λ_{32*} is very close to 0, so the stability status of this equilibrium point may be sensitive to the parameter values chosen. Some experimentation was done within ranges of realistic values for the estimated parameters. In every case, all eigenvalues have negative real parts suggesting with some generality that the equilibrium case where all four variables are present in the system is asymptotically stable. We discuss sensitivity in general in the next section.

Substituting these values into the non-trivial equilibrium and subsequent solution given in the previous section provides the following equilibrium points:

$$\{F = 186.18, A = 19, N = 0.25, P = 300\} \tag{14}$$

The real-world context of these equilibrium points is exciting; it suggests that fish grow to a point well bounded by their carrying capacity, ($K_F = 250$). Ammonia tends towards an amount just below the threshold to which all the fish would die ($K_A = 20$). This suggests that the system is stabilizing in a way which approaches the maximum level of food production for the plants. Nitrate levels appear to be quite low, which is likely due to the superior nutritional uptake of the plants which has been noted in the literature. The plants themselves reach equilibrium at their carrying capacity ($K_P = 300$). This is exactly as anticipated based on the model structure.

Figure 1 shows two plots of the modelled variables over time given the parameter assumptions. As evidenced here, all the variables are approaching their above equilibrium values.The fish and plant follow a logistic trend tending to their carrying capacities. Ammonia also seems to be following this trend, which is due to its

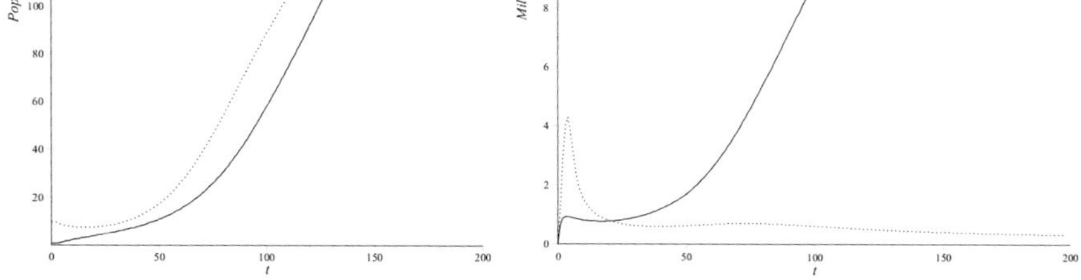

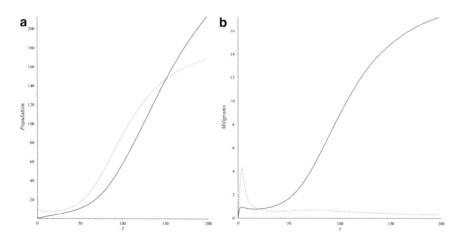

Fig. 1 The population and concentration dynamics of the model over time (**a**) The *solid line* represents the fish population and the *dotted line* represents the plant population (**b**) The *solid line* represents the nitrate population and the *dotted line* represents the ammonia population

growth dependence on fish. Nitrate also shows an interesting trend; as the system is being established, nitrate experiences a brief spike. However, once the plant growth reaches a level which requires significant nitrate sustenance, this spike reverses and the nitrate concentration levels off. These curves appear to loosely represent expected trends based on the aquaponics literature, suggesting the model is reasonably capturing aquaponic behaviour.

5 Sensitivity Analysis

A sensitivity analysis for the parameters of this model was performed with the goals of identifying any unexpected behaviour, guiding any data collection efforts, and most importantly, to give an indication of the importance of accurately estimating the parameter values.

A software toolbox in Matlab was used to perform sensitivity analysis of biological models. SensSB is freely available for academic purposes, and combines a variety of local and global sensitivity methods, both using relative and absolute measures to achieve many of these goals [8, 9].

While locally analysing the sensitivity of parameters is a useful exercise, it has an inherent reliance on the initial numerical estimation of the parameter. Global methods, which test the effect of a parameter while other parameters are varied simultaneously, help avoid this stipulation [9]. Since the initial estimation of the variables from the literature are considered weak points of the model, global sensitivity measures were analysed.

SensSB analyses global sensitivity through three main methods, all of which are discussed in the software documentation (see [9]). In this study, Derivative Based Global Sensitivity Measures (DBGSM), which were introduced in 2009 [5], were selected to optimize accuracy with minimal loss of computational efficiency [9]. DBGSM uses Monte Carlo sampling methods to average local derivatives to a measure $\bar{M}_{ij}$ which averages sensitivity measures S_{ij} over the parameter space.

$$S_{ij} = \frac{\text{\% change in Parameters}}{\text{\% change in Variables}} \tag{15}$$

$$\bar{M}_{ij} = \int_{H^n p} S_{ij} \; dp \tag{16}$$

As seen in Fig. 2, parameter 3 (a_3), or the conversion rate of ammonia to nitrate, has high global absolute sensitivity. In fact, according to the output from SensSB, it accounts for 73.26 % of the total sensitivity in the model. Notably, the relative sensitivity by variable, which shows how sensitive each variable is to the parameters in the model, shows high relative sensitivity for parameters a_2, a_3, a_6, and K_F. In the model, these parameters represent the growth rate of ammonia from fish waste, the conversion rate of ammonia to nitrate, the growth rate of plants and the carrying

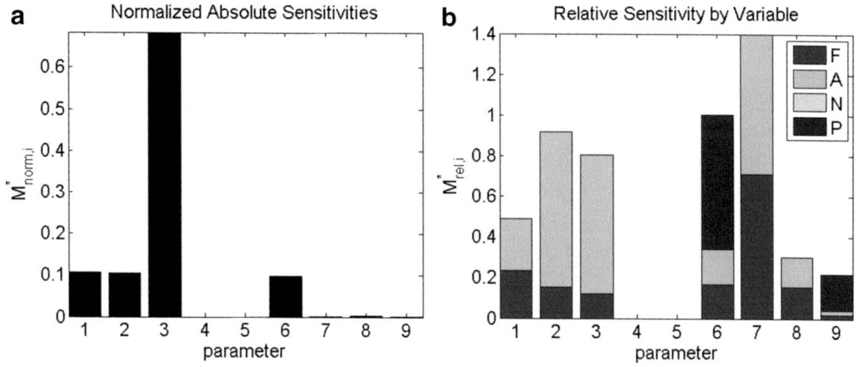

Fig. 2 (**a**) shows the global absolute sensitivity of each of the estimated parameters while (**b**) shows the global relative sensitivity by variable. Note parameters 1–6 are a_1 through a_6, parameter 7 is the carrying capacity K_F, parameter 8 is the carrying capacity K_A, and parameter 9 is the carrying capacity K_P

capacity of fish. This suggests that some more care should be taken in estimating values for these parameters, particularly a_2 and a_3, which do not have reliable values in the literature.

Using values from the previous simulation, both a_2 and a_3 were varied (individually and consecutively) between values of 0 and 100 without loss of stability in the coexistence equilibrium. However, exploring data from an established environment may help clear potential biases associated with this parameter estimation.

6 Future Research Direction

Future models of aquaponic systems should incorporate additional complexity. In particular, removing the assumption that the environment is closed to allow additional harvesting cycle terms for fish and plants would add realism and require additional analysis. Here, the term harvesting refers to the biological process of removing some fish and/or plants, creating a discontinuity in the solution.

As well, there is some indication in the literature that the assumption that the fish population decays linearly with increases in ammonia concentraion is invalid. Instead, it is expected that the fish population would not be affected by the presence of ammonia in the system until the ammonia begins to approach a critical level [1]. Instead, a non-linear polynomial relationship could be used, and may be a better approximation of the real world relationship.

Building from the sensitivity analysis, real data sets could be used to solve the parameter estimation inverse problem to see how well this model, and more complex variations, can accommodate real-world data as an approximation solution.

7 Conclusion

The developed four equation model appears to reasonably simulate biological behaviour. Simulated solutions lead to a few conclusions which may or may not hold in the general case. The most important of these is that the model admits an asymptotically stable equilibrium where fish, ammonia, nitrate, and plants interdependently coexist in the environment. An exploration of parameter space suggests that the stability of such an equilibrium point is robust to changes in the parameters.

Aquaponics shows considerable promise as an ecological agriculture solution. It is the hope that this research can be used to further test the ability of aquaponic solutions as viable practices in both developed and undeveloped countries to aid in concerns over the sustainability of current food practices.

References

1. Wheaton, F.: Recirculating aquaculture systems: an overview of waste management. In: Proceedings of the 4th international conference on recirculating aquaculture, Roanoke (2002)
2. Bhujel, R.C., Little, D.C., Hossain, M.A.: Reproductive performance and growth of stunted and normal Nile Tilapia (Oreochromis niloticus) broodfish at varying feeding rates. IEEE Aquac. **273**, 71–79 (2007)
3. Blidariu, F., Grozea, A.: Increasing the economical efficiency and sustainability of indoor fish farming by means of aquaponics – review. Anim. Sci. Biotechnol. **44**(2), 1–8 (2011)
4. Endut, A., Jusoh, A., Ali, N.: Nitrogen budget and effluent nitrogen components in aquaponics circulation systems. Desalinations Water Treat. **52**, 744–752 (2014)
5. Kucherenko, S., Rodriguez-Fernadez, M., Pantelides, C., Shah, N.: Monte carlo evaluation of derivative based global sensitivity measures. Reliab. Eng. Syst. Saf. **94**, 1135–1148 (2009)
6. Liang, J.L., Chien, Y.H.: Effects of feeding frequency and photoperiod on water quality and crop production in a tilapia-water spinache raft aquaponics system. Int. Biodeterior. Biodegrad. **83**, 693–700 (2013)
7. Pattillo, A.: Aquaponic system design and management. Aquaculture Extension, Iowa State University (2014)
8. Rodriguez-Fernandez, M., Banga, J.R.: SensSB: a software toolbox for the development and sensitivity of systems biology models. Bioinformatics **26**(13), 1, 2 (2009)
9. Rodriguez-Fernandez, M., Banga, J.R.: SensSB – a software toolbox for sensitivity analysis in systems biology models. In: International Conference on Systems Biology (ICSB 2009) Aug 30–Sept 4, Stanford University, Palo Alto (2010)

A New Measure of Robust Stablity for Linear Ordinary Impulsive Differential Equations

Kevin E.M. Church

Abstract A new measure of robust stability for linear ordinary impulsive differential equations with periodic structure is introduced, based on the impulse extension concept. This new stability measure reflects the sensitivity of the model to uncertainty in what we see as the fundamental hypothesis of impulsive models: that the impulse effect occurs quicky enough that its duration can be entirely neglected. The measure, that we call the time-scale tolerance, $\mathscr{E}_t$, has the property that, if the vector of durations of impulse effect, a, satisfies $||a|| < \mathscr{E}_t$, then both the impulsive model and a family of continuous impulse extension equations (a specific functional differential equation) to which it is related, will all be asymptotically stable. We review linear impulse extension equations, state theorems that describe the convergence of their solutions to the associated impulsive solutions, and introduce all the machinery necessary in the development of the time-scale tolerance, stating theoretical results on its existence and how it can be computed in practice. We conclude with two illustrative examples and a discussion of the limitations of the techniques presented, as well as elaborate on the ways they can be improved.

1 Introuction

Impulsive differential equations are frequently employed as models of biological, chemical, and physical systems, among others. The property of these equations that makes them attractive in applications is the impulse effect, which allows for the inclusion of fast dynamics that might otherwise complicate the analysis of the model, were they to be included in a continuous, as opposed to discrete, manner. Most monographs on impulsive differential equations—for example, [1, 5, 6]—state that these equations are reasonable approximations of continuous models with perturbations, if the perturbations themselves occur quickly, relative to the overall dynamics. In practice, this is often taken as an assumption of the model in question. Methods to determine how quickly these perturbations must occur for the model

K.E.M. Church (✉)
Department of Mathematics and Statistics, University of Ottawa, 585 King Edward, Ottawa, ON K1N6N5, Canada
e-mail: church.kevin@gmail.com

© Springer International Publishing Switzerland 2016

J. Bélair et al. (eds.), *Mathematical and Computational Approaches in Advancing Modern Science and Engineering*, DOI 10.1007/978-3-319-30379-6_19

to be a good "fit" to the associated continuous models have yet to be seen in the literature.

In these proceedings, we introduce a quantity that we call the *time-scale tolerance*, denote $\mathscr{E}_t$, for a linear, periodic impulsive differential equation that is asymptotically stable. This quantity has the property, that, if the vector of durations of impulse effect, a, satisfies $||a|| < \mathscr{E}_t$, then both the impulsive model and a family of continuous impulse extension equations (a specific functional differential equation) to which it is related, will all be asymptotically stable.

We review linear impulse extension equations, which were first introduced in [2–4], state theorems that describe the convergence of their solutions to the associated impulsive solutions, and introduce all the machinery necessary in the development of the time-scale tolerance, stating theoretical results on its existence and how it can be computed in practice. We conclude with two illustrative examples and a discussion of the limitations of the present techniques, and how they can be extended to accomodate a larger class of problems.

2 Linear Impulse Extension Equations

To begin, we introduce some notation that will be present throughout this chapter. If $x = \{x_k : k \in \mathbb{Z}\}$ is a real-valued sequence, we denote $\Delta x_k = x_{k+1} - x_k$. The kth element of a real-valued sequence x will always be denoted x_k, and we may abuse notation and identify the sequence x with the symbol x_k. Indexed families of sequences, such as, $\{x^j : j \in U\}$, will always have their index appear in the exponent. In this context, x_k^j denotes the kth element of sequence j from the family U. Finally, our sequences will usually be bi-infinite; that is, indexed by the integers. The symbol $||\cdot||$ will denote a (fixed) Euclidean norm, whenever there is no ambiguity, and if A is a set, its closure will be denoted $\overline{A}$.

Consider a linear, impulsive differential equation with impulses at fixed times

$$\begin{aligned} \frac{dx}{dt} &= A(t)x + g(t), && t \neq \tau_k \\ \Delta x &= B_k x + h_k, && t = \tau_k. \end{aligned} \tag{1}$$

with $t \in \mathbb{R}$, phase space $\Omega \subset \mathbb{R}^n$, $A : \mathbb{R} \to \mathbb{R}^{n\times n}$, $B_k \in \mathbb{R}^{n\times n}$, $h_k \in \mathbb{R}^n$, and sequence of impulses τ_k for $k \in \mathbb{Z}$. We assume that A and g are locally integrable, and that the sequence of impulse times, τ_k, is monotone increasing and unbounded.

An *impulse extension equation for* (1), as described in Church and Smith? [3, 4] and Church [2], is defined as follows.

Definition 1 Consider a linear impulsive differential equation (1).

- A *step sequence over* τ_k is sequence of positive real numbers $a = \{a_k : k \in \mathbb{Z}\}$ such that $a_k \in (0, \Delta\tau_k)$ for all $k \in \mathbb{Z}$. We denote $S_j = S_j(a) \equiv [\tau_j, \tau_j + a_j)$ and

$S = S(a) \equiv \bigcup_{j\in\mathbb{Z}} S_j$. The set of all step sequences will be denoted S^*, and is defined by

$$S^* := \{a : \mathbb{Z} \to \mathbb{R}\ ,\ a_k \in (0, \Delta\tau_k)\}.$$

- The pair $(\varphi_k^B, \varphi_k^h)$, with sequences of functions $\varphi_k^B : \mathbb{R} \times \mathbb{R}_+ \to \mathbb{R}^{n\times n}$ and $\varphi_k^h : \mathbb{R} \times \mathbb{R}_+ \to \mathbb{R}^n$, is a *family of impulse extension for* (1) if for all $a \in S^*$ and all $k \in \mathbb{Z}$, the functions $\varphi_k^B(\cdot, a_k)$ and $\varphi_k^h(\cdot, a_k)$ are integrable on $S_k(a)$ and satisfy the equalities

$$\int_{S_k(a)} \varphi_k^B(t, a_k)dt = B_k, \qquad \int_{S_k(a)} \varphi_k^h(t, a_k)dt = h_k. \tag{2}$$

- Given a step sequence $a \in S^*$ and a family of impulse extensions $\varphi = (\varphi_k^B, \varphi_k^h)$ for (1), the *impulse extension equation associated to* (1) *and induced by* (φ, a) is the (functional) differential equation

$$\frac{dx}{dt} = \begin{cases} A(t)x + g(t), & t \notin S(a), \\ A(t)x + g(t) + \varphi_k^B(t, a_k)x(\tau_k) + \varphi_k^h(t, a_k), & t \in S_k(a). \end{cases} \tag{3}$$

Definition 2 Let a family of impulse extensions, $\varphi = (\varphi_k^B, \varphi_k^h)$, and a step sequence $a \in S^*$ be given. A function $y : I \to \mathbb{R}^n$ defined on an interval $I \subset \mathbb{R}$ is a *classical solution* of the impulse extension equation (3) induced by (φ, a), if y is continuous, the sets $I \cap S_k(a)$ are either empty or contain τ_k, and y satisfies the differential equation (3) almost everywhere on I. Given an *initial condition*

$$x(t_0) = x_0, \tag{4}$$

with $(t_0, x_0) \in \mathbb{R}\times\mathbb{R}^n$, the function $y(t)$ is a solution of the *initial-value problem* (3)–(4) if, in addition, $y(t_0) = x_0$.

Remark 1 Definitions 1 and 2 can be readily modified to accomodate nonlinear ordinary impulsive differential equations; see Church and Smith? [4] and Church [2].

Definition 3 The *predictable set* of the impulse extension equation (3) induced by (φ, a) is

$$\mathscr{P} = \mathbb{R} \setminus \left\{ t \in \overline{S(a)} : \det\left(I + \int_{\max_{\tau_k}\{\tau_k \le t\}}^{t} X^{-1}(s, \tau_k)\varphi_k^B(s, a_k)ds \right) = 0 \right\},$$

where $X(t, s)$ is the Cauchy matrix of the homogeneous system $z' = A(t)z$.

The following theorem states the mode in which solutions of the impulse extension equation, (3), converge to those of the impulsive differential equation, (1).

Theorem 1 *Suppose* $\det(I + B_k) \neq 0$ *for all* $k \in \mathbb{Z}$*, and let* $\varphi = (\varphi_k^B, \varphi_k^h)$ *be a given family of impulse extensions for* (1)*. There exists a positive sequence of real numbers* σ_k*, depending only on* $A(t)$ *and sequence of impulse times* τ_k*, with the following property. Suppose, for* $\xi \in \{B, h\}$ *and each* $k \in \mathbb{Z}$*, there exists* $w_k^\xi : [\tau_k, \tau_{k+1}] \times [0, \Delta\tau_k) \to \mathbb{R}$ *that is continuous and vanishing at* $(\tau_k, 0)$*, for which*

$$\varphi_k^\xi(t, s) - \frac{1}{s}\xi_k = O\left(w_k^\xi(t, s)\frac{1}{e^{\sigma_k s} - 1}\right) \tag{5}$$

for $t \in [\tau_k, \tau_k + s)$ *as* $s \to 0$*. The following are true:*

- *For all* $t_0 \in \mathbb{R}$*, there exists* $\delta > 0$*, such that, for* $a \in S^*$ *with* $||a||_\infty < \delta$ *and all* $x_0 \in \mathbb{R}^n$*, the impulse extension equation* (3) *induced by* (φ, a) *possesses a unique classical solution,* $x(t; a)$*, satisfying the initial condition* $x(t_0) = x_0$.
- *The function* $x(t; a)$ *converges pointwise to* $x(t; 0)$*, the solution the initial value problem* $x(t_0) = x_0$ *for the impulsive differential equation,* (1)*, as* $a \to 0$.
- *If* $N \subset \mathbb{R}$ *is bounded and no strictly decreasing sequence in* N *has an impulse time* τ_k *as its limit, the above convergence is uniform for* $t \in N$.

Proof (Outline) The existence of the sequence σ_k follows from the Generalized Gronwall's inequality: we have $||X(t; \tau_k)|| \leq e^{\sigma_k(t-\tau_k)}$ for $t \in [\tau_k, \tau_{k+1}]$, where $\sigma_k = \int_{\tau_k}^{\tau_{k+1}} ||A(s)||ds$, and $X(t; s)$ is the Cauchy matrix of $x' = A(t)x$. When $||a||_\infty < t_0 - \max\{\tau_k : \tau_k < t_0\}$, the solution of (3)–(4) induced by (φ, a) satisfying $x(t_0; a) = x_0$, can be written as

$$x(t; a) = U(t; a)x_0 + x_p(t; a),$$

for a matrix function $t \mapsto U(t; a)$ satisfying $U(t_0; a) = I$, and $x_p(t_0; a) = 0$ (this follows by Proposition 4.2 of [4]). If $U(t; 0)$ denotes the fundamental matrix solution of the homogeneous equation associated to (1), one can show that the inequality

$$\begin{aligned}||U(t, a) - U(t, 0)|| \leq &||X(t; \tau_k)|| \cdot \left|\left|\left(L_a(t; \tau_k) \prod_{r=k-1}^{0} X(\tau_{r+1}; \tau_r)L_a(\tau_r + a_r; \tau_r)\right)\right.\right.\\ &\left.\left. - (I + B_k)\left(\prod_{r=k-1}^{0} X(\tau_{r+1}; \tau_r)(I + B_r)\right)\right|\right|\end{aligned}$$

holds, where we have assumed $t_0 = \tau_0$ for ease of presentation (other cases follow by similar reasoning, by results from [4]), and

$$L_a(t; \tau_k) = I + \int_{\tau_k}^{\min\{t,\tau_k+a_k\}} X^{-1}(s; \tau_k)\varphi_k^B(s, a_k)ds.$$

It can be shown that the right-hand side of the upper bound converges to zero as $a \to 0$ pointwise, as $||a||_\infty \to 0$ (see the proof of Theorem 3.5.5. from

[2] for the main idea; condition (5) is needed). A similar inequality holds for $||x_p(t; a) - x_p(t; 0)||$, where $x_p(t; 0)$ is the solution of (1) satisfying $x_p(t_0; 0) = 0$, and the convergence result holds for that piece of the solution as well. For uniform convergence, it suffices to consider N to be a finite union of closed intervals with $x_n \downarrow x \in N \Rightarrow x \notin \{\tau_k\}$.

The hypotheses of Theorem 1 are simplified if the equations (1) and (3) are periodic.

Definition 4 The linear impulsive differential equation (1) is *T-periodic with c impulses per period* if $A(t + T) = A(t)$ and $g(t + T) = g(t)$ for all $t \in \mathbb{R}$, and $\tau_{k+c} = \tau_k + T$, $B_{k+c} = B_k$ and $h_{k+c} = h_k$ for all $k \in \mathbb{Z}$. The step sequence $a \in S^*$ is *c-periodic*, and we write $a \in S_c^*$, if $a_{k+c} = a_k$ for all $k \in \mathbb{Z}$. The family of impulse extensions $\varphi = (\varphi_k^B, \varphi_k^h)$ is *(T, c)-periodic* if $\varphi_{k+c}^{\xi}(t + T, s) = \varphi_k^{\xi}(t, s)$ for all $t \in \mathbb{R}$, all $k \in \mathbb{Z}$, all $s \in (0, \Delta\tau_k)$ and $\xi \in \{B, h\}$.

Corollary 1 *Suppose the impulsive differential equation* (1) *is T-periodic with c impulses per period. Let* $\varphi = (\varphi_k^B, \varphi_k^h)$ *be a (T, c)-periodic family of impulse extensions for* (1). *Suppose* $\det(I + B_k) \neq 0$ *for* $k = 0, \ldots, c - 1$. *Let the impulsive differential equation* (1) *have a fundamental matrix* $X(t)$ *with Floquet decomposition* $X(t) = P(t)e^{\Lambda t}$ *satisfying* $X(\tau_0) = I$. *The conclusions of Theorem 1 hold for step sequences* $a \in S_c^*$, *with* $\sigma_k \equiv ||\Lambda||$.

The proof of the above corollary is omitted, since it is simple to prove using Theorem 1. For periodic impulse extension equations, we have an asymptotic Floquet theorem. A proof is available in [2], where it is listed as Theorem 3.6.16.

Theorem 2 *Suppose the impulsive differential equation* (1) *is T-periodic with c impulses per period. Let* $\varphi = (\varphi_k^B, \varphi_k^h)$ *be a (T, c)-periodic family of impulse extensions for* (1). *Let* $\det(I + B_k) \neq 0$ *for* $k = 0, \ldots, c - 1$. *Then, under the hypotheses of Corollary 1 on the asymptotic criterion* (5), *there exists* $\delta > 0$ *such that, if* $a \in S_c^*$ *satisfies* $||a|| < \delta$, *any solution* $x(t)$ *of the* homogeneous impulse extension equation induced by (φ, a),

$$\frac{dx}{dt} = \begin{cases} A(t)x, & t \notin S(a) \\ A(t)x + \varphi_k^B(t, a_k)x(\tau_k), & t \in S_k(a). \end{cases} \tag{6}$$

can be written as a product,

$$x(t) = U_a(t)x_0 = P_a(t)e^{\Lambda_a t}x_0, \tag{7}$$

for some $x_0 \in \mathbb{R}^n$, *T-periodic matrix* $P_a(t)$, *and nonsingular* Λ_a. U_a *can be normalized so that* $U_a(\tau_0) = I$, *and in this case, we have* $\Lambda_a \to \Lambda_0$ *as* $a \to 0$, *where* Λ_0 *is the matrix appearing in the Floquet decomposition,* $X(t) = U_0(t) = P_0(t)e^{\Lambda_0 t}$, *of the homogeneous equation associated to the periodic impulsive differential equation,* (1), *with* $U_0(\tau_0) = I$.

The stability of the periodic linear impulse extension equation induced by some (φ, a) is determined by the spectrum of Λ_a, just as with ordinary and impulsive differential equations. The main difference is that stability (and uniform stability) only holds for initial conditions in particular subsets of the predictable set, $\mathscr{P}$, and such restrictions are in fact, optimal. For details, see [4].

3 The Time-Scale Tolerance for Linear, Periodic Impulsive Differential Equations

Stability of (3) is completely determined by the associated homogeneous equation (6); see [4]. As such, our investigation will now shift to homogeneous, (T, c)-periodic impulsive differential equations,

$$\begin{aligned} \frac{dx}{dt} &= A(t)x, & t &\neq \tau_k \\ \Delta x &= B_k x, & t &= \tau_k. \end{aligned} \tag{8}$$

and impulse extension equations for (8), induced by $(\varphi, a) = (\varphi^B, a)$, with $a \in S_c^*$,

$$\frac{dx}{dt} = \begin{cases} A(t)x, & t \notin S(a) \\ A(t)x + \varphi_k^B(t, a_k)x(\tau_k), & t \in S_k(a). \end{cases} \tag{9}$$

From here onward, M_0 will denote the monodromy matrix for (8) satisfying $M_0 = X(\tau_0 + T, \tau_0)$, where $X(t,s)$ is the Cauchy matrix for (8). We assume $\rho M_0 < 1$ from here onward. We will comment in Sect. 3.3 on what can be done if $\rho M_0 \geq 1$.

Definition 5 Consider a periodic homogeneous impulsive differential equation, (8). Let $\sigma = \{\sigma_k\}$ be a c-element sequence of positive real numbers and $w = \{w_k\}$ be a c-element sequence of functions $w_k : [\tau_k, \tau_{k+1}] \times \overline{S_c^*} \to \mathbb{R}_+$ that are continuous and vanishing at $(\tau_k, 0)$ and such that $w_k(\cdot, a)$ is integrable on $S_k(a)$. A family of periodic impulse extensions, $\varphi = \{\varphi_k\}$, is *uniformly exponentially* (σ, w)*-regulated in the mean* or simply (σ, w)*-regulated* if the inequality

$$\left\| \varphi_k(s, a) - \frac{1}{a_k} B_k \right\| \leq \frac{w_k(s, a)}{e^{\sigma_k a_k} - 1} \tag{10}$$

is satisfied, for all $s \in S_k(a)$ and $k = 0, \ldots, c-1$. A pair (σ, w) that satisfies the above criteria will be referred to as a *uniform exponential regulator*, or simply *exponential regulator*. If φ is uniformly (σ, w)-regulated, we will write $\varphi \in (\sigma, w)$.

Definition 6 If $R = (\sigma, w)$ is an exponential regulator, the (R, a)*-pseudospectral radius of* (8), denoted $\rho(R, a)$, is defined by

$$\rho(R, a) = \sup_{\varphi \in R} \rho M(\varphi, a), \tag{11}$$

and $M(\varphi, a)$ denotes the monodromy matrix of the impulse extension equation for (8) induced by (φ, a).

Definition 7 Suppose (8) is asymptotically stable. Let R be an exponential regulator. The R*-stable set*, denoted $\mathscr{E}_s(R)$, is defined as follows.

$$\mathscr{E}_s(R) = \left\{a \in S_c^* : \forall \varphi \in (\sigma, w), \rho M(\varphi, a) < 1\right\} \tag{12}$$

The R*-time-scale tolerance* is the number

$$\mathscr{E}_t(R) = \sup\{\epsilon : \exists a \in \mathscr{E}_s(R), ||a|| = \epsilon, B_\epsilon(0) \cap S_c^* \subseteq \mathscr{E}_s(R)\}. \tag{13}$$

The time-scale tolerance is defined precisely so that we have the following elementary property, whose proof we omit.

Proposition 1 *Given an exponential regulator* $R = (\sigma, w)$*, the time-scale tolerance behaves as a robust stability threshold for the impulsive system* (1)*; if* $||a|| < \mathscr{E}_t(R)$*, then* $\rho(R, a) < 1$*. In other words, systems* (8) *and the impulse extension equation* (6) *induced by* (φ, a) *are both stable, for all* $\varphi \in R$.

Theorem 3 *Suppose* $\sigma = ||\Lambda||$*, as in Corollary 1. If* $R_\sigma = (\sigma, w)$ *is an exponential regulator and* (8) *is asymptotically stable, then* $\mathscr{E}_t(R_\sigma)$ *is nonzero and the map* $a \mapsto \rho(R_\sigma, a)$ *satisfies*

$$\lim_{a \to 0} \rho(R_\sigma, a) = \rho M_0,$$

where the limit is for $a \in S_c^*$.

Proof (Outline) The monodromy matrix for (6) induced by (φ, a) for $\varphi \in R_\sigma$ can be written as

$$M(\varphi, a) = \prod_{k=c-1}^{0} X(\tau_{k+1}; \tau_k) \left(\int_{S_k(a)} X^{-1}(s; \tau_k) \epsilon_k(s, a) ds \right.$$
$$\left. + \frac{1}{a_k} \int_{S_k(a)} I + X^{-1}(s; \tau_k) B_k ds \right),$$

with $\epsilon_k(s, a) = \varphi_k(s, a) - \frac{1}{a_k} B_k$. Taking norms, each of the ϵ_k terms can be bounded by inequality (10), and the upper bound is independent on the explicit choice of φ, depending only on the regulator R_σ. With the choice of σ given in the theorem, each intergral involving ϵ_K converges to zero, while the other clearly converges to $I + B_k$. Therefore, $M(\varphi, a) \to M_0$ uniformly for $\varphi \in R_\sigma$. The result follows.

In practice, computing the time-scale tolerance is difficult. We can, thankfully, resort to conservative estimates.

Theorem 4 *Let R be an exponential regulator for* (8). *Suppose there exists a continuous function $n : \overline{S_c^*} \to \mathbb{R}_+$ such that $n(0) = 0$ and*

$$||M(\varphi, a) - M_0|| \leq n(a)$$

for all $a \in S_c^*$ and $\varphi \in R$.

1. *Let $\rho_\epsilon M$ denote the ϵ-pseudospectral radius of the matrix M. The following inclusion is valid:*

$$\widehat{\mathscr{E}}_s(R) := \{a \in S_c^* : \rho_{n(a)} M_0 < 1\} \subseteq \mathscr{E}_s(R).$$

2. *Let $h > 0$ denote the unique solution of the equation $\rho_h M_0 = 1$. The inequality*

$$\widehat{\mathscr{E}}_t(R) := \sup\{||a|| : a \in B_{||a||}(0) \cap \widehat{\mathscr{E}}_s(R) \neq \emptyset\} \leq \mathscr{E}_t(R)$$

is valid, and if n is monotone increasing, we have $\widehat{\mathscr{E}}_t(R) = \min\{||a|| : n(a) = h,\ a \in \overline{S_c^*}\}$.

Proof (Outline) By definition of the pseudospectral radius, we have

$$\begin{aligned}\rho_{n(a)} M_0 &= \sup\{\rho(Z) : Z \in \mathbb{R}^{n \times n}, ||Z - M_0|| \leq n(a)\}\\ &\geq \sup\{\rho M(\varphi, a) : \varphi \in R, ||M(\varphi, a) - M_0|| \leq n(a)\}\\ &= \sup\{\rho M(\varphi, a) : \varphi \in R\} = \rho(R, a),\end{aligned}$$

which demonstrates the set inclusion. As for the inequality, that the supremum term is bounded by $\mathscr{E}_t(R)$ is obvious from the set inclusion. That $\widehat{\mathscr{E}}_t(R)$ is achieved at some a for which $n(a) = h$ can be seen by noticing that, as n is continuous and increasing, the set $\widehat{\mathscr{E}}_s(R)$ is star convex with basepoint 0. Consequently, maximizing the radius of a ball in the positive orthant within this set is equivalent to minimizing the distance to the boundary, and the latter is is precisely the level set $n(a) = h$.

Corollary 2 *Denote $X(t) = X(t; \tau_0)$. If $c = 1$, the following inequality holds for all $\varphi \in R = (\sigma, w)$.*

$$\begin{aligned}||M(\varphi, a) - M_0|| \leq ||X(\tau_1)|| \Bigg[&\int_{S_0(a)} ||X^{-1}(s)|| \frac{w_0(s, a)}{e^{\sigma a_0} - 1} ds\\ &+ \left|\left| \frac{1}{a_0} \int_{S_0(a)} (X^{-1}(s) - I) ds B_0 \right|\right| \Bigg]. \end{aligned} \tag{14}$$

Proof (Outline)

$$M(\varphi, a) - M_0 = X(\tau_1)\left[I + \int_{S_0(a)} X^{-1}(s)\left(\varphi(s,a) - \frac{1}{a_0}B_0 + \frac{1}{a_0}B_0\right) ds\right]$$
$$-X(\tau_1)[I + B_0].$$

Re-arranging the above, taking norms and using inequality (10) provides the result.

If $c \neq 1$, a similar estimate to the above holds. However, it is rather cumbersome, and the associated proof is a notationally difficult inductive argument. It is omitted for brevity.

3.1 Example: An Exact Computation for a Scalar Equation

Consider the following scalar impulsive differential equation

$$\begin{aligned} x' &= \sigma x, \quad && t \neq kT \\ \Delta x &= -bx, \quad && t = kT, \end{aligned} \tag{15}$$

with parameters $\sigma > 0$, $b > 0$ and $T > 0$. Assume $M_0 = (1-b)e^{\sigma T} < 1$, so that the trivial solution is asymptotically stable. We choose $w(t,a) = c\left(\frac{a}{T}\right)^{\frac{1}{p}}$ for parameters c and $p > 0$. The bound on the right-hand side of (14), denote $\tilde{n}(a)$, itself has an upper bound:

$$\tilde{n}(a) \leq n(a,p) := \frac{e^{\sigma T}}{\sigma} c \left(\frac{a}{T}\right)^{\frac{1}{p}} + e^{\sigma T} b \left(1 - \frac{1 - e^{-\sigma a}}{\sigma a}\right),$$

and n is strictly increasing in both a and the parameter p. Therefore,

$$\rho_{n(a,p)} \leq \rho_{n(a,\infty)} = \lim_{p\to\infty} \rho_{n(a,p)} M_0 = e^{\sigma T}\left(1 + \frac{c}{\sigma} - \frac{b}{\sigma a}(1 - e^{-\sigma a})\right) \tag{16}$$

for each finite $p > 0$. Solving the equation $\rho_{n(a^*,\infty)} M_0 = 1$ for a^* and applying Theorem 4, the following theorem is proven.

Theorem 5 *Consider the impulsive system* (15). *Define* $u := \frac{1}{b}\left(e^{-\sigma T} - 1 - \frac{c}{\sigma}\right)$. *If* $M_0 := (1-b)e^{\sigma T} < 1$ *and* $c < (1 - M_0)\sigma e^{-\sigma T}$, *then, for all* $a > 0$ *satisfying the inequality*

$$a < \frac{1}{\sigma}\left(W\left(\frac{1}{u}e^{\frac{1}{u}}\right) - \frac{1}{u}\right) := a^*,$$

we have $\rho M(\varphi, a) < 1$, *for all* $\varphi \in (\sigma, w)$, *with* $w(t, a) = c\left(\frac{a}{T}\right)^{\frac{1}{p}}$, *for any* $p > 0$, *where* W *is the principal branch of the Lambert* W *function, or product logarithm function (i.e. the inverse of the map* $x \mapsto xe^x$*).*

3.2 Example: Control of a Pest with Age Structure

Consider the following system of impulsive differential equations.

$$X' = \begin{bmatrix} -10/21 & 5/7 \\ 1/4 & -1/7 \end{bmatrix} X, \qquad t \neq \tau_k \tag{17}$$

$$\Delta X = \begin{bmatrix} -0.7 & 0 \\ 0 & -0.4 \end{bmatrix} X, \qquad t = \tau_{2k} \tag{18}$$

$$\Delta X = \begin{bmatrix} 0 & 0 \\ 0 & -0.7 \end{bmatrix} X, \qquad t = \tau_{2k+1}, \tag{19}$$

with sequence of impulses $\tau_k = 7\lfloor k/2 \rfloor + (k \mod 2)$ and t in units of days. The continuous dynamics, (17), could describe, for example, the population of some pest organism, $X = (X_1, X_2) \geq 0$, with juvenile (X_1) and adult (X_2) life stages. With both impulsive controls included, (17), (18) and (19) is asymptotically stable, with dominant Floquet multiplier equal to 0.4200. If one or both controls are neglected, however, the trivial solution is unstable, so both controls are necessary to control the population.

Figure 1 provides visualizations of the subsets $\widehat{\mathscr{E}_s} \subseteq \mathscr{E}_s$ described in Theorem 4, of the R-stable sets for two uniform exponential regulators for the system (17), (18) and (19). The first regulator, which generates the smaller of the two stable sets (red in the figure), is $R = (\sigma, \sqrt{a_k})$. The second regulator is $R = (\sigma, a_k)$. For both regulators, $\sigma = ||\Lambda||$, as in Corollary 1.

3.3 Limitations

There are two main limitations of the techniques described in these proceedings. First and foremost, only linear systems are treated. The time-scale tolerance can indeed be defined for nonlinear systems of impulsive differential equations in more abstract settings, although the definitions must all be localized around periodic orbits or other stable objects. Some of our current research concerns these problems.

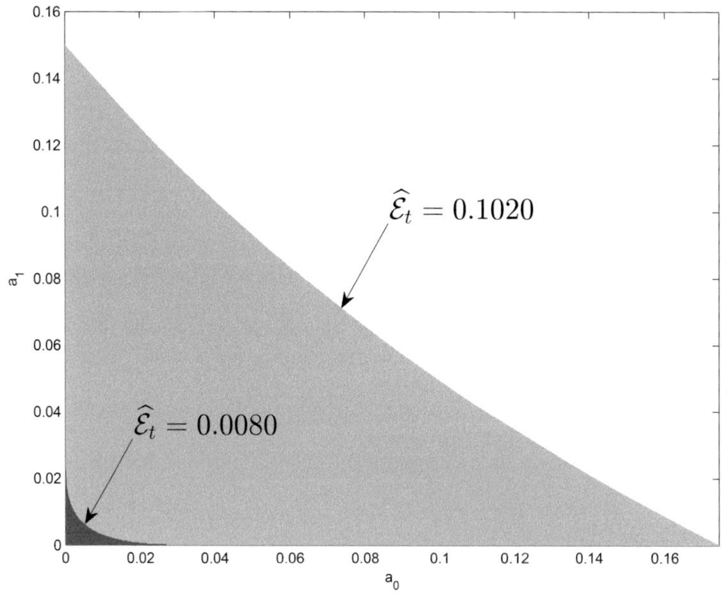

Fig. 1 Conservative approximations, $\widehat{\mathscr{E}}_s$, of the R-stable sets for two uniform exponential regulators. *Arrows* indicate the associated lower bounds for time-scale tolerances

Second, we treated only impulsive systems that are asymptotically stable. These techniques generally fail in the presence of a center subspace; see Example 3.5.6 of [2]. However, a similar approach does work if there is an unstable subspace. For example, if $\rho M_0 > 1$, one might want to know conditions on $a \in S_c^*$ under which $\rho M(\varphi, a) > 1$, for all $\varphi \in R$, with R some suitable set of impulse extensions. If, for some continuous function n satisfying $n(0) = 0$, we have $||M(\varphi, a) - M_0|| \leq n(a)$ for all $\varphi \in R$, one can verify the string of inequalities

$$\begin{aligned}\inf_{\varphi \in R} \rho M(\varphi, a) &\geq \inf\left\{\rho M : ||M - M_0|| \leq \sup_{\varphi \in R} ||M(\varphi, a) - M_0||\right\} \\ &= \inf\{\rho M : ||M - M_0|| \leq \inf\{x : ||M(\varphi, a) - M_0|| \leq x,\ \forall \varphi \in R\}\} \\ &\geq \inf\{\rho M : ||M - M_0|| \leq n(a)\} := \rho^-_{n(a)} M_0\end{aligned}$$

holds. Therefore, an appropriate lower estimate for the analoguous time-scale tolerance can be found by solving the optimization problem

$$\widehat{\mathscr{E}}_t(R) := \sup\{||a|| : a \in B_{||a||}(0) \subseteq \{a \in S_c^* : \rho^-_{n(a)} M_0 > 1\} \neq \emptyset\}.$$

References

1. Bainov, D.D., Simeonov, P.S.: Impulsive Differential Equations: Periodic Solutions and Applications. Longman Scientific & Technical, Burnt Mill (1993)
2. Church, K.: Applications of impulsive differential equations to the control of malaria outbreaks and introduction to impulse extension equations: a general framework to study the validity of ordinary differential equation models with discontinuities in state. M.Sc Thesis, University of Ottawa (2014)
3. Church, K.E.M., Smith?, R.J.: Analysis of piecewise-continuous extensions of periodic linear impulsive differential equations with fixed, strictly inhomogeneous impulses. Dyn. Contin. Discret. Impuls. Syst. Ser. B: Appl. Algorithms **21**, 101–119 (2014)
4. Church, K.E.M., Smith?, R.J.: Existence and uniqueness of solutions of general impulse extension equations with specification to linear equations. Dyn. Contin. Discret. Impuls. Syst. Ser. B: Appl. Algorithms **22**, 163–197 (2015)
5. Lashmikantham, V., Bainov, D.D., Simeonov, P.S.: Theory of Impulsive Differential Equations. World Scientific Publishing, Singapore (1989)
6. Samoilenko, A.M., Perestyuk, N.A.: Impulsive Differential Equations. World Scientific Publishing, Singapore (1995)

Coupled Lattice Boltzmann Modeling of Bidomain Type Models in Cardiac Electrophysiology

S. Corre and A. Belmiloudi

Abstract In this work, a modified coupling Lattice Boltzmann Model (LBM) in simulation of cardiac electrophysiology is developed in order to capture the detailed activities of macro- to micro-scale transport processes. The propagation of electrical activity in the human heart is mathematically modelled by bidomain type systems. As transmembrane potential evolves, we take into account domain anisotropical properties using intracellular and extracellular conductivity, such as in a pacemaker or an electrocardiogram, in both parallel and perpendicular directions to the fibers. The bidomain system represents multi-scale, stiff and strongly nonlinear coupled reaction-diffusion models that consists of a set of ordinary differential equations coupled with a set of partial differential equations. Due to dynamic and geometry complexity, numerical simulation and implementation of bidomain type systems are extremely challenging conceptual and computational problems but are very important in many real-life and biomedical applications. This paper suggests a modified LBM scheme, reliable, efficient, stable and easy to implement in the context of such bidomain systems. The numerical results demonstrate the effectiveness and accuracy of our approach using general methods for bidomain type systems and show good agreement with analytical solutions and numerical results reported in the literature.

1 Introduction

Computational cardiac electrophysiological modeling is now an important field in applied mathematics. Indeed, nowadays, heart and cardiovascular diseases are still the leading cause of death and disability all over the world. That is why we need to improve our knowledge about heart behavior, and more particularly about its electrical behavior. Consequently we want strong methods to compute electrical fluctuations in the myocardium to prevent cardiac disorders (as arrhythmias), or to study interactions between brain and heart. We modelize electrical behavior

S. Corre • A. Belmiloudi (✉)
UEB-IRMAR, Rennes, France
e-mail: Samuel.Corre@insa-rennes.fr; Aziz.Belmiloudi@math.cnrs.fr

© Springer International Publishing Switzerland 2016

J. Bélair et al. (eds.), *Mathematical and Computational Approaches in Advancing Modern Science and Engineering*, DOI 10.1007/978-3-319-30379-6_20

of the heart in the myocardium with the bidomain system, derived from Ohm's law. This biophysical model of electrical cardiac activity links electrophysiological cell models, at small scales, and myocardial tissue mechanics, metabolism and blood flow at large scales. In mathematical viewpoint, bidomain system leads us to compute intracellular and extracellular electrical potentials ρ_i and ρ_e with taking into account the cellular membrane dynamics U. This is a system of non-linear partial differential equations (PDEs) coupled with ordinary differential equations (ODEs). The PDEs describe the propagation of the electrical potentials and ODEs describe the electrochemical processes. During last years a lot of studies about bidomain models have led to results about well-posedness, existence and uniqueness of solutions (see e.g., [2, 3] and the references therein), and several numerical methods based on methods as finite difference method or finite element method are used to solve these models (see e.g., [8] and the references therein). In this paper, we propose a modified Lattice Boltzmann Method (LBM) which is simple to implement, effective, accurate and well suited to bidomain systems which is a coupled nonlinear parabolic/elliptic PDEs. LBM is based on microscopic models and mesoscopic kinetic equations. Indeed, traditional numerical methods as finite difference method or finite element method directly solve governing equations for deriving macroscopic variable, whereas LBM is based on the particle (the discrete) distribution function and numerical solving the continuous Boltzmann transport equation. Then the macroscopic variables of the bidomain system can be recovered from the discrete equations through the multi-scaling Chapman-Enskog expansion procedure. LBM was originated from Boltzmann's kinetic theory of gases (1970s), and attracts more and more attentions for simulating complex fluid flows since 1990s. More recently, LBM has been extended successfully to simulate different types of parabolic reaction-diffusion equation as Keller-Segel chemotaxis model [10] and monodomain model in cardiac electrophysiology [4], or Poisson equation [5].

This paper is organized as follows: in Sect. 2 we recall briefly the derivation of the bidomain model. In Sect. 3 we present and describe the modified LBM method. In Sect. 4, the validity of this method is demonstrated by comparing the numerical solution to the exact solution of bidomain model with a classical FitzHugh-Nagumo model (FHN), and convergence of solution is established. Some interesting numerical simulations to analyze the influence of some parameters on electrical wave propagation, including the bidomain model with a modified FHN model, are also carried out in this section. This paper is ended by a conclusion and some further works.

2 The Bidomain Model

The bidomain model of cardiac tissue is expressed mathematically by the following steady-state of coupled partial differential equations governing the electrical potentials (in the physical region Ω occupied by the excitable cardiac tissue, which is an open, bounded, and connected subset of $\mathbf{R}^d$, $d \leq 3$ and during a time interval $(0, T)$)

$$div(\mathscr{K}_i \nabla \rho_i) = I_m - \kappa f_{is}, \quad div(\mathscr{K}_e \nabla \rho_e) = -I_m - \kappa f_{es}, \tag{1}$$

where ρ_i and ρ_e are the intracellular and extracellular potentials, respectively; $\mathscr{K}_i(x)$ and $\mathscr{K}_e(x)$ are the conductivity tensors describing the anisotropic intracellular and extracellular conductive media; $f_{is}(x,t)$ and $f_{es}(x,t)$ are the respective externally applied current sources. The transmembrane current density is described by I_m and is given by the following expression:

$$I_m = \kappa(c_m \frac{\partial \rho}{\partial t} + \mathscr{I}_{ion}), \tag{2}$$

where ρ is the transmembrane potential, which is defined as $\rho = \rho_i - \rho_e$, κ is the ratio of the membrane surface area to the volume occupied by the tissue, c_m term is the transmembrane capacitance time unit area. The tissue is assumed to be passive, so the capacitance c_m can be assumed to be not a function of the state variables. The nonlinear operator $\mathscr{I}_{ion}(x, t; \rho, U)$ describes the sum of transmembrane ionic currents across the cell membrane with U the electrophysiological ionic state variables (which describe e.g., the dynamics of ion-channel and ion concentrations in different cellular compartments). These variables satisfy the following ODE (with H a nonlinear operator)

$$\frac{\partial U}{\partial t} = H(x, t; \rho, U). \tag{3}$$

From (1), (2) and (3), the bidomain model can be formulated in terms of the state variables ρ, ρ_e and U as follows (in $\mathscr{Q} = \Omega \times (0, T)$)

$$\begin{aligned}
&\kappa(c_m \frac{\partial \rho}{\partial t} + \mathscr{I}_{ion}(.; \rho, U)) - div(\mathscr{K}_i \nabla \rho) = div(\mathscr{K}_i \nabla \rho_e) + \kappa f_{is}, \\
&-div((\mathscr{K}_e + \mathscr{K}_i)\nabla \rho_e) = div(\mathscr{K}_i \nabla \rho) + \kappa(f_{es} + f_{is}), \\
&\frac{\partial U}{\partial t} = H(.; \rho, U).
\end{aligned} \tag{4}$$

The operators $\mathscr{I}_{ion}$ and H which describe electrophysiological behavior of the system have usually the following form (affine functions with respect to U)

$$\mathscr{I}_{ion}(.; \rho, U) = I_1(.; \rho) + I_2(.; \rho)U, \quad H(.; \rho, U) = H_0(.; \rho) + \lambda(.)U. \tag{5}$$

To close the system we impose the following boundary conditions

$$(\mathscr{K}_i \nabla(\rho + \rho_e)).\mathbf{n} = \zeta_i, \quad (\mathscr{K}_e \nabla \rho_e).\mathbf{n} = \zeta_e \quad \text{on } \Sigma = \partial\Omega \times (0, T), \tag{6}$$

where $\mathbf{n}$ being the outward normal to $\Gamma = \partial\Omega$ and ζ_i and ζ_e are the intra- and extracellular currents per unit area applied across the boundary, and the following initial conditions (in Ω)

$$\rho(t = 0) = \rho_0, \;\; U(t = 0) = U_0. \tag{7}$$

Such problems have compatibility conditions determining whether there are any solutions to the PDEs. This is easily found by integrating the second equation of (4) over the domain and using the divergence theorem with the boundary conditions (6) (a.e. in (0, T)). Then (for compatibility reasons), we require the following condition

$$\int_{\Gamma} (\zeta_i + \zeta_e) d\Gamma + \kappa \int_{\Omega} (f_{es} + f_{is}) dx = 0. \tag{8}$$

Moreover, the function ρ_e is defined within a class of equivalence, regardless of a time-dependent function. This function can be fixed, for example by setting the Gauge condition (a.e. in (0, T))

$$\int_{\Omega} \rho_e dx = 0. \tag{9}$$

Under some hypotheses for the data and parameters of the system and some regularity of operators $\mathscr{I}_{ion}$ and H, system (4) with (6)–(7) and under the conditions (8)–(9) is a well-posed problem (for more details see [2]).

3 Numerical Method and Algorithm

In this section, a numerical method is presented for the bidomain system (4) in two space dimensions. For this, we introduce a coupled modified LBM for solving the coupled system of nonlinear parabolic and elliptic equations (i.e. the first and the second equations of (4)). Then we treat the ODE satisfied by ionic state by applying Gronwall Lemma to obtain an integral formulation, and by using a quadrature rule to approximate the obtained integral. In the sequel, without loss of generality, we assume $c_m = 1$ and $\kappa = 1$. Moreover we assume $\mathscr{K}_i = K_i I_d$, $\mathscr{K}_e = K_e I_d$, with I_d identity matrix and K_i, K_e constants.

Remark 1 The developed LBM method has been constructed to take into account the case in which $\mathscr{K}_i = \mathscr{K}_i(\rho, \rho_e)$ and $\mathscr{K}_e = \mathscr{K}_e(\rho, \rho_e)$. In order to simplify the presentation, we have assumed in this paper that these operators are constant matrices.

3.1 LBM for Coupled Parabolic and Elliptic Equations

In this first part, we develop and describe the modified LBM to solve the following system (which corresponds to two first parts of (4))

$$\begin{aligned} &\frac{\partial \rho}{\partial t} - div(K_i \nabla(\rho + \rho_e)) = F(.; \rho, \rho_e), \\ &-div(K_i \nabla \rho + (K_i + K_e) \nabla \rho_e) = G(.; \rho, \rho_e), \end{aligned} \tag{10}$$

where F and G are non linear operators.

3.1.1 LBE for General Reaction-Diffusion Equations

To begin, we introduce the LBM to solve the following reaction-diffusion equation with the macroscopic variable Φ

$$\frac{\partial}{\partial t}\Phi(\mathbf{x},t) - div(K\nabla\Phi(\mathbf{x},t)) = H(\mathbf{x},t;\Phi). \tag{11}$$

The evolution equation of the LBM for (11) is given by

$$\frac{\partial}{\partial t}h(\mathbf{x},t;\mathbf{e}) + \mathbf{e}\cdot\nabla h(\mathbf{x},t;\mathbf{e}) = Q(h(\mathbf{x},t;\mathbf{e})) + P(\mathbf{x},t;\mathbf{e}), \tag{12}$$

where $h(\mathbf{x},t;\mathbf{e})$ is the distribution function of particle moving with velocity $\mathbf{e}$ at position $\mathbf{x}$ and time t, P is the distribution type function of particle of macroscopic external force H moving with velocity $\mathbf{e}$ and Q is the Bhatnagar-Gross-Krook (BGK) collision operator defined by $Q(h) = -\frac{1}{\tau}\left(h(\mathbf{x},t;\mathbf{e}) - h^{eq}(\mathbf{x},t;\mathbf{e})\right)$, where h^{eq} is the equilibrium distribution function and τ is the dimensionless relaxation time. LBM leads us to approximate (12) to recover reaction-diffusion equation (11) with Chapman-Enskog expansion. For that, we discretize $\mathscr{Q}$ in time and space. Then we introduce a lattice size cell Δx and a lattice time step size Δt, and we define streaming lattice speed $c = \Delta x/\Delta t$.

The $D2Q9$ lattice, which involves 9 velocity vectors, is considered for applied lattice scheme, which is the most used scheme for two-dimensional model. Figure 1 shows a typical lattice node of $D2Q9$ model with velocities e_i for various directions defined by

$$\mathbf{e}_0 = \begin{pmatrix}0\\0\end{pmatrix}, \ \mathbf{e}_{i=1,2,3;4} = c\begin{pmatrix}\cos\left((i-1)\frac{\pi}{2}\right)\\ \sin\left((i-1)\frac{\pi}{2}\right)\end{pmatrix}, \ \mathbf{e}_{i=5,6,7,8} = \sqrt{2}c\begin{pmatrix}\cos\left((i-\frac{9}{2})\frac{\pi}{2}\right)\\ \sin\left((i-\frac{9}{2})\frac{\pi}{2}\right)\end{pmatrix}.$$

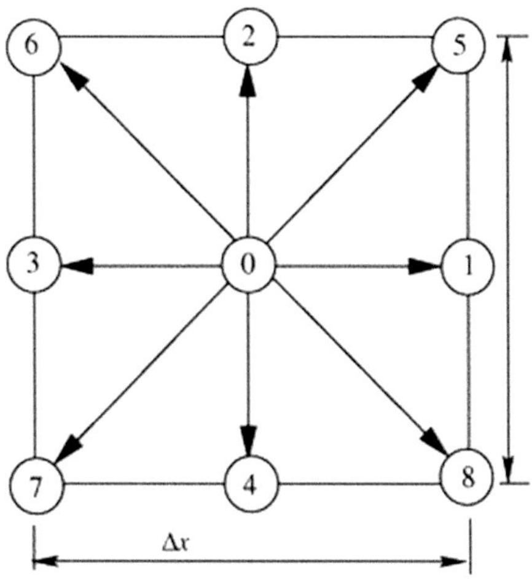

Fig. 1 Particle velocities for D2Q9 LBM

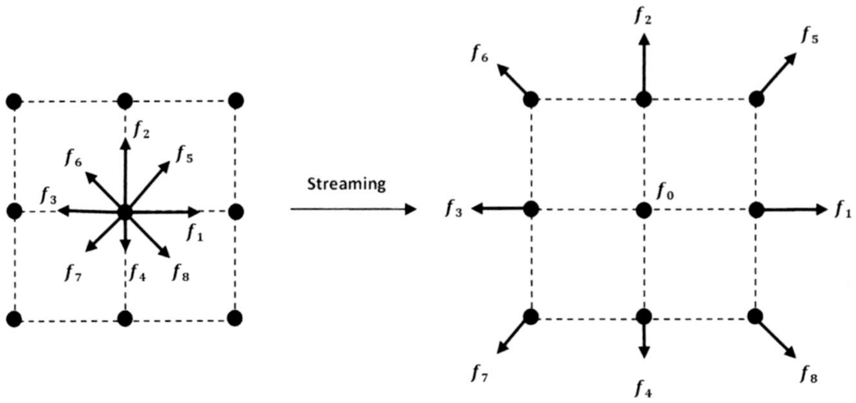

Fig. 2 streaming process of a lattice node

For each particle on the lattice, we associate the discrete distribution functions h_i and h_i^{eq} (in the mesoscopic level), and the discrete operator H_i of H for $i = 0, \ldots, 8$. Then, the form of the Lattice Boltzmann Equation (LBE) with an external force by introducing BGK approximations can be written as follows

$$\begin{aligned} h_i(\mathbf{x} + \mathbf{e}_i \Delta t, t + \Delta t) = h_i(\mathbf{x}, t) - \tfrac{1}{\tau}\left(h_i(\mathbf{x}, t) - h_i^{eq}(\mathbf{x}, t)\right) \\ + \Delta t H_i(\mathbf{x}, t) + \tfrac{\Delta t^2}{2}\tfrac{\partial}{\partial t} H_i(\mathbf{x}, t). \end{aligned} \quad (13)$$

The key steps in LBM, which is directly derived from LBE (13), are the collision and streaming processes (shown on Fig. 2) which are given by

$$h_i^{col}(\mathbf{x}, t) = h_i(\mathbf{x}, t) - \tfrac{1}{\tau}\left(h_i(\mathbf{x}, t) - h_i^{eq}(\mathbf{x}, t)\right) + \Delta t H_i(\mathbf{x}, t) + \tfrac{\Delta t^2}{2}\tfrac{\partial}{\partial t} H_i(\mathbf{x}, t), \quad (14)$$

$$h_i(\mathbf{x} + \mathbf{e}_i \Delta t, t + \Delta t) = h_i^{col}(\mathbf{x}, t). \quad (15)$$

From Chapman-Enskog expansion analysis, the above LBM can recover to the reaction-diffusion equation (11) if we take $h_i^{eq} = w_i \Phi$, $H_i = w_i H$ and the initial distribution at $t = 0$: $h_i(\mathbf{x}, 0) = h_i^{eq}(\mathbf{x}, 0)$. This analysis is based on the following properties:

$$\begin{aligned} &\sum_{i=0}^{8} h_i^{eq}(\mathbf{x}, t) = \sum_{i=0}^{8} h_i(\mathbf{x}, t) = \Phi(\mathbf{x}, t)(\text{macroscopic variable}), \\ &\sum_{i=0}^{8} H_i(\mathbf{x}, t) = H(\mathbf{x}, t), \ \sum_{i=0}^{8} \mathbf{e}_i h_i^{eq}(\mathbf{x}, t) = 0, \ \sum_{i=0}^{8} \mathbf{e}_i H_i(\mathbf{x}, t) = 0, \\ &\sum_{i=0}^{8} \mathbf{e}_i \mathbf{e}_i h_i^{eq}(\mathbf{x}, t) = \frac{c^2}{3}\Phi(\mathbf{x}, t)\mathbf{I}_d, \ \sum_{i=0}^{8} \mathbf{e}_i \mathbf{e}_i H_i(\mathbf{x}, t) = \frac{c^2}{3} H(\mathbf{x}, t)\mathbf{I}_d. \end{aligned} \quad (16)$$

Remark 2 During streaming and collision processes, in order to satisfy boundary conditions, the boundary nodes need special treatments on distribution functions, which are essential to stability and accuracy of the method.

3.1.2 Coupled Modified LBM

To introduce our modified LBM, we have to take into account the coupled terms which link reaction-diffusion equation and elliptic equation in the system (10). As in [10], in order to take into account this coupling, we introduce two correction terms $S_i = w_i\left(\Psi_1(K_i\nabla\rho) + \Psi_2(K_i\nabla\rho_e)\right)$ and $S_i^e = w_i\left(\Psi_1^e(K_i\nabla\rho) + \Psi_2^e((K_i+K_e)\nabla\rho_e)\right)$, where the functions Ψ_1, Ψ_2, Ψ_1^e and Ψ_2^e are determined by Chapman-Enskog expansions. Then, we can solve the reaction-diffusion equation with a first LBE where the distribution function f leads to recover ρ. We construct exactly the same LBM than developed in Sect. 3.1.1. So we choose $f_i^{eq} = w_i\rho$, $F_i = w_iF$ and $\tau = -\frac{3K_i}{c^2\Delta t} + \frac{1}{2}$ to satisfy previous properties (16). Finally, we add the corrector term S_i as follows

$$f_i(\mathbf{x}+\mathbf{e}_i\Delta t, t+\Delta t) = f_i(\mathbf{x},t) - \frac{1}{\tau}\left(f_i(\mathbf{x},t) - f_i^{eq}(\mathbf{x},t)\right) + \Delta t F_i(\mathbf{x},t) + \frac{\Delta t^2}{2}\frac{\partial}{\partial t}F_i(\mathbf{x},t) + \Delta t S_i(\mathbf{x},t). \tag{17}$$

Hence, the macroscopic variable ρ, defined as: $\sum_{i=0}^{8} f_i(\mathbf{x},t) = \rho(\mathbf{x},t)$.

For the elliptic equation, the LBM developed is based on the LBM employed in [5]. The first step is to introduce a new time variable r as $\lim_{r\to\infty}\tilde{\rho}_e(\mathbf{x},r;t) = \rho_e(\mathbf{x},t)$. Then, the equilibrium distribution function is defined as

$$g_i^{eq}(\mathbf{x},r) = \begin{cases} w_i\tilde{\rho}_e(\mathbf{x},r;t) \text{ for } i \neq 0, \\ (w_0-1)\tilde{\rho}_e(\mathbf{x},r;t) \text{ for } i = 0 \end{cases}$$

and we can deduce that $\sum_{i=0}^{8} g_i^{eq}(\mathbf{x},r) = 0$, $\frac{1}{1-w_0}\sum_{i=1}^{8} g_i^{eq}(\mathbf{x},r) = \tilde{\rho}_e(\mathbf{x},r;t)$. Finally, as for previous LBE (17), we add the corrector term S_i^e and we obtain the following LBE (to recover ρ_e)

$$g_i(\mathbf{x}+\mathbf{e}_i\Delta r, r+\Delta r;t) = g_i(\mathbf{x},r;t) - \frac{1}{\tau_e}\left(g_i(\mathbf{x},r;t) - g_i^{eq}(\mathbf{x},r;t)\right) + \Delta t G_i(\mathbf{x},r;t) + \Delta t S_i^e(\mathbf{x},r;t). \tag{18}$$

Hence, the macroscopic variable $\tilde{\rho}_e$, defined as: $\frac{1}{1-w_0}\sum_{i=0}^{8} g_i(\mathbf{x},r;t) = \tilde{\rho}_e(\mathbf{x},r;t)$.

3.2 Treatment of ODE

Now, we present briefly the method to solve ODE satisfy by ionic state U, with initial condition $U(\mathbf{x},0) = U_0(\mathbf{x})$. According to (4) and the form of H given in (5) by $H(\mathbf{x},t;\rho,U) = H_0(\mathbf{x},t;\rho) + \lambda U(\mathbf{x},t)$, with λ assumed to be a constant, and by using Gronwall Lemma we can deduce:

$$U(\mathbf{x},t+\Delta t) = U(\mathbf{x},t)e^{-\lambda\Delta t} + e^{-\lambda t}\int_t^{t+\Delta t} H_0(\mathbf{x},s;\rho(s))e^{\lambda s}ds.$$

Then, according to approximation of derived integral by trapezoidal method between t and $t+\Delta t$ we obtain the following approximation of U denoted also by U

$$U(.,t+\Delta t)=U(.,t)e^{-\lambda\Delta t}+\frac{\Delta t}{2}\left(H_0(.,t+\Delta t;\rho(t+\Delta t))e^{\lambda\Delta t}+H_0(.,t;\rho(t))\right). \tag{19}$$

Finally, after non-dimentionalization, mesh definition and initialization of initial conditions, parameters and data, the proposed algorithm to solve the bidomain system can be summarized as follows

1. Initialization: t=0.
2. LBE according to time t by using (17) to compute $\rho(\mathbf{x},t+\Delta t)$.
3. Trapezoidal method by using (19) to compute $U(\mathbf{x},t+\Delta t)$.
4. Loop on new time variable r:
   a. LBE according to time r by using (18) to compute $\tilde{\rho}_e(\mathbf{x},r+\Delta r;t+\Delta t)$ an approximation of $\rho_e(\mathbf{x},t+\Delta t)$.
   b. If convergence criteria is not reached, set $r=r+\Delta r$ and go back to 4a.
5. Set $\rho_e(\mathbf{x},t+\Delta t) := \tilde{\rho}_e(\mathbf{x},r;t+\Delta t)$.
6. If $t \neq T$, set $t := t+\Delta t$ and go back to 2.

4 Numerical Simulation and Applications

To validate the capacity of our modified coupled LBM to deal with 2D bidomain systems, several situations are numerically simulated. In the first study, we consider the bidomain system with homogeneous Neumann boundary conditions and with a classical FitzHugh-Nagumo model (FHN), in which non linear operators are defined as

$$\mathscr{I} = -1/\alpha_1(\rho-\rho^3/3-U), \quad H = \alpha_2(\rho-\beta_1 U+\beta_2), \tag{20}$$

where α_1, α_2, β_1, β_2 are positive constant parameters respectively called excitation rate constant, recovery rate constant, recovery decay constant and excitation decay constant. First, we consider the system with known analytical solution in order to validate and verify the accuracy and stability of the method. And second, we consider the system as in [8] and we estimate the influence of both constants α_1 and α_2 on the behaviour of the system. In the last application we solve and analyze the system with the following modified FHN model (see e.g., [8])

$$\mathcal{I} = 0.0004(\rho + 85)\,(U + (\rho + 70)(\rho - 40))\,, \quad H = 0.63(\rho + 85) - 0.013U. \tag{21}$$

Nota Bene: If the exact solution ϕ_{sol} is known, we can measure the efficiency of method with the following L^2 relative error: $Err_\phi = \frac{\|\phi_{sol}-\phi\|_{L^2(\Omega)}}{\|\phi_{sol}\|_{L^2(\Omega)}}$.

For simplicity, we assume that the domain is a square region $\Omega = [0;1] \times [0;1]$, the final time is fixed $T = 1$, and the cell surface to volume ratio $\kappa = 1$.

4.1 Benchmark Problem and Validation

In this first analysis, we investigate the accuracy and spatial convergence rate of the proposed modified LBM for which we postulate that the error estimates of the method is of order 2 in space and of order 1 in time (for sufficiently regular solution). We perform a convergence study on cartesian grid, by taking $\alpha_1 = 1$, $\alpha_2 = 1$, $\beta_1 = 1$, $K_i = 1$ and $K_e = 1$, and by setting $\beta_2(x,y,t) = -\rho^{sol}(x,y,t) + 2e^t\cos(\pi(x+y))$. The initial conditions are $\rho(\mathbf{x},0) = 0$, $\rho_e(\mathbf{x},0) = 0$, $U(\mathbf{x},0) = \cos(\pi(x+y))$ and we choose $\mathcal{I}$, I_i and I_e to close the problem according to solution and compatibility conditions. The exact solution is given by: $\rho^{sol}(x,y,t) = tx^2(x-1)^2y^2(y-1)^2$, $\rho_e^{sol}(x,y,t) = t(\cos(\pi x) + \cos(\pi y))$, and $U^{sol}(x,y,t) = e^t\cos(\pi(x+y))$. To study the convergence, we have constructed a sequence of meshes with decreasing spatial step Δx between $1/30$ and $1/200$ and $\Delta t = \Delta x^2$. We just care about relative error on ρ and U at $t = 0.5$ and $t = 1$ (see Table 1). Indeed, the chosen convergence criteria (for the iterative method to

Table 1 Relation between relative error and lattice spacing for ρ and U

t=0.5	Δx	Err_ρ	Err_U
	1/30	0.1914	0.0005
	1/50	0.0389	0.0002
	1/70	0.0116	7.97e−5
	1/100	0.0050	4.48e−5
	1/150	0.0016	1.99e−5
	1/200	0.0007	1.12e−5

t=1	Δx	Err_ρ	Err_U
	1/30	1.7618	0.0009
	1/50	0.4559	0.0003
	1/70	0.1654	0.0001
	1/100	0.0790	7.92e−5
	1/150	0.0284	3.52e−5
	1/200	0.0138	1.99e−5

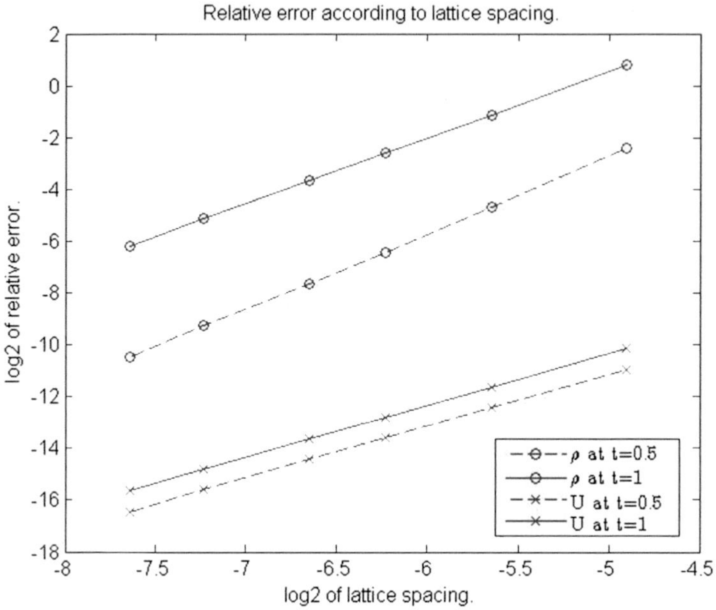

Fig. 3 Error curves for ρ and U

approach ρ_e) involves a constant error because this criteria is not defined in function of lattice size cell. We present on Fig. 3 (at $t = 0.5$ and $t = 1$) the convergence curves, $log(Error)$ versus Δx, for ρ and U. We observe that the slope of error curves for ρ passes approximately from 3 to 2.5 and the slope of error curves for U is approximately equal to 2. This shows that our numerical error estimates are agree with the postulated error estimates, and indicate the good performance of our proposed method.

4.2 *Influence of Some Parameters on Electrical Wave Propagation*

This second computations have been made to test performance of our method by analyzing the influence of some parameters on electrical wave propagation according to FHN models: a classical model (20) and a modified model (21). Here, we take $\Delta x = 1/50$ and $\Delta t = \Delta x^2$.

4.2.1 First Data Setting

In this first application, we consider the classical FHN model (20) with fixed $\beta_1 = 0.5$ and $\beta_2 = 1$ and we study the effect of parameters α_1 and α_2. We assume $K_i = K_e = 1$ and we consider the following initial conditions $\rho(x, y, 0) = -1.28791 + sin(x)$, $\rho_e(x, y, 0) = 0$, $U(x, y, 0) = -0.5758$.

The parameters α_1 and α_2 control the dynamic between transmembrane potential ρ and the ionic state U. In order to study the propagation of an electrical wave through the cardiac tissue, we analyze the relaxation time of ρ. In Table 2, we observe that α_2 has negligible impact on this relaxation time. Conversely, Fig. 4 and Table 2 show that tiny α_1 value leads to near-infinite slope values when potential ρ peaks and relaxes.

Table 2 Relaxation time of ρ according to α_1 and α_2 variations

α_1	α_2	Relaxation time
	0.01	0.4700
0.2	0.05	0.4704
	0.1	0.4716
	0.5	0.4988

α_2	α_1	Relaxation time
	0.01	0.0396
0.2	0.05	0.1568
	0.1	0.2764
	0.5	0.9168

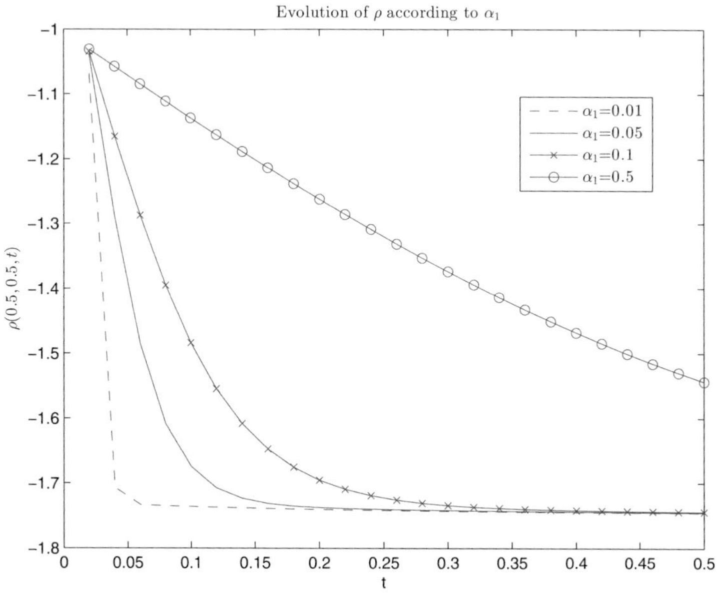

Fig. 4 $\rho(0.5, 0.5, t)$ with $\alpha_1 = 0.01$ to 0.5, $\alpha_2 = 0.2$

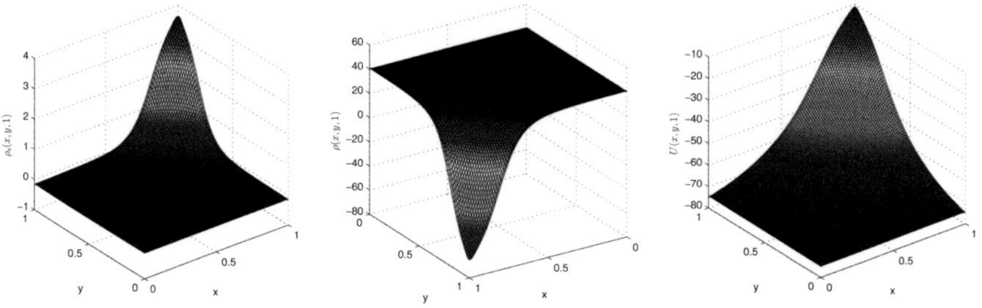

Fig. 5 Evolution of ρ_e, ρ and U at $t = 1$ with $\Delta x = 1/50$

4.2.2 Second Data Setting

In this final application, we consider the modified FHN model (21) which is known to be more adapted than the classical model (20) because of a product between ρ and U. We assume $K_i = 1.75$ and $K_e = 7$ and we consider the following initial conditions: $\rho(x, y, 0) = -85 + 100(1 - sin(xy))$, $\rho_e(x, y, 0) = 0$, $U(x, y, 0) = 0$. Figure 5 shows a color visualization of the simulation for ρ, ρ_e and U at final time $t = T$. The analyze we have done before is still valid.

Remark 3 It is known that the action potential duration (APD) and conduction velocity (CV) could be significantly affected by variation in the values of FHN parameters. So, it will be interesting to perform comparison of the APD and CV for the various FHM ionic-type models (and others) coupling the bidomain model, but also to understand how sensitive APD and CV are to variability in these parameters.

5 Conclusion and Commentary

An efficient and stable coupled LBM to solve a two-dimensional bidomain model of cardiac tissue is developed. From the Chapman-Enskog expansion analysis, the bidomain system which is a coupled of reaction-diffusion, elliptic and EDO equations, can be correctly recovered by our modified LBM. This method is easy to implement and easy to parallelize. It is clear that, due to the multi-scale nature of the system, the cartesian grid used in our preliminary simulations is not very sufficient to compute in a computationally efficient manner real life clinical situation with complex geometry which is in general computationally expensive. Therefore, it is expected to solve the Lattice Boltzmann system on adapted cartesian or triangular unstructured grids as e.g., in [6, 9] and the references therein. Moreover, in order to overcome the limitations of the constraint CFL stability condition, we extend the method to implicit or semi-implicit time schemes, e.g., by using the θ-method (with $\theta \in [0, 1]$) or Runge-kutta methods, coupled with adaptive time stepping strategies, as e.g. in [7] and the references therein. This coupled LBM method will be shown in a forthcoming paper for more general coupled models with realistic complex geometries. It would be interesting to use this developed method with observations coming from experimental data and a more complete description of the biophysical model of electrical cardiac activity. In order to get even closer to a more realistic calculation, it is necessary to study in the future this method coupled with optimization technique and robust control problems by using the approach developed in [1].

Acknowledgements The authors are grateful to the referee for many constructive comments and suggestions which have improved the presentation of this manuscript.

References

1. Belmiloudi, A.: Stabilization, Optimal and Robust Control. Theory and Applications in Biological and Physical Sciences. Springer, London (2008)
2. Belmiloudi, A.: Robust control problem of uncertain bidomain models in cardiac electrophisiology. J. Coupled Syst. Multiscale Dyn. **19**, 332–350 (2013)
3. Bourgault, Y., et al.: Existence and uniqueness of the solution for the bidomain model used in cardiac electrophysiology. Nonlinear Anal-Real **10**, 458–482 (2009)
4. Campos, J.O., et al.: Lattice Boltzmann method for parallel simulations of cardiac electrophysiology using GPUs. J. Comput. Appl. Math. **295**, 70–82 (2016)
5. Chai, Z., Shi, B.C.: Novel Boltzmann model for the Poisson equation. Appl. Math. Model. **32**, 2050–2058 (2008)
6. Fan, Z., et al.: Adapted unstructured LBM for flow simulation on curved surfaces. In: ACM SIGGRAPH, Los Angeles, pp. 245–254 (2005)
7. Huang, J., et al.: A fully implicit method for lattice Boltzmann equations. SIAM J. Sci. Comput. **37**(5), Special Section, S291–S313 (2015)
8. Sharomi, O., Spiteri, R.: Convergence order vs. parallelism in the numerical simulation of the bidomain equations. J. Phys.: Conf. Ser. **385**, 1–6 (2012)
9. Valero-Lara, P., Jansson, J.: A non-uniform Staggered Cartesian grid approach for Lattice-Boltzmann method. Procedia Comput. Sci. **51**, 296–305 (2015)
10. Yang, X., et al.: Coupled lattice Boltzmann method for generalized Keller-Segel chemotaxis model. Comput. Math. Appl. **12**, 1653–1670 (2014)

Dynamics and Bifurcations in Low-Dimensional Models of Intracranial Pressure

D. Evans, C. Drapaca, and J.P. Cusumano

Abstract Intracranial Pressure (ICP) is a physiological parameter of the brain which plays an important role in the diagnosis and treatment of pathologies such as hydrocephalus and traumatic brain injury. Currently, all reliable methods for ICP monitoring involve drilling through the skull to place a pressure probe inside the brain. As a result, ICP is only measured in the most critical cases which require neurosurgical intervention. Mathematical models that relate ICP to physiological parameters whose measurements are minimally invasive could contribute to better diagnostic and treatment protocols for brain disorders. Ideally, such mathematical models should have the capability to predict ICP in real time from non-invasive measurements of other clinically relevant parameters without the need for high-risk procedures. In this paper, we examine in detail the dynamics and stability of a mathematical model proposed by Ursino and Lodi in (J Appl Physiol 82(4):1256–1269, 1997) which predicts ICP from measurements of arterial blood pressure. We study how the equilibria vary with the model parameters and, aided by numerical simulations, we obtain bifurcation diagrams for the system of non-linear ordinary differential equations. Expanding upon the work of Ursino and Lodi, we show that the model exhibits not only one Hopf bifurcation but also a reverse Hopf bifurcation in certain parameter regimes. In addition, we present global phase portraits of the system in interesting parameter configurations.

1 Introduction

Intracranial pressure (ICP) monitoring provides critical physiological information on brain damage in patients suffering from neurological disorders such as hydrocephalus and traumatic brain injury (TBI). The only reliable method of measuring ICP involves drilling a hole directly through the skull, and inserting a catheter with a pressure probe into the brain [5]. Given the invasiveness of this approach, ICP monitoring is only performed on patients in very critical conditions during life-saving neurosurgical interventions. Mathematical models that couple ICP dynamics

D. Evans (✉) • C. Drapaca • J.P. Cusumano
The Pennsylvania State University, University Park, PA, USA
e-mail: dje5104@psu.edu; csd12@psu.edu; jpc3@psu.edu

© Springer International Publishing Switzerland 2016

J. Bélair et al. (eds.), *Mathematical and Computational Approaches in Advancing Modern Science and Engineering*, DOI 10.1007/978-3-319-30379-6_21

and other physiological parameters whose monitoring is less invasive provide a powerful alternative to the direct measurement of ICP. In addition, mathematical models that also incorporate brain's regulatory mechanisms of ICP could prove essential in designing better diagnostic and therapeutic procedures for a wide range of neurological diseases. Although the potential exists to create a model that could be used to estimate ICP in real time, the current state of the art is still far from this goal. A comprehensive review of mathematical models of ICP dynamics can be found by Wakeland and Goldstein [8].

One particular model proposed by Ursino and Lodi [6], predicts ICP dynamics from non-invasive measurements of arterial blood pressure, which appears to be in agreement with the ICP measured in patients with TBI. The model, which we will refer to as the Ursino-Lodi model, is a generalization of the one-compartment mathematical model of ICP dynamics pioneered by Marmarou [4] that includes several blood compartments, each with variable conductances and compliances, and ICP auto-regulation due to blood vessel wall tension, and viscous forces. Since the focus of the Ursino-Lodi model was its clinical applicability, a full mathematical analysis of the model is still missing. Therefore, the aim of this paper is to examine in detail the dynamics and stability of the Ursino-Lodi model. We use numerical simulations to study how the equilibria vary with the model parameters and obtain bifurcation diagrams for the system of non-linear ordinary differential equations. Our in-depth analysis shows that the model exhibits not only one Hopf bifurcation, as reported in [6], but also a reverse Hopf bifurcation in certain parameter regimes. We also present global phase portraits of the system in interesting parameter configurations.

2 Physiological Considerations

The human brain is a multi-phase mixture of many different materials, including cerebrospinal fluid (CSF), brain cells, and cerebral vasculature. The majority of what is considered brain tissue consists of white matter and grey matter, which are both soft and mildy compressible materials [3]. CSF is a colorless liquid made of 99 % water that fills the brain's ventricles, the subarachnoid space (space between the two deeper meninges that envelop the brain tissue: pia mater and the arachnoid mater), and intracellular space. The presence of CSF introduces damping to the brain motion, which cushions and protects it from injury [2]. CSF is constantly formed by the choroid plexuses of the brain's ventricles from the cerebral blood, circulates through the brain tissue and spinal cord, and is constantly reabsorbed into the cerebral veins located in dura mater (above the arachnoid mater) [5]. There exists an approximately constant pressure gradient from the inside of the brain (ventricles containing CSF) to the outside (dura matter). CSF is then forced over this pressure gradient over time, allowing it to be reabsorbed by the venous system. The rate of absorption is linked to the intracranial pressure as follows. Above a certain value of ICP, the relationship is nearly linear while at very low values of ICP, the absorption rate is negligible [1]. The absorption of CSF back into the bloodstream, and hence, its removal from the intracranial compartment, is the primary natural mechanism for lowering the intracranial pressure. As a result, abnormally high intracranial pressure is sometimes caused by high resistances to resorption.

In addition to the brain naturally regulating pressure via the volume of CSF, the brain also contains autoregulatory mechanisms to control the flow of blood in the cerebral arteries. In order to ensure a constant artery-to-vein difference in oxygen concentration, the brain automatically regulates the cerebral blood flow (CBF) in a process known as *cerebral autoregulation*. As oxygen consumption, blood pressure, and blood viscosity change, the cerebral arteries dilate or constrict to regulate the flow of blood to the brain. The volume of the cerebral arteries, the compression of the brain tissue, and the volume of CSF within the skull are all controlled by ICP. The linked mechanisms of feedback and control through all of these processes have the potential to create instability in the steady-state behavior of the ICP, most notably in the case of Lundberg A waves. This pathological phenomenon is characterized by long, sustained increases in the ICP. The ICP, which is normally approximately constant, increases dramatically in oscillations with periods of approximately 5–20 min, and amplitudes of about 50–100 mmHg [5].

3 Ursino-Lodi Model

The Ursino-Lodi mathematical model [6] includes both the ventricular CSF and the cerebral vasculature. The model is made of five compartments: one compartment for the ventricular CSF and four compartments for the blood in the cerebral arteries, capillaries, veins, and venous sinus, respectively. In addition, the model provides a relationship between pressures and volumes of cerebral blood and the pressure and volume of the ventricular CSF. Lastly, the model includes equations which dynamically control the flow of cerebral blood, which account for the effects of cerebral autoregulation. By modeling the effect of cerebral autoregulation, the Ursino-Lodi model has the capability to predict the oscillatory Lundberg A Waves.

Applying the mass conservation principle to the ventricular component yields

$$C_{ic}\frac{dP_{ic}}{dt} = \frac{dV_a}{dt} + \frac{P_c - P_{ic}}{R_f} - \frac{P_{ic} - P_{vs}}{R_o} + I_i, \tag{1}$$

where C_{ic} is the intracranial "compliance", V_a is the blood volume in the cerebral arteries and arterioles, P_c, P_{ic}, and P_{vs} are the capillary, intracranial, and venous sinus pressures, respectively, R_f and R_o are the resistances to CSF formation and, respectively, CSF outflow, and I_i is the rate of externally injected of extracted CSF volume. As in [4], the nonlinear compliance term C_{ic} is assumed to be related to P_{ic} by the following formula:

$$C_{ic} = \frac{1}{k_E P_{ic}}, \tag{2}$$

where k_E is brain's elastance coefficient (a stiffness-like modulus introduced by [4]). The mass preservation principle for the compartment of cerebral capillaries yields the following equation:

$$\frac{P_a - P_c}{R_a} = \frac{P_c - P_{ic}}{R_f} + \frac{P_c - P_{ic}}{R_{pv}} \approx \frac{P_c - P_{ic}}{R_{pv}}, \tag{3}$$

where R_a and R_{pv} denote the resistances of the arterial-arteriolar and respectively venular vessels. The approximation made in Eq. (3) is based on the assumption that the CSF production rate is much less than the cerebral blood flow rate.

By assuming a linear blood volume-pressure relationship $V_a = C_a(P_a - P_{ic})$, where C_a denotes the arterial-arteriolar compliance, the following equation can be obtained through differentiation:

$$\frac{dV_a}{dt} = C_a\left(\frac{dP_a}{dt} - \frac{dP_{ic}}{dt}\right) + \frac{dC_a}{dt}(P_a - P_{ic}). \tag{4}$$

By replacing Eqs. (2) and (4) into Eq. (1), we obtain:

$$\frac{dP_{ic}}{dt} = \frac{k_E P_{ic}}{1 + C_a k_E P_{ic}}\left(C_a\frac{dP_a}{dt} + \frac{dC_a}{dt}(P_a - P_{ic}) + \frac{P_c - P_{ic}}{R_f} - \frac{P_{ic} - P_{vs}}{R_o} + I_i\right). \tag{5}$$

Lastly, the model is completed by assuming that the cerebral autoregulation modifies the compliance C_a as follows:

$$\frac{dC_a}{dt} = \frac{1}{\tau}(-C_a + \sigma(Gx)), \tag{6}$$

where τ is a time constant, G is the maximum autoregulation gain, and

$$\sigma(Gx) = \frac{(C_{an} + \Delta C_a/2) + (C_{an} - \Delta C_a/2)\exp(Gx/k_\sigma)}{1 + \exp(Gx/k_\sigma)}, \tag{7}$$

with $k_\sigma = \Delta C_a/4$.

By definition, the cerebral blood flow q is $q = \dfrac{P_a - P_c}{R_a}$ and thus the normalized deviation of q from its normal value q_n is:

$$x = \frac{q - q_n}{q_n}. \tag{8}$$

The parameter ΔC_a represents the maximum allowed change in the arterial compliance C_a from its basal value C_{an}. Furthermore, it depends on whether the arterial-arteriolar vessels are contracting ($x > 0$) or dilating ($x < 0$) as follows:

$$\Delta C_a = \begin{cases} \Delta C_{a1}, \text{ if } x < 0 \\ \Delta C_{a2}, \text{ if } x > 0\,. \end{cases} \tag{9}$$

By assuming that the arteries are circular cylinders of radius r and the blood is a viscous Newtonian fluid, the Hagen-Poiseuille equation which states that the arterial resistance is inversely proportional to r^4 can be used to obtain the following expression for R_a:

$$R_a = \frac{k_R C_{an}^2}{V_a^2} . \tag{10}$$

The system of non-linear ordinary differential equations (5) and (6) is the Ursino-Lodi model written in terms of the state variables P_{ic} and C_a.

4 ICP Dynamics Predicted by Ursino-Lodi Model

The clinically measured ICP is a pulsatile periodic waveform which has a period equal to one heartbeat. However, the Lundberg A waves happen on a time scale that is much longer than one heartbeat, so in order to observe them we replace the heartbeat with an averaged pressure over one period which greatly simplifies the analysis without loss of applicability. Notably, it ensures that the system of differential equations we work with remains autonomous. The parameters of the Ursino-Lodi model are the variables that are treated as constants with respect to time. A point in the parameter space representing basal parameters accompanies the model, which is presented as Table 1. In our numerical simulations, all parameters are set to these values unless otherwise indicated. Variations from these parameters

Table 1 Ursino-Lodi Model Basal Parameters [6]

Parameter	Value
R_o	526.3 mmHg s ml^{-1}
R_{pv}	1.24 mmHg s ml^{-1}
R_f	2.38×10^3 mmHg s ml^{-1}
ΔC_{a1}	0.75 ml/mmHg
ΔC_{a2}	0.075 ml/mmHg
C_{an}	0.15 ml/mmHg
k_E	0.11 ml^{-1}
k_R	4.91×10^4 mmHg^3
τ	20 s
q_n	12.5 ml/s
G	1.5 ml mmHg^{-1} 100 % CBF change^{-1}
P_a	100 mmHg
P_{ic}	9.5 mmHg
P_c	25 mmHg
P_{vs}	6.0 mmHg
C_a	0.15 ml/mmHg

will be represented as unitless normalized values with respect to the values given in Table 1.

The equilibrium points in the model are found at locations where $\dot{P}_{ic} = \dot{C}_a = 0$, where the dot operator indicates the time derivative. Given positive real input parameters, this system has consistently two equilibrium points in the domain of $P_{ic} \geq 0$, $C_a \geq 0$. These two equilibria are marked with stars in Fig. 2b. What we will refer to as the "primary" equilibrium point in this work is marked "SN" here, for *stable node*, and what we refer to as the "secondary" equilibrium point is marked "SP" here, for *saddle point*, which we will discuss shortly. Although the primary equilibrium point is a stable node for the values of Table 1, we will soon discuss how deviations from these can change its stability.

Nevertheless, we treat both equilibria equally when employing the following techniques. By slowly varying the initial state of the system by one model parameter (here, either R_o, k_E, or G), we obtain a locus of points that correspond to the steady state value of the system. Using the basal case as an initial guess, we use a root-solver (the MATLAB built-in function `fzero`) to track the equilibrium values of P_{ic} and C_a as a function of three of the model parameters: R_o, k_E, and G, for both equilibrium points. Because there is more than one root for the equilibrium state, we kept each solution branch separate by, upon increase of the bifurcation parameter, using the previous solution as the subsequent initial guess for the root-solver. Figures 1 and 2a show how the locations of both equilibria vary as a function of both R_o and G. We found that varying the intracranial elastance, k_E, does not affect the equilibrium values predicted by the model for either equilibrium point. In addition, the location of the secondary equilibrium point does not change as a function of R_o.

We would like to describe how the stability of these equilibria change with respect to the model parameters. We therefore linearize the system of equations (5) and (6) about the equilibria to obtain the linear system $\dot{\mathbf{x}} = \mathbf{J}\mathbf{x}$. Here, $\mathbf{x} = [(P_{ic} - P^*_{ic}),\ (C_a - C^*_a)]^{\mathsf{T}}$ is the perturbation from the equilibrium value, $(P^*_{ic},\ C^*_a)$, and J is the Jacobian matrix of the two-dimensional vector field defined by the right hand sides of Eqs. (5) and (6). Then, the eigenstructure of J determines the stability of the equilibrium point [7].

Confirming what was previously inferred from the phase portrait, the primary equilibrium point is a stable node, having two negative real eigenvalues in the basal condition. Since there are no other stable equilibria in the domain, the basin of attraction is the entire domain $P_{ic} > 0$, $C_a > 0$. In addition, we find that the secondary equilibrium point always occurs at a value of $P_{ic} = 0$ regardless of other input parameters. The eigenvalues of the linearized system near this equilibrium are real with opposite signs. Therefore, this equilibrium can be classified as a *saddle point*.

It is clinically valid that the primary equilibrium is a stable node in the basal condition. That is, given an initial state $(P_{ic},\ C_a)$, the system will autonomously regulate itself to one steady-state value. However, as previously alluded to, there will be instances where the system no longer converges to a constant steady-state value, but instead converges to a stable limit cycle in the phase space, representing

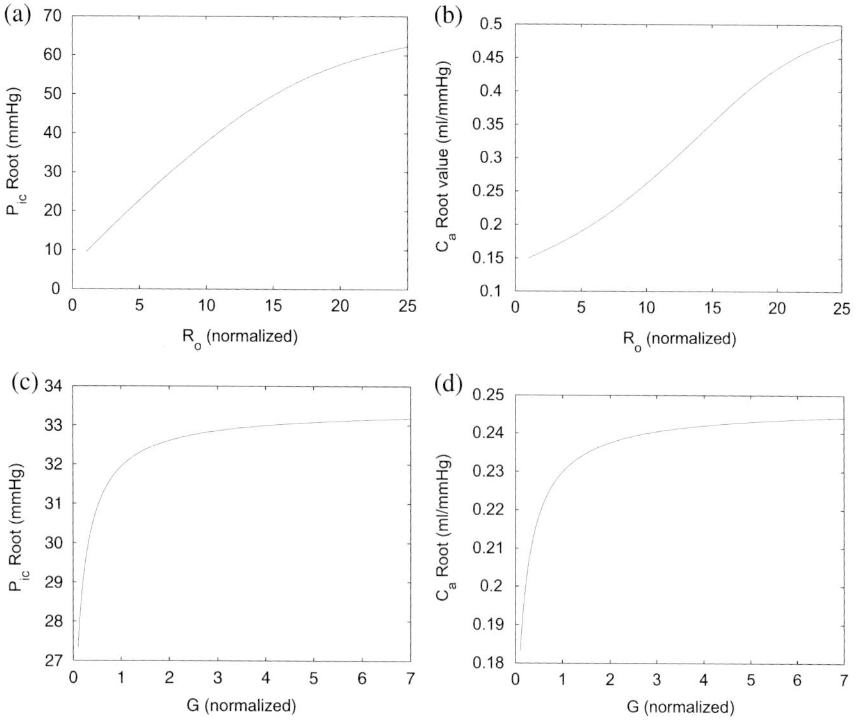

Fig. 1 Location of the primary equilibrium point. (**a**) P_{ic}^* versus R_o, (**b**) C_a^* versus R_o, (**c**) P_{ic}^* versus G, (**d**) C_a^* versus G

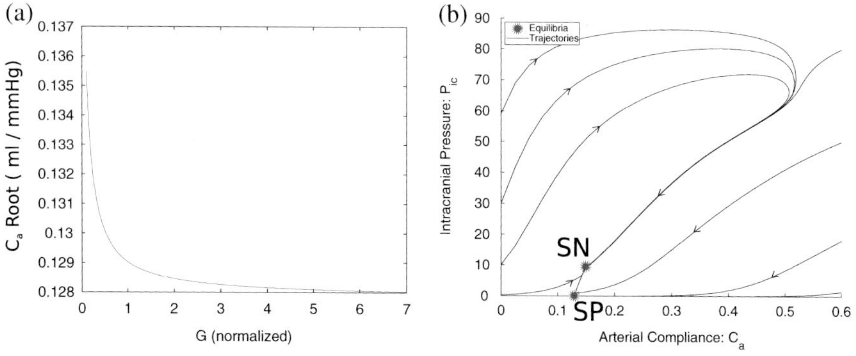

Fig. 2 (**a**) Location of C_a^* on the secondary equilibrium point as a function of G (P_{ic}^* is unaffected). (**b**) Phase portrait of the system in the basal condition as per Table 1. Here, the primary equilibrium point is a stable node (SN), while the secondary equilibrium point is a saddle point (SP)

the clinically observed case of Lundberg A Waves. Mathematically, the transition between the two cases implies the occurrence of a *Hopf bifurcation*, where a stable limit cycle develops around the equilibrium point, which has then lost its stability. The point at which this transition occurs is known as the *Hopf bifurcation point*, which occurs where the pair of eigenvalues are pure imaginary with zero real part. Therefore, in order to find where this bifurcation takes place, we numerically search for cases where the eigenvalues are pure imaginary. We use a fourth-order finite difference approximation of the Jacobian evaluated at the equilibrium point, and evaluate the eigenvalues of the resulting matrix. These relationships are shown in Figs. 3a–c.

If we follow the eigenstructure of the primary equilibrium point as R_o increases (as shown in Fig. 3a), we notice that the equilibrium passes through several stability

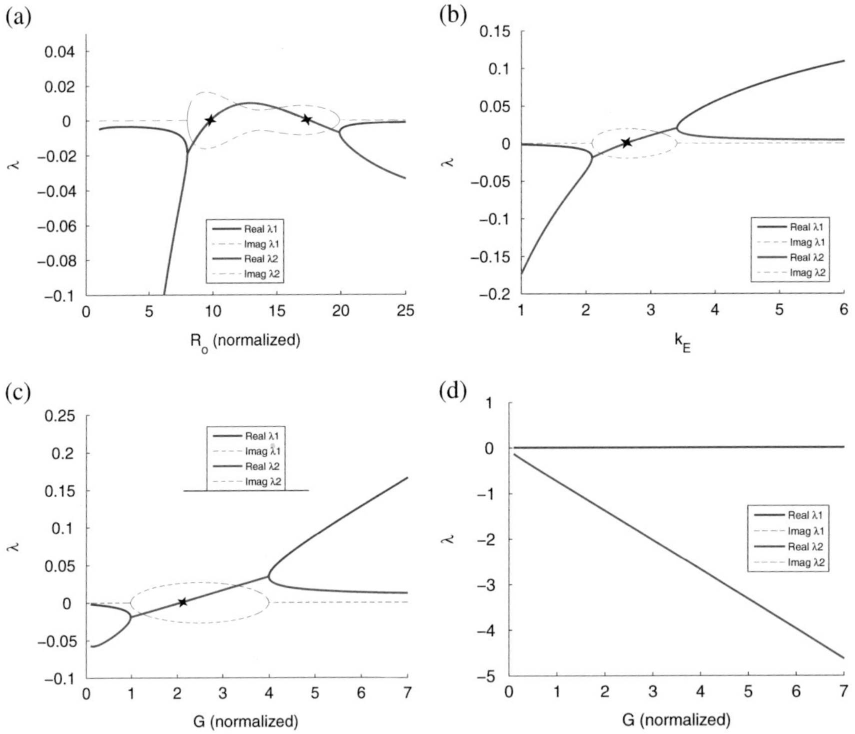

Fig. 3 Eigenvalue plots of equilibrium solution branches, showing changes in stability with parameter variations. The Hopf bifurcation points are marked with stars. (**a**) Primary equilibrium: λ versus R_o. Hopf bifurcation occurs at $R_o \approx 9.793$, and *reverse* Hopf bifurcation occurs at $R_o \approx 17.430$. (**b**) Primary equilibrium: λ versus k_E (with increased $R_o = 8$) Hopf bifurcation occurs at $k_E \approx 2.615$. (**c**) Primary equilibrium: λ versus G (with increased $R_o = 8$ and $k_E = 2.1$). Hopf bifurcation occurs at $G \approx 2.072$. (**d**) Secondary equilibrium: λ versus G. The imaginary parts are not visible in the plot because they are nearly coincident with the real part of λ_1, which takes very small positive values

classifications. The equilibrium begins as a stable node when R_o is unchanged. As R_o increases, the eigenvalues transition from negative and real, to a complex conjugate pair with negative real part, changing the equilibrium to a *stable focus*. Increase R_o further, and we reach the Hopf bifurcation point, where at this instant, the eigenvalues are a pure imaginary complex conjugate pair. The Hopf bifurcation also marks the creation of a limit cycle around the equilibrium point. After the bifurcation point, the eigenvalues are still a complex conjugate pair, but this time they have a positive real part, which means the equilibrium is an *unstable node* in this region. A further increase in R_o causes the reverse to take place, taking the equilibrium back through a reverse Hopf bifurcation, extinguishing the limit cycle, and returning it to a stable focus, and eventually a stable node. It is in the region between the forward and reverse Hopf bifurcation points that we observe a limit cycle in the system. A slightly similar situation occurs when the bifurcation parameters are k_E and G (Fig. 3b, c). While increasing these bifurcation parameters produces the forward Hopf bifurcation in the same way, further increases do not cause a reverse Hopf bifurcation. Instead, the eigenvalues of the unstable focus meet and become a pair of positive real eigenvalues, indicating the equilibrium's change to an *unstable node* (Fig. 4).

For the sake of completeness, we have also analyzed the stability of the saddle point equilibrium. However, we have not observed any changes in the stability of this equilibrium point. Increasing R_o and k_E do not change the eigenvalues at all, and while increasing G changes the values of the eigenvalues, it preserves the overall structure of two real eigenvalues with differing signs (see Fig. 3d).

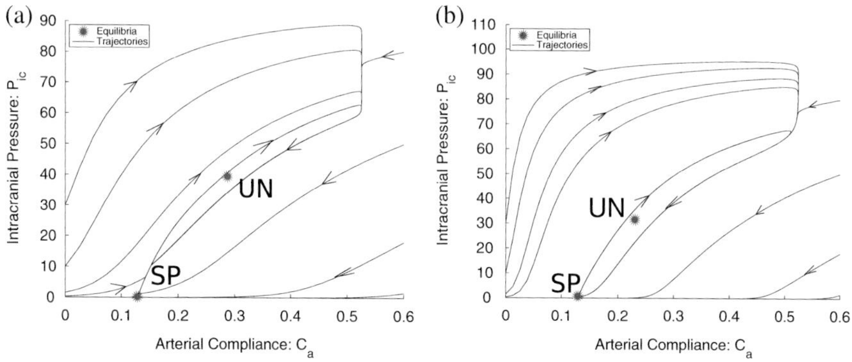

Fig. 4 Phase portraits exhibiting limit cycles, where the primary equilibrium is an unstable node (UN), and the secondary equilibrium is a saddle point (SP). (**a**) Normalized $k_E = 2.1$, $R_o = 10$, and $G = 4$. (**b**) Normalized $k_E = 8$ and $R_o = 8$

5 Conclusions and Further Work

Through the use of numerical simulations, we have shown how varying the parameters G, k_E, and R_o in the Ursino-Lodi model changes the value of the equilibrium intracranial pressure and arterial compliance. We have also shown how varying these parameters can force the primary equilibrium to undergo a Hopf bifurcation, creating a stable limit cycle around the now unstable equilibrium point. In addition to confirming the claims made by Ursino and Lodi, we have also shown the existence of a reverse Hopf bifurcation at very high levels of R_o To the authors' knowledge, no interesting bifurcation phenomena is witnessed by varying the model parameters not mentioned here, hence their exclusion from the work. We intend to use the work presented here when developing new models of intracranial pressure dynamics, to offer additional insight on the stability structure of the steady-state solutions.

References

1. Cutler, R.W.P., Page, L., Galicich, J., Watters, G.V.: Formation and absorption of cerebrospinal fluid in man. Brain **91**(4), 707–720 (1968)
2. Goldsmith, W.: The state of head injury biomechanics: past, present, and future: part 1. Crit. Rev. Biomed. Eng. **29**(5&6), 441–600 (2001)
3. Kyriacou, S.K., Mohamed, A., Miller, K., Neff, S.: Brain mechanics for neurosurgery: modeling issues. Biomech. Model. Mechanobiol. **1**(2), 151–164 (2002)
4. Marmarou, A., Shulman, K., Rosende, R.M.: A nonlinear analysis of the cerebrospinal fluid system and intracranial pressure dynamics. J. Neurosurg. **48**(3), 332–344 (1978)
5. Suarez, J.I.: Critical Care Neurology and Neurosurgery. Springer, New York (2004)
6. Ursino, M., Lodi, C.A.: A simple mathematical model of the interaction between intracranial pressure and cerebral hemodynamics. J. Appl. Physiol. **82**(4), 1256–1269 (1997)
7. Verhulst, F.: Nonlinear Differential Equations and Dynamical Systems, 2nd edn. Springer, New York (1996)
8. Wakeland, W., Goldstein, B.: A review of physiological simulation models of intracranial pressure dynamics. Comput. Biol. Med. **38**(9), 1024–1041 (2008)

Persistent Homology for Analyzing Environmental Lake Monitoring Data

Benjamin A. Fraser, Mark P. Wachowiak, and Renata Wachowiak-Smolíková

Abstract Topological data analysis (TDA) is a new method for analyzing large, high-dimensional, heterogeneous, and noisy data that are characteristic of modern scientific and engineering applications. One major tool in TDA is persistent homology, wherein a filtration of a simplicial complex is generated from point clouds and subsequently analyzed for topological features. Betti numbers are computed across varying spatial resolutions, based on a proximity parameter R, where the n-th Betti number equals the rank of the n-th homology group. In this paper, persistent homology is applied to lake environmental monitoring data collected from a sonde sensor attached to a commercial cruise vessel, and to weather station observations. A modified form of the witness complex described by de Silva is used in an attempt to eliminate the need for persistence and thus to reduce computation time. From preliminary results, witness complexes are very promising in capturing the shape of the data and for detecting patterns. It is therefore proposed that TDA, combined with standard statistical techniques and interactive visualizations, enable insights into observations collected from environmental monitoring sensors.

1 Introduction

Improvements in sensor network technology and sensor mechanisms have facilitated acquisition of vasts amount of data. In the case of biomedical and environmental monitoring, these data tend to be complex, heterogeneous, noisy, unstructured, and high-dimensional [3]. Furthermore, data are frequently acquired as long vectors, where only a small number of coordinates are meaningful to specific questions at hand, but it is often not known a priori which coordinates are relevant [1]. Discovering meaningful structural or temporal patterns in these data – that is, to obtain knowledge about the data's large-scale organization – is very difficult, even for domain experts. New, innovative computational methods and visual data mining tools are therefore needed [3]. Among recently proposed techniques, topological data analysis (TDA) is very promising. TDA provides robust feature definitions,

B.A. Fraser (✉) • M.P. Wachowiak • R. Wachowiak-Smolíková
Nipissing University, North Bay, ON, Canada
e-mail: bfraser826@community.nipissingu.ca; markw@nipissingu.ca; renatas@nipissingu.ca

© Springer International Publishing Switzerland 2016

J. Bélair et al. (eds.), *Mathematical and Computational Approaches in Advancing Modern Science and Engineering*, DOI 10.1007/978-3-319-30379-6_22

emphasizing the global aspects of the data and shape recognition in point clouds of high dimensions. Specifically, homology provides the basis for computing simplicial complexes (a mesh that represents a space in specific ways), in which the topology and geometry of the point clouds are separated [10]. Topological methods for data analysis were described in the seminal paper by Carlsson [1] and subsequently expanded. They are useful in studying connectivity information, such as the classification of loops and high-dimensional surfaces that may implicitly exist in the data. Unlike geometric methods, TDA is not sensitive to properties such as curvature, and, because the focus is on the global structure of the data, numerical values of distance functions are less important. These features allow researchers to study complex data where there is an unclear or incomplete idea of what specific metrics are to be extracted [1]. TDA has been applied to a variety of scientific problems, most notably in biomedicine, including the organization of genomic data from many different sources in the study of breast cancer [6], and for computer-aided diagnosis of pulmonary embolism [7]. TDA is also used for business and social analysis [4].

Analytics and advanced computational methods are increasingly important in environmental data analysis [2, 5]. New institutes and research initiatives focus exclusively on this topic (e.g. the Institute for Environmental Analytics at the University of Reading – www.the-iea.org). In the current paper, we adapt TDA to the analysis of environmental sensor data. Specifically, we modify the witness complex filtration described by de Silva [8] to eliminate the need for persistence, and consequently, to reduce computation time. We analyze a large number of data vectors in $\mathbb{R}^6$ representing various properties (e.g. time, temperature, air pressure, etc.) collected from sensors attached to a commercial cruise vessel, with the goal of obtaining insights into the complex physical phenomena characterizing dynamic lake conditions. In another example, data from weather station sensors are analyzed. After normalization of a Euclidean distance matrix for these vectors, a mesh or triangulation is constructed which accurately represents the topology of this data set. Although the data cannot be characterized as "big data" (very large and heterogeneous data sets that generally cannot be processed or analyzed using standard computational or statistical techniques), the readings are sufficiently numerous and high-dimensional that constructing such a mesh on the full set of points would be prohibitively costly. We must therefore choose a subset that is representative of the full set and retains its topology. The resulting analyses are expected to complement visual and statistical methods to enrich insights and to facilitate discovery of various lake phenomena. The next section describes the TDA algorithm as well as our adaptations, followed by a preliminary investigation of applying TDA to environmental lake monitoring data.

2 Algorithm

We construct a simplicial complex on a representative subset (landmarks) of the full data set using the remaining points as witnesses to decide whether edges will be added between them or not. A different choice of filtration of the witness complex is used than that described by de Silva [8], and the two approaches are compared for their reliance on persistent versus simplicial homology to accurately recover the Betti number profile of the data.

2.1 Simplicial Complexes

A k-simplex is a complete graph on $k - 1$ vertices. Every clique that is a subgraph of a simplex is a face of that simplex.

Definition 2.1 A simplicial complex is a collection K of simplices such that:

- Every face of a simplex in K is in K
- The intersection of any two simplices in K is a face of each of them

In particular, we are interested in the Vietoris-Rips complex [11], the largest simplicial complex on a given set of vertices and edges.

Definition 2.2 Given a vertex set V, the Vietoris-Rips complex $VR(V, R)$ contains the p-simplex $[a_0, \ldots, a_p]$ iff for every edge $[e_j, e_k]$, $0 \leq j \leq k \leq p$, $|e_j - e_k| \leq R$

Constructing a *VR* complex for given V and R then amounts to inserting edges between vertices that are within R of each other, and subsequently determining all cliques in the graph, and including them as simplices.

2.2 Landmarks and Witnesses

Given a set X of N data points, we need to choose a subset of those points consisting of n landmarks that accurately represents the shape of the point cloud. More complex shapes may require more landmarks for accurate representation, whereas a circle could be represented by only four points. Since the remaining $N-n$ points will act as witnesses, it is important not to select too large a proportion of the original N. Ideally, N should be quite large, so that a large number of landmarks can be selected. In practice, the ratio N/n should range from 30 to 100, depending on the size of the original data.

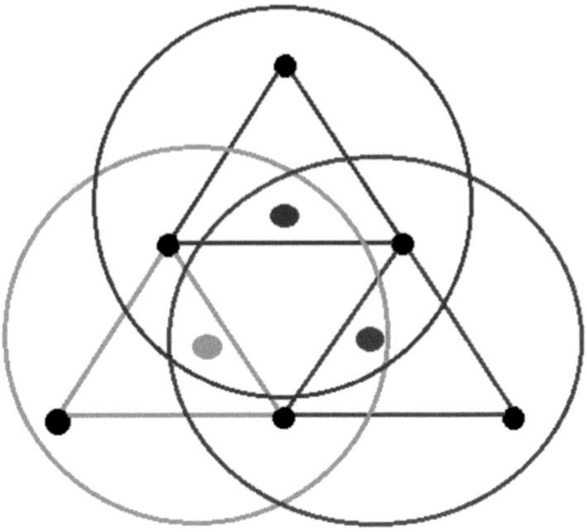

Fig. 1 The *red*, *green* and *blue points* are weak witnesses to the triangles of their respective colour, and to the edges of the inner triangle, but there is no witness to the inner triangle itself

The minmax [8] method of landmark selection is employed, which attempts to choose landmarks as evenly spaced as possible throughout the data. We proceed as follows:

- Choose a random point to be first in a set of landmarks L
- Add each subsequent point $x \in X$ to L to be such that $min\{dist(x, p) \; \forall \; p \in L\}$ is maximized

Let the points in $P = X \setminus L$ be the set of potential witness points.

Definition 2.3 ([8]) A point $w \in P$ is a weak witness to a p-simplex $A = (a_0, a_1, \ldots a_p)$ in L if $|w - a| < |w - b| \; \forall \; a \in A$ and $b \in L \setminus A$. It is a strong witness if $|w - a_0| = |w - a_1| = \ldots = |w - a_p|$ as well.

Example 2.1 Take three vertices of an equilateral triangle, as well as the three midpoints of its edges, as landmarks. Consider the three witnesses which lie just outside the central triangle formed by the midpoints (Fig. 1). These are weak witnesses to its edges, but none of them is a weak witness to the triangle itself; they are weak witnesses to the outer triangles corresponding to their colour.

We may wish to include a simplex, such as the triangle in the preceding example, in the case where all of its subsimplices are included. This motivates the following definition of a witness complex that will be employed, where witnesses are used to determine the 1-skeleton, or neighbourhood graph of edges, and then the higher dimensional simplices will be added by Vietoris-Rips expansion [11].

2.3 Witness Complexes

Two witness complex filtrations are described below. The first is from De Silva [8], the second is the proposed adaptation of this construction. A filtration of a witness complex is a nested sequence of increasing subsets of a simplicial complex. There are three inputs: a parameter v that determines how many landmarks can be witnessed simultaneously by a single point, the distance matrices between data points, and a value R which ranges across an interval and creates the filtration.

Take D to be the $n \times N$ Euclidean distance matrix (on the n landmarks and N full data set). Take E to be the $n \times n$ distance matrix between landmarks, and E' to be the $n \times |P|$ distance matrix (between the n landmarks and $|P|$ potential witnesses). Both witness complexes have the n landmarks as their 0-simplices.

For $W(D; R, v)$ (De Silva, [8]), $R \in [0, \infty)$, and for $W^*(E, E'; R, v)$, $R \in [0, 1]$:

- if $v = 0$, then for $i = 1, 2, \ldots, N$ define $m_i = 0$ and for $i = 1, 2, \ldots, |P|$ define $n_i = 0$
- if $v > 0$, then for $i = 1, 2, \ldots, N$ define m_i to be the v-th smallest entry of the i-th column of D, and for $i = 1, 2, \ldots, |P|$ define n_i to be the v-th smallest entry of the i-th column of E'
  - the edge $\sigma = [ab]$ belongs to $W(D; R, v)$ iff there exists a witness $i \in \{1, 2, \ldots, N\}$ such that $max(D(a, i), D(b, i)) \leq R + m_i$
  - the edge $\sigma = [ab]$ belongs to $W^*(E, E'; R, v)$ iff there exists a witness $i \in \{1, 2, \ldots, |P|\}$ such that $max(E'(a, i), E'(b, i)) \leq n_i$ and $E(a, b) \leq R$
- the p simplex $\sigma = [a_0 a_1 \ldots a_p]$ belongs to the witness complex iff all of its edges also belong to the witness complex

Example 2.2 The reason for splitting the distance matrix D into E and E' in our construction is that landmarks should not act as witnesses. Consider a point cloud where nearly all the vectors are tightly clustered, and there are a few widely spaced outliers (Fig. 2). All of these widely spaced outliers are likely to be chosen as landmarks by minmax, and if they can act as witnesses to each other, they can all be connected to each other by edges, resulting in a complex which does not accurately represent the data's shape. But if landmarks and witnesses are mutually exclusive, then the problem of outlier selection will be mitigated (albeit, each outlier selected as a landmark will still affect b^0).

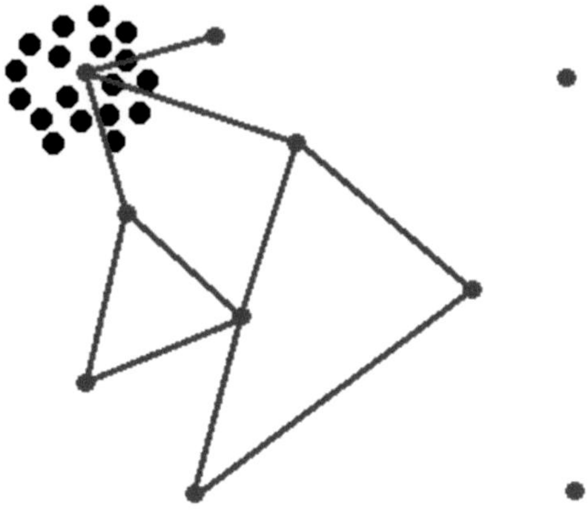

Fig. 2 An example of a witness complex $W(D; 0, 2)$ where outliers were selected as landmarks (*red*), and then acted as witnesses to each other. Features unrepresentative of the data are found

If landmarks are not taken as witnesses, observe the following equality: $W(D; 0, v) = W^*(E, E'; 1, v)$. Further, $W(D; R, v)$ is the same as the Vietoris-Rips complex for $v = 0$, and $W^*(E, E'; R, v)$ is the same for $v \geq n$. Below we explore these relations, and why appropriately chosen values of v can make persistent homology unnecessary.

At the point in the filtrations where $W(D; 0, v) = W^*(E, E'; 1, v)$, each witness point is the center of a smallest possible closed ball which contains at least v landmarks. Any landmarks that all fall within one such ball are joined by edges. This is the intermediate stage, say, $\mathscr{W}$, between completely disconnected vertices and a fully connected graph in which we are interested.

- $W(D; R, v), R \in [0, \infty)$ describes the sequence of simplicial complexes from $\mathscr{W}$ up to the complete graph on n vertices. It does this by expanding the closed balls around each witness point until all of the landmarks are contained in one.
- $W^*(E, E'; R, v), R \in [0, 1]$ describes the sequence of simplicial complexes from the set of disconnected 0-simplices to $\mathscr{W}$. It does this by expanding closed balls centered on the landmarks as an additional condition for connection by an edge.

The rationale behind our definition of $W^*(E, E'; R, v)$ is that we find joining landmarks to each other by considering their own distances from each other rather than from a witness to be a more natural progression, and that we consider $\mathscr{W}$ to be a suitable stopping point, where the simplicial complex accurately represents the shape of the data. (In fact, as will be evident from the examples that follow, there appears to be no need to compute simplicial complexes that are more connected than $\mathscr{W}$.)

Example 2.3 One reason that persistence is necessary for $W(D; R, v)$ is that only v values up to $v = 2$ are considered. Consider the following example: the four corners of a square centered on the origin are chosen as landmark points, and infinitely many witness points are taken uniformly from the interior of the square, excluding the origin (Fig. 3). Then at $R = 0$, $W(D, R, 2)$ is nearly correct, but there exists no witness which has opposite corners across the diagonal as its two nearest landmarks:

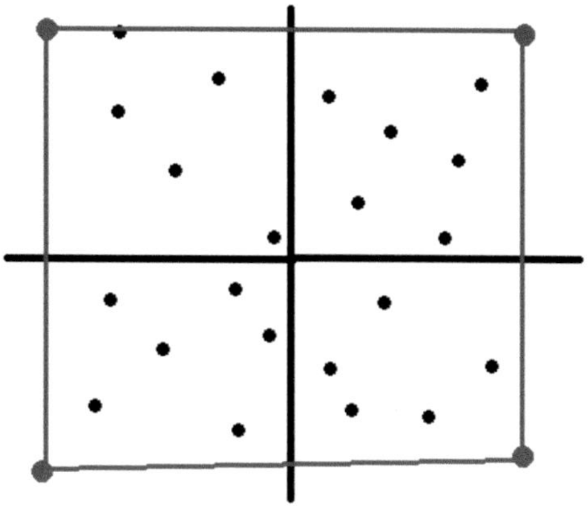

Fig. 3 The 1-skeleton resulting from $W(D, 0, 2)$, with the corners of a square as landmarks. Without a witness precisely at the origin, a 2-dimensional hole is detected. At $R = 0.1$, the hole is filled

as such, they are not joined by edges, and the Betti number b^1 is found to be 1 rather than 0, as it should be. As such, it is necessary to increment R by at least a small amount to fill the hole.

This issue may have been avoided by choosing $v = 3$. In fact, with data of dimension d, choosing $v = d + 1$ generally suffices. Witness complexes are used to reduce the number of cells: in 3 dimensions, a clique on 10 vertices is superfluous. However, a clique on 4 vertices is not. We can recover an accurate Betti profile of our data by increasing v slightly and still keep the number of cells relatively low. And by eliminating the need for persistence (since homology need be computed only once), more landmarks can be selected.

Finally, it remains to be explained why the parameter R should be used to create a filtration at all, if persistence is not used. The answer is that outliers can still be a problem, and $\mathscr{W}$ may not actually be correct.

Example 2.4 Take as our data set a large number of points from $\frac{2}{3}$ of a circle, and two additional points which are very near to each other from the center of the missing arc – these are the outliers. Minmax selects one of these outliers as a landmark, as well as some evenly distributed ones along the $\frac{2}{3}$ arc. An ideal simplicial complex should only connect the landmarks along the $\frac{2}{3}$ arc, but with $v = 3$, the other outlier which was not taken as a landmark witnesses the two ends of that arc and connects them, completing the circle (Fig. 4).

Clearly, this problem could have been avoided by taking the finished complex to be at some $R < 1$, as the edges in the $\frac{2}{3}$ arc would have been formed by $R \geq 0.1$, but the green edges needed $R \geq 0.3$. In general, the more landmarks that are taken, the closer they will be to each other, and the complex should be complete at a lower value of R. Any edges which emerge at higher values are more likely to be artifacts of outliers being witnessed, as in the above example. Therefore, we propose that the filtration be calculated normally, but when the rate of edge addition (relative to

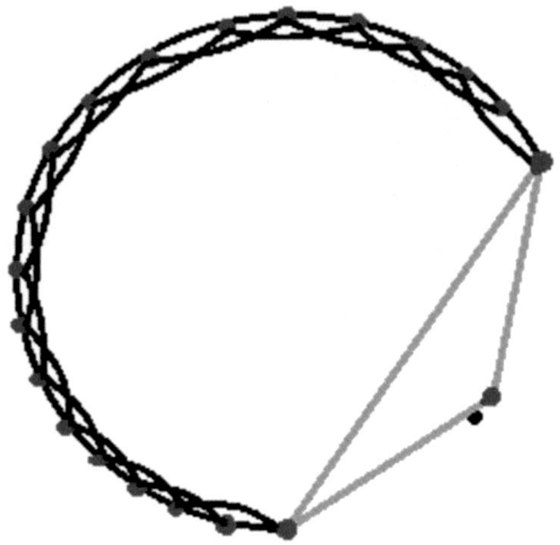

Fig. 4 For $W^*(E, E'; 1, 3)$, the *black point* witnesses the *green edges*, which connect an outlier point and bridge a chasm in the data, which should not have been done

the increase in R) begins to fall off dramatically, that the complex be considered complete and its Betti profile found.

3 Application

The proposed methods are applied to three types of data. The first is an artificially generated sphere point cloud, to mirror the experiments in [8]. The second is lake monitoring data, and the last are vectors of various observed weather properties.

3.1 Sphere Data

The first artificial data set consists of $N = 3200$ points chosen uniformly from the surface of S^2. Using minmax selection, 64 landmarks were chosen. Both $W^*(E, E'; 1, 3)$ and $W^*(E, E'; 1, 4)$ (Fig. 5) recovered the correct Betti profile $(b^0, b^1, b^2) = (1, 0, 1)$ in each of 100 trials.

Then 12,500 points were chosen uniformly from within S^2 having four interior spherical voids, ranging in diameter from 0.2 to 0.5. Again, using minmax, 125 landmarks were chosen. $W^*(E, E'; 1, 3)$ (Fig. 5) recovered the correct Betti profile $(b^0, b^1, b^2) = (1, 0, 4)$ in each of 10 trials.

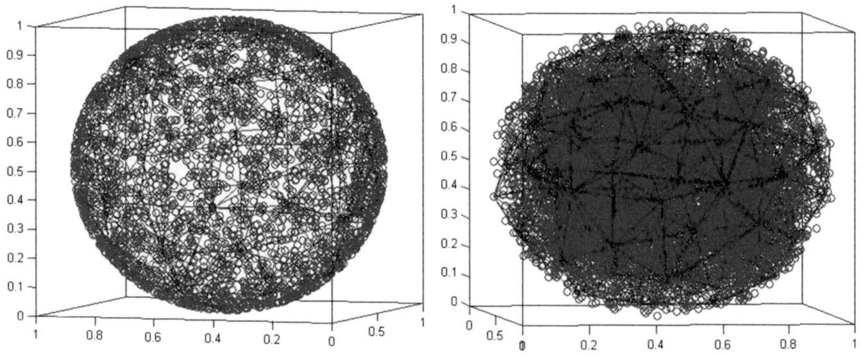

Fig. 5 Simplicial complexes constructed on points taken from S^2 (*left*) and D^2 with 4 spherical voids in it (*right*)

3.2 Lake Monitoring Data

The TDA approach developed in this paper was applied to lake monitoring data to complement existing statistical and visualization analysis tools [9] (see visual.nipissingu.ca/Commanda2). Data were collected from Lake Nipissing (46° 16′ 12″N 79° 47′ 24″W) via a sonde sensor attached to a commercial cruise vessel (the Chief Commanda II). The vectors are in six dimensions: temperature (°C), specific conductivity (μ s/cm), dissolved oxygen concentration (mg/L), pH, chlorophyll RFU (relative fluorescence units), and total algae RFU. A 3-dimensional witness complex $W^*(E, E'; R, 4)$ was constructed on a subseet of 36 landmarks of 216 data vectors from Sept. 4, 2011.

Betti profiles of $(5, 1, 0)$ or $(5, 0, 1)$ were found, but are considered to be unreliable due to the random nature of landmark selection. Removing various dimensions can simplify locating the source of the feature by constructing the same complex on the same data set, minus one of its coordinates. For instance, eliminating chlorophyll RFU made the detection of these features rarer, so we expect it is a contributing factor, while eliminating total algae RFU resulted in these features being detected more frequently. Consequently, the total algae property appears to be irrelevant, as it interferes with the topological feature.

In this case, reducing the coordinates to specific conductivity, pH, and chlorophyll RFU resulted in reliably reproducible 3-dimensional holes appearing in the witness complex $W^*(E, E'; 1, 4)$ (Fig. 6a). Further research is needed to understand these results, and to determine whether these features indicate some interaction between these properties of the lake.

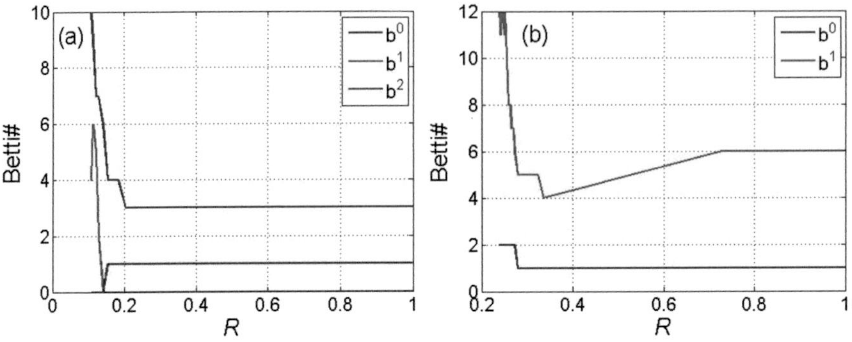

Fig. 6 (**a**) Reliably detected 3-dim hole on dimension-reduced lake data; (**b**) b^1 increasing by 2 at $R = 0.76$ in weather data, a result of outliers

3.3 Weather Data

TDA was applied to data collected from a Nipissing University weather station to complement existing web-based analysis and interactive visualizations (geovisage.nipissingu.ca). Six-dimensional weather data (photosynthetically active radiation (PAR) (μmol/J), rainfall (mm), relative humidity (%), soil moisture (%), soil temperature (°C), and wind speed (m/s)) for Temiskaming Shores, Ontario (approximately 47° 31′N 79° 41′W) were sampled at 5-min intervals, resulting in 8640 vectors for the month of April, 2013. A 2-complex $W^*(E, E'; 1, 3)$ was constructed on a subset of 86 landmarks, and during every such trial there were edges added for large values of R, some even at $R > 0.9$. These edge additions resulted in an increase of the found value of b^1. This is an instance of outliers affecting the result (Fig. 6b), as the majority of the complex was already complete by $R = 0.35$, and the Betti profile found at that point is considered more accurate.

Another strategy for isolating features is to build the complex on different time scales. For instance, a Betti profile of $(b^0, b^1) = (1, 5)$ was most often found for the month of April. It may be helpful to perform the same analysis on individual weeks, or days, to determine whether the appearance of certain features can be related to weather events, such as a thunderstorm, at those particular times.

4 Conclusions

This paper presents an improvement to the persistent homology paradigm for TDA that adapts and builds on the work of de Silva. By employing witness complexes, and with less dependence upon persistence, we are better able to represent the shape of complex data, with improved computational efficiency. Based on its success with artificial data sets and preliminary results from real environmental data, we propose that simplicial homology on this modified version of a witness complex is effective in approximating the shape of complex data, and is potentially useful as an adjunct tool for identifying patterns in these data. Further investigation with methods such as dimension reduction, and determining the parts of the data that give rise to a topological feature would aid in understanding how properties interact. Future work will concentrate on improving algorithm efficiency, and, as described above, in more detailed temporal analysis, and in temporal and spatial comparisons of the features obtained from TDA to determine how these features reflect physical phenomena. Finally, TDA and traditional computational and statistical techniques, as well as interactive visualizations, will be integrated to gain a more complete picture of dynamic environmental conditions.

Acknowledgements M. P. Wachowiak is supported by NSERC Grant #386586-2011.

References

1. Carlsson, G.: Topology and data. Bull. Am. Math. Soc. **46**, 255–308 (2009)
2. Dutta, R., Li, C., Smith, D., Das, A., Aryal, J.: Big data architecture for environmental analytics. In: Denzer, R., Argent, R.M., Schimak, G., Hřebíček, J. (eds.) Environmental Software Systems. Infrastructures, Services and Applications, vol. 448, pp. 578–588. Springer International Publishing, Cham (2015)
3. Holzinger, A.: Extravaganza tutorial on hot ideas for interactive knowledge discovery and data mining in biomedical informatics. In: Ślęzak, D., Tan, A.-H., Peters, J.F., Schwabe, L. (eds.) Brain Informatics and Health. LNCS, vol. 8609, pp. 502–515. Springer, Heidelberg (2014)
4. Lum, P.Y., Singh, G., Lehman, A., Ishkanov, T., Vejdemo-Johansson, M., Alagappan, M., Carlsson, J., Carlsson, G.: Extracting insights from the shape of complex data using topology. Nature **3** (2013). doi:10.1038/srep01236
5. Namieśnik, J., Wardencki, W.: Monitoring and analytics of atmospheric air pollution. Pol. J. Environ. Stud. **11**(3), 211–216 (2002)
6. Nicolau, M., Levine, A.J., Carlsson, G.: Topology based data analysis identifies a subgroup of breast cancers with a unique mutational profile and excellent survival. Proc. Natl. Acad. Sci. **108**(17), 7265–7270 (2011)
7. Rucco, M., Falsetti, L., Herman, D., Petrossian, T., Merelli, E., Nitti, C., Salvi, A.: Using Topological Data Analysis for diagnosis pulmonary embolism. http://arxiv.org/abs/1409.5020 [physics.med-ph]
8. de Silva, V., Carlsson, G.: Topological estimation using witness complexes. In: Proceedings of the Symposium on Point-Based Graphics, pp. 157–166. Eurographics Association, Aire-la-Ville (2004)
9. Wachowiak, M.P., Wachowiak-Smolikova, R., Dobbs, B.T., Abbott, J., Walters, D.: Interactive web-based visualization for lake monitoring in community-based participatory research. Environ. Pollut. **4**(2), 42–54 (2015)
10. Zomorodian, A.: Computational topology. In: Algorithms and Theory of Computation Handbook. Applied Algorithms and Data Structures Series, p. 3. Chapman & Hall/CRC, Boca Raton (2010)
11. Zomorodian, A.: Fast construction of the Vietoris-Rips complex. Comput. Graph. **34**, 263–271 (2010)

Estimating *Escherichia coli* Contamination Spread in Ground Beef Production Using a Discrete Probability Model

Petko M. Kitanov and Allan R. Willms

Abstract Human illness due to contamination of food by pathogenic strains of *Escherichia coli* is a serious public health concern and can cause significant economic losses in the food industry. Undercooked ground beef is the primary means of transmission of pathogenic *E. coli* to humans. In the Western world, most ground beef is produced in large facilities where many carcasses are butchered and various pieces of them are ground together in large batches. Assuming that the source of contamination is a single carcass, the primary determinant of how many batches of ground beef from a particular production cycle are affected is the manner in which pieces of that carcass are spread about in the raw sources that contribute to the ground beef batches. Assuming that ground beef from a particular batch has been identified at the consumer end as contaminated by *E. coli*, we model the probability that previous and subsequent batches generated in the same production cycle are also contaminated. This model may help the beef industry to identify the likelihood of contamination in other batches and potentially save money by not needing to cook or recall unaffected batches of ground beef.

1 Introduction

Pathogenic *Escherichia coli* strains, particularly O157:H7, cause illness in humans. Scallan et al. [8] indicate that a significant part of all cases of acquired food-borne illness in the U.S.A. is caused by the pathogenic strains of *E. coli*. These strains can be transmitted to humans through various food products including produce, dairy, and meat. The primary transmission of these strains from cattle to humans is by consumption of beef, especially under-cooked ground beef.

P.M. Kitanov
University of Ottawa, Ottawa, ON, K1N 6N5, Canada,
e-mail: pkitanov@uottawa.ca

A.R. Willms (✉)
University of Guelph, Guelph, ON, N1G 2W1, Canada,
e-mail: AWillms@uoguelph.ca

© Springer International Publishing Switzerland 2016

J. Bélair et al. (eds.), *Mathematical and Computational Approaches in Advancing Modern Science and Engineering*, DOI 10.1007/978-3-319-30379-6_23

In the western world, since ground beef production is concentrated in large meat processing plants, an *E. coli* contamination event at one of these plants can potentially affect many people over a wide area. Food safety regulations in North America require the removal of all production and raw sources associated with an identified contamination event. This leads to the recalling of large amounts of beef, much of it likely uncontaminated, and consequently large economic losses for the beef industry along with damage to reputation. This paper presents a discrete probability model for estimation of the likelihood that sequential batches of ground beef produced in a large plant are contaminated with pathogenic *E. coli* given that one batch is contaminated.

There have been many studies regarding the transmission of pathogenic *E. coli* and its impact on human health. For example, risk assessment for *E. coli* in ground beef and burgers in different countries has been quantified [4–7, 9]. A risk assessment model for *E. coli* O157:H7 in ground beef and beef cuts in Canada is presented in [10]. One study, [11], used data from a large *E.coli* outbreak in the U.S.A. to identify possible sources of contamination. The differing conditions in production and meat processing plants, distribution networks, and cooking methods, etc., make risk assessment of contamination a very difficult task. The primary method used in the above studies is statistical analysis of empirical data. Stochastic models for outbreak and transmission of *E. coli* O157:H7 infection in cattle, are presented in [12] and [15].

Aslam et al. [1, 2] and Bell [3] have studied the sources of contamination by pathogenic *E. coli* in a beef-packing plant. *E. coli* in meat products, originates mainly from the hides of the incoming animals and is transferred to the trimmings, and subsequently the ground beef, during the dressing of carcasses and carcass breaking. Various decontamination treatments, such as antimicrobial solutions and pasteurization, have been implemented at some beef processing plants. Although it was found in [14] that these treatments significantly reduce the risk of contamination, the hazard can not be completely removed. Also, there are still no effective methods for quickly screening large amounts of ground beef in big production facilities. All of these uncertainties justify using probabilistic methods for control and estimation of *E. coli* contamination in the production of ground beef.

2 The Model

Consider a large ground beef production facility. The ground beef is produced in batches of several tonnes each. The input to these batches are several raw sources, typically a “lean” fresh source and a “fat” fresh source, but also often a frozen source and other sources such as BLBT (Boneless Lean Beef Trimmings). The batch is well-mixed, so that if any contamination is present on any of the raw source material that is input into the batch, the entire batch is deemed to be contaminated.

2.1 Basic Assumptions

The primary assumption is that the contamination is due to a single carcass, often associated with the fat layer on that carcass. Spread of the contaminant in the production process is assumed to be due to division and dispersion of the contaminated carcass portions; transfer via physical contact with other pieces and machinery surfaces is assumed to be negligible. Further, due to the temperature at which production occurs, it is assumed that the contaminant does not grow appreciably.

The manner in which a carcass is spread over some region in a raw source is highly dependent on the production process—the way in which trim is added to the raw source bins, etc. We make a number of simplifying assumptions:

- for each raw source, the number and mass of all pieces from all carcasses is the same, and the manner of spread of the pieces from each carcass throughout the raw source is the same
- material in a raw source bin is not mixed, or what mixing occurs is captured by the carcass distribution function,
- the material in the raw source bins can be ordered and is used as input to the ground beef production batches in that order

This last assumption allows us to define a "mass location" in each raw source; mass from a particular raw source used in batch number b comes from locations just prior to those for the mass from the same source used in batch number $b + 1$. This location is a discrete variable increasing in increments of the mass of a piece, which by the first assumption is always the same in a given raw source. Although these assumptions are all invalid to some degree, they are useful for making a tractable model that we believe to have relevancy to the problem. Further, replacing them with more realistic assumptions would require considerably more information on the production process.

Suppose there are B batches of ground beef produced in a production cycle and S raw sources used as input to these batches, not all of which need be used in any particular batch. It is assumed that a particular batch of ground beef has been identified as being contaminated. This batch is referred to as the "hot" batch, and it is identified as batch number h. The origin of this contamination is due to a single hot raw source and in turn, to a single hot carcass in this raw source. This model computes probabilities of contamination due to the same origin for all other batches in the production cycle.

2.2 Carcass Spread in a Raw Source

Let p_s be the number of pieces, each of mass a_s contributed by each carcass in raw source s. Let C_s be the total number of carcasses in the raw source, then the total number of pieces in the raw source is $N_s = C_s p_s$. Since we are assuming the raw source material is ordered, we assign piece position numbers 1 to N_s for this source. The total mass in the raw source is $M_s = C_s p_s a_s$.

The pieces from each carcass are distributed throughout this raw source in some manner, the same for each carcass. Here we consider a piece-wise linear distribution function that is even around its centre:

$$F(n) = \begin{cases} \frac{(|n|-L_{i-1})H_i^- + (L_i - |n|)H_{i-1}^+}{L_i - L_{i-1}} & \text{if } L_{i-1} < |n| < L_i,\ 1 \le i \le K, \\ \frac{H_i^- + H_i^+}{2} & \text{if } |n| = L_i,\ 0 \le 1 \le K, \\ 0 & \text{if } |n| > L_K. \end{cases} \tag{1}$$

This function's domain has $2K$ pieces, distributed symmetrically around zero, divided by the constants $0 = L_0 < L_1 < L_2 < \cdots < L_K$, where each L_i, $1 \le i \le K$, is a positive multiple of p_s, and $H_i^{\pm} \ge 0$, $0 \le i \le K$, are the limiting values of F at L_i from the left $(-)$ and right $(+)$, respectively. Necessarily, $H_0^- = H_0^+$, $H_K^+ = 0$, and

$$\sum_{i=1}^{K} (L_i - L_{i-1})(H_i^- + H_{i-1}^+) = 1.$$

This class of functions includes the uniform distribution, obtained with $K = 1$, and $H_0^{\pm} = H_1^- = \frac{1}{2L_1}$. If the parameters are chosen such that $H_i^- = H_i^+$, $1 \le i \le K$, then the function F is continuous. Each raw source s may have a different distribution function F_s, and thus a different set of parameters K, L, and H. We assume that the distribution centres, μ_c of each carcass are uniformly spread through the source:

$$\mu_c = \left(c - \frac{1}{2}\right) p_s, \qquad c \in \mathbb{Z}, \quad 1 \le c \le C_s = \frac{N_s}{p_s} = \frac{M_s}{p_s a_s}. \tag{2}$$

Here c is the carcass number and the discrete probability density function for carcass c in this raw source is $F_s(n - \mu_c)$. The expected fraction of a carcass c present in any set of piece positions $R \subseteq \{1, 2, \ldots, N_s\}$ is

$$Q_{sc}(R) = \sum_{n \in R} F_s(n - \mu_c). \tag{3}$$

Equivalently, $F_s(n - \mu_c)$ may be interpreted as the density function for a particular piece of this carcass, and $Q_{sc}(R)$ as the probability that this piece is located in the set R. The piece-wise linear nature of this density function allows it to satisfy the requirement that the fraction of all carcasses at a particular piece position must add to $\frac{1}{p_s}$ so that the expected number of pieces at that location will be one.

2.3 Batch Contamination Probability

Let m_{sb} be the mass input from source s to batch b, and let M_{sb} be the mass input from source s to batches prior to batch b (both assumed to be multiples of a_s). Let n_{sb} and N_{sb} be the number of pieces from source s input to batch b and input to batches prior to b, respectively, that is, $m_{sb} = n_{sb}a_s$ and $M_{sb} = N_{sb}a_s$. The sequential piece positions in raw source s that are input to batch b are then the integers in the interval

$$B_{sb} = [N_{sb} + 1, N_{sb} + n_{sb}].$$

For source s, denote the probability of carcass c being *absent* from the set B_{sb} as $A_{sc}(B_{sb})$. Modelling the selection of pieces as being independent of the other pieces that have already been selected from this carcass and other carcasses, this absence probability is

$$A_{sc}(B_{sb}) = (1 - Q_{sc}(B_{sb}))^{p_s}, \tag{4}$$

where Q_{sc} is defined by (3).

Consider two batches, h and j, which receive input from source s, and a particular carcass with distribution centred at μ_c. The probability that this carcass has at least one piece that is input to batch h and at least one that is input to batch j is

$$\text{Prob}(c \text{ from } s \text{ in } h \,\&\, j) = 1 - \left[A_{sc}(B_{sh}) + A_{sc}(B_{sj}) - A_{sc}(B_{sh} \cup B_{sj})\right]. \tag{5}$$

The last term is present because it is included in both the previous terms but should only be counted once.

Given that batch h is the hot batch and assuming the contamination is due to a hot carcass in raw source s, then the probability that a particular carcass c from source s is the hot carcass is a uniform probability depending on the total number of carcasses, C_{sh} from source s input to h:

$$\text{Prob}(c \text{ from } s \text{ is hot}) = \frac{1}{C_{sh}} \le \frac{p_s}{n_{sh}}. \tag{6}$$

The right hand expression in (6) is an upper bound on the probability since the number of distinct carcasses present, C_{sh}, will likely be more than n_{sh}/p_s, especially if the spread of each carcass is large. A reasonable estimate for C_{sh} might be to include all carcasses c whose expected number of pieces in batch h is at least one, that is $Q_{sc}(B_{sh}) \ge 1/p_s$.

The contamination in batch h may be due to any of the raw sources that were input for this batch and these raw sources may have varying degrees of relative susceptibility to being contaminated. For example, the susceptibility factor, g_s, for raw source number s, might be near zero if the source is frozen material, and might be one if the source is fresh material from the plant. Let f_s be the fraction of fat in raw source s. The probability that raw source s is the origin of the contamination in batch h is the weighted fraction of the input mass, the weights being the product of the susceptibility factors and the fat fractions:

$$\text{Prob}(s \text{ is hot}) = \frac{g_s f_s m_{sh}}{\sum_{k=1}^{S} g_k f_k m_{kh}}. \tag{7}$$

The probability that batch j is also contaminated given that batch h is contaminated is then found by summing over all raw sources:

$$\text{Prob}(j \text{ hot} \mid h) =$$

$$\sum_{s=1}^{S} \text{Prob}(s \text{ is hot}) \sum_{c=1}^{C_s} \text{Prob}(c \text{ from } s \text{ is hot}) \, \text{Prob}(c \text{ from } s \text{ in } h \,\&\, j). \tag{8}$$

This expression can be evaluated using (4), (5), (6), and (7).

One aspect that has not been accounted for is the possibility that one carcass may be present in more than one raw source. This could occur, for example, in a production facility where one fresh lean bin is filled and removed from the trimming line while a particular carcass has only partially been trimmed. In this case pieces from one carcass will occur at the "top" of one raw source bin and the "bottom" of the next, thus potentially spreading the contamination over a much wider area, if this happens to be the hot carcass. The model could be modified to account for this possibility if desired, but would require information about the order and means of filling of the various raw sources.

3 Example

To illustrate the above model we present one example. These data are fictitious but based on typical values one might encounter in a large ground beef production facility. In this example, there are a total of $S = 7$ sources (three frozen lean, two fresh lean, and two fresh fat), and a total of $B = 14$ one-tonne batches of ground beef are produced. Relevant information for the sources is provided in Table 1. The uniform distribution for F ($K = 1$, $H_0^{\pm} = H_1^{-} = 1/(2L_1)$) was used for all raw sources in this example. The mass from each source used in each batch is provided in Table 2.

Using the above model, and letting each batch be the hot batch in turn, probabilities of contamination for each batch were computed. The probability that a particular carcass c in source s is the hot carcass, Prob(c from s is hot), was computed using the suggestion following (6), that is, counting the number of carcasses in source s that are present in batch h as all those carcasses c with

Table 1 Model parameters for the raw sources

Source	g_s	f_s	p_s	a_s (kg)	L_{1s}	C_s	M_s (kg)
I (frozen lean)	0.2	0.05	25	0.6	400	134	2000
II (frozen lean)	0.2	0.09	25	0.6	400	134	2000
III (frozen lean)	0.2	0.07	25	0.6	400	134	2000
IV (fresh lean)	0.8	0.10	20	0.7	350	179	2500
V (fresh lean)	0.8	0.08	20	0.7	350	215	3000
VI (fresh fat)	1.0	0.40	50	0.5	500	80	2000
VII (fresh fat)	1.0	0.45	50	0.5	500	80	2000

Table 2 Source input mass, m_{sb}, (kg) and total fat percentage for each batch

	Source							
	Frozen lean			Fresh lean		Fresh fat		
Batch	I	II	III	IV	V	VI	VII	Fat %
1	312			136		552		25
2	384			52		564		25
3	114			404		260	222	25
4	262			239		231	268	25
5	201	205		89		293	212	25
6	320	180		292		100	108	15
7	407	105		284			204	15
8		390		456			154	15
9		300		205	325		170	15
10		209		211	543		37	10
11		293		132	536		39	10
12		318	94		540		48	10
13			479		454		67	10
14			701		226		73	10

$Q_{sc}(B_{sh}) \geq 1/p_s$. Complete results are given in Table 3 and a selection of these are plotted in Fig. 1.

The likelihood of other batches being contaminated is highly dependent on the source input. In the example, if one of the early batches is the hot batch, then the likelihood of contamination is only nonzero for batches near the hot batch. Conversely, if one of the last batches is the hot batch, then a larger number of other batches have a nonzero probability of being contaminated. This is due to the source configuration shown in Tables 1 and 2. Since the contamination is most likely to be carried in the fatty sources (VI and VII) followed by the lean fresh source (IV and V), the distribution of these sources across the batches is a primary contributor to the contamination probability distribution. The other factor we found to be very important was the values of L_{1s}. If these spread indicators were small, then the contamination was much more confined to nearby batches to the hot batch.

Table 3 Probability of contamination for each batch in percent. Each column corresponds to a different hot batch

	Hot batch													
Batch	1	2	3	4	5	6	7	8	9	10	11	12	13	14
1	99	47	6	0	0	0	0	0	0	0	0	0	0	0
2	62	99	34	13	0	0	0	0	0	0	0	0	0	0
3	5	46	100	65	35	3	0	0	0	0	0	0	0	0
4	0	21	79	100	72	47	19	0	0	0	0	0	0	0
5	0	0	38	66	100	81	47	18	2	0	0	0	0	0
6	0	0	3	34	63	98	69	34	12	0	0	0	0	0
7	0	0	0	12	24	48	100	66	32	6	6	5	2	0
8	0	0	0	0	11	24	66	100	60	24	14	15	15	13
9	0	0	0	0	1	8	37	66	100	63	34	26	29	30
10	0	0	0	0	0	0	17	39	72	100	66	30	32	34
11	0	0	0	0	0	0	13	28	49	60	100	62	35	38
12	0	0	0	0	0	0	8	23	39	20	50	100	64	43
13	0	0	0	0	0	0	3	18	34	17	21	57	100	74
14	0	0	0	0	0	0	0	12	28	14	18	27	67	100

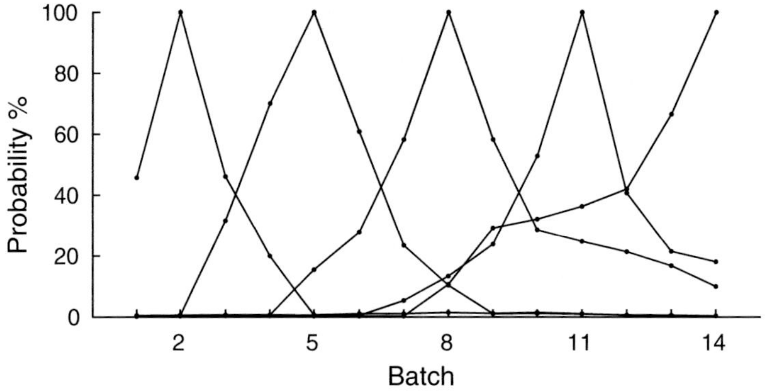

Fig. 1 Probability of contamination for each batch given that a fixed batch is contaminated (hot). The five separate curves correspond to batches 2,5,8,11, and 14 being the hot batch

The overall result then, is the obvious observation that if one wishes to restrict contamination, then one should restrict the spread of each carcass within the raw source, and restrict the spread of the raw source across batches. In other words, try to make it so that one carcass is only present in one batch.

4 Conclusion

We believe that the proposed model may help to reduce economic losses in the beef industry. To our knowledge, this is a novel method for estimating of the probability of *E. coli* contamination in the production of ground beef. Due to the necessarily cautious food safety regulations, in order for this model to be adopted by regulatory agencies, it would need to be further refined to account for plant-specific processes, and its predictions thoroughly tested.

Most of the input to the model is easily obtained from production records, or easily estimated (such as the average size of pieces). The most difficult input values to estimate are the spread parameters, K, L, and $H^{\pm}$, however, some knowledge of these values could be obtained either through detailed observations of the plant processes or by using genetic sampling experiments in the raw sources as well as the final ground beef product. A useful next step would be to use some type of innocuous surrogate for *E. coli* that could be applied to pieces from a particular carcass and then identified in the final product batches. This would allow for better determination of carcass spread within a raw source and would provide data for direct validation of the model.

Further development of the model, allowing for each carcass to have different size pieces, and allowing for carcasses to be present in more than one raw source, is currently being completed [13]. This extended work also incorporates genetic typing data from a production facility to help inform the choice of carcass distribution function.

Acknowledgements The second author was supported by a grant from the Applied Livestock Genomics Program of Genome Alberta.

References

1. Aslam, M., Greer, G.G., Nattress, F.M., Gill, C.O., McMullen, L.M.: Genotypic analysis of *Escherichia coli* recovered from product and equipment at a beef-packing plant. J. Appl. Microbiol. **97**, 78–86 (2004)
2. Aslam, M., Nattress, F.M., Greer, G.G., Yost, C., Gill, C.O., McMullen, L.: Origin of contamination and genetic diversity of *Eschericia coli* in beef cattle. Appl. Environ. Microbiol. **69**(5), 2794–2799 (2003)
3. Bell, R.G.: Distribution and sources of microbial contamination on beef carcasses. J. Appl. Microbiol. **82**, 292–300 (1997)
4. Cassin, M.H., Lammerding, A.M., Todd, E.C.D., Ross, W., McColl, R.S.: Quantitative risk assessment for *Escherichia coli* O157:H7 in ground beef hamburgers. Int. J. Food Microbiol. **41**(1), 21–44 (1998)
5. Duffy, G., Butler, F., Cummins, E., O'Brien, S., Nally, P., Carney, E., Henchion, M., Mahone, D., Cowan, C.: *E. coli* O157:H7 in beef burgers produced in the Republic of Ireland: a quantitative microbial risk assessment. Tech. rep., Ashtown Food Research Centre, Teagasc, Dublin (2006)

6. Duffy, G., Cummins, E., Nally, P., O'Brien, S., Butler, F.: A review of quantitative microbial risk assessment in the management of *Escherichia coli* O157:H7 on beef. Meat Sci. **74**(1), 76–88 (2006)
7. Ebel, E., Schlosser, W., Kause, J., Orloski, K., Roberts, T., Narrod, C., Malcolm, S., Coleman, M., Powell, M.: Draft risk assessment of the public health impact of *Escherichia coli* O157:H7 in ground beef. J. Food Prot. **67**(9), 1991–1999 (2004)
8. Scallan, E., Hoekstra, R.M., Angulo, F.J., Tauxe, R.V., Widdowson, M.A., Roy, S.L., Jones, J.L., Griffin, P.M.: Foodborne illness acquired in the United States – major pathogens. Emerg. Infect. Dis. **17**(1), 7–15 (2011)
9. Signorini, M., Tarabla, H.: Quantitative risk assessment for verocytotoxigenic *Escherichia coli* in ground beef hamburgers in Argentina. Int. J. Food Microbiol. **132**(2–3), 153–161 (2009)
10. Smith, B.A., Fazil, A., Lammerding, A.M.: A risk assessment model for *Escherichia coli* O157:H7 in ground beef and beef cuts in canada: evaluating the effects of interventions. Food Control **29**, 364–381 (2013)
11. Tuttle, J., Gomez, T., Doyle, M.P., Wells, J.G., Zhao, T., Tauxe, R.V., Griffin, P.M.: Lessons from a large outbreak of *Escherichia coli* O157:H7 infections: insights into the infectious dose and method of widespread contamination of hamburger patties. Epidemiol. Infect. **122**(2), 185–192 (1999)
12. Wang, X., Gautam, R., Pinedo, P.J., Allen, L.J.S., Ivanek, R.: A stochastic model for transmission, extinction and outbreak of *Escherichia coli* O157:H7 in cattle as affected by ambient temperature and cleaning practices. J. Math. Biol. (2013). doi:10.1007/s00285-013-0707-1
13. Willms, A.R., Kitanov, P.M.: Probability of *Escherichia coli* contamination spread in ground beef production (2016, submitted)
14. Yang, X., Badoni, M., Youssef, M.K., Gill, C.O.: Enhanced control of microbiological contamination of product at a large beef packing plant. J. Food Prot. **75**(1), 144–149 (2012)
15. Zhang, X.S., Chase-Topping, M.E., McKendrick, I.J., Savill, N.J., Woolhouse, M.E.J.: Spread of *Escherichia coli* O157:H7 infection among Scottish cattle farms: Stochastic models and model selection. Epidemics **2**, 11–20 (2010)

The Impact of Movement on Disease Dynamics in a Multi-city Compartmental Model Including Residency Patch

Diána Knipl

Abstract The impact of population dispersal between two cities on the spread of a disease is investigated analytically. A general SIRS model is presented that tracks the place of residence of individuals, allowing for different movement rates of local residents and visitors in a city. Provided the basic reproduction number is greater than one, we demonstrate in our model that increasing the travel volumes of some infected groups may result in the extinction of a disease, even though the disease cannot be eliminated in each city when the cities are isolated.

1 Introduction

The spatial spread of infectious diseases has been observed many times in history. Most recent examples include the 2002–2003 SARS epidemic in Asia and the global spread of the 2009 pandemic influenza A(H1N1). The Middle East Respiratory Syndrome coronavirus (MERS-CoV) outbreak emerged in 2012, and West Africa is currently witnessing the extensive Ebola virus (EBOV) outbreak, that pose a global threat. There is an increasing interest in the mathematical modelling literature for the spread of epidemics between discrete geographical locations (patches, or cities). Such metapopulation models incorporate single or multiple species occupying multiple spatial patches that are connected by movement dependent or independent of disease status. Such models have been discussed for an array of infectious diseases including measles and influenza, by Arino and coauthors [2–5], Sattenspiel

D. Knipl (✉)
Agent-Based Modelling Laboratory, 331 Lumbers, York University, 4700 Keele St., Toronto, ON, M3J 1P3, Canada

MTA–SZTE Analysis and Stochastic Research Group, University of Szeged, Aradi vértanúk tere 1, Szeged, H-6720, Hungary

Current Institution: Department of Mathematics, University College London, Gower Street, London, WC1E 6BT, UK
e-mail: knipl@yorku.ca; knipl@math.u-szeged.hu; d.knipl@ucl.ac.uk

© Springer International Publishing Switzerland 2016

J. Bélair et al. (eds.), *Mathematical and Computational Approaches in Advancing Modern Science and Engineering*, DOI 10.1007/978-3-319-30379-6_24

and coauthors [14, 15], Wang and coauthors [9, 13, 19, 20]. The work of Arino [1], and Arino and van den Driessche [6] provide a thorough review of the literature.

When considering intervention strategies for epidemic models, our attention is focused on the basic reproduction number $\mathscr{R}_0$, which is the expected number of secondary cases generated by a typical infected host introduced into a susceptible population. This quantity serves as a threshold parameter for disease elimination; if $\mathscr{R}_0 < 1$ then the disease dies out when a small number of infected individuals is introduced whereas if $\mathscr{R}_0 > 1$ then the disease can persist in the population. The above mentioned works illustrate that in metapopulation models $\mathscr{R}_0$ often arises as a complicated formula of the model parameters. Such models include multiple infected classes, and individuals' movement makes it challenging to compute the number of new infections generated by an infected case, and to understand the dependence of $\mathscr{R}_0$ on the movement rates. To calculate $\mathscr{R}_0$ in metapopulation models, the next-generation method is used (see Diekmann et al. [7]).

The models in [4, 5, 14, 15] include residency patch, that is, these models keep track of the patch of origin of an individual as well as where an individual is at a given time (either as resident, or as visitor). There are many reasons why individuals should be distinguished in an epidemic model by their residential statuses; visitors and local residents may have very different contact rates and mixing patterns, but more significantly, these groups are different in their travel rates because in reality, a large part of outbound travels from a city are return trips. In the above works, the basic reproduction number was calculated and its dependence on the movement rates was studied numerically. Some complicated behavior of $\mathscr{R}_0$ in these parameters was highlighted in [1, 2, 4]: numerical simulations suggest that when the infection is present in same patches but absent in others without movement, then travel with small rates can allow for disease persistence in the metapopulation although higher travel rates can drive the disease to extinction.

In this work we present a demographic SIRS epidemic metapopulation model in two cities, and analytically investigate the impact of individuals' movement between the two cities on the disease dynamics. In each city we distinguish residents from visitors, and consider the general situation when individuals with different disease statuses and residential statuses have different movement rates. In our analysis we utilize the concept of the target reproduction number, developed by Shuai et al. in [16, 17]. This quantity measures the effort required to eliminate infectious diseases, when an intervention strategy is targeted at single entries or sets of such entries of a next-generation matrix. Focusing on the control of infected individuals' movement between the two cities—an intervention strategy often applied in pandemic situations—we give conditions and describe how the travel rate of a specific group or some of these groups should be changed to prevent an outbreak.

2 Model Formulation

We formulate a dynamical model to describe the spread of an infectious disease among two cities. We divide the entire populations of the two cities into the disjoint classes S_j^m, I_j^m, R_j^m, $j \in \{1,2\}$, $m \in \{r,v\}$, where the letters S, I, and R represent the compartments of susceptible, infected, and recovered individuals, respectively. Lower index $j \in \{1,2\}$ specifies the current city, upper index $m \in \{r,v\}$ denotes the residential status of the individual in the current city (resident or visitor). An individual who is currently in city j and has residential status v, originally belongs to city k hence we say that this individual has origin in city k ($k \in \{1,2\}, k \neq j$). Let $S_j^m(t)$, $I_j^m(t)$, $R_j^m(t)$, $j \in \{1,2\}$, $m \in \{r,v\}$ be the number of individuals belonging to S_j^m, I_j^m, R_j^m respectively, at time t. The transmission rate between a susceptible individual with residential status m and an infected individual with residential status n in region j ($j \in \{1,2\}, m, n \in \{r,v\}$) is denoted by β_j^{mn}, and disease transmission is modelled by standard incidence. Model parameter γ_j is the recovery rate of infected individuals in city j, and d_j is the natural mortality rate of all individuals with origin in city j. Recovered individuals with residential status m in city j lose disease-induced immunity by rate θ_j^m. For the total population of residents and visitors currently being in city j we use the notations N_j^r and N_j^v, and let N_j^o denote the total population with origin in j. It holds that

$$
\begin{aligned}
N_j^r &= S_j^r + I_j^r + R_j^r, \qquad N_j^v = S_j^v + I_j^v + R_j^v,\\
N_j^o &= S_j^r + I_j^r + R_j^r + S_k^v + I_k^v + R_k^v, \quad k \neq j.
\end{aligned}
$$

For the recruitment term Λ_j into the susceptible resident population we assume that Λ_j is a function of the populations N_j^r and N_k^v ($k \neq j$), that is, the populations with origin in city j. We denote by m_{kj}^{Sm}, m_{kj}^{Im}, and m_{kj}^{Rm} the travel rate of susceptible, infected, and recovered individuals, respectively, with residential status m in city j travelling to city k. Based on the assumptions formulated above, we obtain the following system of differential equations for the disease transmission in city j:

$$
\begin{aligned}
\frac{dS_j^r}{dt} &= \Lambda_j(N_j^r, N_k^v) - \beta_j^{rr}\frac{S_j^r I_j^r}{N_j^r+N_j^v} - \beta_j^{rv}\frac{S_j^r I_j^v}{N_j^r+N_j^v} - d_j S_j^r + \theta_j^r R_j^r - m_{kj}^{Sr} S_j^r + m_{jk}^{Sv} S_k^v,\\
\frac{dI_j^r}{dt} &= \beta_j^{rr}\frac{S_j^r I_j^r}{N_j^r+N_j^v} + \beta_j^{rv}\frac{S_j^r I_j^v}{N_j^r+N_j^v} - (\gamma_j + d_j) I_j^r - m_{kj}^{Ir} I_j^r + m_{jk}^{Iv} I_k^v,\\
\frac{dR_j^r}{dt} &= \gamma_j I_j^r - (\theta_j^r + d_j) R_j^r - m_{kj}^{Rr} R_j^r + m_{jk}^{Rv} R_k^v,\\
\frac{dS_j^v}{dt} &= -\beta_j^{vr}\frac{S_j^v I_j^r}{N_j^r+N_j^v} - \beta_j^{vv}\frac{S_j^v I_j^v}{N_j^r+N_j^v} - d_k S_j^v + \theta_j^v R_j^v - m_{kj}^{Sv} S_j^v + m_{jk}^{Sr} S_k^r,\\
\frac{dI_j^v}{dt} &= \beta_j^{vr}\frac{S_j^v I_j^r}{N_j^r+N_j^v} + \beta_j^{vv}\frac{S_j^v I_j^v}{N_j^r+N_j^v} - (\gamma_j + d_k) I_j^v - m_{kj}^{Iv} I_j^v + m_{jk}^{Ir} I_k^r,\\
\frac{dR_j^v}{dt} &= \gamma_j I_j^v - (\theta_j^v + d_k) R_j^v - m_{kj}^{Rv} R_j^v + m_{jk}^{Rr} R_k^r.
\end{aligned}
\tag{1}
$$

Standard arguments from the theory of differential equations guarantee that the system (1) is well posed. The function forming the right hand side of the system is Lipschitz continuous, which implies the existence of a unique solution. The derivative of each system variable is nonnegative when the variable is zero, hence solutions remain nonnegative for nonnegative initial data. For the dynamics of the total population with origin in city j, we obtain the equation

$$\frac{\mathrm{d}N_j^o}{\mathrm{d}t} = \Lambda_j(N_j^r, N_k^v) - d_j(N_j^r + N_k^v), \quad k \neq j.$$

If $\Lambda_j(N_j^r, N_k^v) = d_j(N_j^r + N_k^v)$ then the population with origin in j is constant. For constant recruitment term Λ_j it is easy to derive that $\hat{N}_j^o = \Lambda_j/d_j$ gives the unique equilibrium of N_j^o. With fixed N_1^o and N_2^o it is obvious from nonnegativity that the solutions of the system (1) are bounded. The model is at an equilibrium if the time derivatives in the system (1) are zero. At a disease-free equilibrium it holds that $I_1^r = I_1^v = I_2^r = I_2^v = 0$ that implies $R_1^r = R_1^v = R_2^r = R_2^v = 0$. Thus at a DFE S_1^r, S_1^v, S_2^r, S_2^v satisfy

$$\begin{aligned}
\Lambda_1(N_1^r, N_2^v) - d_1 S_1^r - m_{21}^{Sr} S_1^r + m_{12}^{Sv} S_2^v &= 0,\\
-d_2 S_1^v - m_{21}^{Sv} S_1^v + m_{12}^{Sr} S_2^r &= 0,\\
\Lambda_2(N_2^r, N_1^v) - d_2 S_2^r - m_{12}^{Sr} S_2^r + m_{21}^{Sv} S_1^v &= 0,\\
-d_1 S_2^v - m_{12}^{Sv} S_2^v + m_{21}^{Sr} S_1^r &= 0.
\end{aligned}$$

Hence if N_1^o and N_2^o are fixed then using that $\Lambda_1 = d_1 N_1^o$ and $\Lambda_2 = d_2 N_2^o$, it follows that

$$\begin{bmatrix} S_1^r \\ S_2^v \end{bmatrix} = \begin{bmatrix} d_1 + m_{21}^{Sr} & -m_{12}^{Sv} \\ -m_{21}^{Sr} & d_1 + m_{12}^{Sv} \end{bmatrix}^{-1} \begin{bmatrix} d_1 N_1^o \\ 0 \end{bmatrix},$$
$$\begin{bmatrix} S_2^r \\ S_1^v \end{bmatrix} = \begin{bmatrix} d_2 + m_{12}^{Sr} & -m_{21}^{Sv} \\ -m_{12}^{Sr} & d_2 + m_{21}^{Sv} \end{bmatrix}^{-1} \begin{bmatrix} d_2 N_2^o \\ 0 \end{bmatrix}.$$

The following result is proved.

Proposition 1 *Assume that the total populations with origin in city 1 and with origin in city 2 are constant. Then there is a unique DFE in the model* (1) *where*

$$\begin{aligned}
S_j^r &= \frac{(d_j + m_{jk}^{Sv}) d_j N_j^o}{(d_j + m_{kj}^{Sr})(d_j + m_{jk}^{Sv}) - m_{jk}^{Sv} m_{kj}^{Sr}},\\
S_j^v &= \frac{m_{jk}^{Sr} d_k N_k^o}{(d_k + m_{jk}^{Sr})(d_k + m_{kj}^{Sv}) - m_{kj}^{Sv} m_{jk}^{Sr}}, \qquad j, k \in \{1, 2\}, k \neq j,\\
I_j^r &= I_j^v = 0,\ R_j^r = R_j^v = 0,\ \textit{and}\ N_j^r = S_j^r,\ N_j^v = S_j^v.
\end{aligned}$$

For the stability of the DFE in the full model (1) we linearize the subsystem of (1) that consists of the equations for I_1^r, I_1^v, I_2^r, and I_2^v—the infected subsystem—about the DFE, and give the Jacobian J, as

$$J = B - G - M.$$

$$B = \begin{bmatrix} \frac{\beta_1^{rr} N_1^r}{N_1^r+N_1^v} & \frac{\beta_1^{rv} N_1^r}{N_1^r+N_1^v} & 0 & 0 \\ \frac{\beta_1^{vr} N_1^v}{N_1^r+N_1^v} & \frac{\beta_1^{vv} N_1^v}{N_1^r+N_1^v} & 0 & 0 \\ 0 & 0 & \frac{\beta_2^{rr} N_2^r}{N_2^r+N_2^v} & \frac{\beta_2^{rv} N_2^r}{N_2^r+N_2^v} \\ 0 & 0 & \frac{\beta_2^{vr} N_2^v}{N_2^r+N_2^v} & \frac{\beta_2^{vv} N_2^v}{N_2^r+N_2^v} \end{bmatrix}, \; M = \begin{bmatrix} m_{21}^{Ir} & 0 & 0 & -m_{12}^{Iv} \\ 0 & m_{21}^{Iv} & -m_{12}^{Ir} & 0 \\ 0 & -m_{21}^{Iv} & m_{12}^{Ir} & 0 \\ -m_{21}^{Ir} & 0 & 0 & m_{12}^{Iv} \end{bmatrix},$$

and $G = \operatorname{diag}(\gamma_1 + d_1, \gamma_1 + d_2, \gamma_2 + d_2, \gamma_2 + d_1) =: \operatorname{diag}(g_1^r, g_1^v, g_2^r, g_2^v)$. Let $s(A)$ denote the maximum real part of all eigenvalues of any square matrix A, and $\rho(A)$ denote the dominant eigenvalue of any square matrix A. We say that a square matrix A has the Z-sign pattern if all entries of A are nonpositive except possibly those in the diagonal. If $A^{-1} \geq 0$ holds then A is a non-singular M-matrix (several definitions exist for M-matrices, see [8, Theorem 5.1]). By [18, Lemma 1] the stability of the DFE is determined by the eigenvalues of J; more precisely, the DFE is locally asymptotically stable if $s(J) < 0$, meaning that all eigenvalues have negative real part, and the DFE is unstable if $s(J) > 0$, when there is an eigenvalue with positive real part. The proof of the next proposition follows by similar arguments as those in the proof of [18, Theorem 2].

Proposition 2 *Consider a splitting $F - V$ of the Jacobian of the infected subsystem about the DFE, where F is a nonnegative matrix and V is a non-singular M-matrix. Then, it holds that $s(J) < 0$ if and only if $\rho(FV^{-1}) < 1$, $s(J) = 0$ if and only if $\rho(FV^{-1}) = 1$, and $s(J) > 0$ if and only if $\rho(FV^{-1}) > 1$.*

The stability of the DFE is often characterized through the basic reproduction number $\mathscr{R}_0$, that is defined as the dominant eigenvalue of the next-generation matrix (NGM). The concept of the NGM was initially introduced by Diekmann et al. [7]. This matrix is computed as $K_0 := F_0 V_0^{-1}$, where F_0 equals B, the transmission matrix describing new infections, and V_0 is defined as $G + M$, the transition matrix for the transitions between and out of infected classes. F_0 and V_0 satisfy the conditions of Proposition 2, hence $\mathscr{R}_0 = \rho(K_0) = \rho(F_0 V_0^{-1})$ is a threshold quantity for the stability of the DFE. We obtain the following corollary from Proposition 2.

Corollary 1 *Consider a splitting $F - V$ of the Jacobian of the infected subsystem about the DFE, where F is a nonnegative matrix and V is a non-singular M-matrix. Then, it holds that $\rho(FV^{-1}) < 1$ if and only if $\mathscr{R}_0 < 1$, $\rho(FV^{-1}) = 1$ if and only if $\mathscr{R}_0 = 1$, and $\rho(FV^{-1}) > 1$ if and only if $\mathscr{R}_0 > 1$.*

With other words, for any splitting $F - V$ of the Jacobian where F is nonnegative and V is a non-singular M-matrix, there arises an alternative NGM by FV^{-1}; moreover, $\rho(FV^{-1})$ and $\mathscr{R}_0$ agree at the threshold value for the stability of the DFE. In the next section we will investigate the impact of movement on the disease dynamics by constructing some alternative next-generation matrices and utilizing the method of Shuai et al. [16] to measure the effort required to control the disease.

3 Main Results

Using the definition of G and the transmission matrix F_0, we introduce the quantities

$$\mathcal{R}_1^r = \frac{\beta_1^{rr}N_1^r+\beta_1^{vr}N_1^v}{g_1^r(N_1^r+N_1^v)},\quad \mathcal{R}_1^v = \frac{\beta_1^{rv}N_1^r+\beta_1^{vv}N_1^v}{g_1^v(N_1^r+N_1^v)},$$
$$\mathcal{R}_2^r = \frac{\beta_2^{rr}N_2^r+\beta_2^{vr}N_2^v}{g_2^r(N_2^r+N_2^v)},\quad \mathcal{R}_2^v = \frac{\beta_2^{rv}N_2^r+\beta_2^{vv}N_2^v}{g_2^v(N_2^r+N_2^v)},$$

where $\mathcal{R}_j^m$ denotes the expected number of new cases in city j when a single infected individual with residential status m who doesn't travel is introduced into city j.

Consider the matrices $F_1 = B - M + \mathrm{diag}(m_{21}^{Ir}, m_{21}^{Iv}, m_{12}^{Ir}, m_{12}^{Iv})$ and $V_1 = G + \mathrm{diag}(m_{21}^{Ir}, m_{21}^{Iv}, m_{12}^{Ir}, m_{12}^{Iv})$. Then $J = F_1 - V_1$ gives another splitting of the Jacobian, moreover $F_1 \geq 0$ and V_1 is a non-singular M-matrix. We obtain the following theorem.

Theorem 1 *If $\mathcal{R}_1^r > 1$, $\mathcal{R}_1^v > 1$, $\mathcal{R}_2^r > 1$, and $\mathcal{R}_2^v > 1$ then the DFE is unstable when the cities are isolated, and movement cannot stabilize the DFE. If the inequalities are reversed then the DFE is stable when the cities are isolated, and movement cannot destabilize the DFE.*

Proof Consider the splitting $J = F_1 - V_1$. As V_1 is a diagonal matrix, one easily computes the alternative NGM

$$K_1 = F_1V_1^{-1} = \begin{bmatrix} \frac{\beta_1^{rr}N_1^r}{(m_{21}^{Ir}+g_1^r)(N_1^r+N_1^v)} & \frac{\beta_1^{rv}N_1^r}{(m_{21}^{Iv}+g_1^v)(N_1^r+N_1^v)} & 0 & \frac{m_{12}^{Iv}}{(m_{12}^{Iv}+g_2^v)} \\ \frac{\beta_1^{vr}N_1^v}{(m_{21}^{Ir}+g_1^r)(N_1^r+N_1^v)} & \frac{\beta_1^{vv}N_1^v}{(m_{21}^{Iv}+g_1^v)(N_1^r+N_1^v)} & \frac{m_{12}^{Ir}}{(m_{12}^{Ir}+g_2^r)} & 0 \\ 0 & \frac{m_{21}^{Iv}}{(m_{21}^{Iv}+g_1^v)} & \frac{\beta_2^{rr}N_2^r}{(m_{12}^{Ir}+g_2^r)(N_2^r+N_2^v)} & \frac{\beta_2^{rv}N_2^r}{(m_{12}^{Iv}+g_2^v)(N_2^r+N_2^v)} \\ \frac{m_{21}^{Ir}}{(m_{21}^{Ir}+g_1^r)} & 0 & \frac{\beta_2^{vr}N_2^v}{(m_{12}^{Ir}+g_2^r)(N_2^r+N_2^v)} & \frac{\beta_2^{vv}N_2^v}{(m_{12}^{Iv}+g_2^v)(N_2^r+N_2^v)} \end{bmatrix}.$$

A standard result for nonnegative matrices (see, e.g., [12, Theorem 1.1]) says that the dominant eigenvalue of a nonnegative matrix is bounded below and above by the minimum and maximum of its column sums. We look at the column sums of K_1 to give upper and lower bounds on the dominant eigenvalue. The column sum in the first column is $\frac{\beta_1^{rr}N_1^r+\beta_1^{vr}N_1^v+m_{21}^{Ir}}{(m_{21}^{Ir}+g_1^r)(N_1^r+N_1^v)}$, and using basic calculus we derive that

$$1 < \frac{\beta_1^{rr}N_1^r+\beta_1^{vr}N_1^v+m_{21}^{Ir}}{(m_{21}^{Ir}+g_1^r)(N_1^r+N_1^v)} \leq \frac{\beta_1^{rr}N_1^r+\beta_1^{vr}N_1^v}{g_1^r(N_1^r+N_1^v)} \qquad \text{if } \frac{\beta_1^{rr}N_1^r+\beta_1^{vr}N_1^v}{(N_1^r+N_1^v)} - g_1^r > 0 \Leftrightarrow \mathcal{R}_1^r > 1,$$

$$\frac{\beta_1^{rr}N_1^r+\beta_1^{vr}N_1^v}{g_1^r(N_1^r+N_1^v)} \leq \frac{\beta_1^{rr}N_1^r+\beta_1^{vr}N_1^v+m_{21}^{Ir}}{(m_{21}^{Ir}+g_1^r)(N_1^r+N_1^v)} < 1 \qquad \text{if } \frac{\beta_1^{rr}N_1^r+\beta_1^{vr}N_1^v}{(N_1^r+N_1^v)} - g_1^r < 0 \Leftrightarrow \mathcal{R}_1^r < 1.$$

Similar results follow for the second, third, and fourth columns. Thus if $\mathscr{R}_1^r > 1$, $\mathscr{R}_1^v > 1$, $\mathscr{R}_2^r > 1$, and $\mathscr{R}_2^v > 1$ hold then all column sums are greater than 1 for any m_{21}^{Ir}, m_{21}^{Iv}, m_{12}^{Ir}, and m_{12}^{Iv}, that implies by Proposition 2 that the dominant eigenvalue of K_1 is greater than 1 and the DFE is unstable. On the other hand, if the above inequalities are reversed then the column sums are less than 1 for any movement rates and the DFE is stable by $\rho(K_1) < 1$. □

Next, we investigate some cases when changing the movement rates of some groups can stabilize the DFE. We construct the matrix $K_2 := F_2V_2^{-1}$, where F_2 is formed as we let $[F_2]_{1,1} = [F_1]_{1,1} - g_1^r$ and $[F_2]_{i,j} = [F_1]_{i,j}$ if $(i,j) \neq (1,1)$, and $V_2 = \operatorname{diag}(m_{21}^{Ir}, g_1^v + m_{21}^{Iv}, g_2^r + m_{12}^{Ir}, g_2^v + m_{12}^{Iv})$. V_2 is a non-singular M-matrix and F_2 is nonnegative if $\frac{\beta_1^{rr}N_1^r}{N_1^r+N_1^v} > g_1^r$. This condition is equivalent to when the number of new infections amongst residents of city 1 is less than 1, when an infected resident who doesn't travel is introduced into city 1. The alternative NGM is computed as

$$K_2 = F_2V_2^{-1} = \begin{bmatrix} \frac{\beta_1^{rr}N_1^r}{m_{21}^{Ir}(N_1^r+N_1^v)} - \frac{g_1^r}{m_{21}^{Ir}} & \frac{\beta_1^{rv}N_1^r}{(m_{21}^{Iv}+g_1^v)(N_1^r+N_1^v)} & 0 & \frac{m_{12}^{Iv}}{(m_{12}^{Iv}+g_2^v)} \\ \frac{\beta_1^{vr}N_1^v}{m_{21}^{Ir}(N_1^r+N_1^v)} & \frac{\beta_1^{vv}N_1^v}{(m_{21}^{Iv}+g_1^v)(N_1^r+N_1^v)} & \frac{m_{12}^{Ir}}{(m_{12}^{Ir}+g_2^r)} & 0 \\ 0 & \frac{m_{21}^{Iv}}{(m_{21}^{Iv}+g_1^v)} & \frac{\beta_2^{rr}N_2^r}{(m_{12}^{Ir}+g_2^r)(N_2^r+N_2^v)} & \frac{\beta_2^{rv}N_2^r}{(m_{12}^{Iv}+g_2^v)(N_2^r+N_2^v)} \\ 1 & 0 & \frac{\beta_2^{vr}N_2^v}{(m_{12}^{Ir}+g_2^r)(N_2^r+N_2^v)} & \frac{\beta_2^{vv}N_2^v}{(m_{12}^{Iv}+g_2^v)(N_2^r+N_2^v)} \end{bmatrix},$$

which is irreducible. Denote by L_2 the matrix that is formed by replacing $[K_2]_{1,1}$ and $[K_2]_{2,1}$ in K_2 by 0. Observe that K_2 converges to L_2 as m_{21}^{Ir} goes to infinity. We show that the disease can be eliminated by controlling only the travel rate of the residents of a single city.

Theorem 2 *Assume that $\mathscr{R}_0 > 1$, that is, the DFE is unstable. If $\frac{\beta_1^{rr}N_1^r}{N_1^r+N_1^v} > g_1^r$ and $\rho(L_2) < 1$ then increasing m_{21}^{Ir} can stabilize the DFE. In particular, if $\frac{\beta_1^{rr}N_1^r}{N_1^r+N_1^v} > g_1^r$ and $\mathscr{R}_1^v < 1$, $\mathscr{R}_2^r < 1$, and $\mathscr{R}_2^v < 1$, then increasing m_{21}^{Ir} can stabilize the DFE.*

Proof We utilize some terminology and results from [16, 17]. Let $S = \{(1,1),(2,1)\}$, and define the 4×4 matrix K_2^S as $[K_2^S]_{i,j} = [K_2]_{i,j}$ if $(i,j) \in S$ and 0 otherwise. Note that S identifies the set of elements in K_2 that depend on m_{21}^{Ir}, and K_2^S contains elements of K_2 that are subject to change when m_{21}^{Ir} is targeted. Following the terminology of [16] it is thus meaningful to refer to S as the target set and to K_2^S as the target matrix. Note that $L_2 = K_2 - K_2^S$, hence the condition $\rho(L_2) < 1$ implies that $\rho(K_2 - K_2^S) < 1$, that is, the controllability condition holds and it is possible to stabilize the DFE by controlling only the elements in S [16].

We compute $\mathscr{T}_S = \rho(K_2^S(I - K_2 + K_2^S)^{-1})$, the number referred to as the target reproduction number in [16]. Here I denotes the 4×4 identity matrix. Let $(m_{21}^{Ir})^c = m_{21}^{Ir}\mathscr{T}_S$, where we denote by $(m_{21}^{Ir})^c$ the controlled travel rate of infected residents of city 1 travelling to city 2. It follows from Corollary 1 and [16, Theorem 2.1] by $\mathscr{R}_0 > 1$ that $(m_{21}^{Ir})^c > m_{21}^{Ir}$. The matrix K_2^c, constructed as we replace m_{21}^{Ir} in K_2 by

$(m_{21}^{Ir})^c$, satisfies $\rho(K_2^c) = 1$ by [16, Theorem 2.2], which means that the disease can be eradicated by increasing m_{21}^{Ir}.

Note that the conditions $\mathscr{R}_1^v < 1$, $\mathscr{R}_2^r < 1$, and $\mathscr{R}_2^v < 1$ ensure that $\rho(L_2) < 1$. Indeed, it is easy to see that the column sums of the second, third, and fourth columns in L_2 are less than 1 for any travel rates, and the column sum in the first column is 1. We now show that 1 is not an eigenvalue of L_2, which together with [12, Theorem 1.1] implies that the dominant eigenvalue of L_2 is less than 1. Assume that 1 is an eigenvalue of L_2, and consider a left eigenvector $\mathbf{v} = [v_1, v_2, v_3, v_4]$ corresponding to 1. It holds that

$$\mathbf{v} \times L_2 = 1 \times \mathbf{v},$$

and we deduce that

$$v_4 = v_1, \quad \max(v_1, v_2, v_3) > v_2, \quad \max(v_2, v_3, v_4) > v_3, \quad \max(v_1, v_3, v_4) > v_4.$$

From the fourth inequality and $v_4 = v_1$ it follows that $v_3 > v_4$, which together with the third inequality implies $v_2 > v_3 > v_4$, but $v_1 > v_2$ by the second inequality, a contradiction to $v_1 = v_4$. The proof is complete. □

To reveal the impact of visitors' travel, a result analogous to Theorem 2 can be formulated. The proof of the following theorem follows by similar arguments to those in Theorem 2.

Theorem 3 *Assume that $\mathscr{R}_0 > 1$, that is, the DFE is unstable. If $\frac{\beta_1^{vv} N_1^v}{N_1^r + N_1^v} > g_1^v$, and $\mathscr{R}_1^r < 1$, $\mathscr{R}_2^r < 1$, and $\mathscr{R}_2^v < 1$, then increasing m_{21}^{Iv} can stabilize the DFE.*

Lastly, we give conditions under which controlling outbound travel from one city is sufficient for disease elimination. Consider two matrices F_3 and V_3, defined as $[F_3]_{1,1} = [F_1]_{1,1} - g_1^r$, $[F_3]_{2,2} = [F_1]_{2,2} - g_1^v$, and $[F_3]_{i,j} = [F_1]_{i,j}$ otherwise, and $V_3 = \text{diag}(m_{21}^{Ir}, m_{21}^{Iv}, g_2^r + m_{12}^{Ir}, g_2^v + m_{12}^{Iv})$. V_3 is a non-singular M-matrix and F_3 is nonnegative if $\frac{\beta_1^{vv} N_1^v}{N_1^r + N_1^v} > g_1^v$ and $\frac{\beta_1^{rr} N_1^r}{N_1^r + N_1^v} > g_1^r$. The following theorem concerns about whether changing the movement rates of the current population of one city can lead to disease eradication.

Theorem 4 *Assume that $\mathscr{R}_0 > 1$, that is, the DFE is unstable. If $\frac{\beta_1^{rr} N_1^r}{N_1^r + N_1^v} > g_1^r$ and $\frac{\beta_1^{vv} N_1^v}{N_1^r + N_1^v} > g_1^v$ but $\mathscr{R}_2^r < 1$ and $\mathscr{R}_2^v < 1$, then increasing m_{21}^{Ir} and m_{21}^{Iv} can stabilize the DFE.*

Proof The proof is similar to the proof of Theorem 2. We compute the alternative NGM

$$K_3 = F_3 V_3^{-1} = \begin{bmatrix} \frac{\beta_1^{rr} N_1^r}{m_{21}^{Ir}(N_1^r + N_1^v)} - \frac{g_1^r}{m_{21}^{Ir}} & \frac{\beta_1^{rv} N_1^r}{m_{21}^{Iv}(N_1^r + N_1^v)} & 0 & \frac{m_{12}^{Iv}}{(m_{12}^{Iv} + g_2^v)} \\ \frac{\beta_1^{vr} N_1^v}{m_{21}^{Ir}(N_1^r + N_1^v)} & \frac{\beta_1^{vv} N_1^v}{m_{21}^{Iv}(N_1^r + N_1^v)} - \frac{g_1^v}{m_{21}^{Iv}} & \frac{m_{12}^{Ir}}{(m_{12}^{Ir} + g_2^r)} & 0 \\ 0 & 1 & \frac{\beta_2^{rr} N_2^r}{(m_{12}^{Ir} + g_2^r)(N_2^r + N_2^v)} & \frac{\beta_2^{rv} N_2^r}{(m_{12}^{Iv} + g_2^v)(N_2^r + N_2^v)} \\ 1 & 0 & \frac{\beta_2^{vr} N_2^v}{(m_{12}^{Ir} + g_2^r)(N_2^r + N_2^v)} & \frac{\beta_2^{vv} N_2^v}{(m_{12}^{Iv} + g_2^v)(N_2^r + N_2^v)} \end{bmatrix},$$

which is irreducible, and define the target set U by identifying the entries of K_3 that depend on m_{21}^{Ir} and/or m_{21}^{Iv}. We let $U = \{(1,1),(1,2),(2,1),(2,2)\}$, and define the 4×4 target matrix K_3^U as $[K_3^U]_{i,j} = [K_3]_{i,j}$ if $(i,j) \in U$ and 0 otherwise. Note that $\rho(K_3) > 1$ holds by $\mathscr{R}_0 > 1$. However, the result in [12, Theorem 1.1] on the upper bound of the dominant eigenvalue implies that $\rho(K_3 - K_3^U) \leq 1$.

Assume that $\rho(K_3 - K_3^U) = 1$, that is, 1 is an eigenvalue of $K_3 - K_3^U$. Then there is a left eigenvector $\mathbf{v} = [v_1, v_2, v_3, v_4]$ such that

$$\mathbf{v} \times (K_3 - K_3^U) = 1 \times \mathbf{v}$$

holds. We derive that

$$v_4 = v_1, \quad v_3 = v_2, \quad \max(v_2, v_3, v_4) > v_3, \quad \max(v_1, v_3, v_4) > v_4,$$

so $v_4 > v_3$ and $v_3 > v_4$ hold by the third and fourth inequalities, a contradiction. We showed that $\rho(K_3 - K_3^U) < 1$, which means that there is a potential to control m_{21}^{Ir} and m_{21}^{Iv} in a way such that the dominant eigenvalue of the controlled matrix drops below 1 (by decreasing targeted entries of K_3 to values close to 0). This condition also allows us to compute the target reproduction number $\mathscr{T}_U = \rho(K_3^U(I - K_3 + K_3^U)^{-1})$.

By [16, Theorem 2.2], the controlled matrix K_3^c satisfies $\rho(K_3^c) = 1$ where K_3^c is formed by replacing $[K_3]_{i,j}$ by $[K_3]_{i,j}/\mathscr{T}_U$ if $(i,j) \in U$, that is achieved by replacing m_{21}^{Ir} by $(m_{21}^{Ir})^c = m_{21}^{Ir}\mathscr{T}_U$, and m_{21}^{Iv} by $(m_{21}^{Iv})^c = m_{21}^{Ir}\mathscr{T}_U$. Note that $\mathscr{T}_U > 1$ by [16, Theorem 2.2], which means that the disease can be eradicated by increasing m_{21}^{Ir} and m_{21}^{Iv}. □

In the case of transmission coefficients equal for all populations present in a city, recovery rates equal for all populations and death rates equal for all populations, $\mathscr{R}_1^r$ and $\mathscr{R}_1^v$ reduce to $\beta_1/(\gamma + d)$, and $\mathscr{R}_2^r$ and $\mathscr{R}_2^v$ reduce to $\beta_2/(\gamma + d)$. Note that these quantities give the expected number of secondary infections generated by a single infected case in city 1 and city 2, respectively, in the absence of movement between the cities. Hence the local reproduction numbers in city 1 and city 2 can be defined as we consider our model without dispersal:

$$\mathscr{R}_1^{loc} = \frac{\beta_1}{\gamma + d}, \qquad \mathscr{R}_2^{loc} = \frac{\beta_2}{\gamma + d}.$$

We derive the following results from Theorems 1 and 4.

Corollary 2 *Suppose that $\beta_1^{mn} = \beta_1$ and $\beta_2^{mn} = \beta_2$ for all $m, n \in \{r, v\}$, $\gamma_j = \gamma$ and $d_j = d$ for all $j \in \{1, 2\}$. Then, the DFE is unstable when the cities are isolated and $\mathscr{R}_1^{loc} > 1$, $\mathscr{R}_2^{loc} > 1$, and movement cannot stabilize the DFE. In the case when $\mathscr{R}_1^{loc} < 1$ and $\mathscr{R}_2^{loc} < 1$, the DFE is stable when the cities are isolated, and movement cannot destabilize the DFE. If the DFE is unstable and $\frac{\beta_1 N_1^r}{N_1^r + N_1^v} > (\gamma + d)$, $\frac{\beta_1 N_1^v}{N_1^r + N_1^v} > (\gamma + d)$ but $\mathscr{R}_2^{loc} < 1$, then increasing the movement rates of individuals in city 1 can stabilize the DFE.*

4 Discussion

A two-city compartmental epidemic model was considered to reveal the impact of population dispersal on disease persistence. This general SIRS model is applicable for an array of infectious diseases, and it can also be reduced to simpler models (SIS, SIR models) by setting parameters (or their inverses) to zero. In the model setup we distinguish local residents from temporary visitors in each city, that results in four infected classes in the model. We demonstrated that controlling the movement of one or two infected groups can be sufficient for preventing a disease outbreak. It was discussed in [11] that the role of different inflow rates of residents and visitors into a city is not necessarily significant in regards of the total epidemic burden, but it is of particular importance for pandemic preparedness, when it comes to assessing the risk for each group to import the infection to a disease-free city.

Modelling the spatial spread of diseases in metapopulations remains a complex task. This paper does not concern with models that include multiple species, hence more analysis is needed to quantify the effect of movement between patches in such models, which are useful in investigating vector-borne diseases and their control strategies. Combining some intervention measures—like the mutual control of dispersal rates and transmission rates—requires less effort for disease elimination, hence there is a potential to incorporate the results of this work into systematic risk assessment analyses, as described in [10].

Acknowledgements D. Knipl acknowledges the support by the Cimplex project funded by the European Commission in the area "FET Proactive: Global Systems Science" (GSS), as a Research and Innovation Action, under the H2020 Framework programme, Grant agreement number 641191.

References

1. Arino, J.: Diseases in metapopulations. In: Modeling and Dynamics of Infectious Diseases. Series in Contemporary Applied Mathematics, vol. 11, pp. 65–123. Higher Education Press, Beijing (2009)
2. Arino, J., Davis, J.R., Hartley, D., Jordan, R., Miller, J.M., van den Driessche, P.: A multi-species epidemic model with spatial dynamics. Math. Med. Biol. **22**(2), 129–142 (2005)
3. Arino, J., Jordan, R., van den Driessche, P.: Quarantine in a multi-species epidemic model with spatial dynamics. Math. Biosci. **206**(1), 46–60 (2007)
4. Arino, J., van den Driessche, P.: A multi-city epidemic model. Math. Popul. Stud. **10**(3), 175–193 (2003)
5. Arino, J., van den Driessche, P.: The basic reproduction number in a multi-city compartmental epidemic model. In: Positive Systems, pp. 135–142 Springer, Berlin/Heidelberg (2003)
6. Arino, J., van den Driessche, P.: Disease spread in metapopulations. In: Nonlinear Dynamics and Evolution Equations, vol. 48, pp. 1–13. American Mathematical Society, Providence (2006)
7. Diekmann, O., Heesterbeek, J.A.P., Metz, J.A.J.: On the definition and computation of the basic reproduction ratio R_0 in models for infectious diseases in heterogeneous populations. J. Math. Biol. **28**, 365–382 (1990)

8. Fiedler, M.: Special Matrices and Their Applications in Numerical Mathematics. Martinus Nijhoff Publishers, Dodrecht (1986)
9. Jin, Y., Wang, W.: The effect of population dispersal on the spread of a disease. J. Math. Anal. Appl. **308**(1), 343–364 (2005)
10. Knipl, D.: A new approach for designing disease intervention strategies in metapopulation models. J. Biol. Dyn. **10**(1), 71–94 (2016). http://dx.doi.org/10.1080/17513758.2015.1107140
11. Knipl, D.H., Röst, G., Wu, J.: Epidemic spread and variation of peak times in connected regions due to travel related infections—dynamics of an anti-gravity type delay differential model. SIAM J. Appl. Dyn. Syst. **12**(4), 1722–1762 (2013)
12. Minc, H.: Nonnegative Matrices. Wiley Interscience, New York (1988)
13. Ruan, S., Wang, W., Levin, S.A.: The effect of global travel on the spread of SARS. Math. Biosci. Eng. **3**(1), 205 (2006)
14. Sattenspiel, L., Dietz, K.: A structured epidemic model incorporating geographic mobility among regions. Math. Biosci. **128**(1), 71–91 (1995)
15. Sattenspiel, L., Herring, D.A.: Simulating the effect of quarantine on the spread of the 1918–19 flu in central Canada. Bull. Math. Biol. **65**(1), 1–26 (2003)
16. Shuai, Z., Heesterbeek, J.A.P., van den Driessche, P.: Extending the type reproduction number to infectious disease control targeting contacts between types. J. Math. Biol. **67**(5), 1067–1082 (2013)
17. Shuai, Z., Heesterbeek, J.A.P., van den Driessche, P.: Erratum to: extending the type reproduction number to infectious disease control targeting contacts between types. J. Math. Biol. **71**(1), 1–3 (2015)
18. Van den Driessche, P., Watmough, J.: Reproduction numbers and sub-threshold endemic equilibria for compartmental models of disease transmission. Math. Biosci. **180**(1), 29–48 (2002)
19. Wang, W., Mulone, G.: Threshold of disease transmission in a patch environment. J. Math. Anal. Appl. **285**(1), 321–335 (2003)
20. Wang, W., Zhao, X.Q.: An epidemic model in a patchy environment. Math. Biosci. **190**(1), 97–112 (2004)

A Chemostat Model with Wall Attachment: The Effect of Biofilm Detachment Rates on Predicted Reactor Performance

Alma Mašić and Hermann J. Eberl

Abstract We consider a previously introduced mathematical model of chemostat with suspended and wall attached growth and exchange of biomass via biofilm detachment and reattachment. In this study we investigate the role of the specific choice of a biomass detachment criterion. We find that this choice does greatly affect output parameters such as biomass in the system, but it does not affect strongly effluent concentration and hence the prediction of reactor performance.

1 Introduction

Bacterial biofilms are layers on immersed surfaces that form wherever environmental conditions sustain microbial growth. They play an important role in several environmental engineering applications, most notably in wastewater treatment [12], where several technologies have been designed based on biofilm properties. An important part of the biofilm life cycle is detachment, or dispersal, the transfer of biomass from the biofilm into the aqueous phase [3, 10]. Several triggers for this phenomenon have been identified, including external factors, such as shear forces, or internal factors, such as weakening of the EPS matrix and quorum sensing signaling mechanisms [11, 13]. Several attempts to incorporate detachment into mesoscale mathematical models of biofilms have been suggested in the literature, both in the traditional 1D Wanner-Gujer biofilm model and its simplification, e.g. [2, 8, 11, 16], and in two- and three-dimensional biofilm models, e.g. [4–6, 18].

It is clear that the mesoscopic description of biofilm detachment strongly affects the mesoscopic biofilm structure predicted by these models. What is less clear is to which extent the mesoscopic description of biofilm detachment affects global parameters that assess reactor performance. For the setting of biofilms in a porous

A. Masic
EAWAG, Überlandstrasse 133, 8600 Dübendorf, Switzerland
e-mail: alma.masic@eawag.ch

H.J. Eberl (✉)
University of Guelph, 50 Stone Rd E, Guelph, ON N1G2W1, Canada
e-mail: heberl@uoguelph.ca

© Springer International Publishing Switzerland 2016

J. Bélair et al. (eds.), *Mathematical and Computational Approaches in Advancing Modern Science and Engineering*, DOI 10.1007/978-3-319-30379-6_25

medium, a partial answer was given in [1], where a multi-scale model was obtained by upscaling a Wanner-Gujer type biofilm model with four different detachment criteria to the reactor scale; it was found that the particular choice of the mesoscopic detachment rate does not affect the macroscopic description. Here we are interested in the question whether for a continuous stirred tank reactor (CSTR) with wall attached and suspended growth the particular description of biofilm detachment affects the effluent substrate concentration predicted by the model. Models of this type arise in the modeling of wastewater treatment processes where both biofilms and nonsessile bacteria are present and contribute to the substrate degradation process. In the focus of our interest is the case where the reactor per se is designed as a biofilm reactor, in which suspended biomass occurs as a side effect through the exchange of biomass between biofilm and aqueous phase.

2 Mathematical Model

We cast a simple model for a continuous stirred tank reactor with suspended and wall attached growth in terms of the dependent variables substrate concentration S [gm^{-3}], suspended biomass u [g] and biofilm thickness λ [m]. Following [9], it reads

$$\dot{S} = D(S^{in} - S) - \frac{u\mu_u(S)}{V\gamma} - \frac{J(S,\lambda)}{V}, \tag{1}$$

$$\dot{u} = u(\mu_u(S) - D - k_u) + A\rho d(\lambda)\lambda - \alpha u, \tag{2}$$

$$\dot{\lambda} = v(\lambda, t) + \frac{\alpha u}{A\rho} - d(\lambda)\lambda. \tag{3}$$

Here, D [d^{-1}] is the dilution rate, S^{in} [gm^{-3}] the inflow substrate concentration, γ [-] the yield coefficient, k_u [d^{-1}] the cell death rate for suspended bacteria, and α [d^{-1}] is the rate at which suspended bacteria attach to the biofilm, V [m^3] is the reactor volume and A [m^2] the colonizable surface area. The biomass density in the biofilm is ρ [gm^{-3}].

In (3), the function $v = v(z,t)$ [md^{-1}] denotes the growth induced velocity of the biomass at a location z in the biofilm. Due to the incompressibility assumption that the biomass density is constant across the biofilm, biofilm expansion is essentially equivalent to biomass growth. Velocity v is obtained as the integral of the biomass production rate

$$v(z,t) = \int_0^z (\mu_\lambda(C(\zeta)) - k_\lambda)d\zeta, \tag{4}$$

where k_λ [d^{-1}] is the cell death rate for biofilm bacteria.

The substrate dependent bacterial growth rates are defined via Monod kinetics, i.e.

$$\mu_u(S) = \frac{\mu_u^{max} S}{K_u + S}, \quad \mu_\lambda(C(z)) = \frac{\mu_\lambda^{max} C(z)}{K_\lambda + C(z)}, \tag{5}$$

where $\mu_u^{max}, \mu_\lambda^{max}$ [d^{-1}] are the maximum specific growth rates, K_u, K_λ [gm^{-3}] the half-saturation concentrations and $C(z)$ [gm^{-3}] denotes the substrate concentration in the biofilm at thickness z [m] from the substratum. It is obtained as the solution of the two-point boundary value problem

$$D_c C''(z) = \frac{\rho}{\gamma} \mu_\lambda(C(z)), \quad C'(0) = 0, \; C(\lambda) = S. \tag{6}$$

Here D_c [$m^2 d^{-1}$] is the diffusion coefficient. The boundary condition at the substratum, $z = 0$, describes that substrate does not leave the reactor through the walls, while the boundary condition at $z = \lambda$ implies that external mass transfer resistance at the biofilm/water interface is neglected. In (6) we used that substrate diffusion is a much faster process than biofilm growth, i.e. that (6) can be considered in a quasi-steady state.

In (1), the sink J [gd^{-1}] denotes the substrate flux from the aqueous phase into the biofilm, i.e.

$$J(S, \lambda) = A D_c \frac{dC}{dz}(\lambda). \tag{7}$$

Detachment of biomass from the biofilm is described by the volumetric detachment rate $d(\lambda)$ [d^{-1}], which we assume to be differentiable.

A frequently used detachment rate expression in biofilm modeling is to assume that d is proportional to λ,

$$d(\lambda) = E_1 \lambda, \tag{8}$$

leading to a quadratic sink term in (3); E_1 [$d^{-1}m^{-1}$] is the erosion or detachment parameter. This traditional detachment model was the only one used in [9]. Another detachment rate function that is found in the literature is to assume a constant rate, i.e

$$d(\lambda) = E_2, \tag{9}$$

which leads to a first order sink term with erosion parameter E_2 [d^{-1}].

It is often assumed that the detachment rate depends also on the hydraulic conditions in the reactor, which determine shear forces acting on the biofilm. Therefore, one can correlate $E_{1,2}$ with D, see below in Sect. 3.

We assume all model parameters to be positive.

First we formally re-write our model as an ordinary initial value problem. Note that integrating (6) once and using the boundary conditions gives

$$\frac{dC}{dz}(\lambda) = \frac{\rho}{\gamma D_c} \int_0^\lambda \mu_\lambda(C(z))dz. \tag{10}$$

We define

$$j(\lambda, S) := \begin{cases} \frac{\rho}{\gamma D_c} \int_0^\lambda \mu_\lambda(C(z))dz, & \lambda > 0 \\ 0 & \lambda = 0. \end{cases} \tag{11}$$

Note that $C(z)$ is indirectly a function of S due to the boundary condition in (6), therefore also j is a function of S. Then (1), (2), (3), (4), (5), (6), and (7) becomes

$$\dot{S} = D(S^{in} - S) - \frac{1}{V}\Big(\frac{u\mu_u(S)}{\gamma} + AD_c j(\lambda, S)\Big) \tag{12}$$

$$\dot{u} = u(\mu_u(S) - D - k_u) + A\rho d(\lambda)\lambda - \alpha u \tag{13}$$

$$\dot{\lambda} = \frac{\gamma D_c}{\rho} j(\lambda, S) - \lambda k_\lambda + \frac{\alpha u}{A\rho} - d(\lambda)\lambda. \tag{14}$$

While the function $j(\lambda, S)$ is not known explicitly in terms of elementary functions, important qualitative properties are known, which we repeat here from [9]:

Lemma 1 *The function $j(\lambda, S)$ for $\lambda \geq 0$, $S \geq 0$ is well-defined, non-negative and differentiable. It has the following properties:*

(a) $j(\cdot, 0) = 0, j(0, \cdot) = 0,$
(b) $\frac{\partial j}{\partial S}(0, S) = 0,$
(c)

$$\frac{S\theta}{K_\lambda + S} \leq \frac{\partial j}{\partial \lambda}(0, S) \leq \frac{S\theta}{K_\lambda}, \tag{15}$$

where $\theta = \frac{\rho\mu_\lambda^{max}}{\gamma D_c}$ *and* K_λ *is the half-saturation coefficient from (5).*

Proposition 1 *The initial value problem of (12), (13) and (14) with non-negative initial data possesses a unique non-negative solution for all $t > 0$. There is no time interval (t_1, t_2), $0 < t_1 < t_2$, over which a non-trivial solution exists with either $u \equiv 0$ or $\lambda \equiv 0$.*

Proof (sketch) The function $j(\lambda, S)$, and thus the system (12), (13) and (14) is well-defined. The tangent criterion of [15] in the usual way can be used to confirm the positive invariance of the non-negative cone. In the non-negative cone the system satisfies a Lipschitz condition, which implies existence and uniqueness. An upper estimate for the solutions can be constructed by the differential inequality techniques from a linear combination of the model equations, more specifically, an upper bound can be derived for $\gamma VS(t) + u(t) + A\rho\lambda(t)$ which then, using non-negativity, implies upper estimates on each of $S(t)$, $u(t)$, $\lambda(t)$. The last statement in the assertion follows directly from (13) and (14). □

Proposition 2 *The washout equilibrium* $(S^{in}, 0, 0)$ *always exists. It is unstable if at least one of* $\mu_u(S^{in}) - D - k_u - \alpha$ *and* $\frac{\gamma D_c}{\rho}\frac{\partial j}{\partial \lambda}(0, S^{in}) - d(0) - k_\lambda$ *is positive; if both expressions are negative then the washout equilibrium is stable if* $\Delta := \left(\mu_u(S^{in}) - D - k_u - \alpha\right)\left(\frac{\gamma D_c}{\rho}\frac{\partial j}{\partial \lambda}(0, S^{in}) - d(0) - k_\lambda\right) - \alpha d(0) > 0$ *and unstable if the inequality is reversed.*

Proof It is easily verified that the trivial equilibrium $E_0 = (S^{in}, 0, 0)$ always exists. To determine the stability of the equilibrium we calculate the Jacobian $J(S^{in}, 0, 0)$ of the right hand side of (12), (13) and (14). Using Lemma 1(a),(b) we find

$$J(S^{in}, 0, 0) = \begin{pmatrix} -D & -\frac{\mu_u(S^{in})}{V\gamma} & -\frac{AD_c}{V}\frac{\partial j}{\partial \lambda}(0, S^{in}) \\ 0 & \mu_u(S^{in}) - D - k_u - \alpha & A\rho d(0) \\ 0 & \frac{\alpha}{A\rho} & \frac{\gamma D_c}{\rho}\frac{\partial j}{\partial \lambda}(0, S^{in}) - d(0) - k_\lambda \end{pmatrix} \tag{16}$$

The eigenvalues are $-D < 0$ as well as the eigenvalues of the 2×2 sub-matrix

$$\begin{pmatrix} j_{22} & j_{23} \\ j_{32} & j_{33} \end{pmatrix} := \begin{pmatrix} \mu_u(S^{in}) - D - k_u - \alpha & A\rho d(0) \\ \frac{\alpha}{A\rho} & \frac{\gamma D_c}{\rho}\frac{\partial j}{\partial \lambda}(0, S^{in}) - d(0) - k_\lambda \end{pmatrix} \tag{17}$$

We distinguish now between three cases: (i) if $j_{22} > 0, j_{33} > 0$ then the trace of this sub-matrix is positive, hence at least one eigenvalue has positive real part and the equilibrium is unstable. (ii) if $j_{22} < 0, j_{33} > 0$ or *vice versa*, then the sub-matrix has a negative determinant and hence one positive and one negative eigenvalue, implying instability. (iii) if $j_{22} < 0, j_{33} < 0$ then the trace of the sub-matrix is negative and its determinant is obtained as

$$\Delta = \left(\mu_u(S^{in}) - D - k_u - \alpha\right)\left(\frac{\gamma D_c}{\rho}\frac{\partial j}{\partial \lambda}(0, S^{in}) - d(0) - k_\lambda\right) - \alpha d(0), \tag{18}$$

i.e. depends on parameters. Positive Δ implies stability, negative Δ implies instability. While for $d(0) = 0$, as in detachment rate function (8), stability is automatic, this is not necessarily true if $d(0) > 0$, as in detachment rate function (9). □

Remark 1 Weaker, but easier to evaluate and to apply stability criteria can be obtained by replacing $\frac{\partial j}{\partial \lambda}(0, S)$ in Proposition 2 with the estimates in Lemma 1(c): [i] The inequality $\frac{\mu_\lambda^{max} S^{in}}{K_\lambda + S^{in}} - d(0) - k_\lambda > 0$ is sufficient (but not necessary) for instability of the washout equilibrium. [ii] If $\mu_u(S^{in}) - D - k_u - \alpha < 0$ and $\frac{\mu_\lambda^{max} S^{in}}{K_\lambda} - d(0) - k_\lambda < 0$ then the inequality $(\mu_u(S^{in}) - D - k_u - \alpha)(\frac{\mu_\lambda^{max} S^{in}}{K_\lambda + S^{in}} - d(0) - k_\lambda) > \alpha d(0)$ is sufficient (but not necessary) for stability of the washout equilibrium, whereas the inequality $(\mu_u(S^{in}) - D - k_u - \alpha)(\frac{\mu_\lambda^{max} S^{in}}{K_\lambda} - d(0) - k_\lambda) < \alpha d(0)$ is sufficient for its instability.

Remark 2 For detachment rate functions with $d(0) = 0$, such as (8), the above analysis simplifies. The eigenvalues of $J(S^{in}, 0, 0)$ are then its diagonal elements and case (iii) above will always be stable. The stability results for cases (i), (ii) are the same but follow immediately from the sign of the diagonal entries. This implies that the answer to the question whether a biofilm can be established in the CSTR with wall attached and suspended growth can depend on the detachment criterion used in the modeling study. Whereas in models with $d(0) = 0$ the specific detachment rate coefficient does not affect the outcome, it does so in detachment models with $d(0) > 0$. This is also consistent with the results of the upscaling study of [1] for porous medium systems.

Non-trivial steady states, in which both wall attached and suspended biomass are present are much more difficult to analyse, even in the algebraically much simpler Freter model, in which no substrate gradients in the wall depositions are accounted for [7, 14]. Therefore, we do not expect any insightful results in pursuing this line of investigation and turn to a numerical study instead.

3 Simulation Results

Typical simulation results are shown in Fig. 1, using the parameters in Table 1. In one case the bulk substrate concentration S^{in} is chosen low enough so that the bacteria cannot be sustained; in the other case it is high enough such that both a biofilm and a suspended population attain a positive equilibrium value. The vast majority of biomass in the system is sessile. These simulations were conducted with the detachment rate function (8).

In Fig. 2 we show for both detachment rate function (8) and (9) the steady state values for λ, u, S for different detachment parameters E_1 and E_2, which have been correlated with the dilution rate. Motivated by [11] we chose the relationship as

$$\tilde{E}_i = E_i \left(\frac{D}{D_0}\right)^{0.58}, \tag{19}$$

where by D_0 we denote a reference dilution rate such that for $D = D_0$ this flow rate dependent criterion is equivalent to a detachment with a given rate constant E_i and for $D > D_0$ we have $\tilde{E}_i > E_i$, while for $D < D_0$ we have $\tilde{E}_i < E_i$. Note that an increased dilution rate implies a more plentiful substrate supply.

The main observation is that the choice of detachment rate function and erosion constant does not have an effect on the substrate concentration in outflow, i.e. the substrate removal performance of the reactor: For smaller dilution rates $D < 50$, the effluent substrate concentration increases almost linearly with D, for larger $D > 50$,

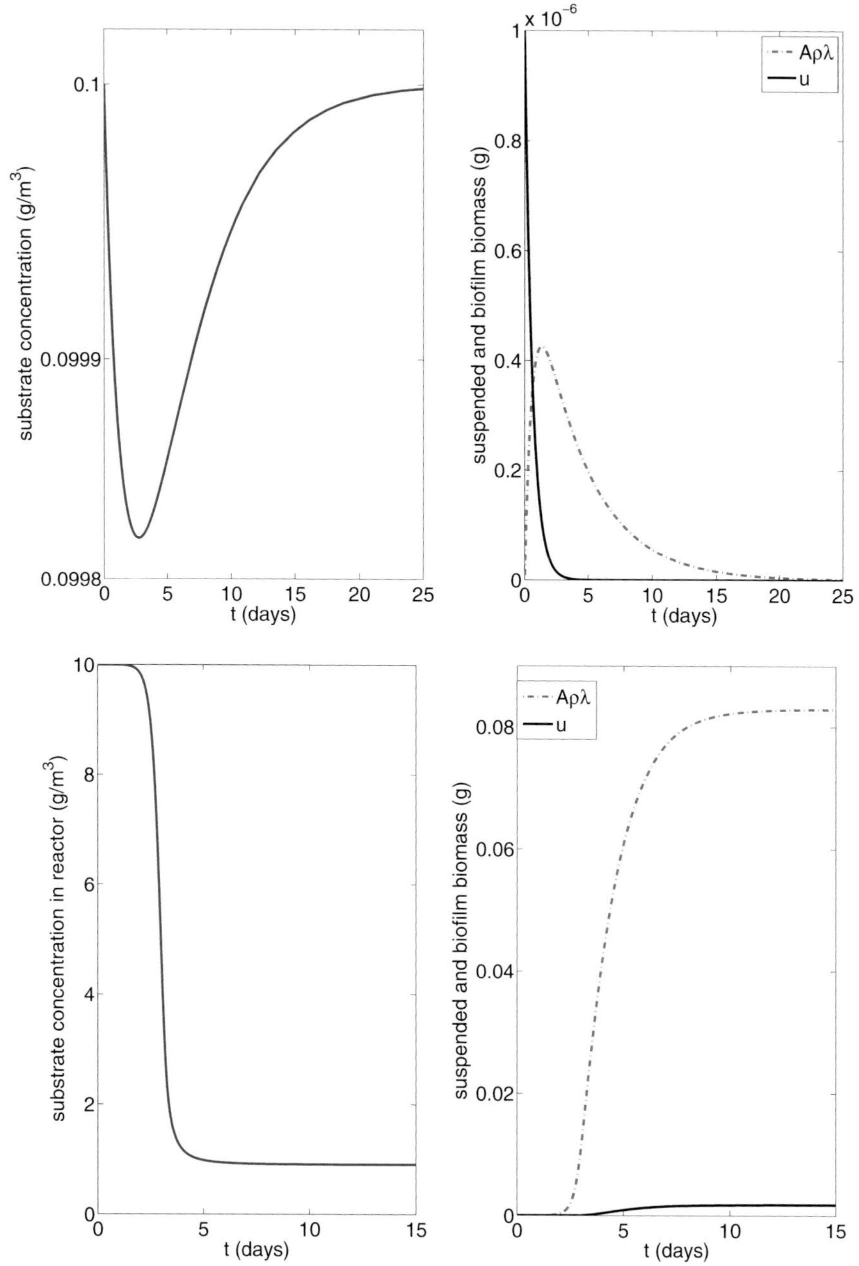

Fig. 1 A typical simulation of model (12), (13) and (14). Case 1 (*top*): The washout equilibrium is stable, case 2 (*bottom*): the washout equilibrium is unstable. In both simulations the detachment criterion (8) was used

Table 1 Model parameters used in the simulations

Symbol	Parameter	Value	Reference
α	Attachment rate	1/day	Assumed
D_c	Diffusion coefficient	10^{-4} m^2/day	[17]
E_1	Erosion parameter	1000/(m·day)	Assumed
γ	Yield of biomass from substrate	0.63 -	[17]
K_λ, K_u	Half-saturation coefficients	4 g/m^3	[17]
k_λ, k_u	Death rates	0.4/day	[17]
$\mu_\lambda^{max}, \mu_u^{max}$	Maximum specific growth rates	6/day	[17]
ρ	Biofilm biomass density	10,000 g/m^3	[17]

the increase becomes sublinear. The results obtained for different base dilution rates D_0, hence different detachment rates E_i, appear indistinguishable for each detachment criterion (8) and (9). Also comparing the results between the simulation experiments for (8) and (9), respectively, shows that for both detachment criteria the same values of S are found.

The biofilm thickness, however, depends strongly on the detachment rate function, and so does the suspended biomass density and by extension the amount of biomass in the effluent of the reactor.

An explanation for this phenomenon is that in a thick enough biofilm, as is typical for many biofilm based wastewater treatment applications, only a small part of the biofilm is active whereas a major part does not contribute to reactor performance due to substrate limitations in the inner layers of the film. An increased detachment force decreases the biofilm thickness but as long as it remains above the thickness of the active layer this does not have an effect on performance. If the detachment is strong enough such that the resulting biofilm becomes too thin, this statement does not hold anymore.

4 Conclusion

We have investigated the question whether the choice of a mesoscale biofilm detachment model affects the prediction of reactor performance in a macroscale biofilm reactor model.

We have found that for the question whether a biofilm can be established or whether biomass is washed out the choice of detachment rate can matter: The detachment coefficient does not affect stability of the washout equilibrium in case of a detachment rate function that vanishes in the absence of biofilm, i.e. if $d(0) = 0$. However, if detachment rate functions with $d(0) > 0$ are used, then the coefficients of the detachment model can affect stability of the washout equilibrium.

In biofilm based reactor technology, the washout equilibrium is of minor importance, because reactors are designed such that sufficiently thick biofilms are

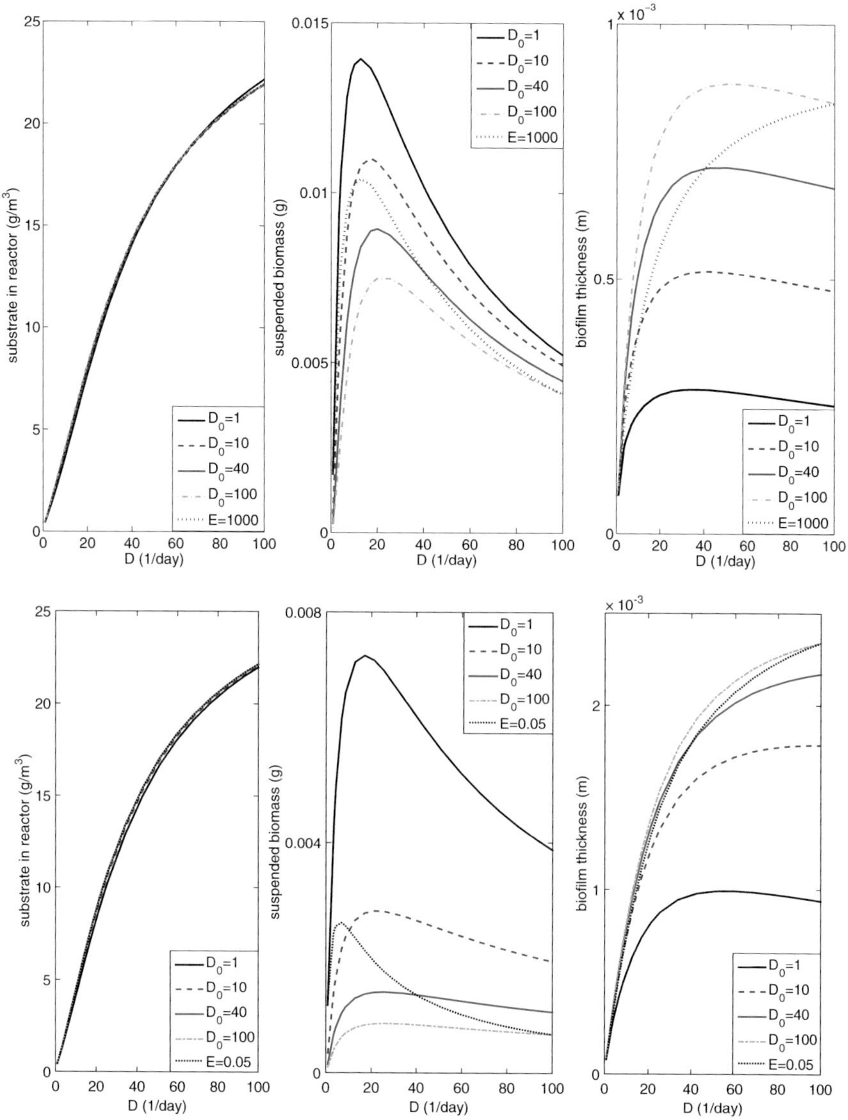

Fig. 2 Steady states attained in dependence of dilution rate *D*. *Top*: detachment criterion (8), reprinted from [9] with permission; *bottom*: detachment criterion (9)

obtained. In our simulations we show that in such situations with well developed, sufficiently thick biofilms the particular choice of a mesoscopic detachment rate function does not quantitatively affect macroscopic reactor performance as measured in terms of effluent substrate concentration, but it does affect the predicted biofilm thickness.

Acknowledgements HJE was supported by NSERC Canada through the Discovery Grant Program and through the Canada Research Chairs Program.

References

1. Abbas, F., Eberl, H.J.: Investigation of the role of mesoscale detachment rate expressions in a macroscale model of a porous medium biofilm Reactor, Int. J. Biomath. Biostat. **2**, 123–143 (2013)
2. Abbas, F., Sudarsan, R., Eberl, H.J.: Longtime behaviour of one-dimensional biofilm models with shear dependent detachment rates. Math. Biosci. Eng. **9**, 215–239 (2012)
3. Bester, E., Edwards, E.A., Wolfaardt, G.M.: Planktonic cell yield is linked to biofilm development. Can. J. Microbiol. **55**, 1195–1206 (2009)
4. Duddu, R., Chopp, D.L., Moran, B: A two-dimensional continuum model of biofilm growth incorporating fluid flow and shear stress based detachment. Biotechnol. Bioeng. **103**, 92–104 (2008)
5. Emerenini, B.O., Hense, B.A., Kuttler, C., Eberl, H.J.: A mathematical model of quorum sensing induced biofilm detachment. PLoS ONE **10**(7), e0132385 (2015)
6. Hunt, S.M., Hamilton, M.A., Sears, J.T., Harkin, G., Reno, J: A computer investigation of chemically mediated detachment in bacterial biofilms. J. Microbiol. **149**, 1155–1163 (2003)
7. Jones, D., Kojouharov, H.V., Le, D., Smith, H.L.: The Freter model: a simple model of biofilm formation. Math. Biol. **47**, 137–152 (2003)
8. Kommendal, R., Bakke, R.: Modeling *Pseudomonas aeruginosa* biofilm detachment. HIT Working Paper no 3/2003 (2003)
9. Masic, A., Eberl, H.J.: Persistence in a single species CSTR model with suspended flocs and wall attached biofilms. Bull. Math. Biol. **74**, 1001–1024 (2012)
10. Morgenroth, E.: Detachment: an often-overlooked phenomenon in biofilm research and modeling. In: Wuertz, S., et al. (eds.) Biofilms in Wastewater Treatment, pp 246–290. IWA Publishing, London (2003)
11. Rittmann, B.E.: The effect of shear stress on biofilm loss rate. Biotechnol. Bioeng. **24**, 501–506 (1982)
12. Rittmann, B.E, McCarty, P.L.: Environmental Biotechnology. McGraw-Hill, Boston (2001)
13. Solano, C., Echeverz, M., Lasa, I.: Biofilm dispersion and quorum sensing. Curr. Opin. Microbiol. **18**, 96–104 (2014)
14. Stemmons, E.D., Smith, H.L.: Competition in a chemostat with wall attachment. SIAM J. Appl. Math. **61**, 567–595 (2000)
15. Walter, W.: Gewöhnliche Differentialgleichungen, 7th edn. Springer-Verlag, Berlin (2000)
16. Wanner, O., Gujer, W.: A multispecies biofilm model. Biotechnol. Bioeng. **28**, 314–328 (1986)
17. Wanner, O., Eberl, H., Morgenroth, E., Noguera, D.R., Picioreanu, C., Rittmann, B., van Loosdrecht, M.: Mathematical modeling of biofilms. Scientific and Technical Report No.18. IWA Publishing, London (2006)
18. Xavier, J.B., Picioreanu, C., van Loosdrecht, M.C.M: A general description of detachment for multidimensional modeling of biofilms. Biotechnol. Bioeng. **91**, 651–669 (2005)

Application of CFD Modelling to the Restoration of Eutrophic Lakes

A. Najafi-Nejad-Nasser, S.S. Li, and C.N. Mulligan

Abstract Eutrophication has been a worldwide lake pollution problem, caused by the presence of excessive nutrients in lakes. The nutrients can come from an external or internal source. The release of phosphorous (P) from resuspended sediments from the lake bottom is a significant internal source. This paper discusses how to effectively control such a release. We considered using artificial circulation technique, and carried out CFD modelling of circulation triggered by air-bubble injection into the lake water. The simulations are based on the RANS equations. We predicted distributed water and air-bubble velocities, as well as air volume fraction. The predictions compare well with experimental data. Turbulent eddy motions cause oxygenated surface water to flow downward and effectively mix with bottom water. The air bubbles directly enhance the dissolved oxygen level. Both mechanisms would inhibit the release of P from bottom sediments. Using proper methods for interphasal forces and turbulence closure is the key to success.

1 Introduction

Lake eutrophication is a nutrient-enrichment scenario where an excess amount of nutrients enters a lake and causes a drastic growth of algae. The subsequent death of algae typically forms a thin greenish layer on the lake surface. Eutrophication reduces light penetration into the lower water column and re-oxygenation of water through air circulation [2]. Dead algae would become food for bacteria. The food consumption process uses oxygen, and thus causes the dissolved oxygen level of the lake water to drop, leading to fish kills [2]. Also, eutrophication causes undesirable odour and taste [6, 9].

Lake eutrophication occurred worldwide. It has been a major environmental issue [10]. Thus, it is worth a while to develop strategies for the effective control of trophic state related to nutrient loading. The aim of this paper is: (1) to characterise water circulation, turbulent mixing, and oxygen concentration, induced by air bubble injection; (2) to determine the accuracy of interphase-force and turbulence

---

A. Najafi-Nejad-Nasser (✉) • S.S. Li • C.N. Mulligan
Concordia University, 1455 de Maisonneuve Blvd. W., Montreal, QC, H3G 1M8, Canada
e-mail: azita.najafi@gmail.com; sam.li@concordia.ca; mulligan@civil.concordia.ca

© Springer International Publishing Switzerland 2016
J. Bélair et al. (eds.), *Mathematical and Computational Approaches in Advancing Modern Science and Engineering*, DOI 10.1007/978-3-319-30379-6_26

closure methods through a comparison between computational results and available laboratory data.

In phytoplankton growth and dynamics, P is a key element. It is the limiting nutrient in fresh water systems. Correspondently, existing strategies for eutrophication control and/or remediation have mostly focused on reducing P load [5, 17]. Various efforts were made to reduce external sources of nutrients loading [12, 21].

However, in the 1970s, studies [13, 20] revealed that P released from lake bottom sediments could continue as internal loading. Internal loading significantly delayed the recovery of eutrophic lakes even after the control of external loading [14]. Under specific conditions, internal loading contributed up to 80 % of total P in lakes [11]. For example, in Lake Ockeechobe in Florida, internal loading was in the same order of magnitude as external sources [19]. Thus, it is of equal importance to be able to effectively control the release of P from bottom sediments to the lake water.

One plausible technique to control internal loading is to introduce bubble plumes to the lake water. Previous studies have identified some relevant parameters associated with this technique [8, 9, 15, 16, 24]. Kim et al. [9] investigated experimentally the effects of bubble size and diffusing area (geometric parameters) on lake destratification. Destratification tends to promote air circulation and improve the dissolved oxygen level. Kim et al. [9] suggested that destratification efficiency is proportional to the bubble diffusing area and inversely proportional to the bubble diameter and overall tank area. Rensen and Roig [15] studied the non-stationary behaviour of flow by conducting experiments of *2D* bubble plumes in a confined tank. According to Imteaz and Aseada [8], the number of ports, air flow rate, and bubbler starting time were important parameters for optimal bubbling operation.

Using the modelling approach, Sahoo and Luktenia [16] studied 1-*D* bubble plumes. They reported that bubbles of close to 1 mm in radius gave a higher oxygen transfer rate and mechanical efficiency than larger bubbles. Yum et al. [24] simulated two-phase bubble plumes, using experimental data for model calibration/verification. Their work led to the derivation of relationships between stratification efficiency, plume spacing, and destratification number. Since the calibration data are from a small tank under controlled environment, uncertainties about the relationships' suitability for application to field conditions exist. More advanced modelling studies are needed to quantify the beneficial effects of bubble plumes on water quality.

This paper tackles the problem by numerically solving the RANS equations on unstructured finite volume mesh. The solution methods entail the use of suitable models for turbulence closure. Previously, a number of investigators have assessed the suitability of turbulence closure models for different applications [7, 18, 22].

2 Computational Model

2.1 Model Domain

The model domain is a cylinder (Fig. 1a) with a height $h = 40$ cm, and a diameter $FG = 50$ cm. The top of the cylinder is open to air. The bottom has a circular hole at the middle. This hole has a diameter $D = 6$ cm. Through this hole (inlet), air bubbles of a given diameter are injected into the otherwise stagnant water contained in the cylinder. This model domain is used because experimental results are available to allow a direct comparison. For efficient computation of water and air bubble motions, we consider three geometric configurations of computational domain: (1) Computational domain A: a simple $2D$ plane cutting through the centreline of the cylinder (Fig. 1c); (2) Computational domain B: an axis-symmetrical domain cutting through the centreline (Fig. 1b); (3) Computational domain C: a 10° wedge of the cylinder (Fig. 1d). Domains A and B are covered by quadrilateral mesh generated using a cell size of 1 mm. The mesh contains 401,802 and 199,562 computing nodes, respectively. Domain C is covered by quadrilateral mesh created using a cell size of 5 mm. The total computing nodes are 33,580.

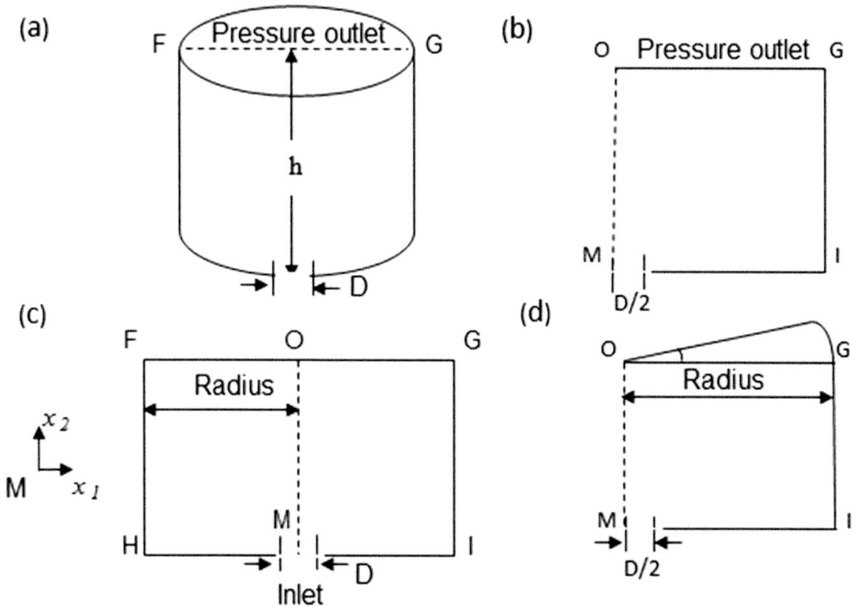

Fig. 1 A diagram of the model cylinder containing water. Water and air-bubble motions are induced by the injection of air bubbles

2.2 Solution Method-Eulerian Approach

The turbulent motions of the liquid phase (water) and gas phase (air bubble) were simulated with the Euler-Euler method using ANSYS Fluent. The motions are described using separate momentum equations. The two phases are related through a momentum exchange term. The interphase momentum transfer is due to interfacial forces acting and interactions between water and air bubbles (such as lift force and drag force) [3, 4, 23]. Hence, a proper solution for the bubble columns dependents on the correct modelling of interphase forces and turbulence models. In this paper, we compare the performance of different interphase force models and two turbulence closure models (the k-ϵ model and the SST k-ω model). All the simulations use unsteady formulation. Phase Coupled SIMPLE (PC-SIMPLE) algorithm is used for pressure velocity coupling. This algorithm is an extension of the SIMPLE algorithm to multiphase problems. The velocity solutions are obtained in a segregated fashion, and coupled by the liquid and gas phases [4].

2.3 Initial and Boundary Conditions

At time $t = 0$, imposed initial conditions are as follows: The free water surface is located at the equilibrium position. The volume fraction of water α_w is equal to one below the free surface (for $x_2 = h$). Water is stagnant or the velocity components u_1 and u_2 are zero in the entire computational domain. Kinematic and dynamic conditions are imposed at the boundaries of the computational domains (Fig. 1), including

(1) Inlet (at the bottom of the cylinder)
(2) Outlet (on the top of the cylinder)
(3) Solid side walls of the cylinder
(4) Axis , OM in Fig. 1b (Runs 2–10, listed in Table 2)
(5) Symmetry (for $3D$ simulation, Run 11)

At the inlet ($-D/2 < x_1 < D/2$ and $x_2 = 0$, Fig. 1), air bubbles of a given diameter d ($d < D$) enter the domain continuously during the simulation time period T. The direction of the entering velocity is upward, and the magnitude is u_o(or $u_2 = u_o$; $u_1 = 0$). At the inlet, the volume fraction of water α_w is taken as zero.

At the outlet (at $x_2 = h$, Fig. 1), fluids are exposed to the atmosphere. Accordingly, the pressure equals to the atmospheric pressure. The volume fraction of water α_w is set to zero.

At the solid walls, no-slip condition is applied for computational domains A, B, and C. For the three cases, the wall distance of the first cell from the walls is $y^+ = 0.33$, $y^+ = 0.45$, and $y^+ = 0.81$, respectively. Thus, no-slip condition is valid. This condition means that both the tangential and normal components of the fluid velocity are set to zero.

2.4 Simulations

Model parameters and their values used in simulations are summarised in Table 1. A total of 11 runs (Runs 1–11, Table 2) were carried out using the Eulerian approach. Runs 1–6 and Runs 10–11 differ from each other in the choice of turbulence dispersion force and lift force; these runs use the k-ϵ model for turbulence closure. Runs 7–9 use the $SSTk$-ω model.

The time period of all the simulations was $t = 10.7$ s. This is long enough since it is more than two times of the advection time scale.

All the 11 runs produced finite volume solutions to the RANS equations. The results are presented and discussed in the next section, along with comparisons with available experimental data.

3 Results and Discussion

Under given conditions (Tables 1 and 2), Runs 1–11 produce water velocities. As an example, velocity vectors and corresponding flow streamlines at a state of equilibrium for Run 3 are plotted in Figs. 2 and 3, respectively. A strong jet is seen to occur in the central region (Fig. 2), as a direct response to bubble injection. Water motions are visible in the entire domain. The jet flow entrains water from both

Table 1 Control parameters and their values

Parameter	Value	Parameter	Value
Time step (s)	0.001	Initial time (s)	0
Cell size for domains A and B (mm)	1	Simulation time period (s)	10.7
Air velocity at the inlet (m/s)	0.085	Number of time step	11,000
Water velocity at the inlet (m/s)	0	Pressure at outlet (Pa)	0
Flow rate at the inlet (m^3/s)	2.4×10^{-4}	Surface diameter (m)	0.5
Bubble size at the inlet (mm)	3	Depth (m)	0.4
Air volume fraction at the inlet	1	Depth of water (m)	0.4
Convergence criteria	10^{-6}	Inlet pipe diameter (m)	0.06

Table 2 Solution domain, water-bubble interaction method, and turbulence closure model

Run	1	2	3	4	5	6	7	8	9	10	11
Domain	A	B	B	B	B	B	B	B	B	B	C
Drag	SN[a]	SN	SN	SN	SN	SN	SN	SN	SN	SN	SN
Lift	(–)	(–)	L[a]	L	T[a]	T	(–)	T	T	T	(–)
Dispersion	(–)	(–)	LDB[a]	S[a]	B[a]	LDB	(–)	LDB	S	S	(–)
Closure	$k-\epsilon$	$k-\epsilon$	$k-\epsilon$	$k-\epsilon$	$k-\epsilon$	$k-\epsilon$	*SST*	*SST*	*SST*	$k-\epsilon$	$k-\epsilon$

[a] *SST* $SSTk-\omega$, *SN* Schiller-Nauman, *L* Legendre, *LDB* Lopez-de-Bertodano, *S* Simonin, *T* Tomiyama, *B* Burns et al. [4]

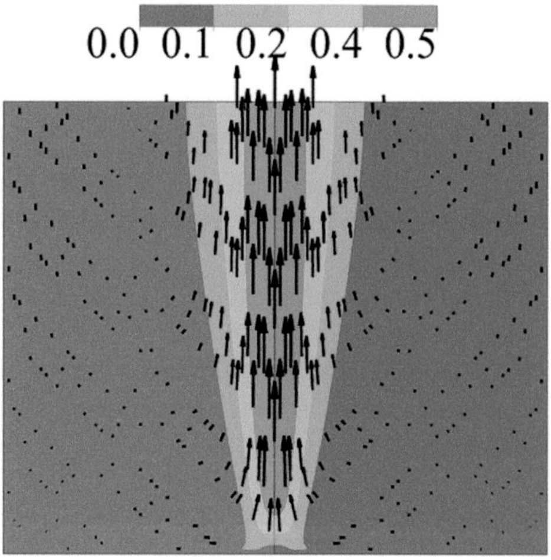

Fig. 2 Water velocity vectors for Run 3 (Table 2)

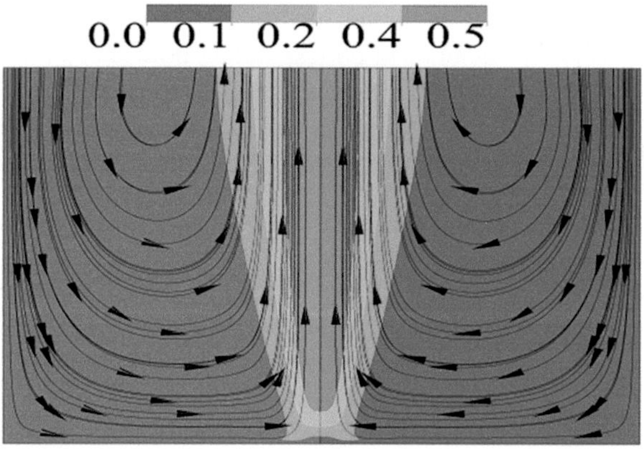

Fig. 3 Water flow streamlines for Run 3

sides in the lower water column and creates eddies (Fig. 3). These eddies produce diverging flows from the centre in the upper water column. Water flow converges to compensate the upward motion at the centre. These flow features are realistic. Also, there are upward and downward motions on both the left and right sides of the water body. These flow patterns would enhance renewal of bottom water with oxygenated surface water.

For Run 3 (Fig. 2), the maximum velocity has a magnitude of nearly 5.6 times of the initial velocity, u_o, of bubbles entering the water column. The maximum velocity has a magnitude of $5.4u_o$ for Run 5, being the lowest among Runs 1–11. For the other runs, the maximum velocity has a magnitude ranging from $5.5u_o$ (for Run 4) to $8.5u_o$ (for Run 7).

The flow streamlines (Fig. 3) show clockwise and counterclockwise eddy motions on the right and left sides of the water body. These eddy motions penetrate the entire water depth, meaning that aeration is effective in producing exchange of water masses. For Runs 1–11, the radius of significant influence is larger than eight times the inlet radius D. Water circulation occurs over virtually the whole width of the water body. In other words, artificial circulation can effectively be created by injecting air bubbles.

The vertical component of water velocities at three selected heights (0.05, 0.1 and 0.2 m) above the bottom varies with radial distance, r, from the centre (Fig. 4a–c). The velocity component has peak values at the centre (or $r = 0$), decreasing rapidly with r. It drops to zero at $r = 0.1$ m for Runs 1, 2, 7 (not shown) and 11 (Fig. 4a–c). The velocity component drops to zero at $r = 0.12$ m, and the flow direction changes from upward to downward for Runs 3–6 and 8–10 (Fig. 4a–c, not shown for Runs 1–7, 9, 10). In summary, Runs 3–6 and 8–10 predict upward flow within $0 < r < 0.12$ m; Runs 1–2, 7 and 11 predict upward flow within $0 < r < 0.1$; this is true at different heights above the bottom. These predictions indicate that the results are sensitive to model setup (Table 2). Experimental data [1] show similar features; water velocity intensifies with increasing height above the inlet but weakens with increasing radial distance.

Bubble rising velocity, u_a, varies with r (Fig. 5a–c). The values are extracted from the model results at model time of $t = 10.7$ s for the same heights as in Fig. 4a–c. u_a decreases from peak values at the centre with increasing r; this feature was observed from experiments [1].

In Table 3, bubble rising velocities at different radial distances are compared with experimental data [1]. Runs 11 produces the smallest relative error.

The results of air and water velocities for Run 3 are better than those for Run 2, especially at larger r distances (not shown). The reason is that the former run considers the effect of interfacial forces. On average (at five different heights above the bottom and different radial distances from the centreline), this consideration has improved the predictions of u_a by 7.3 %, u_w by 6.8 %, and VF_a by 62 %. At the specific height z = 0.3 m, the consideration has reduced the average relative error in the predictions of u_a from 18.8 % to 3.2 %.

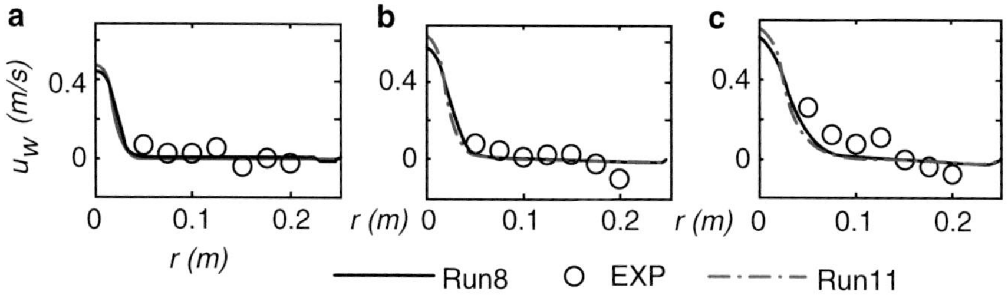

Fig. 4 Water velocity (u_w) distribution with radial distance, r. (**a**) z = 0.05 (m). (**b**) z = 0.1 (m). (**c**) z = 0.2 (m)

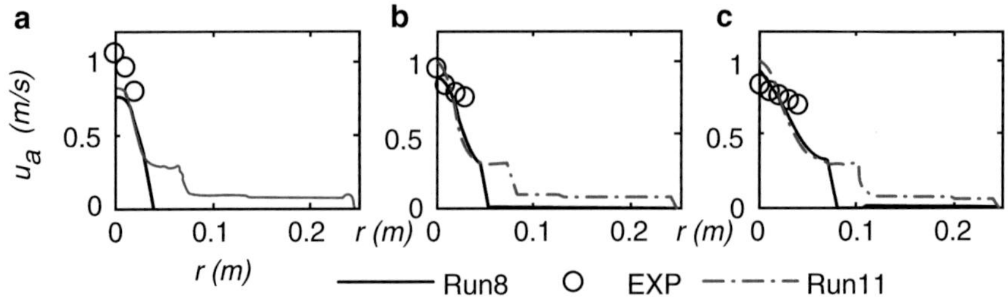

Fig. 5 Distribution of air velocity (u_a) with radial distance, r. (**a**) z = 0.05 (m). (**b**) z = 0.1 (m). (**c**) z = 0.2 (m)

Table 3 Relative errors of computed air velocity (u_a), and average of water turbulent kinetic energy (K)

Run	1	2	3	4	5	6	7	8	9	10	11
u_a (%)	35.3	23.6	19.6	20.1	19.8	17.8	25.9	14.4	20.1	16.2	12.0
K(J/kg)	0.019	0.007	0.006	0.006	0.006	0.006	0.005	0.004	0.004	0.006	0.002

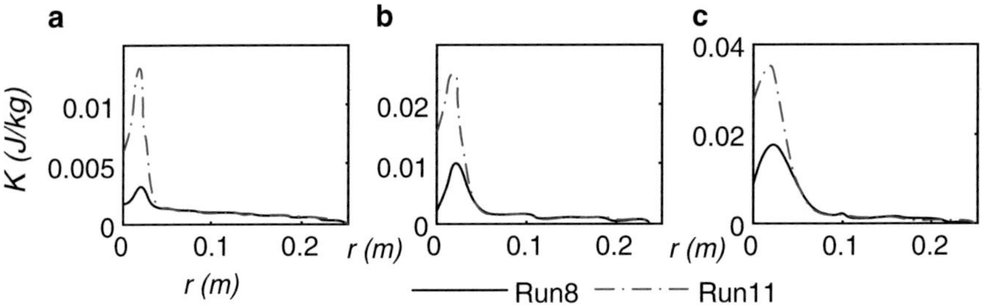

Fig. 6 Distribution of water turbulent kinetic energy (K) with r. (**a**) z = 0.05 (m). (**b**) z = 0.1 (m). (**c**) z = 0.2 (m)

Variations in turbulent kinetic energy, K, of water flow with radial distance r at three different heights from the bottom are illustrated in Fig. 6a–c. As bubbles rise, part of their kinetic energy is converted into turbulent kinetic energy. The maximum K values occur slightly off the centreline. K increases with increasing height. At a given height, it decreases with increasing r. Overall, air bubbles entering into the water body cause water motions and turbulence. Water and bubble motions help prevent sedentary condition, which is known to contribute to water-quality degradation. The averages (over the whole water body) of K values are listed in Table 3.

Distributions of predicted air volume fraction, VF_a, with r are shown in Fig. 7a–c. Values of VF_a are extracted from the model results for Runs 1 to 11 at model time of $t = 10.7s$ for the same heights as in Fig. 4a–c. The snapshots show that at a given height, VF_a decreases from peak values with increasing r. The peak values occur slightly off the centre for Runs 1–11 (Fig. 7a–c, not shown for Runs 1–7, 9, 10).

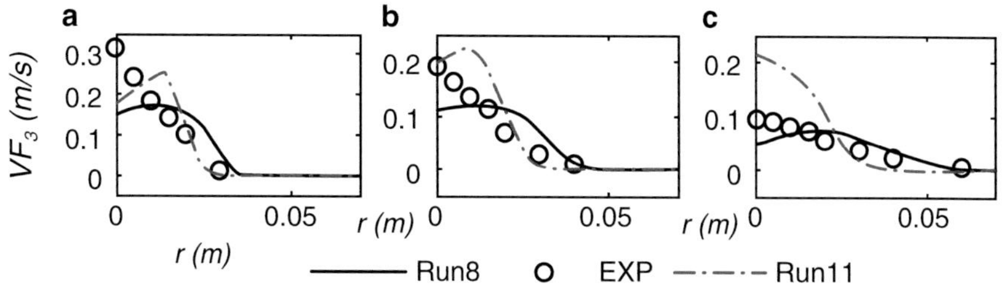

Fig. 7 Distribution of air volume fraction, VF_a, with radial distance r. (**a**) z = 0.05 (m). (**b**) z = 0.1 (m). (**c**) z = 0.2 (m)

A strong lake stratifications known to prohibit fluid motions and mixing in the vertical. However, the consideration of density stratification effects is beyond the scope of this paper.

4 Conclusions

This paper discusses artificial circulation in lakes, induced by injecting air bubbles, for the control of eutrophication. We have simulated artificial circulation using CFD modelling techniques, and reached the following conclusions: (1) The injection triggers turbulent motions of water and bubbles, which feature a strong upward flow above the injection location and energetic turbulent eddies on both sides of the upward flow. (2) These large scale eddies enhance renewal of bottom water with oxygenated surface water, which helps improve the dissolved oxygen level in the lower water column. (3) Air bubbles entering the water column produce turbulent kinetic energy; this source of energy will maintain small scale eddy motions in the lake water, with beneficial mixing effects. (4) The dissolved oxygen level in the lake water is improved as a direct response to air bubbles entering the water column. (5) From the computational perspective, a 2D axisymmetric computational domain is recommended for two reasons: (a) it offers high computational efficiency, relative to a 3D domain; and (b) it appears to be sufficient to capture measured characteristics of air and water velocities. (6) Model predictions of water velocity, air velocity, air volume fraction agree well with experimental data. The best agreement is obtained with the use of the Schiller-Nauman model for drag, the Tomiyama model for lift, the Lopez-de-Bertodano model for turbulent dispersion, and the $k - \epsilon$ model for turbulence closure.

Acknowledgements This study received financial support through Discovery Grants held by S.S. Li and C.N. Mulligan

References

1. Anagbo, P.E., Brimacombe, J.K.: Plume characteristics and liquid circulation in gas injection through a porous plug. Metall. Trans. B **21B**, 637–648 (1990)
2. Ansari, A.A., Gill, S.S., Lanza, G.R., Rast, W.: Eutrophication: Causes, Consequences and Control. Springer, Dordrecht/Heidelberg/London/New York (2011)
3. Azzopardi, B., Zhao, D., Yan, Y., Morvan, H., Mudde, R.F., Lo, S.: Hydrodynamics of Gas-liquid Reactors: Normal Operation and Upset Conditions. Wiley. Chichester, United Kingdom, (2011)
4. ANSYS: Fluent 6.3 User's Guide (2006)
5. Dai, L., Pan, G.: The effects of red soil in removing P from water column and reducing P release from sediment in Lake Taihu. Water Sci. Technol. **65**(5), 1052–1058 (2014)
6. Environment Canada: Canadian guidance framework for the management of P in freshwater systems. Ecosyst. Health: Sci.-Based Solut. **1–8**, 1–114 (2004)
7. Hjarne, J., Chernoray, V., Larsson, J., Lofdahl, L.: Numerical validations of secondary flows and loss development downstream of a highly loaded low pressure turbine outlet guide Vane cascade. ASME Turbo Expo 2007, 723–733 (2007)
8. Imteaz, M.A., Asaeda, T.: Artificial mixing of lake water by bubble plume and effects of bubbling operations on algal bloom. Water Res. **34**(6),1919–1929 (2000)
9. Kim, S.H., Kim, J.Y., Park, H., Park, N.S.: Effects of bubble size and diffusing area on destrafication efficiency in bubble plumes of two-layer stratification. ASCE J. Hydraul. Eng. **136**, 106–115 (2010)
10. Matsui, S., Ide, S., Ando, M.: Lake reservoirs: reflecting waters of sustainable use. Water Sci. Technol. **32**(7),221–224 (1955)
11. Penn, M.R., Auer, M.T., Doerr, S.M., Driscoll, C.T., Brooks, C.M., Effler, S.W.: Seasonality in P release rates from the sediments of a hypereutrophic lake under a matrix Of Ph and redox conditions. Can. J. Fish. Aquat. Sci. **57**, 1033–104 (2000)
12. Phillips, G., Bramwell, A., Pitt, J., Stansfield, J., Perrow, M.: Practical application of 25 Years' research into the management of shallow lakes. Hydrobiologia **395/396** 61–76 (1999)
13. Qunhe, W., Renduo, Z., Shan, H., Hengjun, Z.: Effects of bacteria on nitrogen and P release from river sediment. Environ. Sci. **20**, 404–412 (2008)
14. Reddy, K.R., Q'connor, G.A., Schelske, C.L.: P Biogeochemistry of Subtropical Ecosystems, 1st edn., p. 101. CRC Press, Boca Raton (1999)
15. Rensen, J., Roig, V.: Experimental study of the unsteady structure of a confined bubble plume. Int. J. Multiph. Flow **27**, 1431–1449 (2001)
16. Sahoo, G.B., Luketina, D.: Modeling of bubble plume design and oxygen transfer for reservoir restoration. Water Res. **37**, 393–401 (2003)
17. Schauser, I., Chorus, I.: Assessment of internal and external lake restoration measures for two Berlin lakes. Lake Reserv. Manag. **23**, 366–376 (2007)
18. Schuler, M., Zehnder, F., Weigand, B., von Wolfersdorf, J., Neumann, S.O.: The effect of turning vanes on pressure loss and heat transfer of a Ribbed rectangular two-pass internal cooling channel. ASME J. Turbomach. **133**(2), 0211017–(1–10) (2011)
19. Sharpley, A.N., Chapra, S.C., Wedepohl, R., Sims, J.T., Daniel, T.C., Reddy, K.R.: Managing agricultural P for protecting of surface waters: issues and options. J. Environ. Qual. **23**(3), 437–451 (1994)
20. Sondergaard, M., Jensen, J.P., Jappesen, E.: Role of sediment and internal loading of P in shallow lakes. Hydrobiologia **506–509**, 135–145 (2003)

21. Van Der Molen, D.T., Boers, P.C.M.: Eutrophication Control in the Netherlands. Hydrobiologia **136**, 403–409 (1999)
22. Wang, X., Naji, H., Mezrhab, A.: Computational investigation of different models when predicting airflow in an enclosure. ASME Press. Vessels Piping Div. Conf. **4**, 179–188 (2008)
23. Yeoh, G.H., Tu, J.: Computational Techniques for Multiphase Flows: basics and applications. Elsevier, Butterworth-Heinmann, United Kingdom, (2009)
24. Yum, K., Kim, S.H., Park, H.: Effects of plume spacing and flowrate on destrafication efficiency of air diffusers. Water Res. **42**, 3249–3262 (2008)

On the Co-infection of Malaria and Schistosomiasis

Kazeem O. Okosun and Robert Smith?

Abstract Mathematical models for co-infection of diseases (that is, the simultaneous infection of an individual by multiple diseases) are sorely lacking in the literature. Here we present a mathematical model for the co-infection of malaria and schistosomiasis. We derive reproduction numbers for malaria and schistosomiasis independently, then combine these to determine the effects of disease interactions. Sensitivity indices show that malaria infection may be associated with an increased rate of schistosomiasis infection. However, schistosomiasis infection is not associated with an increased rate of malaria infection. Therefore, whenever there is co-infection of malaria and schistosomiasis in the community, our model suggests that control measures for each disease should be administered concurrently for effective control.

1 Introduction

Malaria and schistosomiasis often overlap in tropical and subtropical countries, imposing tremendous disease burdens [4, 8, 14]. The substantial epidemiological overlap of these two parasitic infections invariably results in frequent co-infections [7, 18]. The challenges facing the development of a highly effective malaria vaccine have generated interest in understanding the interactions between malaria and co-endemic helminth infections, such as those caused by Schistosoma, that could impair vaccine efficacy by modulating host-immune responses to *Plasmodium* infection and treatment [13, 14]. Both malaria and schistosomiasis are endemic to most African nations. However, the extent to which schistosomiasis modifies the rate of febrile malaria remains unclear.

K.O. Okosun
Department of Mathematics, Vaal University of Technology, Vanderbijlpark, South Africa
e-mail: kazeemoare@gmail.com

R. Smith? (✉)
University of Ottawa, Ottawa, ON, Canada
e-mail: rsmith43@uottawa.ca

© Springer International Publishing Switzerland 2016
J. Bélair et al. (eds.), *Mathematical and Computational Approaches in Advancing Modern Science and Engineering*, DOI 10.1007/978-3-319-30379-6_27

Mathematical modelling has been an important tool in understanding the dynamics of disease transmission and also in the decision-making processes regarding intervention mechanisms for disease control. For example, Ross [12] developed the first mathematical models of malaria transmission. His focus was on mosquito control, and he showed that, for the disease to be eliminated, the mosquito population should be brought below a certain threshold. Another classical result is due to Anderson and May [1], who derived a malaria model with the assumption that acquired immunity in malaria is independent of exposure duration.

There is an urgent need for co-infection models for infectious diseases, particularly those that mix neglected tropical diseases with "the big three" (HIV, TB and malaria) [8]. Recently, the authors in [10] proposed a model for schistosomiasis and HIV/AIDS co-dynamics, while the co-infection dynamics of malaria and cholera were studied in [11]. However, few studies have been carried out on the co-infection of schistosomiasis. To the best of our knowledge, no work has been done to investigate the malaria–schistosomiasis co-infection dynamics.

In this paper, we formulate and analyse an SIR (susceptible, infected and recovered) model for malaria–schistosomiasis co-infection, in order to understand the effect that controlling for one disease may have on the other.

2 Model Formulation

Our model subdivides the total human population, denoted by N_h, into subpopulations of susceptible humans S_h, individuals infected only with malaria I_m, individuals infected with only schistosomiasis I_{sc}, individuals infected with both malaria and schistosomiasis C_{ms}, individuals who have recovered from malaria R_m and individuals who have recovered from schistosomiasis R_s. The total mosquito vector population, denoted by N_v, is subdivided into susceptible mosquitoes S_v and mosquitoes infected with malaria I_v. Similarly, the total snail vector population, denoted by N_{sv}, is subdivided into susceptible snails S_{sv} and snails infected with schistosomiasis I_{sv}. Thus $N_h = S_h + I_m + I_s + C_{ms} + R_s + R_m$, $N_v = S_v + I_v$ and $N_{sv} = S_{sv} + I_{sv}$.

The model is given by the following system of ordinary differential equations.

$$\begin{aligned}
S_h' &= \Lambda_h + \varepsilon R_s + \alpha R_m - \beta_1 S_h - \lambda_1 S_h - \mu_h S_h \\
I_m' &= \beta_1 S_h - \lambda_1 I_m - (\psi + \mu_h + \phi) I_m \\
I_{sc}' &= \lambda_1 S_h - \beta_1 I_{sc} - (\omega + \mu_h + \eta) I_{sc} \\
C_{ms}' &= \beta_1 I_{sc} + \lambda_1 I_m - (\delta + \mu_h + \eta + \phi) C_{ms} \\
R_m' &= \psi I_m - (\alpha + \mu_h) R_m + \tau \delta C_{ms} \\
R_s' &= \omega I_{sc} - (\varepsilon + \mu_h) R_s + (1 - \tau) \delta C_{ms}
\end{aligned}$$

$$\begin{aligned} S_v' &= \Lambda_v - \beta_2 S_v - \mu_v S_v \\ I_v' &= \beta_2 S_v - \mu_v I_v \\ S_{sv}' &= \Lambda_s - \lambda_2 S_{sv} - \mu_{sv} S_{sv} \\ I_{sv}' &= \lambda_2 S_{sv} - \mu_{sv} I_{sv}, \end{aligned} \tag{1}$$

with the transmission rates given by

$$\beta_1 = \frac{\beta_h I_v}{N_h}, \quad \lambda_1 = \frac{\lambda I_{sv}}{N_h}, \quad \beta_2 = \frac{\beta_v (I_m + C_{ms})}{N_h}, \quad \lambda_2 = \frac{\lambda_s (I_{sc} + C_{ms})}{N_h}.$$

Birth rates for humans, mosquitoes and snails are, respectively, Λ_h, Λ_v and Λ_{sv}, while the corresponding mortality rates are μ_h, μ_v and μ_{sv}. Here η is the schistosomiasis-related death rate and ϕ is the malaria-related death rate. The immunity-waning rates for malaria and schistosomiasis are α and ε respectively, while the recovery rates from malaria, schistosomiasis and co-infection are ψ, ω and δ respectively. The term $\tau\delta$ accounts for the portion of co-infected individuals who recover from malaria, while $(1 - \tau)\delta$ accounts for co-infected individuals who recover from schistosomiasis; thus τ (satisfying $0 \leq \tau \leq 1$) represents the likelihood of individuals to recover from malaria first. Note that all parameters might in practice vary with time; however, we shall take variations in our critical parameters into account with a sensitivity analysis.

3 Analysis of Malaria–Schistosomiasis Co-infection Model

The malaria–schistosomiasis model (1) has a disease-free equilibrium, given by

$$\mathscr{E}_0 = (S_h^*, I_m^*, I_{sc}^*, C_{ms}^*, R_m^*, R_s^*, S_v^*, I_v^*, S_{sv}^*, I_{sv}^*) = \left(\frac{\Lambda_h}{\mu_h}, 0, 0, 0, 0, 0, \frac{\Lambda_v}{\mu_v}, 0, \frac{\Lambda_s}{\mu_{sv}}, 0\right).$$

The linear stability of $\mathscr{E}_0$ can be established using the next-generation method [17] on the system (1). It follows that the reproduction number of the malaria–schistosomiasis model (1), denoted by $\mathscr{R}_{msc}$, is given by

$$\mathscr{R}_{msc} = \max\{\mathscr{R}_{sc}, \mathscr{R}_{0m}\},$$

where

$$\mathscr{R}_{0m} = \sqrt{\frac{\Lambda_v \beta_h \beta_v \mu_h}{\Lambda_h \mu_v^2 (\psi + \phi + \mu_h)}}$$

$$\mathscr{R}_{sc} = \sqrt{\frac{\lambda \lambda_s \Lambda_s \mu_h}{\Lambda_h (m + \omega + \mu_h) \mu_{sv}^2}}.$$

Note that the reproduction number produced by the next-generation method produces a threshold quantity and not necessarily the average number of secondary infections [9]. We thus have the following theorem.

Theorem 1 *The disease-free equilibrium* $\mathscr{E}_0$ *is locally asymptotically stable whenever* $\mathscr{R}_{msc} < 1$ *and unstable otherwise.*

3.1 Impact of Disease Interactions

To analyse the effects of schistosomiasis on malaria and vice versa, we begin by expressing $\mathscr{R}_{sc}$ in terms of $\mathscr{R}_{0m}$. We solve for μ_h to get

$$\mu_h = \frac{D_1 \mathscr{R}_{0m}^2}{D_2 - D_3 \mathscr{R}_{0m}^2},$$

where

$$D_1 = \Lambda_h \mu_v^2 (\psi + \phi), \qquad D_2 = \Lambda_v \beta_h \beta_v, \qquad D_3 = \Lambda_h \mu_v^2 .$$

Substituting into the expression for $\mathscr{R}_{sc}$, we obtain

$$\mathscr{R}_{sc} = \sqrt{\frac{\lambda \lambda_s \Lambda_s D_1 \mathscr{R}_{0m}^2}{[(\eta + \omega) D_2 + (D_1 - (\eta + \omega) D_3) \mathscr{R}_{0m}^2] \Lambda_h \mu_{sv}^2}} . \tag{2}$$

Differentiating $\mathscr{R}_{sc}$ with respect to $\mathscr{R}_{0m}$ leads to

$$\frac{\partial \mathscr{R}_{sc}}{\partial \mathscr{R}_{0m}} = \frac{(\eta + \omega) D_2 \sqrt{\frac{\lambda \lambda_s \Lambda_s D_1 \mathscr{R}_{0m}^2}{[(\eta+\omega) D_2 + (D_1 - (\eta+\omega) D_3) \mathscr{R}_{0m}^2] \Lambda_h \mu_{sv}^2}}}{[(\eta + \omega) D_2 \mathscr{R}_{0m} + (D_1 - (\eta + \omega) D_3) \mathscr{R}_{0m}^3]} . \tag{3}$$

Similarly, expressing μ_h in terms of $\mathscr{R}_{sc}$, we get

$$\mu_h = \frac{D_4 \mathscr{R}_{sc}^2}{D_5 - D_6 \mathscr{R}_{sc}^2}, \tag{4}$$

where

$$D_4 = \Lambda_h \mu_{sv}^2 (\eta + \omega), \qquad D_5 = \lambda \lambda_s \Lambda_s, \qquad D_6 = \Lambda_h \mu_{sv}^2 .$$

Substituting into the expression for $\mathscr{R}_{0m}$, we obtain

$$\mathscr{R}_{0m} = \sqrt{\frac{D_4 \beta_h \beta_v \Lambda_v \mathscr{R}_{sc}^2}{[(\phi + \psi)D_5 + (D_4 - (\phi + \psi)D_6)\mathscr{R}_{sc}^2]\Lambda_h \mu_{sv}^2}}. \tag{5}$$

3.2 Sensitivity Indices of R_{sc} when Expressed in Terms of R_{0m}

We next derive the sensitivity of R_{sc} in (2) (i.e., when expressed in terms of R_{0m}) to each of the 13 different parameters. However, the expression for the sensitivity indices for some of the parameters are complex, so we evaluate the sensitivity indices of these parameters at the baseline parameter values as given in Table 1. Since the effect of immunity in the control of re-infection is not entirely known [6], we have assumed the schistosomiasis immunity waning rate. Due to a lack of data in the literature, assumptions were made for the recovery rate of co-infected individuals, δ, recovery rate of schistosomiasis-infected individuals, ω, and the rate of recovery from malaria for co-infected individuals, τ.

Table 1 Parameters in the co-infection model

Parameter	Description	value	Ref
ϕ	Malaria-induced death	0.05–0.1 day^{-1}	[16]
β_h	Malaria transmissibility to humans	0.034 day^{-1}	[2]
β_v	Malaria transmissibility to mosquitoes	0.09 day^{-1}	[2]
λ	Schistosomiasis transmissibility to humans	0.406 day^{-1}	[15]
λ_s	Schistosomiasis transmissibility to snails	0.615 day^{-1}	[3]
μ_h	Natural death rate in humans	0.00004 day^{-1}	[2]
μ_v	Natural death rate in mosquitoes	1/15–0.143 day^{-1}	[2]
μ_{sv}	Natural death rate in snails	0.000569 day^{-1}	[3, 15]
α	Malaria immunity waning rate	1/(60*365) day^{-1}	[2]
ε	Schistosomiasis immunity waning rate	0.013 day^{-1}	Assumed
Λ_h	Human birth rate	800 people/day	[3]
Λ_v	Mosquitoes birth rate	1000 mosquitoes/day	[2]
Λ_s	Snail birth rate	100 snails/day	[5]
δ	Recovery rate of co-infected individual	0.35 day^{-1}	Assumed
ω	Recovery rate of schistosomiasis-infected individual	0.0181 day^{-1}	Assumed
ψ	Recovery rate of malaria-infected individual	1/(2*365) day^{-1}	[2]
τ	Co-infected proportion who recover from malaria only	0.1	Assumed
η	Schistosomiasis-induced death	0.0039 day^{-1}	[3]

Table 2 Sensitivity indices of R_{sc} expressed in terms of R_{0m}

	Parameter	Description	Sensitivity index if $R_{0m} < 1$	Sensitivity index if $R_{0m} > 1$
1	μ_{sv}	Snail mortality	-1	-1
2	μ_v	Mosquito mortality	0.56	0.07
3	λ_s	Schistosomiasis transmissibility to snails	0.5	0.5
4	Λ_s	Snail birth rate	0.5	0.5
5	β_h	Malaria transmissibility to humans	-0.28	-0.03
6	β_v	Malaria transmissibility to mosquitoes	-0.28	-0.03
7	Λ_v	Mosquito birth rate	-0.28	-0.03
8	Λ_h	Human birth rate	-0.22	-0.47
9	ϕ	Malaria-induced death	0.12	-0.31
10	ω	Recovery from schistosomiasis	-0.10	0.26
11	m	Schistosomiasis-induced death	-0.02	0.05
12	ψ	Recovery from malaria rate	0.003	-0.0084

The sensitivity index of R_{sc} with respect to λ, for example, is

$$\Upsilon_{\lambda}^{R_{sc}} \equiv \frac{\partial R_{sc}}{\partial \lambda} \times \frac{\lambda}{R_{sc}} = 0.5 . \tag{6}$$

The detailed sensitivity indices of R_{sc} resulting from the evaluation of the other parameters of the model are shown in Table 2.

Table 2 shows the parameters, arranged from the most sensitive to the least. For $R_{0m} < 1$, the most sensitive parameters are the snail mortality rate, the mosquito mortality rate, the transmissibility of schistosomiasis to snails and the snail birth rate (μ_{sv}, μ_v, λ_s and Λ_s, respectively). Since $\Upsilon_{\mu_{sv}}^{R_{sc}} = -1$, increasing (or decreasing) the snail mortality rate μ_{sv} by 10 % decreases (or increases) R_{sc} by 10 %; similarly, increasing (or decreasing) the mosquito mortality rate, μ_v, by 10 % increases (or decreases) R_{sc} by 5.6 %. In the same way, increasing (or decreasing) the transmissibility of schistosomiasis to snails, λ_s, increases (or decreases) R_{sc} by 5 %. As the malaria parameters β_h, β_v and Λ_v increase/decrease by 10 %, the reproduction number of schistosomiasis, R_{sc}, decreases by 2.8 % in all three cases.

For $R_{0m} > 1$, the most sensitive parameters are the snail mortality rate, the rate of a snail getting infected with schistosomiasis, the snail birth rate, the human birth rate, malaria-induced death and recovery from schistosomiasis (μ_{sv}, λ_s, Λ_s, Λ_h, ϕ, ω, respectively). Since $\Upsilon_{\lambda_s}^{R_{sc}} = 0.5$, increasing (or decreasing) by 10 % increases (or decreases) R_{sc} by 5 %; similarly, increasing (or decreasing) the recovery rate, ω, by 10 % increases (or decreases) R_{sc} by 2.6 %. Also, as the malaria parameters β_h, β_v and Λ_v increase/decrease by 10 %, the reproduction number of schistosomiasis, R_{sc}, decreases by only 0.3 % in all three cases.

Table 3 Sensitivity indices of R_{0m} expressed in terms of R_{sc}

	Parameter	Description	Sensitivity index if $R_{sc} < 1$	Sensitivity index if $R_{sc} > 1$
1	β_v	Malaria transmissibility to mosquitoes	0.5	0.5
2	Λ_v	Mosquito birth rate	0.5	0.5
3	λ	Schistosomiasis transmissibility to humans	−0.5	−0.5
4	λ_s	Schistosomiasis transmissibility to snails	−0.5	−0.5
5	Λ_s	Snail birth rate	−0.5	−0.5
6	ϕ	Malaria-induced death	−0.49	−0.49
7	ω	Recovery from schistosomiasis	0.41	0.41
8	m	Schistosomiasis-induced death	0.09	0.09
9	ψ	Recovery from malaria	−0.01	−0.01
10	μ_{sv}	Snail mortality	0.0000002	0.000007
11	Λ_h	Human birth rate	0.0000001	0.000004

It is clear that R_{sc} is sensitive to changes in R_{0m}. That is, the sensitivity of R_{sc} to parameter variations depends on R_{0m}; whenever, $R_{0m} < 1$, R_{sc} is less sensitive to the malaria parameters.

3.3 *Sensitivity Indices of R_{0m} when Expressed in Terms of R_{sc}*

Similar to the previous subsection, we derive the sensitivity of R_{0m} in (5) (i.e. when expressed in terms of R_{sc}) to each of the different parameters. The sensitivity index of R_{0m} with respect to β_h, for example, is

$$\Upsilon_{\beta_h}^{R_{0m}} \equiv \frac{\partial R_{0m}}{\partial \beta_h} \times \frac{\beta_h}{R_{0m}} = 0.5\,. \tag{7}$$

The detailed sensitivity indices of R_{0m} resulting from the evaluation to the other parameters of the model are shown in Table 3. It is clearly seen from Table 3 that the malaria reproduction number, R_{0m}, is not sensitive to any variation in the schistosomiasis reproduction number R_{sc}.

4 Numerical Simulations

Table 1 lists the parameter descriptions and values used in the numerical simulation of the co-infection model.

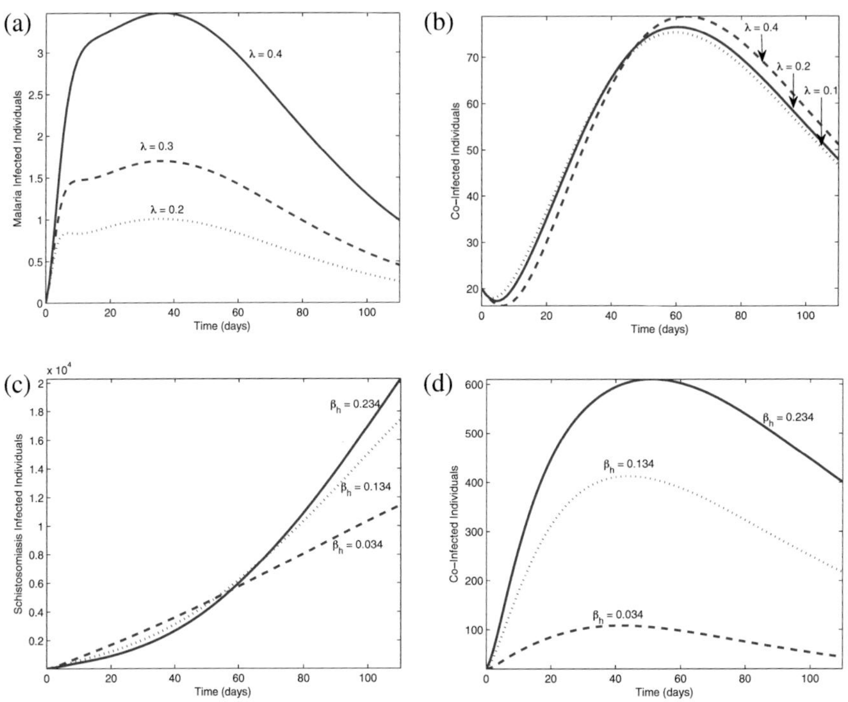

Fig. 1 Simulations of the malaria–schistosomiasis model showing the effect of varying transmission rates

Figure 1a,b shows the effect of varying the schistosomiasis transmission parameter λ on the number of individuals infected with malaria, I_m, and the number of co-infected individuals, C_{ms}. This illustrates that effective control of schistosomiasis would enhance the control of malaria. Conversely, Fig. 1c,d shows the effect of varying the malaria transmission parameter β_h on the number of individuals infected with schistosomiasis, I_{sc}, and the number of co-infected individuals. This illustrates that effective control of malaria would enhance control of co-infection but have only minimal effect on schistosomiasis prevalence.

Figure 2 shows the effect of varying the death rate of mosquitoes μ_v (for example, through spraying) on the number of individuals infected with schistosomiasis and the number of co-infected individuals. As the mosquitoes are controlled, the number of individuals infected with malaria falls dramatically, as does the number of co-infected individuals, while the number of schistosomiasis-infected individuals only decreases slightly.

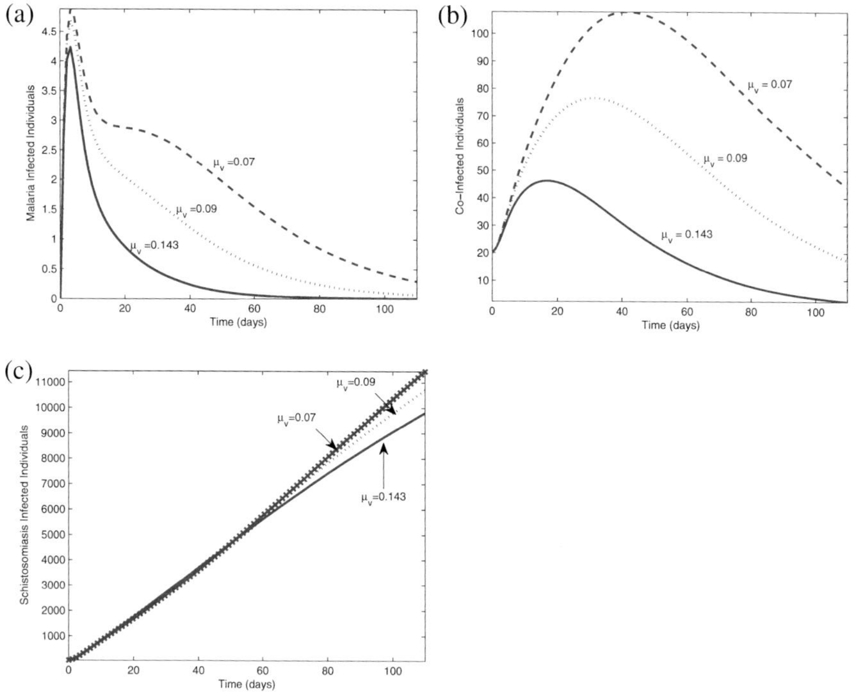

Fig. 2 Simulations of the malaria–schistosomiasis model showing the effect of varying the mosquito death rate

5 Concluding Remarks

In this paper, we formulated and analysed a deterministic model for the transmission of malaria–schistosomiasis co-infection. We derived basic reproduction numbers for each infection and determined the sensitivity of each reproduction number to all parameters. Our analysis shows that malaria infection may be associated with an increased rate of schistosomiasis infection. However, in our model, schistosomiasis infection is not associated with an increased rate of malaria infection. Therefore, whenever there is co-infection of malaria and schistosomiasis in the community, our model suggests that control measures for both diseases should be administered concurrently for effective control.

Acknowledgements The authors are grateful to an anonymous reviewer whose comments enhanced the manuscript. OKO acknowledges the Vaal University of Technology Research Office and the National Research Foundation (NRF), South Africa, through the KIC Grant ID 97192 for the financial support to attend and present this paper at the AMMCS-CAIM 2015 meeting. RS? is supported by an NSERC Discovery Grant. For citation purposes, please note that the question mark in "Smith?" is part of the author's name.

References

1. Anderson, R.M., May, R.M.: Infectious Diseases of Humans: Dynamics and Control. Oxford University Press, Oxford (1991)
2. Blayneh, K.W., Cao, Y., Kwon, H.D.: Optimal control of vector-borne diseases: treatment and prevention. Discret. Contin. Dyn. Syst. Ser B **11(3)**, 587–611 (2009)
3. Chiyaka, E.T., Magombedze, G., Mutimbu, L.: Modelling within host parasite dynamics of schistosomiasis. Comput. Math Methods Med. **11**(3), 255–280 (2010)
4. Doumbo, S., Tran, T.M., Sangala, J., Li, S., Doumtabe, D., et al.: Co-infection of long-term carriers of plasmodium falciparum with schistosoma haematobium enhances protection from febrile malaria: a prospective cohort study in Mali. PLoS Negl. Trop. Dis. **8**(9), e3154 (2014)
5. Feng, Z., Li, C.-C., Milner, F.A.: Schistosomiasis models with two migrating human groups. Math. Comput. Model. **41**(11–12), 1213–1230 (2005)
6. Fulford, A.J.C., Butterworth, A.E., Dunne, D.W., Strurrock, R.F., Ouma, J.H.: Some mathematical and statistical issues in assessing the evidence for acquired immunity to schistosomiasis. In: Isham, V., Medley, G. (eds.) Models for Infectious Human Diseases, pp. 139–159. Cambridge University Press, Cambridge (1996)
7. Hotez, P.J., Molyneux, D.H., Fenwick, A., Ottesen, E., Ehrlich, S.S., et al.: Incorporating a rapid-impact package for neglected tropical diseases with programs for HIV/AIDS, tuberculosis, and malaria. PLoS Med. **3**, e102 (2006)
8. Kealey, A., Smith?, R.J.: Neglected tropical diseases: infection, modelling and control. J. Health Care Poor Underserv. **21**, 53–69 (2010)
9. Li, J., Blakeley, D., Smith?, R.J.: The Failure of R_0. Comput. Math Methods Med. **2011**, Article ID 527610 (2011)
10. Mushayabasa, S., Bhunu, C.P.: Modeling Schistosomiasis and HIV/AIDS co-dynamics. Comput. Math. Methods Med. **2011**, Article ID 846174 (2011)
11. Okosun, K.O., Makinde, O.D.: A co-infection model of malaria and cholera diseases with optimal control. Math. Biosci. **258**, 19–32 (2014)
12. Ross, R.: The Prevention of Malaria, 2nd edn. Murray, London (1911)
13. Salgame, P., Yap, G.S., Gause, W.C.: Effect of helminth-induced immunity on infections with microbial pathogens. Nat. Immunol. **14**, 1118–1126 (2013)
14. Semenya, A.A., Sullivan, J.S., Barnwell, J.W., Secor, W.E.: Schistosoma mansoni Infection impairs antimalaria treatment and immune responses of rhesus macaques infected with mosquito-borne plasmodium coatneyi. Infect. Immun. **80**(11), 3821–3827 (2012)
15. Spear, R.C., Hubbard, A., Liang, S., Seto, E.: Disease transmission models for public health decision making: toward an approach for designing intervention strategies for Schistosomiasis japonica. Environ. Health Perspect. **10**(9), 907–915 (2002)
16. Smith?, R.J., Hove-Musekwa, S.D.: Determining effective spraying periods to control malaria via indoor residual spraying in sub-saharan Africa. Hindawi Publishing Corporation. J. Appl. Math. Dec. Sci. **2008**, Article ID 745463, 19p (2008)
17. van den Driessche, P., Watmough, J.: Reproduction numbers and sub-threshold endemic equilibria for compartmental models of disease transmission. Math. Biosci. **180**, 29–48 (2002)
18. Yapi, R.B., Hürlimann, E., Houngbedji, C.A., Ndri, P.B., Silué, K.D., et al.: Infection and co-infection with helminths and plasmodium among school children in Côte d'Ivoire: results from a national cross-sectional survey. PLoS Negl. Trop. Dis. **8**(6), e2913 (2014)

A Discrete-Continuous Modeling Framework to Study the Role of Swarming in a Honeybee-*Varroa destrutor*-Virus System

Vardayani Ratti, Peter G. Kevan, and Hermann J. Eberl

Abstract In this paper, we present a general discrete-continuous modeling framework to study the effect of swarming on the dynamics of a honeybee colony infested with varroa mite and Acute Bee Paralysis Virus (ABPV) . Two scenarios are studied under which swarming takes place i.e., swarming due to overcrowding and swarming at fixed time intervals. For this purpose, we use an existing mathematical model in the literature. The dependent variables in the model are uninfected bees, infected bees, virus carrying mites and total mites that infest the colony. The model is studied in variable coefficients, in particular, step functions with each season as a constant in time. It is observed that the percentage of healthy bees leaving with the swarm has a great impact on the strength and survival of the parent colony. A colony, that otherwise dies off due to virus, survives as a properly working colony if the percentage of the mites leaving the parent colony is above a critical value.

1 Introduction

A honey bee colony consists of a single reproductive queen, 20,000–60,000 adult workers, 10,000–30,000 individuals at brood stage (eggs, larvae and pupae), and several hundred drones [13]. A large population of workers is needed to carry out the tasks of the bee colony, including cleaning, brood rearing, guarding the hive etc. In order to maintain a high reproduction level in the colony, the bees start preparing for swarming. Swarming is the natural method of reproduction of honeybee colonies. In the process of swarming, the original single colony reproduces to two or more colonies. During swarming, almost 50–70 % of the worker bees leave the parent colony with the queen (old mated queen in case of the first swarm) to a new site [3, 14]. The colony starts preparing for the swarm 1 month before the swarm is issued. Swarming normally takes place in mid-spring [14]. The timing of swarming varies with seasons and location of the hives. Since swarming affects pollination and honey production, it has an impact on agriculture and economy as well. The

V. Ratti (✉) • P.G. Kevan • H.J. Eberl
University of Guelph, 50 Stone Road E, Guelph, ON, N1G 2W1, Canada
e-mail: vratti@uoguelph.ca; pkevan@uoguelph.ca; heberl@uoguelph.ca

© Springer International Publishing Switzerland 2016

J. Bélair et al. (eds.), *Mathematical and Computational Approaches in Advancing Modern Science and Engineering*, DOI 10.1007/978-3-319-30379-6_28

main symptom of swarming is the preparation of brood cups and queen rearing. The causes of swarming could be (i) large colony size, (ii) high proportion of young worker bees, (iii) reduced queen pheromone due to overcrowding, and (iv) abundance of pollen and nectar leading to overcrowding. There is a common cause behind all the symptoms i.e., overcrowding in the colony.

When a swarm issues, the parasites present in the parent colony are divided among the parent colony and the new colony formed after swarming [3]. This may reduce the disease infestation in the parent colony. We investigate the effect of swarming on the colony infested with a parasite *Varroa destructor* and Acute Bee Paralysis Virus. *Varroa destructor* is an ectoparasitic mite that not only feeds on the bees but also carries and transmits fatal viruses in the colony. The mite feeds on bees' haemolymph by piercing their inter-segmental membrane and transmits the virus while feeding on them. When a virus carrying mite feeds on an infected bee, it releases the virus into the bee's haemolymph. Thus, the uninfected bee becomes infected and mite becomes virus free. When a virus free mite feeds on an infected bee, it begins to carry virus. There are 20 bee viruses known so far, out of which at least 14 viruses are reported to be associated with mites [5, 9]. These viruses have different routes of transmission and different levels of virulence. One of the most common virus is the Acute Bee Paralysis Virus. It has also been implicated in colony losses [4, 5, 15]. The bees infected by this virus are unable to fly, loose their body hair and tremble uncontrollably. Unlike the Deformed Wing Virus, where brood infested with the virus develop into an adult sick bee, the brood infected with ABPV does not survive to the adult stage and dies immediately.

There have been several SIR-type mathematical models developed for honeybee-varroa mite-virus systems [2, 11–13]. There have also been models on some aspects of swarming such as the use of bee dance, design of nest selection and decision-making processes [1, 8, 10]. However, none of these models studied the combined effect of swarming and the varroa-virus infestation in a honeybee colony. In this paper, we provide a general framework of difference-differential equations to investigate how swarming affects the fate of the honeybee colony infested with varroa mites and virus. We also use numerical simulations to study (i) swarming due to overcrowding (ii) swarming after a fixed time interval mimicking the natural death cycle of a queen bee. We vary the proportion of healthy bees leaving the parent colony and observe its effect on the dynamics of the colony. We numerically calculate a critical value of the proportion of mites leaving the parent colony below which the parent colony dies off and above which it survives.

2 Model Equations

The underlying model of honeybee-varroa-ABPV disease dynamics from [11] is:

$$\frac{dm}{dt} = \beta_1(M - m)\frac{y}{x + y} - \beta_2 m\frac{x}{x + y} \tag{1}$$

$$\frac{dx}{dt} = \mu g(x)h(m) - \beta_3 m \frac{x}{x+y} - d_1 x - \gamma_1 M x \tag{2}$$

$$\frac{dy}{dt} = \beta_3 m \frac{x}{x+y} - d_2 y - \gamma_2 M y \tag{3}$$

$$\frac{dM}{dt} = rM\left(1 - \frac{M}{\alpha(x+y)}\right) \tag{4}$$

The parameters are assumed to be non-negative. Because the size of the bee colony and the life span of bees vary drastically with seasons, the parameters are assumed to be seasonally varying. In particular, we assume the parameters to be periodic functions of time with a period T; in practice $T = 1$ year.

The parameter μ in (2) is the maximum birth rate, specified as the number of worker bees emerging as adults per day.

The function $g(x)$ expresses that a sufficiently large number of healthy worker bees is required to care for the brood. We think of $g(x)$ as a switch function. If x falls below a critical value, which may seasonally depend on time, essential work in the maintenance of the brood cannot be carried out anymore and no new bees are born. If x is above this value, the birth of bees is not hampered. Thus $g(0,\cdot) = 0$, $\frac{dg(0)}{dx} \geq 0$, $\lim_{x\to\infty} g(x) = 1$. A convenient formulation of such switch like behavior is given by the sigmoidal Hill function

$$g(x) = \frac{x^n}{K^n + x^n} \tag{5}$$

where the parameter K is the size of the bee colony at which the birth rate is half of the maximum possible rate and the integer exponent $n > 1$. If $K = 0$ is chosen, then the bee birth terms of the original model of [13] is recovered. Then the brood is always reared at maximum capacity, independent of the actual bee population size, because $g(x) \equiv 1$.

The function $h(m)$ in (2) indicates that the birth rate is affected by the presence of mites that carry the virus. This is in particular important for viruses like ABPV that kill infected pupae before they develop into bees. The function $h(m)$ is assumed to decrease as m increases, $h(0) = 1$, $\frac{dh}{dm}(m) < 0$ and $\lim_{m\to\infty} h(m) = 0$; [13] suggests that this is an exponential function $h(m) \approx e^{-mk}$, where k is non-negative. We will use this expression in the computer simulations later on.

The parameter β_1 in (1) is the rate at which mites that do not carry the virus acquire it. The rate at which infected mites lose their virus to an uninfected host is β_2. The rate at which uninfected bees become infected is β_3, in bees per virus carrying mite and time.

Finally, d_1 and d_2 are the death rates for uninfected and infected honeybees. We can assume that infected bees live shorter than healthy bees, thus $d_2 > d_1$.

Equation (4) is a logistic growth model for varroa mites. By r we denote the maximum mite birth rate. The carrying capacity for the mites changes with the host population site, $x + y$, and is characterized by the parameter α which indicates how many mites can be sustained per bee on average. The parameters $\gamma_{1,2}$ in Eqs. (2) and (3) represent the mortality rates of bees due to mites feeding on them.

3 Mathematical Formulation of Discrete Interventions

We formulate the process of swarming as discrete interventions in a continuous system in the form of a hybrid system. The discrete interventions (t_i's) may depend on the time or the state variables. For instance, if we assume that swarming takes place after fixed time intervals i.e., mimicking the natural life cycle of a queen bee, the time at which discrete interventions takes place will be given a priori. On the other hand, if we assume that swarming takes place due to overcrowding in the hive, the occurrence of discrete interventions will be state dependent.

The mathematical formulation in terms of hybrid dynamical systems is

$$\dot{z}(t) = f(t, z(t)), \quad t \in (t_i, t_{i+1}), \quad z(t_0) = z_0, \quad z(t_i) = X_i, \tag{6}$$

$$X_{i+1} = F(z_{i+1}), \quad \text{where} \quad z_{i+1} = \lim_{t \nearrow t_{i+1}} z(t). \tag{7}$$

Here t_i are the discrete times at which events take place and $t_0 < \ldots < t_i < t_{i+1}$. Also, $z \in \mathbb{R}^4$, $t \in \mathbb{R}^+$, $i \in \mathbb{N}$, $X \in \mathbb{R}^4$ and $f \in \mathbb{K}$ where

$$\mathbb{K} = \{(m, x, y, M) \in \mathbb{R}^4 : m \geq 0, M \geq 0, x \geq 0, y \geq 0, x + y > 0\}.$$

In the case of swarming, we assume that $F(X) = DX$, where $D \in \mathbb{R}^{4\times4}$ is a diagonal matrix.

Also, $z = (m, x, y, M)^T$, f is the RHS of (1), (2), (3) and (4), and, X and D are given by

$$X = \begin{bmatrix} m \\ x \\ y \\ M \end{bmatrix}, \quad D = \begin{bmatrix} a & 0 & 0 & 0 \\ 0 & b & 0 & 0 \\ 0 & 0 & 1 & 0 \\ 0 & 0 & 0 & a \end{bmatrix}.$$

We assume that the infected bees do not leave the parent colony, because the disease progresses rapidly and sick bees are unable to fly. We also assume that the percentage of total mites and virus carrying mites leaving the parent colony is the same. Thus, two new parameters a and b are introduced. The parameter a is the percentage of mites staying in the parent colony and the parameter b is the percentage of uninfected bees staying in the parent colony after the swarm leaves.

4 Computer Simulations

In the simulation experiments, we study the process of swarming taking place due to overcrowding in the hive and due to events like queen supercedure. We investigate how these two causes of swarming affect the strength and survival of the parent colony infested with mites and virus. In the case where swarming takes place due to overcrowding in the hive, discrete interventions take place when

$$x(t) = cx^* \quad \text{where} \quad 0 < c \leq 1. \tag{8}$$

The quantity x^* is calculated numerically by running the disease free model (without swarming) until the steady state is reached and then taking the maximum of the steady state solution over a 2 year period. For the simulation experiments, we fix $c = 0.95$. In the case where swarming takes place due to queen failure, the time at which discrete interventions (t_i's) take place are fixed a priori as $t_i = t_0 + i(\Delta t)$, where $\Delta t = 2T$, $T = 365$ days and $i = 1, 2, \ldots,$. We assume that $t_0 = 45$ i.e., swarming takes place in mid May [14].

The seasonal averages of the parameters β_i, r, d_1, d_2, k and μ are given in Table 1. Lacking more detailed information about the parameters, we use these values to construct piecewise constant time varying parameter functions, assuming four equally long seasons of 91.25 days [11]. Two sets of seasonal averages of the parameter α, $[0.4784, 0.5, 0.5, 0.4784]$ and $[0.1\ 0.1\ 0.1\ 0.1]$, for the spring, summer, fall and winter respectively, are used depending upon the scenario under study (see [11] for the choice of these values). In order to better present the simulation experiments, we introduce two new variables $a_1 = 1 - a$ and $b_1 = 1 - b$. Here, a_1 is the proportion of mites leaving the parent colony and b_1 is the proportion of uninfected bees leaving the parent colony during swarming.

Table 1 Seasonal averages of model parameters, derived from the data presented in the literature [6, 7, 11, 13]. The parameters included here are kept constant for all simulations; the values of the parameters that are varied are given in the text

Parameter	Spring	Summer	Autumn	Winter	Source
β_1	0.1593	0.1460	0.1489	0.04226	[13]
β_2	0.04959	0.03721	0.04750	0.008460	[13]
β_3	0.1984	0.1460	0.1900	0.03384	[13]
d_1	0.02272	0.04	0.02272	0.005263	[13]
d_2	0.2	0.2	0.2	0.005300	[13]
μ	500	1500	500	0	[13]
k	0.000075	0.00003125	0.000075	N/A	[13]
K	8000	12,000	8000	6000	[11]
r	0.0165	0.0165	0.0045	0.0045	[6, 7]
$\gamma_1 = \gamma_2$	10^{-7}	10^{-7}	10^{-7}	10^{-7}	[11]

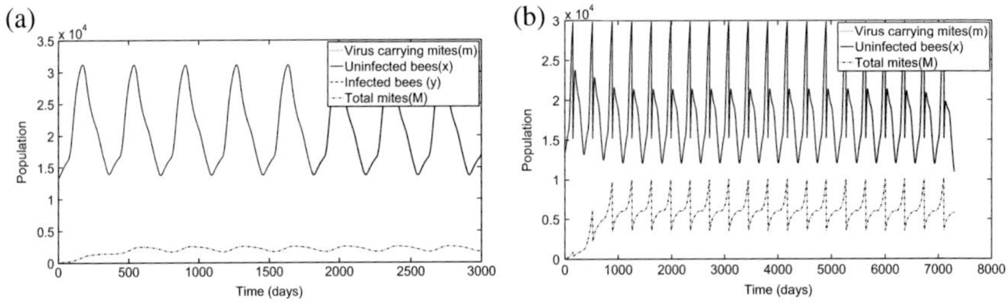

Fig. 1 **(a)** Bee-mite-virus system in the absence of swarming. **(b)** Bee-mite-virus system in the presence of swarming due to overcrowding. Threshold bee population at which swarming takes place is 31,342 and we assume $a_1 = 0.65, b_1 = 0.5$

Illustrative Simulation: The purpose of this simulation is to (i) show a typical simulation of the temporal dynamics of a honeybee colony infested with mites and virus and, (ii) to compare the system with and without the process of swarming. In Fig. 1a, we consider a honeybee-mite-virus system with no swarming taking place. We assume the lower value of the parameter α i.e., [0.1 0.1 0.1 0.1] so that the colony fights off the virus but not the mites [11]. The uninfected bee population increases from 13,350 in the beginning of spring and reaches its maximum of 31,195 in the summer, decreases in the fall and reaches its minimum of 13,350 in the winter. This pattern repeats annually. After an initial transient period of 2 years, the mites establish themselves in the colony and follow the same pattern as the bee population. In Fig. 1b, we consider a honeybee colony infested with mites and virus where swarming takes place due to overcrowding i.e., swarming takes place if the bee population exceeds a certain threshold value x^* (see Eq. 8). We assume the higher value of the parameter α i.e., [0.5 0.4784 0.4784 0.5]. We assume that $a_1 = 0.65$ and $b_1 = 0.5$ which means 50 % of the uninfected bees and 65 % of the mites leave the parent colony during swarming. We choose the parameters, a_1, b_1 and α, in order to present a reference case. The bee population follows the same pattern as in Fig. 1a in the first year. In the second year, the population increases in the spring and summer and then swarming takes place when the bee population reaches the threshold value x^* which in this case is 31,342. The population the drops from 29,670 to 15,207 bees during swarming and then increases to 23,780 due to the model dynamics. The population starts decreasing again in the fall and winter and reaches its minimum of 13,180. This pattern repeats itself annually. After a transient period of 1 year, the mite population also follows the same oscillatory behavior as the bee population.

Simulation Experiment I: In this simulation experiment, we consider the case where swarming takes place due to overcrowding. As in the illustrative simulation, we assume that swarming takes place if the bee population exceeds a certain threshold value, x^*. We use the value of the parameter α to be [0.5 0.4784 0.4784 0.5] for the spring, summer, fall and winter. In Fig. 2a, we compare a disease free colony, a mite infested colony in which only 5 % of the mites leave the parent colony ($a_1 = 0.05$), and, a mite infested colony in which 70 % of the mites leave the parent

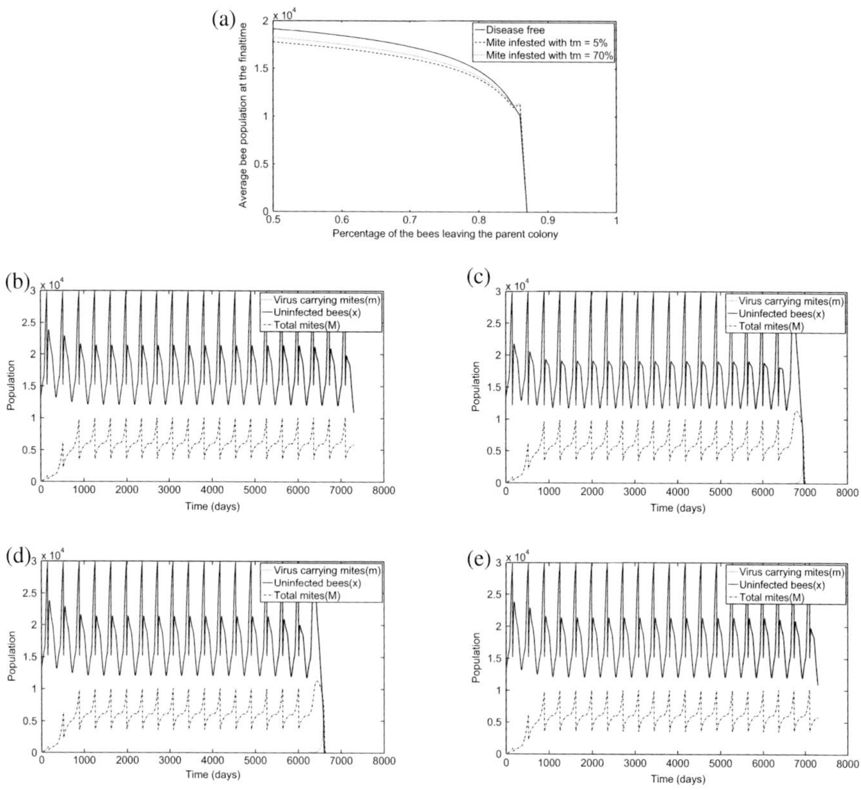

Fig. 2 Swarming due to overcrowding: **(a)** Effect of the percentage of the healthy bees leaving the parent colony on the average healthy bee population of the (i) disease free colony, (ii) mite infested colony with $a_1 = 0.05$, and (iii) mite infested colony with $a_1 = 0.7$. **(b)** The colony fights off the virus when $b_1 = 0.5$ and $a_1 = 0.65$. **(c)** The colony dies off after 7000 days when $b_1 = 0.6$, $a_1 = 0.65$. **(d)** The colony dies off after 6600 days when $a_1 = 0.64$, $b_1 = 0.5$. **(e)** The colony fights off the virus and survives as a properly working colony when $a_1 = 0.65$, $b_1 = 0.5$

colony ($a_1 = 0.7$). We vary the percentage of healthy bees leaving the parent colony over 50–100 % which covers the range (50–70 %) given by [3, 14] i.e., we vary b_1 from 0.5 to 1. In the case of the disease free colony, the average population starts from 19,000 and decreases gradually as the parameter b_1 increases. The average population suddenly drops down when $b_1 = 0.87$. In the case of the mite infested colony with $a_1 = 0.05$, the average bee population starts at a lower level (i.e., at 18,000) than in the disease free case and decreases to 11,000 bees as b_1 increases followed by a sudden drop to 0 when $b_1 = 0.87$. When $a_1 = 0.7$, the average bee population starts at 18,500 which is between the initial average population in the disease free case and the case when $a_1 = 0.05$. The critical value for the percentage of bees leaving the parent colony is the same in all three cases. Figure 2b, c show the effect of the percentage of uninfected bees leaving the parent colony (b_1) on the dynamics of the colony; we fix the parameter $a_1 = 0.65$. In Fig. 2b, the colony fights of the virus and survives as a properly working colony when $b_1 = 0.5$. In Fig. 2c, the colony dies off after 7000 days when $b_1 = 0.6$. Figure 2d, e show how the colony, that otherwise dies off due to virus, survives as a properly working colony if the percentage of mites leaving the parent colony is above a threshold value; we fix the parameter $b_1 = 0.5$. In Fig. 2d, the colony dies off due to virus after 6600 days when $a_1 = 0.64$. Figure 2e shows that when $a_1 = 0.65$, the colony fights off the virus and works as a properly working colony.

Simulation Experiment II: In this simulation experiment, we assume that swarming occurs at fixed time intervals of 2 years mimicking the natural death cycle of the queen bee. We assume that the swarm leaves the colony every 2 years in the mid of May [14]. We use the value of the parameter α to be [0.1 0.1 0.1 0.1] for the spring, summer, fall and winter. In Fig. 3a, we compare a disease free colony, a mite infested colony with $a_1 = 0.05$, and, a mite infested colony with $a_1 = 0.7$. We investigate the effect of the percentage of bees leaving with the swarm (b_1) on the average population of the parent colony. The parameter b_1 is varied from 0.5 to 1. In case of the disease free colony, the average bee population starts from 21,000 and remains constant when the parameter b_1 is varied from 50 % to 76 % and suddenly drops down to 0 when b_1 reaches 0.77. In case of the mite infested colony with $a_1 = 0.05$, the average bee population starts below the disease free population and remains constant until the parameter b_1 reaches 0.75 when it suddenly drops down to 0. In case of the mite infested colony where $a_1 = 0.7$, the average bee population starts at the same level as in case of $a_1 = 0.05$ and remains constant until a_1 reaches 0.76 and then it suddenly drops down to 0. It is interesting to note that the threshold value for the parameter b_1 is the maximum in case of the disease free colony which is followed by the mite infested case with $a_1 = 0.7$ which in turn is followed by the mite infested case with $a_1 = 0.05$.

Figure 3b, c show the effect of the percentage of uninfected bees leaving the parent colony (b_1) on the survival of the colony. The parameter a_1 is fixed to be 0.7. In Fig. 3b, the colony fights off the virus and survives as a properly working colony when $b_1 = 0.76$. In Fig. 3c, the colony dies off after 1000 days when $b_1 = 0.77$. Figure 3d, e show how the colony, that otherwise dies off due to virus, survives as a properly working colony if the parameter a_1 is above a threshold value. The parameter b_1 is fixed to be 0.5. In Fig. 3d, the colony dies off due to virus after 6000 days when $a_1 = 0.91$. Figure 3e shows that when $a_1 = 0.92$, the colony fights off the virus and survives as a properly working colony.

5 Summary and Conclusion

- A general framework is provided to incorporate the discrete interventions into the model using a discrete-continuous model. This framework is applied to the process of swarming which involves two types of discrete interventions: (i) due

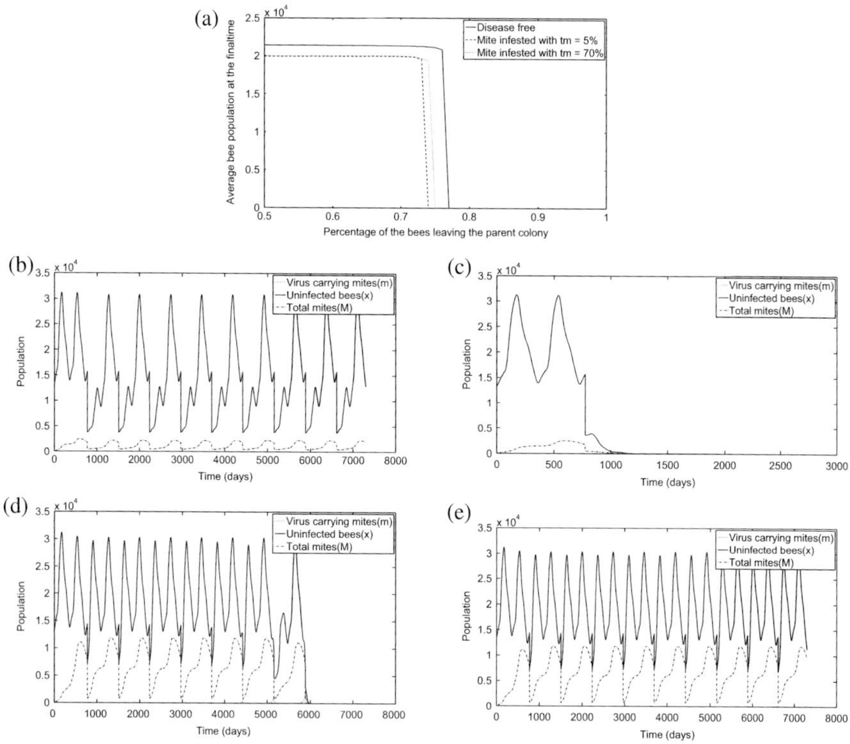

Fig. 3 When swarming takes place every 2 years: **(a)** Effect of the percentage of the healthy bees leaving the parent colony on the average healthy bee population of the colony that is (i) disease free, (ii) mite infested with $a_1 = 0.05$, and (iii) mite infested with $a_1 = 0.7$. **(b)** The colony fights off the virus when $b_1 = 0.76$ and $a_1 = 0.7$. **(c)** The colony dies off after 1000 days when $b_1 = 0.77$ and $a_1 = 0.7$. **(d)** The colony dies off after 6000 days when $a_1 = 0.91$ and $b_1 = 0.5$. **(e)** The colony fights off the virus and survives as a properly working colony when $a_1 = 0.92$ and $b_1 = 0.5$

to queen failure i.e., when discrete events occur at times that are a priori fixed (ii) due to overcrowding i.e., when occurrence of discrete events depend on the state of the system.

- In case of a disease free colony and a mite-infested colony, a critical value of the percentage of bees leaving the colony plays an important role. The colony survives only if the percentage of bees is below this critical value.
- In case of a colony infested with mites and virus, in addition to the percentage of the bees leaving the colony, the percentage of virus carrying mites and virus free mites leaving the colony during swarming also has a huge impact on the survival of the parent colony. Particularly, a colony, that otherwise dies off due to virus, can survive if the percentage of mites leaving the parent colony is above a critical value.

- The critical value of the percentage of bees leaving the colony is lower in case of swarming due to overcrowding as compared to the case where swarming takes place after fixed intervals. This difference could be due to the fact that swarming due to overcrowding takes place every year, however, swarming after fixed intervals is basically every 2 years. The parent colony is able to tolerate greater loss of bees because it gets longer time to establish itself before the next swarming event takes place.

References

1. Britton, N.F., Franks, N.R., Pratt, S. C., Seeley, T.D.: Deciding on a new home: how do honeybees agree? Proc. R. Soc. Biol. Sci. **269**(1498), 1383–1388 (2002)
2. Eberl, H.J., Frederick, M.R., Kevan, P.G.: The importance of brood maintenance terms in simple models of the honeybee – *Varroa destructor* – acute bee paralysis virus complex. Electron. J. Differ. Equ. Conf. Ser. **19**, 85–98 (2010)
3. Fries, I., Hansen, H., Imdorf, A., Rosenkranz, P.: Swarming in honey bees (Apis mellifera) and Varroa destructor population development in Sweden. Apidologie **34**, 389–397 (2003)
4. Genersch, E., von der Ohe, W., Kaatz, H., Schroeder, A., Otten, C.; Büchler, R., Berg, S., Ritter, W., Mühlen, W., Gisder, S., Meixner, M., Liebig, G., Rosenkranz, P.: The German bee monitoring project: a long term study to understand periodically high winter losses of honey bee colonies. Apidologie **41**, 332–352 (2010)
5. Kevan, P.G., Hannan, M., Ostiguy, N., Guzman-Novoa, E.: A summary of the varroa-virus disease complex in honey bees. Am. Bee J. **146**(8), 694–697 (2006)
6. Martin, S.: A population dynamic model of the mite varroa jacobsoni. Ecol. Model. **109**, 267–281 (1998)
7. Martin, S.J.: Varroa destructor reproduction during the winter in apis mellifera colonies in UK. Exp. Appl. Acarol. **25**(4), 321–325 (2001)
8. Myerscough, M.R.: Dancing for a decision: a matrix model for nest-site choice by honey- bees. Proc. R. Soc. Lond. B: Biol. Sci. **270**(1515), 577–582 (2003)
9. Ostiguy, N.: Honey bee viruses: transmission routes and interactions with varroa mites. In: 11 Congreso Internacional De Actualizacion Apicola, vol. 9 al 11De Junio De 2004. Memorias., p. 47 (2004)
10. Passino, K.M., Seeley, T.D.: Modeling and analysis of nest-site selection by honeybee swarms: the speed and accuracy trade-off. Behav. Ecol. Sociobiol. **59**(3), 427–442 (2006)
11. Ratti, V., Kevan, P.G., Eberl, H.J.: A mathematical model for population dynamics in honeybee colonies infested with varroa destructor and the acute bee paralysis virus. Can. Appl. Math. Q. **21**(1), 63–93 (2013)
12. Ratti, V., Kevan, P.G., Eberl, H.J.: A mathematical model for population dynamics in honeybee colonies infested with varroa destructor and the acute bee paralysis virus with seasonal effects. Bull. Math. Biol. **77**(8), 1493–1520 (2015)
13. Sumpter, D.J., Martin, S.J.: The dynamics of virus epidemics in varroa-infested honey bee colonies. J. Anim. Ecol. **73**(1), 51–63 (2004)
14. Winston, M.L.: The biology of the honey bee. Harvard University Press, Cambridge (1991)
15. ZKBS (Zentralkommittee für biologiche Sicherheit des Bundesamts für Verbraucherschutz und Lebensmittelsicherheit); Empfehlung Az.: 45242.0087 - 45242.0094, 2012, (in German: Central Committee for Biological Safety of the Federal Agency for Consumer Protection and Food Safety, Recommendation 45242.0087-45242.0094, 2012)

To a Predictive Model of Pathogen Die-off in Soil Following Manure Application

Andrew Skelton and Allan R. Willms

Abstract The application of manure is an important component of nutrient management in the production of field crops. The regulations governing the safe application of manure are based on laboratory data, which may or may not accurately reflect the environmental fluctuations seen in field conditions.

This study aims to develop a predictive model for pathogen die-off in soil following manure application. An ordinary differential equation model is presented and fit to experimental data. The challenges of modelling field derived data, including detection thresholds, viable but nonculturable bacteria and difficulties in winter data collection, are discussed. The capabilities of the model for predictive purposes and the development of additional experimental trials and data collection methods are discussed.

1 Background

Field production of fruits and vegetables typically includes application of manure as a vital component of nutrient management. In organic farming, manure, compost and other organic materials are the only substances acceptable for use as fertilizer. The Standards Council of Canada 2008 regulations require that

> manure application shall be designed to ensure that manure application a. does not contribute to the contamination of crops by pathogenic bacteria, b. minimizes the potential for run-off into ponds, rivers and streams, c. does not significantly contribute to ground and surface water contamination. The non-composted solid or liquid manure shall be a. incorporated into the soil at least 90 days before the harvesting of crops for human consumption that do not come into contact with soil; b. incorporated into the soil at least 120 days before the harvesting of crops having an edible part that is directly in contact with the surface of the soil or with soil particles.

Pathogen die-off rates depend on environmental and biological conditions and, as such, will differ from season to season and year to year. A number of researchers have studied the effect such conditions have on pathogen die-off, such as the effect

A. Skelton (✉) • A.R. Willms
Department of Mathematics and Statistics, University of Guelph, Guelph, ON, Canada
e-mail: skeltona@uoguelph.ca; awillms@uoguelph.ca

© Springer International Publishing Switzerland 2016

J. Bélair et al. (eds.), *Mathematical and Computational Approaches in Advancing Modern Science and Engineering*, DOI 10.1007/978-3-319-30379-6_29

of temperature [7, 8, 14], soil and manure type [18], manure application methods [2], manure storage [9], intervention strategies [5] and genetic factors [11, 15]. A nice overview of studies conducted on survival of pathogens is given in [17]. A recent study [6] presented results from a meta regression on a large number of potential influencing factors and concluded that the three most important factors affecting pathogen die-off rates were temperature, water and soil types, and whether or not the study was conducted in the lab or in the field. Temperature and soil moisture are environmental factors that are relatively easy to record and predict and, as such, are of use in a predictive model. The third factor is of particular interest as the waiting times required by the regulations are based on data obtained under laboratory conditions.

Data obtained from laboratory studies are subject to highly consistent and controlled environmental conditions and thus do not necessarily accurately reflect the wide fluctuations seen in field conditions. The fixed waiting times required by the regulations may either over- or under-estimate the time required to obtain safe levels of pathogen reduction. It is, therefore, of value to model pathogen die-off in local climatic conditions. Obtaining data in field conditions, however, brings its own set of challenges and concerns. The aim of this study is to develop a predictive model of pathogen die-off and answer questions arising from the study of pathogen die-off in soil. Given measurements of pathogen levels in soil and some knowledge of future environmental conditions, can future pathogen levels be predicted? Can field derived data provide scientific information that can lead to informed policy development which allows for sustainable food production while guarding against food contamination?

2 Experiment

A field study was conducted in 2011–2012 by a team led by Dr. Ann Huber from the Soil Resource Group and Dr. Keith Warriner from the Department of Food Science at the University of Guelph. The project was funded by the Food Safety Research Program facilitated by the Ontario Ministry of Agriculture, Food and Rural Affairs (OMAFRA). A detailed description of the experimental procedure is available in [10] and illustrations and photographs of the experiment are presented in Fig. 1.

Trials were conducted at a farm northwest of Belwood, Ontario at two sites, one with Perth Loam soil and the other with Hillsburg Fine Sandy Loam soil. The two sites were adjacent to each other and thus subject to the same environmental conditions. Manure (either dairy or swine) was spiked with either E.coli O157, Salmonella or Listeria, mixed with the soil, and placed in sentinel vials (see Fig. 1). These vials, which have a membrane impermeable to the contaminant, allow the monitoring of pathogen populations in a confined environment and provide a mechanism for separating the environmental effect on pathogen die-off from the effect due to transport, such as run-off and movement through the soil profile. The sentinel vials were buried either directly under the surface or at a depth of

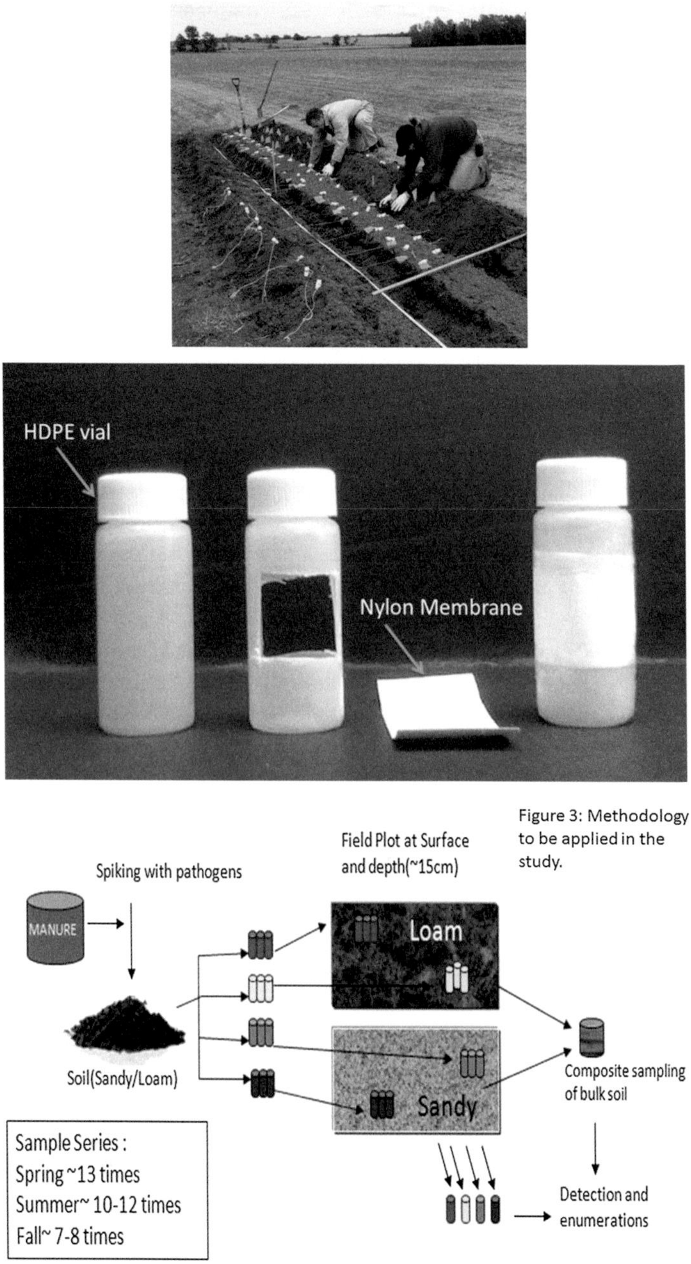

Fig. 1 Illustrations and photographs of the experimental procedure (Images reproduced with permission of Dr. Huber at the Soil Resource Group). *Top*: Introducing the inoculated sentinel vials into the test plot at different depths. *Middle*: Assembly of the sentinel vials into which the pathogen inoculated manure amended soil will be introduced. *Bottom*: Methodology to be applied in the study

approximately 15 cm and additional non-spiked manure was hand applied at a rate equivalent to within the vials. Three trials were conducted: the first from June 16, 2011 to May 29, 2012, the second from September 22, 2011 to June 6, 2012 and the third from November 2, 2011 to June 6, 2012. Each trial included pre-determined sampling dates on which three sentinel vials were extracted and pathogen levels reported. The three trials contained 13, 11 and 8 sampling dates respectively. Our experimental data therefore consists of 72 (3 trials $\times$ 3 pathogens $\times$ 2 depths $\times$ 2 soil types $\times$ 2 manures) time series data sets, with three data replicates at each sampling date.

A typical data set is found in Table 1. Note that there are no sampling dates over the winter months due to the soil becoming frozen and too snow covered to be realistically sampled. When a sentinel vial is extracted from the ground, a soil sample of approximately 10 g is obtained. The sample is processed, diluted and bacteria are cultured and counted. If there are too many bacteria to count after being cultured, the sample is repeatedly diluted by a factor of 10 until the number of bacteria is countable. The data in row four of Table 1, for example, has been diluted twice, so the recordings of 7, 5 and 11, denote true counts of 700, 500 and 1100 bacteria respectively. In each dilution, triplicate counts are obtained. For ease of reading, only the median result is reported in Table 1, but averages are used in the analysis. If there are no bacteria visible at a given dilution, we conclude only that there is less than 1 bacteria at that given dilution level. The data in the last row of Table 1, for example, displays counts of 0 bacteria at a dilution of 0.1. The conclusion that we may draw from this observation is that there is less than 10 bacteria in the sample. Our time series data will, therefore, contain censored data that is known only to have a value beneath a detection threshold. The bacteria counts

Table 1 A typical data set. Data collected for over the September 22, 2011 to June 6, 2012 trial, using dairy manure, in sandy soil, at the surface, measuring levels of E.coli O157. To determine the data values for model fitting, we take the bacterial count, divide by the dilution, scale by the sample weight and take the logarithm of the answer to determine the log bacterial count per 10 g of soil

	Sample weights (g)			Bacterial counts			
Date	Sample 1	Sample 2	Sample 3	Sample 1	Sample 2	Sample 3	Dilution
Sep 22/11	10.000	10.000	10.000	1	1	1	0.001
Sep 25/11	10.063	10.019	10.253	75	73	101	0.01
Oct 3/11	10.247	10.154	10.057	1	4	3	0.001
Oct 13/11	10.297	10.024	10.054	7	5	11	0.01
Oct 19/11	10.328	10.035	10.264	1	2	65	0.1
Nov 10/11	10.254	10.106	10.012	3	0	3	0.1
Nov 25/11	10.107	10.057	10.028	1	2	0	0.1
Mar 27/12	10.039	10.041	10.048	0	0	0	0.1
May 1/12	10.059	10.005	10.140	0	0	0	0.1
May 17/12	10.152	10.128	10.020	0	0	0	0.1
Jun 7/12	9.999	10.002	10.006	0	0	0	0.1

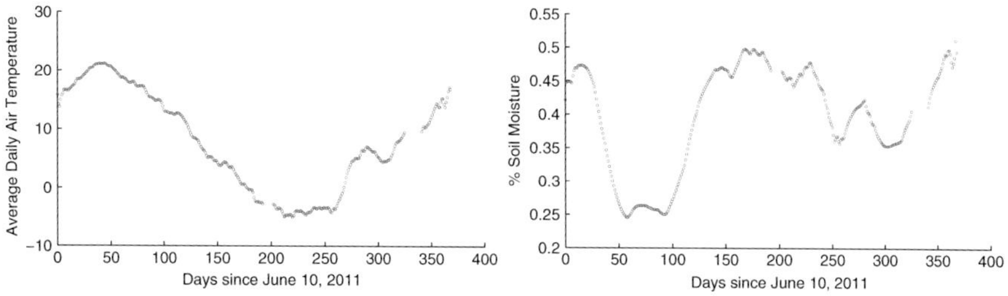

Fig. 2 Environmental data collected. Recordings from the weather station for average daily air temperature and soil moisture over the course of the three trials. Missing data were interpolated from records kept by nearby weather stations

are then scaled by the relative weight of the sample so if, for example, the sample weight was 10.107 g, we would multiply the count by $10/10.107$ to obtain a scaled count. The data with which we will fit our model will be given as a log count of bacteria per 10 g of soil.

Environmental data were collected by a weather station installed at the site. The average daily air temperature and percentage of soil moisture are shown in Fig. 2. There were two periods of missing environmental data. The first was due to strong winds knocking over the weather station, and the second was due to a raccoon consuming the power supply. In both cases, it was possible to obtain replacement weather data spatially averaged between nearby weather stations using data made freely available by Environment Canada.

3 Model and Results

We now fit our experimental data using an ordinary differential equation model. Let $x = x(t)$ denote the logarithm of the bacteria count per 10 g of soil, and $T = T(t), M = M(t)$ denote the average daily air temperature (degrees Celsius) and soil moisture (%) linearly interpolated between adjacent data points from the data shown in Fig. 2.

Pathogen die-off plots typically follow a characteristic logistic curve [2, 3, 7, 11, 14, 18], so we set up a logistic differential equation in which the rate constant is dependent on both temperature and soil moisture. It is common to discuss pathogen die-off in terms of numbers of log reductions, so an empirical model of pathogen die-off in which the dependent variable is given on a logarithmic scale is appropriate. Using a quadratic dependence on temperature and linear dependence on soil moisture, we obtain the following ordinary differential equation

$$x' = -(k_1 + k_2 T + k_3 T^2 + k_4 M)(x - B)(C - x), \tag{1}$$

where the four rate constants k_1, k_2, k_3, k_4, the two shape constants B, C (the baseline and maximum values, respectively, which will depend on the trial, soil type, manure type and sample depth) and the initial condition $x(t_0)$, where t_0 denotes the start of the particular trial of interest and $B < x(t_0) < C$, will be obtained from finding the best model fit to the data. The initial condition must be estimated due to the use of lab bacteria in field trials. When the lab bacteria are exposed to field conditions, a subset of the bacteria will experience shock due to the sudden change in environmental conditions and enter a viable but nonculturable (VBNC) state. Bacteria in a VBNC state cannot be cultured on growth media [13], but are still viable and potentially dangerous, and thus must be accounted for in our analysis. This phenomenon can be seen in the data in Table 1. The first row of data, obtained at the end of the first day of the trial, has an average of 1000 bacteria per 10 g of soil. This observed value is significantly less than the number of bacteria that were originally placed in the spiked soil. The second row of data, obtained 3 days later, has an average of 8406 bacteria per 10 g of soil. This discrepancy reflects the number of bacteria that entered a VBNC state initially, but had recovered after 3 days in the field. In our analysis, we ignore the initial data value and instead estimate the number of bacteria present at the start of the experiment (Table 2).

Owing to the fact that some of the data points are censored due to detection thresholds, we will be required to use the Expectation-Maximization (EM) algorithm to fit our model to the data. The EM algorithm first sets all censored data to exactly half of the appropriate detection threshold (in our case, to either 5, 50, 500, or 5000 bacteria). The following steps are then repeated until convergence: a least squares procedure is used to find the best model fit given the current data values and then the censored data points are adjusted to fit the current parameter set. The mathematical details of the EM algorithm can be found in [4, 12] and the algorithm as applied to censored data at multiple detection thresholds can be found in [1].

Table 2 Best Fit Parameters for the plots shown in Fig. 3. Details of the four trials are as follows. (a) Dairy manure, loam soil, surface, E.Coli, (b) Dairy manure, sand soil, surface, E.Coli, (c) Dairy manure, sand soil, depth, Salmonella, (d) Swine manure, loam soil, depth, Listeria

Parameter	Figure 3a	Figure 3b	Figure 3c	Figure 3d
k_1	5.15×10^{-4}	3.46×10^{-3}	1.62×10^{-3}	7.48×10^{-6}
k_2	-6.51×10^{-5}	3.87×10^{-4}	-4.39×10^{-4}	4.53×10^{-3}
k_3	2.00×10^{-5}	2.89×10^{-5}	1.21×10^{-4}	-1.85×10^{-4}
k_4	1.60×10^{-4}	2.33×10^{-4}	9.95×10^{-4}	3.17×10^{-3}
B	−0.79	−0.39	0.92	1.37
C	9.71	5.25	6.56	6.32
$x(0)$	4.46	4.22	4.18	4.80

To find the best fit parameter set, we will be required to search in a seven-dimensional parameter space. Very little a priori information is available as to the values of many of the parameters. It is therefore useful to reduce the size of the parameter space we will be required to search. We used the method presented in [16] to reduce the size of the search space and improve the speed of the minimization steps of the EM algorithm. This method uses interval analysis and linear multistep discretizations to remove boxes of parameter space that are deemed to be inconsistent with the data. The method is able to quickly remove large regions of parameter space to allow traditional minimization techniques to work more effectively.

Results for representative data sets are presented in Fig. 3. We were, in general, able to obtain good fits to our experimental data. In plots (a) and (d), we can see that there is an expected initial rapid decrease in the bacterial count, followed by a slower die-off (and in fact a small growth in the case of Listeria) over the winter season, followed by another period of more rapid die-off during the spring months. Plots (b) and (c) illustrate the difficulties faced in modelling these data sets. The only viable data exists before the winter freeze, making any dynamics during or after the winter break difficult to model. This lack of information in many data sets has made calibrating a model to the data very difficult. It has, however, directly led to improved experimental design and data collection in a follow-up trial.

4 Conclusion and Future Work

This paper presents the first step in a process by which we hope to develop a predictive model for pathogen die-off in soil following manure application. We fit an ordinary differential equation model to field derived data. Our model was able to reasonably describe the pathogen die-off using soil moisture and air temperature, both commonly available environmental data. To establish a predictive, rather than descriptive, model, we will require additional winter and long-term data with which we can calibrate our model and then use for testing of predictive capabilities. Such data is currently being collected and analysis of the data is ongoing and will be reported elsewhere.

Acknowledgements This work was supported by an NSERC Discovery Grant.

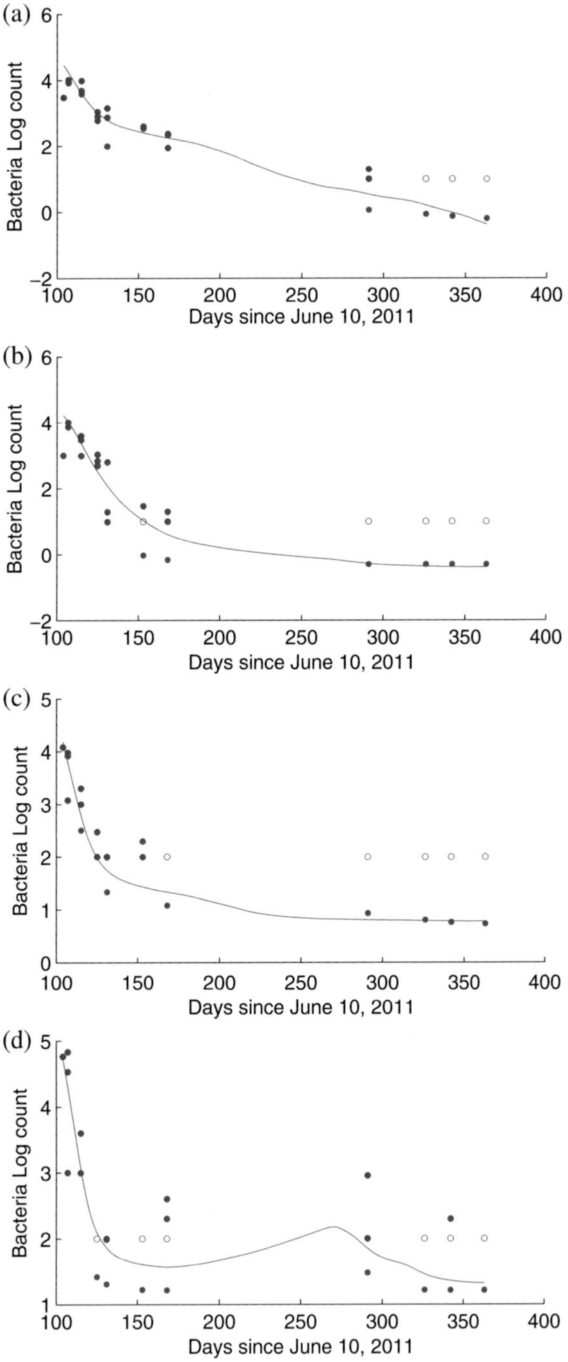

Fig. 3 Best fit plots. The solution to the differential equation model with best fit parameters plotted for four representative data sets from the September 22/11 to June 7/12 trial. The *open circles* represent data points whose values are known only to exist beneath the detection threshold. The *closed dots* represent the predicted data as outputted from the EM algorithm. (**a**) Dairy manure, loam soil, surface, E.Coli, (**b**) Dairy manure, sand soil, surface, E.Coli, (**c**) Dairy manure, sand soil, depth, Salmonella, (**d**) Swine manure, loam soil, depth, Listeria

References

1. Banks, H.T., Davidian, M., Hu, S., Kepler, G.M., Rosenberg, E.S.: Modelling HIV immune response and validation with clinical data. J. Biol. Dyn. **2**(4), 357–385 (2008)
2. Bech, T., Dalsgaard, A., Jacobsen, O.S., Jacobsen, C.S.: Leaching of Salmonella enterica in clay columns comparing two manure application methods. Ground Water **49**(1), 32–42 (2011)
3. Coroller, L., Leguerinel, I., Mettler, E., Savy, N., Mafart, P.: General model, based on two mixed Weibull distributions of bacterial resistance, for describing various shapes of inactivation curves. Appl. Environ. Microbiol. **72**(10), 6493–6502 (2006)
4. Dempster, A.P., Laird, N.M., Rubin, D.B.: Maximum likelihood from incomplete data via the EM algorithm. J. R. Stat. Soc.: Ser. B **39**, 1–38 (1977)
5. Franz, E., Semenov, A.V., van Bruggen, A.H.C.: Modelling the contamination of lettuce with Escherichia coli O157:H7 from manure-amended soil and the effect of intervention strategies. J. Appl. Microbiol. **105**, 1569–1584 (2008)
6. Franz, E., Schijven, J., de Roda Husman, A.M., Blaak, H.: Meta-regression analysis of commensal and pathogenic Escherichia coli survival in soil and water. Environ. Sci. Technol. **48**, 6763–6771 (2014)
7. Fremaux, B., Prigent-Combaret, C., Delignette-Muller, M.L., Mallen, B., Dothal, M., Gleizal, A., Vernozy-Rozand, C.: Persistence of Shiga toxin-producing Escherichia coli O26 in various manure-amended soil types. J. Appl. Microbiol. **104**, 296–304 (2008)
8. Holley, R.A., Arrus, K.M., Ominski, K.H., Tenuta, M., Blank, G.: Salmonella survival in manure-treated soils during simulated seasonal temperature exposure. J. Environ. Q. **35**, 1170–1180 (2006)
9. Huber, A.: Survival of pathogens during storage of livestock manures. OMAFRA Final Report. SR9182 (2009)
10. Huber, A., Warriner, K.: Pathogen die-off rates following manure application under Ontario field conditions – relationship to pre-harvest wait times. OMAFRA Final Report. SF6088 (2012)
11. Ma, J., Ibekwe, A.M., Yi, X., Wang, H., Yamazaki, A., Crowley, D.E., Yang, C.-H.: Persistence of Escherichia coli O157:H7 and its mutants in soils. PLoS ONE **6**(8), 1–9 (2011)
12. McLachlan, G.J., Krishnan, T.: The EM Algorithm and Extensions. Wiley, New York (1997)
13. Oliver, J.D.: The viable but nonculturable state in bacteria. J. Microbiol. **43**(S), 93–100 (2005)
14. Ongeng, D., Muyanja, C., Ryckeboer, J., Springael, D., Geeraerd, A.H.: Kinetic model-based prediction of the persistence of Salmonella enterica serovar Typhimurium under tropical agricultural field conditions. J. Appl. Microbiol. **110**, 995–1006 (2011)
15. Saint-Ruf, C., Garfa-Traore, M., Collin, V., Cordier, C., Franceschi, C., Matica, I.: Massive diversification in aging colonies of Escherichia coli. J. Bacteriol. **196**(17), 3059–3073 (2014)
16. Skelton, A., Willms, A.R.: Parameter range reduction in ordinary differential equation models. J. Sci. Comput. **62**(2), 517–531 (2015)
17. van Elsas, J.D., Semenov, A.V., Costa, R., Trevors, J.T.: Survival of Escherichia coli in the environment: fundamental and public health aspects. ISME J. **5**, 173–183 (2011)
18. You, Y., Rankin, S.C., Aceto, H.W., Benson, C.E., Toth, J.D., Dou, Z.: Survival of Salmonella enterica Serovar newport in manure and manure-amended soils. Appl. Environ. Microbiol. **72**(9), 5777–5783 (2006)

Mathematical Modeling of VEGF Binding, Production, and Release in Angiogenesis

Nicoleta Tarfulea

Abstract This paper presents a new mathematical model for the transduction of the extracellular vascular endothelial growth factor (VEGF) signal into the intracellular VEGF signal. It is based on a signal transduction pathway that accounts for VEGF binding, production, and release. The resulting mathematical model for the evolution of the chemical species concentrations is analyzed analytically and numerically to address qualitative aspects (including positivity, stability and robustness with respect to variations of the secretion rate parameter). Various secretion rate functions are investigated as well.

1 Introduction

In recent years, tumor-induced angiogenesis has become an important field of research because it represents a crucial step in the development of malignant tumors [18]. When a cancerous tumor initially forms, it acquires the nutrients needed for survival from the surrounding tissue. When it grows to a large enough size [17], it needs another source of food and channel for depositing waste. As an early response of tumor cells to hypoxia, they start releasing molecules called tumor angiogenic factors, such as VEGF, into the surrounding tissue to initiate a process called angiogenesis (growth of new blood vessels from preexisting ones). Secreted VEGF then diffuses into the surrounding tissue and binds to specific endothelial cell (EC) receptors. As a response, ECs start to secrete proteolytic enzymes (such as PA-plasmin, MMPs) which dissolve the basal lamina of the parent vessel and the surrounding extracellular matrix (ECM). ECs respond to the stimulus chemotactically and begin to migrate toward it, and so small sprouts which grow toward the tumor are formed [1, 23, 32, 35]. Over the last years it has become evident that an effective way of treating cancer is inhibition of angiogenesis [15, 19, 20, 29, 41], which could be done by targeting endothelial cells rather than tumor cells since: (i) ECs are more accessible than tumor cells; (ii) death of a single

N. Tarfulea (✉)
Purdue University Calumet, 2200 169th Street, Hammond, IN, 46323, USA
e-mail: ntarfule@purduecal.edu

© Springer International Publishing Switzerland 2016

J. Bélair et al. (eds.), *Mathematical and Computational Approaches in Advancing Modern Science and Engineering*, DOI 10.1007/978-3-319-30379-6_30

cell will cause death of approximate 500 tumor cells; (iii) ECs are relatively stable and will not mutate easily into drug resistant variants [50].

The existing mathematical models in the literature [35] have been able to reproduce some characteristics of angiogenesis. These include continuous approaches [2–5, 10–12, 24, 27, 28, 31, 38–40, 44, 52–54], random walk models [2, 6, 13, 22, 38, 39, 46, 47, 51], and cell-based formulations [7–9, 25, 30, 33, 36, 43, 50]. Reviews of such models can be found in [35, 37, 42]. However, it has been discovered that the activated ECs secrete detectable amounts of VEGF [16, 45]. Thus, in this paper we will develop a new mathematical model that account for VEGF binding, production, and release.

2 Description of the Signal Transduction Pathway

The dynamic interaction between angiogenic factors, the uPA-plasmin system, and the presumed activation of matrix-degrading metalloproteinases during angiogenesis is shown in Fig. 1. It can be described as follows [50]: (1) hypoxic tumor cells secrete VEGF; (2) VEGF is sequestered in the ECM by binding to ECM heparin binding sites; (3) VEGF binds to VEGFR (VEGF receptor) on EC cell (VEGFR-2 is considered the main signaling receptor in ECs); (4) inactive uPA (urokinase plasminogen activator) pro-uPA, binds to specific cell surface receptor uPAR; (5) plasminogen (pgn) binds to pgnR; (6) uPAR-bound uPA reacts with bound-plasminogen to generate Plasmin; (7) uPAR occupied by uPA forms complexes with PAI-1 (plasminogen activator inhibitor); (8) Plasmin degrades ECM; (9) Plasmin mobilizes VEGF from ECM reservoir; (10) Plasmin induces the activation of LTGF-

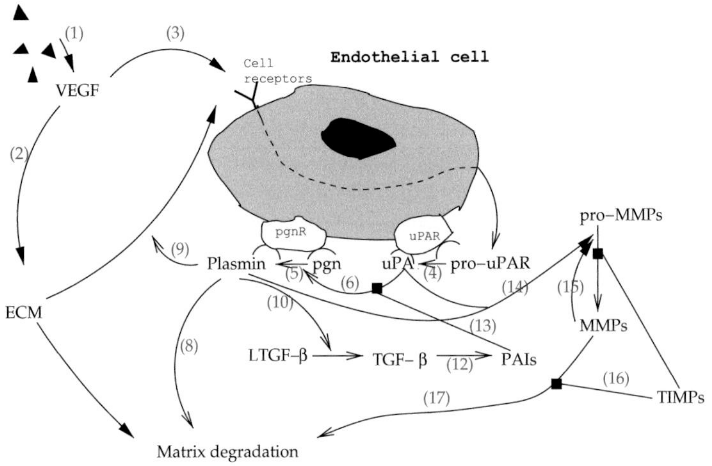

Fig. 1 The biochemical pathway of angiogenesis

β (latent type beta transforming growth factor) from ECM; (11) Plasmin can convert LTGF-β into TGF-β (type beta transforming growth factor); (12) active TGF-β induces synthesis of PAI-1; (13) PAI-1 blocks the conversion of plasminogen into Plasmin; (14) Plasmin, uPA activate some pro-MMPs (matrix metalloproteinase); (15) some MMPs can activate other MMPs; (16) all MMPs are inhibited by TIMPs (tissue inhibitor of MMP enzyme) by binding to pro-MMPs or MMPs; (17) MMPs degrades ECM. The signal transduction part incorporates known biology more completely than any other model [35] and involves 18 variables and 14 biochemical reactions. Using the law of mass action, these lead to a system of ten differential equations and auxiliary algebraic equations for the time evolution of the intracellular species.

3 The Simplified Pathway and the Corresponding Model

3.1 *Assumptions for the Simplified Model Signal Pathway*

Since biological and biochemical processes in angiogenesis are very complex, we seek reductions in the number of variables that do not alter the dynamics of the model.

After VEGF is been secreted into the surrounding tissue by tumor cells and has encountered endothelial cells, it binds to specific proteins (called receptors) that are sitting on the outer surface of the cell. This binding activates a series of relay proteins which transmit a signal that starts a series of cascades inside the cell [50]. As a result the motile machinery of the endothelial cell is activated, and also some VEGF is secreted [26, 45]. Using these facts, we have reduced the scheme to four primary species for the intracellular dynamics. A schematic diagram of the interactions in the proposed model is shown in Fig. 2. The definition of the symbols used in these reactions as well as the mathematical symbol used later are given in Table 1. Here k_i represent the reaction rates, h_i represent the degradation rates, and bsr is the basal secretion rate; all of them are given in Table 2.

Two classes of species are represented in this model: volumetric (growth factor – VEGF) and surface (receptors and ligand-receptor complexes). No factors in the

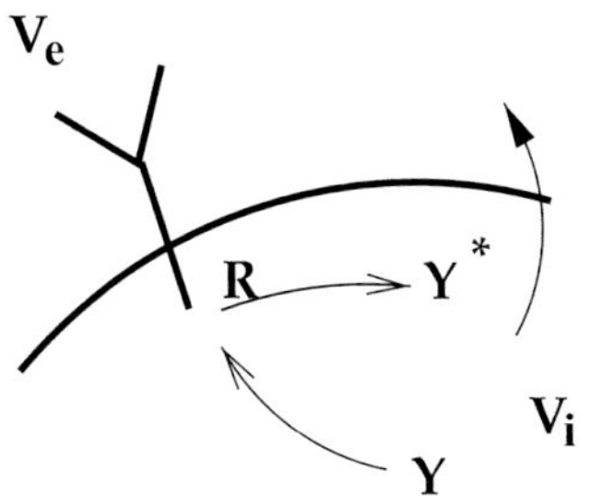

Fig. 2 Simplified pathway for angiogenesis

Table 1 The variables, their symbols, and non-dimensional form

Species	Conc.	Definition	Dimensionless form
V_e	$y_1(t)$	Extracellular *VEGF*	$u_1 = \frac{k_1}{k_5} y_1$
R	$y_2(t)$	*VEGFR*	$u_2 = \frac{y_2}{[R]_T}$
V_eR	$y_3(t)$	*VEGFR* bound with *VEGF*	$u_3 = \frac{y_3}{[R]_T}$
V_eRY	$y_4(t)$	V_eRY complex bound with Y	$u_4 = \frac{k_{-2}+k_3}{k_2[R]_T[Y]_T} y_4$
Y^*	$y_5(t)$	Activated enzyme	$u_5 = \frac{y_5}{[Y]_T}$
S	$y_6(t)$	Substrate	$u_6 = \frac{k_4}{k_3} y_6$
Y^*S	$y_7(t)$	Activated enzyme Y^* bound with S	$u_7 = \frac{y_7}{[Y]_T}$
V_i	$y_8(t)$	Intracellular *VEGF*	$u_8 = \frac{k_1}{k_5} y_8$
Y	$z(t)$	Free Y	–
t	–	Time	$\tau = k_5 t$

Table 2 Estimates for the kinetic parameters – for numerical simulations

Par.	Estimated value	Source	Par.	Estimated value	Source
k_1	$3.8 \cdot 10^6\ M^{-1}s^{-1}$	[14, 21]	k_4	$10^4 m^2/(\mu mol \cdot s)$	[48, 49]
k_{-1}	$95 \cdot 10^{-6} s^{-1}$	[14, 21]	k_{-4}	$0\ s^{-1}$	[48, 49]
k_2	$10^3 m^2/(\mu mol \cdot s)$	[14, 21]	k_5	$6 \cdot 10^{-2} s^{-1}$	[48, 49]
k_{-2}	$0\ s^{-1}$	[48, 49]	k_6	$10 s^{-1}$	[48, 49]
k_3	$10^3 s^{-1}$	[48, 49]	bsr	$3 \cdot 10^{-2} s^{-1}$	[49]
s	$3.8 \cdot 10^{-9} \mu mol/(m^2 s)$	[48, 49]	l_3	$2.8 \cdot 10^{-4} s^{-1}$	[48, 49]
l_2	$10^{-5} s^{-1}$	[21, 49]	l_4	$2.8 \cdot 10^{-4} s^{-1}$	[48, 49]

fluid that could bind the ligands, such as secreted ECM, are included. Also, we assume that receptor concentration is uniform over the cell surface (with no receptor clusters being formed). The detailed biochemical reactions in this mechanism are as follows.

$$V_e + R \underset{k_{-1}}{\overset{k_1}{\rightleftarrows}} V_eR \tag{1}$$

$$V_eR + Y \underset{k_{-2}}{\overset{k_2}{\rightleftarrows}} V_eRY \tag{2}$$

$$V_eRY \xrightarrow{k_3} Y^* + V_eR \tag{3}$$

$$Y^* + S \underset{k_{-4}}{\overset{k_4}{\rightleftarrows}} Y^*S \tag{4}$$

$$Y^*S \xrightarrow{k_5} V_i + Y^* \tag{5}$$

$$Y^* \xrightarrow{k_6} Y \tag{6}$$

$$V_i \xrightarrow{bsr} V_e \tag{7}$$

3.2 The Mathematical Model for the Intracellular Dynamics

We consider the time evolution of species concentrations deduced according to the mechanism described above. A rectangular box was used having dimensions 1 mm× 1 mm × 1.5R, where $R = 10\mu$m is the radius of EC when spherical geometry is assumed. Furthermore, N represents the number of cells, A_c and V_c represent the surface area and the volume of each cell. The reactions that are considered in the mechanism then lead to the following system of differential equations:

$$V_0\frac{dy_1}{dt} = N \cdot A_c k_{-1} y_3 - N \cdot A_c k_1 y_1 y_2 + N \cdot V_c \cdot sr(y_8) - N \cdot V_c h_1 y_1 \tag{8}$$

$$\frac{dy_2}{dt} = k_{-1} y_3 - k_1 y_1 y_2 + s - l_2 y_2 \tag{9}$$

$$\frac{dy_3}{dt} = k_1 y_1 y_2 - k_{-1} y_3 - k_2 y_3 z + (k_{-2} + k_3) y_4 - l_3 y_3 \tag{10}$$

$$\frac{dy_4}{dt} = k_2 y_3 z - (k_{-2} + k_3) y_4 - l_4 y_4 \tag{11}$$

$$\frac{dy_5}{dt} = -k_4 y_5 y_6 + (k_{-4} + k_5) y_7 + k_3 y_4 - k_6 y_5 \tag{12}$$

$$\frac{dy_6}{dt} = -k_4 y_5 y_6 + k_{-4} y_7 \tag{13}$$

$$\frac{dy_7}{dt} = k_4 y_5 y_6 - (k_{-4} + k_5) y_7 \tag{14}$$

$$V_c\frac{dy_8}{dt} = A_c k_5 y_7 - V_c \cdot sr(y_8) - V_c h_2 y_8. \tag{15}$$

Here s is the insertion rate of surface species into endothelial cell membrane and l_2, l_3, and l_4 represent the internalization rates of surface receptors and complexes. Their values are listed in Table 2. We introduce the secretion function $sr(y_8)$ which occurs in the secretion step (7). First, we consider it to be a linear function of the intracellular VEGF (V_i) of the form $sr(y_8) = bsr{\cdot}y_8$. This does not affect the validity of our model since the true secretion rate function is not known. Also, we consider the case where bsr is piecewise and periodic function. Moreover, the secretion function can be easily modified and the model amended as new experimental data becomes available.

We proved the nonnegativity and boundedness of the solution to the initial-value problem corresponding to the system (8), (9), (10), (11), (12), (13), (14) and (15). The initial data is nonnegative.

3.3 Insertion of Surface Species into EC Membrane

In this section, we consider two separated sub-cases, namely, when insertion of surface species into endothelial cell membrane and internalization of surface receptors and complexes are not taken into consideration (Case A), and when they are considered in the system (Case B). A comparison between the two systems will be done.

In **Case A**, we consider the system of differential equations (8), (9), (10), (11), (12), (13), (14) and (15), with the values of the parameters s, l_2, l_3, and l_4 being zero. The conservation relations are $[V_eR]+[R]+[V_eRY]=[R]_T$, $[Y^*]+[Y^*S]+[Y]=[Y]_T$, where subscript T denotes the initial value of a quantity. When these conditions are solved for $y_2(t)$ and $z(t)$ $z(t)=[Y]_T-y_5(t)-y_7(t)$ $y_2(t)=[R]_T-y_3(t)-y_4(t)$, and the result is used in Eqs. (8), (9), (10), (11), (12), (13), (14) and (15), we obtain the system of differential equations for $y_1, y_3, \ldots, y_8$. In the interest of space, this will not be included here. The values used for the kinetic parameters are in Table 2 and were found from experimentally based estimates [14, 21, 48, 49]. We also use $V_e(0)=2.22\,\text{nM}$, $R_T=3.3\cdot 10^{-5}\,\mu\text{mol/m}^2$, and $Y_T=2.50\cdot 10^{-4}\,\mu\text{mol/m}^2$. In **Case B** we take into consideration insertion of surface species into the endothelial cell membrane and internalization of surface receptors and complexes. The reactions that are considered in the mechanism then lead to system of differential equations (8), (9), (10), (11), (12), (13), (14) and (15), which describes the time evolution of species concentrations. The conservation relation is $[Y^*]+[Y^*S]+[Y]=[Y]_T$ and the quantity that is determined by it is $z(t)$ (corresponding to Y – free enzyme) is $z(t)=[Y]_T-y_5(t)-y_7(t)$, where subscript T denotes the initial value of a quantity. As before, using this conservation condition, we obtain a system of differential equations for $y_1, \ldots, y_8$.

3.4 Nondimensionalization and Analysis of the Systems

We introduce dimensionless variables for the independent concentrations (Table 1). We find that they are of two categories: those who reach steady-state rapidly, which are called fast variables and the remaining ones which are called slow variables. In our case the fast variables are denoted by u_4 and u_7. The nondimensionalization gives rise to two small parameters $\epsilon_1=\dfrac{k_5}{k_{-2}+k_3}$ and $\epsilon_2=\dfrac{k_5}{k_3}$. On the τ time scale, the singular equations, corresponding to the fast variables u_4 and u_7, can be reduced to the algebraic system of equations $\frac{du_4}{d\tau} = 0$ and $\frac{du_7}{d\tau} = 0$. By replacing u_4 and u_7 and by dropping the $O(\epsilon)$ terms, after some simplifications, we obtain a system of differential equations for the species concentrations. In case A, the variables are u_1, u_3, u_5, u_6, and u_8. In addition, in Case B we have u_2. Having investigated the dynamics of species concentrations in the described cases, several observations can be made. The analysis in each case was made using the same set of parameters and initial conditions. The reduced system in Case A agrees with the full system only on short time, however we see that the reduced system in Case B and the full system agree on both short and long time period.

3.5 Further Assumptions for Intracellular Dynamics Model

After VEGF is been secreted and released into nearby tissues by the hypoxic tumor, it binds to endothelial cell receptors, process that activates a series of relay proteins which in turn transmit a signal that starts a series of cascades inside the cell. So far the mechanism was reduced to four primary species for the intracellular dynamics. Next, we make the biologically realistic assumption that the substrate is not only consumed, but also produced during the entire process Thus, we can consider that the rate of change of substrate concentration is approximately zero. Moreover, we investigate Case A and Case B too. A similar analysis as well as a comparison between the two systems is performed and we obtain that Case B agrees with the full system on both short and long time period. Thus, we consider the following system to represent the governing equations for the intracellular dynamics.

$$\frac{du_1}{d\tau} = N\gamma_1\alpha_1 u_3 - N\gamma_1 u_1 u_2 + \rho\beta_1 u_8 - \rho\beta_2 u_1 \tag{16}$$

$$\frac{du_2}{d\tau} = \alpha_1 u_3 - u_1 u_2 + \delta_1 - \delta_2 u_2 \tag{17}$$

$$\frac{du_3}{d\tau} = u_1 u_2 - (\alpha_1 + \delta_3) u_3 \tag{18}$$

$$\frac{du_5}{d\tau} = \alpha_4\alpha_2\gamma_2(1 - (1 + \alpha_3 u_6) u_5) u_3 - \alpha_5 u_5 \tag{19}$$

$$\frac{du_8}{d\tau} = \frac{N}{\rho}\frac{\gamma_1}{\gamma_2}\alpha_3 u_5 u_6 - (\beta_1 + \beta_3) u_8, \tag{20}$$

where u_6 is constant and initial condition

$$(u_1(0), u_2(0), u_3(0), u_5(0), u_8(0))^T = (u_{10}, u_{20}, 0, 0, 0)^T.$$

We performed a stability and bifurcation analysis for the above system considering it to be on the form

$$\frac{d\mathbf{u}}{dt} = f(\mathbf{u}, p), \mathbf{u}(0) = \mathbf{u}_0 \tag{21}$$

where $u \in M \subseteq \mathbf{R}^5$ and $p \in I \subseteq \mathbf{R}^m$ [34]. We consider $m = 1$ and the $p = \beta_1$ which is the nondimensional form of the VEGF secretion rate. Furthermore, the subset $I \subseteq [0, 1]$, according with data from biological experiments. We showed that the system has a steady state $(0, 1, 0, 0, 0)$ and a unique steady state in the interior of $\mathbf{R}^5_+$. Moreover, the unique positive steady state is stable for any $\beta_1 \in [0, 1]$ and, since the Jacobian is nonsingular for any $\beta_1 \in [0, 1]$, there are no possible bifurcation points.

4 Conclusions

We have reduced the complicated biochemical pathway of angiogenesis (Fig. 1) to a simplified one (Fig. 2) involving only four primary species for the intracellular dynamics (namely, VEGF, Receptor, Enzyme, and Substrate). Two separated cases have been considered for describing this mechanism (rate of change of substrate concentration is not zero or approximately zero). The behavior of time evolution of species concentrations have been investigated in each case and compared with the expected biological phenomenon. As a conclusion, we have been able to develop a new mathematical model that describes the intracellular dynamics which incorporates a realistic model for signal transduction, VEGF production and release.

Considering the fact that it has only recently been established that the activated endothelial cells secrete detectable amounts of VEGF, quantitative data relative to this fact is almost entirely absent. Therefore, the model was tested for different sets of parameters. A qualitative analysis revealed the fact that the system has a unique positive, asymptotically stable steady state solution. Finally, a bifurcation analysis of the ODE kinetics of the model showed that there are no possible bifurcation points with respect to the coefficient of the secretion rate function, and that the system is robust to variations of this parameter.

First, we considered that the VEGF secretion rate function has a linear dependency on the intracellular VEGF. Next, we investigated the cases when *bsr* is a piecewise function and a periodic function in time, with the same average as the constant value. Interestingly, we obtained the same behavior for short and long time periods for our system (Fig. 3). We note that this analysis could be extended further by investigating different parameter variations, for example the ratio of the total cell volume to the extracellular volume. Additionally, the secretion rate function can be further modified and the model amended as new experimental data becomes available.

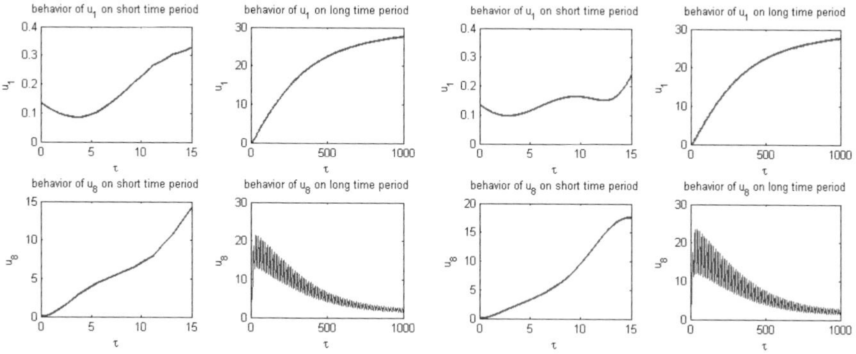

Fig. 3 Piecewise (*left*) and periodic (*right*) secretion rate-*bsr*

References

1. Adams, R.H., Alitalo, K.: Molecular regulation of angiogenesis and lymphangiogenesis. Nat. Rev. Mol. Cell Biol. **8**, 464–478 (2007)
2. Anderson, A.R.A., Chaplain, M.A.J.: Continuous and discrete mathematical models of tumor-induced angiogenesis. Bull. Math. Biol. **60**, 857–900 (1998)
3. Anderson, A.R.A., Chaplain, M.A.J.: A mathematical model for capillary network formation in the absence of endothelial cell proliferation. Appl. Math. Lett. **11**, 109–114 (1998)
4. Arakelyan, L., Vainstein, V., Agur, Z.: A computer algorithm describing the process of vessel formation and maturation, and its use for predicting the effects of anti-angiogenic and anti-maturation therapy on vascular tumor growth. Angiogenesis **5**, 203–214 (2002)
5. Balding, D., McElwain, D.L.S.: A mathematical model of tumor-induced capillary growth. J. Theor. Biol. **114**, 53–73 (1985)
6. Bartha, K., Rieger, H.: Vascular network remodeling via vessel cooption, regression and growth in tumors. J. Theor. Biol. **21**, 903–918 (2006)
7. Bauer, A., Jackson, T., Jiang, Y.: A cell-based model exhibiting branching and anastomosis during tumor-induced angiogenesis. Biophys. J. **92**, 3105–3121 (2007)
8. Bentley, K., Gerhardt, H., Bates, P.A.: Agent-based simulation of notch-mediated tip cell selection in angiogenic sprout initialisation. J. Theor. Biol. **250**, 25–36 (2008)
9. Bentley, K., Mariggi, G., Gerhardt, H., Bates, P.A.: Tipping the balance: robustness of tip cell selection, migration and fusion in angiogenesis. PLoS Comput. Biol. **5**, e1000549 (2009)
10. Billy, F., Ribba, B., Saut, O., et al.: A pharmacologically based multiscale mathematical model of angiogenesis and its use in investigating the efficacy of a new cancer treatment strategy. J. Theor. Biol. **260**, 545–562 (2009)
11. Byrne, H.M., Chaplain, M.A.J.: Mathematical models for tumour angiogenesis: numerical simulations and nonlinear wave solutions. Bull. Math. Biol. **57**, 461–486 (1995)
12. Byrne, H.M., Chaplain, M.A.J.: Explicit solutions of a simplified model of capillary sprout growth during tumor angiogenesis. Appl. Math. Lett. **9**, 69–74 (1996)
13. Capasso, V., Morale, D.: Stochastic modelling of tumour-induced angiogenesis. J. Math. Biol. **58**, 219–233 (2009)
14. Claesson-Welsh, L.: Signal transduction by vascular endothelial growth factor receptors. Biochem. Soc. Trans. **31**, part 1 (2003)
15. Delgado, M., Morales-Rodrigo, C., SuÃąrez, A.: Anti-angiogenic therapy based on the binding to receptors. DCDS-A **32**(11), 3871–3894 (2012)
16. Di Talia, S., Gamba, A., Lamberti, F., Serini, G.: Role of repulsive factors in vascularization dynamics. Phys. Rev. E **73**, 041917:1–11 (2006)

17. Folkman, J.: The vascularization of tumors. In: Cancer Biology: Readings from Scientific American, pp. 115–124. Freeman, New York (1996)
18. Folkman, J. What is evidence that tumors are angiogenesis dependent? J. Natl. Cancer Inst. **82**, 4–6 (1990)
19. Folkman, J.: Clinical applications of research on angiogenesis. N. Engl. J. Med. **333**, 1757–1763 (1995)
20. Folkman, F.: Fighting cancer by attaching its blood supply. Sci. Am. **275**, 150–154 (1996)
21. Gabhann, F.M., Popel, A.S.: Model of competitive binding of vascular endothelial growth factor and placental growth factor to VEGF receptors on endothelial cells. Am. J. Physiol. Heart Circ. Physiol. **286**, H153–H164 (2004)
22. Goede, V., Schmidt, T., Kimmina, S., Kozian, D., Augustin, H.G.: Analysis of blood vessel maturation processes during cyclic ovarian angiogenesis. Lab. Invest. **78**, 1385 (1998)
23. Hegen, A., Koidl, S., Weindel, K., et al.: Expression of angiopoietin-2 in endothelial cells is controlled by positive and negative regulatory promoter elements. Arterioscler. Thromb. Vasc. Biol. **24**, 1803–1809 (2004)
24. Holmes, M.J., Sleeman, D.: A mathematical model of tumour angiogenesis incorporating cellular traction and viscoelastic effects. J. Theor. Biol. **202**, 95–112, 2000
25. Jackson, T.L., Zheng, X.: A cell-based model of endothelial cell elongation, proliferation and maturation during corneal angiogenesis. Bull. Math. Biol. **72**, 830–868 (2010)
26. Jekunen, A., Kairemo, K.: Inhibition of angiogenesis at endothelial cell level. Microsc. Res. Tech. **60**, 85–97 (2003)
27. Levine, H.A., Nilsen-Hamilton, M.: Angiogenesis – a biochemial/mathematical perspective. In: Friedman, A. (ed.) Tutorials in Mathematical Biosciences III, chapter 2. Springer (2006)
28. Levine, H.A., Pamuk, S., Sleeman, B.D., Nilsen-Hamilton, M.: Mathematical modeling of capillary formation and development in tumor angiogenesis: penetration into the stroma. Bull. Math. Biol. **63**, 801–863 (2001)
29. Lignet, F., Benzekry, S., Wilson, S., Billy, F., Saut, O., Tod, M., You, B., Berkane, A.A., Kassour, S., Wei, M.X., Grenier, E., Ribba, B.: Theoretical investigation of the efficacy of antiangiogenic drugs combined to chemotherapy in xenografted mice. J. Theor. Biol. **320**, 86–99 (2013)
30. Liu, G., Qutub, A.A., Vempati, P., Mac Gabhann, F., Popel, A.: Module-based multiscale simulation of angiogenesis in skeletal muscle. Theor. Biol. Med. Model. **8**, 6 (2011)
31. Manoussaki, D.: A mechanochemical model of angiogenesis and vasculogenesis. ESAIM: Math. Model. Numer. Anal. **37**, 581–599 (2003)
32. Matsushita, K., Yamakuchi, M., Morrell, C.N., et al.: Vascular endothelial growth factor regulation of Weibel-Palade-body exocytosis. Blood **105**, 207 (2005)
33. Milde, F., Bergdorf, M., Koumoutsakos, P.: A hybrid model for three-dimensional simulations of sprouting angiogenesis. Biophys. J. **95**, 3146–3160 (2008)
34. Murray, J.D.: Mathematical Biology, 3rd edn. Springer-Verlag, Berlin/Heidelberg/New York (2004)
35. Othmer, H.G., Mantzaris, N., Webb, S.: Mathematical modeling of tumor-induced angiogenesis. J. Math. Biol. **49**(2), 111–187 (2004)
36. Peirce, S.M., Van Gieson, E.J., Skalak, T.C.: Multicellular simulation predicts microvascular patterning and in silico tissue assembly. FASEB J. **18**, 731–733 (2004)
37. Peirce, S.M.: Computational and mathematical modeling of angiogenesis. Microcirculation **15**, 739–751 (2008)
38. Plank, M.J., Sleeman, B.D.: A reinforced random walk model of tumor angiogenesis and anti-angiogenesis strategies. IMA J. Math. Med. Biol. **20**, 135–181 (2003)
39. Plank, M.J., Sleeman, B.D.: Lattice and non-lattice models of tumour angiogenesis. Bull. Math. Biol. **66**, 1785–1819 (2004)

40. Plank, M.J., Sleeman, B.D., Jones, P.F.: A mathematical model of tumour angiogenesis, regulated by vascular endothelial growth factor and the angiopoietins. J. Theor. Biol. **229**, 435–454 (2004)
41. Poleszczuk, J., Bodnar, M., Forys, U.: New approach to modeling of antiangiogenic treatment on the basis of Hahnfeldt et al. model. MBE **8**(2), 591–603 (2011)
42. Qutub, A.A., Mac Gabhann, F., Karagiannis, E.D., Vempat, P., Popel, A.: Multiscale models of angiogenesis. IEEE Eng. Med. Biol. Mag. **28**, 14–31 (2009)
43. Qutub, A.A., Popel, A.: Elongation, proliferation & migration differentiate endothelial cell phenotypes and determine capillary sprouting. BMC Syst. Biol. **3**, 13 (2009)
44. Schugart, R.C., Friedman, A., Zhao, R., Sen, C.K.: Wound angiogenesis as a function of tissue oxygen tension: a mathematical model. PNAS **105**, 2628–2633 (2008)
45. Serini, G., Ambrosi, D., Giraudo, E., Gamba, A., Preziosi, L., Bussolino, F.: Modeling the early stages of vascular network assembly. EMBO J. **8**, 1771–1779 (2003)
46. Stokes, C.L., Lauffenburger, D.A.: Analysis of the roles of microvessel endothelial cell random mobility and chemotaxis in angiogenesis. J. Theor. Biol. **152**, 377–403 (1991)
47. Sun, S., Wheeler, M.F., Obeyesekere, M., Patrick, C.: A deterministic model of growth factor-induced angiogenesis. Bull. Math. Biol. **67**, 313–337 (2005)
48. Tang, Y., Othmer, H.G.: A G-protein-based model for cAMP dynamics in Dictyostelium discoideum. Math. Biosci. **120**, 25–76 (1994)
49. Tang, Y., Othmer, H.G.: Excitation, oscillations and wave propagation in a G-protein-based model of signal transduction in Dictyostelium discoideum. Philos. Trans. R. Soc. **349**, 179–195 (1996)
50. Tarfulea, N.: Mathematical Modeling of Signal Transduction and Cell Motility in Tumor Angiogenesis. University of Minnesota Ph.D. thesis, ProQuest (2006)
51. Tong, S., Yuan, F.: Numerical simulations of angiogenesis in the cornea. Microvasc. Res. **61**, 14–27 (2001)
52. Travasso, R.D.M., Corvera Poir, E., Castro, M., et al.: Tumor angiogenesis and vascular patterning: a mathematical model. PLoS ONE **6**, e19989 (2011)
53. Xue, C., Friedman, A., Sen, C.K.: A mathematical model of ischemic cutaneous wounds. PNAS **106**, 16782–16787 (2009)
54. Zheng, X., Koh, G.Y., Jackson, T.: A continuous model of angiogenesis: initiation, extension, and maturation of new blood vessels modulated by vascular endothelial growth factor, angiopoietins, platelet-derived growth factor-B, and pericytes. DCDS-B **18**, 1109–1154 (2013)

A Mathematical Model of Cytokine Dynamics During a Cytokine Storm

Marianne Waito, Scott R. Walsh, Alexandra Rasiuk, Byram W. Bridle, and Allan R. Willms

Abstract Cytokine storms are a potentially fatal exaggerated immune response consisting of an uncontrolled positive feedback loop between immune cells and cytokines. The dynamics of cytokines are highly complex and little is known about specific interactions. Researchers at the Ontario Veterinary College have encountered cytokine storms during virotherapy. Multiple mouse trials were conducted where a virus was injected into mice whose leukocytes lacked expression of the type I interferon receptor. In each case a rapid, fatal cytokine storm occurred. A nonlinear differential equation model of the recorded cytokine amounts was produced to obtain some information on their mutual interactions. Results provide insight into the complex mechanism that drives the storm and possible ways to prevent such immune responses.

1 Introduction

Overly robust cytokine responses are responsible for a broad array of very challenging and often fatal clinical conditions. These include infectious diseases such as influenza [16], severe acute respiratory syndrome (SARS) [11], bacteria-induced toxic shock syndrome [20] and sepsis [9], as well as multiple sclerosis [13], graft-versus-host disease [7] and sometime adverse effects of therapies [22]. Cytokines are released by cells to coordinate an immune response to help protect against foreign and/or dangerous matter. They are produced in response to infection and inflammation [5]. There are five major classifications of cytokines; interferons (IFN), interleukins (IL), chemokines, colony-stimulating factors (CSF), and tumour necrosis factors (TNF) [23]. Together they function in order to stimulate a response that will control cellular stress as well as minimize the amount of damage to a particular cell or group of cells [5].

M. Waito (✉) • A.R. Willms
Department of Mathematics and Statistics, University of Guelph, Guelph, ON, Canada
e-mail: wilcoxm@uoguelph.ca; awillms@uoguelph.ca

S.R. Walsh • A. Rasiuk • B.W. Bridle
Department of Pathobiology, University of Guelph, Guelph, ON, Canada
e-mail: scott.walsh22@gmail.com; arasiuk@uoguelph.ca; bbridle@uoguelph.ca

© Springer International Publishing Switzerland 2016
J. Bélair et al. (eds.), *Mathematical and Computational Approaches in Advancing Modern Science and Engineering*, DOI 10.1007/978-3-319-30379-6_31

Although the goal of each cytokine is to ultimately protect the host against dangerous insults, it is not uncommon for this process to become unbalanced. For reasons unknown, positive feedback loops can become up-regulated during an immune response involving complex interaction between cytokines and immune cells. The fine balance that is typical of healthy individuals is lost and gives rise to a cascade which is termed a cytokine storm. The dynamics of cytokines during normal immune responses and even more so during storms are highly complex and little is known about specific interactions [5, 27]. Cytokine storms not only have very negative effects on the immune response, they often have life threatening effects including a decrease in blood pressure, increase in heart rate and can often lead to death of the host [27].

As more biological data is collected, the immense complexities of the immune system is becoming increasingly apparent [26]. It is clear that mathematical modelling has become an essential tool to complement experimental techniques both in vitro and in vivo. Combining these approaches can result in major advancements in both the understanding of cancer and immune system dynamics including cytokine interactions, with the potential to identify strategies to control toxic cytokine storms [26].

Previous mathematical models include at most two cytokines and do not focus the research on the interactions and function of the cytokines. Typical cytokines included in models are IL-2, IL-10, IL-12 and TNF. Unfortunately, even the models that contain cytokines do not investigate the specific interactions that we are interested in. These models exclude many of the cytokines that are of greatest interest to us [6, 12, 15, 19].

Since our focus is to model only the dynamics of cytokines, we turn to ordinary differential equation (ODE) models to provide a good framework for exploring interactions between different cytokine populations. To keep the model relatively simple, we ignore the effects of effector cells (activated immune cells) due to the complexities it brings to the model. To date, the only model that is of a similar nature is produced by [27] called *Dynamics of a Cytokine Storm*. In this paper, the interactions of nine cytokines were modelled using data from a human clinical trial where six volunteers took part in a study that accidentally led them to undergo the effects of a cytokine storm [27]. This model was primarily constructed to look at the effects of the antibody responsible for the cytokine storm using a linear ODE model where parameter estimation determined coupling parameters between cytokines. The coupling parameters identified which cytokine was responsible for enhancing or inhibiting each other cytokine as well as self-regulation [27].

In this paper, we determine how cytokines interact with one another based on a set of time series data provided by Dr. Byram Bridle and his lab members Dr. Scott Walsh and Alexandra Rasiuk at the Ontario Veterinary College (OVC). Following administration of a highly attenuated virus to mice with leukocytes lacking the type I IFN receptor, a deadly cytokine storm developed leading to death in only 24 h. Down-regulation of anti-viral IFN signaling is a common mechanism used by viruses during infection, including those associated with cytokine storms (e.g. influenza virus [14], SARS-coronavirus [3] and Ebola virus [2]). Thus, a natural question arises: what are the specific dynamics of cytokines with respect to both initiating and exacerbating a cytokine storm when type I interferon signalling is impaired?

2 Mathematical Model

Most biological systems are innately non-linear and thus a non-linear ODE model is essential to obtain the proper dynamics [26]. The model we use is given by:

$$\dot{x}_i = -\mu_i x_i + \frac{M_i}{1 + e^{-y_i}} \qquad \text{for } i,j = 1, 2, \ldots, 13. \tag{1}$$

The first and most common assumption throughout the literature is that the concentration of a particular cytokine, x_i, continues to decrease linearly when there is no outside stimulus [10, 17]. Since cytokines are secreted by other cells it makes sense that in the absence of these producers, the number of cytokines will rapidly decline at a rate μ_i (>0) [10, 17, 25]. The second assumption is that the rate of production of cytokines is dependent on interactions with other cytokines and is sigmoidal in shape of the form

$$\frac{M_i}{1 + e^{-y_i}},$$

where M_i (>0) is the maximum production rate. The interaction with other cytokines, y_i, determines the slope of the function, and thus how much of a cytokine is produced, potentially offsetting some or all of the decay [10, 17, 25]. The interaction factor y_i is given by

$$y_i = \alpha_{0,i} + \sum_{j=1}^{n} \alpha_{i,j} x_j + S_i, \tag{2}$$

meaning that y_i is composed of effects from the presence of other cytokines, $\alpha_{i,j}$; effects from components for which there is no data, $\alpha_{0,i}$; and the stimulus, S_i, by the virus. The stimulus is of the form

$$S_i = be^{-t/\tau},$$

where there is an initial dose of the drug, b, which decays exponentially with characteristic time τ.

3 Data

Data was collected and provided by Dr. Byram Bridle, Dr. Scott Walsh and Alexandra Rasiuk who study the role of type I IFN signalling in the regulation of cytokine responses at the OVC. Chimeric mice were made by lethal irradiation of the bone marrow of C57BL/6 mice (Charles River Laboratories) followed by reconstitution with bone marrow from either wild-type or type I IFN receptor (IFNAR)-knockout donors (the latter provided by Laurel Lenz, University of Colorado School of Medicine). These mouse-based experiments were approved by the institutional Animal Care Committee and complied with the standard of the Canadian Council of Animal Care. These mice were infected intravenously with recombinant Vesicular Stomatitis Virus with a deletion of methionine at position 51 of the matrix protein (VSVΔm51) [21]. The matrix protein of VSV suppresses antiviral type I IFN responses. Therefore, this mutant virus renders the already attenuated laboratory strain of VSV even safer and it is being developed as an oncolytic virus for the treatment of cancers via intravenous infusion [24]. Surprisingly, mice lacking the IFNAR on their leukocytes experienced a profound cytokine storm, ultimately leading to death in only 24 h. The resulting time series data provided concentrations of 13 different cytokines in plasma, measured using a multiplex array (BioRad).

Concentrations of 13 cytokines were recorded at times 0, 2.5, 5, 10 and 24 h for both the wild-type mice and mice lacking the IFNAR on 20% of their leukocytes. Raw data was then normalized to account for the vast variability in the concentrations. This was done by dividing the concentration at each time point by the sum of the concentrations across all time points for a particular cytokine. In order to produce an accurate fit to the model it was essential to group cytokines to reduce the number of parameters. Grouping was based on the inflammatory classification of each cytokine as well as similarity of cytokine profile. The time at which the peak concentration occurred was recorded along with either the pro- or anti-inflammatory classification. Groupings can be seen in Table 1. Parameter estimation techniques, specifically *fmincon* in MATLAB, were used to fit the model to the data. The cost

Table 1 Grouping of 13 cytokines based on raw data and inflammatory classification

Group	Cytokines	Inflammatory classification	WT data peak (h)	KO data peak (h)
1	IL-13, IL-5	Anti	2.5	24
2	MIP-1β, TNF-α, IL-12	Pro	2.5	24
3	IL-1β, eotaxin, IFN-γ	Pro	5	24
4	IL-4, IL-10	Anti	5	24
5	IL-6	Pro/Anti	2.5	10
6	MIP-1α	Pro	5	24
7	Rantes	Pro	10	24

function is the typical least squares function with a normalization matrix to offset discrepancies between groups.

4 Results

Mathematical analysis is an important tool for biological predictions and for producing more questions and further areas of research both in biology and mathematics [1]. Model results from parameter optimization of IFNAR-knockout mice data are shown in Fig. 1. The model provides an accurate fit to the cytokine time series data for both IFNAR-knockout and wild-type mice (not shown). The model parameters that provide the most interesting information are the $\alpha_{i,j}$ values, as they relate how each cytokine group interacts with each of the other groups. The sign and magnitude of these values indicate the type of effect a particular cytokine has; a negative parameter value indicates an inhibitory effect, a positive value an enhancing effect and the largest magnitude produces the most significant coupling. The three largest $\alpha_{i,j}$ values for each group of the IFNAR-knockout mice are displayed in Fig. 2 where the lines indicate either enhancing (solid line with an arrowhead) or inhibiting (dotted line with a 'T') and the line thickness displays the magnitude (the thicker the line, the greater the magnitude).

Preliminary results from Fig. 2 show that Groups 2, 4, 5 and 6 are the most significant. Each of them have a large number of interactions, multiple two-way paths, exhibit definite enhancement or inhibition, and fit well with what is known biologically.

The most significant group, Group 2, consists of three pro-inflammatory cytokines: MIP-1β, TNF-α and IL-12. Of the three, TNF-α is assumed to be the dominating cytokine. TNF has been thoroughly studied and is one of the most well known cytokines as it plays an important role in the outcome of immune function

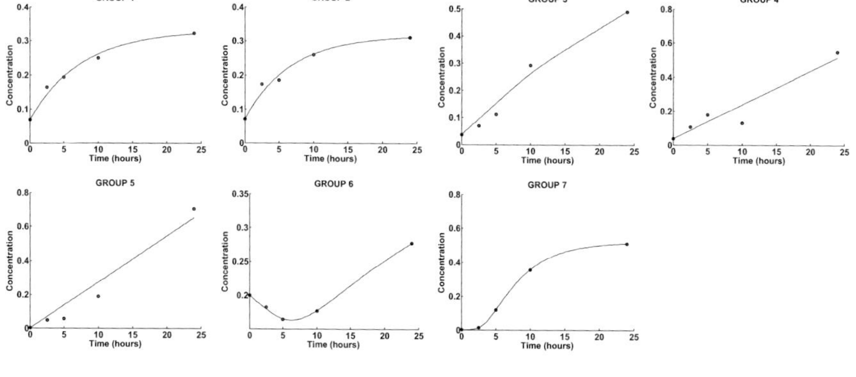

Fig. 1 Model results from parameter estimation of IFNAR-knockout mice. *Dots* represent raw time series data, *lines* show the model prediction

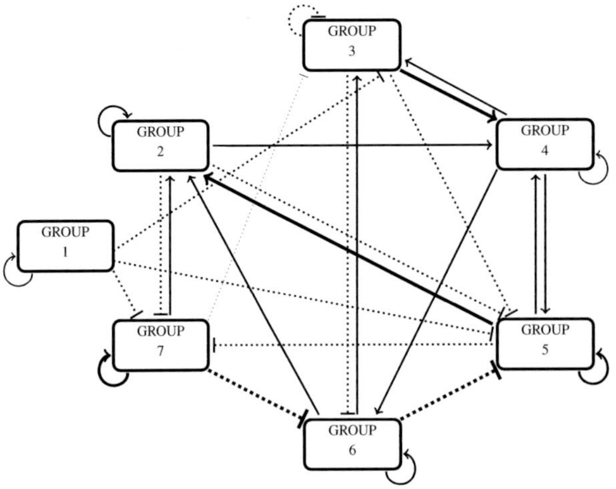

GROUP	CYTOKINES
1	IL-13, IL-5
2	MIP-β, TNF-α, IL-12
3	IL-1β, eotaxin, IFN-γ
4	IL-4, IL-10
5	IL-6
6	MIP-1α
7	Rantes

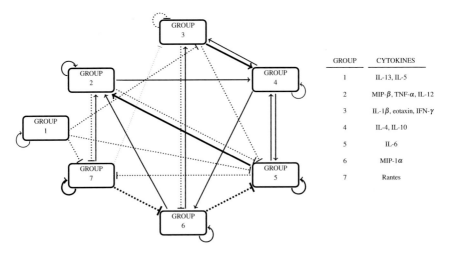

Fig. 2 Largest three interactions for each cytokine group of IFNAR-knockout mice are represented with a line. *Solid lines* with an arrow indicate enhancing effects while *dotted lines* with a 'T' indicate inhibitory effects

[18, 23]. Model results for Group 2 show that it is an integral part of six significant interactions and has multiple two-way paths, one with Group 5 and another with Group 7. All three interactions that affect Group 2 are enhancing, indicating that as the storm progresses, the amount of cytokine will continue to rise. TNF-α, along with IL-1β (found in Group 3), are considered early-response cytokines, occurring soon after an immune response is triggered. It is known biologically that TNF-α promotes the generation of IL-1β [23]. Although the results are not shown for the wild-type case, the model results indicate that in fact Group 2 does enhance Group 3. In the case of IFNAR-knockout mice however, Group 2 inhibits Group 3 (the relatively small interaction is not shown in Fig. 2). This difference begins to shed light on why a storm occurred in mice lacking IFNAR on leukocytes.

Another central group is Group 4 which is made up of two anti-inflammatory cytokines, IL-4 and IL-10, with the latter being a very prominent inhibitor. IL-10 is often produced once a cytokine storm has begun in an attempt to return the balance that has been lost, termed immunoparalysis. Although overproduction can often allow the host to survive the cytokine storm longer, it is not likely it will survive long term [8, 18, 23]. Model results for Group 4 show that there are six interactions that are significant as well as two primarily enhancing two-way interactions with Groups 3 and 5. The three interactions that affect Group 4 are all enhancing, meaning that the amount of Group 4 will likely increase as the storm continues. Biologically it is known that IL-10 plays a role in the down regulation of both TNF-α (Group 2) and IL-1β (Group 3) [18]. Referring to Fig. 2, the model verifies that in fact Group 4 does inhibit Group 2, however Group 4 enhances Group 3. This could be due to the grouping of cytokines, since Group 2 contains MIP-1β and IL-12 as well. On the contrary, in the wild-type case, Group 4 inhibits Group 3 while it enhances Group 2.

A well-studied cytokine that is known to be a key component of a cytokine storm is IL-6, Group 5, which has both pro- and anti-inflammatory properties and is a central cytokine used to assess cytokine responses in the host [23]. Figure 2 displays the importance of this group with the eight interactions and two primarily enhancing two-way paths. It is known biologically that the production of IL-6 is stimulated by TNF-α and IL-1β [23]. Results from the model using wild-type data verify that indeed both Groups 2 and 3 enhance Group 5, however the interaction between Group 3 and 5 is relatively small. IFNAR-knockout results shown in Fig. 2 imply that Groups 2 and 3 inhibit Group 5 and that those interactions are significant. Again, this can provide some insight into how a storm was able to occur.

Group 6, MIP-1α, is required for a typical inflammatory response to viruses [4]. It is a pro-inflammatory chemokine that inhibits proliferation of hematopoietic stem cells in vitro and in vivo [4]. Model results for MIP-1α show that six interactions are significant as well as a two-way interaction with Group 3. The three interactions that affect MIP-1α are both inhibiting and enhancing whereas in the wild-type case the interactions are purely enhancing, causing an increase in the amount of MIP-1α as the storm continues.

It has been noted that the fine balance of pro- and anti-inflammatory mechanisms is critical in maintaining stability, and if these mechanisms become unbalanced, the outcome may contribute to a cytokine storm [23]. Groups 1 and 4 are anti-inflammatory cytokines, while the remainder act primarily as pro-inflammatory cytokines. For wild-type populations, of the three most significant groups (2, 4, and 6), the anti-inflammatory cytokines are being inhibited, while the pro-inflammatory cytokines are being enhanced. This balance in this system becomes lost in mice lacking the IFNAR on their leukocytes, as shown in Fig. 2. Instead of the anti-inflammatory cytokines (Group 4) being inhibited, they are instead enhanced. Immediately it becomes apparent that the fine balance that is typical of healthy individuals has become unstable.

Future work including sensitivity analysis of cytokine groupings could provide further information on the significance of the groupings and individual cytokines. In conclusion, cytokines belonging to Groups 2, 4, 5 and 6, particularly TNF-α, IL-10, IL-6 and MIP-1β, have the largest effects on the dynamics of this particular cytokine storm. Changes introduced into the system by knocking out IFNAR cause key interactions to swap from enhancing to inhibiting and vice versa. It is possible that reducing the alterations in the effects of Groups 2, 4, 5 and 6 could lead to the reduction in severity and possibly even the entire storm.

Acknowledgements This research was supported in part by the Government of Ontario and the University of Guelph.

References

1. Bellomo, N., Li, N.K., Maini, P.K.: On the foundations of cancer modelling: selected topics, speculations, and perspectives. World Sci. **18**(4), 593–646 (2008)
2. Chang, T.-H., Kubota, T., Matsuoka, M., Jones, S., Bradfute, S.B., Bray, M., Ozato, K.: Ebola zaire virus blocks type i interferon production by exploiting the host sumo modification machinery. PLoS Pathog. **5**(6), e1000493 (2009)
3. Chen, X., Yang, X., Zheng, Y., Yang, Y., Xing, Y., Chen, Z.: Sars coronavirus papain-like protease inhibits the type i interferon signaling pathway through interaction with the sting-traf3-tbk1 complex. Protein Cell **5**(5), 369–381 (2014)
4. Cook, D.N.: The role of mip-1α in inflammation and hematopoiesis. J. Leucoc. Biol. **59**(1), 61–66 (1996)
5. Dranoff, G.: Cytokines in cancer pathogenesis and cancer therapy. Nature **4**(1), 11–22 (2004)
6. Eftimie, R., Bramson, J.L., Earn, D.J.D.: Interactions between the immune system and cancer: a brief review of non-spatial mathematical models. Bull. Math. Biol. **73**(1), 2–32 (2011)
7. Ferrara, J.L.: Cytokine dysregulation as a mechanism of graft versus host disease. Curr. Opin. Immunol. **5**(5), 794–799 (1993)
8. Frazier, W.J., Hall, M.W.: Immunoparalysis and adverse outcomes from critical illness. Pediatr. Clin. N. Am. **55**(3), 647–668 (2008)
9. Harrison, C.: Sepsis: calming the cytokine storm. Nat. Rev. Drug Discov. **9**(5), 360–361 (2005)
10. Hone, A., van den Berg, H.: Mathematical analysis of artificial immune system dynamics and performance. In: Flower, D., Timmis, J. (eds.) In Silico Immunology, pp. 351–374. Springer, New York (2007)
11. Huang, J., Wu, S., Barrera, J., Matthews, K., Pan, D.: The Hippo signaling pathway coordinately regulates cell proliferation and apoptosis by inactivating Yorkie, the drosophila homolog of YAP. Cell **122**(3), 421–434 (2005)
12. Kirschner, D., Panetta, J.C.: Modeling immunotherapy of the tumor-immune interaction. J. Math. Biol. **37**, 235–252 (1998)
13. Link, H.: The cytokine storm in multiple sclerosis. Mult. Scler. **4**(1), 12–15 (1998)
14. Medina, R.A., Garcia-Sastre, A.: Influenza a virus: new research developments. Nat. Rev. Microbiol. **9**, 590–603 (2011)
15. Müller, J., Tjardes, T.: Modeling the cytokine network in vitro and in vivo. J. Theor. Med. **5**(2), 93–110 (2003)
16. Pothlichet, J., Meunier, I., Davis, B.K., Ting, J.P.-Y., Skamene, E., von Messling, V., Vidal, S.M.: Type I IFN triggers RIG-I/TLR3/NLRP3-dependent inflammasome activation in influenza A virus infected cells. PLoS Pathog. **9**(4), e1003256 (2013)
17. Read, M., Timmis, J., Andrews, P.S.: Empirical investigation of an artificial cytokine network. In: Bentley, P.J., Lee, D., Jung, S. (eds.) Artifical Immune Systems, pp. 340–351. Springer, Berlin/Heidelberg (2008)
18. Reis, E.A.G., Hagan, J.E., Ribeiro, G.S., Teixeira-Carvalho, A., Martins-Filho, O.A., Montgomery, R.R., Shaw, A.C., Ko, A.I., Reis, M.G.: Cytokine response signatures in disease progression and development of severe clinical outcomes for leptospirosis. PLoS Negl. Trop. Dis. **7**(9), e2457 (2013)
19. Rodríguez-Pérez, D., Sotolongo-Grau, O., Riquelme, R.E., Sotolongo-Costa, O., Antonio Santos Miranda, J., Antoranz, J.C.: Assessment of cancer immunotherapy outcome in terms of the immune response time features. Math. Med. Biol. **25**, 287–300 (2008)
20. Silversides, J.A., Lappin, E., Ferguson, A.J.: Staphylococcal toxic shock syndrome: mechanisms and management. Curr. Infect. Dis. Rep. **12**(5), 392–400 (2010)
21. Stojdl, D.F., Lichty, B.D., tenOever, B.R., Paterson, J.M., Power, A.T., Knowles, S., Marius, R., Reynard, J., Poliquin, L., Atkins, H., Brown, E.G., Durbin, R.K., Durbin, J.E., Hiscott, J., Bell, J.C.: Vsv strains with defects in their ability to shutdown innate immunity are potent systemic anti-cancer agents. Cancer Cell **4**(4), 263–275 (2003)

22. Suntharalingam, G., Perry, M.R., Ward, S., Brett, S.J., Castello-Cortes, A., Brunner, M.D., Panoskaltsis, N.: Cytokine storm in a phase 1 trial of the anti-cd28 monoclonal antibody tgn1412. N. Engl. J. Med. **355**(10), 1018–1028 (2006)
23. Tisoncik, J.R., Korth, M.J., Simmons, C.P., Farrar, J., Martin, T.R., Katze, M.G.: Into the eye of the cytokine storm. Microbiol. Mol. Biol. Rev. **76**(1), 16–32 (2012)
24. Wollmann, G., Rogulin, V., Simon, I., Rose, J.K., van den Pol, A.N.: Some attenuate variants of vesicular stomatitis virus show enhanced oncolytic activity against human glioblastoma cells relative to normal brain cells. J. Virol. **84**(3), 1563–1573 (2010)
25. Wu, J.T., Kirn, D.H., Wein, L.M.: Analysis of a three-way race between tumor growth, a replication-competent virus and an immune response. Bull. Math. Biol. **66**, 605–625 (2004)
26. Yates, A., Chan, C.C.W., Callard, R.E., George, A.J.T., Stark, J.: An approach to modelling in immunology. Brief. Bioinform. **2**(3), 245–257 (2001)
27. Yiu, H.H., Graham, A.L., Stengel, R.F.: Dynamics of a cytokine storm. PLOS One **7**(10), e45027 (2012)

Examining the Role of Social Feedbacks and Misperception in a Model of Fish-Borne Pollution Illness

Michael Yodzis

Abstract Pollution-induced illnesses are caused by toxicants that result from human activity and should be entirely preventable. However, social pressures and misperceptions can undermine the efforts to limit pollution, and vulnerable populations can remain exposed for decades. We present a human-environment system model for the effects of water pollution on the health and livelihood of a fishing community. The model includes dynamic social feedbacks that determine how effectively the population recognizes the injured and acts to reduce its pollution exposure. Our work is motivated by a historical incident from 1949 to 1968 in Minamata, Japan where methylmercury effluent from a local factory poisoned fish populations and humans who ate them. We will discuss the conditions that allow for the outbreak of a pollution-induced epidemic, and explore the role that misperception plays in allowing it to persist.

1 Introduction

A human-environment system (HES) is characterized by the interaction of human activities and natural processes [12, 13]. The HES model that we present in this paper is developed in [20] in the context of the methylmercury-poisoning incident in Minamata, Japan. It represents the transmission of pollution to humans through

The present contribution draws heavily from previous joint work with Chris T. Bauch and Madhur Anand that was published under the title 'Coupling fishery dynamics, human health and social learning in a model of fish-borne pollution exposure'. The results are extended to include a more complete explanation of the model's dependence on its initial conditions, and the sensitivity analysis is enlarged to study the influence of the ecological parameters on the model's qualitative outcomes.
The original version of this chapter was revised. An erratum to this chapter can be found at DOI 10.1007/978-3-319-30379-6_71.

M. Yodzis (✉)
University of Guelph, Guelph, ON, Canada
e-mail: myodzis@uoguelph.ca

© Springer International Publishing Switzerland 2016

J. Bélair et al. (eds.), *Mathematical and Computational Approaches in Advancing Modern Science and Engineering*, DOI 10.1007/978-3-319-30379-6_32

fish-eating, and the social feedbacks to limit exposure through the boycott of fish-eating and the demand for pollution control.

This model is motivated by several key observations. First is that the epidemiological data from Minamata show that the community was unwilling to recognize new pollution victims from 1960 to 1971 [7, 8]. Historical accounts describe the dependence of the community on the polluting industry, and how this dependence fed a widespread stigmatization of the pollution victims and underestimation of the health damages [5, 9, 17].

Our model depicts the voluntary actions that citizens must take to protect themselves when proactive leadership from the government or industry is lacking. Legal theorists who study the regulatory compliance of polluting industries observe that government policy may not be enforced effectively without pressure from citizens [19]. Social misperception can be the single largest obstacle to resolving pollution problems. Sometimes this is fuelled by an inconclusive scientific understanding, given that the symptoms of pollution exposure can be slow to appear, and difficult to separate from other causes or confounding variables [6, 7, 19]. Misperception may also be fuelled by economic expediency and deliberate foul play. Therefore, citizen action and changing social perception play a large role in helping to resolve pollution problems.

Our objectives are to examine the ecological and social conditions that cause the outbreak of a pollution-induced epidemic, and to study the role of social feedbacks and misperception effects that allow the epidemic to persist.

2 Model

To focus on numerical simplicity and qualitative behaviour, in [20] we build a system of normalized equations whose state variables are proportions rather than absolute quantities. The pollution is driven by an input $n(t)$, which is the baseline emissions loading rate from an industry into a water source. Whereas $n(t)$ is a prescribed function, the net emissions loading $E(t)$ is affected by feedback from the social demand for abatement. The fishery consists of fish $F(t)$ and boats $B(t)$, whose interactions are governed by Lotka-Volterra type equations and emissions mortality.

Human injuries are driven by the fish-borne pollution exposure. This exposure depends on the net emissions level $E(t)$, the fish catch $HF(t)B(t)$ with harvest rate H, and a factor of pollution availability to humans, ε_I.

To model human perception, we introduce $P(t)$ for the perceived injured and $I(t)$ for the actual injured. The parameter s is the stigma rate, or the rate at which the community denies new victims as the perceived injured $P(t)$ increases. This stigma effect causes a disparity between the actual and perceived number of injured. If there is no stigma, $s = 0$, then $I(t)$ and $P(t)$ grow at the same rate. However, when $s > 0$, then a distinct negative feedback is introduced in P': the rate at which the community is willing to recognize new victims decreases as the number of perceived victims increases. Then $P(t)$ is an underestimate of the actual health damages $I(t)$.

Observe that in the model, all social feedbacks and decisions are driven by $P(t)$ but are blind to $I(t)$. $P(t)$ affects the level of fish-eating, $1 - bP(t)$, where b is the boycott rate. $P(t)$ also influences the public demand for abatement $X(t)$, which grows through social learning in the form of an imitation game [10]. We assume that citizens sample the preferences of others and aim to minimize a cost function, $hP(t) - \beta X(t)$, by choosing either to favour abatement or non-abatement. κ is a sampling and imitation rate, h measures the health costs from pollution, and β reflects costs to productivity and job security that is claimed by the industry and its supporters in response to growing X. In turn, the population preference $X(t)$ feeds back to influence the net emissions $E(t)$, thus closing the loop.

Altogether, the equations for the human-environment system are:

$$\left.\begin{aligned} F'(t) &= rF(t)(1 - F(t)) - HF(t)B(t) - \varepsilon_F E(t)F(t) \\ B'(t) &= (1 - bP(t))HF(t)B(t) - cB(t) \\ I'(t) &= \varepsilon_I E(t)HF(t)B(t)\,[1 - bP(t) - I(t)] \\ P'(t) &= \varepsilon_I E(t)HF(t)B(t)\,[1 - bP(t) - sP(t) - P(t)] \\ X'(t) &= \kappa X(t)(1 - X(t))\,[hP(t) - \beta X(t)] \end{aligned}\right\} \tag{1}$$

where the net emissions loading $E(t)$ is given by:

$$E(t) = n(t)\,(1 - f(X(t))) \tag{2}$$

Note that the form of the abatement function $f(X)$ is important. It should be monotone increasing in X, and have a social concern threshold $\breve{x}$ above which there is complete abatement: i.e. whenever $X \geq \breve{x}$, then $f(X) = 1$ and $E(t) = 0$. Table 1 gives a summary of the model variables and parameters, including the values selected in [20] to simulate the historic pollution epidemic in Minamata. In these simulations, f is chosen to be a nonlinear function with $\breve{x} = 0.4$,

With the appropriate choice of initial conditions, we can constrain the solutions of (1) to an invariant region Ω in the non-negative orthant of the *FBIPX* state space,

$$\Omega := \left\{ \begin{aligned} (F, B, I, P, X) : F, X \in [0, 1], \quad & B \geq 0, \\ I \in \left[0, \tfrac{1+s}{1+b+s}\right], \; & P \in \left[0, \tfrac{1}{1+b+s}\right] \end{aligned} \right\} \tag{3}$$

This result is proved in the supplementary appendix of [20] using standard invariance theorems for systems of ordinary differential equations.

Pollution Epidemic The system (1) has nine equilibria, with a single nontrivial equilibrium:

Table 1 Model variables and parameters

Variable	Meaning	Range	I.C.[a]
E	Net emissions loading	[0, 1]	0.0
F	Fish	[0, 1]	1.0
B	Boats	≥ 0	0.1
I	Cumulative injured	[0, 1]	0.0
P	Cumulative perceived injured	[0, 1]	0.0
X	Demand for pollution abatement	[0, 1]	0.01
$f(X)$	Abatement level	[0, 1]	0.0
Parameter	Meaning	Range	Baseline[b]
H	Harvesting rate	[0, 1]	0.7
r	Fecundity	(0, 1]	1.0
c	Normalized boat costs	(0, 1]	0.35
ε_F	Fish pollution mortality	[0, 1]	0.7
ε_I	Pollution availability to humans	[0, 1]	0.009
b	Rate of fish boycott per unit injury	≥ 0	100
s	Rate of stigmatization per unit injury	≥ 0	600
κ	Sampling and imitation rate	[0, 1]	0.5
h	Rate of health concern per unit injury	≥ 0	1000
β	Pushback to demand for abatement	≥ 0	4.0
$\breve{x}$	Social concern threshold	[0, 1]	0.4

[a]Initial Condition values selected for simulation in [20]
[b]Baseline parameter values selected for simulation in [20]

$$\begin{pmatrix} F^* \\ B^* \\ I^* \\ P^* \\ X^* \end{pmatrix} = \begin{pmatrix} \frac{c}{H(1-bP^*)} \\ \frac{1}{H}\left(r\,(1-F^*) - \varepsilon_F\,(1-f(X^*))\right) \\ \frac{1+s}{1+b+s} \\ \frac{1}{1+b+s} \\ \frac{h}{\beta(1+b+s)} \end{pmatrix} \tag{4}$$

This equilibrium corresponds to a pollution epidemic steady-state, as it occurs when the emissions and fish catch coexist.

Existence and Stability Conditions This equilibrium cannot occur in Ω unless the following three inequality conditions hold: the social concern is below the threshold needed to abate the emissions,

$$0 \leq \frac{h}{\beta(1+b+s)} < \breve{x} \tag{5}$$

the fish reproduction rate exceeds the pollution mortality,

$$r > \varepsilon_F(1-f(X^*)) \tag{6}$$

and the returns from the harvest exceed the cost of fishing,

$$c < H(1 - bP^*)\left(1 - \frac{\varepsilon_F(1 - f(X^*))}{r}\right). \tag{7}$$

Using these same inequalities, it is shown in [20] that when the nontrivial equilibrium exists in Ω it is locally attractive there.

Convergence of Solutions to the Pollution Epidemic Equilibrium Even if a pollution epidemic equilibrium exists and is locally attractive in Ω, it will not necessarily be reached. The existence and local stability conditions alone are not enough to guarantee convergence to steady-state, since the system (1) is highly coupled and nonlinear. The convergence behaviour of this system is sensitive to the relative growth rates of each variable, particularly those that affect the viability of the fishery.

Solutions converge to the pollution epidemic equilibrium if the injuries $P(t)$ and social concern $X(t)$ grow fast enough that the fish catch $F(t)B(t)$ attains its steady-state. Whereas when they grow too slowly, the fish catch $F(t)B(t)$ will collapse. These observations are illustrated in the following schematic:

$$\begin{array}{rlcrl}
P(t) & \xrightarrow[FAST]{} P^* & \qquad & P(t) & \xrightarrow[SLOW]{} P^* \\
X(t) & \xrightarrow[FAST]{} X^* & & X(t) & \xrightarrow[SLOW]{} X^* \\
F(t)B(t) & \xrightarrow[converges]{} F^*B^* & & F(t)B(t) & \xrightarrow[collapses]{} 0
\end{array}$$

Due to the nonlinearity of the system, we are not able to characterize the convergence behaviour more precisely in terms of the model parameters, as we were in (5), (6) and (7). However, there are insights to be gained from writing each variable in terms of its associated integral equation, evaluated up to some time T:

$$P(T) = P(t_0) + \varepsilon_I \int_{t_0}^{T} H\,F(t)\,B(t)\,(1 - f(X(t)))\;[1 - (1 + b + s)\,P(t)]\;dt$$

$$X(T) = X(t_0) + \kappa \int_{t_0}^{T} X(t)\,(1 - X(t))\;[h\,P(t) - \beta\,X(t)]\;dt$$

$$F(T) = F(t_0) + r \int_{t_0}^{T} F(t)\,(1 - F(t)) - H\,F(t)\,B(t)\,dt - \varepsilon_F \int_{t_0}^{T} F(t)\,(1 - f(X(t)))\,dt$$

$$B(T) = B(t_0) + \int_{t_0}^{T} (1 - b\,P(t))H\,F(t)\,B(t) - c\,B(t)\,dt$$

Observe that these integral equations depend on the initial conditions and the parameter values ε_I and κ, whereas the inequalities (5), (6) and (7) do not. We see that the fishery is sensitive to the time evolution of $P(t)$ and $X(t)$, because $P(t)$ affects both the demand for fish $1 - bP(t)$ and the demand for abatement $X(t)$, while $X(t)$ affects the emissions. In general, during the time that $P(t)$ and $X(t)$ are small, both the number of boats and the level of emissions are high. If the fish population is driven below the break-even cost required for the boats, the fish catch collapses and an epidemic is averted.

3 Analysis

In [20] our model is applied to simulate the pollution epidemic that occurred from 1949 to 1968 in Minamata, Japan. Given that some of the datasets from Minamata are incomplete or sampled at irregular time intervals, the simulations are meant to agree qualitatively with the available data, and to identify a set of baseline parameter values that are physically plausible. These baseline values displayed in Table 1. The simulations run from 1945 to 1976.

We are interested in examining how the social feedbacks, driven by the parameters s, b and h, affect the resulting steady-state. The system (1) has nine equilibria that we class into four types: (I) No fish catch, no emissions; (II) Fish catch, no emissions; (III) No fish catch, emissions; (IV) Fish catch and emissions. Type IV is a nontrivial equilibrium, a pollution epidemic steady-state.

Figure 1 depicts the steady-states of the system solved in s-b parameter space, with all other parameters held at their baseline values. We numerically solve F, B and X for large time $t = 1{,}000{,}000$ to approximate the steady-state, and then plot it in s-b space coloured according to its equilibrium type.

We are also able to test our theoretical expectations against the numerical results. In Fig. 1, the hatched region demarcates where there is potential for a pollution

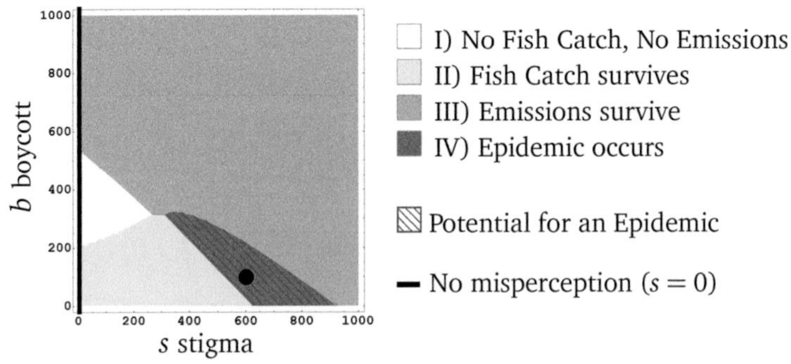

Fig. 1 Changing values for the social feedbacks s (stigma) and b (boycott) yield alternative qualitative outcomes. The parameter plane shows the dynamical outcomes defined by the equilibrium-types I, II, III, and IV that result for various values of s and b. The hatched region signifies where a potential epidemic is possible, and the *solid point* represents the baseline values used in the simulation of Minamata; $s = 600$, $b = 100$

epidemic according to the inequalities (5), (6) and (7). We find that this region overlaps exactly with the numerically solved steady-state region in dark grey, which indicates that a pollution epidemic has occurred.

As the parameter plane in Fig. 1 shows us, an epidemic does not occur unless the stigma s is sufficiently large. It is very important to recognize how unlikely the conditions for a pollution epidemic (type-IV) equilibrium are without social stigma/misperception.

We want to know what parameters are required for us to have a pollution epidemic when $s = 0$, and how physically plausible they are compared to the baseline parameter values used in the simulation of Minamata. Figure 1 depicts the axis $s = 0$ in black. To find when a pollution epidemic becomes possible without stigma, we must vary our parameters so that the inequalities (5), (6) and (7) are satisfied on the axis $s = 0$. To find when this happens, we can consider the boundary case of (5) with $X^* = \breve{x}$,

$$\frac{h}{\beta(1+b)} = \breve{x} \tag{8}$$

so that (6) reduces to $r > 0$ and (7) reduces to

$$c < H\left(\frac{1}{1+b}\right) \tag{9}$$

Combining (8) and (9) yields

$$\frac{h}{\beta} < \frac{H\breve{x}}{c} \tag{10}$$

Varying each of these parameters individually while holding the others at their baseline values from Table 1, we find that the potential for a pollution epidemic exists for some b if either the health concern is decreased to $0 < h < 3.2$, the boat costs are decreased to $0 < c < 0.00112$, or the fish harvesting rate is increased to $H > 218.75$. Note that in the ratio h/β, it suffices to consider changing h only. As the numerical results in Fig. 2 show, the time evolution of the solutions is sensitive to the collapse of the fishery, and they do not necessarily converge to the pollution epidemic equilibrium.

When the health concern is decreased to $h = 3$ a pollution epidemic becomes possible for $0.25 < b < 0.55$. However, as Fig. 2a demonstrates, small values of h and b make the social response to the pollution negligible. The fish catch collapses before the injuries grow large enough to trigger any abatement, and a pollution epidemic is averted.

Decreasing the boat costs to $c = 0.001$, a pollution epidemic becomes possible for values of $b > 624$. The numerical results in Fig. 2b indicate that the pollution epidemic occurs. However, is this plausible? The accompanying plot in Fig. 2b suggests that this operating cost for the boats is unrealistically low. The boats grow

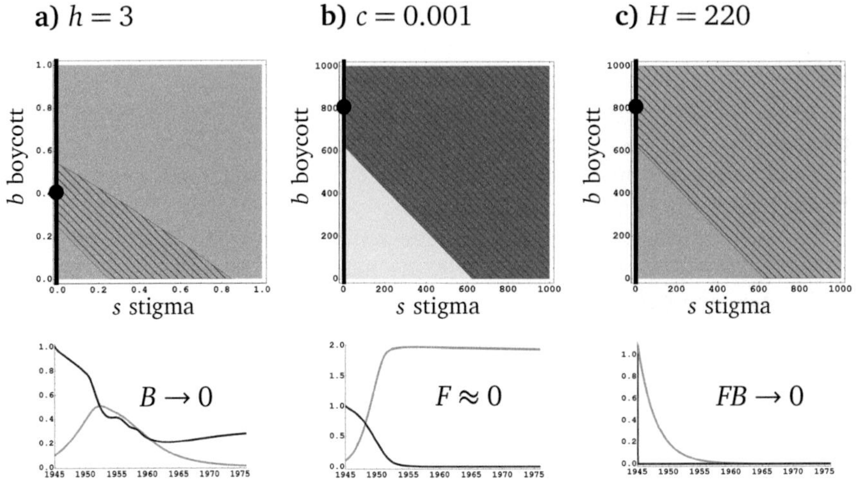

Fig. 2 We seek the conditions for which a pollution epidemic is possible with no stigma. The panels depict the s-b parameter planes that result when (**a**) $h = 3$, (**b**) $c = 0.001$ and (**c**) $H = 220$. Each is coloured according to the legend in Fig. 1. The *solid dot* in each plane represents the values of s and b used to generate each of the plots below, which show the time evolution of the fish (in *dark grey*) and boats (in *light grey*)

to an unrealistically high level and push the fish population toward zero. For the short time-scale we are interested in, the potential for a pollution epidemic is averted by the decline of the fish catch to near-zero levels. If we solve numerically over a very large time scale as $t \to \infty$, it turns out that the fish and boats exhibit damped oscillations, and the injuries grow.

Another option is to increase the harvest rate to $H = 220$ with $b > 624$. However, as the plots in Fig. 2c show, this unsustainable harvest pressure causes the fish catch to collapse before the variables are able to reach a pollution epidemic steady-state.

To examine the effects of the parameters r and ε_F on the occurrence of a pollution epidemic, we return to the inequalities (5), (6) and (7). Allowing r, ε_F and b to be free while holding $s = 0$ and all other parameters at baseline, we find that the inequalities are satisfied when $r > 0$ and $\varepsilon_F < 0$. However, negative values for the pollution mortality rate of fish violate our parameter requirements and are biologically implausible.

Instead, we explore the case $\varepsilon_F = 0$ so that the fish do not die from pollution, while freeing r and another parameter, h. Holding $s = 0$ and all other parameters at baseline, we find that the inequalities (5), (6) and (7) are satisfied if $r > 0$, $0 < h < 3.2$, and $0 < b < 1$. Figure 3a shows that an epidemic does occur for the values $r = 1$, $h = 1$ and $b = 0.9$. To determine whether this is realistic, we look to the time series plots in Figs. 3b–e. The emissions and fish catch coexist and the injuries grow steadily. Since $s = 0$, all of these injuries are fully perceived. However, the growth of the social concern X and the reduction of fish-eating $-bP$ remain negligible, because the social feedbacks h and b are very small. Assuming that the nature of the

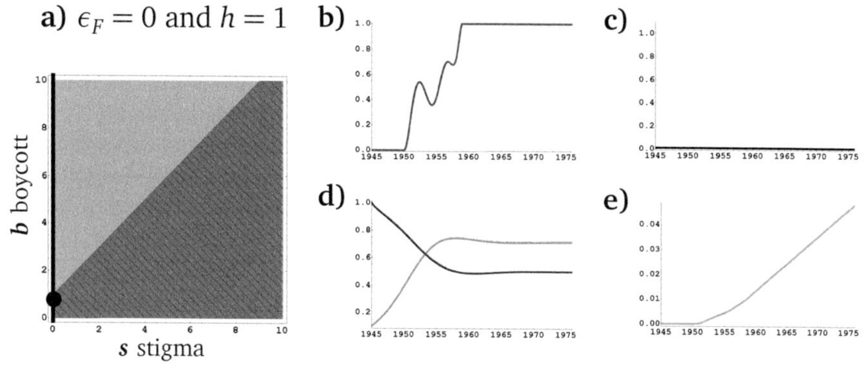

Fig. 3 We seek the conditions for which a pollution epidemic is possible with no stigma. Panel (**a**) depicts the *s*-*b* parameter plane that results when $\varepsilon_F = 0$ and $h = 1$. The plane is coloured according to the legend in Fig. 1, and the black point gives the values $s = 0$ and $b = 1$ used to generate the accompanying time series plots: (**b**) emissions, (**c**) social concern, (**d**) fish (in *dark grey*), boats (in *light grey*), (**e**) perceived and actual injuries

pollution injuries is serious, like those experienced from methylmercury-poisoning in Minamata, this situation of full perception without response is socially unrealistic.

4 Discussion and Conclusion

The human-environment system model studied in this paper allows us to understand how social feedbacks that protect people from pollution rely on accurate information. Our model is capable of reaching a pollution epidemic steady-state where the pollution exposure continues indefinitely. We conclude that the pollution epidemic equilibrium is unlikely to occur in the absence of stigma/misperception, except under conditions that are socially or biologically implausible, or which violate our parameter constraints. This highlights the importance of including misperception as a factor in the model.

Our HES model is based on interactions that are endogenous and localized; there are no external shocks, and no corrective mechanisms to move the system away from a pollution equilibrium once it settles there. The very existence of a pollution epidemic equilibrium, and the role played by misperception in augmenting its occurrence, reveals that our model can fail to be self-regulating. This reflects real-world circumstances in which citizen stakeholders have information that is inaccurate or different to that held by industry, rendering citizens' voluntary protective actions inadequate to resolve the problem. Our HES model represents the conditions that might have persisted for decades in Minamata if a second methylmercury-poisoning incident had not occurred in Niigata, Japan, to increase citizens' awareness and stimulate relief for the victims [1, 17].

In models where human activity interacts with natural processes, we believe there is more room to study the extent to which human decisions can be made from inaccurate information, and to study how this affects the qualitative outcome of the model. For example, although medical researchers recognize the role of stigma in prolonging epidemics, especially when it shames people from seeking medical attention [4, 18], it is not clear that these insights have been readily taken up in epidemiological models. Risk perception and behavioural strategies to avoid infectious disease are more widely modelled [14], and some agent-based models explicitly distinguish between injured and perceived injured [15]. Social learning models are increasingly used in mathematical epidemiology to study vaccine scares [3], and in ecology to study resource management and conservation [2, 11, 16].

Up to now, these methods have not been widely applied to understand pollution illnesses, in spite of the fact that pollution is intimately bound to social activity. We have worked to incorporate these insights into our model, and the feedbacks involving stigma and fish-eating boycott represent a break from existing approaches.

References

1. Almeida, P., Stearns, L.B.: Political opportunities and local grassroots environmental movements: the case of minamata. Soc. Probl. **45**(1), 37–60 (1998)
2. Barlow, L.-A., Cecile, J., Bauch, C.T.. Anand, M.: Modelling interactions between forest pest invasions and human decisions regarding firewood transport restrictions. PLoS ONE **9**(4), e90511 (2014)
3. Bauch, C.T., Bhattacharyya, S.: Evolutionary game theory and social learning can determine how vaccine scares unfold. PLoS Comput. Biol. **8**(4), 1–12 (2012)
4. Des Jarlais, D.C., et al.: Effects of toxicants on populations: a qualitative approach ii. First order kinetics. Am. J. Public Health **96**(3), 561–567 (2006)
5. George, T.S.: Minamata: Pollution and the Struggle for Democracy in Postwar Japan. Number 194 in Harvard East Asian Monographs. Harvard University Asia Center, Cambridge/London (2001)
6. Grandjean, P., et al.: Adverse effects of methylmercury: environmental health research implications. Environ. Health Perspect. **118**(8), 1137–1145 (2010)
7. Harada, M.: Sachie, T., George, T.S. (English Trans.) (2004). Minamata Disease. Patients Alliance, Tokyo, (1972)
8. Harada, M.: Smith, A. (English Trans.). Minamta disease: a medical report. In: Smith, E., Smith, A. (eds.) Minamata, pp. 180–192. Holt, Rinehart and Winston, New York (1975)
9. Harada, M.: Minamata disease: methylmercury poisoning in Japan caused by environmental pollution. Crit. Rev. Toxicol. **25**(1), 1–24 (1995)
10. Hofbauer, J., Sigmund, K.: 8.1. Imitation dynamics. In: Evolutionary Games and Population Dynamics, pp. 86–86. Cambridge University Press, Cambridge (1998)
11. Innes, C., et al.: The impact of human-environmental interactions on the stability of forest-grassland mosaic ecosystems. Nat. Sci. Rep. **3**(2689), 1–10 (2013)
12. Levin, S., Clark, W.C.: Toward a science of sustainability. In: Center for International Development Working Papers 196. John F. Kennedy School of Government, Harvard University (2010)
13. Liu, J., et al.: Coupled human and natural systems. AMBIO: J. Hum. Environ. – BioOne **36**(8), 639–649 (2007)
14. Perisic, A., Bauch, C.: Social contact networks and disease eradicability under voluntary vaccination. PLoS Comput. Biol. **5**(2), 1–8 (2009)

15. Poletti, P.: Risk perception and effectiveness of uncoordinated behavioral responses in an emerging epidemic. Math. Biosci. **238**, 80–89 (2012)
16. Satake, A., et al.: Synchronized deforestation induced by social learning under uncertainty of forest-use value. Ecol. Econ. **63**(2), 452–462 (2007)
17. Smith, E., Smith, A.: Minamata. Holt, Rinehart and Winston, New York (1975)
18. Van Brakel, W.H.: Measuring health-related stigma – a literature review. Psychol. Health Med. **11**(3), 307–334 (2007)
19. Van Rooij, B.: The people vs. pollution: understanding citizen action against pollution in China. J. Contemp. China **19**(63), 56–77 (2010)
20. Yodzis, M., Bauch, C.T., Anand, M.: Coupling fishery dynamics, human health and social learning in a model of fish-borne pollution exposure. Sustain. Sci. **11**, 179–192 (2016)

Part III
Computational Engineering and Mathematical Foundation, Numerical Methods and Algorithms

Stability Properties of Switched Singular Systems Subject to Impulsive Effects

Mohamad S. Alwan, Humeyra Kiyak, and Xinzhi Liu

Abstract This paper addresses impulsive switched singular systems with nonlinear perturbation term. The main theme is to establish exponential stability of the systems where the impulses are of fixed time type and treated as perturbation. We first establish the exponential stability of a single-mode impulsive systems using the Lyapunov method. We have observed that if the underlying continuous system is stable and the impulses are applied slowly, then it is guaranteed that the impulsive system maintains the stability property. Later, a switched system with impulsive effects is considered. The method of multiple Lyapunov function and average dwell time switching signal are used. We have noticed that if all subsystems are exponentially stable and the average dwell time is sufficiently large, then the impulsive switched system is exponentially stable. Numerical examples with simulations are given to illustrate the effectiveness of the proposed theoretical results.

1 Introduction

Singular systems (or differential algebraic systems) have received considerable attentions due their extensive applications in the areas of control systems, electrical power systems, mechanical systems, and robotics. Readers may refer to [4, 4–6, 9, 12]. The theory of singular system is well studied in the literature; see for instance [3, 4, 6, 7, 12, 15], and [5].

On the other hand, the dynamics of many natural or engineering systems are subject to abrupt changes, where the durations of these changes are sufficiently small that can be reasonably approximated by instantaneous changes of states or impulses. The resulting systems can be suitably modeled as an impulsive systems. Nowadays, the applications of impulsive system have been found in many different real world or man-made systems, such as mechanical and engineering systems including mass-spring systems, robotics, or electrical circuits. It can also be found in biological systems including the function of the heart and biological neural network,

M.S. Alwan (✉) • H. Kiyak • X. Liu
Department of Applied Mathematics, University of Waterloo, Waterloo, ON, Canada
e-mail: malwan@uwaterloo.ca; hkiyak@uwaterloo.ca; xinzhi.liu@uwaterloo.ca

© Springer International Publishing Switzerland 2016

J. Bélair et al. (eds.), *Mathematical and Computational Approaches in Advancing Modern Science and Engineering*, DOI 10.1007/978-3-319-30379-6_33

and epidemic disease models including pulse vaccination. The theory of impulsive system is well presented and more application can be found in [1, 2, 11, 14, 16].

When applying the impulsive effects to the singular system, the resulting system is called impulsive singular system. Many singular systems exhibit impulsive and switching behaviors, which are characterized by switches of states and abrupt changes at switching time; that is, the systems switch with impulse effects [1, 2, 11, 13].

The objective of the paper is to establish the exponential stability of a nonlinear impulsive switched singular system. The nonlinearity is linearly bounded, and both the impulsive effects and switching are applied to the system at fixed times. Using the method of Lyapunov function for impulsive singular system, some sufficient conditions are given to establish the stability property. On the other hand, multiple Lyapunov function and the average dwell-time approach are used to determine some sufficient conditions for impulsive switched singular system.

The rest of the paper is organized as follows: In Sect. 2, the problem formulation and some background are given. In Sect. 3, we state and prove the main results. Numerical examples are presented in Sect. 4 to clarify the proposed approach. Finally, a conclusion of these results is given in Sect. 5.

2 Problem Formulation

Denote by $\mathbb{R}^n$ the n-dimensional Euclidean space with the norm $\|\cdot\|$, $\mathbb{R}^{nxn}$ the set of all nxn square matrices and $\mathbb{N}$ the set of natural numbers. If $A \in \mathbb{R}^{nxn}$, denote by λ_{min} and λ_{max} the minimum and maximum eigenvalues of A, respectively, and its norm is $\|A\| = \sqrt{\lambda_{max}(A^T A)}$, where the superscript T represents the matrix transpose. If A is a positive definite matrix, then we write $A > 0$.

Consider the impulsive switched singular systems of the form

$$\begin{aligned} E\dot{x}(t) &= A_{\sigma(t)}x(t) + g_{\sigma(t)}(t,x), && t \neq t_k, \\ \Delta x(t) &= B_k x(t), && t = t_k, \\ x(t_0) &= x_0, \end{aligned} \tag{1}$$

where $x \in \mathbb{R}^n$ is the system state variable, and $A_{\sigma(t)}, B_k, E \in \mathbb{R}^{nxn}$ are system coefficient matrices with E being singular with $\text{rank}(E) = r < n$, the matrix pairs $(E, A_{\sigma(t)})$ being regular, and B_k being constant matrices. The switching signal $\sigma : [t_0, \infty) \to \mathscr{S}$ is a piecewise constant function taking values in a finite compact set $\mathscr{S} = \{1, 2, \ldots, N\}$ for some $N \in \mathbb{N}$. $\{t_k\}_{k=1}^{\infty}$ are the impulsive times that form an increasing sequence satisfying $t_{k-1} < t_k$ and $\lim_{k\to\infty} t_k = \infty$. $\Delta x = x(t^+) - x(t^-)$ where $x(t^-)$ (and $x(t^+)$) is the state just before (and just after) the impulsive action with $x(t^+) = \lim_{s\to t^+} x(s)$. The solution x is assumed to be left-continuous, i.e., $x(t_k^-) = x(t_k)$. For $i \in \mathscr{S}$, $g_i(t,x) : \mathbb{R}^+ \times \mathbb{R}^n \to \mathbb{R}^n$ is piecewise continuous vector-valued functions with $g_i(t,0) \equiv 0,\ t \in \mathbb{R}^+$ to ensure the existence of solutions for system (1).

When $\sigma(t) = \sigma$ for all $t > t_0$, system (1) is reduced to the following single-mode impulsive singular system:

$$
\begin{aligned}
E\dot{x}(t) &= Ax(t) + g(t,x), && t \neq t_k \\
\Delta x(t) &= B_k x(t), && t = t_k, \quad k \in \mathbb{N} \\
x(t_0) &= x_0,
\end{aligned}
\tag{2}
$$

where the subscript σ is dropped for simplicity of notation.

In the following, some definitions and theorems are introduced.

Definition 1 ([7, 8]) Matrix pair (E,A) is *regular* if there exists a constant scalar $\gamma \in \mathbb{C}$ such that $det(\gamma E - A) \neq 0$. The matrix pair (E,A) is said to be *impulse free* if $\deg(det(\gamma E - A)) = rank(E)$.

Definition 2 ([7, 17]) Regular system (1) is said to be exponentially stable if there exist $\alpha,\ \beta > 0$ such that its state $x(t)$ satisfies $\|x(t)\| \leq \alpha\|x(t_0)\|e^{-\beta t}$, for all $t > t_0$.

Definition 3 ([17]) System (1) is said to be E-exponentially stable if there exist constants $\alpha,\ \beta > 0$ such that $\|Ex(t)\| \leq \alpha\|Ex(t_0)\|e^{-\beta t}$, for all $t > t_0$.

Definition 4 ([7]) System (1) is admissible if it is stable and impulse-free.

Theorem 1 ([8]) *Matrix pair (E,A) is regular if and only if there exist two nonsingular matrices Q, P such that $Q = \begin{bmatrix} Q^1 \\ Q^2 \end{bmatrix}$ and $P = \begin{bmatrix} P^1 & P^2 \end{bmatrix}$ where $Q^1 \in \mathbb{R}^{r\times n}$, $Q^2 \in \mathbb{R}^{(n-r)\times n}$, $P^1 \in \mathbb{R}^{n\times r}$, $P^2 \in \mathbb{R}^{n\times(n-r)}$ and the following standard decomposition holds:*

$$
QEP = diag\,(I_r, N)\,, \quad QAP = diag\,(A_1, I_{n-r})\,, \quad Qg(t,x) = \begin{bmatrix} g_1(t,x_1,x_2) \\ g_2(t,x_1,x_2) \end{bmatrix}, \tag{3}
$$

where $r = rank(E)$, $A_1 \in \mathbb{R}^{r\times r}$, $N \in \mathbb{R}^{(n-r)\times(n-r)}$ is a nilpotent matrix with nilpotent index h.

Theorem 2 ([7]) *The singular system $E\dot{x}(t) = Ax(t)$ is stable if and only if (E,A) has finite eigenvalues with negative real parts.*

Theorem 3 ([7]) *If system (1) is admissible, then for any $Y > 0$ there exists $X > 0$ satisfying $E^TXA + A^TXE = -E^TYE$.*

3 Main Results

In this section, exponential stability of system (1) and (2) are discussed.

3.1 Singular Systems Subject to Impulsive Effects

Theorem 4 *Assume that system (2) is impulse free, the eigenvalues of the matrix pair (E, A) have negative real parts, the singular matrix E and the matrix $I + B_k$ for $I \in \mathbb{R}^{nxn}$ identity matrix are commutative, and $\|g(t,x)\| \leq \gamma^* \|Ex\|$ for $\gamma^* < \frac{\lambda_{\min}(Y)}{2\|X\|}$ positive constant where $X > 0$ satisfying*

$$A^T XE + E^T XA = -E^T YE \tag{4}$$

for any $Y > 0$. Then, the trivial solution of the nonlinear singular impulsive system (2) is exponentially stable if the following inequality holds:

$$\ln \alpha_k - \nu(t_k - t_{k-1}) \leq 0, \quad k = 1, 2, \ldots \tag{5}$$

where $\alpha_k = \frac{\lambda_{\max}[(I+B_k)^T X(I+B_k)]}{\lambda_{\min}(X)}$, $0 < \nu < \xi$, and $\xi = \frac{\lambda_{\min}(Y) - 2\|X\|\gamma^*}{\lambda_{\max}(X)}$.

Proof Let $x(t) = x(t, t_0, x_0)$ be the solution of the system (2). Define

$$\upsilon(t) = V(x(t)) = x^T(t)E^T XEx(t), \qquad t \neq t_k$$

as a Lyapunov function candidate. Then, along the trajectory of (2),

$$\begin{aligned} \dot{\upsilon}(t) &= x^T(t)\left(A^T XE + E^T XA\right)x(t) + g^T(t,x)XEx(t) + x^T(t)E^T Xg(t,x) \\ &= -x^T(t)E^T YEx(t) + 2x^T(t)E^T Xg(t,x) \end{aligned} \tag{6}$$

where $-E^T YE = A^T XE + E^T XA$. Since $X > 0$ and $Y > 0$, we have [10]

$$\lambda_{min}(X)\|Ex\|^2 \leq x^T E^T XEx \leq \lambda_{max}(X)\|Ex\|^2 \tag{7}$$

$$\lambda_{min}(Y)\|Ex\|^2 \leq x^T E^T YEx \leq \lambda_{max}(Y)\|Ex\|^2. \tag{8}$$

Thus, using (7), (8), and $\|g(t,x)\| \leq \gamma^* \|Ex(t)\|$ in (6) lead to

$$\dot{\upsilon}(t) \leq -\left(\lambda_{\min}(Y) - 2\|X\|\gamma^*\right)\|Ex(t)\|^2 \leq -\xi\upsilon(t), \quad t \in (t_{k-1}, t_k],$$

where $\xi = \frac{\lambda_{\min}(Y) - 2\|X\|\gamma^*}{\lambda_{\max}(X)}$ is a positive constant if $\gamma^* < \frac{\lambda_{\min}(Y)}{2\|X\|}$. Then,

$$\upsilon(t) \leq \upsilon(t_{k-1}^+)e^{-\xi(t-t_{k-1})}, \quad t \in (t_{k-1}, t_k]. \tag{9}$$

At $t = t_k^+$, we have

$$\upsilon(t_k^+) = x^T(t_k)(I + B_k)^T E^T XE(I + B_k)x(t_k) \leq \alpha_k \upsilon(t_k), \tag{10}$$

where $\alpha_k = \frac{\lambda_{\max}[(I+B_k)^T X(I+B_k)]}{\lambda_{\min}(X)}$. Using (9) and (10), one may get

$$\begin{aligned}\upsilon(t) &\leq \alpha_1\alpha_2 \ldots \alpha_k \upsilon(t_0^+)e^{-\xi(t-t_0)} \\ &= \upsilon(t_0^+)\alpha_1 e^{-\nu(t_1-t_0)}\alpha_2 e^{-\nu(t_2-t_1)} \ldots \alpha_k e^{-\nu(t_k-t_{k-1})}e^{-\nu(t-t_k)}e^{-(\xi-\nu)(t-t_0)}.\end{aligned}$$

By assumption (5), we have

$$\upsilon(t) \leq \upsilon(t_0^+)e^{-(\xi-\nu)(t-t_0)}, \quad t \geq t_0,$$

which implies that

$$\begin{aligned}&\lambda_{\min}(X)\|Ex(t)\|^2 \leq \upsilon(t) \leq \upsilon(t_0^+)e^{-(\xi-\nu)(t-t_0)} \leq \lambda_{\max}(X)\|Ex(t_0^+)\|^2 e^{-(\xi-\nu)(t-t_0)} \\ &\Rightarrow \|Ex(t)\| \leq K\|Ex(t_0^+)\|e^{-(\xi-\nu)(t-t_0)/2}, \quad t \geq t_0, \end{aligned} \tag{11}$$

where $K = \sqrt{\lambda_{\max}(X)/\lambda_{\min}(X)}$. Thus, the trivial solution of the system (2) is E-exponentially stable.

Let

$$P^{-1}x(t) = \begin{bmatrix} x_1(t) \\ x_2(t) \end{bmatrix}. \tag{12}$$

Then, it follows from the standard decomposition form that system (2) is equivalent to

$$\dot{x}_1 = A_1 x_1 + Q^1 g(t,x) \tag{13}$$

$$0 = x_2 + Q^2 g(t,x) \tag{14}$$

where $x_1 \in \mathbb{R}^r$ and $x_2 \in \mathbb{R}^{n-r}$. By (12), we have

$$QEx(t) = QEP\begin{bmatrix} x_1(t) \\ x_2(t) \end{bmatrix} = \begin{bmatrix} I_r & 0 \\ 0 & 0 \end{bmatrix}\begin{bmatrix} x_1(t) \\ x_2(t) \end{bmatrix} = \begin{bmatrix} x_1(t) \\ 0 \end{bmatrix}. \tag{15}$$

From (11) and (15), we have

$$\|x_1\| = \|QEx(t)\| \leq K\|Q\|\|Ex(t_0^+)\|e^{-(\xi-\nu)(t-t_0)/2} \tag{16}$$

which implies that x_1 is exponentially stable.

We need to show that x_2 is also exponentially stable. It follows from (14) and $\|g(t,x)\| \leq \gamma^* \|Ex(t)\|$ that

$$\|x_2\| = \|Q^2 g(t,x)\| \leq \gamma^* K \|Q^2\| \|Ex(t_0^+)\| e^{-(\xi-\nu)(t-t_0)/2}. \tag{17}$$

This means x_2 is exponentially stable. From (16) and (17), the trivial solution of (2) is exponentially stable. This completes the proof.

Remark 1 The validity of the matrix equation (4) guarantees that the Lyapunov function is decreasing for all $t \neq t_k$. While the inequality in (5) is made to ensure that the impulses are applied slowly in order to maintain the stability of the impulsive system in (2).

3.2 Switched Singular Systems Subject to Impulsive Effects

Theorem 5 *For any $i \in \mathscr{S}$, assume that system (1) is impulse free, the eigenvalues of the matrix pairs (E, A_i) have negative real parts, and $\|g_i(t,x)\| \leq \gamma_i \|Ex\|$ for $\gamma_i < \frac{\lambda_{\min}(Y_i)}{2\|X_i\|}$ positive constant where $X_i > 0$ satisfying $A_i^T X_i E + E^T X_i A_i = -E^T Y_i E$ for any $Y_i > 0$. Then, the trivial solution of (1) is exponentially stable if the following assumptions hold:*

(i) For any $i, j \in \mathscr{S}$ there exists $\alpha_k > 1$ such that

$$(I + B_k)^T E^T X_j E (I + B_k) \leq \alpha_k E^T X_i E. \tag{18}$$

(ii) For any t_0, the switching law satisfies

$$N(t, t_0) \leq N_0 + \frac{t - t_0}{T_a}$$

where $N(t, t_0)$ represents the number of switchings in (t, t_0), and N_0 and T_a are the chatter bound and average dwell time to be defined, respectively.

Proof Let $x(t) = x(t, t_0, x_0)$ be the solution of the system (1). Define

$$\upsilon_i(t) = V_i(x(t)) = x^T(t) E^T X_i E x(t), \qquad t \neq t_k,$$

as a Lyapunov function candidate for ith subsystem. Then, derivative of υ_i along the trajectory of (1) is given by

$$\begin{aligned} \dot{\upsilon}_i(t) &= -x^T(t) E^T Y_i E x(t) + 2x^T(t) E^T X_i g_i(t,x) \\ &\leq -\lambda_{\min}(Y_i) \|Ex(t)\|^2 + 2\|x^T(t) E^T\| \|X_i\| \|g_i(t,x)\| \\ &\leq -(\lambda_{\min}(Y_i) - 2\|X_i\| \gamma_i) \|Ex(t)\|^2 \leq -\xi_i \upsilon_i(t), \quad t \in (t_{k-1}, t_k], \end{aligned} \tag{19}$$

where $\xi_i = \frac{\lambda_{\min}(Y_i) - 2\|X_i\|\gamma_i}{\lambda_{\max}(X_i)} > 0$ if $\gamma_i < \frac{\lambda_{\min}(Y_i)}{2\|X_i\|}$. Then,

$$\upsilon_i(t) \leq \upsilon_i(t_{k-1}^+)e^{-\xi_i(t-t_{k-1})}, \quad t \in (t_{k-1}, t_k]. \tag{20}$$

At $t = t_k^+, k \in \mathbb{N}$, suppose $\sigma(t_k^+) = j$, it follows from (1) and (18) that

$$\upsilon_j(t_k^+) = x^T(t_k)(I + B_k)^T E^T X_j E(I + B_k)x(t_k) \leq \alpha_k \upsilon_i(t_k). \tag{21}$$

Using (20) and (21) leads to

$$\upsilon_i(t) \leq \alpha_1\alpha_2 \ldots \alpha_{i-1}\upsilon_1(t_0^+)e^{-\xi_1(t_1-t_0)}e^{-\xi_2(t_2-t_1)} \ldots e^{-\xi_i(t-t_k)}, \quad t \geq t_0.$$

Let $\xi = \min\{\xi_i;\ i \in \mathscr{S}\}$. Then, we have

$$\upsilon_i(t) \leq \alpha_1\alpha_2 \ldots \alpha_{i-1}\upsilon_1(t_0^+)e^{-\xi(t-t_0)}. \tag{22}$$

To use assumption (ii), let $\alpha = \max\{\alpha_k; \quad k = 1, 2, \ldots, i-1\}$. Then, (22) becomes

$$\upsilon_i(t) \leq \alpha^{i-1}\upsilon_1(t_0^+)e^{-\xi(t-t_0)} = \upsilon_1(t_0^+)e^{(i-1)\ln\alpha - \xi(t-t_0)}.$$

Applying (ii) with $N_0 = \frac{\eta}{\ln\alpha}$, where η is an arbitrary constant, and $T_a = \frac{\ln\alpha}{\xi - \xi^*}$, $(\xi > \xi^*)$ leads to

$$\upsilon_i(t) \leq \upsilon_1(t_0^+)e^{\eta - \xi^*(t-t_0)},$$

which implies that

$$\|Ex(t)\| \leq K\|Ex(t_0^+)\|e^{(\eta - \xi^*(t-t_0))/2}, \quad t \geq t_0, \tag{23}$$

where $K = \sqrt{\lambda_{\max}(X_1)/\lambda_{\min}(X_i)}$. Thus, the trivial solution of the system (1) is E-exponentially stable.

As done in Theorem 4, we can show that

$$\|x_1\| \leq K\|Q_i\|\|Ex(t_0^+)\|e^{(\eta-\xi^*(t-t_0))/2} \text{ and } \|x_2\| \leq \gamma_i K\|Q_i^2\|\|Ex(t_0^+)\|e^{(\eta-\xi^*(t-t_0))/2}.$$

That is, the trivial solution of (1) is exponentially stable.

4 Numerical Example

In this section, we present some examples to illustrate the effectiveness of the theoretical results.

Example 1 Consider the singular impulsive system in (2) where

$$E = \begin{bmatrix} 1 & 0 & 0 & 0 \\ 0 & 0 & 1 & 0 \\ 0 & 0 & 0 & 0 \\ 0 & 0 & 0 & 0 \end{bmatrix}, A = \begin{bmatrix} 0 & 1 & 0 & 0 \\ 1 & 0 & 0 & 0 \\ -1 & 0 & 0 & 1 \\ 0 & 1 & 1 & 1 \end{bmatrix}, g(t,x) = \begin{bmatrix} 0 \\ 0 \\ 0 \\ -x_1(t)\sin(t) \end{bmatrix},$$

$$B_k = \begin{bmatrix} 0.01 & 0 & 0 & 0 \\ 0 & 0.01 & 0 & 0 \\ 0 & 0 & 0.01 & 0 \\ 0 & 0 & 0 & 0.01 \end{bmatrix}$$

and the admissible initial value is $x(0) = [2 \;\; -1 \;\; 0 \;\; 0]^T$.

Since $\deg(\det(sE - A)) = rank(E) = 2$, the system is impulse free. Also, the system is stable because the eigenvalues of (E, A) are $\frac{-1 \pm i\sqrt{3}}{2}$. Given

$$Y = \begin{bmatrix} 2 & 0 & 0 & 0 \\ 0 & 2 & 0 & 0 \\ 0 & 0 & 1 & 0 \\ 0 & 0 & 0 & 1 \end{bmatrix},$$

the equation $E^T XA + A^T XE = -E^T YE$ is satisfied where

$$X = \begin{bmatrix} 2 & 1 & 2 & -2 \\ 1 & 3 & 1 & -1 \\ 2 & 1 & 3 & 0 \\ -2 & -1 & 0 & 8 \end{bmatrix}.$$

We also get $\xi = 0.1112$, $\gamma^* = 0.000001$ and $\alpha_k = 84.5806$, and from (5) $t_k - t_{k-1} \geq 39.91$. The simulation result is shown in Fig. 1, where $t_k - t_{k-1} = 40$.

Example 2 Consider the impulsive switched singular system given by (1) where $x = \begin{bmatrix} x_1(t) \\ x_2(t) \end{bmatrix}$, $\sigma(t) \in \mathscr{S} = \{1, 2\}$, $E = \begin{bmatrix} 4 & 0 \\ 2 & 0 \end{bmatrix}$ with $rank(E) = 1$, $B_k = 0.1I$ and

$$A_1 = \begin{bmatrix} -2 & 1 \\ 1 & -2 \end{bmatrix}, \quad g_1(t,x) = \left[\tfrac{1}{15}\tanh(x_1(t)) \quad \tfrac{1}{15}\tanh(x_2(t))\right]^T,$$

$$A_2 = \begin{bmatrix} -4 & 1 \\ 1 & -4 \end{bmatrix}, \quad g_2(t,x) = \left[\tfrac{1}{2}\tanh(x_1(t)) \quad \tfrac{1}{2}\tanh(x_2(t))\right]^T.$$

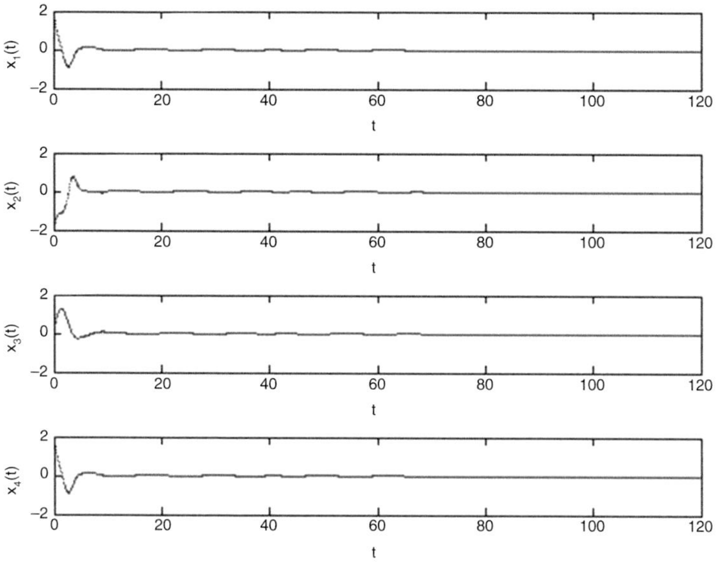

Fig. 1 Exponential stability result of Example 1

From the Jordan canonical form of $(\gamma E - A_i)^{-1}E$ for $\gamma \in \mathbb{C}$, we find that

$$Q_1 = \begin{bmatrix} \frac{1}{5} & \frac{1}{10} \\ -\frac{2}{5} & \frac{4}{5} \end{bmatrix}, \quad P_1 = \begin{bmatrix} 1 & 0 \\ \frac{4}{5} & -\frac{1}{2} \end{bmatrix}, \quad Q_2 = \begin{bmatrix} \frac{4}{9} & \frac{1}{9} \\ -\frac{1}{12} & \frac{1}{6} \end{bmatrix}, \quad P_2 = \begin{bmatrix} \frac{1}{2} & 0 \\ \frac{1}{3} & -\frac{4}{3} \end{bmatrix},$$

such that $Q_1EP_1 = Q_2EP_2 = \begin{bmatrix} 1 & 0 \\ 0 & 0 \end{bmatrix}$, $Q_1A_1P_1 = \begin{bmatrix} -\frac{3}{10} & 0 \\ 0 & 1 \end{bmatrix}$, $Q_2A_2P_2 = \begin{bmatrix} -\frac{5}{6} & 0 \\ 0 & 1 \end{bmatrix}$. From the decomposition form it is clear that the systems are impulse free. Moreover, the eigenvalues of (E, A_1) and (E, A_2) are negative.

$X_1 = \begin{bmatrix} \frac{5}{3} & 0 \\ 0 & \frac{5}{3} \end{bmatrix} > 0$ satisfies $A_1^T X_1 E + E^T X_1 A_1 = -E^T Y_1 E$ for any $Y_1 = I > 0$. Using X_1 and Y_1, we can find that $\gamma_1 < 0.3$ where γ_1 satisfies $\|g_1(t,x)\| \leq \gamma_1 \|Ex\|$. Similarly, $X_2 = \begin{bmatrix} \frac{2}{3} & 0 \\ 0 & \frac{1}{3} \end{bmatrix} > 0$ satisfying $A_2^T X_2 E + E^T X_2 A_2 = -E^T Y_2 E$ for any $Y_2 = I > 0$. Thus, by X_2 and Y_2, $\gamma_2 < 0.75$ is found where γ_2 satisfies $\|g_2(t,x)\| \leq \gamma_2 \|Ex\|$. We also get $\xi_1 = 0.62$ and $\xi_2 = 1.5020$. Thus, $\xi = 0.62$. The inequality $(I + B_k)^T E^T X_j E (I + B_k) \leq \alpha_k E^T X_i E$ holds where $\alpha_k = 1.21$. If we choose $\eta = 0$ and $\xi^* = 0.02$, then the chatter bound $N_0 = 0$, and the average dwell time $T_a = 0.3177$. The simulation result is shown in Fig. 2, where $t_k - t_{k-1} = 1$.

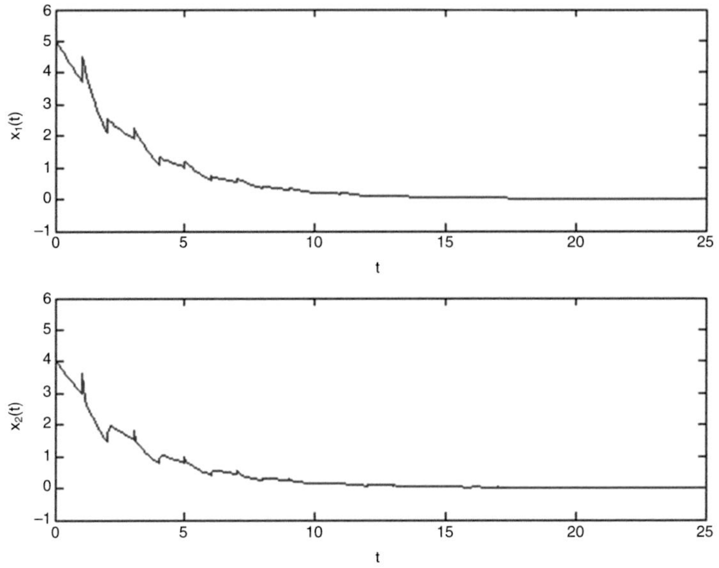

Fig. 2 Exponential stability result of Example 2

5 Conclusion

A nonlinear impulsive singular system has been studied. In Sect. 3.1, the exponential stability has been established for the impulsive system when we have used the Lyapunov method. It has been noticed that if the continuous system is stable and the time between every successive impulses is sufficiently large, then the impulsive system is also exponentially stable. In Sect. 3.2, we have addressed the impulsive switched system and developed new sufficient condition to guarantee the exponential stability using multiple Lyapunov function method and average dwell time switching signal to organize the jumps among the subsystems. It has been noticed that when all subsystems are stable, if the average dwell time is sufficiently large, then the entire system is also exponentially stable.

References

1. Bainov, D.D.: Systems with Impulse Effect: Stability, Theory, and Applications. Ellis Horwood, Chichester/Halsted Press, New York, Toronto (1989)
2. Bainov, D.D., Simeonov, P.S.: Impulsive Differential Equations: Periodic Solutions and Applications. Longman Scientific & Technical, Harlow/Wiley, New York/Burnt Mill, Harlow (1993)
3. Brenan, K.E., Campbell, S.L., Petzold, L.R.: Numerical Solution of Initial-Value Problems in Differential-Algebraic Equations. North-Holland, New York (1989)

4. Campbell, S.L.: Singular Systems of Differential Equations. Pitman Advanced Pub. Program, San Francisco (1980)
5. Campbell, S.L.: Singular Systems of Differential Equations II. Pitman, San Francisco (1982)
6. Dai, L.: Singular Control Systems. Springer-Verlag, Berlin (1989)
7. Duan, G.: Analysis and Design of Descriptor Linear Systems. Springer, New York (2010).
8. Feng, G., Cao, J.: Stability analysis of impulsive switched singular systems. IET Control Theory Appl. **9**(6), 863–870 (2015)
9. Hill, D.J., Mareels, I.M.Y.: Stability theory for differential/algebraic systems with application to power systems. IEEE Trans. Circuits Syst. **37**(11), 1416–1423 (1990)
10. Khalil, H.K.: Nonlinear Systems. Prentice Hall, Upper Saddle River (2002)
11. Lakshmikantham, V.: Theory of Impulsive Differential Equations. World Scientific, Singapore (1989)
12. Lewis, F.: A survey of linear singular systems. Circuits Syst. Signal Process **5**(1), 3–36 (1986)
13. Li, Z.: Switched and Impulsive Systems: Analysis, Design, and Applications. Springer, Berlin/New York (2005)
14. Sun, J.T., Zhang, Y.P.: Stability analysis of impulsive control systems. IEE Proc. Control Theory Appl. **150**(4), 331–334 (2003)
15. Wang, C.J.: Controllability and observability of linear time varying singular systems. IEEE Trans. Autom. Control **44**(10), 1901–1905 (1999)
16. Yang, T.: Impulsive control. IEEE Trans. Autom. Control **44**(5), 1081–1083 (1999)
17. Yao, J., Guan, Z.H., Chen, G., Ho, D.W.C.: Stability, robust stabilization and control of singular-impulsive systems via switching control. Syst. Control Lett. **55**(11), 879–886 (2006)

Input-to-State Stability and H_∞ Performance for Stochastic Control Systems with Piecewise Constant Arguments

Mohamad S. Alwan and Xinzhi Liu

Abstract This paper addresses stochastic control system of differential equations with piecewise constant arguments (SEPCA). The piecewise constant arguments are of delay type. The system is viewed as a hybrid (or particularly switched) system. This approach motivates the applicability of the classical theory of ordinary differential equations, but not of functional differential equations, and the design of a switching law. The main theme of this work is to establish the problems of input-to-state stabilization (ISS), and H_∞ performance for a class of an uncertain control SEPCA. To analyze these result, a common Lyapunov function together with the techniques of differential inequalities and Razumikhin condition is used. A numerical example with simulations is presented to clarify the validity of the proposed theoretical approaches.

1 Introduction

A nonlinear deterministic differential equations with piecewise constant arguments (EPCA) may have the form

$$\dot{x}(t) = f(t, x(t), x(\gamma(t))), \tag{1}$$

where the argument γ is a piecewise constant function defined on intervals with a certain length, and it may be defined by $\gamma(t) = [t], [t-n], t-n[t]$, or $[t+1]$, for all t and a positive integer n, where $[\cdot]$ is the greatest-integer function [8, 22].

EPCA appeared in some sequential-continuous disease models [6], control systems [7], population growth models [1, 4, 10], chaotic synchronization–including delayed secure communication [3], and in approximating the solutions of some delay differential equations [9, 11, 22]. The dynamics of these equations contain both continuous and discrete components, which enable this type of systems of equations to be adequately embedded under the hybrid system umbrella [4]. EPCA

M.S. Alwan (✉) • X. Liu
Department of Applied Mathematics, University of Waterloo, Waterloo, ON, N2L 3G1, Canada
e-mail: malwan@uwaterloo.ca; xzliu@uwaterloo.ca

© Springer International Publishing Switzerland 2016

J. Bélair et al. (eds.), *Mathematical and Computational Approaches in Advancing Modern Science and Engineering*, DOI 10.1007/978-3-319-30379-6_34

can also be a special type of functional differential equations, where the state history is given at certain individual points, rather than intervals.

The theory of EPCA is well discussed in [8, 22]. Further properties and applications of these equations were considered in many other works; readers may refer to [2, 5, 7, 14, 15, 21, 23].

The notion of ISS deals with the response of asymptotically stable systems to bounded, small input or disturbance regardless of the magnitude of system initial state. During the last two decays, ISS has become a central foundation of modern nonlinear feedback and design. It has been now playing a key role in systems with recursive design and co-prime factorizations, and a connection between the input-output (or external) stability and state (or internal) stability. For more properties, applications, and implications of the ISS, readers may consult [13] and [16–20].

In this paper, stochastic EPCA (or SEPCA) is studied. As stated earlier, the system will be viewed as a switched system. The main objective is to establish the problem of input-to-state stabilization and H_∞ performance. To analyze these results, the Lyapunov function method and Razumikhin condition are used. We also present exponential stability as a special case of the aforementioned results. Finally, to enhance the findings of this paper, some numerical examples and simulations are presented. We believe that these results will have a great contribution to many research fields in sciences and engineering, particularly in the fields of modern nonlinear control design and recursive system design and in dynamical systems subject to external disturbances.

The paper is organized as follows: In Sect. 2, the system is described, and some notations and definitions that will be used in the sequel are given. The main contributions of the paper are presented in Sect. 3, and a conclusion is given in Sect. 4.

2 Problem Formulation and Background

Denote by $\mathbb{R}_+$ the set of all non-negative real numbers, $\mathbb{R}^n$ the n-dimensional Euclidean space and its norm $\|x\| = \sqrt{\sum_{i=1}^n x_i^2}$ for every $x \in \mathbb{R}^n$, $\mathbb{R}^{n\times m}$ the set of all $n \times m$ real matrices, and Ω a sample space. Let $\mathscr{C}([a,b],\mathbb{D})$ be the space of continuous functions mapping $[a,b]$, with $a < b$ for any $a, b \in \mathbb{R}_+$, into $\mathbb{D}$, for some open set $\mathbb{D} \subseteq \mathbb{R}^n$, and $\mathscr{C}^{1,2}(\mathbb{R}_+ \times \mathbb{R}^n; \mathbb{R}_+)$ be the space of all functions $V(t,x)$ defined on $\mathbb{R}_+ \times \mathbb{R}^n$ such that they are continuously differentiable once in t and twice in x. For instance, if $V(t,x) \in \mathscr{C}^{1,2}(\mathbb{R}_+ \times \mathbb{R}^n; \mathbb{R}_+)$, then we have

$$V_t = \frac{\partial V}{\partial t}, \quad V_x = \left(\frac{\partial V}{\partial x_1}, \cdots, \frac{\partial V}{\partial x_n}\right), \quad V_{xx} = \left(\frac{\partial^2 V}{\partial x_i \partial x_j}\right)_{n\times n}.$$

Let $(\Omega, \mathscr{F}, \{\mathscr{F}_t\}_{t\geq 0}, \mathbb{P})$ be a complete probability space with filtration $\{\mathscr{F}_t\}_{t\geq 0}$ satisfying the usual condition (i.e., it is right continuous and $\mathscr{F}_0$ contains all $\mathbb{P}$-

null set of $\mathscr{F}$). For all $t \in \mathbb{R}_+$ and $\omega \in \Omega$, let $W(t,\omega) \in \mathbb{R}^m$ be an m-dimensional Wiener (or Brownian motion) process. We also assume that, for a given W and filtration $\mathscr{F}_t$ for all $t \in [a,b] \subseteq \mathbb{R}_+$, the process W is $\mathscr{F}_t$-measurable and, for all $s \leq t$, the random variable $W(t,\omega) - W(s,\omega)$ is independent of the σ-algebra $\mathscr{F}_s$. Denote by $\mathscr{L}_{ad}(\Omega, L^p[a,b])$ the class of all random processes f which are $\mathscr{F}_t$-adapted and almost all their sample paths are integrable in the Riemann sense. For simplicity of notation, we will drop Ω and ω when writing random processes.

A nonlinear system with SEPCA may have the following form

$$dx(t) = f\big(t, x(t), x(\gamma(t))\big)dt + g\big(t, x(t), x(\gamma(t))\big)dW(t), \quad x(t_0) = x_0, \tag{2}$$

where $x \in \mathbb{R}^n$ is the system state, and for all $t \geq t_0$, the function $\gamma(t)$ is a piecewise constant function taking values in the set $\Xi = \{\xi_k\}_{k=0}^{\infty}$. So that one may redefine this function as follows: for a non-negative integer k, $\gamma : [t_k, t_{k+1}) \to \Xi$. This piecewise constant function represents the switching signal of the system which has the roles of switching between the values of system state argument x. Accordingly, one may define system (2) as follows: for all $t \in [t_k, t_{k+1})$,

$$dx(t) = f\big(t, x(t), x(\xi_k)\big)dt + g\big(t, x(t), x(\xi_k)\big)dW(t), \quad x(t_0) = x_0, \tag{3}$$

or, equivalently,

$$x(t) = x_0 + \int_{t_0}^{t} f\big(s, x(s), x(\xi_k)\big)ds + \int_{t_0}^{t} g\big(s, x(s), x(\xi_k)\big)dW(s), \tag{4}$$

where the first integral is a Riemann integral almost surely (a.s.), and the second one is an Itô integral satisfying

$$\mathbb{E}\Big[\int_{t_0}^{t} g\big(s, x(s), x(\xi_k)\big)dW(s)\Big] = 0, \text{ and}$$

$$\mathbb{E}\Big\|\int_{t_0}^{t} g\big(s, x(s), x(\xi_k)\big)dW(s)\Big\|^2 = \int_{t_0}^{t} \mathbb{E}\big\|g\big(s, x(s), x(\xi_k)\big)\big\|^2 ds.$$

The following definitions and assumption will be needed throughout this paper.

Definition 1 For any $\alpha,\ \beta \in \mathbb{R}$, an $\mathbb{R}^n$-valued stochastic process $x : (\alpha, \beta) \to \mathbb{R}$ is said to be a *solution* of (2) (or (3)) if the following hold:

1. $x(t)$ is continuous and $\mathscr{F}_t$-adapted for all $t \in (\alpha, \beta)$;
2. $f(t, x(t), x(\xi_k)) \in \mathscr{L}_{ad}(\Omega, L^1(\alpha, \beta))$ and $g(t, x(t), x(\xi_k)) \in \mathscr{L}_{ad}(\Omega, L^2(\alpha, \beta))$;
3. the stochastic integral equation (4) holds (a.s.).

Itô Lemma. For any $t_0 \in \mathbb{R}_+$ and $t \geq t_0$, let $x(t)$ be an $\mathbb{R}^n$-dimensional Itô stochastic process, i.e., it is $\mathscr{F}_t$-adapted and satisfying,

$$x(t) = x(t_0) + \int_{t_0}^{t} f(s, x(s))ds + \int_{t_0}^{t} \sigma(s, x(s))dW(s), \qquad \text{(a.s.)}.$$

Suppose that $V \in \mathscr{C}^{1,2}(\mathbb{R}_+ \times \mathbb{R}^n; \mathbb{R})$. Then, for any $t \geq t_0$, $V(t, x(t))$ is an Itô stochastic process satisfying

$$V(t, x(t)) = V(t_0, x(t_0)) + \int_{t_0}^{t} \mathscr{L}V(s, x(s))f(s, x(s))ds + \int_{t_0}^{t} V_x(s, x(s))\sigma(s, x(s))dW(s),$$

(a.s.), where

$$\mathscr{L}V(t, x(t)) = V_t(t, x(t)) + V_x(t, x(t))f(t, x(t)) + \frac{1}{2}\text{tr}[\sigma^T(t, x(t))V_{xx}(t, x(t))\sigma(t, x(t))]$$

is the infinitesimal operator acting on the process $V(t, x(t))$ with $V_t(t, x(t))$, $V_x(t, x(t))$, and $V_{xx}(t, x(t))$ being the partial differentials of the process $V(t, x(t))$ with respect to t, x, and twice with respect to x, respectively.

Having defined SEPCA, we consider the following uncertain control SEPCA

$$dx(t) = \Big((A + \Delta A)x(t) + Bu(t) + Gw(t) + f(x_\xi)\Big)dt + g(x_\xi)\, dW(t), \tag{5a}$$

$$z(t) = Cx(t) + Fu(t), \tag{5b}$$

$$x(t_0) = x_0, \tag{5c}$$

where $x \in \mathbb{R}^n$ is the system state, $u \in \mathbb{R}^p$ is the control input of the form Kx, where $K \in \mathbb{R}^{p\times n}$ is a control gain matrix, $w \in \mathbb{R}^q$ is a disturbance input, which is assumed to be in $L_2[0, \infty)$ (i.e., $\|w\|_2^2 = \int_0^\infty \|w(t)\|^2 dt < \infty$), $f \in \mathbb{R}^n$ and $g \in \mathbb{R}^{n\times m}$ represent lumped uncertainties, $z \in \mathbb{R}^r$ is the controlled measured output, A, B, G, C, and F are constant matrices that describe the nominal system, and $\Delta A(\cdot)$ is real-valued matrix, which is piecewise continuous function representing parameter uncertainties. We also assume that $f(0) = 0 \in \mathbb{R}^n$ and $g(0) = 0 \in \mathbb{R}^{n\times m}$ to ensure that the system admits a trivial solution. A symmetric matrix P is said to be positive definite if the scalar $x^T Px > 0$ for all nonzero $x \in \mathbb{R}^n$ and $x^T Px = 0$ for $x = 0$.

Definition 2 A function $a \in \mathscr{C}(\mathbb{R}_+; \mathbb{R}_+)$ is said to belong to class $\mathscr{K}$ if $a(0) = 0$ and it is strictly increasing [12]; it said to belong to class $\mathscr{K}_1$ if $a \in \mathscr{K}$ and it is convex; it said to belong to class $\mathscr{K}_2$ if $a \in \mathscr{K}$ and it is concave.

Definition 3 For any $t_0 \in \mathbb{R}_+$, $t \geq t_0$, and $x_0 \in \mathbb{R}^n$, let $x(t) = x(t, t_0, x_0)$ be a solution of uncertain SEPCA (5). Then, the system is said to be robustly globally input-to-state stable (ISS) in the m.s. if there exist functions $\beta \in \mathscr{KL}$ and $\rho \in \mathscr{K}$

such that, for any input disturbance w, the solution satisfies

$$\mathbb{E}[\|x(t)\|^2] \leq \beta\big(\mathbb{E}[\|x_0\|^2], t - t_0\big) + \rho(\|w(t)\|), \qquad \forall\, t \geq t_0,$$

for any solution $x(t) = x(t, t_0, x_0)$ of (5), and any $x_0 \in \mathbb{R}^n$ with $\mathbb{E}[\|x_0\|^2] < \infty$. Particularly, if $\beta(s,t) = se^{-\lambda t}$, for some positive λ, then the system is said to be input-to-state exponentially stable in the m.s.

Definition 4 Given a constant $\bar{\gamma} > 0$, uncertain SEPCA (5) is said to be input-to-state stabilizable with an H_∞-norm bound $\bar{\gamma}$ if there exists a state feedback law

$$u(t) = Kx(t),$$

where $K = -\frac{1}{2}\varepsilon B^T P$ for some a constant $\varepsilon > 0$ and positive-definite matrix P, such that, for any admissible parameter uncertainty $\Delta A(\cdot)$, the corresponding closed-loop system is uniformly asymptotically (or exponentially) ISS in the (m.s.) and the controlled output z satisfies

$$\|z\|^2_{\mathbb{E}} := \mathbb{E}\Big[\int_0^\infty \|z(t)\|^2 dt\Big] \leq \bar{\gamma}^2 \|w\|_2^2 + m_0,$$

for some positive constant m_0.

Assumption A For all $t \in \mathbb{R}_+$, the admissible parameter uncertainty of the system is defined by $\Delta A = D\mathcal{U}(t)H$, where D and H are known real constant matrices with appropriate dimensions that give the structure of the uncertainty, and $\mathcal{U}$ is an unknown real time-varying matrix containing the uncertain parameters in the linear part and satisfying $\|\mathcal{U}(t)\| \leq 1$.

3 Main Results

Theorem 1 *For all $t \in [t_k, t_{k+1})$, let the controller gain K and disturbance level $\bar{\gamma} > 0$ be given. Assume that Assumption A holds, and there exist positive constants $\varepsilon_1, \varepsilon_2$, and a positive-definite matrix P such that the following inequality*

$$\Big(A + BK\Big)^T P + P\Big(A + BK\Big) + P\Big(\varepsilon_1 DD^T + \frac{1}{\varepsilon_2} I + \bar{\gamma}^{-2} GG^T\Big)P + \zeta\bar{q}P$$
$$+ \frac{1}{\varepsilon_1} H^T H + \varepsilon_2 \bar{q} \|U\|^2 I + C_c^T C_c + \alpha P < 0 \tag{6}$$

holds, where $\bar{q} > 1$, $\alpha > 0$, $\zeta > 0$ is such that

$$tr[g^T(y)Pg(y)] \leq 2\zeta\bar{q}x^T Px, \qquad (a.s.), \tag{7}$$

U is an $n \times n$ matrix such that $\|f(y)\|^2 \le \bar{q}\|U\|^2\|x\|^2$, and $C_c = C + FK$. Suppose further that for any $k \in \mathbb{N}$, the following dwell-time-type condition

$$t_{k+1} - t_k \ge \frac{1}{\bar{\alpha}} \ln\Big(\frac{ad_k}{bd_{k+1}}\Big) \tag{8}$$

holds, where $\bar{\alpha} > 0$ and d_k is a constant satisfying $d_{k+1} < d_k < 1$ and $\lim_{k\to\infty} d_k = 0$. Then, system (5) is m.s. robustly globally exponentially input-to-state stabilized by the feedback control $u = Kx$ with an H_∞-norm bound $\bar{\gamma} > 0$.

Proof For all $t \in [t_k, t_{k+1})$, let $x(t) = x(t, t_0, x_0)$ be the solution of (5) and define $V(x) = x^T Px$ as Lyapunov function candidate. Then,

$$\begin{aligned}\mathscr{L}V(x) &\le x^T[A^TP + PA + 2K^TB^TP]x + 2x^TP(\Delta A)x + 2x^TPf(x_\xi)\\ &\quad + \zeta\bar{q}x^TPx + 2x^TPGw, \qquad \text{(a.s.)},\end{aligned}$$

where $\bar{q}$ is provided by using the Razumikhin technique.

Claim There exits $\epsilon_2 > 0$ such that $2x^TPf(x_\xi) \le x^T\Big(\varepsilon_2\bar{q}\|U\|^2I + \dfrac{1}{\varepsilon_2}P^2\Big)x$.

Proof of the Claim Since

$$\begin{aligned}0 &\le \Big(\sqrt{\varepsilon_2}f^T(x_\xi) - \frac{1}{\sqrt{\varepsilon_2}}x^TP\Big)\Big(\sqrt{\varepsilon_2}f^T(x_\xi) - \frac{1}{\sqrt{\varepsilon_2}}x^TP\Big)^T\\ &= \varepsilon_2 f^T(x_\xi)f(x_\xi) + \frac{1}{\varepsilon_2}x^TP^2x - 2x^TPf(x_\xi)\\ &\le \varepsilon_2\|U\|^2\|x_\xi\|^2 + \frac{1}{\varepsilon_2}x^TP^2x - 2x^TPf(x_\xi)\end{aligned}$$

which implies the desired result. Where $\bar{q} = q/\lambda_{\min}(P) > 1$, with $q > 1$.

By the facts that $2x^TP(\Delta A)x \le x^T\Big(\varepsilon_1PDD^TP + \frac{1}{\varepsilon_1}H^TH\Big)x$, we get

$$\begin{aligned}\mathscr{L}V(x) &\le x^T\Big((A+BK)^TP + P(A+BK)\Big)x + x^T\Big(\varepsilon_1PDD^TP + \frac{1}{\varepsilon_1}H^TH\Big)x\\ &\quad + x^T\Big(\varepsilon_2\bar{q}\|U\|^2I + \frac{1}{\varepsilon_2}P^2\Big)x + \zeta\bar{q}x^TPx + \varepsilon_3x^TPGG^TPx + \frac{1}{\varepsilon_3}w^Tw, \text{(a.s.)}.\end{aligned}$$

Making use of inequality (6) with $\varepsilon_3 = \bar{\gamma}^{-2}$, we get, for all $t \in [t_k, t_{k+1})$,

$$\begin{aligned}\mathscr{L}V(x) &\le -(\alpha - \theta)x^TPx - \theta x^TPx + \frac{1}{\varepsilon_3}w^Tw\\ &= -\bar{\alpha}V(x), \qquad \bar{\alpha} = \alpha - \theta, \qquad \text{(a.s.)},\end{aligned}$$

provided that $V(x) \geq \frac{1}{\theta\varepsilon_3}\|w\|^2$ or $\|x\| \geq \sqrt{\frac{1}{a\theta\varepsilon_3}}\|w\|$, for some positive constant $\theta < \alpha$. Applying the Itô formula to process $V(x)$ and taking the mathematical expectation give $D^+m(t) \leq -\bar{\alpha}m(t)$, which implies that

$$m(t) \leq m(t_k)e^{-\bar{\alpha}(t-t_k)}, \qquad \text{whenever} \qquad \|x\| \geq \sqrt{\frac{1}{a\theta\varepsilon_3}}\|w\|,$$

where $m(t) = \mathbb{E}[V(x(t))],\ \forall t \in [t_k, t_{k+1})$. By the dwell-time condition (8), we get

$$\mathbb{E}[\|x(t_{k+1})\|^2] \leq d_{k+1}\mathbb{E}[\|x(t_k)\|^2],$$

where we have used the fact $b\|x\|^2 \leq x^TPx \leq a\|x\|^2$. Since $\lim_{k\to\infty} d_k = 0$, the limit of $x(t_{k+1})$ will eventually converge to the limit set of radius $\sqrt{\frac{1}{a\theta\varepsilon_3}}\|w\|$ in the m.s.; that is the solution x is robustly globally input-to-state stabilized in the m.s.

To prove the upper bound on the output magnitude $\|z\|$, we introduce the performance function

$$J = \mathbb{E}\int_{t_0}^{\infty}(z^Tz - \bar{\gamma}^2w^Tw)dt.$$

It follows that

$$\begin{aligned}
J &= \mathbb{E}\int_{t_0}^{\infty}(z^Tz - \bar{\gamma}^2w^Tw)dt + \mathbb{E}\int_{t_0}^{\infty}dV(x)dt - \mathbb{E}[V(x(\infty))] + \mathbb{E}[V(x_0)] \\
&\leq \mathbb{E}\int_{t_0}^{\infty}(z^Tz - \bar{\gamma}^2w^Tw)dt + \mathbb{E}\int_{t_0}^{\infty}\Big[x^T\Big((A+BK)^TP + P(A+BK) + P(\varepsilon_1DD^T + \frac{1}{\varepsilon_2}I)P \\
&\quad + \bar{q}\varepsilon_2\|U\|^2I + \frac{1}{\varepsilon_1}H^TH + \bar{q}\zeta P - \bar{\gamma}^{-2}PGG^TP + \bar{\gamma}^{-2}PGG^TP\Big)x + 2x^TPGw\Big]dt + \mathbb{E}[V(x_0)] \\
&= \mathbb{E}\int_{t_0}^{\infty}x^T\Big((A+BK)^TP + P(A+BK) + P(\varepsilon_1DD^T + \frac{1}{\varepsilon_2}I + \bar{\gamma}^{-2}GG^T)P \\
&\quad + \bar{q}\varepsilon_2\|U\|^2I + \frac{1}{\varepsilon_1}H^TH + \bar{q}\zeta P + C_c^TC_c\Big)x\,dt \\
&\quad - \mathbb{E}\int_{t_0}^{\infty}\Big(w - \bar{\gamma}^{-2}G^TPx\Big)^T\bar{\gamma}^2\Big(w - \bar{\gamma}^{-2}G^TPx\Big)dt + \mathbb{E}[V(x_0)].
\end{aligned}$$

By the theorem assumption in (6) and strict negativeness of the second term, we get

$$\|z\|_{\mathbb{E}}^2 \leq \bar{\gamma}^2\|w\|_2^2 + m_0, \qquad \text{where } m_0 = \mathbb{E}[V(x_0)] \qquad \square$$

Remarks

1. The conditions on f and g mean that the system perturbations are bounded by linear growth bounds.

2. The dwell-time-type condition in (8) is made to generate a sequence of solution trajectories evaluated at the switching times that is convergent in the m.s. to a limit set with a radius depending on the system disturbance.
3. If the controlled measured output is available at the time instances ξ, that is to say $z_\xi = C_c x_\xi$, where C_c is as defined before, then assumption (6) becomes

$$\Big(A+BK\Big)^T P + P\Big(A+BK\Big) + P\Big(\varepsilon_1 DD^T + \frac{1}{\varepsilon_2} I + \bar{\gamma}^{-2} GG^T\Big)P + \zeta\bar{q}P$$
$$+ \frac{1}{\varepsilon_1} H^T H + \varepsilon_2 \bar{q} \|U\|^2 I + \bar{q} C_c^T C_c + \alpha P < 0,$$

where in this case $J = \mathbb{E}\int_{t_0}^{\infty} (z_\xi^T z_\xi - \bar{\gamma}^2 w^T w)dt$.

Corollary *In Theorem 1, if $w(t) \equiv 0$ for all $t \geq t_0$, and the inequality*

$$\Big(A+BK\Big)^T P + P\Big(A+BK\Big) + P\Big(\varepsilon_1 DD^T + \frac{1}{\varepsilon_2} I\Big)P$$
$$+ \zeta\bar{q}P + \frac{1}{\varepsilon_1} H^T H + \varepsilon_2 \bar{q} \|U\|^2 I + C_c^T C_c + \alpha P < 0$$

holds, then $x \equiv 0$ is robustly globally exponentially stable in the m.s.

Example Consider uncertain control SEPCA (5), where

$$A = \begin{bmatrix} -7 & 0 \\ 0 & 0.1 \end{bmatrix}, \quad B = \begin{bmatrix} 0.1 & 0.4 \\ 0 & -9 \end{bmatrix}, \quad G = \begin{bmatrix} 0 & 1 \\ 1 & 0 \end{bmatrix}, \quad w(t) = \sin(t)\begin{bmatrix} 1 \\ 1 \end{bmatrix}, \quad C = I_{2\times 2},$$

$$F = \begin{bmatrix} 0.1 & -0.1 \\ -2 & 0 \end{bmatrix}, \quad D = \begin{bmatrix} 1 \\ 0 \end{bmatrix}, \quad H = \begin{bmatrix} 0 & 1 \end{bmatrix}, \quad \mathscr{U}(t) = \sin(t), \quad f(x_\xi) = \frac{1}{2}\begin{bmatrix} x_{1_\xi} \\ x_{2_\xi} \end{bmatrix},$$

$$g(x_\xi) = \begin{bmatrix} x_{1_\xi} & 0 \\ 0 & x_{2_\xi} \end{bmatrix}, \quad \varepsilon_1 = 1, \varepsilon_2 = 0.1, \bar{\gamma} = 0.1, \alpha = 2, \bar{q} = 2, \text{ and } \zeta = 0.6386.$$

By the theorem assumptions, we have, after some algebraic manipulations, $U = 0.5 I_{2\times 2}$, and from the algebraic Riccati-like inequality (6), we have

$$P = \begin{bmatrix} 0.0593 & -0.0125 \\ -0.0125 & 0.1748 \end{bmatrix}, \quad \text{with } b = \lambda_{\min}(P) = 0.0579 \text{ and } a = \lambda_{\max}(P) = 0.1762,$$

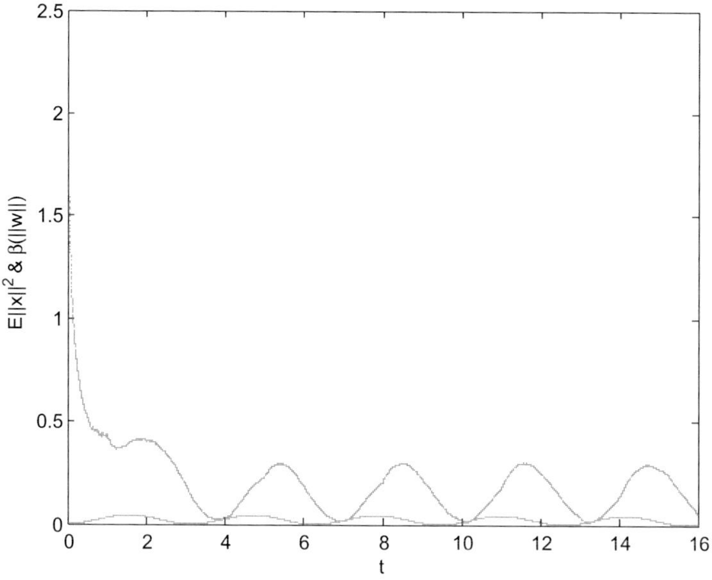

Fig. 1 Input-to-state stability in the m.s.

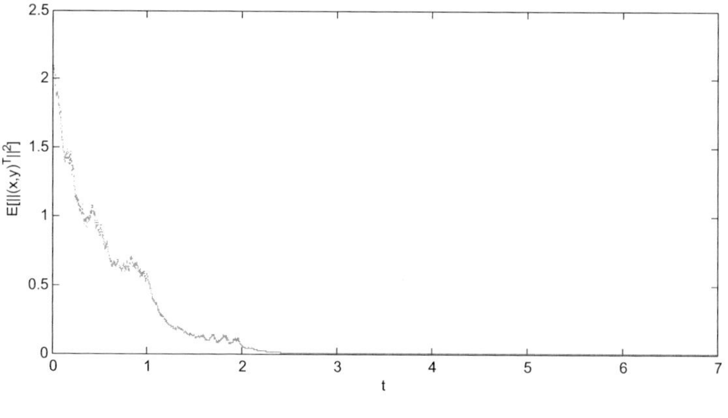

Fig. 2 Exponential stability in the m.s. ($w(t) \equiv 0$)

and the control gain matrix is $K = \begin{bmatrix} -0.0009 \; 0.0002 \\ -0.0204 \; 0.2308 \end{bmatrix}$. The simulation results of $\mathbb{E}[\|x(t)\|^2]$ (above) and $\beta(\|w(t)\|) = \frac{1}{a\theta\varepsilon_3}\sin^2(t)$ (below) is shown in Fig. 1, where $\theta = 1.3$, $\varepsilon_3 = \bar{\gamma}^{-2} = 1/0.01$, and, by the dwell-time condition in (8), we have chosen $t_{k+1} - t_k = 1$, where $d_k = 1/2^k$, and $\xi_k = t_k$, for all $k = 0, 1, \cdots$. If $w(t) \equiv 0$, $x(t) \equiv 0$ is exponentially stable in the m.s., as shown in Fig. 2.

4 Conclusions

In this paper, we have considered a control system with uncertain SEPCA, which has been viewed as a switched system. The focus was on establishing input-to-state stabilization and H_∞ performance, where the method of Lyapunov function and Razumikhin techniques have been used to write some sufficient conditions.

Acknowledgements This research was financially supported by Natural Sciences and Engineering Research Council of Canada (NSERC).

References

1. Akhmet, M.U., Aruğtaslan, D., Liu, X.Z.: Permanence of nonautonomous ratio-dependent predator-prey systems with piecewise constant argument of generalized type. DCDIS-A: Math. Anal. **15**, 37–51 (2008)
2. Akhmet, M.U., Aruğaslan, D.: Lyapunov-Razumikhin method for differential equations with piecewise constant argument. DCDS **25**(2), 457–466 (2009)
3. Alwan, M.S., Liu, X.Z., Akhmet, M.U.: On chaotic synchronization via impulsive control and piecewise constant arguments. DCDIS-Ser. B: Appl. Algorithms **22**, 53–67 (2015)
4. Alwan, M.S., Liu, X.Z., Xie, W.-C.: Comparison principle and stability results for nonlinear differential equations with piecewise constant arguments. JFI **350**, 211–230 (2013)
5. Bao, G., Wen, S., Zeng, Z.: Robust stability analysis of interval fuzzy Cohen-Grossberg neural networks with piecewise constant argument of generalized type. Neural Netw. **33**, 32–41 (2012)
6. Busenberg, S., Cooke, K.L.: Models of vertically transmitted with sequential-continuous dynamics. In: Lakshmikantham, V. (ed.) Nonlinear Phenomena: In Mathematical Sciences. Academic, New York, pp. 179–187 (1982)
7. Chen, L., Leng, Y., Guo, H., Shi, P., Zhang, L.: H_∞ control of a class of discrete-time Markov jump linear systems with piecewise-constant TPs subject to average dwell time switching. J. Frankl. Inst. **349**, 1989–2003 (2012)
9. Cooke, K.L., Györi, I.: Numerical approximation of the solutions of delay differential equations on an infinite interval using piecewise constant arguments. Comput. Math. Appl. **28**, 81–92 (1994)
8. Cooke, K.L., Wiener, J.: A survey of differential equations with piecewise constants arguments. In: Busenberg, S., Martelli, M. (eds.) Delay Differential Equations and Dynamical Systems. Lecture Notes in Mathematics, vol. 1475, pp. 1–15. Springer-Verlag, Heidelberg/New York (1991)
10. Gopalsamy, K.: Stability and Oscillations in Delay Differential Equations of Population Dynamics. Kluwer Academic, Dordrecht (1992)
11. Györi, I., Hartung, F.: On numerical approximation using differential equations with piecewise-constant arguments. Period. Math. Hung. **56**(1), 55–69 (2008)
12. Khalil, H.K.: Nonlinear Systems. Prentice Hall, Upper Saddle River (2002)
13. Kokotovic, P., Arcak, M.: Constructive nonlinear control: Progress in the $90's$. In: Proceedings of the 14th IFAC World Congress, the Plenary and Index Volume, Beijing, pp. 49–77 (1999)
14. Nieto, J.J., Rodríguez-López, R.: Second-order linear differential equations with piecewise constant arguments subject to nonlocal boundary conditions. AMC **218**, 9647–9656 (2012)
15. Nieto, J.J., Rodríguez-López, R.: Study of solutions to some functional differential equations with piecewise constant arguments. Abstr. Appl. Anal. art. no. 851691 (2012)

21. Ozturk, I., Bozkurt, F.: Stability analysis of a population model with piecewise constant arguments. Nonlinear Anal.: Real World Appl. **12**, 1532–1545 (2011)
19. Sontag, E.D.: Comments on integral variants of ISS. Syst. Control Lett. **34**, 93–100 (1998)
16. Sontag, E.D.: Smooth stabilization implies coprime factorization. IEEE Trans. Autom. Control **34**(4), 435–443 (1989)
18. Sontag, E.D., Wang, Y.: New characterization of input-to-state stability property. IEEE Trans. Autom. Control **34**(41) 1283–1294 (1996)
17. Sontag, E.D., Wang, Y.: On characterization of input-to-state stability property. Syst. Control Lett. **24**, 351–359 (1995)
20. Teel, A.R., Moreau, L., Nešic, D.: A unified framework for input-to-state stability in systems with two time scales. IEEE Trans. Autom. Control **48**(9), 1526–1544 (2003)
22. Wiener, J.: Generalized Solutions of Functional Differential Equations. World Scientific, Singapore (1993)
23. Wu, Z.-G., Park, J.H., Su, H., Chu, J.: Stochastic stability analysis of piecewise homogeneous Markovian jump neural networks with mixed time-delays. JFI **349**, 2136–2150 (2012)

Switched Singularly Perturbed Systems with Reliable Controllers

Mohamad S. Alwan, Xinzhi Liu, and Taghreed G. Sugati

Abstract This paper addresses the problem of exponential stability for a class of switched control singularly perturbed systems (SCSPS) not only when all the control actuators are operational, but also when some of them experience failures. Multiple Lyapunov functions and average dwell-time switching signal approach are used to establish the stability criteria for the proposed systems. In this paper, we assume that a full access to all the system modes is available, though the mode-dependent, slow-state feedback controllers experience faulty actuators of an outage type. In the stability analysis, the system under study is viewed as an interconnected system that has been decomposed into isolated, lower order, slow and fast subsystems, and the interconnection between them. It has been observed that if the degree of stability of each isolated mode is greater than the interconnection between them, the interconnected mode is exponentially stable, and, then, the full order SCSPS is also exponentially stable for all admissible switching signals with average dwell-time. A numerical example with simulations is introduced to illustrate the validity of the proposed theoretical results.

1 Introduction

A switched system is a special class of hybrid systems that consists of a family of continuous- or discrete-time dynamical subsystems, and a switching rule that controls the switchings amongst them. In reality, many applications in different fields such as aircraft, automotive industry, robotics, control systems, biological, epidemic disease models, etc., are operating based on switchings between multiple dynamical modes. The stability of such systems has received great attention in the last few decades. For more readings, one may refer to [3, 9, 10, 14, 17] and the references therein.

It is well known that even if the individual modes are stable, the switched system may not be stable [9]. However, it has been shown that the stability of the switched system can be achieved if the dwell-time, the time between any two consecutive

M.S. Alwan (✉) • X. Liu • T.G. Sugati
Department of Applied Mathematics, University of Waterloo, Waterloo, ON, N2L 3G1, Canada
e-mail: malwan@uwaterloo.ca; xinzhi.liu@uwaterloo.ca; tsugati@uwaterloo.ca

© Springer International Publishing Switzerland 2016

J. Bélair et al. (eds.), *Mathematical and Computational Approaches in Advancing Modern Science and Engineering*, DOI 10.1007/978-3-319-30379-6_35

switchings, is sufficiently large [12]. Practically, in some situations, such as aging systems or systems with finite escape time, the dwell-time condition may not hold, and yet one can get the same stability result if the average dwell-time is satisfied [5].

The reliable control is the controller that tolerates actuator and/or sensor failures. The failure of control components is frequently encountered in practice, yet the immediate repair may be impossible in some critical cases. Therefore, designing a reliable controller to guarantee stability is necessary. This problem has drawn researcher attention for many years; see for instance [4, 13, 15, 16].

Systems involving multiple time-scale dynamics (known as singularly perturbed systems) arise in a large class of applications in science and engineering such as celestial mechanics, many-particle dynamics, and climate systems, etc. They are characterized by small parameters multiplied by the highest derivatives. The stability problem of these systems has attracted many researchers; see [1, 2, 6–8, 11] and some references therein.

In this paper, we address SCSPS where the controllers are subject to faulty actuators. The continuous states are viewed as an interconnected system with two-time scale (slow and fast) subsystems. Moreover, due to dominant behaviour of the reduced systems, the stabilization of the full order systems is achieved through the controller of the slow reduced order subsystem. This in turn results in lessening some unnecessary sufficient conditions imposed on the fast subsystem. The stability analysis is obtained by multiple Lyapunov functions method after decomposing the system into isolated, lower order, slow and fast subsystems, and the interconnection between them.

It has been observed that if the degree of stability of each isolated mode is greater than the interconnection between them, the underlying interconnected mode of the switched system is exponentially stable. Moreover, if switching among the system modes follows the average dwell-time rule, then the SCSPS is also exponentially stable. The relationship between the stability degrees of the isolated subsystems and the interconnection strength is usually formulated by the so-called M-matrix. Finally, a numerical example and simulations are provided to justify the proposed theoretical results.

The rest of this paper is organized as follows. Section 2 involves the problem description and some definitions. The main results and proofs are stated in Sect. 3. A numerical example with simulations is presented in Sect. 4. The conclusion is given in Sect. 5.

2 Problem Formulation and Preliminaries

Throughout this paper, $\mathbb{R}^n$ denotes the n-dimensional Euclidean space; $\mathbb{R}_+$ refers to the nonnegative real numbers; $\mathbb{R}^{n\times m}$ is the class of all $n \times m$ real matrices. A symmetric matrix P is said to be positive definite if all its eigenvalues are positive. Moreover, If $P \in \mathbb{R}^{n\times n}$, denote by $\lambda_{max}(P)(\lambda_{min}(P))$ the maximum (minimum) eigenvalue of P. If $V(x) = x^T Px$, the inequalities $\lambda_{min}(P)||x||^2 \leq V(x) \leq$

$\lambda_{max}(P)||x||^2$ are true. If $x \in \mathbb{R}^n$, then $||x||$ refers to the Euclidean vector norm of x. Consider the following system

$$\dot{x} = A_{11_{\rho(t)}}x + A_{12_{\rho(t)}}z + B_{1_{\rho(t)}}u, \tag{1a}$$

$$\varepsilon_{\rho(t)}\dot{z} = A_{21_{\rho(t)}}x + A_{22_{\rho(t)}}z + B_{2_{\rho(t)}}u, \tag{1b}$$

$$x(t_0) = x_0, \;\; z(t_0) = z_0, \tag{1c}$$

where $x \in \mathbb{R}^m$, $z \in \mathbb{R}^n$ are the system slow and fast states respectively, $u \in \mathbb{R}^l$ is the control input of the form $u = Kx$ for some control gain $K \in \mathbb{R}^{l\times m}$, $\rho : [t_0, \infty) \rightarrow \mathscr{S} = \{1, 2, \cdots, N\}$ is a piecewise constant function known as the switching signal (or law). For each $i \in \mathscr{S}$, $A_{11_i} \in \mathbb{R}^{m\times m}$, $A_{12_i} \in \mathbb{R}^{m\times n}$, $A_{21_i} \in \mathbb{R}^{n\times m}$, $A_{22_i} \in \mathbb{R}^{n\times n}$, are known real constant matrices with A_{22_i} is a nonsingular Hurwitz matrix, $B_{1_i} \in \mathbb{R}^{m\times l}$, $B_{2_i} \in \mathbb{R}^{n\times l}$, and $0 < \varepsilon_i \ll 1$. Setting $\varepsilon_i = 0$ implies that $z = h_i(x) = -A_{22_i}^{-1}[A_{21_i}x + B_{2_i}u]$. Plug z into (1a) gives *the slow reduced subsystem* $\dot{x}_s = A_{0_i}x_s + B_{0_i}u$ where $A_{0_i} = A_{11_i} - A_{12_i}A_{22_i}^{-1}A_{21_i}$, and $B_{0_i} = B_{1_i} - A_{12_i}A_{22_i}^{-1}B_{2_i}$. Choose $u = Kx_s$ such that (A_{0_i}, B_{0_i}) is stabilizable.

For simplicity of notation, we use x instead of x_s to refer to the slow reduced system.

Definition 1 The trivial solution of system (1) is said to be globally exponentially stable (g.e.s.) if there exist positive constants L and λ such that

$$||x(t)|| + ||z(t)|| \leq L(||x(t_0)|| + ||z(t_0)||)e^{-\lambda(t-t_0)}, \quad t \geq t_0 \in \mathbb{R}_+,$$

for all $x(t)$ and $z(t)$, the solutions of system (1), and any $x_0 \in \mathbb{R}^m$, $z_0 \in \mathbb{R}^n$.

Definition 2 An $n \times n$ matrix $M = [m_{ij}]$ with $m_{ij} \leq 0$, for all $i \neq j$, is said to be an M-matrix if all its leading successive principle minors are positive, i.e.,

$$det \begin{bmatrix} m_{11} & m_{12} & \cdots & m_{1k} \\ m_{21} & m_{22} & \cdots & m_{2k} \\ \cdots & \cdots & \cdots & \cdots \\ m_{k1} & m_{k2} & \cdots & m_{kk} \end{bmatrix} > 0, \quad k = 1, 2, \cdots, n.$$

Average dwell-time Condition (ADTC) [5]. The number of switchings $N(t_0, t)$ in the interval (t_0, t) for a finite t satisfies $N(t_0, t) \leq N_0 + \frac{t-t_0}{\tau_a}$, where N_0 is the chatter bound, and τ_a is the average dwell-time.

3 The Main Results

In this section, we present our main results.

3.1 Normal Case

For any $i \in \mathscr{S}$, the closed-loop system becomes

$$\begin{cases} \dot{x} = (A_{11_i} + B_{1_i}K_i)x + A_{12_i}z, \\ \varepsilon_i\dot{z} = (A_{21_i} + B_{2_i}K_i)x + A_{22_i}z, \\ x(t_0) = x_0, \quad z(t_0) = z_0. \end{cases} \tag{2}$$

Theorem 1 *The trivial solution of system* (1) *is g.e.s. if ADTC holds, and the following assumptions hold*

(i) $Re[\lambda(A_{22_i})] < 0,$ *and* (A_{0_i}, B_{0_i}) *is stabilizable;*
(ii) there exist positive constants a_{ji}, $j = 1, \cdots, 6$ *such that*

$$2x^T P_{1_i} A_{12_i} h_i(x) \le a_{1i} x^T x, \tag{3}$$

$$2x^T P_{1_i} A_{12_i}(z - h_i(x)) \le a_{2i} x^T x + a_{3i}(z - h_i(x))^T (z - h_i(x)), \tag{4}$$

$$2(z - h_i(x))^T P_{2_i} \mathscr{R}_{1_i} x \le a_{4i} x^T x + a_{5i}(z - h_i(x))^T (z - h_i(x)), \tag{5}$$

$$(z - h_i(x))^T \mathscr{R}_{2_i}(z - h_i(x)) \le a_{6i}(z - h_i(x))^T (z - h_i(x)) \tag{6}$$

where $h_i(x) = -A_{22_i}^{-1}(A_{21_i} + B_{2_i}K_i)x$, P_{2_i} *is the solution of the Lyapunov equation* $A_{22_i}^T P_{2_i} + P_{2_i} A_{22_i} = -I_{i_n}$, *where* I_{i_n} *is an identity matrix,* $\mathscr{R}_{1_i} = A_{22_i}^{-1}(A_{21_i} + B_{2_i}K_i)[A_{11_i} + B_{1_i}K_i - A_{12_i}A_{22_i}^{-1}(A_{21_i} + B_{2_i}K_i)]$, *and* $\mathscr{R}_{2_i} = 2P_{2_i}A_{22_i}^{-1}(A_{21_i} + B_{2_i}K_i)A_{12_i}$*;*
(iii) there exist a positive constant ε_i *such that* $-\bar{A}_i$ *is an M-matrix where*

$$\bar{A}_i = \begin{bmatrix} \frac{\lambda_{\max}(N_i)}{\lambda_{\max}(P_{1_i})} & \frac{a_{3i}}{\lambda_{\min}(P_{2_i})} \\ \frac{a_{4i}}{\lambda_{\min}(P_{1_i})} & \frac{a_{6i}}{\lambda_{\min}(P_{2_i})} - \frac{(1 - a_{5i}\varepsilon_i)}{\varepsilon_i \lambda_{\max}(P_{2_i})} \end{bmatrix},$$

where $N_i = -Q_i + (a_{1i} + a_{2i})I + M^T P_i + P_i M^T$ *such that* $M = A_{12_i}A_{22_i}^{-1}(A_{21_i} + B_{2_i}K_i)$ *and* $(A_{0_i} + B_{0_i}K_i)^T P_{1_i} + P_{1_i}(A_{0_i} + B_{0_i}K_i) = -Q_i$ *for a given* K_i.

Proof Let $V_i(x) = x^T P_{1_i} x$ and $W_i((z - h_i(x))(t)) = (z - h_i(x))^T P_{2_i}(z - h_i(x))$ be Lyapunov function candidates for the slow and the fast subsystem, respectively. Then,

$$\begin{aligned} \dot{V}_i(x) &\le x^T(-Q_i + a_{1i}I + a_{2i}I)x + a_{3i}(z - h_i(x))^T (z - h_i(x)) \\ &\le \frac{\lambda_{\max}(N_i)}{\lambda_{\max}(P_{1_i})} V_i(x) + \frac{a_{3i}}{\lambda_{\min}(P_{2_i})} W_i((z - h_i(x))(t)), \end{aligned} \tag{7}$$

where $N_i = -Q_i + (a_{1i} + a_{2i})I + M^T P_i + P_i M^T$ such that $M = A_{12_i} A_{22_i}^{-1}(A_{21_i} + B_{2_i} K_i)$ is negative definite. We also have

$$\begin{aligned}\dot{W}_i((z-h_i(x))(t)) =& -\frac{1}{\varepsilon_i}(z-h_i(x))^T(z-h_i(x)) - 2(z-h_i(x))^T P_{2_i}\dot{h}_i(x)\\ &\leq (a_{5i} - \frac{1}{\varepsilon_i})(z-h_i(x))^T(z-h_i(x)) + (z-h_i(x))^T \mathscr{R}_{2_i}(z-h_i(x))\\ &+ a_{4i}x^T x\\ &\leq \frac{a_{4i}}{\lambda_{\min}(P_{1_i})}V_i(x) + \Big[\frac{a_{6i}}{\lambda_{\min}(P_{2_i})} - \frac{(1-a_{5i}\varepsilon_i)}{\varepsilon_i\lambda_{\max}(P_{2_i})}\Big]W_i((z-h_i(x))(t)).\end{aligned} \tag{8}$$

where $\mathscr{R}_{2_i} = 2P_{2_i}A_{22_i}^{-1}(A_{21_i} + B_{2_i}K_i)A_{12_i}$. Combining (7) and (8), we get

$$\begin{bmatrix} \dot{V}_i(x) \\ \dot{W}_i\big((z-h_i(x))(t)\big)\end{bmatrix} \leq \begin{bmatrix} \frac{\lambda_{\max}(N_i)}{\lambda_{\max}(P_{1_i})} & \frac{a_{3i}}{\lambda_{\min}(P_{2_i})} \\ \frac{a_{4i}}{\lambda_{\min}(P_{1_i})} & \frac{a_{6i}}{\lambda_{\min}(P_{2_i})} - \frac{(1-a_{5i}\varepsilon_i)}{\varepsilon_i\lambda_{\max}(P_{2_i})}\end{bmatrix}\begin{bmatrix} V_i(x) \\ W_i\big((z-h_i(x))(t)\big)\end{bmatrix}.$$

Then, we have

$$\bar{A}_i = \begin{bmatrix} \frac{\lambda_{\max}(N_i)}{\lambda_{\max}(P_{1_i})} & \frac{a_{3i}}{\lambda_{\min}(P_{2_i})} \\ \frac{a_{4i}}{\lambda_{\min}(P_{1_i})} & \frac{a_{6i}}{\lambda_{\min}(P_{2_i})} - \frac{(1-a_{5i}\varepsilon_i)}{\varepsilon_i\lambda_{\max}(P_{2_i})}\end{bmatrix}.$$

Then there exists $\eta_i = -\lambda_{\max}(\bar{A}_i) > 0$ such that for $t \in [t_{k-1}, t_k)$,

$$V_i(x) \leq \Big(V_i(x(t_{k-1})) + W_i\big((z-h_i(x))(t_{k-1})\big)\Big)e^{-\eta_i(t-t_{k-1})},$$

and

$$W_i\big((z-h_i(x))(t)\big) \leq \Big(V_i(x(t_{k-1})) + W_i\big((z-h_i(x))(t_{k-1})\big)\Big)e^{-\eta_i(t-t_{k-1})},$$

For any $i, j \in \mathscr{S},\ M > 1$, we have

$$\begin{aligned}V_j(x(t)) &\leq \mu V_i(x(t)),\\ W_j\big((z-h_j(x))(t)\big) &\leq \mu W_i\big((z-h_i(x))(t)\big).\end{aligned}$$

If the system switches among its modes, one may get for all $t \geq t_0$,

$$\begin{aligned}V_i(x(t)) \leq& 2\mu e^{-\eta_1(t_1-t_0)} \cdot 2\mu e^{-\eta_2(t_2-t_1)} \cdots 2\mu e^{-\eta_{k-1}(t_{k-1}-t_{k-2})}\\ &\times \Big[V_1(x(t_0)) + W_1\big((z-h_1(x))(t_0)\big)\Big]e^{-\eta_k(t-t_{k-1})}.\end{aligned}$$

Let $\eta = \min\{\eta_j : j = 1, 2, \cdots, k\}$. Then

$$V_i(x(t)) \leq (2\mu)^{k-1}\Big[V_1(x(t_0)) + W_1\big((z - h_1(x))(t_0)\big)\Big]e^{-\eta(t-t_0)}$$
$$\leq \Big[V_1(x(t_0)) + W_1\big((z - h_1(x))(t_0)\big)\Big]e^{(k-1)\ln\rho - \eta(t-t_0)},$$

where $\rho = 2\mu$. Applying the ADTC with $N_0 = \frac{\gamma}{\ln\rho}$, γ is an arbitrary constant, $\tau_a = \frac{\ln\rho}{(\eta-\eta^*)}$ with $\eta > \eta^*$ leads to

$$V_i(x(t)) \leq \Big[V_1(x(t_0)) + W_1\big((z - h_1(x))(t_0)\big)\Big]e^{\rho - \eta(t-t_0)}$$

and

$$W_i\big((z - h_i(x))(t)\big) \leq \Big[V_1(x(t_0)) + W_1\big((z - h_1(x))(t_0)\big)\Big]e^{\rho - \eta(t-t_0)}$$

This implies that there exists $L > 0$ such that

$$||x(t)|| + ||z(t)|| \leq L(||x(t_0)|| + ||z(t_0)||)e^{-\eta^*(t-t_0)/2}. \qquad \square$$

3.2 Faulty Case

To analyze the reliable stabilization with respect to actuator failures, for any $i \in \mathscr{S}$, consider the decomposition of the control matrix $B_i = B_{i\Sigma} + B_{i\bar{\Sigma}}$, where Σ the set of actuators that are susceptible to failure, and $\bar{\Sigma} \subseteq \{1, 2, \ldots, l\} - \Sigma$ the set of actuators which are robust to failures and essential to stabilize the given system, moreover, the matrices $B_{i\Sigma}$, $B_{i\bar{\Sigma}}$ are the control matrices associated with Σ, $\bar{\Sigma}$ respectively, and are generated by zeroing out the columns corresponding to $\bar{\Sigma}$ and Σ, respectively. The pair $(A_i, B_{i\bar{\Sigma}})$ is assumed to be stabilizable. For a fixed i, let $\sigma \subseteq \Sigma$ corresponds to some of the actuators that experience failure, and assume that the output of faulty actuators is zero. Then, the decomposition becomes $B_i = B_{i\sigma} + B_{i\bar{\sigma}}$, where $B_{i\sigma}$ and $B_{i\bar{\sigma}}$ have the same definition of $B_{i\Sigma}$ and $B_{i\bar{\Sigma}}$, respectively. Since the control input u is applied to the system through the normal actuators, the closed-loop system becomes

$$\dot{x} = (A_{11_i} + B_{1\bar{\sigma}_i}K_{i\bar{\sigma}})x + A_{12_i}z, \tag{9a}$$
$$\varepsilon_i\dot{z} = (A_{21_i} + B_{2\bar{\sigma}_i}K_{i\bar{\sigma}})x + A_{22_i}z, \tag{9b}$$
$$x(t_0) = x_0, \quad z(t_0) = z_0. \tag{9c}$$

where $K_{i\bar{\sigma}} = -\frac{1}{2}\beta_i B_{0i\bar{\sigma}}^T P_{i\bar{\sigma}}$, with $B_{0i\bar{\sigma}} = B_{1\bar{\sigma}_i} - A_{12_i}A_{22_i}^{-1}B_{2\bar{\sigma}_i}$, and $P_{i\bar{\sigma}}$ is a positive definite matrix such that $(A_{0_i} + B_{0i\bar{\sigma}}K_{i\bar{\sigma}})^T P_{i\bar{\sigma}} + P_{i\bar{\sigma}}(A_{0_i} + B_{0i\bar{\sigma}}K_{i\bar{\sigma}}) = -I$. Setting

$\varepsilon_i = 0$, one may get $z = h_{i\bar{\sigma}}(x) = -A_{22_i}^{-1}(A_{21_i} + B_{2\bar{\sigma}_i}K_{i\bar{\sigma}})x$. In the following theorem, we assume that $\bar{\sigma}_i = \bar{\Sigma}_i$.

Theorem 2 *The trivial solution of system* (9) *is g.e.s. if ADTC and the following assumptions hold for any $i \in \mathcal{S}$*

(i) $Re[\lambda(A_{22_i})] < 0,$ *and* $A_{11_i}^T P_{1_i} + P_{1_i}A_{11_i} + \beta_i P_{1_i}\big(A_{12_i}A_{22_i}^{-1}B_{2\bar{\Sigma}_i}B_{1\bar{\Sigma}_i}^T - B_{1\bar{\Sigma}_i}B_{1\bar{\Sigma}_i}^T\big)P_{1_i} + \alpha_i I = 0;$

(ii) there exist positive constants a_{ji}, $j = 1, \cdots, 6$ *such that*

$$2x^T P_{1_i}A_{12_i}h_{i\bar{\Sigma}}(x) \le a_{1i}x^T x, \tag{10}$$

$$2x^T P_{1_i}A_{12_i}(z - h_{i\bar{\Sigma}}(x)) \le a_{2i}x^T x + a_{3i}(z - h_{i\bar{\Sigma}}(x))^T(z - h_{i\bar{\Sigma}}(x)), \tag{11}$$

$$2(z - h_{i\bar{\Sigma}}(x))^T P_{2_i}\mathcal{R}_{1_{i\bar{\Sigma}}}x \le a_{4i}x^T x + a_{5i}(z - h_{i\bar{\Sigma}}(x))^T(z - h_{i\bar{\Sigma}}(x)), \tag{12}$$

$$(z - h_{i\bar{\Sigma}}(x))^T\mathcal{R}_{2\bar{\Sigma}_i}(z - h_{i\bar{\Sigma}}(x)) \le a_{6i}(z - h_{i\bar{\Sigma}}(x))^T(z - h_{i\bar{\Sigma}}(x)), \tag{13}$$

where $h_{i\bar{\Sigma}}(x) = -A_{22_i}^{-1}(A_{21_i} + B_{2\bar{\Sigma}_i}K_{i\bar{\Sigma}})x$, P_{2_i} *is the solution of* $A_{22_i}^T P_{2_i} + P_{2_i}A_{22_i} = -I_{i_n}$, $\mathcal{R}_{1_{i\bar{\Sigma}}} = A_{22_i}^{-1}(A_{21_i} + B_{2\bar{\Sigma}_i}K_{i\bar{\Sigma}})[A_{11_i} + B_{1_i}K_i - A_{12_i}A_{22_i}^{-1}(A_{21_i} + B_{2\bar{\Sigma}_i}K_{i\bar{\Sigma}})]$ *where* $K_{i\bar{\Sigma}} = -\frac{1}{2}\beta_i B_{0i\bar{\Sigma}}^T P_{i\bar{\Sigma}}$, *and* $\mathcal{R}_{2\bar{\Sigma}_i} = 2P_{2_i}A_{22_i}^{-1}(A_{21_i} + \frac{1}{2}\beta_i B_{2\bar{\Sigma}_i}B_{2\bar{\Sigma}_i}^T(A_{12_i}A_{22_i}^{-1})^T P_{1_i})A_{12_i} - \frac{1}{2}\beta_i B_{2\bar{\Sigma}_i}B_{1\bar{\Sigma}_i}^T P_{1_i}$;

(iii) there exist a positive constant ε_i *such that* $-\bar{A}_{i\bar{\Sigma}}$ *is an M-matrix where*

$$\bar{A}_{i\bar{\Sigma}} = \begin{bmatrix} \frac{\lambda_{\max}(N_{i\bar{\Sigma}})}{\lambda_{\max}(P_{1_i})} & \frac{a_{3i}}{\lambda_{\min}(P_{2_i})} \\ \frac{a_{4i}}{\lambda_{\min}(P_{1_i})} & \frac{\varepsilon_i(a_{5i}+a_{6i})-1}{\varepsilon_i\lambda_{\max}(P_{2_i})} \end{bmatrix}.$$

Proof Let $V_i(x) = x^T P_{1_i}x$ and $W_i\big((z - h_{\bar{\Sigma}_i}(x))(t)\big) = (z - h_{\bar{\Sigma}_i}(x))^T P_{2_i}(z - h_{\bar{\Sigma}_i}(x))$ be Lyapunov function candidates. Then, we have

$$\begin{aligned}\dot{V}_i(x) &\le x^T(-\alpha_i + a_{1i} + a_{2i})Ix + a_{3i}(z - h_{i\bar{\Sigma}}(x))^T(z - h_{i\bar{\Sigma}}(x)) \\ &\le \frac{-\alpha_i + a_{1i} + a_{2i}}{\lambda_{\max}(P_{1_i})}V_i(x) + \frac{a_{3i}}{\lambda_{\min}(P_{2_i})}W_i\big((z - h_{\bar{\Sigma}_i}(x))(t)\big)\end{aligned} \tag{14}$$

We also have

$$\begin{aligned}\dot{W}_i\big((z - h_{\bar{\Sigma}_i}(x))(t)\big) &= -\frac{1}{\varepsilon_i}(z - h_{i\bar{\Sigma}}(x))^T(z - h_{i\bar{\Sigma}}(x)) - 2(z - h_{i\bar{\Sigma}}(x))^T P_{2_i}\dot{h}_{i\bar{\Sigma}}(x) \\ &\le (a_{5i} - \frac{1}{\varepsilon_i})(z - h_{i\bar{\Sigma}}(x))^T(z - h_{i\bar{\Sigma}}(x)) + a_{4i}x^T x \\ &\quad + (z - h_{i\bar{\Sigma}}(x))^T\mathcal{R}_{2\bar{\Sigma}_i}(z - h_{i\bar{\Sigma}}(x)) \\ &\le \frac{a_{4i}}{\lambda_{\min}(P_{1_i})}V_i(x) + \Big[\frac{\varepsilon_i(a_{5i} + a_{6i}) - 1}{\varepsilon_i\lambda_{\max}(P_{2_i})}\Big]W_i((z - h_{i\bar{\Sigma}}(x))(t)),\end{aligned} \tag{15}$$

where $\mathscr{R}_{2\bar{\Sigma}_i} = 2P_{2_i}A_{22_i}^{-1}(A_{21_i} - \frac{1}{2}\beta_i B_{2\bar{\Sigma}_i}B_{1\bar{\Sigma}_i}^T P_{1_i} + \frac{1}{2}\beta_i B_{2\bar{\Sigma}_i}B_{2\bar{\Sigma}_i}^T (A_{12_i}A_{22_i}^{-1})^T P_{1_i})A_{12_i}$. Combining (14) and (15), we get the M-matrix $-\bar{A}_{i\bar{\Sigma}}$ with

$$\bar{A}_{i\bar{\Sigma}} = \begin{bmatrix} \frac{\lambda_{\max}(N_{i\bar{\Sigma}})}{\lambda_{\max}(P_{1_i})} & \frac{a_{3i}}{\lambda_{\min}(P_{2_i})} \\ \frac{a_{4i}}{\lambda_{\min}(P_{1_i})} & \frac{\varepsilon_i(a_{5i}+a_{6i})-1}{\varepsilon_i\lambda_{\max}(P_{2_i})} \end{bmatrix}.$$

Proceeding as done in the proof of Theorem 1, we get the desired result. □

4 Numerical Example

Example 1 Consider system (1) with $\mathscr{S} = \{1, 2\}$,

$$A_{11_1} = \begin{bmatrix} -5 & 0 \\ 0 & -10 \end{bmatrix}, A_{12_1} = \begin{bmatrix} 0.1 & 2 \\ 0.1 & 0 \end{bmatrix}, A_{21_1} = \begin{bmatrix} 1 & 3 \\ 2 & 1 \end{bmatrix}, A_{22_1} = \begin{bmatrix} 1 & -2 \\ 3 & -2 \end{bmatrix},$$

$$A11_2 = \begin{bmatrix} -3 & 1 \\ 0 & -6 \end{bmatrix}, A_{12_2} = \begin{bmatrix} 1 & 0 \\ 0.1 & 0.3 \end{bmatrix}, A_{21_2} = \begin{bmatrix} 2 & 3 \\ 1 & 1 \end{bmatrix}, A_{22_2} = \begin{bmatrix} -2 & 1 \\ 1 & -1 \end{bmatrix},$$

$$B_{1_1} = \begin{bmatrix} -5 & 0.5 \\ 0.1 & 0.15 \end{bmatrix}, B_{2_1} = \begin{bmatrix} 3 & -1 \\ 1 & 4 \end{bmatrix}, B_{1_2} = \begin{bmatrix} 4 & 5 \\ 0.5 & 1 \end{bmatrix}, B_{2_2} = \begin{bmatrix} 2 & -2 \\ 1 & 3 \end{bmatrix},$$

$\varepsilon_1 = 0.01,\ \beta_1 = 0.5,\ a_{11} = 0.1,\ a_{21} = 0.15,\ a_{31} = 0.02,\ a_{41} = 0.01,\ a_{51} = 70,\ Q_1 = -4I$,

$\varepsilon_2 = 0.02,\ \beta_2 = 0.25,\ a_{12} = 0.3,\ a_{22} = 0.2,\ a_{32} = 0.2,\ a_{42} = 0.02,\ a_{52} = 30$, and $Q_2 = -I$.

Case 1. When all actuators are operational, we have $P_{1_1} = \begin{bmatrix} 0.1025 & 0.0274 \\ 0.0274 & 0.0615 \end{bmatrix}$, $P_{1_2} = \begin{bmatrix} 0.1697 & 0.0616 \\ 0.0616 & 0.1322 \end{bmatrix}$, $P_{2_1} = \begin{bmatrix} 1.5 & 1 \\ 1 & 1.75 \end{bmatrix}$, $P_{2_1} = \begin{bmatrix} 0.5 & 0.5 \\ 0.5 & 1 \end{bmatrix}$, and $K_1 = \begin{bmatrix} 0.0217 & 0.0031 \\ 0.0840 & 0.0238 \end{bmatrix}$, $K_2 = \begin{bmatrix} -0.1638 & -0.0869 \\ -0.1449 & -0.0842 \end{bmatrix}$. Thus, the matrices $A_{0_i} + B_{0_i}K_i$ $(i = 1, 2)$ are Hurwitz and $\tau_a = \frac{\ln\mu}{\alpha^* - \nu} = 1.8330$.

Case 2. When there are failures in the first actuator of B_{1_i}, and the second actuator of B_{2_i} for both modes, i.e.,

$$B_{1\bar{\Sigma}_1} = \begin{bmatrix} 0 & 0.5 \\ 0 & 0.15 \end{bmatrix}, B_{2\bar{\Sigma}_1} = \begin{bmatrix} 3 & 0 \\ 1 & 0 \end{bmatrix}, B_{1\bar{\Sigma}_2} = \begin{bmatrix} 0 & 5 \\ 0 & 1 \end{bmatrix}, B_{2\bar{\Sigma}_2} = \begin{bmatrix} 2 & 0 \\ 1 & 0 \end{bmatrix},$$

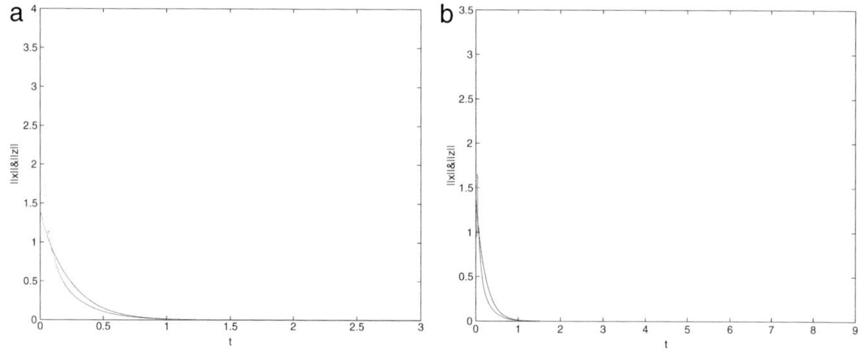

Fig. 1 Singularly perturbed switched system. (**a**) Operational actuators. (**b**) Faulty actuators

we have $P_{1_1} = \begin{bmatrix} 0.0993 & 0.0257 \\ 0.0257 & 0.0606 \end{bmatrix}, P_{1_2} = \begin{bmatrix} 0.2828 & 0.1570 \\ 0.1570 & 0.2145 \end{bmatrix}$, P_{2_1} and P_{2_2} are the same as for the normal case, and $K_1 = \begin{bmatrix} -0.1024 & -0.0278 \\ -0.0134 & -0.0055 \end{bmatrix}, K_2 = \begin{bmatrix} -0.1355 & -0.0991 \\ -0.1964 & -0.1249 \end{bmatrix}$. Thus, the matrices $A_{0_i} + B_{0_i}K_i$ $(i = 1, 2)$ are Hurwitz and $\tau_a = \frac{\ln \mu}{\alpha^* - \nu} = 4.1498$.

Figure 1a,b show the simulation results of $||x||$ (top) and $||z||$ (bottom) for the normal and the faulty cases respectively.

5 Conclusion

This paper has established new sufficient conditions that guaranteed the global exponential stability of SCSPS. The output of the faulty actuators has been treated as an outage. So that, as a future work, one may consider nonzero output which can be viewed as an external disturbance to the system. We have shown that, using ADTC with multiple Lyapunov functions, the full order switched system has been exponentially stabilized by $u = K_i x$ where in the faulty case, $K_{i\bar{\sigma}} = -\frac{1}{2}\beta_i B_{0i\bar{\sigma}}^T P_{i\bar{\sigma}}$. A numerical example has been introduced to clarify the proposed results.

Acknowledgements This work was partially supported by NSERC Canada. The third author acknowledges the sponsorship of King Abdulaziz University, Saudi Arabia.

References

1. Alwan, M.S., Liu, X.Z.: Stability of singularly perturbed switched systems with time delay and impulsive effects. Nonlinear Anal. **17**, 4297–4308 (2009)
2. Alwan, M.S., Liu, X.Z., Ingalls, B.: Exponential stability of singularly perturbed switched systems with time delay. Nonlinear Anal.: Hybrid Syst. **2**(3), 913–921 (2008)
3. Branicky, M.S.: Multiple Lyapunov functions and other analysis tools for switched hybrid systems. IEEE Trans. Autom. Control **43**(4), 475–482 (1998)
4. Cheng, X.M, Gui, W.H., Gan, Z.J.: Robust reliable control for a class of time-varying uncertain impulsive systems. J. Cent. S. Univ. Technol. **12**(1), 199–202 (2005)
5. Hespanha, J.P., Morse, A.S.: Stability of switched systems with average dwell-time. In: Proceedings of the 38th IEEE Conference on Decision and Control, vol. 3, Phoenix, pp. 2655–2660 (1999)
6. Kang, K.-I., Park, K.-S., Lim, J.-T.: Exponential stability of singularly perturbed systems with time delay and uncertainties. Int. J. Syst. Sci. **46**(1), 170–178 (2015)
7. Khalil, H.K.: Nonlinear Systems, 3rd edn. Prentice-Hall, Upper Saddle River (2002)
8. Kokotovic, P.V., Khalil, H.K., O'Reilly, J.: Singular Perturbation Methods in Control: Analysis and Design. Academic, London (1986)
9. Liberzon, D.: Switching in Systems and Control. Birkhäuser, Boston (2003)
10. Liberzon, D., Morse, A.S.: Basic Problems is Stability and Design of Switched Systems. IEEE Control Syst. Mag. **19**(5), 59–70 (1999)
11. Liu, X., Shen, X., Zhang, Y.: Exponential stability of singularly perturbed systems with time delay. Appl. Anal. **82**(2), 117–130 (2003)
12. Morse, S.: Supervisory control of families of linear set-point controllers, part 1: exact matching. IEEE Trans. Autom. Control **41**(10), 1413–1431 (1996)
13. Seo, C.J., Kim, B.K.: Robust and reliable H_∞ control for linear systems with parameter uncertainty and actuator failure. Automatica **32**(3), 465–467 (1996)
14. Shorten, R., Wirth, F., Mason, O., Wulff, K., King, C.: Stability criteria for switched and hybrid systems. SIAM Rev. **49**(4), 545–592 (2007)
15. Veillette, R.J.: Reliable state feedback and reliable observers. In: Proceedings of the 31st Conference on Decision and Control, Tucson, pp. 2898–2903 (1992)
16. Veillette, R.J., Medanic, J.V., Perkins, W.R.: Design of reliable control systems. IEEE Trans. Autom. Control **37**(3), 290–304 (1992)
17. Zhai, G., Hu, B., Yasuda, K., Michel, A.N.: Stability analysis of switched systems with stable and unstable subsystems: an average dwell time approach. Proc. Am. Control Conf. **1**(6), 200–204 (2000)

Application of an Optimized SLW Model in CFD Simulation of a Furnace

Masoud Darbandi, Bagher Abrar, and Gerry E. Schneider

Abstract Radiation is the most important part of heat transfer in combustion processes in applications such as furnaces. A proper radiation model for CFD simulations is the one, which provides the required accuracy with minimum computational cost. In this work, we focus on radiation modeling in CFD simulation of a laboratory scaled furnace. We use an optimized spectral line-based weighted-sum-of-gray-gases (SLW) model, which only needs four radiation transfer equations (RTE) solution for accurate prediction of radiative heat transfer in non-gray combustion fields. This is while the classic non-optimized SLW model needs at least 10–20 RTEs solution for the same case. We apply both the optimized and non-optimized SLW model to CFD simulation of the furnace. To evaluate the achieved results, we compare them with available measured data. We further compare the results of optimized SLW model with those of non-optimized SLW model. The comparisons demonstrate that the optimized SLW model provides the same accuracy of non-optimized SLW mode, while it requires less than 80% computational time.

1 Introduction

The fully coupled computational fluid dynamic (CFD) simulation is widely used for design and optimization proposes in different industrial furnaces, boilers, and fire-heaters. Radiation is the most important part of heat transfer in such combustion devices. The process of turbulent combustion and heat transfer strongly depend on each other and the small influence of thermal radiation may be magnified by the non-linear processes of turbulent combustion. Moreover, radiation can affect flame temperature and species concentration calculations. The inclusion of radiative heat transfer reduces the size of flame region, where the maximum temperatures occur.

M. Darbandi (✉) • B. Abrar
Department of Aerospace Engineering, Center of Excellence in Aerospace Systems, Sharif University of Technology, Tehran, P.O. Box 11365-8639, Iran
e-mail: darbandi@sharif.edu

G.E. Schneider
Department of Mechanical and Mechatronics Engineering, University of Waterloo, Waterloo, ON, N2L 3GI, Canada

© Springer International Publishing Switzerland 2016

J. Bélair et al. (eds.), *Mathematical and Computational Approaches in Advancing Modern Science and Engineering*, DOI 10.1007/978-3-319-30379-6_36

Therefore, the effects of thermal radiation should take into account even in case of non-luminous flames [1–3]. This requires the solution of radiative transfer equation (RTE), which determines the variation of radiation intensity in spatial and angular space. However, accurate solution of RTE in combustion fields is a very complicated task. It is mainly because of non-gray behavior of combustion gases. The absorption coefficient of non-gray gases, which appears in RTE, varies very rapidly and strongly with the wave number. Cumber and Fairweather [4] recommended that non-gray radiation models should be used for accurate simulations of combustion processes.

Among different non-gray radiation models, the spectral line-based weighted-sum-of-gray-gases (SLW) model has been the subject of many researches in the last two decades [5]. The SLW is categorized in the modern class of global models. The global models can be considered as improved versions of the weighted-sum-of-gray-gases (WSGG) model, which benefit form definition of absorption-line-black-body-distribution-function (ALBDF) to obtain the radiation parameters. Goutiere, et al. [6] evaluated different radiation models to calculate the radiative heat transfer in two dimensional enclosures. They showed that the SLW model can accurately predict the radiation heat transfer in their test cases. Similar investigations were also conducted by Coelho [7] in three-dimensional enclosures. He also confirmed the accuracy of SLW model. Johansson, et al. [8] evaluated WSGG and SLW models in several 1D tests mimic the conditions in oxy-fired boilers. Their results showed that the SLW model usually yields more accurate predictions than the WSGG model. Recently, Modest [9] reviewed historical development of nongray models. He proposed to include global models in simulations of complicated combustion systems. However, one should consider the computational cost of global models, such as the SLW model, before applying it in CFD simulations. The classic SLW model approximates the non-gray gases by sum-up of 10–20 gray gases. Therefore, it requires 10–20 RTEs solution for these gray gases, which could be very time consuming in real scale problems. It is well known that reducing the number of SLW's gray gases and therefore reducing the computational cost of RTEs solution is not readily possible, because it could introduce some errors in the predicted results [5].

In this work, we conduct a CFD simulation for turbulent reacting flow in a laboratory scaled gas fired furnace. We apply an optimized SLW model for efficient calculation of radiation heat transfer. The optimized SLW model approximates non-gray combustion gas mixture by sum-up of only 3 gray gases plus 1 clear gas. However, the accuracy of the optimized SLW model is preserved through the use of an optimization procedure. This would make the optimized SLW model computationally more efficient than the classic non-optimized SLW model.

2 Governing Equations and Physical Models

2.1 Turbulent Flow Field and Energy Transport Equations

The mass and momentum governing equations for a turbulent flow can be expressed as

$$\nabla \cdot (\rho \mathbf{V}) = 0 \tag{1}$$

$$\nabla \cdot (\rho \mathbf{V}\mathbf{V}) = -p\mathbf{I} + \nabla \cdot \bar{\tau} \tag{2}$$

in which, ρ is the density, $\mathbf{V}$ is the velocity vector, p is the pressure, and $\bar{\tau}$ is the stress tensor. Here, we use the κ-ε model along with the standard wall functions [10], which is commonly accepted for turbulence modeling in industrial applications such as furnaces and boilers [11, 12].

The governing equation for turbulent energy transport is given by

$$\nabla \cdot (\rho \mathbf{V} H) = \nabla \cdot (\frac{k_{eff}}{c_p} \nabla H) + S_H \tag{3}$$

in which, H is enthalpy. The viscous heating terms are neglected here. The effective thermal conductivity is given by $k_{eff} = k + (c_p \mu_t)/Pr_t$, with $Pr_t = 0.85$. k, c_p , μ_t are thermal conductivity coefficient, specific heat at constant pressure, and fluid viscosity coefficient, respectively. S_H also represents source term due to radiation heat transfer.

2.2 Combustion Model

Here, we use the mixture fraction theory for combustion modeling. The mixture fraction is defined as $f = (Y_i - Y_{i,o})/(Y_{i,f} - Y_{i,o})$, in which Y_i is the mass fraction of species i. The subscripts f and o also refer to fuel and oxidizer at inlet streams. In this theory, a transport equation is solved for each of mean mixture fraction and mixture fraction variance. One advantage of using mixture fraction theory for turbulent non-premixed combustion modeling is that, under the assumption of chemical equilibrium, instantaneous values of species concentration, density, and temperature can uniquely relate to mixture fraction via $\varphi_i = \varphi_i(f_i, H_i)$. However, only averaged values of these scalars, i.e. φ, f, and H are predicted in turbulent flow modeling. The relationship between the averaged values and the instantaneous values can be achieved through the use of presumed β-shaped probability density function (PDF)[13].

Here, we perform the chemical equilibrium calculations by means of Gibbs' free energy minimization. We assume there are 20 species and radicals in the equilibrium mixture including CH_4, C_2H_6, C_3H_8, C_4H_{10}, CO_2, N_2, O_2, H_2O, CO, H_2, OH, O,

O_3, HOCO, HCO, CHO, H_2O_2, HONO, HO_2, and H. More details on governing equations and combustion modeling in turbulent reacting flow can be found in Ref. [14].

2.3 Radiation Model

Radiation effects are taken into account in the CFD simulation of combustion processes via addition of radiation source term in the right hand side of energy transport equation, i.e., Eq. (3). In the SLW model, this source term is given by

$$S_H = \sum_{j=1}^{J} \kappa_j (4\pi a_j I_b - \int_{4\pi} I_j d\Omega) \tag{4}$$

The calculation of above source term requires the solution of j RTE s as follows:

$$\frac{dI_j}{ds} = \kappa_j (a_j I_b - I_j) \tag{5}$$

in which, I is the radiation intensity, κ_j is the absorption coefficient, and a_j is the emissivity weighting factor. κ_j is related to the absorption cross section as described by $\kappa_j = NC_j$. To achieve C_j, the entire range of absorption cross section between C_{max} and C_{min} is divided into J logarithmically spaced intervals. The interval borders named as supplemental absorption cross section are obtained using $\tilde{C}_j = C_{min}(C_{max}/C_{min})^{(j/J)}$, where C_j is defined as $C_j = \sqrt{\tilde{C}_j \tilde{C}_{j+1}}$. The emissivity weighting factor a_j is also calculated by employing the ALBDF definition as follows:

$$F(C_\eta, T_g, T_b, Y) = \frac{1}{I_b(T_b)} \int_{\{\eta: C_\eta(T_g, Y) \le C\}} I_{b\eta}(T_b, \eta) d\eta \tag{6}$$

Using the above definition, a_j is written as $a_j = F(\tilde{C}_{j+1}) - F(\tilde{C}_j)$. To obtain the values of ALBDFs, here we use the tabulated values, which are presented by Pearson and Webb [15] for H_2O, CO_2, and CO gases based on the recent HITEMP2010 spectroscopic database [16].

2.4 The Optimized SLW Model

As outlined in the previous sections, the SLW model is involved with $J + 1$ RTEs solutions selections. About $J = 10 - 20$ gray gases are usually required to achieve the required accuracy in the SLW model. However, 10–20 RTEs solutions may need much computational efforts in real scale combustion application problems. This is a disadvantage, which makes the use of classic SLW model prohibitive in CFD simulations. On the other hand, choosing a value of J bellow 10 may deteriorate the accuracy of SLW model [5]. Here, we propose the use of optimized SLW model to reduce the number of gray gases. Instead of logarithmic discretization in the classic non-optimized SLW model, we benefit from the optimization procedure to obtain gray gas parameters a_j and κ_j in the optimized SLW model. The optimization procedure minimizes the difference between the calculated values of total emissivity based on only three gray gases parameters and the true values of total emissivity over the path lengths L_i. Mathematically, it is required to minimize the following objective function:

$$\text{error} = \sum_i [\varepsilon_{\text{optimized SLW}}(L_i) - \varepsilon_{\text{true}}(L_i)]^2 \tag{7}$$

where the values of total emissivity, based on the three gray gases parameters, are calculated from

$$\varepsilon_{\text{optimized SLW}}(L_i) = \sum_{j=1}^{J=3} a_j(1 - e^{-\kappa_j L_i}) \tag{8}$$

and the true values of total emissivity ε_{true} can be obtained using the detail spectroscopic data in the HITEMP2010 database and taking the following integration:

$$\varepsilon_{\text{true}}(L_i) = \frac{1}{I_b} \int_\eta I_{b\eta}(1 - e^{-\kappa_\eta L_i}) d\eta \tag{9}$$

The path lengths used in Eqs. (7), (8) and (9) should cover a range of typical lengths consistent with the problem in hand. Here, we use a typical range from 0.01 to 10 times of the characteristic length of the problem including about 10 path lengths in each decade. The minimization of the objective function in Eq. (7) is in fact a non-linear curve-fit problem, in which the values of a_j and κ_j parameters are obtained for each of the three gray gases as of the results. We use the trust region algorithm for solving this problem. In this way, we can accurately approximate the non-gray gases mixture by sum-up of only 3 gray and 1 clear gases. In the optimization procedure, the true values of ε are first calculated over a range of distances L typical of problem under investigation. Then, the optimized parameters a_j and κ_j for each of 3 gray gases are directly achieved using the least square curve fit to these ε data.

3 Numerical Methods and Computational Procedure

We use the finite-volume method to treat the flow governing equations and to derive the sets of linear algebraic equations. In this method, the solution domain is discretized to a number of control volumes. The governing equations are integrated over the entire faces of each control volume. We also use the finite-volume method to derive a conservative statement for the RTE. For this purpose, each octant of the angular space 4π is discretized into $N_\theta \times N_\varphi$ non-overlapping control angles Ω_l. In this work, we use the SIMPLE algorithm to solve the flow governing equations. Details of the employed finite volume method for the flow and RTE governing equations and the SIMPLE algorithm are given in our previous works [17, 18].

4 Problem Description

We consider a laboratory scale furnace with available measured data in this study [19]. The furnace is a circumferentially symmetric 300 kW BERL combustor, with properly insulated octagonal cross section, conical hood and cylindrical exhaust duct. It is equipped with a vertically fired burner in the bottom. The burner has 24 radial fuel injection holes and a bluff center body. Natural gas is radially introduced through these holes. Air is introduced through an annular inlet equipped with a proper swirler. Therefore, a non-premixed turbulent swirl stabilized flame forms in the furnace. We model this problem as axisymmetric by appropriate adjustment with the real 3D furnace. Figure 1 presents a schematic of the furnace and a close up of burner's head along with main dimensions.

The natural gas is assumed to be composed of 96.5 % CH_4, 1.7 % C_2H_6, 0.1 % C_3H_8, 0.1 % C_4H_{10}, 0.3 % CO_2, and 1.3 % N_2. Fuel jet is considered to have a mean radial velocity of $v = 157.77$ m/s at fuel inlet boundary. Air stream is also considered to have a mean axial velocity of $u = 31.35$ m/s and a mean swirl velocity of $w = 20.97$ m/s. The fuel and air inlet temperatures are 312 and

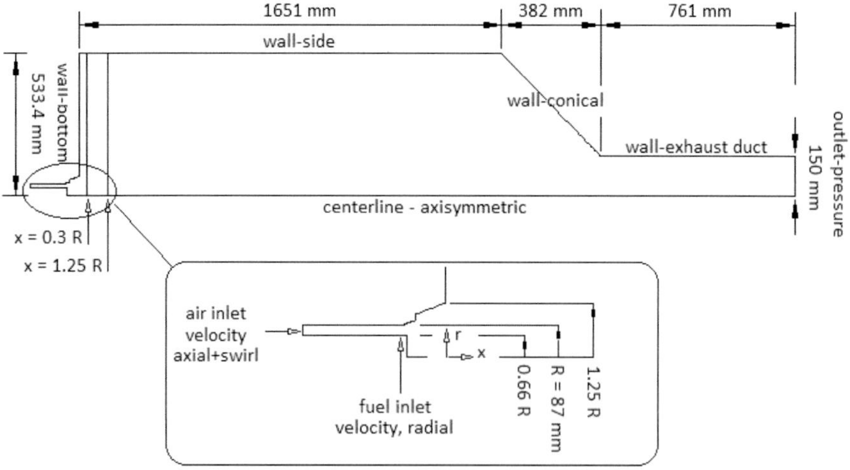

Fig. 1 Geometry of the furnace and close-up of the burner, Sayre et al. [19]

308 K, respectively. The temperature of the bottom wall, side wall, conical wall, and exhaust duct wall are set to be constant at 1100, 1220, 1305, and 1370 K, respectively. All walls are treated to be diffuse with an internal emissivity of 0.5. The gauge pressure is also set to be zero at the furnace outlet.

5 Results and Discussion

Here we solve the problem using both the optimized and non-optimized SLW models. In the non-optimized SLW model, $j = 20$ gray gases are used. In this section, we first provide the results, which are achieved using the optimized SLW model and compare them with the available measured data of furnace. Then we further compare the achieved radiative quantities of the optimized SLW model with those of the non-optimized SLW model. We obtain our results using a non-uniform structured grid of about 10,000 control volumes and $N_\theta \times N_\varphi = 4 \times 4$ control angles.

Figure 2 gives qualitative representations of the achieved flame and using the optimized SLW model. Figure 2a displays the temperature contour, Fig. 2b and c displays the H_2O and CO_2 mole fraction contours, and Fig. 2d displays the CO mole fraction contour. All the presented contours in Fig. 2 have important role in radiation calculations.

Figure 3 presents radial temperature profiles in different axial locations of the furnace. We compare the results of optimized SLW model with the available measured data of Sayre, et al. [19]. It seems that our results present a thinner flame than the measured data. This may be partially because of the axisymmetric modeling of a three-dimensional furnace. Also the sharp gradients of temperature variations in our results can be partially due to ignoring the finite rate of reactions in our equilibrium chemistry calculations. Moving downstream in the furnace, these are better agreement between our predictions and the measured data.

Figure 4 compares the results of the optimized and non-optimized SLW models with each other. Figure 4a presents the contours of radiation source term. For a better comparison, we limit the range of depicted source term. Comparison in this figure reveals that the results of optimized SLW model are in complete agreement with those of the non-optimized SLW model. Figure 4b displays the variation of incident radiation heat flux along the furnace walls and the variation of temperature along the furnace centerline. Again, there are complete agreement between the results of optimized and non-optimized SLW models. These comparisons reveal that the optimized SLW model can provide the same accuracy as it is provided by the non-optimized SLW model.

Table 1 compares the number of equations and the required CPU times for 1 iteration of coupled CFD solution using different SLW models. The optimized SLW model only needs 4 RTEs to be solved, while the non-optimized SLW model needs 21 RTE solutions. This would indicate a great computational advantage for the optimized SLW model. Here, all the computations are conducted on a laptop equipped with an Intel CORE i5 CPU and 4 GB RAM. As is seen in Table 1,

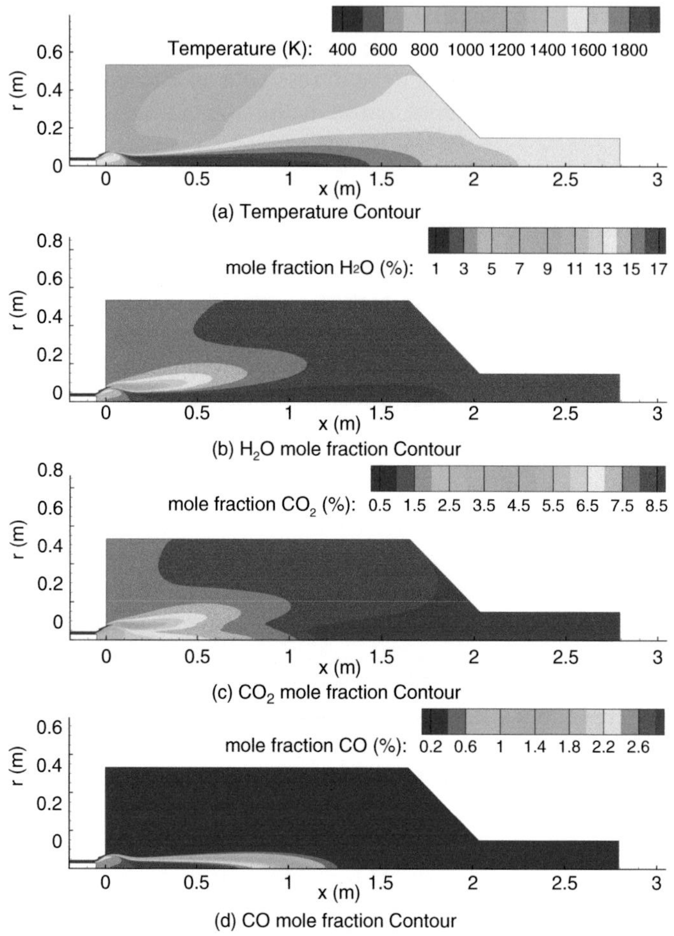

Fig. 2 Contours of **a**: temperature and **b**: H_2O, **c**: CO_2, and **d**: CO mole fractions in the furnace

each solution iteration only lasts 0.8 s with the optimized SLW model. However, the required time for each solution iteration is 4.1 s with the non-optimized SLW model. In other words, using the optimized SLW model makes over 80 % reductions in the required computational time in comparison with using the non-optimized SLW model. This is where both models provide the same level of accuracy, see Fig. 4.

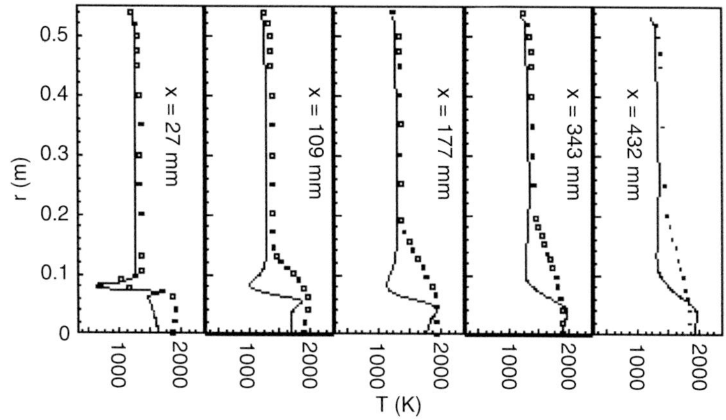

Fig. 3 Radial temperature profiles at different axial distances in furnace and comparison with the measured data of Sayre, et al. [19]

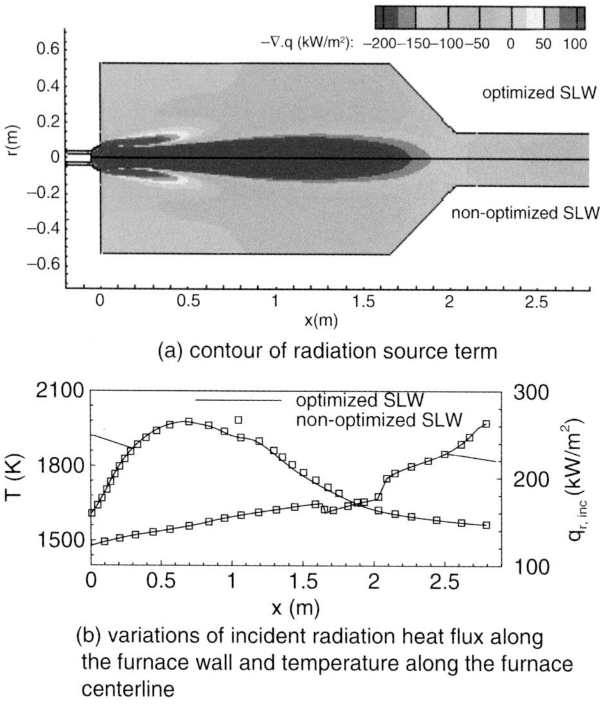

Fig. 4 Comparison between the results of optimized and non-optimized SLW models, **a**: contours of radiation source term, **b**: variations of incident radiation heat flux along the furnace wall and temperature along the furnace centerline

Table 1 Number of equations and required CPU times for each coupled CFD solution iteration using different SLW models

Model	Number of RTEs	CPU time (s)
Optimized SLW	4	0.8
Non-optimized SLW	21	4.1

6 Conclusion

In this work we introduced the optimized SLW model, which used only 3 gray gases plus 1 clear gas to approximate the radiation in non-gray gas mixtures. Therefore, it reduces the number of radiative transfer equatous to only 4 RTEs solution. This is while the classic non-optimized SLW model requires about 10–20 RTEs solution to support the same accuracy. We applied both the optimized and non-optimized SLW models in the CFD simulation of turbulent reacting flow in a furnace. The achieved results were in good agreements with available measured data in furnace. We further compared the results of optimized SLW model with those of non-optimized SLW model. The comparisons demonstrated that the accuracy of the optimized SLW model was as good as the non-optimized SLW model. This is while the use of optimized SLW model required 80 % less computational time. This improvement is of great value for intensive CFD simulations. Therefore, we suggest the optimized SLW model for radiation heat transfer calculation in CFD simulation of combustion processes.

Acknowledgements The authors would like to thank the research deputy of Sharif University of Technology for the financial support received during this research work.

References

1. Keramida, E.P., Liakos, H.H., Founti, M.A., Boudouvis, A.G., Markatos, N.C.: Radiative heat transfer in natural gas-fired furnaces. IJHMT **43**(10), 1801–1809 (2000)
2. Xu, X., Chen, Y., Wang, H.: Detailed numerical simulation of thermal radiation influence in Sandia flame D. IJHMT **49**(13), 2347–2355 (2006)
3. Kontogeorgos, D.A., Keramida, E.P., Founti, M.A.: Assessment of simplified thermal radiation models for engineering calculations in natural gas-fired furnace. IJHMT **50**(25), 5260–5268 (2007)
4. Cumber, P.S., Fairweather, M.: Evaluation of flame emission models combined with the discrete transfer method for combustion system simulation. IJHMT **48**(25), 5221–5239 (2005)
5. Denison, M.K., Webb, B.W.: A spectral line-based weighted-sum-of-gray-gases model for arbitrary RTE solvers. J. Heat Transf. **115**(4), 1004–1012 (1993)
6. Goutiere, V., Liu, F., Charrette, A.: An assessment of real-gas modelling in 2D enclosures. JQSRT **64**(3), 299–326 (2000)
7. Coelho, P.J.: Numerical simulation of radiative heat transfer from non-gray gases in three-dimensional enclosures. JQSRT **74**(3), 307–328 (2002)
8. Johansson, R., Andersson, K., Leckner, B., Thunman, H.: Models for gaseous radiative heat transfer applied to oxy-fuel conditions in boilers. IJHMT **53**(1), 220–230 (2010)

9. Modest, M.F.: The treatment of nongray properties in radiative heat transfer: from past to present. J. Heat Transf. **135**(6), 061801 (2013)
10. Launder, B.E., Spalding, D.B.: The numerical computation of turbulent flows. Comput. Methods Appl. Mech. Eng. **3**(2), 269–289 (1974)
11. Belosevic, S., Sijercic, M., Oka, S., Tucakovic, D.: Three-dimensional modeling of utility boiler pulverized coal tangentially fired furnace. IJHMT **49**(19), 3371–3378 (2006)
12. Crnomarkovic, N., Sijercic, M., Belosevic, S., Tucakovic, D., Zivanovic, T.: Numerical investigation of processes in the lignite-fired furnace when simple gray gas and weighted sum of gray gases models are used. IJHMT **56**(1), 197–205 (2013)
13. Cumber, P.S., Onokpe, O.: Turbulent radiation interaction in jet flames: sensitivity to the PDF. IJHMT **57**(1), 250–264 (2013)
14. Kuo, K.K.: Principles of Combustion. Wiley, Hoboken (2005)
15. Pearson, J.T., Webb, B.W., Solovjov, V.P., Ma, J.: Updated correlation of the absorption line blackbody distribution function for H2O based on the HITEMP2010 database. JQSRT 128:10–17 (2013)
16. Rothman, L.S., Gordon, I.E., Barber, R.J., Dothe, H., Gamache, R.R., Goldman, A., Tennyson, J.: HITEMP, the high-temperature molecular spectroscopic database. JQSRT **111**(15), 2139–2150 (2010)
17. Darbandi, M., Abrar, B., Schneider, G.E.: Solving combined natural convection-radiation in participating media considering the compressibility effects. In: Proceeding of the 52nd Aerospace Sciences Meeting, AIAA, Maryland (2014)
18. Darbandi, M., Abrar, B.: A compressible approach to solve combined natural convection-radiation heat transfer in participating media. Numer. Heat Transf. B **66**(5), 446–469 (2014)
19. Sayre, A., Lallemant, N.D.J., Weber, R.: Scaling Characteristics of Aerodynamics and Low-NOx Properties of Industrial Natural Gas Burners, The SCALING 400 Study, Part IV: The 300 kW BERL Test Results. International Flame Research Foundation (1994)

Numerical Investigation on Periodic Simulation of Flow Through Ducted Axial Fan

Seyedali Sabzpoushan, Masoud Darbandi, Mohsen Mohammadi, and Gerry E. Schneider

Abstract In this paper, the flow through an axial fan is suitably simulated considering relatively low mesh sizes and benefiting from the periodic boundary condition. The current periodic boundary condition implementation has major differences with the past classical ones, which were routinely used in literature. In this regard, we first discuss the ambiguities behind the current periodic geometry and the proposed mesh generation and possible boundary condition choices. Then, proper remedies are proposed to resolve them. One remedy returns to the proper choice of pitch ratio magnitude. After practicing various pitch ratio magnitudes, we eventually arrive to an optimum one, which provides suitable numerical accuracy despite using sufficiently low number of mesh elements. In order to validate the achieved numerical solutions, they are compared with the experiments and other available numerical results. The proposed approach can be also used in simulations of other turbomachinary cases, where there are serious computer memory and computational time limitations. For instance, using the current periodic boundary condition approach, one can readily simulate very huge wind tunnels considering a full 3D model of its fans, i.e., including its rotor and stator instead of considering simple fan pressure jump or actuator disc models.

S. Sabzpoushan • M. Mohammadi
Department of Aerospace Engineering, Sharif University of Technology, Tehran, P.O. Box 11365-8639, Iran
e-mail: s_sabzpoushan@ae.sharif.edu; mohammadi@ae.sharif.edu

M. Darbandi (✉)
Department of Aerospace Engineering, Center of Excellence in Aerospace Systems, Sharif University of Technology, Tehran, P.O. Box 11365-8639, Iran
e-mail: darbandi@sharif.edu

G.E. Schneider
Department of Mechanical and Mechatronics Engineering, University of Waterloo, Waterloo, ON, N2L 3G1, Canada

© Springer International Publishing Switzerland 2016

J. Bélair et al. (eds.), *Mathematical and Computational Approaches in Advancing Modern Science and Engineering*, DOI 10.1007/978-3-319-30379-6_37

1 Introduction

Fan has various applications in different industrial equipment, e.g., heat exchangers, air conditioning systems, and wind tunnels. In case of applications with large volume of air, the industries may have to consider using parallel fans. However, this may lead to a difficulty that the flow at upstream and/or downstream of fans may not perform axisymmetric pattern. In such cases, one may not simply apply the periodic boundary condition at the boundaries of a slice of flow passage. It is because if one wishes to simulate such a flow field, this would cause pitch ratio at the interfaces between the moving rotor domain and the stationary ducts domains, which in turn will cause critical regions at such interfaces. We should be careful on two key points at these interfaces. First, one should know that the face mesh at these interfaces must be as similar as possible to avoid missing data from one zone to the next one [1]. Second, there is an important constraint indicating that the pitch ratio cannot be further up than a certain limit [2]. As a rule of thumb, the pitch ratio must not exceed the upper limit of about 10 if one expects an almost reliable simulation [3]. Literature shows that there have been several efforts to simulate axial fans assuming both full rotor disc consideration and periodic blade boundary condition implementation. Fidalgo [1] simulated a three-dimensional unsteady flow through a full-annulus rotor disc with distortion in its inlet total pressure. Additionally, Denton [2] raised some limitations in turbomachinary CFD indicating that if the engineers do not care those limitations, they may not get solutions with sufficient accuracies. He specially focused on specific geometries such as the tip clearance and leading edge shapes. Cevik [4] performed a simulation on the full geometry of an axial fan with a diameter of 3.13 m placed in a duct with 12.5 m length. Using the k-epsilon turbulence model, the total number of mesh elements were about 2,500,000. There are some other publications, which use the true periodic boundary condition. For example, Le Roax [5] simulated the rotor disc and its upstream/downstream ducts completely and all together using the periodic boundary condition.

At this stage, we need to clarify the "pitch ratio" importance in axial fan modelling by introducing a simple example (see Fig. 1). Prior to any explanation, we should mention that this simple example is so helpful to examine the validity and reliability of current periodic simulation at next stage. This can be done for more complicated cases by comparing its solution with the fan experimental data and full blade rotor simulation results. As it is shown in Fig. 2(a), the target fan has 12 blades located in a long duct. For this geometry, the pitch ratio is $360°/360° = 1$ at the interface of rotational and stationary zones. Now, suppose a periodic domain in which, we consider only two blades of the fan with $(2/12)\times 360° = 60°$ annulus slice cross section. This slice would interact with the full 360° domains at its downstream and upstream zones. For this geometry, the pitch ratio becomes $360°/60° = 6$. In this article, we focus on mass flow rate and flow uniformity rather than other parameters. In case of using parallel set of fans, interactions between the inflow and outflow streams would also become important.

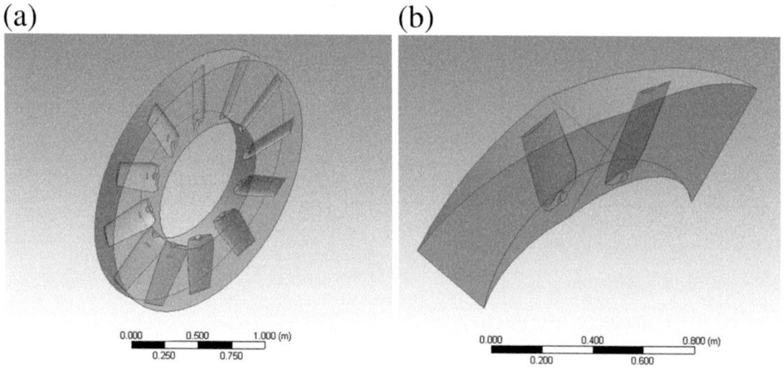

Fig. 1 Two types of rotor disc geometry. (**a**) Full rotor disc. (**b**) Sliced rotor disc

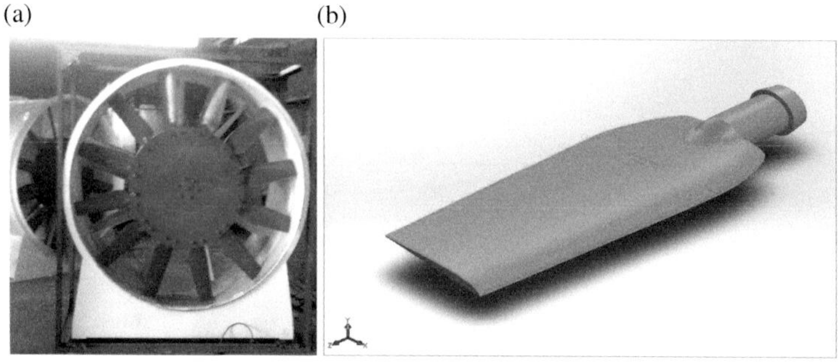

Fig. 2 The fan and its rotor blade geometries. (**a**) The assembled fan. (**b**) The result of rotor blade CMM

2 Geometry and the Solution Domain

The length of rotor blade shown in Fig. 2 is 0.5 m from root to tip and the diameter of its casing is 2 m. The maximum allowable rotational speed is about 900 rpm. In order to eliminate the swirl of flow through the rotor, a stator has been embedded downstream of the rotor, if this swirl can have considerable effects. Fortunately, in case of using periodic rotor disc, the flow passage in sliced rotor domain does not need to be coincident with the streamlines. It is because any amount of air exiting from one of the two periodic sides, would enter from the other side.

Before applying the current approach in complicated geometries with complex conditions, it is better to examine it in a simple geometry test case. Figure 3 presents this simple test case, which is constructed based on a complex wind tunnel air driver geometry. According to the standards related to the methods of fan testing [4, 6, 7], the minimum lengths for the upstream and downstream ducts of an axial fan are assumed to be 10D and 5D respectively, where D is fan casing diameter. We should satisfy these recommended minimum standard values to let the flow become fully developed before passing through the rotor and also at the end of downstream duct.

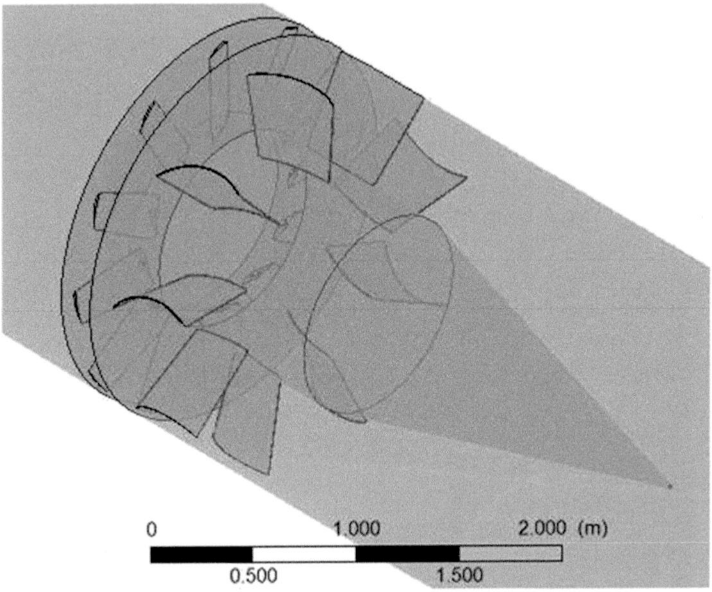

Fig. 3 The fan between forward and aft ducts in a simple test geometry

3 Grid Distribution and Boundary Conditions

Figure 4 shows rotor and stator blade sections accompanied with the generated mesh. The boundary layer mesh and mesh concentration have been used wherever needed. The boundary layer mesh is controlled suitably to achieve a maximum thickness of about 10 cm.

Evidently, it is unavoidable facing with some limitations in mesh generation. The sources can be due to the geometric constrains. For example, we are faced with very tiny gap at the blade tip clearance, which has considerable effects on the achieved fan performance; especially the fan noise generation [8, 9]. This tiny gap makes it so difficult to connect the boundary layers generated on the upper and lower surfaces to the blade tip. However, there are two possible approaches to resolve this problem. One is to reduce the number of layers at the blade's tip and casing. The other one is to ignore the layer's growth rate and to compress them. This sometimes requires to change the first layer height. The first approach needs a sudden detraction of layers, which is harmful for the numerical solution, especially for capturing the separation

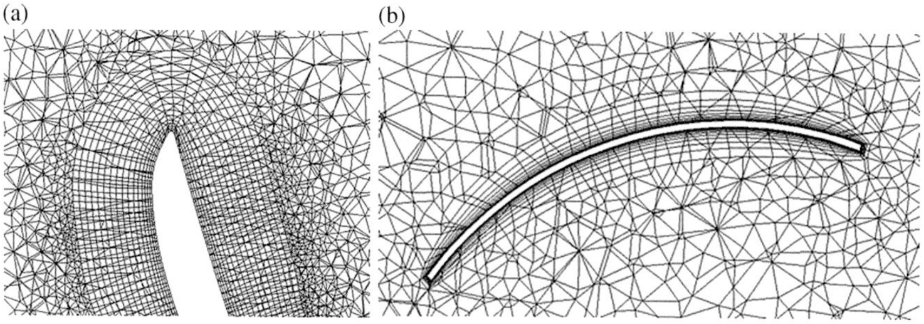

Fig. 4 Two sample sections of mesh distributions around the rotor blade and the stator vane. (**a**) Rotor blade leading edge. (**b**) Stator vane

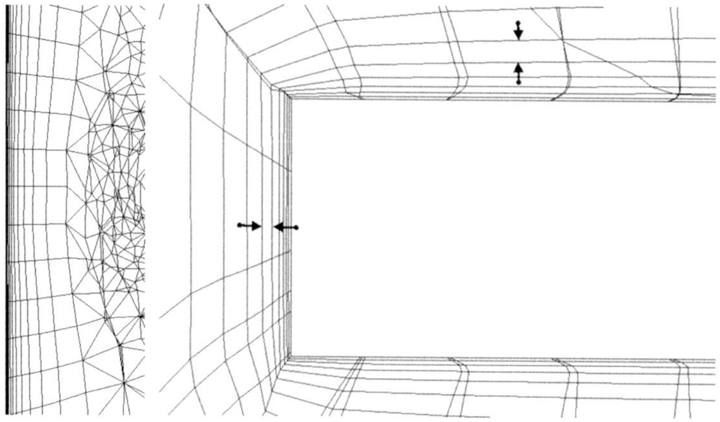

Fig. 5 The mesh cluster between the rotor blade tip and its casing

and tip vortices. Hence, the second approach is chosen, see Fig. 5. Controlling y^+ in the range of 1–100, we can employ *k-ω* SST turbulence model. As another concern, it is important for the surface mesh on the blade to preserve the real shape of leading edge [5, 10]. Finally, Fig. 6 enforces this point that the surface mesh on both sides of an interface with any pitch ratio must be as similar as possible.

If we choose a slice of rotor disc (and that of the stator disc if it exists) instead of the entire rotor and stator discs, the number of elements (and consequently the required computer memory) will decrease considerably. This needs to apply suitable periodic conditions. After performing some mesh independency efforts, we achieved to a grid resolution of approximately 600,000 elements. This is where the entire rotor disc solution domain needed about 1,700,000 elements to result in mesh-independent solution. Such mesh studies were conducted based on monitoring both

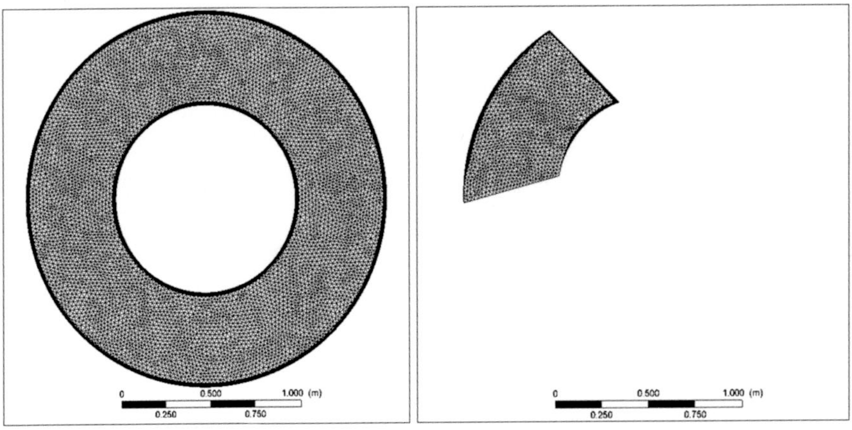

Fig. 6 Similar mesh generations at the stationary and rotational domains' interface

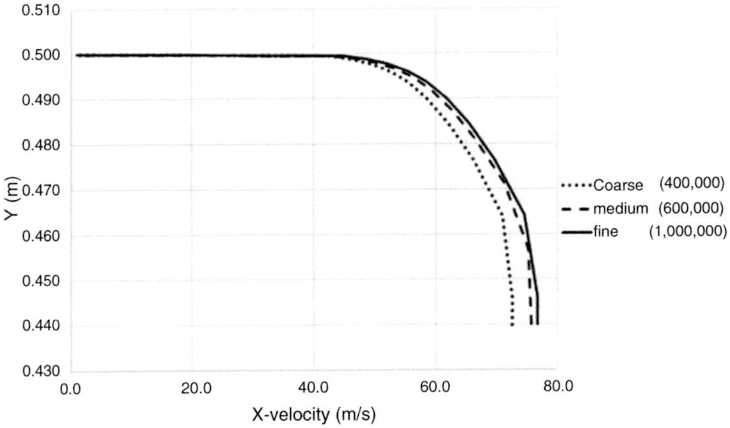

Fig. 7 The mesh refinement study via monitoring the boundary layer velocity profile in three different grid resolutions

the mass flow rate magnitude and the boundary layer velocity profile. Figure 7 shows the results of mesh refinement study for the sliced rotor considering the boundary layer velocity profile inside the duct wall. This figure also verifies our estimation on the boundary layer thickness which is about 6 cm, whereas the boundary layer mesh had been generated with a total thickness of 10 cm.

Here, we would like to mention some points on the chosen boundary conditions. The boundary conditions at the inlet and outlet sections of the duct are not fixed by neither mass flow rate, nor velocity magnitude and direction. So, the solution can be gradually converged during the numerical procedure. We determine the turbulence intensity of 1 % at the inlet section and the static temperature of 298 K at this boundary. Determining the total pressure at the inlet and the average static pressure at the outlet could be other possible choices as the boundary conditions. These two alternative choices require more flow parameters and constraints to be defined at these boundaries and hence lower the problem generality [11]. The interface connections are chosen in a manner to enforce equal area-weighted average of normal velocities at both sides of the interface. To achieve more similarity with the real conditions, the rotor casing (shroud) is forced to have a counter-rotating RPM relative to the rotational domain in order to be motionless in stationary reference frame [10, 12, 13]. The ambient pressure is equal to the value reported in standard atmosphere table for Tehran altitude, which is about 87,000 Pa.

4 The Results and Discussion

According to the experimental tests performed at the atmospheric pressure of 87,000 Pa under the standard conditions, the air mass flow rate for a single fan at the maximum rotational speed is about 53~54 kg/s. As seen in Fig. 8, a high pitch ratio would lead to invalid solution with about 60 % deviation from the exact value. However, it promptly drops below 10 % after this point and approaches rapidly to the exact solution. Thus, this can not only help us to find the proper pitch ratio magnitude, but also can be used as a type of validation for the proposed simulation case.

Figure 9 shows the tangential velocity contours just after the exit of rotor (before entering the stator) and right after the stator. To elaborate the swirl and the effect of stator on its decay, Fig. 10 presents the streamlines passing the fan and compares the flow swirl in the absence and presence of stator. The symmetry in these contours

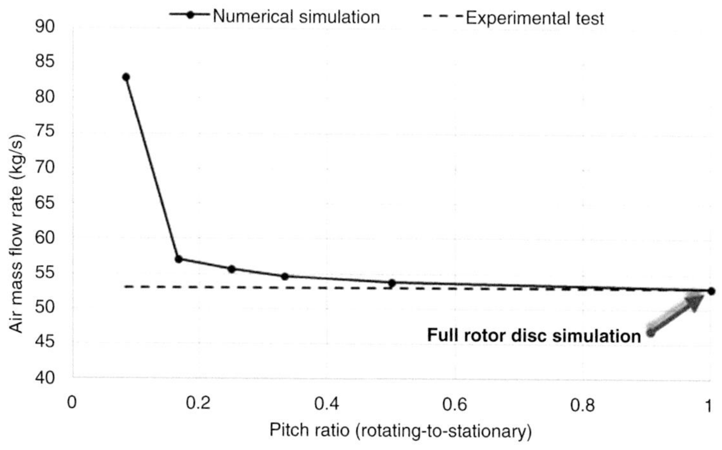

Fig. 8 Effect of pitch ratio on the calculated mass flow rate through the axial fan

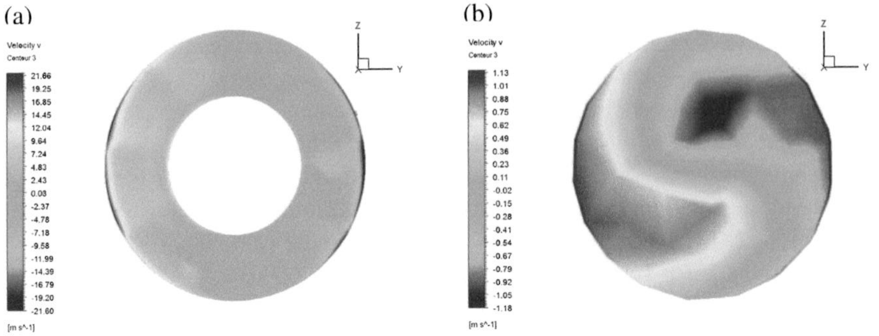

Fig. 9 Tangential velocity contours, indicating the swirl strength. (**a**) Just at the downstream of rotor. (**b**) right after the stator

and non-swirling flow at the upstream of the rotor (see streamlines in Fig. 10) also imply that the current periodic simulation performed reliably.

To provide more qualitative validations, one can check the pressure change over the fan. The current calculations showed that the area-weighted average of static pressure (gauge) over two sections somewhere at middle lengths of upstream and downstream ducts would be about −156 and 23 Pa, respectively. This value is about −180 Pa at inlet and 2 Pa at outlet. They indicate that the static head of fan would be approximately 180 Pa. Figure 11 shows the static pressure contours at two sections just before and after the rotor disc. According to available fan performance curves, this is exactly what we expect for the static head of such a fan with a relatively low incidence angle (about 17°) of rotor blades.

Figures 12 and 13 show two important detected phenomena, which support the current periodic boundary condition approach in axial fan simulation. The first one is small vortices, which can potentially cause flow to separate from the blade surface. The second one is the flow leakage from the blade pressure side to its suction side through the gap at tip clearance. It is also known as downwash phenomenon.

5 Conclusion

We introduced a new periodic boundary condition implementation, which was relatively different from the past classical ones. This new method can be readily used to simulate a single fan or a set of several parallel fans despite facing with asymmetric flow conditions. The major obstacle is the pitch ratio, which occurs at the interfaces between the rotor domain and the stationary upstream and downstream ducts' domains. After practicing various pitch ratios, we eventually found an optimum magnitude, which would provide suitable mesh sizes with guaranteed numerical accuracy. Our study showed that the number of mesh elements required to

(a)

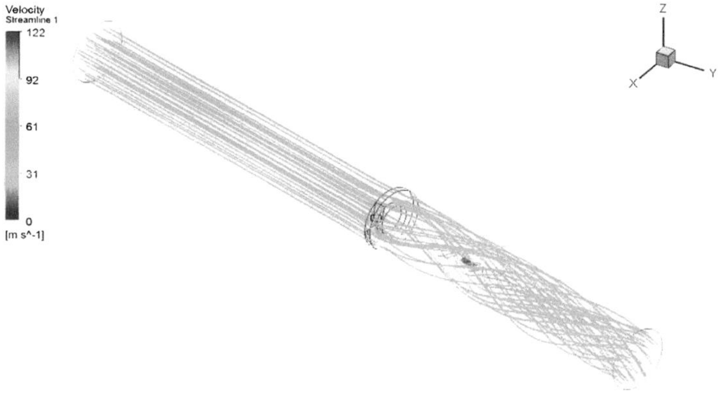

(b)

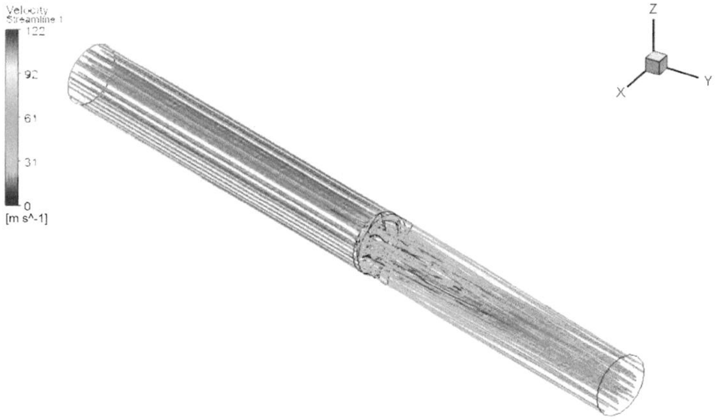

Fig. 10 The streamlines inside the duct at a rotor speed of 900 rpm (**a**) without stator (**b**) with stator

simulate a full-geometry fan in such applications is more than 11 times the number of elements required using the proposed periodic approach. Indeed, this reduction in the number of elements is so essential in many real industrial applications. For example, it is important when one wishes to simulate the set of several combined fans, e.g., a set of four parallel fans, utilized in a huge air supply ventilation system.

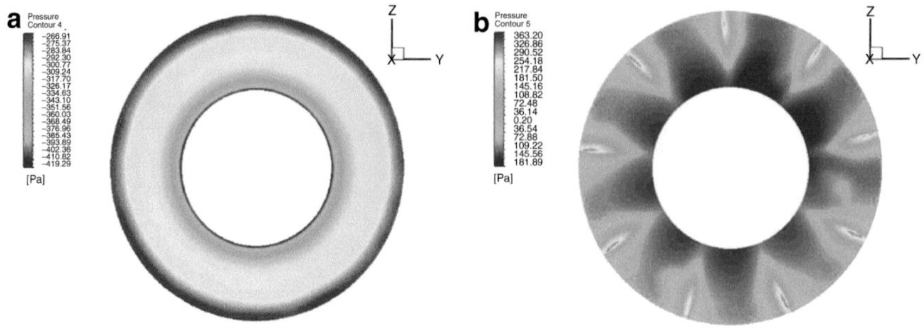

Fig. 11 Contours of gauge static pressure. (**a**) Just before passing the rotor. (**b**) Just after passing the rotor

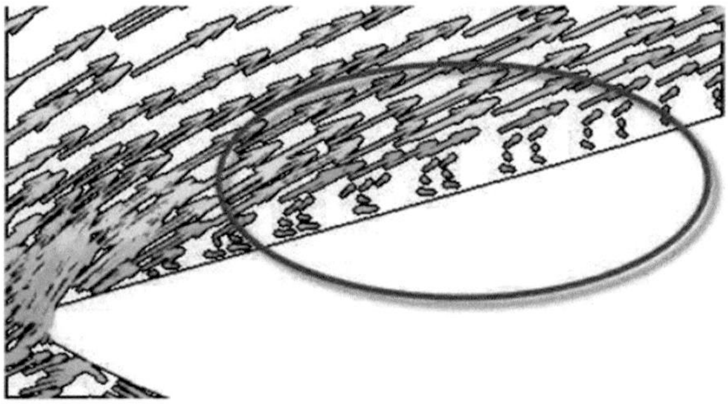

Fig. 12 Demonstration of small vortices appearing close to the rotor blade face

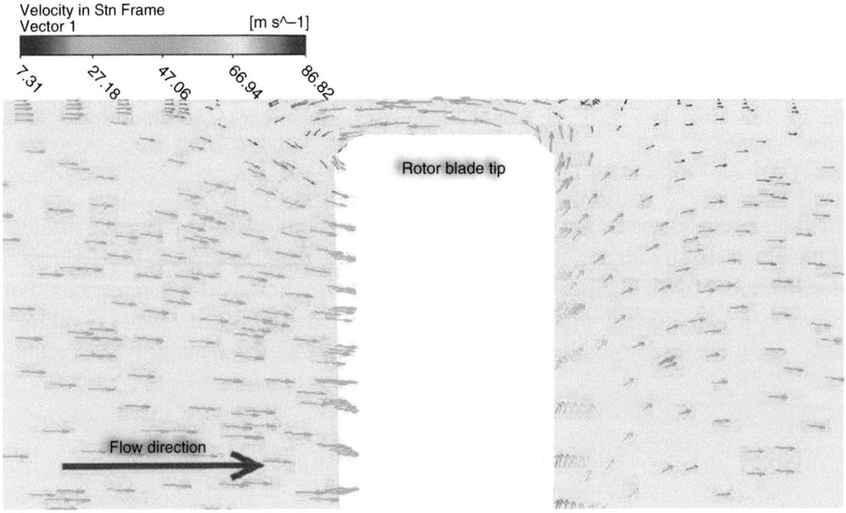

Fig. 13 Demonstration of flow leakage from the high pressure side to the low pressure side through the tip clearance gap

Acknowledgements The authors would like to greatly thank the financial supports received from the Deputy of Research and Technology of Sharif University of Technology (SUT).

References

1. Fidalgo, J.: A Study of Fan-Distortion Interaction with the NASA Rotor-67 Transonic Stage. In: ASME Turbo Expo, June 2010, Glasgow (2010)
2. Denton, J.D.: Some Limitations of Turbomachinery CFD. In: ASME Turbo Expo, June 2010, Glasgow (2010)
3. Bhasker, C.: Simulation of three dimensional flows in industrial components using CFD techniques. In: Minin, I. (ed.) Computational Fluid Dynamics Technologies and Applications. InTech (2011). ISBN:978-953-307-169-5, doi:10.5772/19909
4. Cevik, F.: Design of an axial flow fan for a vertical wind tunnel for paratroopers. Master thesis, Mechanical Engineering Department, Middle East Technical University (2010)
5. le Roux, F.N.: The CFD simulation of an axial flow fan. Master thesis, Department of Mechanical and Mechatronic Engineering, University of Stellenbosch (2010)
6. AMCA 210: Laboratory Methods of Testing Fans for Aerodynamic Performance Rating. Air Movement and Control Association International, Inc., Arlington Heights, IL (1999)
7. Cory, W.B.: Fans and Ventilation: A Practical Guide. Elsevier-Roles Ltd, Amsterdam (2005)
8. Srinivas, G., Srinivasa Rao., P.: Numerical simulation of axial flow fan using Gambit and Fluent. IJRET **3**(3), 586–590 (2014)
9. Raj, S.A., Pandian, P.P.: Effect of tip injection on an axial flow fan under distorted inflow. IJASER **3**(1), 302–309 (2014)
10. Dwivedi, D., Dandotiya, D.S.: CFD analysis of axial flow fans with skewed blades. IJETAE **3**(10), 741–752 (2013)
11. Augustyn, O.P.H.: Experimental and numerical analysis of axial flow fans. Master thesis, Stellenbosch University (2013)
12. Raj, A.S., Pandian, P.P.: Numerical simulation of static inflow distortion on an axial flow fan. IJMERR **3**(2), 20–25 (2014)
13. Sarmiento, A.L.E., Gamboa, Y.F.Q., Oliveira, W., Camacho, R.G.R.: Performance analysis through computational fluid dynamics of axial rotor with symmetric blades used in tunnel ventilation. HIDRO and HYDRO, PCH NOTICIAS and SHP NEWS **60**(1), 22–25 (2014)

Numerical Analysis of Turbulent Convective Heat Transfer in a Rotor-Stator Configuration

D.-D. Dang and X.-T. Pham

Abstract This paper presents the numerical analysis of convective heat transfer of a rotor-stator configuration, which is typically found in hydro-generators. The Reynolds Averaged Navier Stokes (RANS) turbulence models based on the eddy-viscosity approximation were employed. Different steady and unsteady multiple frames of reference models were used to deal with the flow interaction in the rotor-stator system. The fluid flow and heat transfer analysis were performed using conjugate heat transfer methodology, in which the governing equations for the fluid dynamics, heat conduction with additional constraints on the fluid-solid interface were simultaneously solved. The computed convective heat transfer coefficient was compared against available experimental data to assess the suitability of turbulence models.

1 Introduction

During the operation of hydro-generators, the heat generated by the electromagnetic and electrical loses causes the internal heating in the solid components. The solid temperatures that are too high might reduce the lifetime or at worst result in the breakdown of the machine.

In order to maintain the temperature of the solid parts within a safe margin, the heat generated is removed by a circulation of cooling air. The most important parameter governing the thermal performance of the hydro-generators is therefore the convective heat transfer coefficient (CHTC) on the boundary between the cooling fluid and the solid surface. Traditionally, CHTCs were approximated by the convective empirical correlations which are available in literature and widely

D.-D. Dang (✉) • X.-T. Pham
Ecole de technologie superieure, Montreal, QC H3C 1K3, Canada
e-mail: ak58860@ens.etsmtl.ca,tan.pham@etsmtl.ca

© Springer International Publishing Switzerland 2016

J. Bélair et al. (eds.), *Mathematical and Computational Approaches in Advancing Modern Science and Engineering*, DOI 10.1007/978-3-319-30379-6_38

used for thermal analysis. However, those correlations were originally established for simple configurations, and might not be applicable or insufficiently accurate for the case of complex geometries.

Prediction of convective heat transfer coefficient in hydro-generators is crucially important but not a trivial task due to the complexity of flow dynamics in the machine. The flow field in the rotor-stator system exhibits a number of phenomena which are challenging for numerical modelling such as rotation, turbulence and unsteady nature.

With respect to methodology, Lumped-Parameter Thermal Network (LPTN), Finite Element Analysis(FEA) and Computational Fluid Dynamics(CFD) are the most common approaches for thermal analysis of electrical machines [2]. Although the analytical LPTN and FEA have advantage of being very fast and efficient, these approaches require the knowledge of CHTCs in prior to the computation, which are challenging to obtain and therefore not always available. These methods are beyond the scope of the current research that focuses mainly on CFD application.

Pickering et al. [11] are considered as the first authors who applied CFD to predict the temperature in the solid of electrical machines. The numerical analysis of an air-cooled, 4-pole generator case concluded that CFD showed to be a potential tool for thermal analysis of electrical machines. However, because of limited computing resources at the time, only simulations with a coarse mesh near the pole surface could be performed. Depraz et al. [3] studied the cooling system of a large hydro-generator using the three-dimensional flow simulation. The comparison between the numerical results by CFD and traditional two-dimensional network based flow calculation showed a good agreement. Li [7] employed the conjugate heat transfer (CHT) methodology to predict the temperature distribution in the solid pole of the large hydro-generators. The CHT calculations used a coarse mesh on the wall surface for wall-functions utilization. However, the comparison with the experimental data was not carried out in his work. Recently Toriano et al. [12] proposed a hybrid method combining the numerical and experimental techniques to evaluate the heat transfer coefficient on the pole-face of a scaled rotating prototype.

Although numerous factors might affect the heat transfer prediction by conjugate heat transfer approach, the aim of the current study is limited to investigate the sensitivity of implemented numerical parameters on the prediction of fluid flow and heat transfer in the rotor-stator system. In particular the influence of turbulence models and steady-state Multiple Frames of Reference models are analyzed.

2 Computational Domain and Meshes

Because of the limited access on the real generator for experimental measurement, a scaled prototype was built. Since it is not realizable from the computational perspective to perform simulations of the entire scaled rotating model with an

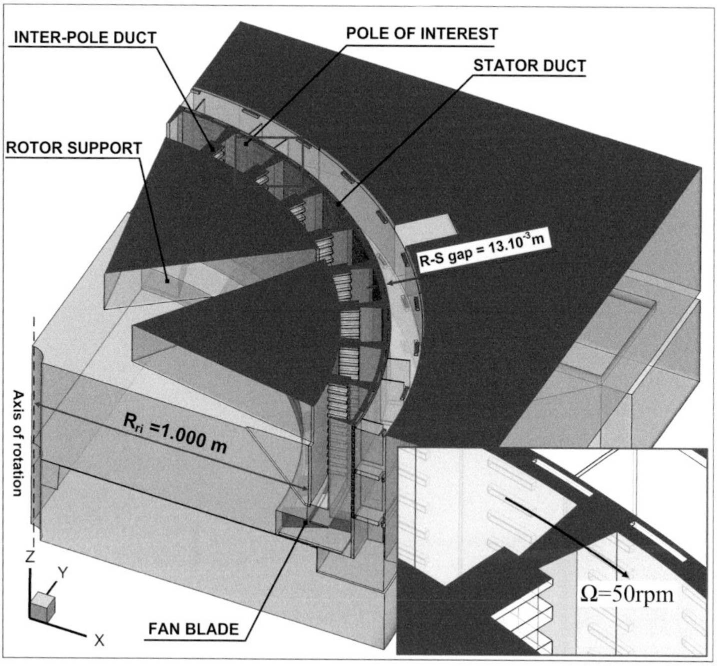

Fig. 1 Computational domain of scaled rotating model

adequately fine grid; calculations were carried out in two steps. First, the simulation with the full scaled model (Fig. 1) has been computed with a coarse 86 M hexahedral cells grid using the standard $k-\epsilon$ turbulence model. This calculation was then used to generate the conditions at the inlet boundary for the following second step of the calculation. The second configuration was compromised to only one section of 10 degree single-pole in (r, θ) plane which allowed to perform with a finer mesh up to 1.6 M hexahedra cells (Fig. 2). The extensive parametric analysis will be performed on this simplifed 2-D configuration using various turbulence and multiple frame of reference models.

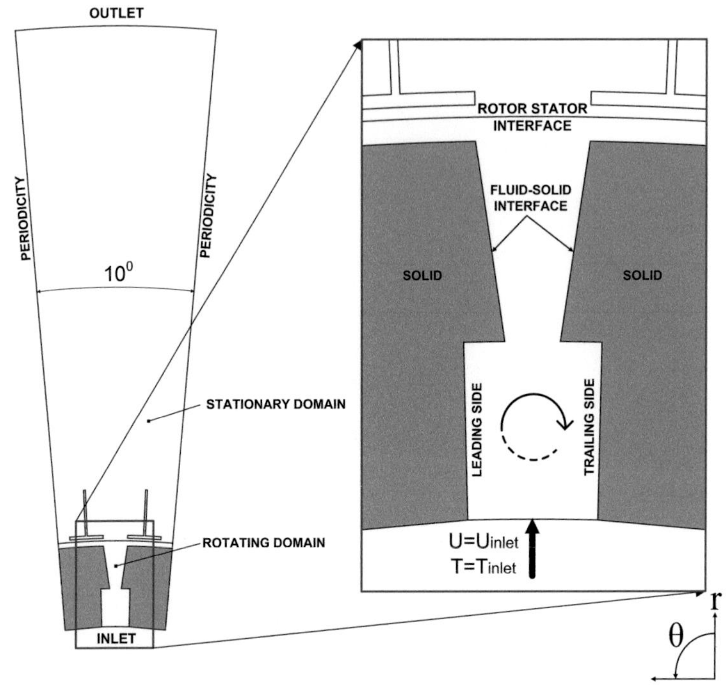

Fig. 2 Computational domain of two-dimensional simplified configuration

3 Mathematical Model and Numerical Method

3.1 Governing Equations

The governing equations for incompressible steady state flow [4] are expressed in the Cartesian tensor notation for a rotating coordinate system as follows.

Continuity equations:

$$\frac{\partial}{\partial x_i}(\rho U_i) = 0; \tag{1}$$

Momentum transport equation:

$$\frac{\partial}{\partial x_j}(\rho U_i U_j) = -\frac{\partial P}{\partial x_i} + \frac{\partial}{\partial x_j}\left[\mu(\frac{\partial U_i}{\partial x_j} + \frac{\partial U_j}{\partial x_i}) - \rho\overline{u_i u_j}\right] - \underbrace{2\rho\varepsilon_{ijp}\Omega_p U_j}_{centrifugalforce} - \underbrace{\rho\left[\Omega_j X_j \Omega_i - \Omega_j X_i \Omega_j\right]}_{Coriolisforce} \tag{2}$$

The two last terms in momentum equations were added to account for the Coriolis and centrifugal forces in the rotating frame of reference. These contributions however are eliminated in the stationary frame of reference.

Energy equation:

$$\frac{\partial}{\partial x_j}\left(\rho U_j h_{tot}\right) = \frac{\partial}{\partial x_j}\left(\lambda_f \frac{\partial T}{\partial x_j} - \rho\overline{u_j h}\right) + \frac{\partial}{\partial x_j}\left[U_i\left(\tau_{ij} - \rho\overline{u_i u_j}\right)\right] + S_E, \tag{3}$$

where the specific total enthalpy for ideal gas is given by:

$$h_{total} = C_p T + \frac{P}{\rho} + \frac{U_i^2}{2} + \frac{1}{2}\overline{u_i^2} \tag{4}$$

The heat conduction equation in the solid has the same form as the energy equation for fluid Eq. 3, except that the velocity components are set to be zero, and the solid thermal conductivity λ_s was substituted.

On the fluid-solid interface, additional constraints need to be added to ensure the equilibrium of heat flux and the temperature continuity through the two mediums.

$$\begin{aligned} T_{solid} &= T_{fluid} \\ \lambda_s \frac{\partial T}{\partial n}\Big|_{solid} &= \lambda_f \frac{\partial T}{\partial n}\Big|_{fluid} \end{aligned} \tag{5}$$

3.2 Turbulence Models and Boundary Layer Modelling

The presence of Reynolds stresses $-\rho\overline{u_i u_j}$ in Eqs. 2–3 means that the derived equations are not closed, which necessitates additional equations for closure. The turbulent stresses and turbulent heat flux are obtained using the effective viscosity approximation:

$$\rho\overline{u_i u_j} = -\frac{2}{3}k\delta_{ij} - \mu_t\left(\frac{\partial U_i}{\partial x_j} + \frac{\partial U_j}{\partial x_i}\right) \tag{6}$$

$$-\rho\overline{u_j h} = \frac{\mu_t}{Pr_t}\frac{\partial h}{\partial x_j} \tag{7}$$

The turbulent eddy-viscosity μ_t in Eqs. (6) and (7) is calculated by the turbulence models. Several turbulence model bases on the eddy viscosity approximation are implemented to calculate the flow dynamics and heat transfer. Standard $k-\epsilon$ (SKE) [6] model was developed for high-Re flow and has been widely used in industrial simulation because it compromises between the robustness and accuracy. However, SKE was consistently found inappropriate for the low-Re flows or flows with separation and reattachment of boundary layer. To improve the prediction, the

Shear Stress Transport (SST) model [8] was also implemented. Menter [9] presented several test cases to show that the SST $k - \omega$ predicts flows with separation in the presence of adverse pressure gradients more accurately than the SKE model.

Boundary layer modelling Modelling of the near-wall region is of major importance for CFD simulations, especially for the cases in which the prediction of the flow quantities on the wall (skin friction, convective heat transfer) is important. In principle, low-Reynolds number (LRN) and wall-functions are two common approaches for the near-wall modelling. The LRN model refers to an approach that resolves the entire boundary layer using a very fine mesh in the near wall regions. The grids used for the low-Re number model require typically a dimensionless wall distance of the wall-adjacent cell about unity, i.e. $y^+ < 1$. The grid for the low-Re number is shown in Fig. 3 (right), in which the wall-space of the first cell was set as 6.10^{-3} mm. Because of high computational cost associated with LRN model, the wall-functions are often used instead. The idea behind the wall-functions approach is to place the first computational node into the logarithmic layer and employ a semi-empirical dimensionless profile to obtain the wall shear stress. The advantage of this approach is that the boundary layer can be resolved with an adequately small number of grid points. The wall functions approach requires the y^+ value of the first node between 30 and 300, i.e., $30 < y^+ < 300$. In the present CFD code, the LRN and wall-functions models are employed for the $\omega-$base and $\epsilon-$based turbulence models, respectively.

Rotor-stator interaction The fluid flows in a rotor-stator system having non-axisymmetric components are always unsteady, which is a challenge for CFD simulations [10]. In an effort to make the rotor-stator analysis practical with limited computing resource, several steady-state Multiple Frames of Reference (MFR) models have been developed. In these models, an interface is defined between rotating and stationary components in a manner so that the steady-state calculations in each frame are supported while maintaining as much interaction between the components as possible [5]. There are two steady-state interface techniques are

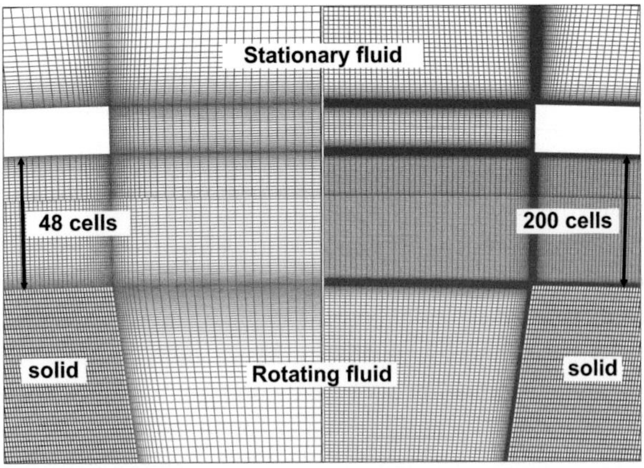

Fig. 3 Grids for different boundary layer models: wall-functions (*left*) and low-Re number model (*right*)

available in the present CFD code. The first model is called Mixing Plane, where the upstream flow velocity profile is first averages circumferentially before transferring to the downstream region. In this context, any non-uniformity in the circumferential direction will not be preserved in the next region. The second type of steady-state interface is called Frozen Rotor, where the flow profile variation is now preserved across the interface, however, the relative position between the rotor and stator is fixed in time and space. Also, the transient sliding interface is also available to model the unsteady flow due to the relative motion between the rotating and stationary components. In this model, the flow field variation is fully taken into account.

3.3 Boundary Conditions and Convergence Criteria

The boundary conditions for the configuration illustrated in Fig. 2 are defined as follows. The fluid velocity at the inlet boundary was set to be uniform, and normal to the inlet surface. The average value of the inlet profile was derived from the 3D simulation on the coarse grid with SKE turbulence model. For the case with rotational speed 50 rpm, $U_{inlet} = 2.10\,\text{m/s}$. The inlet temperature was set to 298 K (25 °C). For all wall boundaries, no-slip conditions with zero velocity were assumed. The circumferential periodicity feature of the geometry was specified in the numerical model by defining rotational periodic interfaces. On the rotor-stator interface, the general grid interface (GGI) was defined, which allows performing calculations of non-conforming meshes in a conservative manner. The governing equations presented in Sect. 2 were solved by CFD code ANSYS CFX-15.0. Considering the recommendation of ASME V&V numerical accuracy guideline [1], all equations were solved using second-order accuracy schemes.

Convergence was judged by examining the residual levels as well as by monitoring the relevant variables at critical locations. The used criterion requires that the maximum residual of all equations dropped at least by the order of 10^{-4} and the variables of interest at considered locations keeps constant at the steady-state value.

4 Results and Discussions

The flow and turbulence structure in the inter-pole duct calculated using mixing-plane model with the SST $k - \omega$ is illustrated in Fig. 4. The flow structure observed are typical for the flow in the duct with forward facing-step, in which flow with uniform velocity at the inlet develops in upstream region until it impinges on the step; a circulation bubble is formed in the step corner due to the adverse pressure

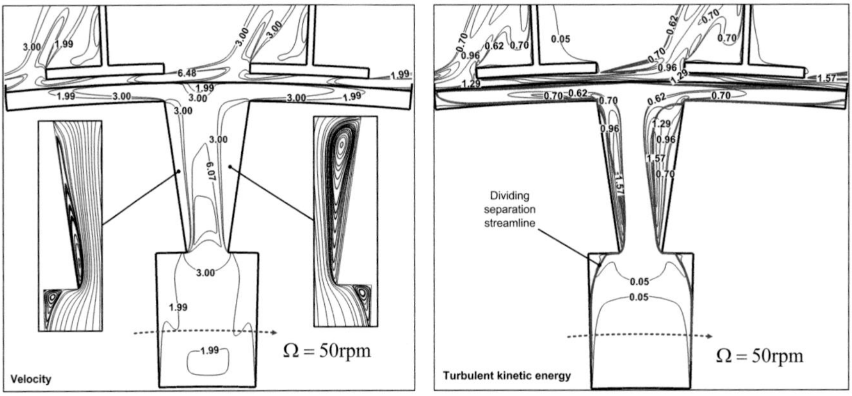

Fig. 4 Velocity contour (*left*) and turbulent kinetic energy contour (*right*) predicted by mixing-plane

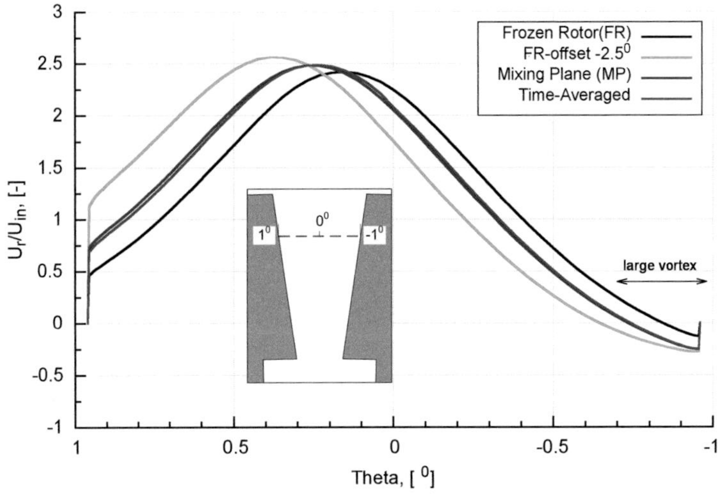

Fig. 5 Normalized radial velocity in the inter-pole duct predicted by different MFR models

gradient caused by step blockage. The boundary layer separates over the step and reattaches to the pole surface thereafter. Due to the system rotation the flow structure is asymmetric in the inter-pole duct; the Coriolis force have the effect of shrinking the recirculation size on the leading edge and enlarging the bubble size on the trailing edge.

The steady state models frozen rotor and mixing-plane was employed for an inherently unsteady flow in rotor-stator interaction, which is necessary to verify. The calculated result using these models are compared with the time-averaged transient simulation as reference result. Figure 5 shows the normalized radial velocity in the

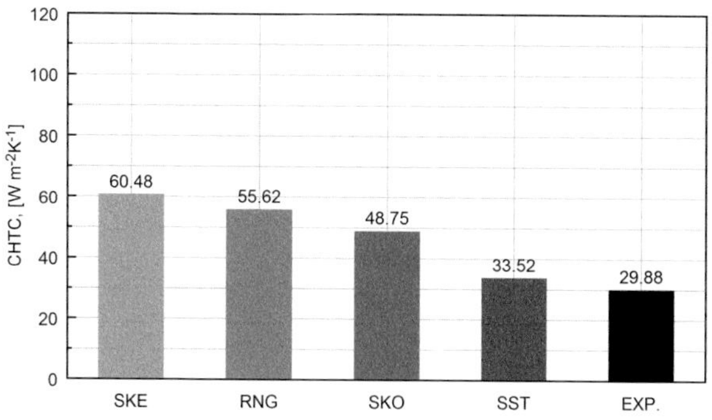

Fig. 6 Average heat transfer coefficient on the pole-face computed by different turbulence models

duct at r = 1.12 m for different MFR models. It is observed that the mixing-plane model shows a better agreement with the time-average transient model meanwhile the result computed by FR model significantly varies depending the relative position of rotor and stator. This results implies that the circumferential flow variation that each passages rotates during a full revolution is small at the rotor-stator interface. Since the simulation using mixing-plane model requires less computational effort than transient simulations, this model will be employed for the further parametric analysis.

Figure 6 reports the comparision of average CHTC on the pole face predicted by different turbulence models closure, including the standard $k - \epsilon$ (SKE), the Re-Normalisation Group $k - \epsilon$ (RNG), the standard $k - \omega$ (SKO) and SST $k - \omega$ (SST) models. The averaged value is calculated by Eq. 8.

$$\bar{h} = \frac{\int_A q_w dA}{\int_A \left(T_w - T_{ref}\right) dA} \tag{8}$$

where the T_{ref} is the reference temperature, which is fluid temeprature at 5 mm away from the pole face for both numerical and experimental approach. The experimental data is extracted from Toriano et al. [12]. It can be seen that all turbulence models predict higher CHTC in comparison with experimental data. The results calculated by SST $k - \omega$ shows the best agreement with the experiments, the relative error is calculated as 11 %.

In Fig. 7, the dimensionless temperature T^* profiles are compared for different positions at the pole-face along lines normal to the surface, as functions of y^*,

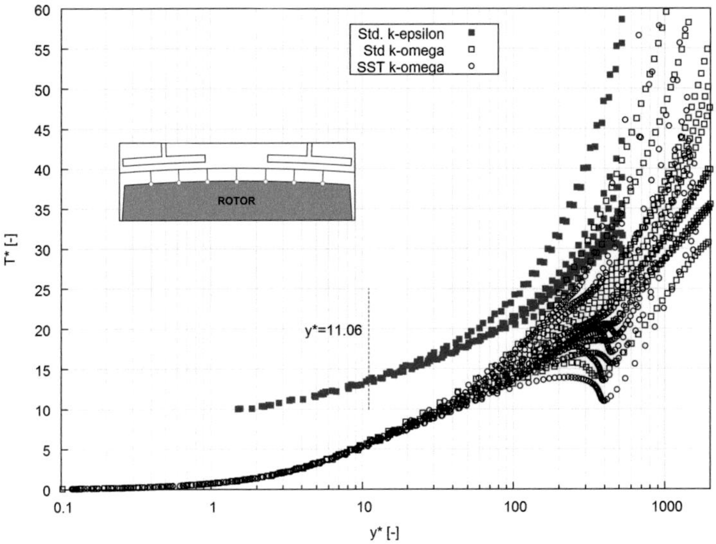

Fig. 7 Profiles of dimensionless temperature T^* as functions of y^* along lines normal to the pole face

defined by:

$$T^* = \frac{\rho C_\mu^{1/4} k^{1/2} \left(T_w - T\right) C_p}{q_w} \tag{9}$$

$$y^* = \frac{\rho C_\mu^{1/4} k^{1/2} y}{\mu} \tag{10}$$

According to Eq. 9, the higher T^* values results in a lower wall heat flux in case of the same level of predicted turbulence kinetic energy(TKE). However, the TKE predicted by wall-functions approach are found much higher than the low-Re number model. The example of the TKE profiles at two positions 2° and 3° are illustrated in Fig. 8. Overall, this results in the significant over-estimation of the CHTC by the wall function, as showed in Fig. 6. The maximum discrepancy between the CHTC predicted by wall-functions and experimental data is up to 50 %.

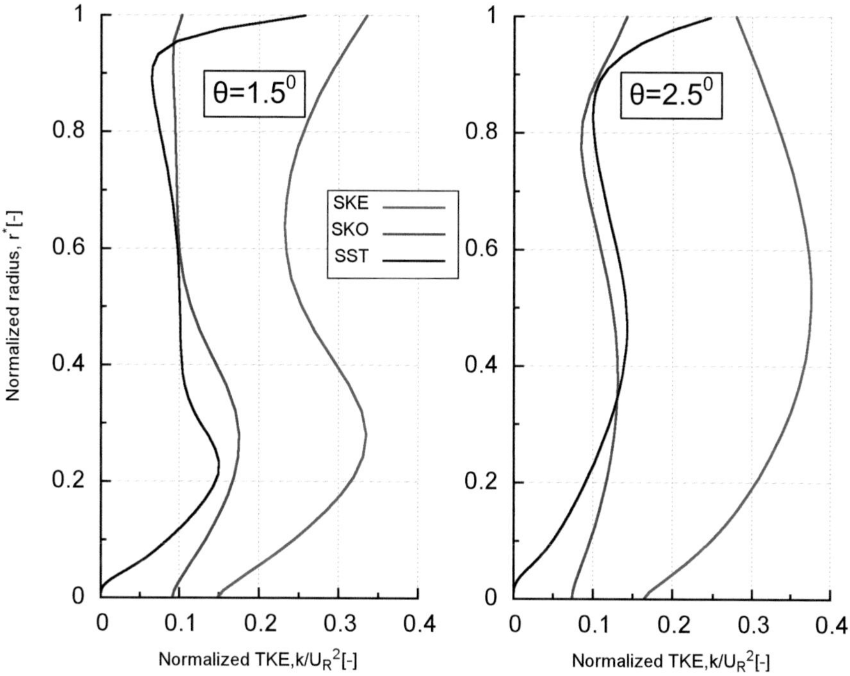

Fig. 8 Normalized turbulence kinetic energy in the air-gap predicted by different turbulence models

5 Conclusion

High resolution steady-state RANS CFD simulations of conjugate heat transfer in a rotor-stator configuration was performed. The focus was the sensitivity analysis of numerical parameters on the fluid flow and convective heat transfer prediction on the pole-face. The results showed that the mixing-plane model demonstrated a good agreement with the transient model than the frozen rotor, whose results significantly varies depending on the relative position of the rotor and stator. The heat transfer coefficient prediction by different turbulence models shows that the SST with low-Re model for the near-wall region is more appropriate than the wall-functions, which over-estimates up to 50 % this coefficient on the pole face on average.

Acknowledgements The support of the Natural Sciences and Engineering Research Council of Canada(NSERC) and Fonds de recherche du Quebec Nature et technologies(FRQNT) are gratefully acknowledged. We are also grateful for the funding and computing infrastructure provided by Institut de recherche d'Hydro-Quebec(IREQ).

References

1. ASME Standard for verification and validation in computational fluid dynamics and heat transfer.: ASME V&V 20-2009, American Society of Mechanical Engineers (2009)
2. Boglietti, A., Cavagnino, A., et al.: Evolution and modern approaches for thermal analysis of electrical machines. IEEE Trans. Ind. Electron. **56**(3), 872–882 (2009)
3. Depraz, R., Zickermann, R., Schwery, A., Avellan, F.: CFD validation and air cooling design methodology for large hydro generator. In: Proceedings of 17th International Conference on Electrical Machines ICEM, Crete Island, Greece (2006)
4. Ferziger, J.H., Peric, M.: Computational Methods for Fluid Dynamics. Springer, Berlin/Heidelberg (2001)
5. Galpin, P., Broberg, R., Hutchinson, B.: Three-dimensional Navier-Stokes predictions of steady state rotor-stator interaction with pitch change. In: Proceedings of 3rd Annual Conference of the CFD Society of Canada, Banff, Alberta (1995)
6. Launder, B.E., Spalding, D.B.: The numerical computation of turbulent flows. Comput. Methods Appl. Mech. Eng. **3**(2), 269–289 (1974)
7. Li W, Guan C, Chen Y (2014) Influence of rotation on rotor fluid and temperature distribution in a large air-cooled hydrogenerator. IEEE Trans. Energy Convers. **28**, 117–24
8. Menter, F.R.: Improved Two-Equation k-omega Turbulence Models for Aerodynamic Flows. NASA TM-103975 (1992)
9. Menter, F.R., Kuntz, M., Langtry, R.: Ten years of industrial experience with the SST turbulence model. Turbul. Heat Mass Transf. **4**, 625–32 (2003)
10. Moradnia, P., Chernoray, V., Nilsson, H.: Experimental assessment of a fully predictive CFD approach, for flow of cooling air in an electric generator. Appl. Energy **124**, 223–30 (2014)
11. Pickering, S.J., Lampard, D., Shanel, M.: Modelling ventilation and cooling of the rotors of salient pole machines. In: IEEE International Electric Machines and Drives Conference, Cambridge, MA, pp. 806–808 (2001)
12. Torriano, F., Lancial, N. et al.: Heat transfer coefficient distribution on the pole face of a hydrogenerator scale model. Appl. Therm. Eng. **70**(1), 153–162 (2014)

Determining Sparse Jacobian Matrices Using Two-Sided Compression: An Algorithm and Lower Bounds

Daya R. Gaur, Shahadat Hossain, and Anik Saha

Abstract We study the determination of large and sparse derivative matrices using row and column compression. This sparse matrix determination problem has rich combinatorial structure which must be exploited to effectively solve any reasonably sized problem. We present a new algorithm for computing a two-sided compression of a sparse matrix. We give new lower bounds on the number of matrix-vector products needed to determine the matrix. The effectiveness of our algorithm is demonstrated by numerical testing on a set of practical test instances drawn from the literature.

1 Introduction

The determination of the Jacobian matrix $F'(x)$ of a mapping $F : \Re^n \rightarrow \Re^m$ is a key computation in Newton's method for solving nonlinear optimization problems. Using forward difference (FD), the product of the Jacobian matrix with a vector s may be approximated as

$$\left.\frac{\partial F(x+ts)}{\partial t}\right|_{t=0} = F'(x)s \equiv As \approx \frac{1}{\epsilon}[F(x+\epsilon s) - F(x)] \equiv b, \qquad (1)$$

with one extra evaluation of F at $(x + \epsilon s)$, where $\epsilon > 0$ is a small increment, assuming $F(x)$ has already been evaluated. Algorithmic (or Automatic) Differentiation (AD) [7] gives $b = F'(x)s$ and $c = w^\top F'(x)$ in *forward mode* and *reverse mode*, respectively. The numerical values are accurate up to the machine round-off and the cost of the matrix-vector product is a small multiple of the cost of one function evaluation. The key observation is that the product of the Jacobian matrix of function F, evaluated at a given point x, with a given vector s can be obtained as As or $A^\top s$ at a cost that is a small multiple of the cost of evaluating the function F at x [7]. Therefore, in what follows, we express the computational

D.R. Gaur • S. Hossain (✉) • A. Saha
Department of Mathematics and Computer Science, University of Lethbridge, Lethbridge, AB, Canada
e-mail: daya.gaur@uleth.ca; shahadat.hossain@uleth.ca; anik.saha@uleth.ca

© Springer International Publishing Switzerland 2016

J. Bélair et al. (eds.), *Mathematical and Computational Approaches in Advancing Modern Science and Engineering*, DOI 10.1007/978-3-319-30379-6_39

Fig. 1 Algorithm for FD estimation of the Jacobian matrix of F at x

```
for j ← 1 to n
    do
        B(:, j) ← (1/ε)[F(x + εI(:, j)) − F(x)]
```

Fig. 2 A row compression matrix S

$$B = \underbrace{\begin{pmatrix} a_{11} & 0 & 0 & 0 & 0 & a_{16} \\ a_{21} & 0 & a_{23} & a_{24} & a_{25} & 0 \\ a_{31} & 0 & a_{33} & 0 & 0 & 0 \\ 0 & a_{42} & 0 & a_{44} & 0 & 0 \\ a_{51} & 0 & 0 & 0 & a_{55} & 0 \\ 0 & a_{62} & 0 & 0 & 0 & a_{66} \end{pmatrix}}_{A} \underbrace{\begin{pmatrix} 1 & 0 & 0 & 0 \\ 1 & 0 & 0 & 0 \\ 0 & 1 & 0 & 0 \\ 0 & 0 & 1 & 0 \\ 0 & 0 & 0 & 1 \\ 0 & 1 & 0 & 0 \end{pmatrix}}_{S}$$

$$= \begin{pmatrix} a_{11} & a_{16} & 0 & 0 \\ a_{21} & a_{23} & a_{24} & a_{25} \\ a_{31} & a_{33} & 0 & 0 \\ a_{42} & 0 & a_{44} & 0 \\ a_{51} & 0 & 0 & a_{55} \\ a_{62} & a_{66} & 0 & 0 \end{pmatrix}$$

cost of determining a Jacobian matrix in terms of the number of matrix-vector products (MVPs) of the kind As or $A^\top s$. The complete Jacobian can be determined, for example, by differencing along n coordinate directions as shown in Fig. 1 where I denotes the identity matrix. We use colon notation of [5] to specify submatrices of a matrix. In AD reverse mode, the Jacobian matrix can be determined row-by-row as vector-matrix products $s^\top A$. Many large-scale practical optimization problems often contain useful structure. For example, the component functions F_i nonlinearly depend only on a small subset of the n independent variables giving rise to partially separable structure [6]. It is usually the case that zero-nonzero pattern of the Jacobian is a priori known or can be computed easily. Consider the matrix A in Fig. 2. Define matrix S where $S(:,1) = e_1 + e_2$, $S(:,2) = e_3 + e_6$, $S(:,3) = e_4$, $S(:,4) = e_5$ and compute the product $B = AS$ using, for example, the algorithm in Fig. 1. It is clear from Fig. 2 that the FD estimation of the unknown elements of matrix A can simply be read-off, i.e., without incurring any floating-point operations, from the *compressed matrix* B at a cost of only 4 (instead of 6) additional evaluations of function F. Alternatively, we may define a column compression matrix W where $W(:,1) = e_2 + e_6$, $W(:,2) = e_4 + e_5$, $W(:,3) = e_1$, $W(:,4) = e_3$ and compute the compressed matrix $C^\top = W^\top A$ in reverse mode AD. In either case, 4 matrix-vector products are required to completely determine the matrix. We say that matrix A is *directly determined* if all the nonzero unknowns can be read-off from the compressed matrices B or C as defined above. The Jacobian matrix determination problem (JMDP) based on matrix-vector product calculation can be stated as below (See [10]).

Given the specification of the sparsity pattern for matrix A, obtain matrices $S \in \{0,1\}^{n \times p}$ and $W \in \{0,1\}^{m \times q}$ such that matrix A is directly determined and $p + q$ is minimum.

A group of columns satisfying the property that no two of them contain nonzero entries in the same row position is called *structurally orthogonal*. The observation that a group of structurally orthogonal columns can be approximated with only one extra function evaluation was reported first in a seminal paper by Curtis, Powell, and Reid [3]. Their row compression algorithm[1] was further analyzed by Coleman and Moré [1] who gave the first graph coloring interpretation of the compression problem. The classical arrow-head matrix example demonstrates that one-sided compression (i.e., either column or row) may not yield full exploitation of sparsity [9]. Hossain and Steihaug [9], and Coleman and Verma [2] independently proposed techniques for two-sided compression (i.e., combined row and column compression) for the sparse Jacobian determination problem. Unfortunately, sparse Jacobian determination using compressions (one-sided or two-sided) is NP-hard [2] implying that exact methods are impractical except for small problem instances. Greedy heuristics *complete direct cover (CDC)* by Hossain and Steihaug [9] and *minimum nonzero count ordering (MNCO)* by Coleman and Verma [2] produce matrices S and W by considering the matrix rows and columns in specified orders. Recently, Juedes and Jones [11] proposed *an approximation algorithm for minimum star bi-coloring (ASBC)* with an approximation guarantee of $O(n^{\frac{2}{3}})$ of the optimal. An easy to compute "good" lower bound on the number of groups allows one to measure the effectiveness of such algorithms. For row compression (of A or $A^\top$) a satisfactory lower bound on the number of matrix-vector products (or AD passes) is given by $\rho_{\max}(A) = \max_i \rho_i$ where ρ_i denotes the number of nonzero entries in row i. For two-sided compressions, [8] proposed the expression $\lceil \frac{\text{nnz}(A)}{\max\{m,n\}} \rceil$ where $\text{nnz}(A)$ denotes the number of nonzero entries in matrix A. The bound is derived using a consistency argument in solving the system of linear equations defined by the products $B = JS$ and $C = W^\top J$. Juedes and Jones [11] derive the same bound using a graph theoretic argument. In this paper we propose a new and more effective lower bound which generalizes the above lower bound. We propose a new heuristic algorithm for two-sided compression of sparse Jacobian matrices. We demonstrate the effectiveness of our lower bound and algorithm by extensive computational experiments on a standard set of test instances. The remainder of the paper is organized as follows. In Sect. 2, we derive the new lower bound on the number of matrix-vector products which is followed by the description of the new two-sided compression algorithm. Section 3 contains results of numerical experiments from the lower bound computation and the two-sided compression algorithm on a set of test matrices. Test results indicate that the new lower bound is superior to the existing one. On many of the test matrices, our algorithm produced better results. The paper is concluded in Sect. 4 with comments on future research.

[1] A row compression is a one-sided compression where only seed matrix S is defined.

2 Two-Sided Compression

As discussed in Sect. 1 the example in Fig. 2 can be determined using 4 MVPs using either row or column compression with one-sided compression. This is also the minimum possible since the maximum number of nonzero entries in any row of A or $A^\top$, $\rho_{\max}(A) = \rho_{\max}(A^\top)$, is 4. However, a closer examination of Fig. 2 reveals that the last two columns of the compressed matrix B are quite sparse – only two of the entries are nonzero. In other words, there are available sparsity which are not exploited by one-sided compressions. On the other hand, with the following two-sided compression

$$S = \begin{pmatrix} 1 & 0 \\ 1 & 0 \\ 0 & 1 \\ 0 & 1 \\ 0 & 1 \\ 0 & 1 \end{pmatrix}, \quad W = \begin{pmatrix} 1 \\ 1 \\ 0 \\ 0 \\ 0 \\ 0 \end{pmatrix}$$

we can compute the MVPs

$$AS = \begin{pmatrix} a_{11} & a_{16} \\ a_{21} & a_{23} + a_{24} + a_{25} \\ a_{31} & a_{33} \\ a_{42} & a_{44} \\ a_{51} & a_{55} \\ a_{62} & a_{66} \end{pmatrix}, \quad W^\top A = \begin{pmatrix} a_{11} + a_{21} & 0 & a_{23} & a_{24} & a_{25} & a_{16} \end{pmatrix}$$

such that all the nonzero entries are determined uniquely. A lower bound on the number of MVPs for one-sided compressions is given by the easily evaluated expression $\min\{\rho_{\max}(A),\ \rho_{\max}(A^\top)\}$ which has been found to be a good approximation to the chromatic number of graphs associated with the matrices on an extensive set of test instances [1]. On the other hand, only a handful of published literature consider two-sided compressions. The Ph.D. dissertation [9] derived the bound $LB2S = \lceil \frac{\mathrm{nnz}(A)}{\max\{m,n\}} \rceil$ on the number of MVPs in a two-sided compression which is also independently derived and analyzed experimentally in [11]. In this paper we generalize this lower bound and show that the new lower bound is always at least as good and in many test instances it yields better result.

2.1 A New Lower Bound

Following [9] the bound *LB2S* can be derived intuitively in the following way. Consider a two-sided compression $S \in \{0,1\}^{n\times p}, W \in \{0,1\}^{m\times q}$. With an MVP of type $As, s \in \{0,1\}^n$, at most m nonzero entries can be uniquely determined. Similarly, with an MVP of type $w^\top A, w \in \{0,1\}^m$, at most n nonzero entries can be uniquely determined. If there are nnz(A) nonzero entries in matrix A, then its follows that

$$\text{nnz}(A) \le pm + qn \le \max\{m,n\}(p+q) \text{ implying } \left\lceil \frac{\text{nnz}(A)}{\max\{m,n\}} \right\rceil \le (p+q).$$

In one-sided compression a row with the maximum number of nonzero determines a lower bound for the number of MVPs. In graph-theoretic terminology this represents a subgraph where each pair of vertices are connected by an edge. Our proposal is to extend this key observation to the two-sided scenario. In other words, we claim that a dense submatrix of a given matrix A yields a lower bound on the number of MVPs to uniquely determine matrix A. Clearly, a bound of the type *LB2S* on a submatrix of A is also a bound on the whole matrix A. Now consider a submatrix $A' \in \Re^{m'\times n'}$ sufficiently dense such that neither the rows nor the columns of the submatrix are structurally orthogonal. A trivial example is a completely dense submatrix. Then, we need at least $\min\{m', n'\}$ MVPs to uniquely determine matrix A. In graph-theoretic terminology, this completely dense submatrix corresponds to a complete bipartite subgraph $K_{m',n'}$ of the bipartite graph associated with matrix A. We want to find the largest dense submatrix of the kind described above. Unfortunately, the complexity of deciding whether or not a bipartite graph contains a complete bipartite subgraph of size k is NP-complete [4]. Instead of trying to find the largest dense submatrix of matrix A we therefore settle for a reasonably dense submatrix which is easy to compute. The above discussion can be succinctly stated as: find permutation matrices P and Q such that the largest dense submatrix of A is included in some *leading submatrix* of the permuted matrix PAQ. The k-th leading submatrix of matrix A is denoted by $A(1:k, 1:k)$. To move dense rows and columns toward the top left corner of matrix A we use the following simple procedure.

Order the rows in non increasing order of the number of nonzero entries followed by the same ordering procedure applied to the columns. Matrices P and Q represent the row and column ordering, respectively.

We then apply the formula *LB2S* on each leading submatrix of permuted matrix PAQ. The largest *LB2S* value is our lower bound *LB2SX* on the number of MVPs to determine matrix A. Thus,

$$LB2SX = \max_i\{LB2S_i\}$$

and

$$LB2S_i = \left\lceil \frac{\text{nnz}_i(PAQ)}{i} \right\rceil,$$

where $\text{nnz}_i(.)$, $i = 1, \ldots, \min\{m, n\}$ denotes the number of nonzero entries in the i-th leading submatrix of the argument.

The above discussion can be summarized in the following lemma.

Lemma 1 *A lower bound on the number of matrix-vector products to directly determine* $A \in \Re^{m \times n}$ *is given by*

$$LB2SX = \max_i \{LB2S_i\}$$

with

$$LB2S_i = \left\lceil \frac{\text{nnz}_i(PAQ)}{i} \right\rceil,$$

where $\text{nnz}_i(.)$, $i = 1, \ldots, \min\{m, n\}$ *denotes the number of nonzero entries in the* i*-th leading submatrix of the argument and* P *and* Q *are permutation matrices. If* $m = n$*, then we have*

$$LB2SX \geq LB2S = \left\lceil \frac{nnz(A)}{\max\{m, n\}} \right\rceil.$$

2.2 A Heuristic for Two-Sided Compression

The heuristic algorithm that we outline here is inspired by the Recursive Largest First (RLF) coloring heuristic due to [12]. The main idea of our compression algorithm is to group columns and rows such that for each nonzero a_{ij},

1. there is a column group k containing column j such that there is no column $j' \neq j$ also in group k for which $a_{ij'} \neq 0$, or
2. there is a row group l containing row i such that there is no row $i' \neq i$ also in group l for which $a_{i'j} \neq 0$

This condition is termed as *direct cover condition* in [8] and the nonzero a_{ij} meeting this is said to be *directly covered* or simply *covered*, for brevity. The *complete direct cover (CDC)* algorithm of [9] groups the columns and rows while respecting the cover condition until all the nonzero entries are covered. The resulting column groups and row groups define the matrices S and W such that the Jacobian matrix is directly determined. The columns and rows are assigned to groups according to the number of nonzero entries that are yet to be covered: higher the number of entries earlier they are grouped. In our algorithm 2SIDEDCOMPRESSION, as depicted in

Fig. 3, we too take into account the number of entries that are yet to be covered in forming the groups. First, we introduce terminology that will enable us to describe the algorithm. Given a sparse matrix A let $I = \{1, \ldots, m\}$ and $J = \{1, \ldots, n\}$ denote the set of row indices and the set of column indices, respectively. For row index i and $\mathscr{I} \subseteq I$ define,

$$fbdn_{\mathscr{I}}(i) = \big\{i' | i' \in \mathscr{I} \text{ and there is an index } j \in J \text{ for which } a_{ij} \neq 0, \\ a_{i'j} \neq 0 \text{ are not covered}\big\}.$$

Analogously, for column index j and $\mathscr{J} \subseteq J$ define,

$$fbdn_{\mathscr{J}}(j) = \big\{j' | j' \in \mathscr{J} \text{ and there is an index } i \in I \text{ for which } a_{ij} \neq 0, \\ a_{ij'} \neq 0 \text{ are not covered}\big\}.$$

Also, let $deg_{\mathscr{I}}(i) = |fbdn_{\mathscr{I}}(i)|$ and $deg_{\mathscr{J}}(j) = \big|fbdn_{\mathscr{J}}(j)\big|$ denote the cardinality of the respective index sets.

Thus, $deg_{\mathscr{I}}(i)$ denotes the number of rows corresponding to set $\mathscr{I}$ that cannot be grouped with row i. Similarly, $deg_{\mathscr{J}}(j)$ denotes the number of columns corresponding to set $\mathscr{J}$ that cannot be grouped with column j.

On return from 2SIDEDCOMPRESSION each column is given a label from the set $\{0, 1, \ldots p\}$ indicating the column group it is included in and each row is given label from the set $\{0, 1, \ldots q\}$ indicating the row group it is included in. A column (or row) that is labeled with group number 0 indicates that its nonzero entries are directly covered by other group(s). Variable *ncovered* represents the number of nonzero entries that have been directly covered. Initially, each column/row is labeled with group number 0. The **while** -loop in line 5 is iterated until all the nonzero entries of matrix A are directly covered. Variables I_u and J_u contain the indices of rows and columns, respectively, that are yet to be included in a group other than group 0. The pseudocode in lines 7–16 and lines 17–25 compute the next row group and column group, respectively. The **if** -construct in lines 26–34 chooses the group which covers the most nonzero entries (Fig. 3).

3 Numerical Experiments

In this section we provide numerical evidence to demonstrate the effectiveness of the methods developed in Sect. 2. For numerical experiments we choose the set of test instances used in ([11] Table 2). Table 1 compares the lower bound on the number of MVPs given by *LB2SX* and *LB2S*. In Tables 1 and 2, $\rho_{\max}(.)$ represents the maximum number of nonzero entries in any row of the matrix given in the argument. Table 1 includes the test matrices of ([11] Table 2) where the two bounds differ. Out of 32 instances, *LB2SX* is strictly better on 14 of them; on the remaining 16 instances *LB2SX* and *LB2S* yield the same value. On problem eris1176 algorithm

```
2SIDEDCOMPRESSION
p ← 0; q ← 0            ▷ p: the number of column groups; q: the number of row groups
I_u ← I; J_u ← J        ▷ I = {1,2,...,n}, J = {1,2,...,m}
label columns corresponding to J_u with group number p and label rows corresponding to I_u with group number q
ncovered ← 0            ▷ no nonzero unknown is directly covered
while ncovered ≠ nnz
    do I_* ← Ø                    ▷ I_*: next row group
       let i ∈ I_u be a row with the maximum number of uncovered nonzero entries
       I_* ← I_* ∪ {i}
       F ← fbdn_{I_u}(i)
       X ← I_u \ ({i} ∪ F)
       while X ≠ Ø
           do let i' ∈ X be such that deg_F(i') is maximum
              I_* ← I_* ∪ {i'}
              F ← F ∪ fbdn_X(i')
              X ← X \ ({i'} ∪ fbdn_X(i'))
       J_* ← Ø                    ▷ J_*: next column group
       let j ∈ J_u be a column with the maximum number of uncovered nonzero entries
       J_* ← J_* ∪ {j}
       F ← fbdn_{J_u}(j)
       X ← J_u \ ({j} ∪ F)
       while X ≠ Ø
           do Let j' ∈ X be such that deg_F(j') is maximum
              J_* ← J_* ∪ {j'}
              F ← F ∪ fbdn_X(j')
              X ← X \ ({j'} ∪ fbdn_X(j'))
       if the number nonzero entries covered by rows corresponding to I_* is larger than that of J_*
         then q ← q + 1
              label rows corresponding to I_* with group number q
              I_u ← I_u \ I_*
              ncovered ← ncovered + number of nonzero entries covered by rows corresponding to I_*
       else p ← p + 1
              label columns corresponding to J_* with group number p
              J_u ← J_u \ J_*
              ncovered ← ncovered + number of nonzero entries covered by columns corresponding to J_*
```

Fig. 3 An algorithm for two-sided compression of a sparse Jacobian matrix

Table 1 Lower bound on the number of MVPs in two-sided compression of sparse Jacobian

Matrix	m	n	$\rho_{\max}(A)$	$\rho_{\max}(A^\top)$	*LB2S*	*LB2SX*
abb313	313	176	6	26	5	6
arc130	130	130	2	12	10	16
bp0	822	822	266	20	4	7
bp200	822	822	283	21	5	7
bp400	822	822	295	21	5	7
bp800	822	822	304	21	6	8
bp1000	822	822	308	21	6	8
eris1176	1176	1176	99	99	16	80
lund_a	147	147	21	21	17	18
lund_b	147	147	21	21	17	18
str_0	363	363	34	34	7	17
str_200	363	363	30	26	9	17
str_400	363	363	33	34	9	18
will_57	57	57	11	11	5	6

Table 2 Number of MVPs in two-sided compression heuristics

Matrix	m	n	$\rho_{\max}(A)$	$\rho_{\max}(A^\top)$	ASBC	2SIDEDCOMPRESSION
abb313	313	176	6	26	17	**10**
ash219	219	85	2	9	8	**4**
ash292	292	292	14	14	19	**14**
ash331	331	104	2	12	10	**6**
ash608	608	188	2	12	11	**6**
ash958	958	292	2	13	12	**6**
bp0	822	822	266	20	**16**	17
bp200	822	822	283	21	**17**	19
bp400	822	822	295	21	**19**	20
curtis54	54	54	12	16	12	**11**
eris1176	1176	1176	99	99	92	**80**
fs_541_1	541	541	11	541	18	**13**
fs_541_2	541	541	11	541	18	**13**
ibm32	32	32	8	7	9	**8**
lund_a	147	147	21	21	26	**24**
lund_b	147	147	21	21	27	**23**
shl_0	663	663	422	4	7	**4**
shl_200	663	663	440	4	7	**4**
shl_400	663	663	426	4	6	**4**
str_0	363	363	34	34	**25**	27
str_200	363	363	30	26	**31**	32
will199	199	199	6	9	9	**7**

2SIDEDCOMPRESSION is optimal (see Table 2) since the number of MVPs needed to directly determine the matrix is same as the lower bound given by *LB2SX*. The matrix is pattern symmetric and will require at least 99 MVPs to determine it by any one-sided compression method. With a two-sided compression, however, at least 80 MVPs will be needed to determine the matrix. The *LB2S* gives a lower bound of 16 MVPs which is much lower than the optimum number of MVPs (= 80) required to directly determine the matrix. Table 2 compares the number of MVPs needed by algorithm ASBC and our heuristic algorithm 2SIDEDCOMPRESSION. The instances are chosen from ([11] Table 2) where the number of MVPs required by the two methods are different. Out of 22 instances, algorithm 2SIDEDCOMPRESSION yields better result than ASBC on 17 (marked in bold face) of them. For the instances where ASBC is better, the difference is at most 2.

4 Conclusion

In this paper we have considered the problem of sparse Jacobian matrix determination with two-sided compressions. The new lower bound on the number of MVPs generalizes the lower bound of [9, 11]. The results from numerical experiments on a standard set of test instances provide strong evidence that the proposed lower bound is superior to that of [9, 11]. The new lower bound is easy to compute, never smaller than the bound proposed in [9, 11], and strictly larger on about 50 % of the test instances. The Heuristic algorithm 2SIDEDCOMPRESSION saves, on average, 13 % MVPs compared with the ASBC algorithm and, on average it is within approximately 1.5 times the optimal.

There are a number of possible extensions to this research that we plan to pursue in future. In our lower bound calculation we use the simple strategy of reordering of rows and columns according to the nonzero density. More sophisticated ordering techniques can be considered here. For example, lining up the nonzero entries together in rows and columns may reveal the dense submatrices which can be identified more easily. This will yield the added benefit of utilizing data locality to improve cache memory performance in a computer implementation. Also, the current computer implementation of the compression algorithm can be improved by employing more advanced data structures for manipulating sparse matrices.

Acknowledgements This research is supported in part by Natural Sciences and Engineering Research Council of Canada (NSERC) Discovery Grant (Individual).

References

1. Coleman, T.F., Moré, J.J.: Estimation of sparse Jacobian matrices and graph coloring problems. SIAM J. Numer. Anal. **20**(1), 187–209 (1983)
2. Coleman, T.F., Verma, A.: The efficient computation of sparse Jacobian matrices using automatic differentiation. SIAM J. Sci. Comput. **19**(4), 1210–1233 (1998)
3. Curtis, A.R., Powell, M.J.D., Reid, J.K.: On the estimation of sparse Jacobian matrices. IMA J. Appl. Math. **13**(1), 117–119 (1974)
4. Garey, M.R., Johnson, D.S.: Computers and Intractability: A Guide to the Theory of NP-Completeness. Freeman, San Francisco (1979)
5. Golub, G.H., Van Loan, C.F.: Matrix Computations, 3rd edn. Johns Hopkins University Press, Baltimore (1996)
6. Griewank, A., Toint, Ph.L.: On the unconstrained optimization of partially separable objective functions. In: Powell, M.J.D. (ed.), Nonlinear Optimization, pp. 301–312. Academic, London (1982)
7. Griewank, A., Walther, A.: Evaluating Derivatives: Principles and Techniques of Algorithmic Differentiation, 2nd edn. Society for Industrial and Applied Mathematics, Philadelphia (2008)
8. Hossain, A.K.M.S.: On the computation of sparse Jacobian matrices and newton steps. Ph.D. Dissertation, Department of Informatics, University of Bergen (1998)
9. Hossain, A.K.M.S., Steihaug, T.: Computing a sparse Jacobian matrix by rows and columns. Optim. Methods Softw. **10**, 33–48 (1998)
10. Hossain, S., Steihaug, T.: Graph models and their efficient implementation for sparse Jacobian matrix determination. Discret. Appl. Math. **161**(2), 1747–1754 (2013)
11. Juedes, D., Jones, J.: Coloring Jacobians revisited: a new algorithm for star and acyclic bicoloring. Optim. Methods Softw. **27**, 295–309 (2012)
12. Leighton, F.T.: A graph coloring algorithm for large scheduling problems. J. Res. Natl. Bur. Stand. **84**, 489–505 (1979)

An h-Adaptive Implementation of the Discontinuous Galerkin Method for Nonlinear Hyperbolic Conservation Laws on Unstructured Meshes for Graphics Processing Units

Andrew Giuliani and Lilia Krivodonova

Abstract For computationally difficult problems, mesh adaptivity becomes a necessity in order to efficiently use computing resources and resolve fine solution features. The discontinuous Galerkin (DG) method for hyperbolic conservation laws is a numerical method adapted to execution on graphics processing units (GPUs). In this work, we give the framework of an efficient h-adaptive implementation of the modal DG method on NVIDIA GPUs, outlining implementation considerations in the context of GPU computing. Finally, we demonstrate the effectiveness of our implementation with a computed example.

1 Introduction

A relatively popular approach to numerical simulation is mesh adaptivity. Notably, in general it is not possible to know a priori how to optimally distribute on a domain the numerical degrees of freedom (DOFs) e.g. the number of elements and local order of approximation, for finite element methods. This is especially true for time-dependent simulations. For many difficult problems in computational fluid dynamics (CFD), computing resources will quickly be expended if DOFs are not added satisfactorily. Fortunately, mesh adaptation strategies exist that automatically add or remove DOFs solely in areas of the solution where they are needed [12]. One approach is spatial refinement, or h-adaptivity, which attempts to selectively subdivide or merge mesh cells thereby locally increasing or decreasing solution resolution. H-adaptivity can be implemented in varying fashions; we discuss three possible ways: component grids, and block or cell-based refinement.

Berger considered component grids in detail on structured regular meshes [1, 2]. This method achieves h-adaptivity through a hierarchy of independent meshes that

---

A. Giuliani (✉) • L. Krivodonova
University of Waterloo, 200 University Avenue West, Waterloo, ON N2L 3G1, Canada
e-mail: agiulian@uwaterloo.ca,lgk@uwaterloo.ca

© Springer International Publishing Switzerland 2016

J. Bélair et al. (eds.), *Mathematical and Computational Approaches in Advancing Modern Science and Engineering*, DOI 10.1007/978-3-319-30379-6_40

are overlay one another. An error indicator first flags elements that do not satisfy some criterion. These elements are then clustered into groups based on spatial proximity and other criteria. Then, additional rotated rectangular grids with finer mesh spacing are created and overlay these groups. This is done multiple times until an error tolerance is reached, creating a series of embedded grids After an appropriate number of timesteps, the mesh hierarchy is redetermined through a process called 'regridding'. Since Berger [1, 2], a simplified approach has been applied to GPUs in [3, 10] for the shallow water equations. Other approaches include block and cell-based refinement strategies.

In block-based refinement, cells are grouped into blocks of a predefined size. If any elements in a group are flagged for refinement, then all elements in that group are refined and replaced by a number of refined blocks. This allows connectivity information to be stored with respect to blocks in a quadtree data structure for 2D and octree for 3D [9, 11]. Block-based implementations have been discussed in [14].

In cell-based refinement, elements are refined individually rather than in groups which maintains high-resolution locality. This strategy can be implemented both on structured and unstructured meshes, though many cell-based approaches to date deal with Cartesian grids [7]. A trade off of refinement locality is that more connectivity information is required than in block-based refinement. This is because data must be stored on the level of elements, rather than blocks. A quadtree and octree data structure can be used again to represent parent-child relationships for determining mesh connectivity [5, 13]. The advantage of this strategy is that irregular boundaries can be dealt with very easily through cut-cells, higher order elements or curved boundary conditions.

GPUs are useful tools in scientific computing applications. The compute capacity of NVIDIA GPUs is quite impressive; single and double precision arithmetic throughput can exceed 1 TFLOP/s depending on the model. When this is considered in conjunction with the price of such devices, GPUs are an attractive scientific computing platform.

We now focus on the main subject of this work: an efficient cell-based h-adaptive DG-GPU algorithm in NVIDIA's CUDA C on unstructured triangular meshes for nonlinear hyperbolic conservation laws. We begin with the derivation of a DG method for these partial differential equations (PDEs). We then outline the implementation details and present a computed example, examining the solver's performance characteristics.

2 The Discontinuous Galerkin Method

In two dimensions, hyperbolic conservation laws are PDEs of the form

$$\frac{d}{dt}\mathbf{u} + \nabla_{xy} \cdot \mathbf{F}(\mathbf{u}) = 0, \tag{1}$$

with the solution $\mathbf{u}(\mathbf{x}, t) = (u_1, u_2, \ldots, u_M)^{\mathsf{T}}$ defined on $\Omega \times [0, T]$ such that $\mathbf{x} = (x, y)$, $\mathbf{x} \in \Omega \subset \mathbb{R}^2$, T is a final time and $\mathbf{F}(\mathbf{u}) = (\mathbf{F}_1(\mathbf{u}), \mathbf{F}_2(\mathbf{u}))$ is the flux function.

Additionally, the initial condition

$$\mathbf{u}(\mathbf{x}, 0) = \mathbf{u}_0(\mathbf{x}),$$

along with appropriate boundary conditions are applied. The DG method is a high-order method without an extensive stencil that can successfully capture shocks and discontinuities in the numerical solution of conservation laws. As opposed to the standard finite element method, continuity at the interface between adjacent elements is not imposed.

This method can be formulated by first dividing the domain Ω into an unstructured mesh of N triangles such that $\Omega = \bigcup_{i=1}^{N} \Omega_i$. We note that nonconforming tessellations are permitted. In our implementation, the difference in refinement level between adjacent cells must not exceed 1, see Fig. 1.

We define $S^p(\Omega_i)$ to be the space of polynomials of degree at most p on Ω_i. Additionally, $\{\widetilde{v_j}\}_{j=1..N_p}$ is a set of orthonormal basis functions for $S^p(\Omega_i)$, where N_p represents the number of basis functions.

The weak form of the conservation law is obtained by multiplying Eq. (1) by a test function $\widetilde{v_j} \in S^p(\Omega_i)$ and integrating on an element Ω_i. After integrating by parts, we obtain

$$\int_{\Omega_i} \mathbf{u}_t \widetilde{v_j} d\Omega_i - \int_{\Omega_i} \mathbf{F}(\mathbf{u}) \nabla_{xy} \widetilde{v_j} d\Omega_i + \int_{\partial\Omega_i} \widetilde{v_j} \mathbf{F}(\mathbf{u}) \cdot \mathbf{n} dl = 0, \tag{2}$$

where $\mathbf{n}$ is the outward facing normal on $\partial\Omega_i$.

Each element Ω_i is mapped to the canonical triangle Ω_c, having vertices at $(0, 0)$, $(1, 0)$, $(0, 1)$. We use the transformation

$$\begin{pmatrix} x \\ y \\ 1 \end{pmatrix} = \begin{pmatrix} x_{i,1} & x_{i,2} & x_{i,3} \\ y_{i,1} & y_{i,2} & y_{i,3} \\ 1 & 1 & 1 \end{pmatrix} \begin{pmatrix} 1 - r - s \\ r \\ s \end{pmatrix}, \tag{3}$$

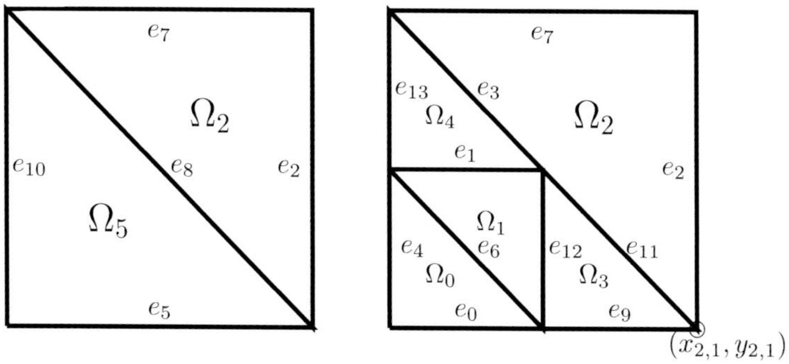

Fig. 1 Examples of admissible meshes: (*left*) two element mesh (*right*) Ω_5 is refined yielding a five element mesh

where $(x_{i,z}, y_{i,z})_{z=1,2,3}$ are the three original vertices of element Ω_i in physical space and (r, s) are the transformed coordinates in the computational space. The Jacobian of the transformation is denoted by J_i and we require that $\det J_i$ be positive, i.e. that the vertices be ordered counter-clockwise. Vertex ordering also defines the side numbers $q = 1, 2 \ldots N_i^{\text{sides}}$ of each cell, where N_i^{sides} is the number of sides an element has; for example $s_{i,q}$ refers to the qth side of Ω_i beginning clockwise from $(x_{i,1}, y_{i,1})$. In the refined mesh of Fig. 1, $s_{2,4}$ is edge e_{11} with the indicated $(x_{2,1}, y_{2,1})$.

Additionally, we map each edge of the mesh e_k, i.e. $s_{i,q}$, to the canonical interval $I_c = [-1, 1]$ with the transformation

$$\begin{pmatrix} x \\ y \end{pmatrix} = \begin{pmatrix} x_{k,1} & x_{k,2} \\ y_{k,1} & y_{k,2} \end{pmatrix} \begin{pmatrix} \frac{1}{2}(1-\xi) \\ \frac{1}{2}(1+\xi) \end{pmatrix}, \tag{4}$$

where ξ is the transformed coordinate in the computational space and $(x_{k,z}, y_{k,z})_{z=1,2}$ are the two original vertices of e_k. The determinant of the Jacobian of (4) is

$$l_{i,q} = \frac{1}{2}\sqrt{(x_{k,1} - x_{k,2})^2 + (y_{k,1} - y_{k,2})^2}.$$

As for other Galerkin methods, the solution on element Ω_i is approximated by a linear combination of the basis functions $\widetilde{v_j}$, i.e. $\mathbf{U}_i = \sum_{j=1}^{N_p} \mathbf{c}_{i,j}\widetilde{v_j}$ with $\mathbf{c}_{i,j} = [c_{i,j}^1, c_{i,j}^2, \ldots, c_{i,j}^m, \ldots, c_{i,j}^M]^\mathsf{T}$, the modal degrees of freedom. Finally, we separate the second term in (2) into line integrals along the edges of the element. It follows that equation (2) becomes

$$\frac{d}{dt}\mathbf{c}_{i,j} = \frac{1}{\det J_i}\left(\int_{\Omega_c} \mathbf{F}(\mathbf{U}_i) \cdot (J_i^{-1}\nabla_{rs} v_j) \det J_i d\Omega_c \right.$$
$$\left. - \sum_{q=1}^{N_i^{\text{sides}}} \int_{I_0} v_{j,q}\mathbf{F}(\mathbf{U}_i) \cdot \mathbf{n}_{i,q} l_{i,q}\, d\xi \right), \tag{5}$$

where v_j and $v_{j,q}$ are values of the basis functions mapped from the physical element Ω_i to the canonical element Ω_c and from side $s_{i,q}$ to the canonical side I_0, respectively. Finally, $\mathbf{n}_{i,q}$ is the outward facing normal of $s_{i,q}$. The right-hand side of equation (5) is composed of two terms: a volume and surface integral. As continuity between elements is not imposed, the solution $\mathbf{U}_i$ is multivalued in the surface integral. We therefore introduce a numerical flux $\mathbf{F}(\mathbf{U}_i, \mathbf{U}_k)$ to allow information to be exchanged between adjacent cells Ω_i and Ω_k. We can evaluate these terms exactly for a linear flux and approximately for a nonlinear flux with numerical quadrature. The equation now becomes

$$\frac{d}{dt}\mathbf{c}_{i,j} = \frac{1}{\det J_i}\Bigg(\int_{\Omega_c} \mathbf{F}(\mathbf{U}_i)\cdot(J_i^{-1}\nabla_{rs}v_j)\,\det J_i d\Omega_c - \sum_{q=1}^{N_i^{\text{sides}}}\int_{I_c} v_{j,q}\mathbf{F}(\mathbf{U}_i,\mathbf{U}_{k_q})\cdot\mathbf{n}_{i,q}l_{i,q}\,d\xi\Bigg). \tag{6}$$

Because equation (6) is a system of ordinary differential equations (ODEs) for the degrees of freedom $\mathbf{c}_{i,j}$, it can be solved in time with a standard ODE solver, e.g. Runge-Kutta (RK) method.

As the volume and surface integral contributions can be computed independently of one another, the method is predisposed to applications on highly parallel GPU architectures [6]. Developing algorithms for this computing platform is an active area of research. An efficient h-adaptive, block-based structured mesh DG-GPU implementation has been presented in [3, 10]. Cell-based h-adaptivity on unstructured meshes in the context of GPUs can be more difficult due to the inevitable irregularity of memory accesses and operations both in time stepping and mesh adaptivity operations. We will now describe the implementation details of our efficient h-adaptive DG-GPU algorithm in NVIDIA's CUDA C; this implementation consists of a time stepping and mesh adaptation module.

3 Time Stepping Details

With the time stepping module, we calculate the right-hand side of (6), without mesh modification. Every n regular timesteps, the adaptive module is executed which refines and coarsens select elements. The time stepping aspects of the solver are outlined in more detail in [6]. This module consists of three compute kernels that efficiently evaluate the right-hand side of (6) for a standard RK ODE time integrator. The first kernel, `eval_volume`, launches one thread per element Ω_i; each thread evaluates the volume integral terms for its allocated element. Likewise, the second kernel, `eval_surface`, launches one thread per edge e_k; each thread evaluates the surface integral terms for its allocated edge. Finally, `eval_rhs` launches one thread per element and appropriately sums the volume and surface terms over the mesh. An essential aspect of these final two kernels is the use of connectivity data on the GPU.

DOF storage In order fully utilize the compute capacity of the GPU, the modal DOFs must be stored in a specific fashion. See [6] for further information.

Connectivity We store connectivity information in GPU DRAM memory. We store simple pointers in linear arrays as integers, which encode the element and edge IDs Ω_i and e_k, respectively. The bidirectional pointers are displayed in Fig. 2. Elements

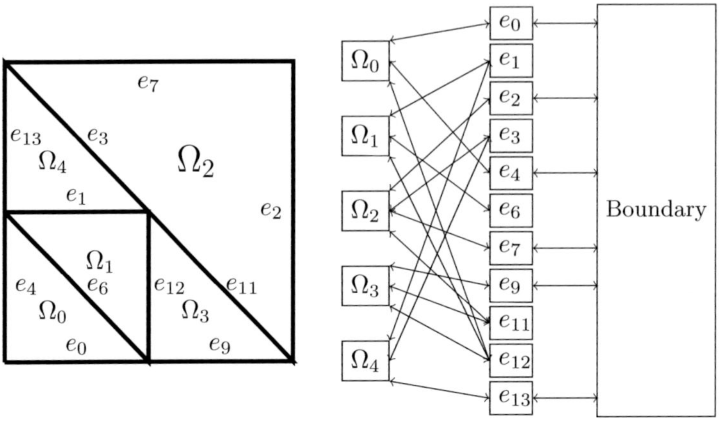

Fig. 2 (*Left*) nonconforming mesh, (*right*) connectivity data

point to their edges and edges point to their two elements; note that ghost cells enforce boundary conditions. Due to the nonconforming nature of our meshes, each element requires N_i^{sides} pointers to its respective edges, see Ω_2 in Fig. 2 ($N_2^{\text{sides}} = 4 \neq 3$).

4 Mesh Adaptation

The mesh adaptation module of our implementation works to optimally distribute the spatial DOFs on the computational domain. We note that work is completed directly in GPU video memory with minimal memory transfer between host and device.

Tree structures A tree structure for both elements and edges is implemented, keeping track of parent-child relationships. These data structures, along with other information, allow us to redetermine the mesh connectivity described in Sect. 3 after elements and edges are refined and coarsened. In the tree, each element and edge point to their children; likewise, each element and edge also point to their parents.

In Fig. 1, element Ω_5 is refined into four new elements. We denote Ω_5 as the parent of the four child elements Ω_3, Ω_0, Ω_4 and Ω_1. Similar terminology is used for the parent edges: e_5, e_8 and e_{10} and their children. The bidirectional pointers that would be stored in GPU DRAM for this example are presented in Fig. 3. We highlight the elements and edges in the trees without children, corresponding to those components which are active in the current mesh.

Error indicator In order to determine which elements need to be refined and which need to be coarsened, an error indicator is used. In the literature, there are a substantial number of indicators available. The number of times an element

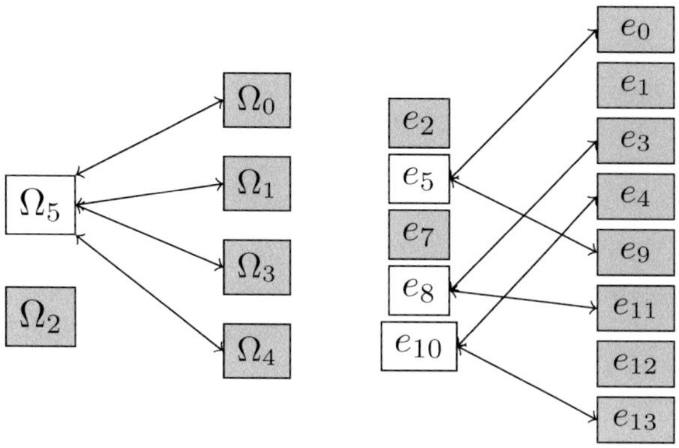

Fig. 3 Tree data stored for the two sequence of meshes in Fig. 1. Elements and edges highlighted in *blue* are active in the current mesh

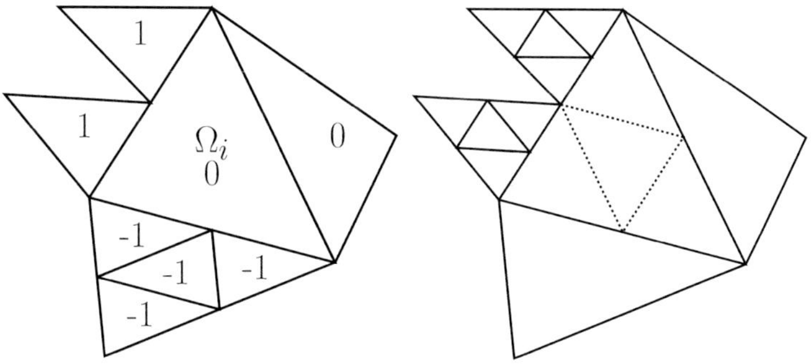

Fig. 4 (*Left*) initial configuration around Ω_i along with the values of `aflag` for each element, (*right*) mesh after refinement and smoothing; *dotted line* indicates the elements added

is refined and coarsened as determined from the error indicator is stored as an integer in an array `aflag`. A positive integer indicates refinement, whereas a negative integer indicates coarsening. Our implementation allows a maximum level of refinement of $l_{max} = 5$.

Flag smoothing Flag smoothing is an operation which enforces that the flags in `aflag` do not create elements that have an interelement difference in refinement level greater than the maximum allowed, i.e. $|l_i - l_j| \leq 1$ where l_i and l_j are the refinement levels of the adjacent elements Ω_i and Ω_j, respectively. We must therefore modify `aflag` so that after the refinement and coarsening operation, an incompatible mesh is not created.

For example, in Fig. 4 on the left we have the mesh configuration around an arbitrary element Ω_i with value of `aflag` on each element. The two elements being refined once will induce an interelement jump in refinement level of 2 with Ω_i. The smoothing operation will supress this jump by refining element Ω_i once as shown in the mesh on the right.

Refinement and coarsening After `aflag` is smoothed, the elements and edges may now be refined and coarsened. We implement refinement and coarsening kernels which modify the element and edge trees. Solution coefficients are redetermined using an L_2 projection and geometric data is adapted accordingly. The final step in the mesh adaptivity algorithm is redetermining the mesh connectivity detailed in Sect. 3. This is done in part with the element and edge trees presented in Sect. 4.

Memory management In refining and coarsening the mesh, we will naturally have to add or remove data from lists, e.g. the modal degrees of freedom **C**. A naive solution would be to simply replace the undesired element by a `null` placeholder. Unfortunately, this would lead to 'holes' in our arrays which is not an efficient use of memory and further would limit the effectiveness of coalesced memory accesses. We must therefore not only remove elements from arrays but also shift those that remain so that the reduced array is dense.

On GPUs, this problem does not have a simple solution. Indeed, parallelizing this operation, called a 'reduction', while avoiding race conditions must be done with great care. There are application programming interfaces (APIs) that have this task optimized; we use CUDPP in our implementation [4]. The details of a reduction are not presented here for brevity, an involved examination of this issue can be found in [8].

5 Computed Example

We now present an example demonstrating the effectiveness of this implementation. Consider the two dimensional advection equation

$$\frac{d}{dt}u + \nabla_{xy} \cdot [-2\pi y u, 2\pi x u] = 0, \tag{7}$$

on the domain $\Omega = [-1, 1]^2$, subject to the initial condition

$$u_0(x, y) = 5 \exp\left(\frac{-(x-0.2)^2 - y^2}{2 \cdot 0.15^2}\right). \tag{8}$$

This is called the rotating hill test problem, whereby a Gaussian pulse is advected around the origin; the exact solution is expressed as

$$u(x, y, t) = u_0(x\cos(2\pi t) + y\sin(2\pi t), -x\sin(2\pi t) + y\cos(2\pi t)). \tag{9}$$

This problem would benefit from h-adaptivity as much of the variation in the solution is spatially concentrated in a small, moving region of the domain.

We apply our adaptive code to this problem and advance the solution until a final time of $T = 1$, i.e. one full rotation around the origin. The DG spatial discretization uses the upwind numerical flux and is integrated in time with an RK time integrator of appropriate order. The adaptive routines are executed every 100 regular timesteps, allowing an element to be refined a maximum of 3 times. The initial mesh was composed of 1080 elements; if uniformly refined three times, it would comprise 69,120 elements. The numerical solution and final adaptive mesh are plotted in Fig. 5 for $p = 1$. In Table 1, we present the computed L_2 error of the final solution for both the adaptive and uniformly refined meshes. The advantage of h-adaptivity is apparent as only a fraction of the computational work is required to achieve an error comparable to that of a uniformly refined mesh. In all simulations, the refinement and coarsening subroutines comprise at most ~2 % of the total runtime.

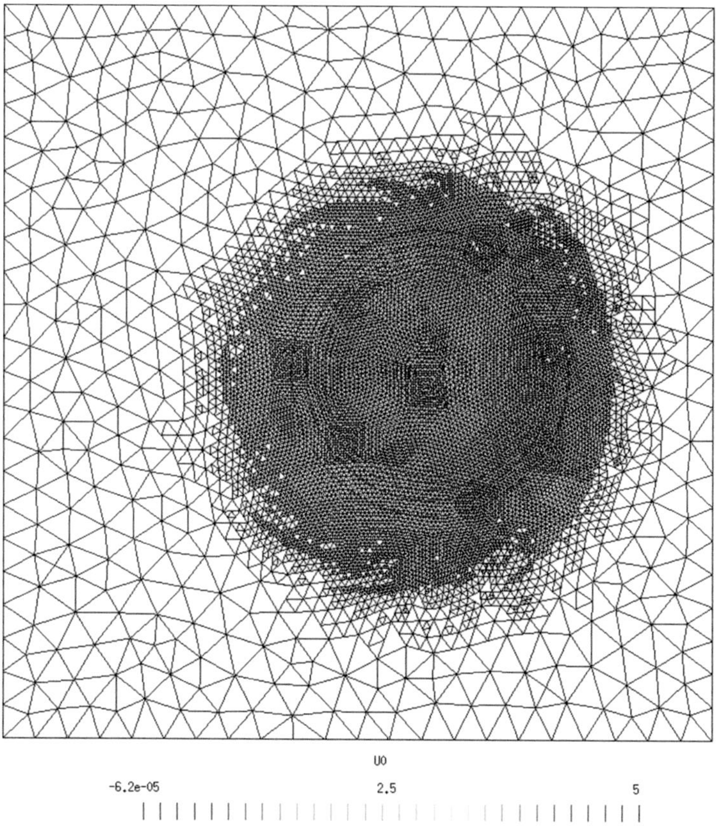

Fig. 5 Final refined mesh at $T = 1$ with the solution isolines for $p = 1$

Table 1 L_2 error for $p = 1, 2, 3$ of the h-adaptive solution, and uniformly refined solution

p	Average number of elements	Adaptive mesh L_2 error	Uniformly refined mesh L_2 error
1	18,195	4.684E-4	4.453E-4
2	24,906	4.373E-6	4.226E-6
3	34,785	4.514E-8	4.274E-8

6 Conclusion

In conclusion, we have presented the framework of an efficient h-adaptive implementation of the DG method for hyperbolic conservation laws on GPUs. We plan on implementing a combined hp-adaptive solver, and extending support to 3 dimensional unstructured tetrahedral meshes with local time stepping.

Acknowledgements This research was supported in part by the Natural Sciences and Engineering Research Council of Canada (NSERC) grant 341373-07 and the NSERC CGS-M grant.

References

1. Berger, M.: Adaptive mesh refinement for hyperbolic partial differential equations. Ph.D. thesis, Stanford University (1982)
2. Berger, M.J., Oliger, J.: Adaptive mesh refinement for hyperbolic partial differential equations. J. Comput. Phys. **53**(3), 484–512 (1984)
3. Brodtkorb, A.R., Sætra, M.L., Altinakar, M.: Efficient shallow water simulations on GPUs: implementation, visualization, verification, and validation. Comput. Fluids **55**, 1–12 (2012)
4. CUDA Data Parallel Primitives Library. http://cudpp.github.io/. CUDPP 2.2
5. Flaherty, J.E., Loy, R.M., Shephard, M.S., Szymanski, B.K., Teresco, J.D., Ziantz, L.H.: Adaptive local refinement with octree load balancing for the parallel solution of three-dimensional conservation laws. J. Parallel Distrib. Comput. **47**(2), 139–152 (1997)
6. Fuhry, M., Giuliani, A., Krivodonova, L.: Discontinuous Galerkin methods on graphics processing units for nonlinear hyperbolic conservation laws. Int. J. Numer. Methods Fluids **76**(12), 982–1003 (2014)
7. Gao, X.: A parallel solution-adaptive method for turbulent non-premixed combusting flows. Ph.D. thesis, University of Toronto (2008)
8. Harris, M., Sengupta, S., Owens, J.D.: Parallel prefix sum (scan) with CUDA. In: Nguyen, H., NVIDIA Corporation (eds.) GPU Gems 3. Addison-Wesley, Upper Saddle River (2008)
9. Ivan, L., Sterck, H.D., Northrup, S.A., and Groth, C.: Multi-dimensional finite-volume scheme for hyperbolic conservation laws on three-dimensional solution-adaptive cubed-sphere grids. J. Comput. Phys. **255**(0), 205–227 (2013)
10. Sætra, M.L., Brodtkorb, A.R., Lie, K.-A.: Efficient GPU-implementation of adaptive mesh refinement for the shallow-water equations. J. Sci. Comput. **63**(1), 23–48 (2014). doi:10.1007/s10915-014-9883-4. ISSN:1573-7691
11. Susanto, A.: High-order finite-volume schemes for magnetohydrodynamics. Ph.D. thesis, University of Waterloo (2014)

12. Wang, Z.: Adaptive High-order Methods in Computational Fluid Dynamics. Advances in Computational Fluid Dynamics. World Scientific, Singapore/Hackensack (2011)
13. Zeeuw, D.L.D.: A quadtree-based adaptively-refined cartesian-grid algorithm for solution of the Euler equations. Ph.D. thesis, The University of Michigan, Ann Arbor (1993)
14. Zhang, J.Z.: Parallel anisotropic block-based adaptive mesh refinement finite-volume scheme. Ph.D. thesis, University of Toronto (2011)

Extending BACOLI to Solve the Monodomain Model

Elham Mirshekari and Raymond J. Spiteri

Abstract BACOLI is a numerical software package that solves systems of parabolic partial differential equations (PDEs) in one spatial dimension. It is based on high-order B-spline collocation and features adaptivity in space and time. The monodomain model of cardiac electrophysiology is a multi-scale model that couples electrical activity in myocardial tissue at the tissue scale with that at the cellular scale. This leads to a (parabolic) reaction-diffusion PDE coupled with a set of nonlinear (non-parabolic) PDEs that do not involve spatial derivatives. In this paper, we extend BACOLI to solve this more general class of problem, of which the monodomain model is one example. We demonstrate that the extended BACOLI software package outperforms the Chaste software package, which is a powerful, widely used, and well-respected software package for heart simulation, in terms of execution time on the monodomain equation with two cell models of varying stiffness.

1 Introduction

BACOLI is a numerical software package that solves systems of parabolic partial differential equations (PDEs) of the form

$$\boldsymbol{u}_t(x,t) = \boldsymbol{f}(t,x,\boldsymbol{u}(x,t),\boldsymbol{u}_x(x,t),\boldsymbol{u}_{xx}(x,t)), \quad a \le x \le b, \quad t_0 \le t \le t_f, \tag{1a}$$

where $\boldsymbol{u}(x,t) \in \mathbb{R}^{M_u}$ is the unknown vector function, subject to separated boundary conditions of the form

$$\mathfrak{B}_a(t,\boldsymbol{u}(a,t),\boldsymbol{u}_x(a,t)) = \boldsymbol{0}, \quad \mathfrak{B}_b(t,\boldsymbol{u}(b,t),\boldsymbol{u}_x(b,t)) = \boldsymbol{0}, \quad t_0 \le t \le t_f, \tag{1b}$$

E. Mirshekari
Department of Mathematics and Statistics, University of Saskatchewan, Saskatoon, SK, Canada
e-mail: elm662@mail.usask.ca

R.J. Spiteri (✉)
Department of Computer Science, University of Saskatchewan, Saskatoon, SK, Canada
e-mail: spiteri@cs.usask.ca

© Springer International Publishing Switzerland 2016

J. Bélair et al. (eds.), *Mathematical and Computational Approaches in Advancing Modern Science and Engineering*, DOI 10.1007/978-3-319-30379-6_41

and the initial condition

$$u(x, t_0) = u_0(x), \quad a \le x \le b. \tag{1c}$$

BACOLI is one of only a few software packages to feature adaptive error control in both space and time [18]. The software package is designed to produce an approximate solution such that the error is estimated to be less than a user-supplied tolerance [9]. Adaptive error control increases both the efficiency as well as the reliability of the simulation.

The monodomain model is a multi-scale mathematical model for the evolution of the electrical potential in myocardial tissue that relates the propagation of ionic currents at the tissue scale with their generation at the cellular scale. Because of their global nature, the mathematical models of the propagation at the tissue level depend on spatial derivatives. On the other hand, because of their local nature, the models of a single cell do not.

The monodomain model can be written as

$$\begin{aligned} \chi C_m \frac{\partial v}{\partial t} + \chi I_{\text{ion}}(s, v, t) &= \frac{\lambda}{1+\lambda} \nabla \cdot (\sigma_i \nabla v), \\ \frac{\partial s}{\partial t} &= f(s, v, t), \end{aligned} \tag{2a}$$

with boundary condition

$$\hat{n}.(\sigma_i \nabla v) = 0, \tag{2b}$$

where v is the transmembrane potential, s is a (model-dependent) vector of cellular states such as gating variables and ionic concentrations, I_{ion} is the ionic current across the membrane per unit cell membrane area, χ is the area of cell membrane per unit volume, σ_i is the conductivity, C_m is the capacitance of the cell membrane per unit area, and $\hat{n}$ is the unit normal vector to the boundary [14]. The system (2) is supplemented by appropriate initial conditions.

The cell models considered in this paper are those of Luo–Rudy [10] (considered to be mildly stiff) and the epicardial variant of ten Tusscher and Panfilov [15] (considered to be stiff). The cell models do not have any dependency on the spatial derivatives and thus cannot be solved directly using BACOLI. We extend BACOLI so that it can solve PDEs of the form (2a).

The remainder of this paper unfolds as follows. A review of the BACOLI software package is provided in Sect. 2. Some basic information on the of the heart is provided in Sect. 3. The performance in terms of execution time of the extended version of BACOLI is demonstrated on the monodomain model with two different cell models in Sect. 4.

2 Review of BACOLI

This section describes some basic concepts behind the BACOLI software package, including its structure and the modifications to extend it to solve multi-scale systems like the monodomain model.

2.1 *Basic Concepts*

BACOLI uses the following discretization strategy. Consider the interval $[a, b]$ divided into N_x subintervals $[x_{i-1}, x_i]$, where $i = 1, 2, \ldots, N_x$, with a partition π of the form

$$\pi : a = x_0 < x_1 < \ldots < x_{N_x} = b.$$

On subinterval i, $i = 1, 2, \ldots, N_x$, the unknown function is approximated by a local polynomial. The spline $s(x)$ obtained from joining the local polynomials is a piecewise polynomial with certain properties; e.g., it can have several continuous derivatives [6].

The standard B-representation of a spline is as a linear combination of B-splines. A B-spline is piecewise polynomial that has minimal support with respect to its degree, continuity, and knot sequence. For the knot sequence $X_1 \leq X_2 \leq \ldots \leq X_{D+M+1}$, we have

$$s(x) = \sum_{j=1}^{D} B_{j,M}(x) a_j,$$

where $B_{j,M}(x)$ is B-spline j of degree M, D is the number of the B-spline basis functions, and the coefficients a_j are the unknown coefficients that are defined according to the choice of method, e.g., collocation . There are different types of knots mentioned in [7]. The type that is considered in BACOLI is *open uniform*, defined as

$$\begin{cases} X_k = X_1, & \text{if } k \leq M + 1, \\ X_{k+1} - X_k = \texttt{constant}, & M + 1 \leq k < D + 1, \\ X_k = X_{D+M+1}, & k \geq D + 1, \end{cases}$$

where $k = 1, 2, \ldots, D + M + 1$; i.e., at each boundary, the multiplicity of the knots is $(M + 1)$.

With the choice of open uniform knots, the associated B-splines have two properties at the left boundary:

$$B_{1,M}(x_1) = 1, \quad \text{and} \quad B'_{1,M}(x_1) = -B'_{2,M}(x_1).$$

Similar properties hold at the right boundary.

2.2 Structure of the BACOLI Software Package

The exact solution of (1), $\boldsymbol{u}(x,t)$, is approximated by a linear combination of these B-spline functions as

$$\boldsymbol{U}(x,t) = \sum_{j=1}^{D} B_{j,M}(x)\boldsymbol{y}_j(t),$$

where $\boldsymbol{y}_j(t)$ is the vector of unknown coefficients of the B-spline j. In order to determine $\boldsymbol{y}_j(t)$, the BACOLI software package requires $\boldsymbol{U}(x,t)$ to satisfy (1a) at each of the $M-1$ Gaussian collocation points of each subinterval [18].

The collocation equations defining $\{\boldsymbol{y}_k(t)\}_{k=1}^{D}$ are of the form

$$\frac{d}{dt}\boldsymbol{U}(\xi_l,t) = \boldsymbol{f}(t,\xi_l,\boldsymbol{U}(\xi_l,t),\boldsymbol{U}_x(\xi_l,t),\boldsymbol{U}_{xx}(\xi_l,t)), \tag{3}$$

where $l = 1+(i-1)(M-1)+j$, $i = 1,2,\ldots,N_x$, and $j = 1,2,\ldots,M-1$. There are at most $(M+1)$ B-spline basis functions that do not vanish at each internal collocation point [5]. Generally, on subinterval i, only the B-splines $\{B_{j,M}\}_{j=j_0}^{j_f}$ do not vanish at the internal ξ_l, where $j_0 = (i-1)(M-1)+1$ and $j_f = i(M-1)+2$. Taking advantage of this property,

$$\boldsymbol{U}(\xi_l,t) = \sum_{j=j_0}^{j_f} B_{j,M}(\xi_l)\boldsymbol{y}_j(t). \tag{4}$$

Inserting (4) into (3) results in the system of ODEs

$$\sum_{j=j_0}^{j_f} B_{j,M}(\xi_m)\boldsymbol{y}'_j(t) = \boldsymbol{f}(t,\xi_m,\boldsymbol{U}(\xi_m,t),\boldsymbol{U}_x(\xi_m,t),\boldsymbol{U}_{xx}(\xi_m,t)), \tag{5}$$

where $m = 1,2,\ldots,M-1$. Once the system of ODEs is constructed, the discretized boundary conditions are added to this system of ODEs to form a system of initial-value differential algebraic equations (DAEs). In other words, BACOLI treats the boundary conditions directly. For simplicity, on the spatial interval [0, 1], at the left boundary,

$$\boldsymbol{U}(0,t) = B_{1,M}(0)\boldsymbol{y}_1(t), \quad \boldsymbol{U}_x(0,t) = B'_{1,M}(0)\boldsymbol{y}_1(t) + B'_{2,M}(0)\boldsymbol{y}_2(t).$$

Similarly, at the right boundary,

$$\boldsymbol{U}(1,t) = B_{D,M}(1)\boldsymbol{y}_D(t), \quad \boldsymbol{U}_x(1,t) = B'_{D,M}(1)\boldsymbol{y}_D(t) + B'_{D-1,M}(0)\boldsymbol{y}_{D-1}(t).$$

Coupling the ODEs in (5) with the discretized boundary conditions gives an index-1 initial-value system of DAEs of the form

$$\begin{aligned}
&\mathfrak{B}_a(t,\boldsymbol{U}(0,t),\boldsymbol{U}_x(0,t)) = \boldsymbol{0}, \\
&\sum_{j=j_0}^{j_f} B_{j,M}(\xi_l)\boldsymbol{y}'_j(t) = \boldsymbol{f}(t,\xi_l,\boldsymbol{U}(\xi_l,t),\boldsymbol{U}_x(\xi_l,t),\boldsymbol{U}_{xx}(\xi_l,t)), \qquad (6)\\
&\mathfrak{B}_b(t,\boldsymbol{U}(1,t),\boldsymbol{U}_x(1,t)) = \boldsymbol{0},
\end{aligned}$$

where ξ_l, $l = 1, 2, \ldots, M-1$, is collocation point l in subinterval i, $i = 1, 2, \ldots, N_x$. This is a system of DM_u equations in DM_u unknowns. The system (6) is also supplemented by appropriate initial conditions.

This system of DAEs is integrated with respect to the independent variable t in order to determine the unknown coefficients $\boldsymbol{y}_j(t)$. The variable-stepsize, variable-order DAE solver DASSL [13] is applied to integrate the resulting initial-value DAEs with respect to time. After solving (6) with DASSL, an approximate solution is available. The BACOLI software package has two options for the estimation of the spatial error. One is through the use of a super-convergent interpolant (SCI) [1], and the other is a low-order interpolant (LOI) [2]. The SCI option generates a numerical solution in the space of polynomials of degree M and an error estimate for that solution. On the other hand, the LOI option generates a numerical solution in the space of polynomials of degree $M + 1$ and an error estimate for that solution. Because one can consider the SCI option to be a standard error control mode, we use SCI in this paper.

More specifically, in the SCI option, at any time step, based on the solution computed through the solver DASSL, $\boldsymbol{U}(x,t)$, which is of degree M, an interpolant, $\tilde{\boldsymbol{U}}(x,t)$, of degree $(M + 1)$ is constructed using some super-convergent values. The difference between $\boldsymbol{U}(x,t)$ and $\tilde{\boldsymbol{U}}(x,t)$ gives an asymptotic estimate of the error for the approximate solution $\boldsymbol{U}(x,t)$. Finally, BACOLI computes a spatial error estimate, and if necessary, performs a spatial adaptation in the form of remeshing.

2.3 The Extended `BACOLI` Software Package

The `BACOLI` software package has been extended to solve system of equations given by

$$\begin{aligned} u_t(x,t) &= f(t,x,\boldsymbol{v}(x,t),u(x,t),u_x(x,t),u_{xx}(x,t)), \\ \boldsymbol{v}_t(x,t) &= \boldsymbol{g}(t,x,\boldsymbol{v}(x,t),u(x,t)), \\ & a \le x \le b, \quad t \ge t_0, \end{aligned}$$

subject to the separated boundary conditions

$$\mathfrak{B}_a(t,\boldsymbol{v}(a,t),u(a,t),u_x(a,t)) = 0, \quad \mathfrak{B}_b(t,\boldsymbol{v}(b,t),u(b,t),u_x(b,t)) = 0,$$

and the initial conditions

$$u(x,t_0) = u_0(x), \qquad \boldsymbol{v}(x,t_0) = \boldsymbol{v}_0(x),$$

where $u(x,t)$ and $\boldsymbol{v}(x,t)$ are the unknown scalar and vector functions of sizes 1 and M_v, respectively. Full details on the modifications and a detailed list and description of the modified subroutines may be found in [12].

3 Electrophysiology Background

The heart can be modelled as an electric dipole with separated positive and negative charges that create an electrical field. The body behaves like a conductor when it is exposed to such an electrical field. In each heart cycle, an electrical current runs through the body. This electrical current causes an electrical potential at each point of the body, giving rise to a time-varying potential difference throughout the body. The potential variations during a heart cycle represent the electrical activation of the heart muscle cells. The electrocardiogram (ECG) is a time-dependent reading of specific potential differences with respect to a reference potential [14]. Figure 1 represents recorded variations of the transmembrane potential as a function of time for the cell model of Bondarenko et al. [3, 4]. A *control volume* is a fixed region considered in space in order to study the energies or fluids that pass its boundary [8]. *Continuum modelling* of a tissue is a method of modelling based on the concept of the control volume, in which specific sample points are considered that relate to a quantity that is the average of its neighbouring cells. In continuum modelling, instead of considering individual cells, only those sample points are considered.

The monodomain model is a continuum-based, multi-scale mathematical model for the electrical activity in heart tissue [14]. The standard formulation of this model is (2). In this paper, the constants χ, σ_i, and C_m are taken to be $1400\,\text{cm}^{-1}$,

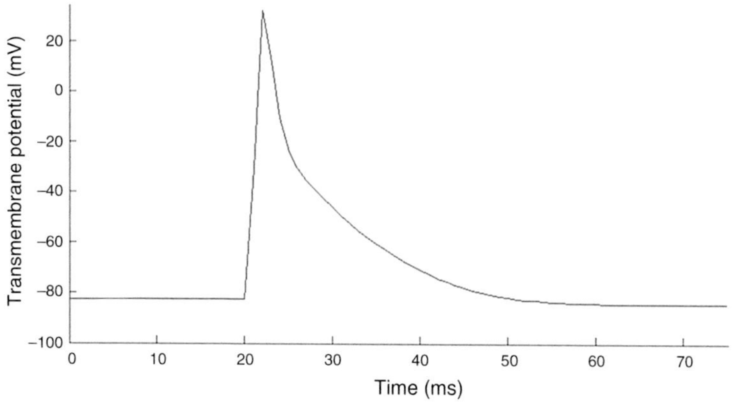

Fig. 1 Transmembrane potential as a function of time steps for the cell model of Bondarenko et al. [3, 4]

1.75 mS/cm, and 1 μF/cm^2, respectively, as per the default values from the Chaste software package (see below).

4 Numerical Results

This section presents the numerical results of solving the monodomain equation coupled with the Luo–Rudy I cell model [10] and the epicardial variant of the 2006 cell model of ten Tusscher and Panfilov [15]. BACOLI accepts two tolerances, a relative and an absolute tolerance.

The boundary conditions for both cases are insulating boundary conditions on a domain chosen as [0, 1]

$$v_x(0,t) = 0, \qquad v_x(1,t) = 0.$$

In order to measure the errors of the numerical solution obtained with the extended BACOLI software package, a *reference solution* is generated with Chaste, a powerful, widely used, and well-respected software package for heart simulation. This software applies a semi-implicit method [19] to the monodomain model (2a), and the cell models (which reduce to ordinary differential equations at each discrete point in space) are solved with Heun's method. The reference solution is generated by comparing increasingly accurate solutions for which the time step is halved and the number of spatial mesh points is doubled. The process of generating these increasingly accurate solutions continues until a desired number of matching digits are obtained. In our numerical experiments, the comparison is made at 21 equally spaced time steps in the temporal domain and 101 equally spaced spatial points.

In order to measure the accuracy and efficiency of any numerical method, one can compute an average of the error at N_t points in the temporal domain $t \in [t_0, t_f]$ and at N_x points in the spatial domain, i.e., the average is over $N = N_t N_x$ space-time points. After generating the reference solution for all N points, the *mixed root mean square* (MRMS) [11] error is computed as

$$e_{\text{MRMS}} = \sqrt{\frac{1}{N}\sum_{i=1}^{N}\left(\frac{\hat{v}_i - v_i}{1 + |\hat{v}_i|}\right)^2},$$

where $\hat{v}_i$ and v_i, respectively, denote the reference solution and the numerical solution for the voltage at space-time point i.

4.1 Monodomain Equation with Luo–Rudy I Cell Model

The Luo–Rudy I cell model consists of 8 variables and simulates the action potential of a guinea pig ventricular cell [10]. The initial values are taken from [16]. Table 1 shows some accuracy results. The first column contains the spatial points for the output. The second column represents the reference solution generated by `Chaste`. This reference solution has 4 matching digits and MRMS error 4.86×10^{-4} % between two consecutive phases. The third column represents the solution generated with the extended `BACOLI` at tolerance 1×10^{-8} and SCI spatial error estimate. The results are reported at the time $t_f = 5$ ms. The matching digits are 7 at $x = 0.0$ and 6 elsewhere.

For the extended `BACOLI` software package, the absolute and relative tolerances are 1.25×10^{-2}, and the number of collocation points is 3. These tolerances are roughly the largest that gives a numerical solution with MRMS error of about 5 %. Because it is possible to achieve MRMS error of about 5 % with relatively large tolerances, it is most efficient for the extended `BACOLI` to use a relatively small number of collocation points. The initial mesh is set to have 4 uniform subintervals in order to minimize any bias in the extended `BACOLI` software package determining a suitable mesh. The resulting numerical solution has MRMS error 4.15 %, and that for `Chaste` is almost the same. The results show that with

Table 1 Solution of the Luo–Rudy I for the voltage (mV) after 5 ms, obtained with `Chaste` and the extended `BACOLI` software package (SCI scheme and tolerance 1×10^{-8})

x	Reference solution	Extended `BACOLI` solution
0.0	**19.72929**206746164	**19.7292917**671303982
0.2	**18.86929**313940373	**18.869273**4805216737
0.4	**18.85464**593756643	**18.854628**0990272841
0.6	**19.29529**722142660	**19.295242**7151648926
0.8	**16.61051**308307339	**16.610495**7169274741
1.0	**20.52934**860573765	**20.529334**7833165818

the extended BACOLI software package it takes 0.18 s and with Chaste it takes 0.50 s. Therefore, in this case there is a speed-up of a factor of around three in the extended BACOLI software package. Such a saving becomes significant when such simulations are run many times sequentially as may be required for parameter or other types of optimization; for example, running the Luo–Rudy I model a million times consecutively with the extended BACOLI software package would theoretically save about $320,000$ s or 3.7 days on a six-day run.

4.2 Monodomain Equation with Ten Tusscher Cell Model

The epicardial variant of the model of ten Tusscher and Panfilov [15] consists of 19 variables and simulates the human ventricle [15]. The initial values are taken from [17]. Table 2 shows some accuracy results. The first column contains the spatial points for the output. The second column represents the reference solution generated by Chaste. This reference solution has 3 matching digits and MRMS error 3.92×10^{-3} % between two consecutive phases. The third column represents the solution generated with the extended BACOLI at tolerance 1×10^{-8} and SCI spatial error estimate. The results are reported at the time $t_f = 5$ ms. The matching digits vary between 3 and 6.

For the extended BACOLI software package, the absolute and relative tolerances are 1×10^{-2}, and the number of collocation points is 3. These tolerances are roughly the largest that gives a numerical solution with MRMS error of about 5 %. For consistency with the previous experiment, the initial mesh size is taken to be uniform with 4 subintervals. The resulting numerical solution has MRMS error 4.53 % and that for Chaste is 5.60 %. The results show that with the extended BACOLI software package it takes 1.16 s and with Chaste it takes 2.58 s. Therefore, in this case there is a speed-up of a factor of around two in the extended BACOLI software package.

Table 2 Solution of the epicardial variant of the model of ten Tusscher and Panfilov [15] for the voltage (mV) after 5 ms, obtained with Chaste and the extended BACOLI software package (SCI scheme and tolerance 1×10^{-8})

x	Reference solution	Extended BACOLI solution
0.0	**15.7367**1511358962	**15.7366**969196549213
0.2	**15.8815**3751011188	**15.8815**217931186137
0.4	**16.0789**0988375955	**16.0788**979751286440
0.6	**16.401**15130209244	**16.401**0866279204457
0.8	**17.36**927102074231	**17.36**68857822675768
1.0	**20.7**3937330189026	**20.7**486443492494281

5 Conclusions

Modifications to the BACOLI software package were made to extend it to solve a more general class of problems. The extended BACOLI software package can solve problems in one dimension that consist of a scalar parabolic PDE coupled with any number of PDEs that have no dependency on spatial derivatives. The practical application of such problems described in this paper is the mathematical model of the electrical activity of the heart known as the monodomain model. The examples considered in this paper are the monodomain equation coupled with the Luo–Rudy I cell model and the monodomain equation coupled with the epicardial variant of the cell model of ten Tusscher and Panfilov [15]. For these examples, we demonstrated a speed-up of factors of 2–3 in execution time of the extended BACOLI software package in comparison with the Chaste software package for accuracies typically required in practice.

References

1. Arsenault, T., Muir, P.H., Smith, T.: Superconvergent interpolants for efficient spatial error estimation in 1D PDE collocation solvers. Can. Appl. Math. Q. **17**(3), 409–431 (2009)
2. Arsenault, T., Smith, T., Muir, P.H., Pew, J.: Asymptotically correct interpolation-based spatial error estimation for 1D PDE solvers. Can. Appl. Math. Q. **20**(3), 307–328 (2012)
3. Auckland Bioengineering Institute.: The CellML project. http://www.cellml.org/
4. Bondarenko, V.E., Szigeti, G.P., Bett, G.C.L., Kim, S.J., Rasmusson, R.L.: Computer model of action potential of mouse ventricular myocytes. Am. J. Physiol.-Heart C. **287**(3), H1378–H1403 (2004)
5. de Boor, C.: A Practical Guide to Splines. Springer, New York (1978)
6. de Boor, C., The MathWorks Inc.: Spline Toolbox User's Guide. The MathWorks, Inc., Natick (1999)
7. Dodgson, N.: B-splines. http://www.cl.cam.ac.uk/teaching/2000/AGraphHCI/SMEG/node4.html (Aug 2013)
8. Engineers Edge LLC.: Control volume – fluid flow. http://www.engineersedge.com/fluid_flow/control_volume.htm (Feb 2014)
9. Holder, D., Huo, L., Martin, C.F.: The Control of Error in Numerical Methods. Springer, New York (2007)
10. Luo, C., Rudy, Y.: A model of ventricular cardiac action potential. Circ. Res. **68**(6), 1501–1526 (1991)
11. Marsh, M.E., Torabi Ziaratgahi, S., Spiteri, R.J.: The secrets to the success of the Rush – Larsen method and its generalizations. IEEE Trans. Bio-Med. Eng. **59**(9), 2506–2515 (2012)
12. Mirshekari, E.: Extending BACOLI to solve multi-scale problems. Master's thesis, Department of Mathematics and Statistics, University of Saskatchewan (2014)
13. Petzold, L.R.: A description of DASSL: a differential/algebraic system solver. Technical report, Sandia National Labs., Livermore (1982)
14. Sundnes, J., Lines, G.T., Cai, X., Nielsen, B.F., Mardal, K.A., Tveito, A.: Computing the Electrical Activity in the Heart. Springer, Berlin (2006)
15. ten Tusscher, K.H.W.J., Panfilov, A.V.: Alternans and spiral breakup in a human ventricular tissue model. Am. J. Physiol. Heart Circ. Physiol. **291**(3), 1088–1100 (2006)

16. The CellML Project.: A model of the ventricular cardiac action potential. http://models.cellml.org/exposure/2d2ce7737b42a4f72d6bf8b67f6eb5a2/luo_rudy_1991.cellml/@@cellml_codegen/F77 (Feb 2014)
17. The CellML Project: Alternans and spiral breakup in a human ventricular tissue model (epicardial model). http://models.cellml.org/exposure/a7179d94365ff0c9c0e6eb7c6a787d3d/ten_tusscher_model_2006_IK1Ko_epi_units.cellml/@@cellml_codegen/F77 (Apr 2014)
18. Wang, R., Keast, P., Muir, P.H.: A high-order global spatially adaptive collocation method for 1-D parabolic PDEs. Appl. Numer. Math. **50**(2), 239–260 (2004)
19. Whiteley, J.P.: An efficient numerical technique for the solution of the monodomain and bidomain equations. IEEE Trans. Bio-Med. Eng. **53**(11), 2139–2147 (2006)

An Analysis of the Reliability of Error Control B-Spline Gaussian Collocation PDE Software

Paul Muir and Jack Pew

Abstract B-spline Gaussian collocation software for the numerical solution of PDEs has been widely used for several decades. BACOL and BACOLI are recently developed packages of this class that provide control of estimates of the temporal and spatial errors of the numerical solution through the use of adaptive time-stepping/adaptive time integration method order selection and adaptive spatial mesh refinement. Previous studies have investigated the performance of the BACOL and BACOLI packages, primarily with respect to work-accuracy measures. In this paper, we investigate the reliability of the BACOL and BACOLI packages, focusing on the relationship between the requested tolerance and the accuracy achieved. In particular, we consider the effect, on the reliability of the software, of (i) the degree of the piecewise polynomials employed in the representation of the spatial dependence of the approximate solution, (ii) the type of spatial error control employed, and (iii) the type of spatial error estimate computed.

1 Introduction/Background

B-spline Gaussian collocation software for the numerical solution of PDEs has been widely used for several decades. Although there has been some development of software of this type for 2D PDEs, see, e.g., [9] and references within, software for the 1D case is at a more advanced state, and it is this case that we focus on in this paper. (However, since the approach for the 2D case considered in [9] builds on the numerical algorithms used in the 1D case, the questions we consider in this paper will be relevant for the 2D case.) A characterizing feature of such software is the representation of the numerical solution as a linear combination of known B-spline [4] basis functions (piecewise polynomials of a given degree p) in space within unknown time-dependent coefficients. The B-spline basis is chosen to have

P. Muir (✉) • J. Pew
Saint Mary's University, Halifax, NS, B3H 3C3, Canada
e-mail: muir@smu.ca; jack.pew@gmail.com

© Springer International Publishing Switzerland 2016
J. Bélair et al. (eds.), *Mathematical and Computational Approaches in Advancing Modern Science and Engineering*, DOI 10.1007/978-3-319-30379-6_42

C^1-continuity and depends on a spatial mesh, $\{x_i\}$, $i = 0, \ldots, NINT$, that partitions the spatial domain. The approximate solution has the form

$$\underline{U}(x,t) = \sum_{i=1}^{NC} \underline{y}_i(t) B_i(x), \tag{1}$$

where $\underline{y}_i(t)$ is the time-dependent coefficient of the i-th B-spline basis function, $B_i(x)$, and $NC = NINT(p-1)+2$. The B-spline coefficients are determined by solving collocation conditions, obtained by requiring the approximate solution to exactly satisfy the PDEs at the images of the Gauss points of a given order on each subinterval of the spatial mesh, and additional equations that depend on the boundary conditions. Assuming a system of PDEs of the form

$$\underline{u}_t(x,t) = \underline{f}\left(t, x, \underline{u}(x,t), \underline{u}_x(x,t), \underline{u}_{xx}(x,t)\right), \qquad a \le x \le b, \quad t \ge t_0, \tag{2}$$

with initial conditions

$$\underline{u}(x,t_0) = \underline{u}_0(x), \qquad a \le x \le b, \tag{3}$$

and boundary conditions

$$\underline{b}_L\left(t, \underline{u}(a,t), \underline{u}_x(a,t)\right) = \underline{0}, \qquad \underline{b}_R\left(t, \underline{u}(b,t), \underline{u}_x(b,t)\right) = \underline{0}, \quad t \ge t_0, \tag{4}$$

the collocation conditions are a system of ODEs the form

$$\underline{U}_t(\xi_l,t) = \underline{f}\left(t, \xi_l, \underline{U}(\xi_l,t), \underline{U}_x(\xi_l,t), \underline{U}_{xx}(\xi_l,t)\right), \tag{5}$$

for $l = 2, \ldots, NC-1$. The collocation points are $\xi_l = x_{i-1} + h_i\rho_j, l = 1 + (i-1)(p-1)+j, i = 1, \ldots, NINT, j = 1, \ldots, p-1$, $\{\rho_i\}_{i=1}^{p-1}$ are the images of the Gauss points on [0, 1], and $h_i = x_i - x_{i-1}$. (Thus, *kcol*, the number of collocation points per subinterval, is $p-1$.)

About 35 years ago, the B-spline Gaussian collocation package, PDECOL [10] was developed. The B-spline basis was implemented using the de Boor B-spline package [4] and the ODE solver, GEARIB [6], was used to solve the collocation conditions (5) together with the time differentiated boundary conditions. Temporal error control was provided by the ODE solver. PDECOL used a fixed spatial mesh which meant that no control of the spatial error was provided.

About a decade later, a modification of PDECOL, called EPDCOL [8] was released; EPDCOL replaced the band matrix solver employed by PDECOL with an almost block diagonal (ABD) solver, COLROW [5], more appropriate for the structure of the linear systems arising from the B-spline collocation process. Savings in execution time of about 50 % were obtained.

The next update to this family of PDE solvers, BACOL [18], released a little over a decade later, was a new implementation of the B-spline Gaussian collocation algorithm, with several additional features. BACOL employed a Differential-Algebraic Equation (DAE) solver, DASSL [3], so that the boundary conditions (4) could be treated directly together with the ODEs (5) arising from the collocation process. Temporal error control was provided by DASSL. The most significant new feature was the implementation of spatial error estimation and control. The spatial error estimates were obtained by computing a second (higher order) collocation solution, $\underline{\bar{U}}(x, t)$, based on B-splines of degree $p + 1$, which was then used together with $\underline{U}(x, t)$ to compute a set of spatial error estimates of the form

$$E_j(t) = \sqrt{\int_a^b \left(\frac{U_j(x,t) - \bar{U}_j(x,t)}{ATOL_j + RTOL_j |U_j(x,t)|} \right)^2 dx}, \quad j = 1, \ldots, NPDE, \tag{6}$$

where $U_j(x, t)$ is the jth component of $\underline{U}(x, t)$, $\bar{U}_j(x, t)$ is the jth component of $\underline{\bar{U}}(x, t)$, $ATOL_j$ and $RTOL_j$ are absolute and relative tolerances corresponding to the jth component of the solution, and the number of PDEs is $NPDE$. At the end of each time step, t, the collocation solution, $\underline{U}(x, t)$ was accepted if

$$\max_{1 \le j \le NPDE} E_j(t) \le 1. \tag{7}$$

Otherwise, the time step was rejected and BACOL computed a second set of spatial error estimates of the form,

$$\hat{E}_i(t) = \sqrt{\sum_{j=1}^{NPDE} \int_{x_{i-1}}^{x_i} \left(\frac{U_j(x,t) - \bar{U}_j(x,t)}{ATOL_j + RTOL_j |U_j(x,t)|} \right)^2 dx}, \quad i = 1, \ldots, NINT, \tag{8}$$

which were then used within an adaptive mesh refinement (AMR) process for which the goal was to determine a spatial mesh for the current time such that the estimated spatial errors were (i) approximately equidistributed over the mesh subintervals and (ii) less than the user tolerance [20]. BACOL was shown to have superior performance compared to a number of comparable packages, especially for problems with sharp moving layers and higher accuracy requirements [19].

In BACOL, the spatial error estimate is used to control the error in $\underline{U}(x, t)$, the approximate solution returned by the code. We refer to this as Standard (ST) error control. A minor modification of BACOL would allow it to return $\underline{\bar{U}}(x, t)$ and in this case, the computation of the *returned* solution would be controlled based on an error estimate for a collocation solution that is of one lower order. We refer to this as Local Extrapolation (LE) error control; see, e.g., [7], in the context of Runge-Kutta formula pairs for the numerical integration of ODEs.

The most recent update to this family of solvers is a modification of BACOL known as BACOLI [12]. The key feature of BACOLI is the avoidance of the computation of $\underline{\bar{U}}(x, t)$, which represented a substantial cost within BACOL. BACOLI instead computes only one collocation solution and one low cost interpolant that is used in the computation of the spatial error estimate. There are two options available for the interpolant. One involves a superconvergent interpolant (SCI) [1] which is based on the presence of certain points within the spatial domain where $\underline{U}(x,t)$ is of at least one order of accuracy higher. The SCI is of the same order of accuracy as $\bar{\underline{U}}(x,t)$ and leads to a spatial error estimate for $\underline{U}(x,t)$, similar to the situation in BACOL. This option therefore provides ST error control. The other option involves a lower order interpolant (LOI) [2], of one order of accuracy lower than $\underline{U}(x,t)$. The LOI is based on interpolation of $\underline{U}(x,t)$ at certain points such that the interpolation error of the LOI agrees asymptotically with the error of a collocation solution of one order lower than $\underline{U}(x,t)$. This option therefore provides LE error control. BACOLI has been shown to be about twice as fast as BACOL [12].

When considering the effectiveness of a numerical software package, the most common analysis involves work-precision (i.e., work-accuracy) diagrams that show the relationship between the CPU time and the accuracy. A number of such studies involving BACOL and BACOLI have been performed; see, e.g., [12, 19]. *However, an equally important measure of the quality of a software package is its reliability , and a key component of this measure involves assessing the relationship between the tolerance requested and the accuracy achieved.* This depends on the quality of the error estimates and the algorithms that adapt the computation to attempt to control the error estimates with respect to the user-specified tolerances.

The purpose of this paper is to present an experimental analysis of the reliability of the error control PDE solvers BACOL and BACOLI on a standard test problem chosen from the literature. (Additional results of this type for several other test problems are reported in [11] and some preliminary results of this type for BACOL were reported in [17].) In particular, we are interested in how the choice of the degree, p, of the B-spline basis and the choice of error control mode (ST or LE) impact on the reliability of the solvers. We also examine the effectiveness of the interpolation-based spatial error estimates provided by the SCI and LOI schemes.

2 Numerical Results

In this section we present numerical results for four code combinations identified in the previous section:

- BACOL with ST error control : **BAC/ST**
- BACOL with LE error control : **BAC/LE**
- BACOLI with the SCI scheme/ST error control : **SCI/ST**
- BACOLI with the LOI scheme/LE error control : **LOI/LE**

We consider the One Layer Burgers Equation (**OLBE**) with $\epsilon = 10^{-4}$:

$$u_t = \epsilon u_{xx} - u u_x, \tag{9}$$

with boundary conditions at $x = 0$ and $x = 1$ ($t > 0$) and an initial condition at $t = 0$ ($0 \leq x \leq 1$) such that the exact solution is

$$u(x,t) = \frac{1}{2} - \frac{1}{2}\tanh\left(\frac{x - \frac{t}{2} - \frac{1}{4}}{4\epsilon}\right).$$

The solution has a sharp layer region around $x = 0.25$ when $t = 0$. As t goes from 0 to 1, the layer moves to the right and is located around $x = 0.75$ when $t = 1$. A plot of the solution for $\epsilon = 10^{-4}$ is given in [12].

We apply each of the four code combinations identified above to the OLBE with $\epsilon = 10^{-4}$, for a range of *kcol* values (recall that the degree of the piecewise polynomials in space is $p = kcol + 1$) and for 91 tolerance values (with $ATOL_1 = RTOL_1$) uniformly distributed from 10^{-1} to 10^{-10}. For each experiment, we report the L^2-norm of the error at the final time, $T_{out} = 1$. Additional results, including results for other test problems, are given in [11].

2.1 Reliability vs. kcol

In this subsection, we consider, for the LOI/LE code, the relationship between the error and the tolerance, for $kcol = 3, 5, 7, 9$. We provide plots of the tolerance vs. the error and plots of the tolerance vs. the error tolerance ratio (ETR), i.e., the error divided by the tolerance. See Figs. 1, 2, 3, and 4. In Fig. 1a, we see that there is good correlation between the requested tolerance and the achieved accuracy; the points are clustered around the reference line corresponding to equal values of the tolerance and error. In Fig. 1b, considering the horizontal reference line corresponding to an ETR value of 1, we see that the ETR values vary over the range of 91 tolerance values, with some errors greater than the tolerance and some less than the tolerance. The other horizontal line represents the average of the ETR values. This figure also

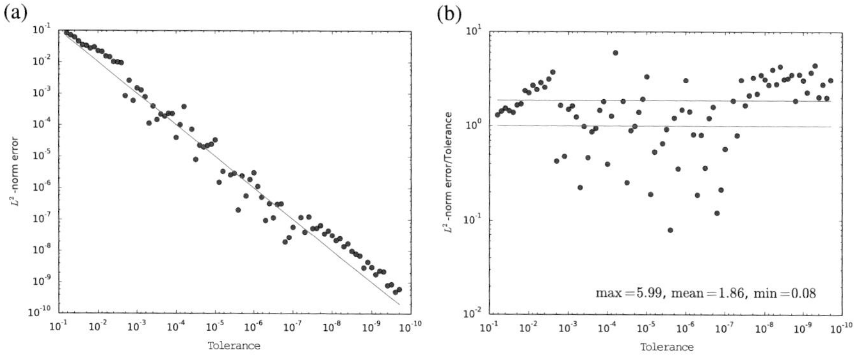

Fig. 1 (**a**) Tolerance vs. Error, LOI/LE, $kcol = 3$, (**b**) Tolerance vs. ETR, LOI/LE, $kcol = 3$

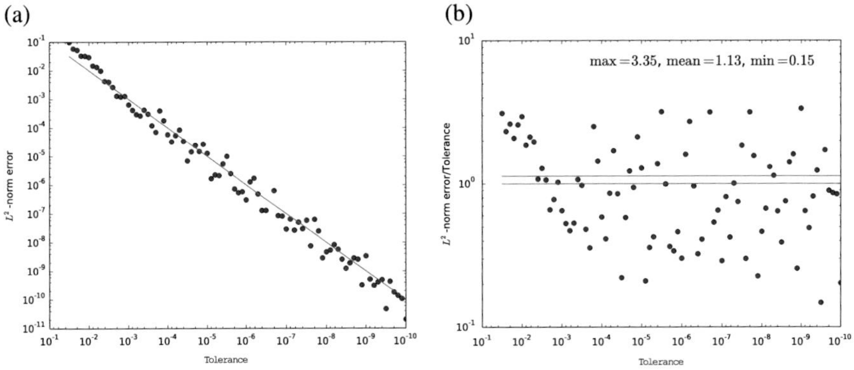

Fig. 2 (**a**) Tolerance vs. Error, LOI/LE, *kcol* = 5, (**b**) Tolerance vs. ETR, LOI/LE, *kcol* = 5

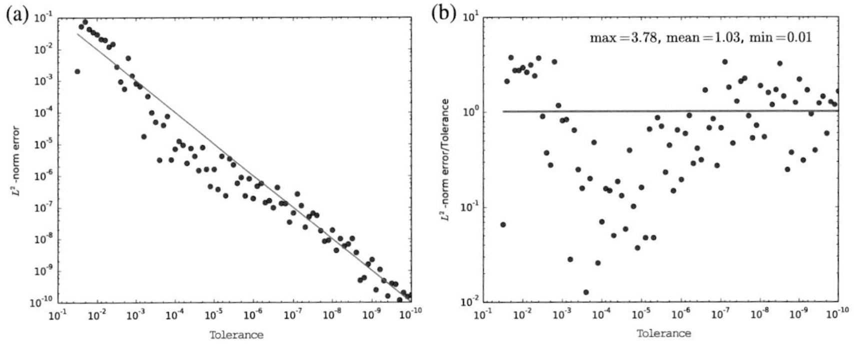

Fig. 3 (**a**) Tolerance vs. Error, LOI/LE, *kcol* = 7, (**b**) Tolerance vs. ETR, LOI/LE, *kcol* = 7

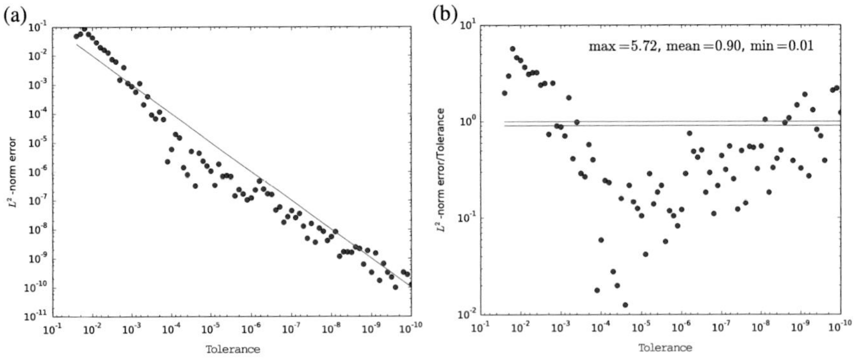

Fig. 4 (**a**) Tolerance vs. Error, LOI/LE, *kcol* = 9, (**b**) Tolerance vs. ETR, LOI/LE, *kcol* = 9

shows that maximum ETR is 5.99, the minimum is 0.08, and that on average the error is 1.86 times the tolerance. Figures 2, 3, and 4, give similar results for the $kcol = 5, 7, 9$ cases, and show maximum, minimum, and average ETR values of $\{3.35, 0.15, 1.13\}$, $\{3.78, 0.01, 1.03\}$, and $\{5.72, 0.01, 0.90\}$, respectively. Similar results are reported in [11] for the other code combinations and test problems.

2.2 *Reliability vs. Error Control Mode*

In this subsection we consider the BAC/ST and BAC/LE codes for the $kcol = 5$ case. From the results for these codes we can assess the effect that the error control mode has on reliability since this is the only difference between these codes. From Figs. 5 and 6 we can see that the two codes appear to have quite similar performance. For the BAC/ST code, the maximum, minimum, and average ETR values are 4.49, 0.28, and 1.79, while for the BAC/LE code, the corresponding values are 4.49,

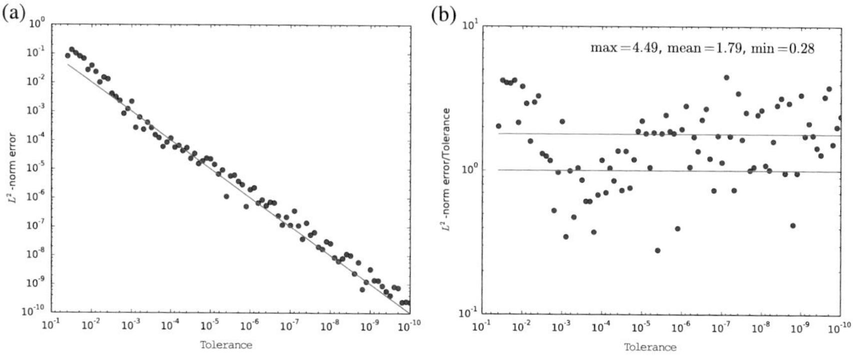

Fig. 5 (**a**) Tolerance vs. Error, BAC/ST, $kcol = 5$, (**b**) Tolerance vs. ETR, BAC/ST, $kcol = 5$

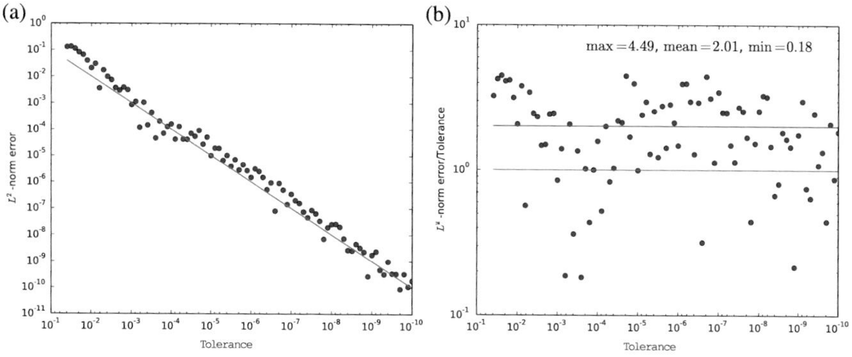

Fig. 6 (**a**) Tolerance vs. Error, BAC/LE, $kcol = 5$, (**b**) Tolerance vs. ETR, BAC/LE, $kcol = 5$

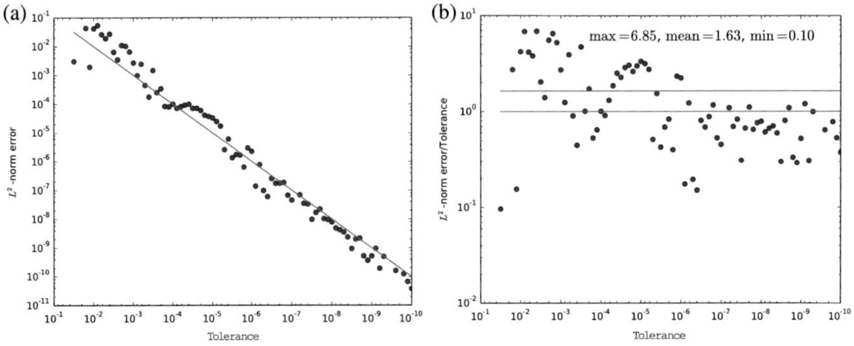

Fig. 7 (**a**) Tolerance vs. Error, SCI/ST, $kcol = 5$, (**b**) Tolerance vs. ETR, SCI/ST, $kcol = 5$

0.18, and 2.01. Similar results are reported in [11] for the other *kcol* values and test problems.

2.3 *Reliability vs. Error Estimation Scheme*

In this section we compare BAC/ST with SCI/ST and BAC/LE with LOI/LE, for $kcol = 5$. This will allow us to assess the effectiveness of the interpolation-based error estimation schemes. The plots for the SCI/ST code are given in Fig. 7. A comparison of Figs. 5 and 7 show that the BAC/ST and SCI/ST codes have comparable reliability, while a comparison of Figs. 6 and 2 show that the BAC/LE and LOI/LE codes have comparable reliability. From Fig. 7b, we see that for the SCI/ST code the maximum ETR is 6.85, the minimum ETR is 0.10, and the average is 1.63. Generally similar results for the other *kcol* values and test problems are reported in [11]. (The only exception, as documented in [12] and [11], is for the SCI/ST, $kcol = 3$ case, where issues associated with the quality of its error estimates have been observed.)

3 Discussion/Conclusions/Future Work

Several conclusions can be drawn from the results presented in the previous section:

- The codes considered in this paper appear to be reliable; the error is generally well correlated with the requested tolerance. We see that in almost all cases the error is within the correct order of magnitude of the requested tolerance. *Furthermore the error is, on average, a small multiple of the requested tolerance.* This suggests that the error estimates computed by the codes are generally of good quality and that the new SCI and LOI error estimation schemes are of comparable quality to the original error estimation scheme employed by BACOL.

- The relationship between the error and tolerance for each of the two versions of BACOL that employ different error control modes (ST vs. LE) is quite similar. This suggests that the choice of error control mode does not have a strong impact on the reliability of the solver.
- The degree p of the B-spline basis as specified by the choice of *kcol* ($p = kcol+1$) does appear to have an impact on the reliability. A comparison of the results for the $kcol = 3, 5, 7$, and 9 cases shows that as *kcol* is increased, the reliability of the codes tends to increase; that is, the error is, on average, almost equal to the tolerance for higher *kcol* values. For the $kcol = 9$ case, the error is often significantly less than the tolerance. (Of course, while it is desirable for the error to be about the same size as or even slightly less than the tolerance, if the error is too much less than the tolerance this can lead to inefficiency in the computation.)

While the entire family of software packages discussed in this paper leave *kcol* as a parameter that must be chosen by the user, it is in fact not clear how to make an appropriate choice of this parameter for a given problem. The correlation between higher *kcol* values and improved reliability that can be seen from the results presented in this paper, as well as earlier results investigating the relationship between the choice of *kcol*, the tolerance, and the efficiency of the solvers [12], will provide a basis for further analysis that will allow us to develop a new release of BACOLI in which the code itself will choose *kcol* appropriately to improve the efficiency and reliability of the computation.

Since BACOL and BACOLI employ DASSL to compute the time-dependent B-spline coefficients, the overall reliability of the computations performed by these codes is dependent on the reliability of DASSL. An example of the relationship between the tolerance and the error for DASSL can be seen in Figure 1 of [16]; it shows generally good correlation between the error and the tolerance, but with variations in the ETR of approximately one order of magnitude in either direction, similar to those reported here for BACOL and BACOLI.

Furthermore, for DASSL and the codes considered in this paper, the relationship between the tolerance and the error is not particularly smooth. That is, while the correlation between a given tolerance and the corresponding error is generally good (i.e., the error is, on average, within a small multiple of the tolerance), a small change in the tolerance does not necessarily lead to a similar small change in the error. This is related to what is referred to as the "conditioning" or "stability" of the software [16]. The papers [14, 15], and [13] describe how control theoretic techniques can be employed within ODE or DAE software to provide an improved condition number for the software. We plan to introduce these type of techniques into the BACOL and BACOLI packages.

Finally, as mentioned earlier, since the approach for the 2D case considered in [9] builds on the 1D case, we expect the results of this paper to be relevant for the numerical treatment of 2D PDEs.

References

1. Arsenault, T., Smith, T., Muir, P.H.: Superconvergent interpolants for efficient spatial error estimation in 1D PDE collocation solvers. Can. Appl. Math. Q. **17**, 409–431 (2009)
2. Arsenault, T., Smith, T., Muir, P.H., Pew, J.: Asymptotically correct interpolation-based spatial error estimation for 1D PDE solvers. Can. Appl. Math. Q. **20**, 307–328 (2012)
3. Brenan, K.E., Campbell, S.L., Petzold, L.R.: Numerical Solution of Initial-Value Problems in Differential-Algebraic Equations. Society for Industrial and Applied Mathematics (SIAM), Philadelphia (1989)
4. de Boor, C.: A Practical Guide to Splines. Volume 27 of Applied Mathematical Sciences. Springer, New York (1978)
5. Díaz, J.C., Fairweather, G., Keast, P.: Algorithm 603. COLROW and ARCECO: FORTRAN packages for solving certain almost block diagonal linear systems by modified alternate row and column elimination. ACM Trans. Math. Softw. **9**(3), 376–380 (1983)
6. Gear, C.W.: Numerical Initial Value Problems in Ordinary Differential Equations. Prentice-Hall, Englewood Cliffs (1971)
7. Hairer, E., Nørsett, S.P., Wanner, G.: Solving Ordinary Differential Equations. I. Volume 8 of Springer Series in Computational Mathematics, 2nd edn. Springer, Berlin (1993)
8. Keast, P., Muir, P.H.: Algorithm 688: EPDCOL: a more efficient PDECOL code. ACM Trans. Math. Softw. **17**(2), 153–166 (1991)
9. Li, Z., Muir, P.H.: B-spline Gaussian collocation software for two-dimensional parabolic PDEs. Adv. Appl. Math. Mech. **5**, 528–547 (2013)
10. Madsen, N.K., Sincovec, R.F.: Algorithm 540: PDECOL, general collocation software for partial differential equations. ACM Trans. Math. Softw. **5**(3), 326–351 (1979)
11. Muir, P.H., Pew, J.: Tolerance vs. error results for a class of error control B-spline Gaussian collocation PDE solvers. Saint Mary's University, Department of Mathematics and Computing Science Technical Report Series. http://cs.smu.ca/tech_reports (2015)
12. Pew, J., Li, Z., Muir, P.H.: A computational study of the efficiency of collocation software for 1D parabolic PDEs with interpolation-based spatial error estimation. Saint Mary's University, Dept. of Mathematics and Computing Science Technical Report Series. http://cs.smu.ca/tech_reports (2013)
13. Pulverer, G., Söderlind, G., Weinmüller, E.: Automatic grid control in adaptive BVP solvers. Numer. Algorithms **56**(1), 61–92 (2011)
14. Söderlind, G.: Digital filters in adaptive time-stepping. ACM Trans. Math. Softw. **29**(1), 1–26 (2003)
15. Söderlind, G., Wang, L.: Adaptive time-stepping and computational stability. J. Comput. Appl. Math. **185**(2), 225–243 (2006)
16. Söderlind, G., Wang, L.: Evaluating numerical ODE/DAE methods, algorithms and software. J. Comput. Appl. Math. **185**(2), 244–260 (2006)
17. Wang, R.: High order adaptive collocation software for 1-D parabolic PDEs. Ph.D. thesis, Dalhousie University (2002)
18. Wang, R., Keast, P., Muir, P.H.: BACOL: B-spline Adaptive COLlocation software for 1D parabolic PDEs. ACM Trans. Math. Softw. **30**(4), 454–470 (2004)
19. Wang, R., Keast, P., Muir, P.H.: A comparison of adaptive software for 1D parabolic PDEs. J. Comput. Appl. Math. **169**(1), 127–150 (2004)
20. Wang, R., Keast, P., Muir, P.H.: A high-order global spatially adaptive collocation method for 1-D parabolic PDEs. Appl. Numer. Math. **50**(2), 239–260 (2004)

On the Simulation of Porous Media Flow Using a New Meshless Lattice Boltzmann Method

S. Hossein Musavi and Mahmud Ashrafizaadeh

Abstract We propose a new meshless lattice Boltzmann method (MLLBM) for the simulation of nearly incompressible flows in porous media. As for the standard lattice Boltzmann method, the collision and the streaming operators are split. While the collision equation remains unaltered, the streaming equation is discretized using the Lax-Wendroff scheme in time, and the meshless local Petrov-Galerkin scheme in space. The Poiseuille flow is solved to validate the numerical scheme. The method is then used to simulate a flow in a porous medium. Although in general, meshless methods suffer from high computational costs, the present method shows promising accuracy and performance at a lower memory and computational costs for complex geometries, when compared with standard lattice Boltzmann methods.

1 Introduction

Since the beginning of the development of the lattice Boltzmann method (LBM), a great number of studies have been conducted to extend the capabilities of the standard lattice Boltzmann method to handle nonuniform and unstructured grids [1–6]. A major trend in these studies is to apply some standard numerical techniques, such as the finite difference (FD), the finite volume (FV), and the finite element (FE) methods, for the discretization of the Boltzmann equation. However, for geometrically complex problems, such as flows in porous media, where the time and computational requirements of creating good quality meshes are significant, the idea of developing a meshless lattice Boltzmann solver naturally arises.

In this study, we implement our newly proposed meshless lattice Boltzmann method [7] for the simulation of nearly incompressible flows in porous media. In our formulation, we split the collision and the advection equations following the standard lattice Boltzmann method [8]. The collision equation is the same as for the standard LBM, however, the advection equation is discretized using the

S.H. Musavi • M. Ashrafizaadeh (✉)
Department of Mechanical Engineering, Isfahan University of Technology, Isfahan 8415683111, Iran
e-mail: hmusavi@me.iut.ac.ir; mahmud@cc.iut.ac.ir

© Springer International Publishing Switzerland 2016

J. Bélair et al. (eds.), *Mathematical and Computational Approaches in Advancing Modern Science and Engineering*, DOI 10.1007/978-3-319-30379-6_43

Lax-Wendroff scheme in time and the meshless local Petrov-Galerkin (MLPG) method in space.

The meshless feature of the proposed method makes it a more flexible lattice Boltzmann solver, especially for cases where using the usual mesh generation techniques introduces significant numerical errors into the solution, or where improving the mesh quality is a complex and time consuming process. These features of the present method are well illustrated in a number of test cases described in the next sections.

2 Formulation

The discrete Boltzmann equation (DBE) with the BGK collision approximation is

$$\frac{\partial f_i}{\partial t} + c_{i,\alpha}\frac{\partial f_i}{\partial x_\alpha} = -\frac{1}{\tau}(f_i - f_i^{\mathrm{eq}}),\ i = 1,\ldots,nQ\,, \tag{1}$$

where f_i is the particle distribution function, τ is the relaxation time towards equilibrium, nQ is the number of discrete microscopic velocities and $\mathbf{c}_i$ is the discrete microscopic velocity. For the D2Q9 lattice we have

$$\begin{aligned}
\mathbf{c}_0 &= \mathbf{0}\,,\\
\mathbf{c}_i &= \cos(i-1)\frac{\pi}{4}\mathbf{e}_x + \sin(i-1)\frac{\pi}{4}\mathbf{e}_y,\ i = 1,3,5,7\,,\\
\mathbf{c}_i &= \sqrt{2}[\cos(i-1)\frac{\pi}{4}\mathbf{e}_x + \sin(i-1)\frac{\pi}{4}\mathbf{e}_y],\ i = 2,4,6,8\,,
\end{aligned} \tag{2}$$

and f_i^{eq} is the equilibrium distribution, that is

$$f_i^{\mathrm{eq}} = \rho t_i\left(1 + \frac{c_{i,\alpha}u_\alpha}{c_s^2} + \frac{(c_{i,\alpha}u_\alpha)^2}{2c_s^4} - \frac{u_\alpha^2}{2c_s^2}\right). \tag{3}$$

As usual, we break Eq. (1) into two steps, namely the collision step:

$$\tilde{f}_i = f_i - \frac{1}{\tau}(f_i - f_i^{\mathrm{eq}}), \tag{4}$$

and the advection step:

$$\frac{\partial f_i}{\partial t} + c_{i,\alpha}\frac{\partial f_i}{\partial x_\alpha} = 0. \tag{5}$$

We discreatize Eq. (5) in time using the Lax-Wendroff scheme, which reads

$$f_i^{n+1} = f_i^n - \delta t c_{i,\alpha} \frac{\partial f_i^n}{\partial x_\alpha} + \frac{\delta t^2}{2} c_{i,\alpha} c_{i,\beta} \frac{\partial^2 f_i^n}{\partial x_\alpha \partial x_\beta}. \quad (6)$$

where δt is the time step size and the superscript n is the time step number.

In order to apply the meshless local Petrov-Galerkin scheme to discretize Eq. (6) in space, first, the local weak form of Eq. (6) on the control volume Ω_I of point I is derived by taking its inner product with a local test function W_I over Ω_I, and using integration by parts, so that we obtain

$$\int_{\Omega_I} W_I f_i^{n+1} d\Omega = \int_{\Omega_I} W_I f_i^n d\Omega - \int_{\Omega_I} \left(\delta t W_I c_{i,\alpha} \frac{\partial f_i^n}{\partial x_\alpha} + \frac{\delta t^2}{2} c_{i,\alpha} c_{i,\beta} \frac{\partial W_I}{\partial x_\beta} \frac{\partial f_i^n}{\partial x_\alpha} \right) d\Omega$$
$$+ \frac{\delta t^2}{2} \int_{\Gamma_I} W_I c_{i,\alpha} c_{i,\beta} \frac{\partial f_i^n}{\partial x_\alpha} n_\beta d\Gamma, \quad (7)$$

where Γ_I is the boundary of the control volume Ω_I and n_β is the unit outward normal vector of Γ_I. Equation (7) is the local weak form of Eq. (6).

In the next step, the field variable f_i is to be expressed in terms of nodal values $f_{i,J}$ by a local interpolation scheme, that is

$$f_i(\mathbf{x}, t) = \sum_{J=1}^{N_s} \phi_J(\mathbf{x}) f_{i,J}(t) = \boldsymbol{\Phi}^{\mathrm{T}}(\mathbf{x}) \mathbf{f}_s(t), \quad (8)$$

where N_s is the number of nodal points in a local interpolation domain of point $\mathbf{x}$ called the support domain, $\boldsymbol{\Phi}^{\mathrm{T}}(\mathbf{x}) = \{\phi_1(\mathbf{x})\ \phi_2(\mathbf{x})\ \ldots\ \phi_{N_s}(\mathbf{x})\}$ is the transpose of the vector of shape functions, and $\mathbf{f}_s(t) = \{f_{i,1}(t)\ f_{i,2}(t)\ \ldots\ f_{i,N_s}(t)\}^{\mathrm{T}}$ is the vector of the nodal values of f_i in the support domain. In this study, we make use of the local radial point interpolation method (LRPIM) [9] which uses the local radial basis functions (RBF) augmented with polynomials as the basis function; thus $\boldsymbol{\Phi}^{\mathrm{T}}(\mathbf{x})$ in the interpolation equation (8) is the transpose of the vector of LRPIM shape functions given as [9]

$$\tilde{\boldsymbol{\Phi}}^{\mathrm{T}}(\mathbf{x}) = \{\mathbf{R}^{\mathrm{T}}(\mathbf{x})\ \mathbf{p}^{\mathrm{T}}(\mathbf{x})\} \mathbf{G}^{-1}, \quad (9)$$

where $\mathbf{R}^{\mathrm{T}}(\mathbf{x}) = \{R_1(\mathbf{x})\ R_2(\mathbf{x})\ \ldots\ R_{N_s}(\mathbf{x})\}$ is the transpose of the vector of radial basis functions (RBF), $\mathbf{p}^{\mathrm{T}}(\mathbf{x}) = \{1\ x\ y\ \ldots\ p_m(\mathbf{x})\}$ is the transpose of the vector of monomial basis functions, m is the number of monomial basis functions, $\tilde{\boldsymbol{\Phi}}^{\mathrm{T}}(\mathbf{x}) = \{\boldsymbol{\Phi}^{\mathrm{T}}(\mathbf{x})\ \phi_{N_s+1}(\mathbf{x})\ \ldots\ \phi_{N_s+m}(\mathbf{x})\}$ is the transpose of the extended vector of the shape functions, and $\mathbf{G}$ is a symmetric matrix defined in Ref. [9]. Substituting Eq. (8) in

Eq. (7) we write,

$$\sum_{J=1}^{N_I}\Big[\int_{\Omega_I} W_I\phi_J d\Omega\Big] f_{i,J}^{n+1} = \sum_{J=1}^{N_I}\Big[\int_{\Omega_I} W_I\phi_J d\Omega$$
$$-\int_{\Omega_I}\Big(\delta t W_I + \frac{\delta t^2}{2} c_{i,\beta}\frac{\partial W_I}{\partial x_\beta}\Big) c_{i,\alpha}\frac{\partial \phi_J}{\partial x_\alpha} d\Omega$$
$$+\frac{\delta t^2}{2}\int_{\Gamma_I} W_I c_{i,\alpha} c_{i,\beta}\frac{\partial \phi_J}{\partial x_\alpha} n_\beta d\Gamma\Big] f_{i,J}^{n}, \tag{10}$$

where N_I is the number of nodal points involved in the interpolation of the field variable on the inner and the boundary points of the control volume Ω_I. By introducing the mass matrix as

$$M_{IJ} = \int_{\Omega_I} W_I\phi_J d\Omega, \tag{11}$$

and the stiffness matrix as

$$K_{i,IJ} = -\int_{\Omega_I}\Big(\delta t W_I + \frac{\delta t^2}{2} c_{i,\beta}\frac{\partial W_I}{\partial x_\beta}\Big) c_{i,\alpha}\frac{\partial \phi_J}{\partial x_\alpha} d\Omega$$
$$+\frac{\delta t^2}{2}\int_{\Gamma_I} W_I c_{i,\alpha}\frac{\partial \phi_J}{\partial x_\alpha} c_{i,\beta} n_\beta d\Gamma, \tag{12}$$

we rewrite Eq. (10) as follows,

$$\sum_{J=1}^{N_I} M_{IJ} f_{i,J}^{n+1} = \sum_{J=1}^{N_I}[M_{IJ} + K_{i,IJ}] f_{i,J}^{n}. \tag{13}$$

To complete the discretization process, the integrals of equations (11) and (12) are to be evaluated numerically. The Gauss quadrature scheme is employed for this purpose. We have

$$M_{IJ} = \sum_{k=1}^{N_G} \xi_k W_I(\mathbf{x}_k)\phi_J(\mathbf{x}_k)|\mathbf{J}^{\Omega_I}|, \tag{14}$$

and

$$
\begin{aligned}
K_{i,IJ} = & -\sum_{k=1}^{N_G} \xi_k \Big(\delta t W_I(\mathbf{x}_k) + \frac{\delta t^2}{2} c_{i,\beta} \frac{\partial W_I}{\partial x_\beta}\Big|_{\mathbf{x}_k} \Big) \Big(c_{i,\alpha} \frac{\partial \phi_J}{\partial x_\alpha}\Big|_{\mathbf{x}_k} \Big) |\mathbf{J}^{\Omega_I}| \\
& + \frac{\delta t^2}{2} \sum_{k=1}^{N_G^b} \xi_k W_I(\mathbf{x}_k) \Big(c_{i,\alpha} \frac{\partial \phi_J}{\partial x_\alpha}\Big|_{\mathbf{x}_k} \Big) (c_{i,\beta} n_\beta) |\mathbf{J}^{\Gamma_I}|,
\end{aligned} \tag{15}
$$

where ξ_k is the Gauss weighting factor for the Gauss quadrature point $\mathbf{x}_k$, $\mathbf{J}^{\Omega_I}$ and $\mathbf{J}^{\Gamma_I}$ are the mapping Jacobian matrices for the domain and the boundary integrations, respectively, and N_G and N_G^b are the number of Gauss points used for the domain and the boundary integrations, respectively.

Now, Eq. (13) becomes the fully discretized equation for the nodal point I. Writing this equation for all of the nodal points in the computational domain ($I = 1, \ldots, N$), and assembling the resulting equations in a global system of equations, we can write

$$
\mathbf{M}\mathbf{f}_i^{n+1} = [\mathbf{M} + \mathbf{K}_i]\mathbf{f}_i^n, \; i = 1, \ldots, nQ, \tag{16}
$$

where $\mathbf{M}$, $\mathbf{K}$, and $\mathbf{f}_i$ are the global mass matrix, stiffness matrix, and particle distribution vector, respectively. Equation (16) is a system of N equations with N unknowns which should be solved separately for each direction i after imposing the boundary conditions.

The advection equation of the particle distributions is a hyperbolic equation, which requires boundary conditions for the incoming particles at the boundary ($c_{i,\beta} n_\beta < 0$). In this study, we impose boundary conditions using the bounce-back scheme of non-equilibrium distributions, i.e.

$$
f_i - f_i^{\mathrm{eq}} = f_{i^*} - f_{i^*}^{\mathrm{eq}}, \tag{17}
$$

where f_{i^*} is the outgoing particle distribution along the opposite direction of the incoming distribution f_i. Substituting the equilibrium distribution of Eq. (3) in the above equation, we obtain

$$
f_i = f_{i^*} + 2\rho_b t_i (c_{i,\alpha} u_{b,\alpha}) / c_s^2, \tag{18}
$$

where ρ_b and $u_{b,\alpha}$ are the macroscopic density and velocity at the boundary. If f_{i^*} in Eq. (18) is considered to be the post-collision (pre-streaming) distribution, then Eq. (18) becomes an explicit essential boundary condition for the discretized system of Eq. (16).

The coefficient matrix in Eq. (16) is a sparse matrix which can be efficiently solved using sparse iterative solvers such as BiCGStab. However, the explicit nature of the standard lattice Boltzmann method, and the diagonally dominant character of the mass matrix, motivated us to find rational ways of diagonalizing the mass matrix, and thus save much of the computational time. In this study, we use row-sum lumping, in which the sum of the elements of each row of the mass matrix is used as the diagonal element. As a result, our meshless lattice Boltzmann method becomes an explicit solver for the fluid flow problems. The maximum value for the time step leading to a stable solution is determined using the Courant number, CFL $=\max\{|\mathbf{c}_i|\}\,\delta t/\delta x_{\min}$, where $\delta x_{\min}$ is the minimum point spacing in the domain. For a stable solution the CFL number should be smaller than 1.0.

3 Results

3.1 Poiseuille Flow

The first test case considered in this study is the pressure driven fully developed flow between two parallel plates. In solving this test case using the proposed method, we consider a square computational domain in the *xy* plane and discretize it using 9×9, 17×17, 23×23, and 65×65 uniform point distributions. The constant pressure boundary condition is imposed in the inlet and outlet and the no-slip and the impermeability boundary conditions are imposed at the walls. For each point distribution, time iterations continue until a steady state is reached. The result of our method for the velocity distribution is depicted and compared with the analytical solution in Fig. 1.

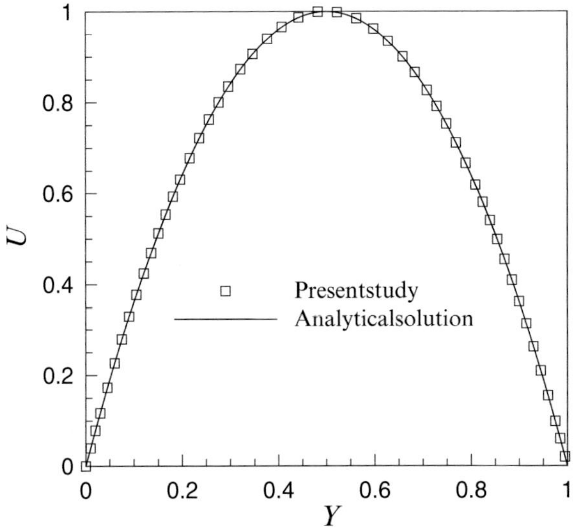

Fig. 1 The velocity distributions for the Poiseuille flow; *line*: analytical solution, *symbols*: present study

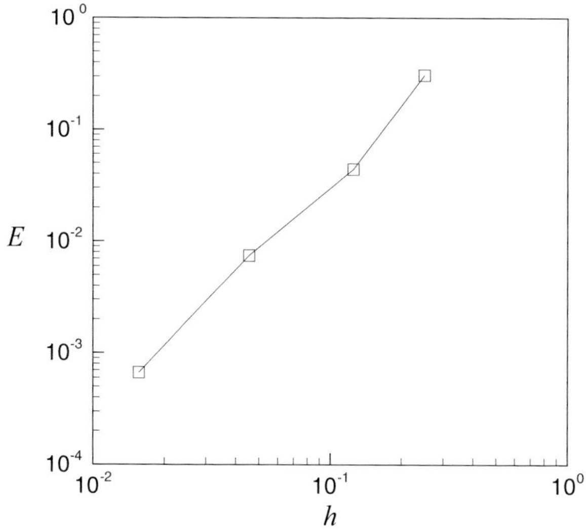

Fig. 2 Numerical convergence of the meshless Lattice Boltzmann method in L^2 error norm with respect to the point spacing for the Poiseuille flow

In order to determine the order of accuracy of the numerical scheme, we employ the following relative L^2 error norm,

$$E = \left(\frac{\sum_{I=1}^{N_e} (U_{aI} - U_{nI})^2}{\sum_{I=1}^{N_e} U_{aI}^2} \right)^{1/2}, \tag{19}$$

where U_{aI} and U_{nI} are the analytical and numerical solutions of the velocity at point I, respectively, and N_e is the fixed number of points used for the error analysis. For this test case, a 10×10 uniform point distribution is used for the error analysis. The variation of the above error norm with respect to the point spacing h is sketched in the logarithmic diagram of Fig. 2. The rate of the convergence of our numerical method, computed by the linear regression of the data in Fig. 2, is $R = 2.15$.

3.2 *Flow in a Porous Medium*

One of the most common applications of the standard lattice Boltzmann method is the simulation of the flow in porous media. In order to illustrate the ability of the present meshless lattice Boltzmann method to deal with complex geometries, we simulate a pressure-driven fluid flow in a typical two-dimensional porous medium shown in Fig. 3 and compute the permeability of the medium. One of the common definitions of the Reynolds number for the porous media flow is

$$Re = \frac{\rho v D_p}{\mu}, \tag{20}$$

where ρ and μ are the density and viscosity of the fluid, v is the superficial flow velocity, and D_p is a representative grain size for the porous media, which is usually chosen to be the average grain diameter. Experimental studies have illustrated that the flow regimes with low Reynolds numbers (typically less than 10) are Darcian, meaning that the superficial velocity-pressure gradient relation obeys the Darcy's law

$$v = -\frac{k}{\mu}\nabla p, \tag{21}$$

where k is the permeability of the medium measured in m^2 or *Darcy*, where $1\,\mathrm{Darcy} = 9.869233 \times 10^{-13}\,\mathrm{m}^2$.

In our simulation, the no-slip boundary condition is imposed on the top and the bottom walls and on the grain surfaces using the bounce-back scheme and the constant pressure boundary condition is imposed on the left and right boundaries. The pressure differences between inlet and outlet boundaries in our simulations are

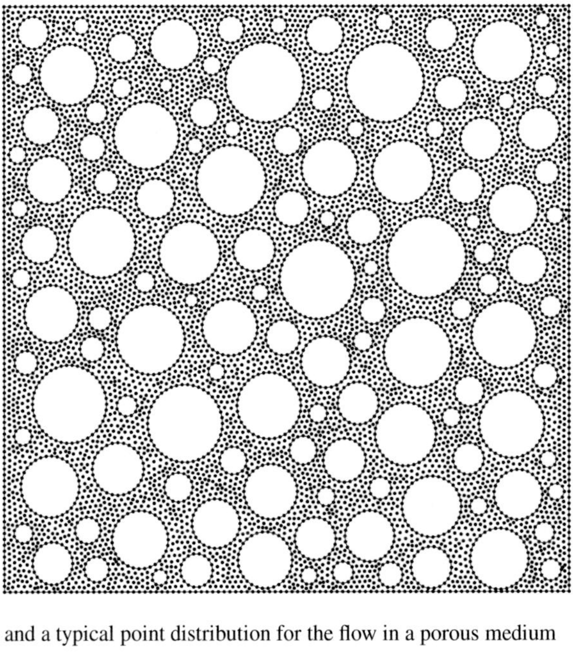

Fig. 3 Domain and a typical point distribution for the flow in a porous medium

set so that the Reynolds number of the flow is low and the flow remains in the Darcian regime.

The domain is discretized using arbitrary sets of nodal points as shown in Fig. 3. First, to investigate the convergence of the numerical method with respect to the domain discretization, we discretize the domain with 2601, 7364, 25284, and 90043 arbitrary distributed nodal points and compute the superficial velocity for each case. The results are depicted in Fig. 4. As shown in this figure, the difference between the cases with 25284 and 90043 points is less than 1 % and therefore we use the case with 25284 points in all the following simulations. However, as observed from the figure, the standard LBM requires 640000 grid points for the accurate simulation of the flow. It means that the required number of points in our method is about 1/25th of that of the standard LBM.

Next, to obtain the permeability of the medium, we compute the superficial velocity for different pressure gradients and perform the linear interpolation of the results as is depicted in Fig. 5. From this figure, the permeability is obtained as $k = 1.53 \times 10^{-5}\,\mathrm{m}^2 = 1.55 \times 10^7$ Darcy.

The pressure and velocity magnitude distributions of the porous medium flow are depicted in Figs. 6 and 7, respectively, for the dimensionless pressure gradient of 0.1.

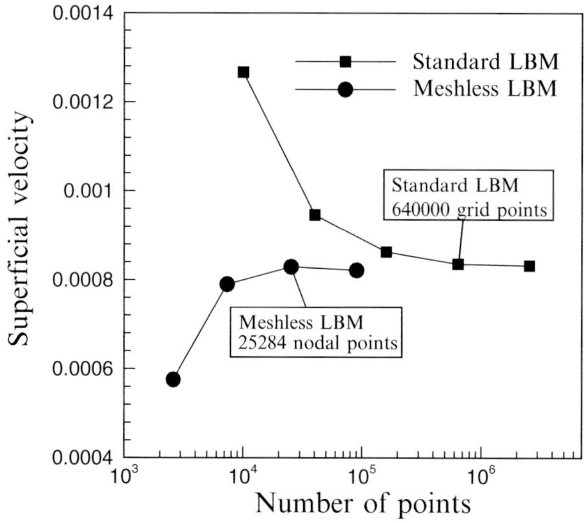

Fig. 4 The superficial velocity versus the number of nodal points used in the meshless LBM

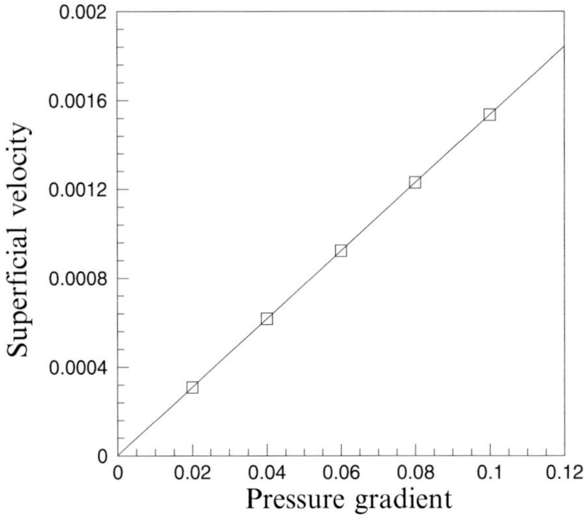

Fig. 5 The superficial velocity with respect to the pressure gradient for the flow in the porous medium

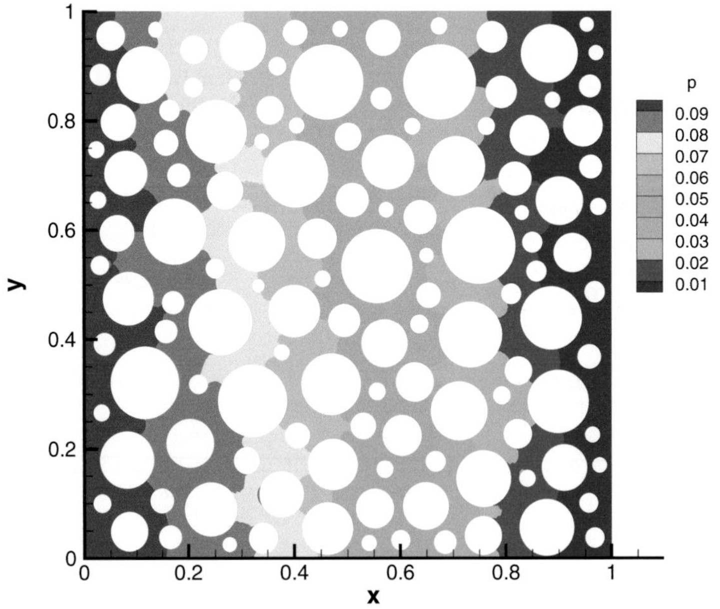

Fig. 6 Pressure distribution for the flow in the porous medium

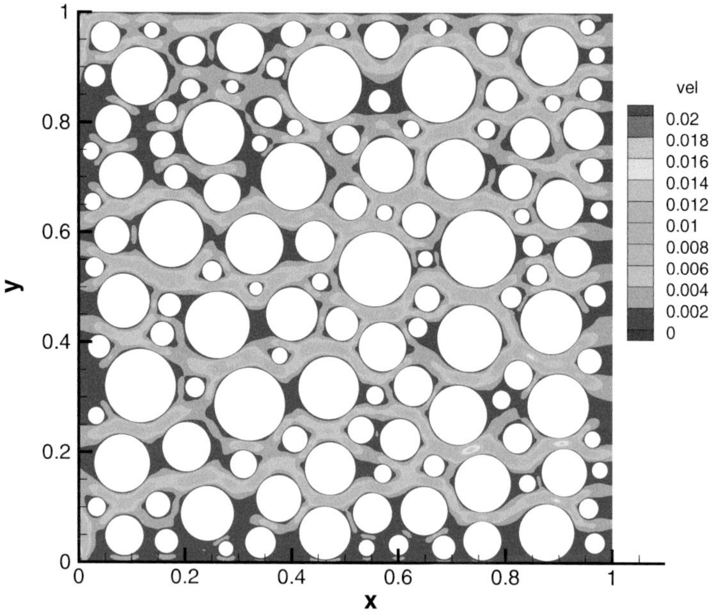

Fig. 7 Velocity magnitude distribution for the flow in the porous medium

4 Conclusions

A new meshless lattice Boltzmann method (MLLBM) has been developed for the simulation of the nearly incompressible fluid flows. The main advantage of our method with respect to the previous extensions of the lattice Boltzmann method is to eliminate the need for any meshes. A feature that shows its superiority over the standard lattice Boltzmann method in complex geometries, such as porous media.

Two test cases have been considered in this study. First, the Poiseuille flow has been solved and compared with the analytical solution, showing the second order of accuracy of our method. Next, the flow in a porous medium has been simulated to illustrate the capability of the method in dealing with domains with a very complex geometry. Although the results presented in this study are for two-dimensional cases, the extension for three-dimensional problems is straightforward and remains for our future works. However, our preliminary 3D developments show that the time and memory saving for the meshless LBM method should be even higher than that of the 2D model.

References

1. Bardow, A., Karlin, I.V., Gusev, A.A.: General characteristic-based algorithm for off-lattice Boltzmann simulations. EPL (Europhys. Lett.) **75**(3), 434 (2006)
2. Patil, D.V., Lakshmisha, K.N.: Finite volume TVD formulation of lattice Boltzmann simulation on unstructured mesh. J. Comput. Phys. **228**(14), 5262–5279 (2009)
3. Li, Y., LeBoeuf, E.J., Basu, P.K.: Least-squares finite-element scheme for the lattice Boltzmann method on an unstructured mesh. Phys. Rev. E **72**(4), 046711 (2005)
4. Shi, X., Lin, J., Yu, Z.: Discontinuous Galerkin spectral element lattice Boltzmann method on triangular element. Int. J. Numer. Methods Fluids **42**(11), 1249–1261 (2003)
5. Düster, A., Demkowicz, L., Rank, E.: High-order finite elements applied to the discrete Boltzmann equation. Int. J. Numer. Methods Eng. **67**(8), 1094–1121 (2006)
6. Min, M., Lee, T.: A spectral-element discontinuous Galerkin lattice Boltzmann method for nearly incompressible flows. J. Comput. Phys. **230**(1), 245–259 (2011)
7. Musavi, S.H., Ashrafizaadeh, M.: Meshless lattice Boltzmann method for the simulation of fluid flows. Phys. Rev. E **91**(2), 023310 (2015)
8. Wolf-Gladrow, D.A.: Lattice-Gas Cellular Automata and Lattice Boltzmann Models: An Introduction, vol. 1725. Springer, Berlin (2000)
9. Liu, G.R., Gu, Y.T.: An Introduction to Meshfree Methods and Their Programming. Springer Science and Business Media, Dordrecht/New York (2005)

A Comparison Between Two and Three-Dimensional Simulations of Finite Amplitude Sound Waves in a Trumpet

Janelle Resch, Lilia Krivodonova, and John Vanderkooy

Abstract Simplifying a three-dimensional problem of simulating sound propagation in musical instruments is frequently done by exploiting axial symmetry and reducing the problem to one or two dimensions. We examine if such dimension reduction is valid. We numerically solve the equations of motion of compressible gases using the discontinuous Galerkin method to model nonlinear sound propagation inside a trumpet. The numerical results in two and three dimensions are then compared with experimental data. Experiments were carried out on a trumpet in which the sound pressure waves of the B_3^b and B_4^b notes played at forte were recorded. We found that it is crucial to consider the problem with all three spatial dimensions to ensure reflections in the bell region are properly modelled. Additionally, the shape of flare must be carefully approximated to compute the propagating waves accurately.

1 Introduction

In order to accurately model sound propagation in brass musical instruments, higher amplitude propagating waves must be considered. For musical instruments such as the trombone or trumpet, pressure variations inside the instrument can be a significant fraction of atmospheric pressure if loud, high frequency notes are played. Assuming the pressure disturbance entering the narrow tubing of the bore is large enough, nonlinear behaviour can cause waveform distortion. In particular, the crest of the wave will travel faster than the trough causing the wave to steepen. Wave steepening will excite higher harmonic components of the sound pressure waves which influences the timbre of the sound by giving it a more 'brassy' effect [3]. The

J. Resch (✉) • L. Krivodonova
Department of Applied Mathematics, University of Waterloo, 200 University Avenue W., Waterloo, ON N2L 3G1, Canada
e-mail: jresch@uwaterloo.ca; lgk@uwaterloo.ca

J. Vanderkooy
Department of Physics and Astronomy, University of Waterloo, 200 University Avenue W., Waterloo, ON N2L 3G1, Canada
e-mail: jv@uwaterloo.ca

© Springer International Publishing Switzerland 2016
J. Bélair et al. (eds.), *Mathematical and Computational Approaches in Advancing Modern Science and Engineering*, DOI 10.1007/978-3-319-30379-6_44

linear acoustic equations used to model sound propagation outside an instrument are not suitable to model the sound propagation inside. Instead, wave propagation inside of the instrument should take compressibility into account which can be modelled using the Euler equations. We solve these equations numerically using the discontinuous Galerkin method (DGM). This method is particularly useful because it can handle unstructured meshes and has excellent dissipative and dispersive properties. In certain regimes (not considered here) shock waves can be produced inside an instrument and this method can handle such nonlinear effects [2].

Previously, in [5], we modelled nonlinear wave propagation in a 2D trumpet. In addition, we considered the consequences of neglecting the third spatial dimension since the spreading of the waves in 2D and 3D differs. We calculated that the 2D – 3D dimensionality difference (which we call the *dimension factor*) in our results would be approximately 14 dB. We arrived at this value by assuming that the axial pressure in both 2D and 3D is a good measure of the total energy leaving the bell. A full derivation of the dimension factor value can be found in [5]. After taking this amplitude difference into account, we obtained a good match between the experimental and numerical data, for the frequency components that propagate out of the trumpet bell.

Here we present some of our preliminary results for extending the model to 3D. In particular, we will be able to verify and further investigate the influence that spatial dimensions have on such models. This is an important aspect to consider since many mathematical descriptions of sound wave propagation in the literature are reduced to 2D or even 1D.

2 Experimental Data and Numerical Setup

The experimental data that we collected from a B^b Barcelona BTR-200LQ trumpet shown in Fig. 1 is discussed in detail in [5]. We took one period of the sound pressure measurements obtained from a quarter-inch microphone carefully mounted

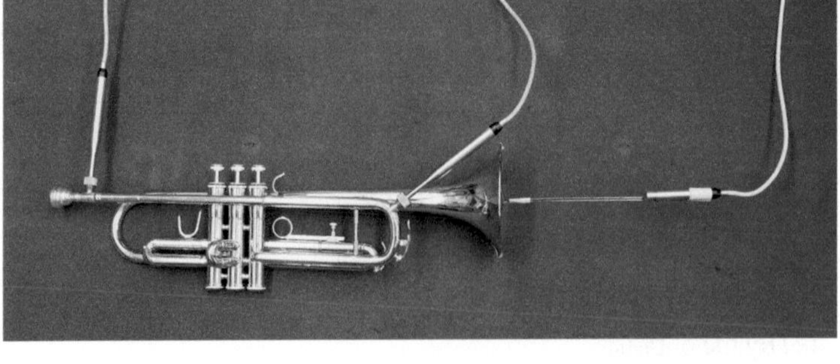

Fig. 1 Placement of microphones on the Barcelona BTR-200LQ trumpet

on the shank of the trumpet mouthpiece, and applied Fourier analysis to the data to reconstruct the waveform as a sum of 30 cosine waves. This experimental data obtained at the mouthpiece was then used as a boundary condition on the pressure for our simulations. The musical notes recorded and then simulated were the B_3^b and B_4^b played at forte. A half-inch microphone was also placed outside the trumpet on the central axis about 17 cm away from the bell. We compared our simulation results with the pressure measurements obtained from this microphone outside the bell.

For all simulations, we assumed that the initial flow was at rest and related pressure and velocity through the 1D expression derived from linear acoustic theory. Since there is little return from the reflections at the bell, at least for the higher frequencies, the plane wave velocity expression is a reasonable approximation. For the boundary conditions, we prescribed reflecting boundary conditions on the inner and outer walls of the trumpet mesh (excluding the mouthpiece). Since we modelled a trumpet in a box, we used pass-through boundary conditions on the computational domain.

3 The Influence of the Geometry and Dimension of the Bell

Several numerical simulations have been carried out to determine if the model considered is adequate. Some such computational experiments are discussed in [5]. We have discovered that considering the proper geometry of brass musical instruments is critical to simulate wave propagation with any accuracy. We have also investigated the importance of the bell geometry. One of the most significant properties of the bell is how it influences the reflections of the sound pressure waves. The location at which the harmonic waves reflect in the bell is dependent on their frequency. For higher frequency waves, a larger portion of the energy will be lost or be completely transmitted from the bell [1].

In the literature, it is common for the trumpet flare to be approximated by several different functions or combination of functions, e.g., exponential functions, hyperbolic functions, etc. However, we were uncertain if such geometric descriptions would be sufficient to accurately model wave propagation through the trumpet bell. If the propagating sound pressure waves are particularly sensitive to the bell's curvature, simulations may produce exaggerated discrepancies if the bell shape is poorly approximated. Simulations performed in 2D should be relevant for the frequency components that are transmitted from the bell. Therefore, to reduce run time, we first examined our problem by neglecting the third spatial dimension and focused on the flare shape.

The first mesh, which will be called *mesh 1*, represented a 2D trumpet where the bell was approximated by an elliptic function. To improve this approximation, bell measurements were taken at several locations along the flare, which was then interpolated by the obtained set of points and cubic splines. The measurements were done at the university's machine shop by sampling the diameter of the bore near expansion, 1.02 m from the mouthpiece. This flare shape was used to construct a 2D trumpet mesh called *mesh 2*. To obtain more precise points to approximate the trumpet shape, a photo of the trumpet flare was taken. The *grabit* software from Math Works Inc. was used to accurately trace out the shape of the trumpet bell. This third bell shape was used to create a 2D mesh which we will call *mesh 3*. Since this representation of the bell was most accurate, it was further used to extend mesh 3 into a full 3D mesh. This 3D mesh will be referred to as *mesh 4*. The total number of cells is 8190, 8467, 8038 and 14,362 for mesh 1, mesh 2, mesh 3 and mesh 4, respectively. The meshes consist of triangular or tetrahedral elements with adaptive sizes to accurately resolve the geometric features of the trumpet. More details of the mesh construction can be found in [5]. The computational domain and a sample mesh are shown in Fig. 2.

Once these flare shapes were obtained, simulations were carried out on the meshes of a trumpet that is 1.48 m in total length. The diameter of the bore is constant except near the end of the instrument where it slowly increases and then flares rapidly to give the bell shape. This expansion for all meshes begins 1.02 m from the mouthpiece end. In [5], we justified that the bends of the instrument do not

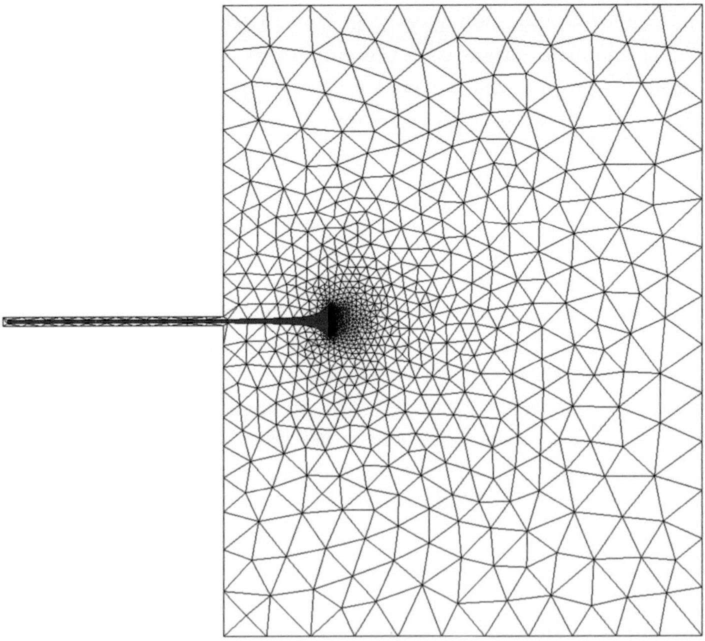

Fig. 2 Mesh 3: Two-dimensional computational domain and trumpet mesh constructed by tracing out the shape of the bell in Matlab

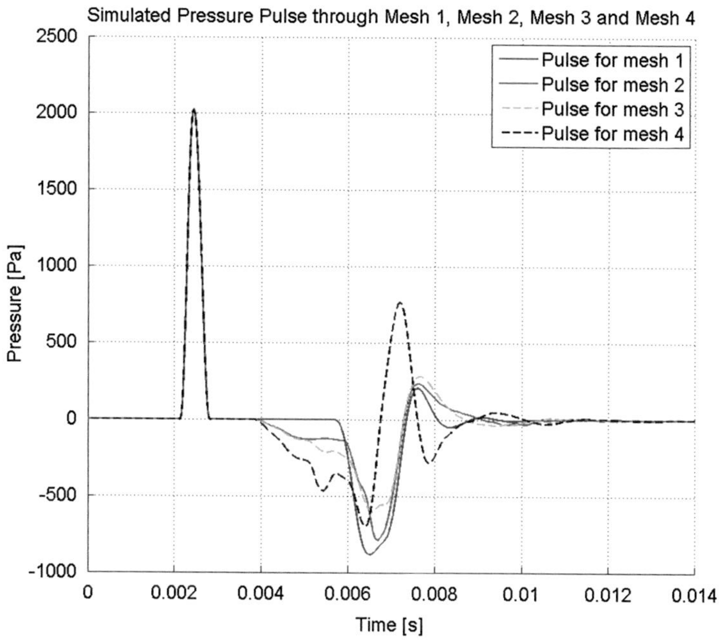

Fig. 3 Initial and reflected wave of simulated pressure pulse through mesh 1, mesh 2, mesh 3 and mesh 4

greatly influence the wave propagation in the instrument, especially in comparison to the bell. Therefore, for all the simulations discussed, the bends will not be modelled, i.e., the bends and coils of the trumpet will be unwrapped and the tubing will be straightened out.

Before simulating the musical notes however, we considered a simpler problem. We sent a pressure pulse down the trumpet meshes from the mouthpiece to examine their acoustic behaviour. As expected, the variations in the bell geometry greatly influenced wave propagation. The pulse generated at the mouthpiece was given by

$$p = \begin{cases} 1.0 + (0.01 - 0.01\cos(1500\mathrm{t})), & \text{if t} < \frac{2\pi}{1500} \\ 1.0, & \text{else} \end{cases}$$

which corresponds to a unipolar pressure pulse with an amplitude of approximately 2000 Pa. In Fig. 3, we show the pressure simulated at a point located at the mid-length of the cylindrical bore. The first peak located at approximately $t = 0.0025$ corresponds to the initial pulse moving from the mouthpiece towards the bell. The second, inverted peak seen at about $t = 0.007$ corresponds to the signal travelling back to the mouthpiece after it has been reflected by the bell. In Fig. 4, we also plotted the *reflected transfer* data which is calculated from the frequency content of the reflected pulse divided by that of the incident pulse. This curve represents the power reflected by the bell meshes.

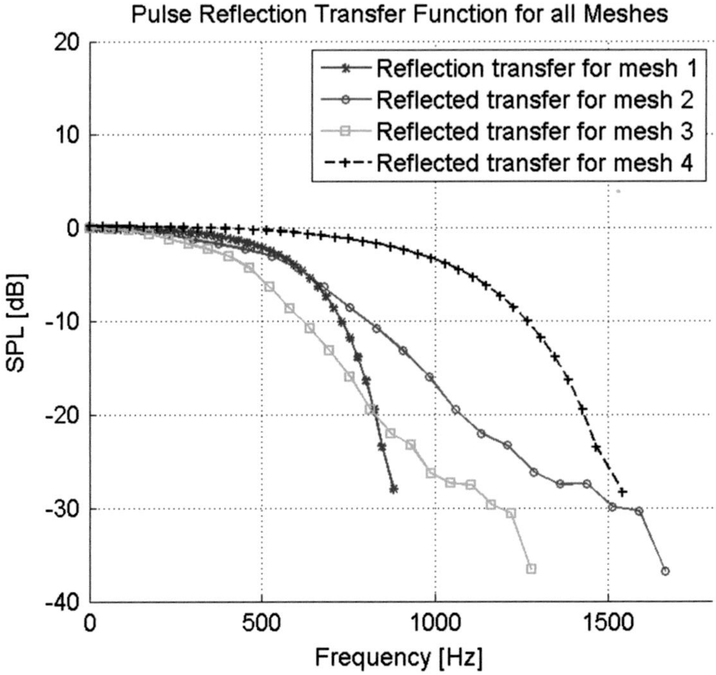

Fig. 4 Corresponding reflected transfer function of pressure pulses depicted in Fig. 3

Table 1 The number of cells, and the frequency components that are mostly confined to the instrument or transmitted from the bell are summarized for each mesh

Mesh name	Cell count	Reflected freq. (Hz)	Transmitted freq. (Hz)
Mesh 1	8190	600	800
Mesh 2	8467	600	1000
Mesh 3	8038	400	1000
Mesh 4	14,362	900	1200

We observe in Figs. 3 and 4 a rather significant difference in the shape of the returning pressure pulse for each mesh considered, especially when comparing the 2D and 3D results. We found that different frequency components were confined to the instrument and transmitted from the bell for each mesh. This was accomplished by examining the relationship between the reflected transfer function and its corresponding compliment, the transmission transfer function. Frequency components less than approximately 550 Hz for mesh 1, 600 Hz for mesh 2, 400 Hz for mesh 3, and 900 Hz for mesh 4, are mostly reflected before or within the bell [5]. For frequency components greater than 800 Hz for mesh 1, 1000 Hz for mesh 2 and mesh 3, and 1200 Hz for mesh 4, the waves are being mostly transmitted from the bell. The properties and number of cells for each mesh is summarized in Table 1.

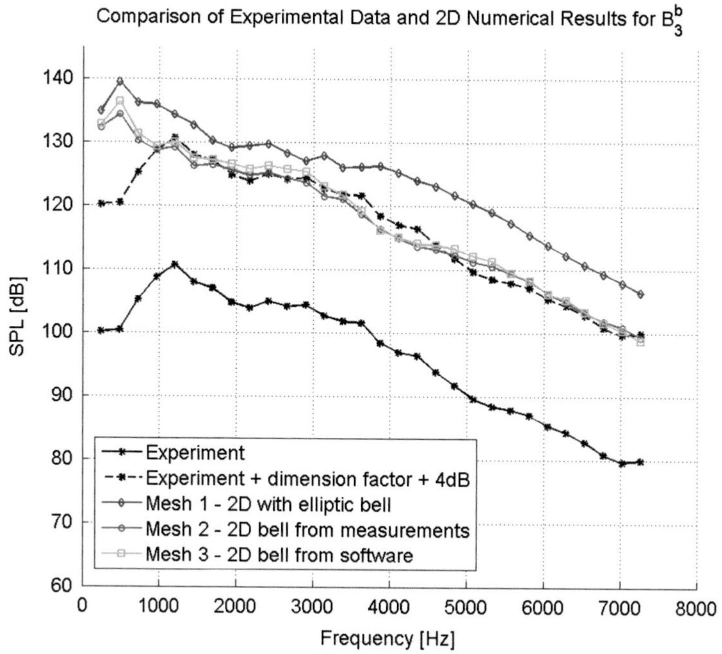

Fig. 5 Two-dimensional numerical and experimental results for the B_3^b

Further simulations were then carried out using the measured B_3^b and B_4^b pressure waveforms. In particular, we solved the full 2D and 3D sets of compressible Euler equations with the initial and boundary conditions mentioned above on the meshes. Each note was initialized with 30 harmonics.

The frequency spectra of the 2D simulation results for the B_3^b and B_4^b notes outside of the trumpet bell can be seen in Figs. 5 and 6, respectively. Figures 5 and 6 also depict the frequency spectra of the experimental data. A comparison of the numerical and experimental data shows that the numerical amplitude is approximately 19+ dB off, where 14 dB are expected since the third spatial dimension is neglected (see analysis in [5]). However, we are not certain why we obtained an additional decibel difference of roughly 4+ dB. As we will discuss in the next section, a possible explanation is that the subtleties of the bore geometry near the mouthpiece were not considered. Energy losses were also neglected and some conjecture that this could make a difference of several decibels (up to 6 dB) [4]. Nonetheless, to easily compare the harmonic distribution of the experimental and 2D numerical data presented in Figs. 5 and 6, we shifted the experimental data by the decibel difference stated in the plots.

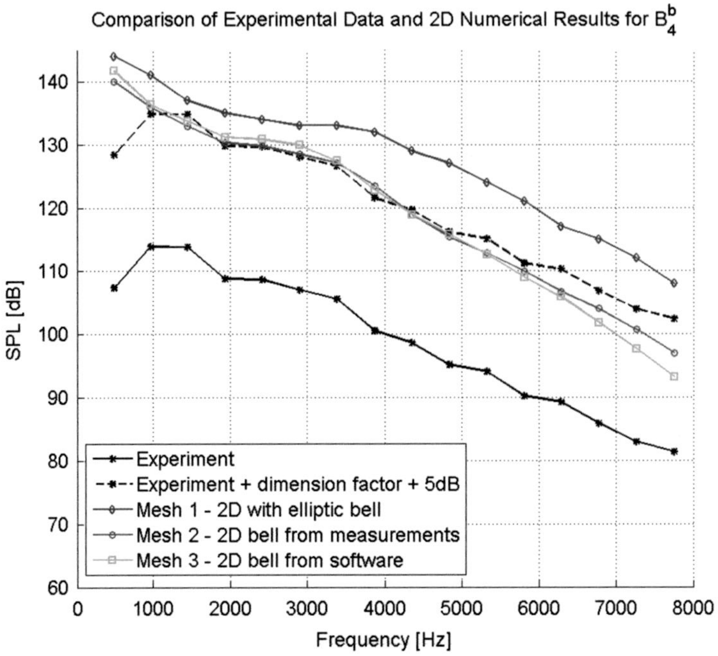

Fig. 6 Two-dimensional numerical and experimental results for the B_4^b

We can see in Figs. 5 and 6 that the performed simulations on mesh 1, i.e., the 2D trumpet mesh where the bell is approximated by an elliptic function, gives the poorest results in shape and amplitude. The 2D simulation results carried out on mesh 2 and mesh 3 however show improvement in both aspects, i.e., in amplitude and harmonic distribution. In particular, the spectral data aligns relatively well with the shifted experimental curve, with the exception of the lowest frequency components. The first three harmonics for the B_3^b and the fundamental frequency for the B_4^b deviate most from the experimental data. However, we postulate that the lowest frequencies are mostly confined within the instrument and consequentially, are greatly influenced by the reflections that take place near the bell. We also predicted that we would observe some amplitude differences and harmonic distribution discrepancies from examining the pulse results. We further hypothesized that the experimental and numerical data will better match (specifically for the lowest harmonic components of the notes) if the equivalent simulations were carried out in 3D, i.e., on mesh 4.

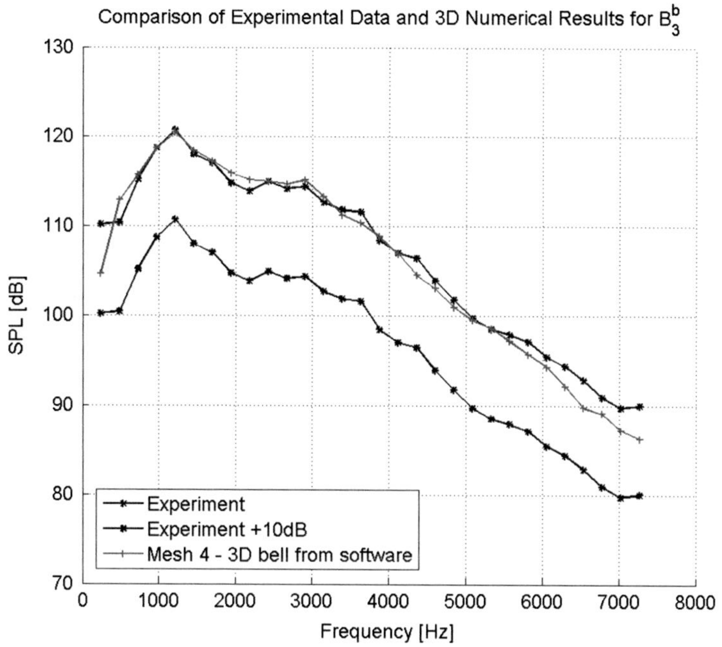

Fig. 7 Three-dimensional numerical and experimental results for the B_3^b

The frequency spectra of the measured B_3^b and B_4^b pressure waveforms outside the trumpet bell are plotted in Figs. 7 and 8, respectively. The result of solving the full 3D system on mesh 4 is also plotted for the B_3^b and B_4^b notes in Figs. 7 and 8, respectively. Overall, we see that the 3D simulations show greater similarity to experimental data. There are however still some discrepancies, particularly in the amplitude of the 3D solution curves. The B_3^b and B_4^b notes are overestimated by approximately 10 dB. For comparative purposes, we again shifted the experimental curves by the amplitude difference. When this shift is considered, the shape of the experimental data and numerical solutions are in better agreement compared to the results in Figs. 5 and 6. More specifically, the resulting lower frequency components for both simulated notes demonstrate that the bell reflections are more accurately modelled when the third spatial dimension is considered. In the next section, we conjecture on possible sources for the 10 dB difference obtained in our 3D numerical simulations.

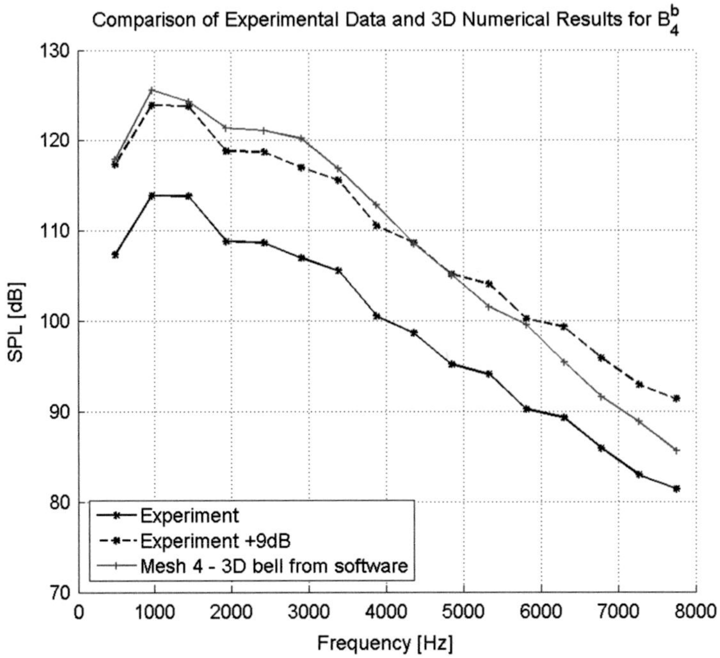

Fig. 8 Three-dimensional numerical and experimental results for the B_4^b

4 Discussion Regarding the Geometry of the Bore Near the Mouthpiece

Neglecting subtleties of the geometry near the mouthpiece of the trumpet may be an explanation for the observed amplitude discrepancies in our simulation results. In reality, the tube of the trumpet near the mouthpiece does not maintain constant radius. The radius of the trumpet bore from the mouthpiece slowly increases for approximately 22 cm. The tubing then remains roughly at a constant radius until the bore begins to widen at 102 cm from the mouthpiece. Furthermore, the geometry of the trumpet mouthpiece itself is also quite complex and varies in shape and dynamics for each unique mouthpiece model. We tried to avoid some of these potential effects by measuring the sound pressure waveforms at the shank of the mouthpiece. However, the mouthpiece throat's radius is approximately 0.4× the shank's radius; and the radius of the trumpet bore at 20 cm from the shank is approximately 1.42× the shank's radius. It is thus possible that neglecting these geometric attributes of the trumpet may cause an overestimation in the simulated amplitude.

We are currently investigating how the geometry of the bore between the mouthpiece and the first bend of the trumpet influences the wave propagation being modelled. We speculate that in addition to the geometric shape of the tube near the mouthpiece, the rate at which the radius increases may be important to consider. Just as the rapid flare of the trumpet influences the wave propagation, specifically the harmonic reflections, we hypothesize that the increase in radius at the mouthpiece may have a similar effect. We are also currently studying the influence of such reflections near the mouthpiece.

5 Conclusion

We have presented our 2D and 3D simulation results of modelling sound production in a trumpet. For comparative purposes, a plot of the experimental data and all simulation results presented here for the B_3^b and B_4^b notes can be found in Fig. 9. Although neglecting the third spatial dimension is frequently done, it is evident that such an assumption will not accurately model the spreading of the waves in all spatial directions. Furthermore, in Figs. 3, 4, 5, and 6, we can see that wave propagation is very sensitive to the geometry of the flare, specifically where the reflections at the bell are sensitive even to minor changes to its shape. We have shown that it is not sufficient to merely approximate the bell geometry by a smooth curve, it must be constructed with maximal agreement to physical geometry to obtain good results.

Our 3D simulation results presented in Figs. 7 and 8 are significantly more accurate than our 2D simulations. In addition, these findings have initiated an investigation to determine the importance of the tube profile near the mouthpiece of the trumpet on accuracy of the model. While the tube appears to be straight from the outside, it has a complex geometry inside that cannot be inferred. We would need to cut open an instrument to measure the true geometry and are currently looking for a less invasive solution. Preliminary results indicate that better match of the experimental and numerical data can be achieved.

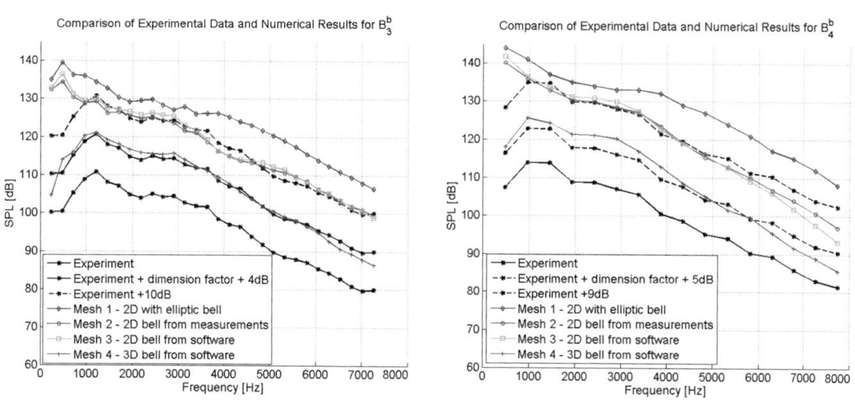

Fig. 9 Numerical and experimental results for the B_3^b and B_4^b

References

1. Benade, A.H.: Fundamentals of Musical Acoustics. Dover Publications, New York (1900)
2. Flaherty, J.E., Krivodonova, L., Remacle, J.F., Shephard, M.S.: Some aspects of discontinuous Galerkin methods for hyperbolic conservation laws. J. Finite Elem. Anal. Des. **38**, 889–908 (2002)
3. Hirschberg, A.J., Gilbert, J., Msallam, R., Wijnands, A.P.J.: Shock waves in trombones. J. Acoust. Soc. Am. **99**, 1754–1758 (1995)
4. Kausel, W., Moore, T.: Influence of wall vibrations on the sound of brass wind instruments. J. Acoust. Soc. Am. **128**, 3161–3174 (2010)
5. Resch, J., Krivodonova, L., Vanderkooy, J.: A two-dimensional study of finite amplitude sound waves in a trumpet using the discontinuous Galerkin method. J. Comput. Acous. (2012). doi:10.1142/S0218396X14500076

A Dual-Rotor Horizontal Axis Wind Turbine In-House Code (DR_HAWT)

K. Lee Slew, M. Miller, A. Fereidooni, P. Tawagi, G. El-Hage, M. Hou, and E. Matida

Abstract This paper describes the DR_HAWT (Dual-Rotor Horizontal Axis Wind Turbine) in-house code developed by the present authors, the National Renewable Energy Laboratory (NREL) Phase VI validation test case, and a dual-rotor configuration simulation. DR_HAWT uses a combination of the Blade Element Momentum theory (BEM) and the vortex filament method to accurately predict the aerodynamic performance of horizontal axis wind turbines for a single or dual-rotor configuration. Using two-dimensional (2D) airfoil coefficients obtained from XFOIL, the aerodynamic power prediction produced by DR_HAWT resulted in an average error of less than 20 % when compared to the NREL Phase VI experiment. Upon correcting the 2D wind tunnel aerodynamic coefficient data for 3D flow effects, good agreement was obtained with the exception of the deep-stall region. In addition, a dual-rotor test case consisting of a combination of a half-scaled and full-sized NREL Phase VI rotor, is presented. The dual-rotor configuration resulted in a 16 % average increase in power generation as compared to the full-sized rotor alone.

1 Introduction

With recent advancements in wind energy technology, and the increase in oil prices, wind energy is now considered to be a favourable green-energy alternative [1]. In an attempt to maximize energy extraction while minimizing costs, a Contra-Rotating Dual-Rotor Horizontal Axis Wind Turbine configuration has been proposed. It is hypothesized that by having both an upwind rotor and a downwind rotor, more power can be produced on a single wind turbine tower.

K. Lee Slew (✉) • M. Miller • E. Matida
Carleton University, 1125 Colonel By Drive, Ottawa, ON K1S 5B6, Canada
e-mail: KennyLeeSlew@cmail.carleton.ca; MikeMiller@cmail.carleton.ca

A. Fereidooni
National Research Council Canada, 2320 Lester Road, Ottawa, ON K1V 1S2, Canada

P. Tawagi • G. El-Hage • M. Hou
ZEC Wind Power, 1800 Woodward Drive, Ottawa, ON K2C 0P7, Canada

© Springer International Publishing Switzerland 2016

J. Bélair et al. (eds.), *Mathematical and Computational Approaches in Advancing Modern Science and Engineering*, DOI 10.1007/978-3-319-30379-6_45

Due to significant improvements in computational technology and knowledge of Horizontal Axis Wind Turbine (HAWT) aerodynamics, computational simulation tools have become a preferred method in the aerodynamic performance prediction of wind turbines. Commercially available CFD packages offer accurate aerodynamic predictions whilst capturing the complex flow phenomena that occur with HAWTs, however, these programs are typically computationally expensive. For this reason, Blade Element Momentum theory (BEM) is often used by the wind energy industry as a preliminary aerodynamic performance prediction tool.

BEM is computationally inexpensive and yields accurate results provided acceptable airfoil aerodynamic data is available, however, BEM cannot capture all the flow phenomena which occurs in the wake of a HAWT. To reduce runtimes compared to CFD, while providing a more accurate representation of the wake, an in-house code, utilizing a grid-less technique, for Dual-Rotor Horizontal Axis Wind Turbines (DR_HAWT) was created by the present authors based on a combination of BEM [1] and the vortex filament method [2, 3]. DR_HAWT possesses the ability to predict the aerodynamic performance of a single or dual-rotor configuration as well as parallel processing capabilities.

To verify and validate DR_HAWT, the National Renewable Energy Laboratory (NREL) Phase VI wind turbine was simulated. The experiment was performed in the NASA-Ames 24.4 m × 36.6 m wind tunnel using a two-bladed, 10-m-diameter HAWT. The tapered and twisted blades utilize the S809 airfoil; further geometric details can be found in Ref. [4]. All comparisons in this paper, unless stated otherwise, are to the aerodynamic power output of the Sequence S experiment [4].

This paper will briefly describe the BEM and vortex filament methods implemented in DR_HAWT including the 3D correction factors used to account for the effect of rotation. The NREL validation test case will be presented as well as a dual-rotor HAWT configuration.

2 Methodology

2.1 Blade Element Momentum Theory

The well documented BEM theory [1] is a combination of Blade Element theory and the Momentum theory. The wind turbine blade is discretized into elements along the blade span, where geometric data are known. At a specified Reynolds number (Re), velocity and angle of attack (AOA), the coefficient of lift and drag (C_L and C_D, respectively) can be obtained from look-up tables for each specific airfoil. Using the aerodynamic coefficients, the tangential and normal forces acting on the blade elements can be determined. These forces can then be integrated along the span of the blade to determine the rotor torque, thrust and power.

The wake of the wind turbine is commonly accounted for in the BEM method by induction factors. Due to the extraction of energy from the wind flow, these induction factors decrease the velocity of the flow in the axial and tangential direction by means of a semi-empirical model. The BEM method, combined with various other correction factors such as Prandtl's tip loss factor, provides an adequate aerodynamic prediction of single rotor configuration of HAWT [1]. However, for a dual-rotor configuration, where the influence of the upwind and downwind rotor wake is an intrinsic aspect of the aerodynamic performance prediction, a better means of representing the wake is necessary.

2.2 Vortex Filament Method

In order to determine the influence of the wake on the aerodynamic performance of a HAWT, particularly in a dual-rotor configuration, a vortex filament method has been adapted from Strickland et al. [2, 3]. In this three-dimensional (3D), grid-less free vortex model, vortex filaments are used in place of the induction factors in the BEM method. These vortex filaments are free to convect in the fluid domain thereby providing a suitable spatial and temporal representation of the wake.

Similar to the BEM method, the blade is discretized into elements where the geometric characteristics are known. Each element is defined by unit vectors in the spanwise (**s**), chordwise (**c**), and normal (**n**) directions, located at the aerodynamic center of each element. In addition, the position of the aerodynamic center of each element and the shed vortices are defined in a global coordinate system (**i**, **j**, and **k** axes) located at the center of the upwind hub. Figures 1 and 2 illustrate the coordinate systems used in DR_HAWT.

With respect to these coordinate systems, the relative velocity, $\mathbf{V_{rel}}$, seen by a blade element as shown in Fig. 2, can be evaluated by vector sum of the blade rotational velocity, $\mathbf{V_T}$, free stream velocity, $\mathbf{V_o}$, and the induced velocity caused by the vortex filaments, $\mathbf{V_{ind}}$. The local flow conditions (Re, AOA) are calculated and,

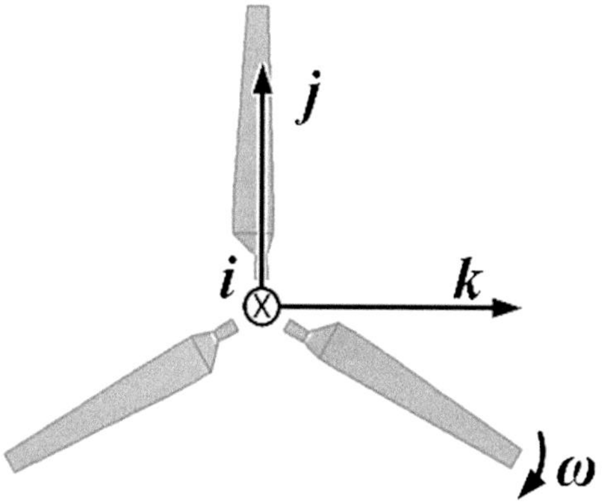

Fig. 1 Global coordinate system used in DR_HAWT

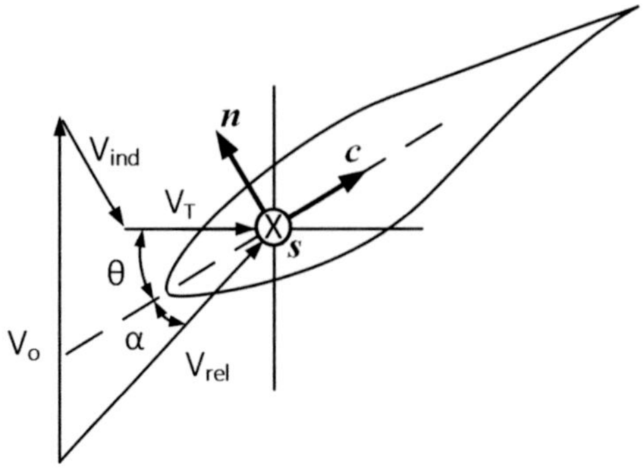

Fig. 2 Velocity components acting on a blade element

the lift and drag coefficients are linearly interpolated from look-up tables consisting of experimental or numerical airfoil aerodynamic data.

Based on the Lifting Line Theory, each blade element is associated with a bound vortex filament of strength, Γ_B. This strength can be expressed using the Kutta-Joukowski theorem as follows:

$$\Gamma_B = \frac{1}{2} C_L c |\mathbf{V_{rel}}| \tag{1}$$

The bound vortex strength is a function of the C_L, the chord length (c) and the magnitude of the relative velocity. Since the relative velocity takes into account the induced velocity, a predictor-corrector method is employed for which the bound vortex strength is predicted based on the previous time step induced velocities. The predicted strength is then used to correct the induced velocities which in turn corrects the bound vortex strength. This process is repeated until the differences between the predicted and corrected values are below 1×10^{-3}. Once the bound vorticity and the induced velocities have been corrected, the strength of the vortices that constitute the wake can be computed.

Figure 3 illustrates the representation of blade elements with the wake which is made up of vortex filaments. For a time step *NT*, the bound vortex strength of an element *i* is written as Γ_{NT}^{i}. As the blade rotates, vortices are shed in the form of spanwise and trailing vortex filaments parallel to the span and chord of the blade respectively. The strength of these vortex filaments are calculated based on Kelvin's Theorem for which any change in the bound vortex strength over time must be accompanied by an equal and opposite vortex strength in the wake. This forms the spanwise vortices (bold lines) whereas the trailing vortices (thin lines) result from the spanwise variation in bound vortex strength. The tip vortex is assumed to have the same magnitude and opposite direction as the bound vortex closest to the tip.

Each vortex filament in the wake convects with a local velocity. As the vortex filament first leaves the blade, its initial velocity is assumed to be equal to the sum of the induced velocities at the blade element end and the oncoming freestream

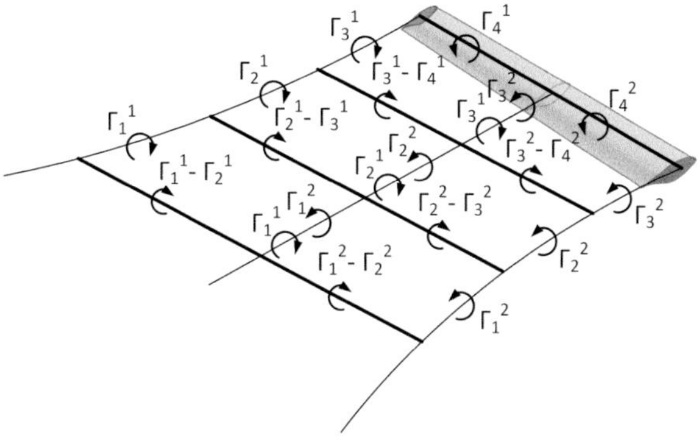

Fig. 3 Spanwise and trailing vortices being shed from a blade

velocity. In order to calculate the distance travelled by any vortex filament in the wake for a given time step, a second order Adams-Bashforth explicit integration formula is used. Knowing the strength and position of the shed vortices, the induced velocity at any point can be calculated using the Biot-Savart Law given by:

$$\mathbf{V}_{\mathbf{ind}} = \frac{\Gamma}{4\pi} \int_l \frac{\mathbf{r} \times \mathbf{dl}}{|\mathbf{r}|^3} \tag{2}$$

where $\mathbf{r}$ is the position vector from the point of interest in the fluid domain to the incremental length (**dl**) along the vortex filament. The total perturbation velocity induced by all the vortex filaments represents the effect of the wake on the blades. In order to calculate the torque, thrust and power generated by the blades, the normal and tangential forces are integrated along the span of the blade by assuming linear variation along each element [1]. Finally, the aerodynamic power produced by the wind turbine is the product of the rotational velocity and the total torque of the rotor.

2.3 Aerodynamic Coefficients

This study uses the 2D (non-rotating) airfoil coefficient data (C_L and C_D) obtained either by using experimental wind tunnel data, acquired from Ref. [4] or XFOIL which numerically calculates the lift and drag coefficients for a given airfoil profile based on pressure distributions obtained using a panel-method [5]. For comparison purposes, the same numeric values of Reynolds numbers are used in both cases for the aerodynamic coefficient input files. In addition, the experimental data was corrected to obtain a better understanding of 3D flow effects due to rotation.

In an attempt to model these 3D effects, several numerical correction factors have been formulated by the likes of Du and Selig [6], Snel et al. [7], Bak et al. [8], Lindenburg [9], and more. These various models all simply correct the 2D C_L and

C_D wind tunnel data using the following equations.

$$C_{L,3D} = C_{L,2D} + f(c/r, \ldots)\Delta C_L \tag{3}$$

$$C_{D,3D} = C_{D,2D} + f(c/r, \ldots)\Delta C_D \tag{4}$$

where f is the correction factor which in all cases is a function of at least the chord to radial (c/r) location ratio. ΔC_L and ΔC_D is the difference between the 2D C_L and C_D wind tunnel data and the 2D C_L and C_D if hypothetically the flow remained attached at all AOA. Some 3D correction formulations, such as that of Du and Selig, have empirical correction factors which are built into f. The fine tuning of the 3D corrections is only viable when experimental 3D data, like the data obtained from the NREL experiment, is available. Obtaining a suitable universal empirical correction factor for the 3D corrections should be performed by simulating several different experiments.

The 3D correction factors of Snel et al. and Du and Selig were implemented in DR_HAWT for AOA up to 30°, at which point the corrected values linearly decay to the 2D value at 55° as per Ref. [10]. Figure 4 illustrates an example of the Snel et al. 3D corrections applied to the C_L of the S809 airfoil at a Reynolds number of 6.5×10^5. It should be noted that the Snel et al. correction only corrects the C_L whereas Du and Selig corrects both the C_L and C_D.

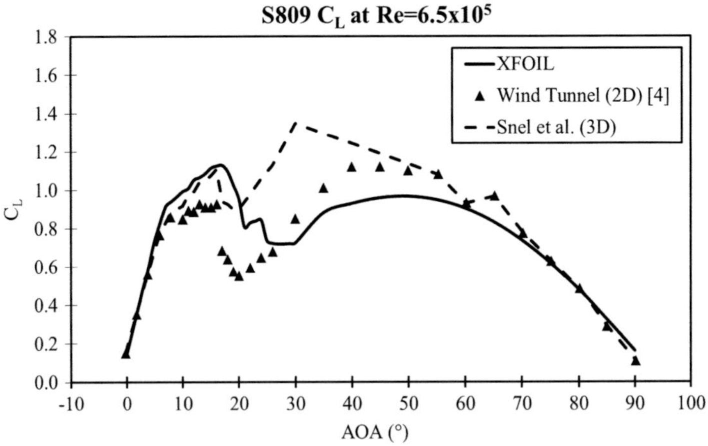

Fig. 4 C_L curves obtained by XFOIL, 2D wind tunnel data and 3D corrections (Snel et al.) for a given Reynolds number

2.4 Convergence Analyses

Prior to simulating the Sequence S of the NREL Phase VI experiment, a sensitivity analysis was performed to determine the effect of changing the number of elements, time increment, and number of time steps in DR_HAWT for the performance prediction of the NREL Phase VI wind turbine. Default values of 26 discretized blade elements used to define the geometry of each blade in Ref. [4], a time increment of 0.03145 s and 200 time steps were both halved and doubled. The outcome of this study revealed that the default values were adequate for this simulation.

Typically, converged values of power, torque and thrust for a given oncoming wind speed are obtained after the wake has travelled 3–4 diameters downstream. Correspondingly, a single-rotor configuration for 200 time steps with a time increment of 0.03145 s, has a simulation runtime of approximately 60 min using an i7 3.60 GHz processor with 16 GB of RAM; even faster runtimes can be obtained if the code is run with parallel processing. This highlights DR_HAWT's ability to run parametric studies of a single or dual-rotor configuration in a timely manner compared to existing commercially available CFD packages.

3 Results and Discussion

Figure 5 illustrates the output power predicted by DR_HAWT using aerodynamic coefficients from XFOIL, wind tunnel (2D) experiments, wind tunnel data corrected for 3D flow effects using Snel et al. and Du and Selig, as well as the UAE experimental data. Based on XFOIL data, DR_HAWT resulted in an average

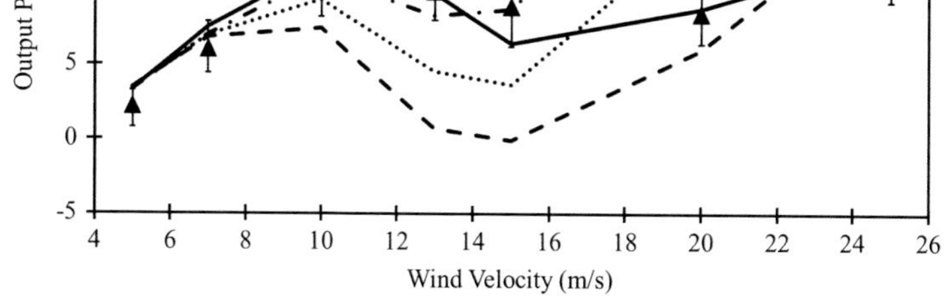

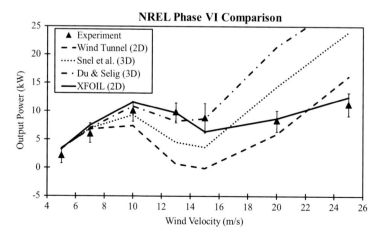

Fig. 5 Aerodynamic power prediction produced by DR_HAWT

error below 20 % when compared with the experimental data, which is within the measurement error bars [11]. With the 2D wind tunnel data, however, large differences in excess of 100 % were obtained in the stall region (10–25 m/s).

The large variation in aerodynamic power curves indicate that the input aerodynamic coefficient is an influencing factor in the output results produced by DR_HAWT. As seen in Fig. 5, this mainly affects the outcome in the stall region (10–25 m/s) where separation and 3D flow effects are present on the blade.

Due to the rotational nature of the wind turbine, the flow over the wind turbine blade is not 2D but it is in fact 3D which has been found to delay stall. The pressure gradient due to rotation of the blade invokes flow in the radial direction which alters the boundary layer, hence resulting in higher lift and lower drag values. For this reason, wind turbines generally will obtain higher power values than predicted when using 2D wind tunnel airfoil data alone [10]. Although closer to the experimentally obtained aerodynamic power, particularly in the pre-stall (below 7 m/s) and stall regimes (7–15 m/s), the 3D corrections still do not produce favourable results when deep-stall is prominent (above 15 m/s). According to Vermeer et al., the use of 3D corrections at high wind velocity is questionable [12].

No matter if XFOIL, 2D wind tunnel or 3D corrected aerodynamic coefficient data is used for the simulations, an over prediction in power up to 7 m/s wind speeds can be observed. The consistency in power values obtained by the various simulations is due to the fact that the majority of the blade is in the linear, pre-stall, region of the lift and drag coefficient curves. As illustrated in Fig. 4, this linear region does not vary substantially between the different methods of obtaining the aerodynamic airfoil data since the flow is attached. DR_HAWT was found to consistently over predict the tangential force compared to experimental values [11]. Under the assumption that the C_L and C_D are not contributing factors in the pre-stall region, this discrepancy must result from an inaccuracy in the method used to determine the AOA. The challenge of predicting the AOA was also identified by Lindenburg [10], Sant et al. [13] and Jonkman [14] suggesting that further investigation into the methods used by DR_HAWT is required.

4 Dual-Rotor Configuration

In order to examine the effects of adding a second rotor to the HAWT, a half-sized, geometrically scaled NREL Phase VI rotor and an unmodified NREL UAE Phase VI rotor were simulated in the upwind and downwind positions, respectively. The upwind rotor had a rotational velocity twice that of the unmodified rotor in order to maintain similar tip speeds. The rotors were separated coaxially by half the diameter of the unmodified rotor. The contra-rotating dual-rotor configuration was simulated using XFOIL 2D data for wind speeds ranging from 5 to 25 m/s. As can be seen in Fig. 6, the addition of the upwind rotor resulted in an average increase in power of 16 % over the single rotor configuration. The power generated by the dual-rotor configuration is lower than that of the combined individual single rotors by approximately 6 %. This is as expected since the presence of the wake from one rotor onto another is typically detrimental to their performance. A visualization of the dual-rotor configuration post-processed in ParaView [15] can be seen in Fig. 7.

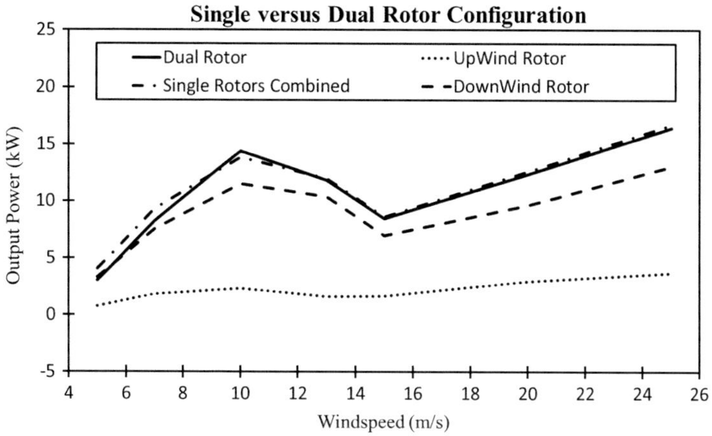

Fig. 6 Aerodynamic power comparison of a single and dual-rotor HAWT configuration

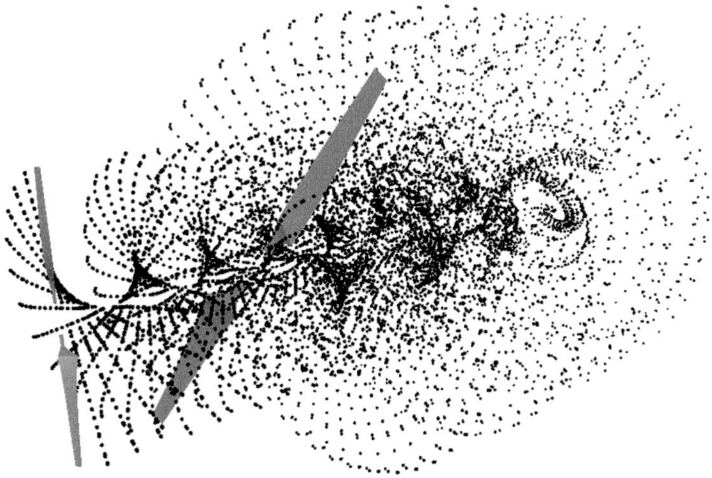

Fig. 7 Wake visualization of a dual-rotor configuration where the dots represent the end points of a vortex filament

5 Conclusion

An in-house code, named DR_HAWT was created by the present authors for the aerodynamic performance prediction of a single or dual-rotor HAWT configuration. Converged values of aerodynamic loads are obtained within approximately 60 min. DR_HAWT was validated against the NREL UAE Phase VI experiment with relatively good agreement. Power predicted by DR_HAWT using numerically obtained data from XFOIL, fell within the experimental error bars and had an overall error of less than 20 %. This adequate aerodynamic performance prediction and fast computational run times facilitate parametric studies of a single or dual-rotor HAWT configuration.

Further analyses revealed that the input lift and drag coefficient data have a large influence on the predicted aerodynamic forces. 3D flow effects were found to have a profound significance particularly in the stall region. 3D corrections formulated by Snel et al. and Du and Selig were applied to 2D coefficient data and found to show improvement with the exception of the deep stall region (above 15 m/s). A contra-rotating dual-rotor HAWT configuration based on the NREL Phase VI rotor, was found to produce 16 % more power than the single NREL Phase VI rotor alone, demonstrating the potential benefits of a dual-rotor wind turbine configuration.

Acknowledgements The authors would like to thank NSERC (National Sciences and Engineering Research Council of Canada) and OCE (Ontario Centres of Excellence) for the financial support as well as Scott Schreck of the National Renewable Energy Laboratory for providing experimental data for validation purposes.

References

1. Hansen, M.: Aerodynamics of Wind Turbines. EarthScan, London (2008)
2. Strickland, J., Webster, B., Nguyen, T.: A Vortex Model of the Darrieus Turbine: An Analytical and Experimental Study. Sandia National Laboratories, Lubbock (1980)
3. Fereidooni, A.: Numerical Study of Aeroelastic Behaviour of a Troposkien Shape Vertical Axis Wind Turbine – MASc Thesis. Carleton University, Ottawa (2013)
4. Hand, M., Simms, D., Fingersh, L., Jager, D., Cotrell, J., Schreck, S., Larwood, S.: Unsteady Aerodynamics Experiment Phase VI: Wind Tunnel Test Configurations and Available Data Campaigns. National Renewable Energy Laboratory, Golden (2001)
5. Drela, M.: XFOIL: An Analysis and Design System for Low Reynolds Number Airfoils. Dept. of Aeronautics and Astronautics, MIT, Cambridge (1989)
6. Du, Z., Selig, M.S.: A 3-D Stall-Delay Model for Horizontal Axis Wind Turbine Performance Prediction. Am. Inst. Aeronaut. Astronaut. **21**, 9–19 (1998)
7. Snel, H., Houwink, R., Bosschers, J.: Sectional Predicition of Lift Coefficients on Rotating Wind Turbine Blades in Stall. Energy Research Centre of the Netherlands, Petten (1994)
8. Bak, C., Johansen, J., Anderson, P.B.: Three-Dimensional Corrections of Airfoil Characteristics Based on Pressure Distributions. In: European Wind Energy Conference & Exhibition, Athens (2006)
9. Lindenburg, C.: Modelling of Rotational Augmentation Based on Engineering Considerations and Measurements. In: European Wind Energy Conference, London (2004)

10. Lindenburg, C.: Investigation into Rotor Blade Aerodynamics: Analysis of the Stationary Measurements on the UAE Phase-VI Rotor in the NASA-Ames Wind Tunnel. Energy research Centre of the Netherlands, Petten (2003)
11. Schreck, S.: Sequence S Data. National Renewable Energy Laboratory via Private Communications, Golden (2015)
12. Vermeer, L., Sørensen, J., Crespo, A.: Wind Turbine Wake Aerodynamics. Prog. Aerosp. Sci. **39**, 467–510 (2003)
13. Sant, T., van Kuik, G., van Bussel, G.: Estimating the Unsteady Angle of Attack from Blade Pressure Measurements on the NREL Phase VI Rotor in Yaw using a Free Wake Vortex Model. In: 44th AIAA Aerospace Sciences Meeting and Exhibition, Reno (2006)
14. Jonkman, J.: Modeling of the UAE Wind Turbine for Refinement of FAST_AD. National Renewable Energy Laboratory, Golden (2003)
15. Ayachit, U.: The ParaView Guide: A Parallel Visualization Application. Kitware, Inc., Clifton Park, NY (2015)

Numerical Study of the Installed Controlled Diffusion Airfoil at Transitional Reynolds Number

Hao Wu, Paul Laffay, Alexandre Idier, Prateek Jaiswal, Marlène Sanjosé, and Stéphane Moreau

Abstract Reynolds-Averaged Navier-Stokes simulations have been carried out for the self-noise study of a Controlled Diffusion airfoil at chord Reynolds number of 1.5×10^5. Two numerical setups of the anechoic open-jet facilities where experimental data on aerodynamics have been collected using hot-wire and particle image velocimetry measurements are investigated to capture installation effects and to compare different turbulent models. The installation effects are observed between different wind tunnel nozzle dimensions. Different turbulent models in simulations have been compared for such flow case. The $k\omega-SST$ model shows better agreement with the experimental data at a reasonable computational cost, providing a reliable initialisation field and boundary conditions for future direct numerical simulation studies.

1 Introduction

In the design process of a rotating machine such as an automotive engine cooling fan, a wind turbine or an air-conditioning unit, one major evaluation index is the noise level for a given loading. Where other noise sources can be reduced or avoided by a careful design, the trailing-edge (TE) noise, or airfoil self-noise, is the only remaining noise source when an airfoil encounters a homogeneous stationary flow. Such a noise is generated by the interaction between the TE with pressure fluctuations convecting in a boundary layer. If the boundary layer is laminar, tonal noise will be introduced; if the boundary layer is turbulent, pressure fluctuations are present over a wide range of frequencies, which leads to broadband noise radiation. The study of TE noise received much attention mainly in the 1970s and early 1980s. Most of such studies during that time involved experimental measurements of wall-pressure fluctuations and far field acoustic on various airfoils in open-jet anechoic wind tunnels [1, 2]. In recent years, Computational Fluid Dynamics (CFD)

H. Wu (✉) • P. Laffay • A. Idier • P. Jaiswal • M. Sanjosé • S. Moreau
Department of Mechanical Engineering, Université de Sherbrooke, Sherbrooke, QC J1K2R1, Canada
e-mail: hao.wu@usherbrooke.ca

© Springer International Publishing Switzerland 2016

J. Bélair et al. (eds.), *Mathematical and Computational Approaches in Advancing Modern Science and Engineering*, DOI 10.1007/978-3-319-30379-6_46

methods have become an important complementary part aside from the experimental methods. In practice however, simplified flow configurations are often employed in simulations whereas most trailing-edge aeroacoustics experiments have been conducted in open-jet wind-tunnel facilities, where the airfoil is immersed in a jet downstream of the nozzle exit. It has been seen that the proximity of the airfoil to the jet nozzle exit and the limited jet width relative to the airfoil thickness and chord length can cause the airfoil loading and flow characteristics to deviate significantly from those measured in free air and hence, alter the radiated noise field [3]. Installation effects thus take place for different jet configurations and require simulations to model it.

As aerodynamics has direct influence on aeroacoustics, the present work aims at presenting and quantifying the installation effects as well as studying the influence of different turbulent models on such flow case simulation. Experimental and numerical setups are firstly introduced with associated technical methods. Results on a systematic CFD study, based on Reynolds-Averaged Navier-Stokes (RANS) are then presented and are compared with flows data over the Controlled Diffusion (CD) airfoil (a cambered airfoil originally developed at Valeo Motors and Actuators) installed in the anechoic wind tunnels of Ecole Centrale Lyon (ECL) and of Université de Sherbrooke (UdS). The evaluation of different turbulence models and installation effects observed in simulations will be discussed. The results give a synthesis of the previous RANS study and provide guidance for the appropriate initial field needed in future direct numerical simulation (DNS) studies [4, 5].

2 Experimental Setup

All the measurements were done in an open-jet anechoic wind tunnel shown in Fig. 1a. The airfoil mock-up (Fig. 1b) is held between two side plates in a $30 \times 30\,\text{cm}^2$ test section, to keep the flow two-dimensional. This setup is very similar to that used in [3, 6] at ECL, of which, the jet nozzle dimension is $50 \times 25\,\text{cm}^2$. The self-noise is the noise radiated by the turbulent eddies coming from the turbulent boundary layer. The setup should be free of any additional noise source that would make the self-noise. In particular, very low turbulence intensity is achieved by a very high convergent ratio of about 1:25. The measured turbulence intensity is 0.3–0.4 %. The wind tunnel is also acoustically treated to achieve low background noise. According to the above mentioned criteria, it becomes clear that open-jet anechoic wind tunnel is best suited for the study of airfoil self-noise [7]. The hot wire and PIV measurements have been performed in the wake and pressure sensor probes on the airfoil to detect airfoil loading, as described below.

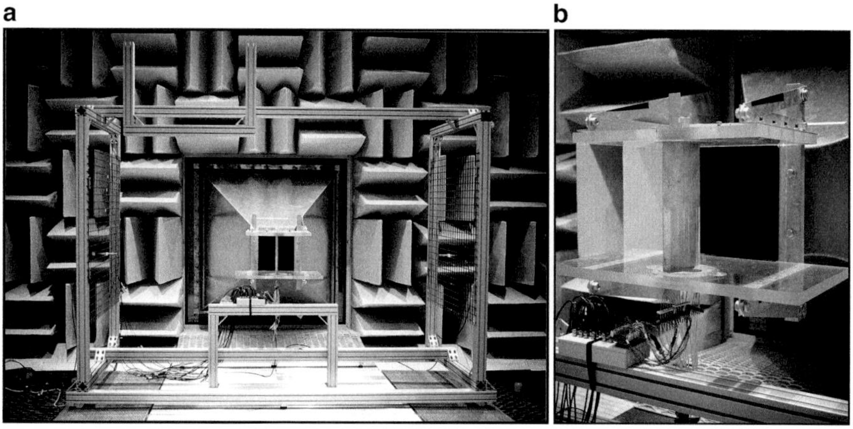

Fig. 1 (**a**) Front view of the CD airfoil in the anechoic wind tunnel and (**b**) Close side-view

2.1 Wall Pressure Measurements

The mean wall pressure on the CD airfoil was measured using a Baratron capacitance manometer which is connected to a pin hole on the surface of the airfoil using a capillary tube. There are in total 21 probes, 18 of them are placed in streamwise direction and rest 3 in spanwise direction. All the streamwise probes are placed at the mockup mid-span as shown in Fig. 1b.

2.2 Velocity Measurements

The streamwise velocity in the wake was obtained using a TSI 1210-T1.5 single hot-wire probe The length of the probe is 1.27 mm and a diameter of 3.8 μm. The displacement of the hot wire in the wake was realized using a Superior Electric M062-FD03 Slo-Syn Stepping Motor. The minimum displacement of the hot wire (the highest spatial resolution) in the near wake was 0.05 mm. The hot-wire probe is connected to a Constant Temperature Anemometer System IFA 300. The data acquisition was achieved using a National Instrument BNC 2090 system controlled with Labview at the sampling frequency of 20 kHz. The relative measurement error was calculated taking into account parameters listed in [8]. It was found to equal to 2.215 %.

Planar PIV (Particle Image Velocimetry) measurements have been performed using a single LaVision sCMOS 5.5 megapixel camera with a pixel pitch of 6.5 μm and a Evergreen 70 mj ND:YAG laser [9]. The laser-sheet thickness was measured to be about 2 mm. 700 images in double frame were recorded using sCMOS camera fitted with Nikon 50 mm lens at an acquisition rate of 15 Hz. The image magnification was about 0.07. The particle image diameter and depth of field were adjusted using lens aperture [9], chosen to be 11 for the current experiments. This is done so that particle image diameter is greater than 2 pixels (calculated by neglecting any lens aberrations). At this image diameter under sampling of particle image which can cause peak locking can be avoided [10]. The depth of field calculated was about 60 mm, which is much larger than the measured laser sheet thickness so all the particles can be assumed in focus. Time between frames was selected based on many considerations like minimization of loss-of-pair due out of plane motion, truncation error due to constant velocity assumption of the particle between the two frames and minimization of relative error on the displacement estimate[11]. Finally it must be remembered that the dynamic velocity range of PIV increases with time separation between the two frames [12]. Taking all these considerations into effect, time between two frames was chosen to be about 45 μs. The free stream displacement particle between the two frames was measured to be roughly about 7.5 pixels in the object plane. Taking the smallest resolvable displacement fluctuation to be 0.1 pixel, the dynamic velocity range was calculated to be about 75 [9]. The seeding density during the experiments was kept at about 0.055 ppp to ensure a more that 10 particles were present for the final interrogation domain for a better signal-to-noise ratio in cross correlation [13]. The parameters used for PIV measurements are summarized below in Table 1. All the results were processed using DAVIS 8 software. To improve the accuracy in peak fitting and sub-pixel accuracy, normalized cross correlation option in DAVIS 8 was selected for the calculation of cross correlation. The final interrogation window size was kept at 16 × 16 pixel with a 50 % overlap. An adaptive window shape was selected for image cross correlation to take into account the effect of shear in the near wake. The error analysis in PIV is a topic of active research and depends upon many factors. In absence of detailed error analysis the value of error was chosen to be equal to 0.1 pixel which corresponds to a typical error in measurement of displacement in PIV [9]. On the other hand typical error on pulse separation has an order of magnitude in nanoseconds while time between frames is in the order

Table 1 Parameters used in the PIV measurement

Parameters	Value
Depth of focus	~59 mm
Seeding density	$\geqslant$ 0.055 ppp
Mode	Double frame
Frequency	50 Hz
Number of images	700
Window of interrogation	16 × 16
Focal length	57.41 mm
FOV	240 mm
Particle image diameter	2.33 pixels
Magnification	0.07

of microseconds hence relative error is very small, compared to relative error in displacement.

3 Numerical Setup

3.1 Flow Conditions and Physical Models

The flow conditions for CD airfoil are a freestream velocity U_0 of 16 m/s measured at the wind tunnel nozzle exit away from the airfoil. The Mach number is 0.05 and the Reynolds number is 1.5×10^5 based on the airfoil chord length $c = 0.1356$ m. The flow is therefore turbulent or transitional. To model the turbulence in RANS simulations, 4 turbulence models are tested: $k - \epsilon$ (standard and with low Reynolds number correction), $k\omega - SST$, and $tr - k - kl - \omega$. As the RANS results will serve as the initialization field for future DNS study, the transitional model $tr - k - kl - w$ is chosen for its capacity to capture transition process in the simulation. Air is supposed to be an incompressible perfect gas. The geometric angle of attack with respect to the wind tunnel axis is 8°. To evaluate the airfoil loading and the installation effects, the pressure coefficient C_p and the friction coefficient C_f are introduced as shown respectively in Eq. (1),

$$C_p = \frac{P_s - P_0}{\frac{1}{2}\rho_0 U_0^2} \quad and \quad C_f = \frac{\tau_n}{\frac{1}{2}\rho_0 U_0^2} \tag{1}$$

where P_s stands for the static pressure, τ_n for the local wall-shear stress and all the parameters with the sub-index 0 correspond to the reference values in this flow case. The values of the results shown afterwards in the section *Results* are non-dimensionalized based on c and U_0.

3.2 Numerical Parameters

To introduce the installation effects, the simulation domain includes the complete nozzle geometry to assess the effect of the interaction between the jet shear layer and the airfoil (Fig. 2). A 2D domain which represents the mid-section of the *Experimental Setup* is generated by Centaur. Hybrid grids with a total 69,000 cells are employed to get a balance between simulation accuracy and computational cost according to what was reported in previous simulations [3] Quadrilaterals are used to refine the grid close to the walls in order to capture the boundary layer around the airfoil and the shear layer effect of the wind tunnel exit as shown in Fig. 2b, c. The dimensionless wall-normal grid spacing in wall units Δy^+ is smaller than 1 over most of the chord length on both pressure and suction sides. The mesh for

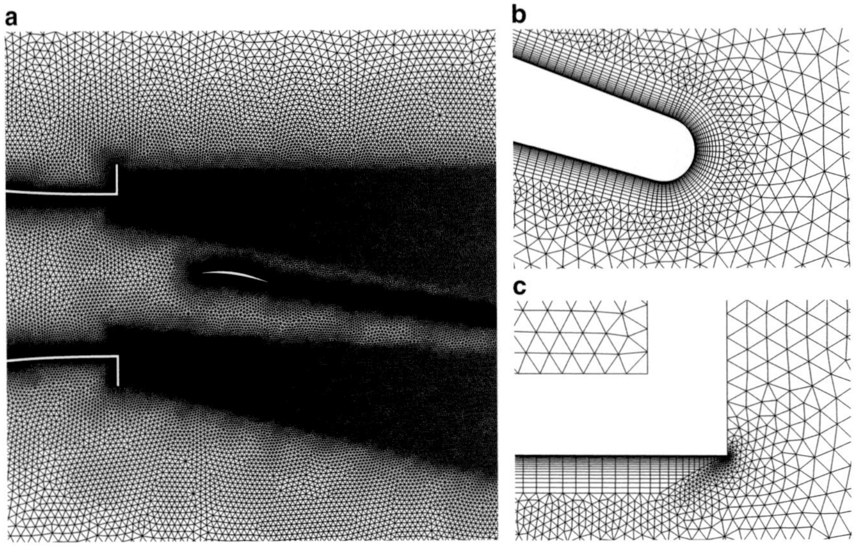

Fig. 2 Hybrid mesh (**a**); Zoom view of airfoil trainling-edge and nozzle exit (**b**) (**c**)

ECL installation is generated using the same methodology as in Fig. 2. The RANS simulations are performed using ANSYS FLUENT-15. The pressure-based solver provides two types of pressure-velocity coupling algorithms: either in a segregated manner (SIMPLE) or in a coupled manner (Coupled). The results demonstrate no difference in the wake velocity profiles or loading of the airfoil. The Coupled scheme takes $2 \sim 3$ times longer than SIMPLE. In the present paper, calculations presented are performed using the SIMPLE scheme only.

4 Results

4.1 Effects of Turbulence Models

4.1.1 Velocity and Turbulence Kinetic Energy Field

All models are able to establish a converged wake zone except for the $tr-k-kl-\omega$ model as shown in Fig. 3. Even from the established field obtained with the $k\omega-SST$ model that can be seen as a reference from previous studies[3], the scaled residuals of the $tr-k-kl-\omega$ model oscillate around much higher levels, two orders of magnitude higher than other models. With regards to turbulence, the $k-\epsilon$ over-predicts the turbulence kinetic energy (TKE) (Fig. 3). By using damping functions, the $k-\epsilon$ model with low Reynolds-number correction has been implemented in ANSYS Fluent and shows better behaviour than the standard $k-\epsilon$ model. Yet it

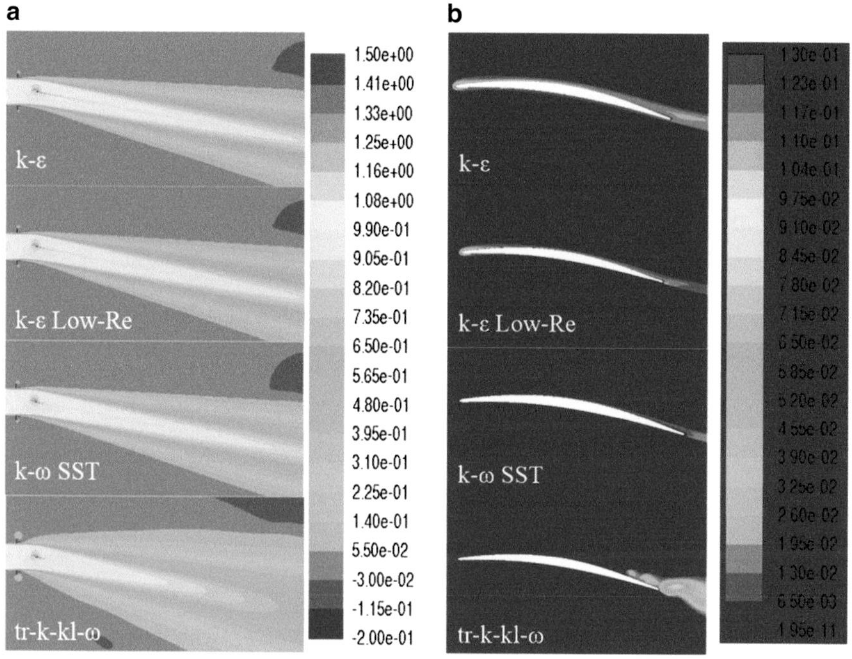

Fig. 3 Velocity (*left*) and turbulence kinetic energy (*right*) iso-contours

still over-predicts the turbulence over the whole profile. The $k\omega - SST$ model only sees the transition process to turbulence close to the trailing edge. The $tr - k - kl - \omega$ has unphysical boundary-layer development at the TE which is caused by high numerical instabilities. To make a fair comparison, the simulations for $tr - k - kl - \omega$ have then been conducted in a URANS (Unsteady Reynolds Averaged Navier-Stokes) mode and then averaged to get a steady solution.

4.1.2 Airfoil Loading

The $tr - k - kl - \omega$ model averaged data is closer to the mean wall-pressure measurement in the TE region and has similar prediction capacity as the $k\omega - SST$ model at other positions along the airfoil (Fig. 4). Compared with the $k\omega - SST$ model or other $k\omega$ based models, the $tr - k - kl - \omega$ model includes three transport equations to solve the turbulent kinetic energy (k_t), the laminar kinetic energy (k_l) and the scale-determining variable ω where $\omega = \epsilon / k_t$, the ratio of isotropic dissipation and the turbulence kinetic energy respectively. In the ω equation, a transitional production term is introduced to produce a reduction in turbulence length scale during the transition breakdown process, which essentially helps predict the magnitude of low-frequency velocity fluctuations in the pre-transitional

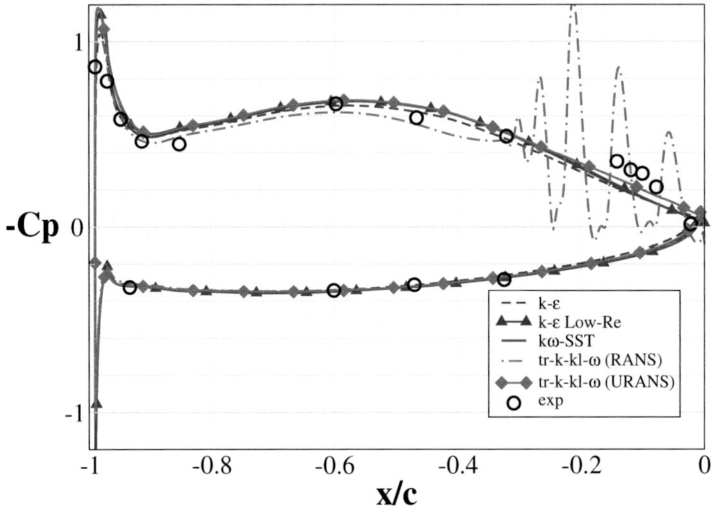

Fig. 4 Comparison of mean wall-pressure coefficient

boundary layer that have been identified as the precursors to transition [14]. As the numerical transitional process at UdS appears around TE, this model has shown its advantage on capturing such fluctuations compared with other tested models. The unsteadiness appearing in the $tr - k - kl - \omega$ RANS simulation at the TE can be potentially verified in the future by measurements of wall-pressure fluctuations.

4.1.3 Velocity Profile in the Wake Region

The wake velocity profile is a consequence of the transition process on the airfoil. Here the velocity magnitude from different models are compared with both hot-wire and PIV measurements in Fig. 5. Simulation results tend to over-predict the velocity deficit at all measuring positions but are consistent with their respective production of turbulence. Some of the discrepancies can be attributed to the isotropic hypothesis on calculating the turbulence kinetic energy made in the RANS simulation. The $k{-}\epsilon$ with low Reynolds number correction model improves the result from the standard $k{-}\epsilon$ model by nearly 10 %. Yet, the $k\omega{-}SST$ model shows better overall comparison with the experimental data.

In summary, for RANS simulations of such a flow case, the $tr - k - kl - \omega$ model must be run in an unsteady mode and its mean solution is similar to that of the $k\omega - SST$ model with an airfoil loading even slightly better compared with experiment. The $k\omega - SST$ model provides the best overall prediction among the tested models considering its reasonable computational time. It is thus used as a reference model for future comparisons. By using damping functions, the

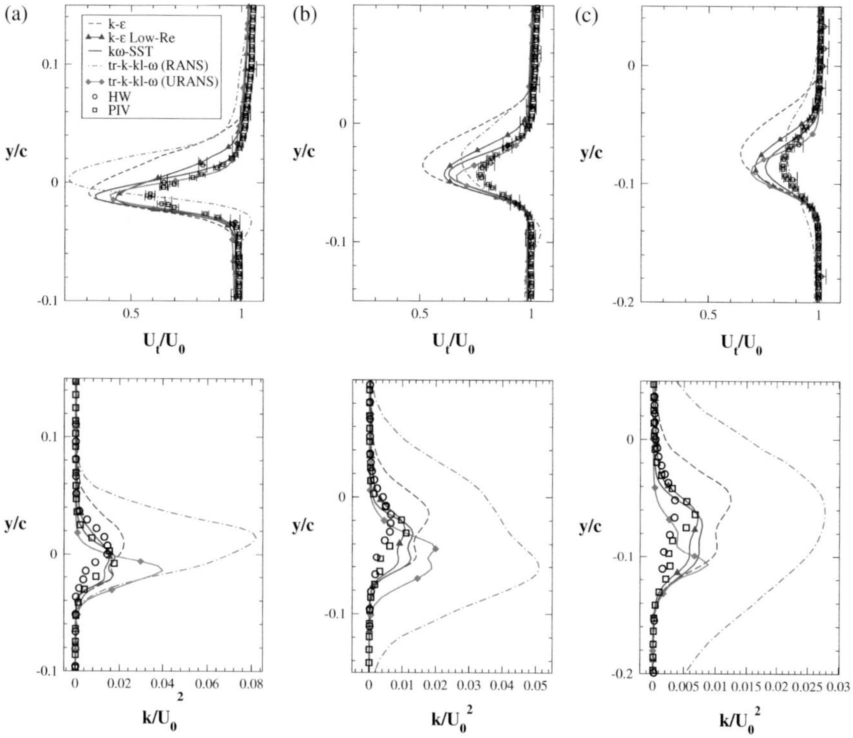

Fig. 5 Wake velocity and TKE profiles. (**a**) $x/c = 0.07$. (**b**) $x/c = 0.22$. (**c**) $x/c = 0.44$

two-equation $k - \epsilon$ model with low Reynolds number correction shows better behaviour in wake velocity profiles than the the standard $k - \epsilon$ model.

4.2 Installation Effects

As firstly mentioned by Moreau et al. [3], installation effects are significant between an isolated airfoil case and one that is installed in an open-jet wind tunnel. In this paper, the effects of nozzle jet width is specifically investigated by comparing results in two different setups at ECL and UdS respectively.

The loading of the airfoil is changed as can be seen from the wall pressure coefficient in Fig. 7. This is mostly caused by a different transition process. At ECL, a separation bubble turns the laminar boundary layer into a turbulent one at the leading edge, which is hardly the case in the UdS set-up for which, the transition process is less severe. Indeed Fig. 6 shows that the friction coefficient C_f is barely negative in the UdeS case and the laminar recirculation bubble is hardly formed. Because of the more limited nozzle jet width at UdS, the boundary layer

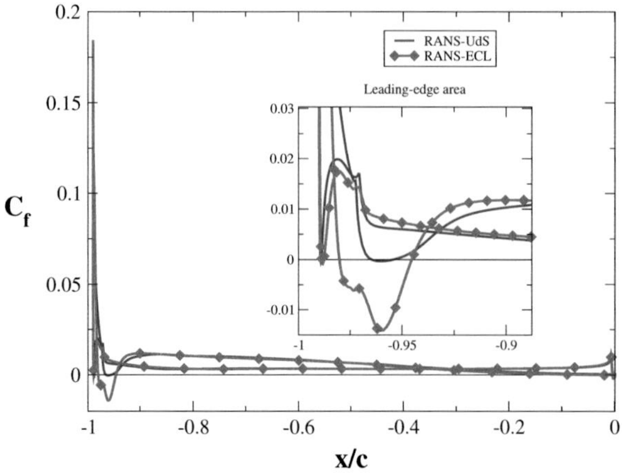

Fig. 6 Mean friction coefficient: global and zoom in the leading-edge area

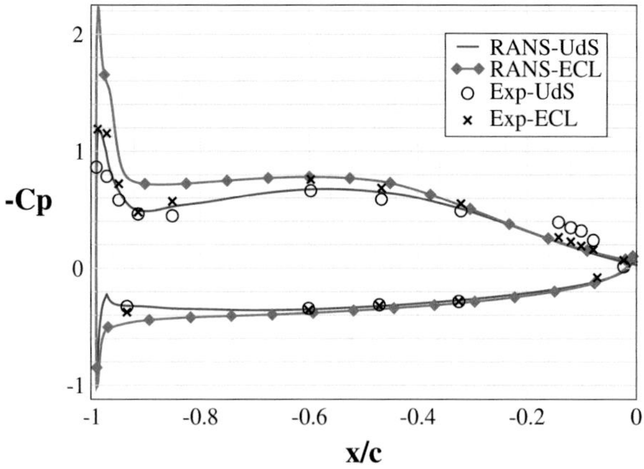

Fig. 7 Mean wall pressure coefficient

at the leading edge develops in a different way as a consequence of the larger flow confinement. Besides, there is a difference of 15 % on C_p at the TE in Fig. 7 between the experimental data and the simulation results at UdS as no flow separation is captured in this simulation. Consequently, the RANS simulations have a hard time predicting the transition process correctly, and the flow separation either at the leading edge (laminar recirculation bubble) or at the TE (driven by the adverse pressure gradient).

5 Conclusion

RANS simulations have yielded the transitional flow around a CD airfoil installed in anechoic wind tunnels for future TE noise study. The results have been systematically compared with experimental data. Different turbulence models have been tested to compare their capacity to capture transitional process. The $k\omega - SST$ model has a better global behaviour considering the reasonable computational cost and is thus considered as the reference model for future RANS simulations. The transitional model $tr{-}k{-}kl{-}\omega$ slightly improves the airfoil mean loading according to the averaged URANS data. Installation effects for the two setups with different jet nozzle size have been studied for flow with same inlet condition. In the larger nozzle case, a laminar recirculation bubble that triggers the boundary layer development on the suction side is observed while in the other case not. RANS simulations show limitations in predicting the laminar transition process and flow separations under adverse pressure gradient.

References

1. Brooks, T.F., Hodgson, T.H.: Trailing edge noise prediction from measured surface pressures. J. Sound Vib. **78**(1), 69–117 (1981)
2. Fink, M.R.: Experimental evaluation of theories for trailing edge and incidence fluctuation noise. AIAA J. **13**(11), 1472–1477 (1975)
3. Moreau, S., Henner, M., Iaccarino, G., Wang, M., Roger, M.: Analysis of flow conditions in freejet experiments for studying airfoil self-noise. AIAA J. **41**(10), 1895–1905 (2003)
4. Wang, M., Moreau, S., Iaccarino, G., Roger, M.: LES prediction of wall-pressure fluctuations and noise of a low-speed airfoil. Int. J. Aeroacou. **8**(3), 177–197 (2009)
5. Winkler, J., Sandberg, R.D., Moreau, S.: Direct numerical simulation of the self-noise radiated by an airfoil in a narrow stream. In: 18th CEAS/AIAA Aeroacoustics Conference, Colorado Springs, pp. 2012–2059 (2012)
6. Neal, D.R.: The effects of rotation on the flow field over a controlled-diffusion airfoil. Ph.D. thesis, Michigan State University (2010)
7. Vathylakis, A., Kim, J.H., Chong, T.P.: Design of a low-noise aeroacoustic wind tunnel facility at Brunel University. In: 20th AIAA/CEAS Aeroacoustic Conference and Exhibit, Atlanta (2014)
8. Jorgenson, F.: How to Measure Turbulence with Hot Wire Anemometers. Dantec Dynamics, Skovlunde (2004)
9. Raffel, M., Willert, C.E., Kompenhans, J.: Particle Image Velocimetry: A Practical Guide. Springer, Heidelberg/New York (2013)
10. Westerweel, J.: Fundamentals of digital particle image velocimetry. Meas. Sci. Technol. **8**(12), 1379 (1997)
11. Adrian, R.J., Westerweel, J.: Particle Image Velocimetry, Number 30. Cambridge University Press, Cambridge/New York (2011)
12. Adrian, R.J.: Dynamic ranges of velocity and spatial resolution of particle image velocimetry. Meas. Sci. Technol. **8**(12), 1393 (1997)
13. Keane, R.D., Adrian, R.J.: Optimization of particle image velocimeters. I. Double pulsed systems. Meas. Sci. Technol. **1**(11), 1202 (1990)
14. Walters, D.K., Cokljat, D.: A three-equation eddy-viscosity model for Reynolds-averaged Navier–Stokes simulations of transitional flow. J. Fluids Eng. **130**(12), 121401 (2008)

Part IV
Mathematics and Computation in Finance, Economics, and Social Sciences

Financial Markets in the Context of the General Theory of Optional Processes

M.N. Abdelghani and A.V. Melnikov

Abstract A probability space is considered "unusual" if the information flow is not right or left continuous or is not complete. On these probability spaces lives certain type of stochastic processes known as optional processes including optional semimartingales. Optional semimartingales have right and left limits but are not necessarily right or left continuous. Here, we present a short summary of the calculus of optional processes and define stochastic logarithms and present some of its properties. Moreover, we develop a financial market model based on optional semimartingales and methods for finding local martingale deflators for this market. Finally, we present some financial examples.

1 Introduction

The assumption that the stochastic basis $\left(\Omega, \mathscr{F}, \mathbf{F} = (\mathscr{F}_t)_{t\geq 0}, \mathbf{P}\right)$ satisfy the usual conditions, where $\mathbf{F}$ is complete and right-continuous, is a foundation concept in the theory of stochastic processes. Stochastic processes that are adapted to this basis form a large class of processes known as semimartingales whose paths are right-continuous with left limits (RCLL). This theory has been instrumental in generating many important results in mathematical finance and in the theory of stochastic processes. It is difficult to conceive of a theory of stochastic processes without the usual conditions and RCLL processes.

However, it turns out that it is not difficult to give some examples of stochastic basis without the usual conditions. In 1975 Dellacherie [4] started to study the theory of stochastic processes without the assumption of the usual conditions, termed "unusual conditions".

Further developments of this theory were carried on by Lepingle [12], Horowitz [8], Lenglart [11], and mostly by Galtchouk [6, 7].

We believe that the theory optional processes will offer a natural foundation and a versatile set of tools for modeling financial markets. We can mention here few research problems that were not treated with the methods of the calculus of

M.N. Abdelghani (✉) • A.V. Melnikov
University of Alberta, Edmonton, AB, Canada
e-mail: mnabdelghani@gmail.com; melnikov@ualberta.ca

© Springer International Publishing Switzerland 2016

J. Bélair et al. (eds.), *Mathematical and Computational Approaches in Advancing Modern Science and Engineering*, DOI 10.1007/978-3-319-30379-6_47

optional processes but possibly should be. The first one, is a recent development in mathematical finance specially in pricing of derivative contracts and hedging under transaction costs (see [2] for details) that hints to the needed application of the calculus of optional processes to price derivatives and hedge under transaction costs. Furthermore, in models [1] with stochastic dividends paid at random times, there is an opportunity to treat these problems in the context of optional semimartingale theory in a natural way. Duffie [5] presented a new approach to modeling term structures of bonds and contingent claims that are subject to default risk. Perhaps Duffie's method could be studied with the methods of optional calculus. The aim of this paper is to present the theory of optional processes on unusual stochastic basis, develop new results and bring its methods to mathematical finance. The paper is organized as follows. Section 2 presents foundation material on optional processes. Section 3 introduces stochastic exponentials and logarithms. Section 4 describes optional semimartingale model of a financial market and two methods for finding local martingale deflators for these markets. Section 5 presents examples of optional semimartingale markets. Finally, we give some concluding remarks.

2 Foundation

Suppose we are given $\left(\Omega, \mathscr{F}, \mathbf{F} = (\mathscr{F}_t)_{t\geq 0}, \mathbf{P}\right)$, $t \in [0,\infty)$, where $\mathscr{F}_t \in \mathbf{F}$, $\mathscr{F}_s \subseteq \mathscr{F}_t$, $s \leq t$, a complete probability space. It is complete because $\mathscr{F}$ contains all $\mathbf{P}$ null sets. But this space is unusual because the family $\mathbf{F}$ is not assumed to be complete, right or left continuous. On this space, we introduce $\mathscr{O}(\mathbf{F})$ and $\mathscr{P}(\mathbf{F})$ the σ-algebras of optional and predictable processes, respectively (see [7]). A random process $X = (X_t)$, $t \in [0,\infty)$, is said to be *optional* if it is $\mathscr{O}(\mathbf{F})$-measurable. In general, optional processes have right and left limits but are not necessarily continuous. For an optional process we can define the following: (a) $X_- = (X_{t-})_{t\geq 0}$, a left continuous version of the process X and $X_+ = (X_{t+})_{t\geq 0}$, the right continuous version of X; (b) The jump processes $\triangle X = (\triangle X_t)_{t\geq 0}$ and $\triangle X_t = X_t - X_{t-}$ and (c) $\triangle^+ X = (\triangle^+ X_t)_{t\geq 0}$, $\triangle^+ X_t = X_{t+} - X_t$. A random process (X_t), $t \in [0,\infty)$, is predictable if $X \in \mathscr{P}(\mathbf{F})$ and *strongly predictable* if $X \in \mathscr{P}(\mathbf{F})$ and $X_+ \in \mathscr{O}(\mathbf{F})$. We denote by $\mathscr{P}_s(\mathbf{F})$ the set of strongly predictable processes. An optional semimartingale $X = (X_t)_{t\geq 0}$ is an optional process that can be decomposed to an optional local martingale $M \in \mathscr{M}_{loc}$ and an optional finite variation processes $A \in \mathscr{V}$, i.e. $X = X_0 + M + A$, [7]. A semimartingale X is called special if the above decomposition exists but with A being a strongly predictable process ($A \in \mathscr{A}_{loc}$ the set of locally integrable finite variation processes [7]). Let $\mathscr{S}$ denote the set of optional semimartingales and $\mathscr{S}p$ the set of special optional semimartingales. If $X \in \mathscr{S}p$ then the semimartingale decomposition is unique. Using the decomposition optional martingales and of finite variation processes we can decompose a semimartingale further to $X = X_0 + X^r + X^g = X_0 + (A^r + M^r) + (A^g + M^g)$ where A^r and A^g are right and left continuous finite variation processes, respectively. $M^r \in \mathscr{M}^r_{loc}$ is a right-continuous local martingale and $M^g \in \mathscr{M}^g_{loc}$ is a left-continuous local martingale.

The stochastic integral with respect to optional semimartingale X is defined as a bilinear form $(\varphi, \phi) \circ X_t$ where

$$Y_t = (\varphi, \phi) \circ X_t = Y_0 + \varphi \cdot X^r_t + \phi \odot X^g_t = \int_{0+}^{t} \varphi_{s-} dX^r_s + \int_0^{t-} \phi_s dX^g_{s+}.$$

where Y_t is again an optional semimartingale $\varphi_- \in \mathscr{P}(\mathbf{F})$, and $\phi \in \mathscr{O}(\mathbf{F})$, such that $\left(\varphi^2 \cdot [X^r, X^r]\right)^{1/2} \in \mathscr{A}_{loc}$ and $\left(\phi^2 \odot [X^g, X^g]\right)^{1/2} \in \mathscr{A}_{loc}$. The integral $\int_{0+}^t \varphi_{s-} dX^r_s$ is our usual stochastic integral with respect to RCLL semimartingale however the integral $\int_0^{t-} \phi_s dX^g_{s+}$ is Galchuk stochastic integral [7] with respect to left-continuous semimartingale.

3 Stochastic Exponentials and Logarithms

Stochastic exponentials and logarithms are indispensable tools of financial mathematics (see, Melnikov et al. [13]). They describe relative returns, link hedging with the calculation of minimal entropy and therefore utility indifference. Moreover, they determine the structure of the Girsanov transformation.

For optional semimartingales the stochastic exponential was defined by Galchuk [7]. If $X \in \mathscr{S}$ then there exists a unique semimartingale $Z \in \mathscr{S}$ such that

$$Z_t = Z_0 + Z \circ X_t = Z_0 + \int_{0+}^{t} Z_{s-} dX^r_s + \int_0^{t-} Z_s dX^g_{s+} = Z_0 \mathscr{E}(X)_t, \tag{1}$$

Two useful properties of stochastic exponentials are the inverse and product formulas. The inverse of stochastic exponential is $\mathscr{E}^{-1}(h) = \mathscr{E}(-h^*)$, such that,

$$h^*_t = h_t - \langle h^c, h^c \rangle_t - \sum_{0<s\leq t} \frac{(\triangle h_s)^2}{1 + \triangle h_s} - \sum_{0\leq s<t} \frac{(\triangle^+ h_s)^2}{1 + \triangle^+ h_s}.$$

And, the product of stochastic exponentials is $\mathscr{E}(U)\mathscr{E}(V) = \mathscr{E}(U + V + [U, V])$ where $[U, V]_t = \langle U^c, V^c \rangle_t + \sum_{0<s\leq t} \triangle U_s \triangle V_s + \sum_{0\leq s<t} \triangle^+ U_s \triangle^+ V_s$.

The stochastic logarithm is defined by the following theorem.

Theorem 1 *Let Y be a real valued optional semimartingale such that the processes Y_- and Y do not vanish then the process*

$$X_t = \frac{1}{Y} \circ Y_t = \int_{0+}^{t} \frac{1}{Y_{s-}} dY^r_s + \int_0^{t-} \frac{1}{Y_s} dY^g_{s+}, \quad X_0 = 0, \tag{2}$$

also denoted by $X = \mathscr{L}\text{og}Y$ *is called the stochastic logarithm of* Y*. The process* X *is a unique semimartingale such that* $Y = Y_0\mathscr{E}(X)$*. Moreover, if* $\Delta X \neq -1$ *and* $\Delta^+ X \neq -1$ *we also have*

$$\mathscr{L}\text{og}Y_t = \log\left|\frac{Y_t}{Y_0}\right| + \frac{1}{2Y^2}\circ\langle Y^c, Y^c\rangle_t - \sum_{0<s\leq t}\left(\log\left|1+\frac{\Delta Y_s}{Y_{s-}}\right| - \frac{\Delta Y_s}{Y_{s-}}\right) \tag{3}$$
$$- \sum_{0\leq s<t}\left(\log\left|1+\frac{\Delta^+ Y_s}{Y_s}\right| - \frac{\Delta^+ Y_s}{Y_s}\right).$$

It is important to note that the process Y *need not be positive for* $\mathscr{L}\text{og}(Y)$ *to exist, in accordance with the fact that the stochastic exponential* $\mathscr{E}(X)$ *may take negative values.*

Proof The assumptions that Y_- and Y don't vanish implies that: $S_n = \inf\left(t : |Y_{t-}| \leq \frac{1}{n}\right) \uparrow \infty$, hence $1/Y_-$ is locally bounded; likewise, $T_n = \inf\left(t : |Y_t| \leq \frac{1}{n}\right) \uparrow \infty$, hence $1/Y$ is also locally bounded. Therefore, the stochastic integral in (2) makes sense. Let $\tilde{Y} = Y/Y_0$ then $\tilde{Y}_0 = 1$. By equation (2) we have that $X = (1/\tilde{Y})\circ\tilde{Y}$. Therefore, $1+\tilde{Y}\circ X = 1+\tilde{Y}\circ\left(\frac{1}{\tilde{Y}}\circ\tilde{Y}\right) = 1+\left(\tilde{Y}\frac{1}{\tilde{Y}}\right)\circ\tilde{Y} = \tilde{Y}$, i.e. $\tilde{Y} = \mathscr{E}(X)$. Furthermore, $\Delta X = \Delta Y/Y_- \neq -1$ and $\Delta^+ X = \Delta^+ Y/Y \neq -1$. To obtain uniqueness let $\tilde{X}$ be any other semimartingale satisfying $Y = \mathscr{E}(\tilde{X})$. Since $\tilde{Y} = Y/Y_0$ then $\tilde{Y} = \mathscr{E}(\tilde{X})$. Therefore $\tilde{Y} = 1+\tilde{Y}\circ\tilde{X}$ and $Y = Y_0 + \tilde{Y}\circ\tilde{X}$. But since $X_0 = 0$ we have $\tilde{X} = \frac{\tilde{Y}}{\tilde{Y}}\circ\tilde{X} = \frac{1}{\tilde{Y}}\circ(Y - Y_0) = \frac{1}{Y}\circ Y = X$ and we obtain uniqueness.

To deduce equation (3) we apply Gal'chuk-Ito's lemma (see [7]) for the optional semimartingale $\log|Y|$. But the log function explodes at 0. To circumvent this problem consider for each n the C^2 functions $f_n(x) = \log|x|$ on $\mathbb{R}$ such that $|x| \geq 1/n$. Consequently for all n, $t < T_n$ and $t < S_n$ we get $\log|Y_t| = \log|Y_0| + \frac{1}{Y}\circ Y_t - \frac{1}{2Y^2}\circ\langle Y^c, Y^c\rangle_t + \sum_{0<s\leq t}\left(\Delta\log|Y_s| - \frac{\Delta Y_s}{Y_{s-}}\right) + \sum_{0\leq s<t}\left(\Delta^+\log|Y_s| - \frac{\Delta^+ Y_s}{Y_s}\right)$. This result together with equation (2) yields equation (3) for $t < T_n$ and $t < S_n$. Since, $T_n \uparrow \infty$ and $S_n \uparrow \infty$ we obtain equation (3) everywhere.

Now, we present some of the properties of stochastic logarithms.

Lemma 1 *(a) If* X *is a semimartingale satisfying* $\Delta X \neq -1$ *and* $\Delta^+ X \neq -1$ *then* $\mathscr{L}\text{og}(\mathscr{E}(X)) = X - X_0$*. (b) If* Y *is a semimartingale such that* Y *and* Y_- *do not vanish, then* $\mathscr{E}(\mathscr{L}\text{og}(Y)) = Y/Y_0$*. (c) For any two optional semimartingales* X *and* Z *we get the following identities:* $\mathscr{L}\text{og}\,(XZ) = \mathscr{L}\text{og}X + \mathscr{L}\text{og}Z + [\mathscr{L}\text{og}X, \mathscr{L}\text{og}Z]$*;* $\mathscr{L}\text{og}\left(\frac{1}{X}\right) = 1 - \mathscr{L}\text{og}\,(X) - \left[X, \frac{1}{X}\right]$*.*

Proof (a) $\mathscr{L}\text{og}(\mathscr{E}(X)) = \mathscr{E}(X)^{-1}\circ\mathscr{E}(X) = \frac{\mathscr{E}(X)}{\mathscr{E}(X)}\circ X = X - X_0$. (b) Let $Z = \mathscr{E}(\mathscr{L}\text{og}(Y))$ then by (a) we find that $\mathscr{L}\text{og}\,(Z) = \mathscr{L}\text{og}\,(\mathscr{E}(\mathscr{L}\text{og}(Y))) = \mathscr{L}\text{og}(Y) - \mathscr{L}\text{og}(Y_0) = \mathscr{L}\text{og}(Y/Y_0)$. Therefore, $Z = Y/Y_0$. (c) By the integral definition of the stochastic logarithm we find $\mathscr{L}\text{og}\,(XZ) = \frac{1}{XZ}\circ(X\circ Z + Z\circ X + [X,Z]) = \mathscr{L}\text{og}X + \mathscr{L}\text{og}Z + [\mathscr{L}\text{og}X, \mathscr{L}\text{og}Z]$. Using integration by parts and the integral definition the stochastic logarithm, $\mathscr{L}\text{og}\left(\frac{1}{X}\right) = X \circ \left(\frac{1}{X}\right) = 1 - \frac{1}{X} \circ X - \left[X, \frac{1}{X}\right] = 1 - \mathscr{L}\text{og}\,(X) - \left[X, \frac{1}{X}\right]$.

4 Markets of Optional Processes

Let $\left(\Omega, \mathscr{F}, \mathbf{F} = (\mathscr{F}_t)_{t\geq 0}, \mathbf{P}\right)$, $t \in [0, \infty)$, be the unusual stochastic basis and that the financial market stays on this space. The market consists of two types of securities x and X. A portfolio $\pi = (\eta, \xi)$, is composed of the optional processes η and ξ. η is the volume of the reference asset x while ξ is the volume of the security X. Suppose $x_t > 0$ and $X_t \geq 0$ for all $t \geq 0$ and write the ratio process $R_t = X_t/x_t$. Then, the value of the portfolio is

$$Y_t = \eta_t + \xi_t R_t. \tag{4}$$

We restrict the portfolio π to be self-financing that is we must have,

$$Y_t = Y_0 + \xi \circ R_t. \tag{5}$$

Reconciling equations (4) and (5) we obtain $C_t = \eta_t + R \circ \xi_t + [\xi, R]_t = Y_0 = C_0$ where C_t is the consumption process with its initial value C_0. Since the ratio process R is optional semimartingale then ξ, evolves in the space $\mathscr{P}(\mathbf{F}) \times \mathscr{O}(\mathbf{F})$ with the predictable part determining the volume of R^r and the optional part determining the volume of R^g. Also, η belongs to the space $\mathscr{O}(\mathbf{F})$. Furthermore, for the integral in equation (5) to be well defined ξ must be R-integrable, $\int_0^\infty \xi_s^2 d[R, R]_s \in \mathscr{A}_{loc}$.

One can interpret the trading strategy ξ in several different ways. First, eventhough ξ gives us the option to trade different parts of the ratio-process, R^r predictably and R^g optionally, we can still trade both parts in the same way, predictably. Alternatively, in a portfolio of a sum of left-continuous and right-continuous assets (see example Sect. 5) we can trade each asset independently. The left-continuous asset will be traded with an optional strategy while the right-continuous asset will be traded by a predictable one. This certainly a possible portfolio in the current structure of financial markets. Moreover, with a theory of optimal portfolio with consumption and endowment streams we can interpret the predictable trades of R^r as the optimal portfolio while the optional trades of R^g as an optimal consumption or endowment streams. Yet another way to think about optional portfolios in financial markets is in terms of market-orders. Market-orders can either be predictable or optional. Optional in the sense that trades can be executed based on a condition, for example the stock price passing some boundary. Conditional trades are optional processes that act on the left-continuous part of the ratio-process R^g.

Finally, note that the ratio process is an optional semimartingale and must be transformed to a local optional martingale for any pricing and hedging theory to be viable. In the next section we show how to find local martingale transforms that change the ratio process to some optional local martingales. A special subset of these martingale deflators can transform RLL optional semimartingales to RCLL local martingale. This special subset of transforms can be useful in cases where markets only allow for predictable trading strategies. We will illustrate this procedure in example Sect. 5.

As in Melnikov et al. [13] for RCLL We suppose that the dynamics of securities in our market follows the stochastic exponential, $X_t = X_0\mathscr{E}_t(H)$ and $x_t = x_0\mathscr{E}_t(h)$ where x_0 and X_0 are $\mathscr{F}_0$-measurable random variables. $h = (h_t)_{t\geq 0}$ and $H = (H_t)_{t\leq 0}$ are optional semimartingales admitting the representations, $h_t = h_0 + a_t + m_t$ and $H_t = H_0 + A_t + M_t$ with respect to (w.r.t) $\mathbf{P}$. $a = (a_t)_{t\geq 0}$ and $A = (A_t)_{t\geq 0}$ are locally bounded variation processes and predictable. $m = (m_t)_{t\geq 0}$ and $M = (M_t)_{t\geq 0}$ are optional local martingales. A local martingale deflator is a strictly positive supermartingale multiplier used in mathematical finance to transform the value process of a portfolio to a supermartingale (i.e. a local martingale). Here we will develop methods for finding local martingale deflators. We can write R as,

$$R_t = \frac{X_t}{x_t} = R_0\mathscr{E}(H)_t\mathscr{E}^{-1}(h)_t = R_0\mathscr{E}(H_t)\mathscr{E}(-h_t^*) = R_0\mathscr{E}(\Psi(h_t, H_t)), \tag{6}$$

$$\Psi_t = \Psi(h_t, H_t) = H_t - h_t^* - [H, h^*]_t = H_t - h_t + \langle h^c, h^c - H^c\rangle_t + J^d + J^g,$$

$$J^d = \sum_{0<s\leq t} \frac{\Delta h_s(\Delta h_s - \Delta H_s)}{1+\Delta h_s}, \quad J^g = \sum_{0\leq s<t} \frac{\Delta^+ h_s\left(\Delta^+ h_s - \Delta^+ H_s\right)}{1+\Delta^+ h_s}.$$

if Ψ is a local optional martingale then R is a local optional martingale and we are done. Otherwise, we have to find a strictly positive transformation $Z \in \mathscr{M}_{loc}$ that will render $ZR \in \mathscr{M}_{loc}$. Z is known as the local martingale deflator. For a strictly positive Z, we can define $N \in \mathscr{M}_{loc}$ with $N = \mathscr{L}\mathrm{og}(Z) = Z^{-1} \circ Z$ or $Z = \mathscr{E}(N)$. To find N we have the following theorem;

Theorem 2 *Given $R = R_0\mathscr{E}(\Psi(h, H))$ where $\Psi(h, H)$ as in equation (6) and $Z = \mathscr{E}(N)$ where Z, $N \in \mathscr{M}_{loc}(\mathbf{P}, \mathbf{F})$ and $Z > 0$ then $ZR \in \mathscr{M}_{loc}$ is a local optional martingale if and only if $(A - a) + \langle m^c - N^c, m^c - M^c\rangle + \tilde{K}^d + \tilde{K}^g = 0$, where $\tilde{K}^d$ and $\tilde{K}^g$ are the compensators of the processes*

$$K^d = \sum_{0<s\leq t} \frac{(\Delta h_s - \Delta N_s)(\Delta h_s - \Delta H_s)}{1+\Delta h_s}, K^g = \sum_{0\leq s<t} \frac{\left(\Delta^+ h_s - \Delta^+ N_s\right)\left(\Delta^+ h_s - \Delta^+ H_s\right)}{1+\Delta^+ h_s}.$$

Proof Suppose $Z_t = \mathscr{E}(N)_t \in \mathscr{M}_{loc}$, $Z_t > 0$ for all t such that $ZR \in \mathscr{M}_{loc}$ then$ZR = R_0\mathscr{E}(N)\mathscr{E}(\Psi(h, H)) = R_0\mathscr{E}(\Psi(h, H, N))$, where $\Psi(h, H, N) = N_t + H_t - h_t + \langle h^c, h^c - H^c\rangle_t + J_t^d + J_t^g + [N, H] - [N, h] + [N, J^d] + [N, J^g]$, hence,

$$\begin{aligned}\Psi(h, H, N) = {} & N_t + H_t - h_t + \langle h^c, h^c - H^c\rangle_t + J_t^d + J_t^g \\ & + \langle N^c, H^c\rangle_t + \sum_{0<s\leq t} \Delta N_s \Delta H_s + \sum_{0\leq s<t} \Delta^+ N_s \Delta^+ H_s\end{aligned}$$

$$
\begin{aligned}
&- \langle N^c, h^c \rangle_t - \sum_{0<s\leq t} \Delta N_s \Delta h_s - \sum_{0\leq s<t} \Delta^+ N_s \Delta^+ h_s \\
&+ \sum_{0<s\leq t} \Delta N_s \frac{\Delta h_s (\Delta h_s - \Delta H_s)}{1+\Delta h_s} \\
&+ \sum_{0\leq s<t} \Delta^+ N_s \frac{\Delta^+ h_s \left(\Delta^+ h_s - \Delta^+ H_s\right)}{1+\Delta^+ h_s}.
\end{aligned}
$$

therefore, $\Psi(h, H, N) = N_t + H_t - h_t + \langle h^c - N^c, h^c - H^c \rangle_t + K^d + K^g$. So, if $\Psi(h, H, N) \in \mathscr{M}_{loc}$ then $ZR \in \mathscr{M}_{loc}$. And if $\Delta^+\Psi(h, H, N) \neq -1$ and $\Delta\Psi(h, H, N) \neq -1$ then $\Psi(h, H, N) \in \mathscr{M}_{loc} \Leftrightarrow ZR \in \mathscr{M}_{loc}$. Now, let us take into consideration the decomposition of H and h and write $\Psi(h, H, N) = (A - a) + (M - m + N) + \langle (m - N)^c, (m - M)^c \rangle + K^d + K^g$. So, $\Psi(h, H, N)$ is a local optional martingale under $\mathbf{P}$ if $(A - a) + \langle m^c - N^c, m^c - M^c \rangle + \tilde{K}^d + \tilde{K}^g = 0$ where $\tilde{K}^d$ and $\tilde{K}^g$ are the compensators of K^d and K^g, respectively.

By finding all $N \in \mathscr{M}_{loc}$ such that the above equation (4) is valid and $\mathscr{E}(N) > 0$ we find the set of all appropriate local optional martingale transforms Z such that ZR is a local optional martingale. Note that if Z is a local martingales transform such that ZR is a local martingale then it is true for all self financing strategies π.

Theorem 3 *If Z is a local martingale transform of R, that is ZR is a local optional martingale, and π is a self financing portfolio which is R-integrable then ZY_t^π is a local optional martingale.*

Proof Z is a local martingale transform of R therefore $Z > 0$. $\pi = (\eta, \xi)$ is self financing and R-integrable, then $Y_t^\pi = Y_0 + \xi \circ R_t$ and $Z_t Y_t^\pi$ can be written as, $d\left(Z_t Y_t^\pi\right) = \xi_t \left[Z_t dR_t + dZ_t R_t + d\left[Z_t, R_t\right]\right] + dZ_t \eta_t = \xi_t d\left(Z_t R_t\right) + dZ_t \eta_t$. This leads us to the following result $Z_t Y_t^\pi = \xi \circ Z_t R_t + \eta \circ Z_t$. $\eta \circ Z_t$ and $\xi \circ Z_t R_t$ are local optional martingales therefore their sum $Z_t Y_t^\pi$ is a well defined local optional martingale. Note, that we have implicitly used the fact that η is bounded, i.e. comes from the fact that π is a self financing and also that, $\int_0^\infty \xi_t^2 d\left[ZR\right]_t \in \mathscr{A}_{loc}$.

On the other hand, if we know that there exist a Z such that ZY^π is a local optional martingale then what can we say about the portfolio π and the product ZR? It is reasonable to suppose that $Z = \mathscr{E}(N) > 0$, π-self-financing, ξ is R-integrable and η is bounded. In this case, $\xi \circ Z_t R_t = Z_t Y_t^\pi - \eta \circ Z_t$ is a sum of two local optional martingales and therefore a local optional martingale it self, for any optional process ξ, in particular for $\xi = 1$; therefore ZR is a local optional martingale.

An alternative approach can be developed using stochastic logarithms. What is interesting about this approach is that we don't have to define the process R as a stochastic exponential of an underlying process Ψ. All that is required is that the ratio process R and its predictable version R_- don't vanish, except on sets of measure zero. This approach is technically based on the following lemmas and results.

Lemma 2 *Suppose* $X = X_0 + A + M$, $x = x_0 + a + m$, $R = X/x$ *and* $R_- \neq 0$ *and* $R \neq 0$ *a.s.* **P**, *then* $\mathscr{L}\text{og}(R)$ *is a local optional martingale if and only if* $\frac{1}{X} \circ A - \frac{1}{x} \circ a + \frac{1}{x^2} [m, m] - \frac{1}{xX} \circ [M, m] = -1$.

Lemma 3 *Suppose that* R_- *and* R *don't vanish then* $\mathscr{L}\text{og}(R) \in \mathscr{M}_{loc} \Leftrightarrow R \in \mathscr{M}_{loc}$.

If R is not a local martingale then there exists a local martingale $Z > 0$ such that ZR is a local martingale. The following lemma helps us with finding Z.

Lemma 4 *Let* $X = X_0 + A + M$, $x = x_0 + a + m$, $R = X/x$ *and suppose that* R_- *and* R *don't vanish then* $\mathscr{L}\text{og}(ZR)$ *is a local optional martingale if and only if* $1+\frac{1}{X}\circ A-\frac{1}{x}\circ a+\frac{1}{x^2}\circ[m,m]-\frac{1}{xX}\circ[M,m]+\frac{1}{ZX}\circ[Z,M]-\frac{1}{xZ}\circ[Z,m] = -1$ *furthermore if* $Z = \mathscr{E}(N) > 0$ *then* $\frac{1}{X}\circ A-\frac{1}{x}\circ a+\frac{1}{x^2}\circ[m,m]-\frac{1}{xX}\circ[M,m]+\frac{1}{X}\circ[N,M]-\frac{1}{x}\circ[N,m] = -1$.

Corollary 1 *Suppose that* R_- *and* R *don't vanish then* $\mathscr{L}\text{og}(ZR) \in \mathscr{M}_{loc} \Leftrightarrow ZR \in \mathscr{M}_{loc}$.

5 Illustrative Examples

Consider a market composed of a bond x and an asset X evolving according to $x_t = x_0\mathscr{E}(h)_t$ and $X_t = X_0\mathscr{E}(H)_t$ where $h_t = rt + bL_t^g$, $h_0 = 0$, $H_t = \mu t + \sigma W_t + aL_t^d$, $H_0 = 0$. $L_t^d = L_t - \lambda t$, $L_t^g = -\bar{L}_{t-} + \gamma t$, and r, μ, σ, a, and b are constants. W is diffusion term and L and $\bar{L}$ are Poisson with constant intensity λ and γ respectively. Let $\mathscr{F}_t$ be the natural filtration that is neither right or left continuous. Here the bond is modeled by a left continuous process for which we have assumed that its jumps don't necessarily avoid the jumps of the asset (see also [5] for bonds that can experience defaults). We believe our model gives a better description of a portfolio of stocks and bonds than models that assume RCLL processes on usual probability space.

Given x and X the ratio process is $R_t = \frac{X_0}{x_0}\mathscr{E}(H_t - h_t^* - [H, h^*]_t)$. We want to find $Z = \mathscr{E}(N)$ such that ZR is a local martingale. In Sect. 4 we showed that associated with the product $ZR = \frac{X_0}{x_0}\mathscr{E}(\Psi(h, H, N))$ is the process $\Psi(h, H, N)$. To compute a reasonable form for $\Psi(h, H, N)$ we suppose that $N_t = \varsigma W_t + cL_t^d + \theta L_t^g$ an optional local martingale for which Z an optional local martingale deflator, hence

$$\begin{aligned}
\Psi(h, H, N) &= \left[\varsigma W_t + cL_t^d + \theta L_t^g\right] + \left[\mu t + \sigma W_t + aL_t^d\right] - \left[rt + bL_t^g\right] \\
&\quad + \left\langle \left[rt + bL_t^g\right]^c - \left[\varsigma W_t + cL_t^d + \theta L_t^g\right]^c, \left[rt + bL_t^g\right]^c - \left[\mu t + \sigma W_t + aL_t^d\right]^c \right\rangle \\
&\quad + \sum_{0<s\le t} ac \left(\triangle L_s^d\right)^2 + \sum_{0\le s<t} \frac{b\,(b-\theta)\left(\triangle^+ L_s^g\right)^2}{1 + b\triangle^+ L_s^g} \\
&= (\mu - r + \varsigma\sigma)\, t + (\varsigma + \sigma)\, W_t + (c + a)\, L_t^d + (\theta - b)\, L_t^g + acL_t
\end{aligned}$$

$$+ (b-\theta) L_t^g + b(\theta-b)\left[L_t^g, L_t^g\right]$$
$$= (\mu - r + \varsigma\sigma) t + (\varsigma + \sigma) W_t + (c+a) L_t^d + acL_t + b(\theta - b) \bar{L}_{t-}.$$

because W_t and L_t^d are martingales, $\Psi(h, H, N)$ is a martingale if and only if $\mu - r + \varsigma\sigma + ca\lambda + \theta b - b^2\gamma = 0$. The solution of this equation leads to infinitely many solutions which means the market is incomplete. Here are some solutions. Let $\theta = 0$ which leads to right continuous local martingale deflator. Another solution is a one which will eliminate the effects of jumps on drift that is by letting $\theta = -1/b$ and $c = 1/a\lambda$, then $\varsigma = \left(r - \mu + b^2\gamma\right)/\sigma$.

This is a simple but important example that we have alluded to in Sect. 4. Suppose that the market participants can't trade the left-continuous part of the market ratio process R with an optional trading strategy. A way around this problem is to transform RLL semimartingles to RCLL local optional martingales. Consider a market ratio process $R = \mathscr{E}(M) > 0$, $M_t = rt + aW_t + bD_t + cG_t$ where W is a diffusion process, D a right-continuous compensated Poisson with intensity λ and G a left continuous compensated Poisson with intensity γ. Let $N = \alpha W + \beta D + \kappa G$. We like to find $Z = \mathscr{E}(N) > 0$ such that ZR is RCLL. $ZR = \mathscr{E}(N)\mathscr{E}(M) = \mathscr{E}(N + M + [N, M])$. Therefore, $ZR = \mathscr{E}((a+\alpha)W + (b+\beta)D + (c+\kappa)G + (r + a\alpha + b\lambda\beta + c\gamma\kappa)t)$. If we choose $\kappa = -c$ we get rid of the left-continuous part of ZR. And, if we choose α and β such that $r + a\alpha + b\lambda\beta - \gamma c^2 = 0$ we obtain the RCLL local optional martingale ZR we are searching for. However, we still need to make sure that our choice $Z = \mathscr{E}(\alpha W + \beta D - cG) > 0$. This is true, if and only if $(1 + \beta \triangle D_t)(1 - c \triangle^+ G_t) > 0$ or $(1+\beta)(1-c) > 0$ for all time t.

6 Conclusion

Optional processes including left continuous ones are a natural occurrence in financial markets. For example, a defaultable bond is modeled by a left continuous process in [5], stochastic dividends in [16] and transaction costs in [2, 3]. Also optional processes appears for optimal consumption from investment [15], for consumption from investment with random endowment [9, 14, 17]. Furthermore, optional processes naturally arise in the context of super-replication in incomplete markets as a result of optional decomposition theorem [10]. Therefore, to study financial markets with optional processes, left-continuous processes and a mixture of right and left continuous processes the calculus of optional processes is going to be an indispensable tool in the future study of mathematical finance. We have introduced the notions of the unusual basis and optional semimartingales, presented a summary of the calculus of optional processes and introduced new results of stochastic logarithms. We have also described the optional semimartingale model of financial market and described a procedure of finding local martingale deflators for this market. Finally we presented illustrative examples. The first example is of a portfolio of a left continuous bond and right continuous stock where we showed how to find local martingale deflators for this market. The second examples demostrates a method for finding a subset of local martingale deflators that transforms a ladlag semimartingale to cadlag local optional martingale.

Acknowledgements The research is supported by the NSERC discovery grant #5901.

References

1. Albrecher, H., Bauerle, N., Thonhauser, S.: Optimal dividend-payout in random discrete time. Stat. Risk Model. Appl. Financ. Insur. **28**(3), 251–276 (2011)
2. Czichowsky, C., Schachermayer, W.: Duality theory for portfolio optimisation under transaction costs. arXiv:1408.5989 [q-fin.MF] (2014)
3. Davis, M.H., Panas, V.G., Zariphopoulou, T.: European option pricing with transaction costs. SIAM J. Control Optim. **31**(2), 470–493 (1993)
4. Dellacherie, C.: Deux remarques sur la separabilite optionelle. Sem. Probabilites XI Univ. Strasbourg, Lecture Notes in Mathematics, vol. 581, pp. 47–50. Springer, Berlin (1977)
5. Duffie, D., Singleton, K.J.: Modeling term structures of defaultable bonds. Rev. Financ. stud. **12**(4), 687–720 (1999)
6. Gal'chuk, L.I.: Optional martingales. Matem. Sb. **4**(8), 483–521 (1980)
7. Gal'chuk, L.I.: Stochastic integrals with respect to optional semimartingales and random measures. Theory Probab. Appl. **XXIX**(1), 93–108 (1985)
8. Horowitz, J.: Optional supermartingales and the Andersen-Jessen theorem. Z. Wahrscheinlichkeitstheorie und Verw Gebiete. **43**(3), 263–272 (1978)
9. Karatzas, I., Zitkovic, G.: Optimal consumption from investment and random endowment in incomplete semimartingale markets. Ann. Probab. **31**(4), 1821–1858 (2003)
10. Kramkov, D.O.: Optional decomposition of supermartingales and hedging contingent claims in incomplete security markets. Probab. Theory Relat. Fields **105**, 459–479 (1994)
11. Lenglart, E.: Tribus de Meyer et théorie des processus. Lecture Notes in Mathematics, vol. 784, pp. 500–546. Springer, Berlin/New York (1980)
12. Lepingle, D.: Sur la représentation des sa uts des martingales. Lectures Notes on Mathematics, vol. 581, pp. 418–434. Springer, Berlin/New York (1977)
13. Melnikov, A.V., Volkov, S.N., Nechaev M.L.: Mathematics of Financial Obligations. Translations of Mathematical Monographs, vol. 212. American Mathematical Society, Providence (2002)
14. Mostovyi, O.: Optimal investment with intermediate consumption and random endowment. Math. Financ. (2014). Published on-line
15. Mostovyi, O.: Necessary and sufficient conditions in the problem of optimal investment with intermediate consumption. Financ. Stoch. **19**(1), 135–159 (2015)
16. Wenshen, X., Ling, Z., Zhin, W.: An option pricing problem with the underlying stock paying dividend. Appl. Math. JCU **12**(B), 447–454 (1997)
17. Zitkovic, G.: Utility maximization with a stochastic clock and an unbounded random endowment. Ann. Appl. Probab. **15**(1B), 748–777 (2005)

A Sufficient Condition for Continuous-Time Finite Skip-Free Markov Chains to Have Real Eigenvalues

Michael C.H. Choi and Pierre Patie

Abstract We provide a sufficient condition for the negative of the infinitesimal generator of a continuous-time finite skip-free Markov chain to have only real and non-negative eigenvalues. The condition includes stochastic monotonicity and certain requirements on the transition rates of the chain. We also give a sample path illustration of Markov chains that satisfy the conditions and its Siegmund dual. We illustrate our result by detailing an example which also reveals that our conditions are not necessary.

1 Introduction

A Markov chain on a denumerable state space is said to be *upward skip-free* if the only upward transition is of unit size, yet it can have downward jump of any arbitrary magnitude. In this paper, we consider a continuous time upward skip-free Markov chain $X = (X_t)_{t \geq 0}$ on a finite state space $E := \{0, 1, \ldots, a\}$ with the infinitesimal generator $G := (G(i,j))_{i,j \in E}$ and transition semigroup $P = (P_t)_{t \geq 0}$, where

$$P_t' = P_t G, \quad t \geq 0.$$

Since X is upward skip-free, $G(i,j) = 0$ for $j > i + 1, i, j \in E$. The generator G necessarily satisfies $0 \leq G(i,j) < \infty$ for $i \neq j$ and for all $j \in E$,

$$-G(j,j) \geq \sum_{k \neq j} G(j,k). \tag{1}$$

In what follows, we assume that X is conservative (i.e. the equality in (1) holds) and $G(j,j) < \infty$ for $j \in E$. We refer the interested readers to Asmussen [1], Norris [8] for a general reference in continuous-time Markov chains.

M.C.H. Choi (✉) • P. Patie
School of Operations Research and Information Engineering, Cornell University, Ithaca, NY, USA
e-mail: cc2373@cornell.edu; pp396@cornell.edu

© Springer International Publishing Switzerland 2016
J. Bélair et al. (eds.), *Mathematical and Computational Approaches in Advancing Modern Science and Engineering*, DOI 10.1007/978-3-319-30379-6_48

The spectrum of a skip-free Markov chain plays an important role in the study of upward hitting time as well as the fastest strong stationary time, see e.g. Diaconis and Fill [3] and Fill [4]. It is shown in Fill [4] that if the eigenvalues of $-G$ are all real and non-negative, then the law of the upward hitting time can be interpreted as a convolution of exponential random variables with parameters given by the aforementioned eigenvalues. However, it is not known under which condition(s) do a skip-free Markov chain admit real and non-negative spectrum. We aim to fill in this gap in the literature by providing a sufficient condition. The main idea is to develop a similarity transformation of G with its so-called Siegmund dual $\hat{G}$ which is the generator of a birth-and-death process.

The rest of the paper is organized as follow. We review Siegmund duality and the concept of stochastic monotonicity in Sect. 2. They serve as important tools for our main result, which is presented in Sect. 3. Finally, we provide a sample path illustration in Sect. 4.

2 Siegmund Duality and Stochastic Monotonicity

Siegmund duality of Markov processes was first introduced by Siegmund [9], which was further developed and applied by Diaconis and Fill [3], Huillet and Martinez [5], Jansen and Kurt [6].

Definition 1 (Siegmund kernel) The Siegmund kernel $H_S = (H_S(i,j))_{i,j\in E}$ is defined to be

$$H_S(i,j) := \mathbb{1}_{\{i\leq j\}}\,,$$

where $\mathbb{1}$ is the indicator function. Its inverse H_S^{-1} is

$$H_S^{-1}(i,j) = \mathbb{1}_{\{i=j\}} - \mathbb{1}_{\{i=j-1\}}\,.$$

Definition 2 (Siegmund dual) We say that $\hat{G}$ is the Siegmund dual (or H_S-dual) of G if $\hat{G}$ is an infinitesimal generator such that

$$\hat{G}(i,j) := (H_S^{-1}GH_S)^T(i,j) \geq 0\,, \quad i \neq j\,.$$

That is, apart from the diagonal entries $(\hat{G}(j,j))_{j\in E}$, $(\hat{G}(i,j))_{i\neq j}$ is elementwise non-negative.

Next, we recall the definition of stochastic monotonicity for Markov chains.

Definition 3 (Stochastic monotone) G is said to be stochastically monotone if

$$\sum_{k\geq j} G(h,k) \leq \sum_{k\geq j} G(i,k)\,, \quad 0 \leq h \leq i < j\,, \tag{2}$$

$$\sum_{k \le l} G(h,k) \ge \sum_{k \le l} G(i,k)\,, \quad 0 \le l < h \le i\,. \tag{3}$$

It is known that the existence of Siegmund dual is equivalent to stochastic monotonicity, which is shown in the following lemma. The proof of a more general case can be found in Asmussen [1, Proposition 4.1].

Lemma 1 *Suppose that $(X_t)_{t\ge 0}$ is a continuous-time Markov chain on E with infinitesimal generator G. The Siegmund dual $\hat{G}$ exists if and only if G is stochastically monotone.*

Proof For any $i, j \in E$, since $H_S\hat{G}^T(i,j) = GH_S(i,j)$, $H_S\hat{G}^T(i,j) = \sum_{k\ge i} \hat{G}(j,k)$ and $GH_S(i,j) = \sum_{k\le j} G(i,k)$, we have

$$\begin{aligned}
\hat{G}(j,i) &= \sum_{k\ge i} \hat{G}(j,k) - \sum_{k>i} \hat{G}(j,k) \\
&= \sum_{k\le j} (G(i,k) - G(i+1,k)) \qquad (4)\\
&= \sum_{k>j} (G(i+1,k) - G(i,k)) \quad \text{(since G is conservative)}\,. \qquad (5)
\end{aligned}$$

In view of (3) and (4), and (2) and (5), $\hat{G}(i,j) \ge 0$ for $i \ne j$ if and only if G is stochastically monotone. □

Finally, we recall a lemma shown in Huillet and Martinez [5] that will be used in the proof of our main result Proposition 1.

Lemma 2 *Suppose that $(X_t)_{t\ge 0}$ is a continuous-time Markov chain on E with infinitesimal generator G. If G is irreducible on E (resp. on $E\backslash\{0\}$, where 0 is an absorbing state) and is stochastically monotone, then*

1. *$\hat{G}$ is conservative except at 0 (resp. on E). That is, $-\hat{G}(0,0) > \sum_{k\ne 0} \hat{G}(0,k)$, and (1) holds (with G replaced by $\hat{G}$) for $j \ne 0$.*
2. *a is the unique absorbing state in $\hat{G}$.*

3 Main Result

As X is upward skip-free, the infinitesimal generator G is non-symmetric and non-reversible (with respect to the stationary distribution of the chain), making the analysis of the spectrum of G difficult. One way to overcome this is to develop conditions such that its Siegmund dual $\hat{G}$ corresponds to a birth-and-death process, which leads us to the following result.

Proposition 1 *Suppose that $(X_t)_{t\geq 0}$ is an upward skip-free Markov chain on $E = \{0, 1, \ldots, a\}$ with infinitesimal generator G (and $a \geq 3$), and assume that $(X_t)_{t\geq 0}$ is either irreducible on E or irreducible on $E\backslash\{0\}$ with 0 being an absorbing boundary. If*

1. *G is stochastically monotone, and for every $i \in [0, a-2]$,*

$$\sum_{k\leq i} G(i+1,k) > \sum_{k\leq i} G(i+2,k)\,, \quad \textit{and}$$

2. *for every $i \in [0, a-3]$,*

$$\sum_{k\leq i} G(j,k) = \sum_{k\leq i} G(j+1,k) \quad \forall j \in [i+2, a-1]\,,$$

then all the eigenvalues of $-G$ are real, distinct and non-negative.

Proof Since G is stochastically monotone, its Siegmund dual $\hat{G}$ exists by Lemma 1. For $i \in [2, a-1]$, $\hat{G}(i, i-1) = G(i, i+1) > 0$ since G is upward skip-free, conservative and irreducible on $E\backslash\{0\}$, and $\hat{G}(i,j) = 0 \quad \forall j \leq i-2$. We also have $\hat{G}(1,0) = G(1,2) > 0$. For $i \in [0, a-2]$, condition 1 gives $\hat{G}(i, i+1) = \sum_{k\leq i} G(i+1,k) - \sum_{k\leq i} G(i+2,k) > 0$. Condition 2 guarantees that $\hat{G}(i,j) = 0$ for each $i \in [0, a-3]$ and $j \in [i+2, a-1]$. That is, $\hat{G}$ is an irreducible birth-and-death process when restricted to the state space $\{0, \ldots, a-1\}$. Moreover, by Lemma 2, a is the unique absorbing state of $\hat{G}$, and as a result the row corresponding to state a is zero. Denote by $\hat{G}^{BD}$ the restriction of $\hat{G}$ to $\{0, \ldots, a-1\}$. By breaking off the last row and last column of $\hat{G}$, we can write

$$\hat{G} = \begin{pmatrix} \hat{G}^{BD} & \mathbf{h} \\ \mathbf{0}^T & 0 \end{pmatrix} = (H_S^{-1} G H_S)^T\,, \tag{6}$$

where $\mathbf{0}$ is a column vector of zero, and $\mathbf{h}$ is a column vector storing $\hat{G}(i, a)$ for $i \in [0, a-1]$. Since $\hat{G}$ is a similarity transformation of G, they share the same eigenvalues. In addition, we observe from (6) that the eigenvalues of G are 0 and that of $\hat{G}^{BD}$. Now, for $i \in [0, a-1]$, let

$$\pi^{BD}(0) = c > 0\,,$$

$$\pi^{BD}(i) = \pi^{BD}(0) \frac{\hat{G}^{BD}(0,1)\ldots\hat{G}^{BD}(i-1,i)}{\hat{G}^{BD}(i,i-1)\ldots\hat{G}^{BD}(1,0)}\,,$$

then $\pi^{BD}(i)\hat{G}^{BD}(i,j) = \pi^{BD}(j)\hat{G}^{BD}(j,i)$ on $E\backslash\{a\}$, where both sides are non-zero only when $|i-j| \leq 1$. Define on $E\backslash\{a\}$

$$A(i,j) := \pi(i)^{1/2}\pi(j)^{-1/2}\hat{G}^{BD}(i,j)\,.$$

If D_π is a diagonal matrix with $D_{\pi^{BD}}(i,i) = \pi^{BD}(i)$, then $A = D_{\pi^{BD}}^{1/2}\hat{G}^{BD}D_{\pi^{BD}}^{-1/2}$. Since $\hat{G}^{BD}$ is reversible, A is symmetric, and the spectral theorem for symmetric matrices tells us that A has real eigenvalues. As a result, $\hat{G}^{BD}$ has real eigenvalues, and hence G has real eigenvalues.

We will now show that G has distinct and non-positive eigenvalues. Since $\hat{G}$ is the infinitesimal generator of a finite Markov chain, $\hat{G}$ is uniformizable. Let $b := \max_{i\in E} -\hat{G}(i,i) < \infty$. Then $\hat{P} := I + \frac{1}{b}\hat{G}$ is a stochastic matrix of a discrete time upward skip-free Markov chain with a being the unique absorbing state. By the result of Kent and Longford [7, Sec. 4 Lemma 1], $\hat{P}$ has distinct eigenvalues, which implies $\hat{G}$ has distinct eigenvalues as well. If G has a positive eigenvalue λ, then $1 + \frac{\lambda}{b} > 1$ is an eigenvalue for $\hat{P}$, which is not possible since a stochastic matrix has eigenvalues less than or equal to 1. Therefore, all the eigenvalues of G are non-positive. □

Condition 1 and 2 are by no means necessary for G to admit a real spectrum. This can be illustrated with the following example:

Example 1 (Condition 1 *and* 2 *are not necessary)*

$$G = \begin{pmatrix} -0.3 & 0.3 & 0 & 0 \\ 0.4 & -0.7 & 0.3 & 0 \\ 0.5 & 0.3 & -0.85 & 0.05 \\ 0.1 & 0.2 & 0.4 & -0.7 \end{pmatrix}$$

has eigenvalues $0, -0.63, -0.89, -1.04$, and does not satisfy condition 1 (since $\hat{G}(0,1) = G(1,0) - G(2,0) = -0.1 < 0$) and 2 (since $\hat{G}(0,2) = G(2,0) - G(3,0) = 0.4 > 0$). Therefore, condition 1 and 2 are not necessary for real, distinct and non-positive eigenvalues. □

4 Pathwise Illustration of the Main Result

We recall the notion of strong pathwise duality for Markov processes. Relying on the fact that Siegmund duality is a strong pathwise duality, we demonstrate Proposition 1 from a sample path perspective in Fig. 2.

Definition 4 (strong pathwise duality) Let $(X_t)_{t\geq 0}$ and $(Y_t)_{t\geq 0}$ be two continuous-time Markov chains on finite state spaces E and F respectively, and $H : E \times F \to \mathbb{R}$ be a measurable and bounded function. Suppose that for every $T > 0$ there are families of processes $\{(X_s^x)_{s\in[0,T]}\}_{x\in E}$ and $\{(Y_s^y)_{s\in[0,T]}\}_{y\in F}$, defined on a *common* probability space $(\Omega, \mathscr{F}, \mathbb{P})$, such that the following holds:

1. For all $x \in E$ and $y \in F$, the finite dimensional distributions of $(X_s^x)_{s\in[0,T]}$ under $\mathbb{P}$ (resp. $(Y_s^y)_{s\in[0,T]}$ under $\mathbb{P}$) agree with that of $(X_t)_{t\in[0,T]}$ under $\mathbb{P}_x$ (resp. $(Y_t)_{t\in[0,T]}$ under $\mathbb{P}^y$).

2. For all $s \in [0, T]$ and all $x \in E$, $y \in F$,

$$H(x, Y_T^y) = H(X_s^x, Y_{T-s}^y) = H(X_T^x, y)\,, \quad \mathbb{P} - a.s..$$

Then $(X_t)_{t \geq 0}$ and $(Y_t)_{t \geq 0}$ are said to be *strongly pathwise dual* with respect to H.

Theorem 1 (Siegmund duality is strong pathwise duality (Clifford and Sudbury [2])) *Let (X_t) and (Y_t) be two continuous-time Markov chains on a finite state space E, which are dual with respect to the Siegmund kernel $H_S(i,j) = \mathbb{1}_{\{i \leq j\}}$. Then they are strongly pathwise dual.*

We now discuss a procedure to simulate (X_t^x) with generator G and its Siegmund dual (Y_t^y) with generator $\hat{G}$ over the time frame $[0, T]$. We summarize the procedure in Clifford and Sudbury [2] as follows: since the state space E is finite, both (X_t^x) and (Y_t^y) are uniformizable. Suppose that $X_0^x = x$. Let us write $c = \max\{\max_i -G(i,i), \max_i -\hat{G}(i,i)\} > 0$. Then the inter-arrival time (or holding time) $(t_i)_{i=1}^n$ with $t_0 = 0$ of the transition of (X_t^x) is independent and exponentially distributed with mean c^{-1}, and the transition follows that of the discrete time embedded Markov chain with transition matrix $P = I + \frac{G}{c}$, which is driven by n i.i.d. $(U_i)_{i=1}^n$ uniformly distributed numbers on $(0, 1)$.

To simulate (Y_t^y), we reverse in both jump directions and jump time of that of (X_t^x). Precisely, suppose that $Y_0^y = y$. The holding time of (Y_t^y) is $(T - t_{n-k})_{k=0}^n$, and its transition is governed by the transition matrix $\hat{P} = I + \frac{\hat{G}}{c}$ driven by $(V_i)_{i=1}^n$, where $V_i = 1 - U_{n+1-i}$. Let $\tilde{Y}_t^y := Y_{T-t}^y$. The above procedure is summarized into Algorithm 1:

To illustrate Algorithm 1, we use it to simulate a continuous time birth and death processes (X_t^3) on $E = \{0, \ldots, 7\}$ with birth rate and death rate both equal to 0.5 for all states apart from 0 and 7, where we have an absorbing boundary at 0,and a reflecting boundary at 7 with $G(7, 6) = 0.5$. The sample paths are plotted in Fig. 1.

Algorithm 1 Simulate a stochastically monotone CTMC and its Siegmund dual

---

Require: $x, y, T, G, \hat{G}$
  $c \leftarrow \max\{\max_i -G(i,i), \max_i -\hat{G}(i,i)\}$
  $P \leftarrow I + \dfrac{G}{c}$; $\hat{P} \leftarrow I + \dfrac{\hat{G}}{c}$
  $i \leftarrow 1$; $t_0 \leftarrow 0$; $X_0^x \leftarrow x$; $Y_0^y \leftarrow y$
  repeat
    Sample $t_i \sim$ Exponential(mean $= c^{-1}$)
    $i \leftarrow i + 1$
  until $\sum_i t_i > T$
  $n \leftarrow length(t_i) - 1$
  Sample $(U_i)_{i=1}^n \sim$ Uniform(0, 1); Set $V_i \leftarrow 1 - U_{n+1-i}$ for $i = 1, \ldots, n$
  Sample n steps of P that starts at x and is driven by $(U_i)_{i=1}^n$, say $(X_i^x)_{i=1}^n$
  Sample n steps of $\hat{P}$ that starts at y and is driven by $(V_i)_{i=1}^n$, say $(Y_i^y)_{i=1}^n$
  return $(t_i)_{i=0}^n$, $(X_i^x)_{i=0}^n$, $(Y_i^y)_{i=0}^n$

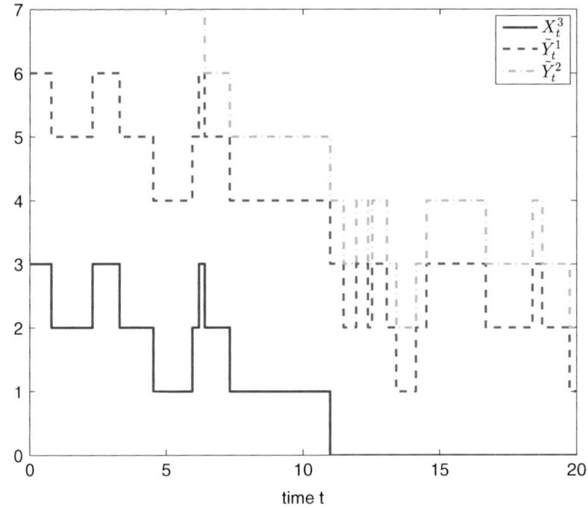

Fig. 1 Sample paths of birth-death (X_t^3) with an absorbing boundary at 0, $G(i, i+1) = G(i, i-1) = 0.5$, ($\tilde{Y}_t^4$) and ($\tilde{Y}_t^5$) with $T = 20$, $a = 7$

This figure should be compared with Jansen and Kurt [6, Fig. 1 Sec. 4.1]. We can observe that strong pathwise duality (cf. Definition 4) holds, that is, $X_{20}^3 \leq y$ if and only if $\tilde{Y}_0^y \geq 3$.

Next, to demonstrate our main result Proposition 1, we apply Algorithm 1 to simulate an upward skip-free process (X_t^4) with generator G in (7). It can be checked that G fulfills condition (1) and (2) in Proposition 1. Strong pathwise duality holds in Fig. 2, that is, $X_{20}^4 \leq y$ if and only if $\tilde{Y}_0^y \geq 4$. In essence, while (X_t^4) is upward skip-free, its Siegmund dual can be considered as a birth-and-death process before it gets absorbed to state 7.

$$G = \begin{pmatrix} 0 & 0 & 0 & 0 & 0 & 0 & 0 & 0 \\ 0.3 & -0.5 & 0.2 & 0 & 0 & 0 & 0 & 0 \\ 0.1 & 0.25 & -0.65 & 0.3 & 0 & 0 & 0 & 0 \\ 0.1 & 0.2 & 0.4 & -1.1 & 0.4 & 0 & 0 & 0 \\ 0.1 & 0.2 & 0.3 & 0.35 & -1.45 & 0.5 & 0 & 0 \\ 0.1 & 0.2 & 0.3 & 0.3 & 0.6 & -2.1 & 0.6 & 0 \\ 0.1 & 0.2 & 0.3 & 0.3 & 0.5 & 1.0 & -3.1 & 0.7 \\ 0.1 & 0.2 & 0.3 & 0.3 & 0.5 & 0.7 & 1.0 & -3.1 \end{pmatrix} \tag{7}$$

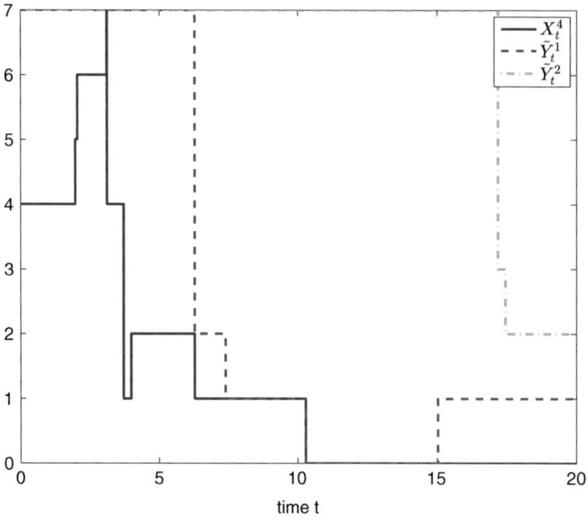

Fig. 2 Sample paths of (X_t^4) with G given by (7), $(\tilde{Y}_t^1)$ and $(\tilde{Y}_t^2)$ with $T = 20$, $a = 7$

Acknowledgements The authors would like to thank an anonymous referee for insightful suggestions which improved an earlier version of the paper.

References

1. Asmussen, S.: Applied Probability and Queues. Volume 51 of Applications of Mathematics (New York), 2nd edn. Springer, New York (2003). ISBN: 0-387-00211-1. Stochastic Modelling and Applied Probability
2. Clifford, P., Sudbury, A.: A sample path proof of the duality for stochastically monotone Markov processes. Ann. Probab. **13**(2), 558–565 (1985). ISSN: 0091-1798
3. Diaconis, P., Fill, J.A.: Strong stationary times via a new form of duality. Ann. Probab. **18**(4), 1483–1522 (1990). ISSN: 0091-1798
4. Fill, J.A.: On hitting times and fastest strong stationary times for skip-free and more general chains. J. Theor. Probab. **22**(3), 587–600 (2009). ISSN: 0894-9840, doi:10.1007/s10959-009-0233-7
5. Huillet, T., Martinez, S.: Duality and intertwining for discrete Markov kernels: relations and examples. Adv. Appl. Probab. **43**(2), 437–460 (2011). ISSN: 0001-8678, doi:10.1239/aap/1308662487
6. Jansen, S., Kurt, N.: On the notion(s) of duality for Markov processes. Probab. Surv. **11**, 59–120 (2014). ISSN: 1549-5787, doi:10.1214/12-PS206
7. Kent, J.T., Longford, N.T.: An eigenvalue decomposition for first hitting times in random walks. Z. Wahrsch. Verw. Gebiete **63**(1), 71–84 (1983). ISSN: 0044-3719, doi:10.1007/BF00534178
8. Norris, J.R.: Markov Chains. Volume 2 of Cambridge Series in Statistical and Probabilistic Mathematics. Cambridge University Press, Cambridge (1998). ISBN: 0-521-48181-3. Reprint of 1997 original
9. Siegmund, D.: The equivalence of absorbing and reflecting barrier problems for stochastically monotone Markov processes. Ann. Probab. **4**(6), 914–924 (1976)

Bifurcations in the Solution Structure of Market Equilibrium Problems

F. Etbaigha and M. Cojocaru

Abstract In this study, the well-known market disequilibrium model with excess supply and demand is investigated to determine if it exhibits changes in the structure and the number of equilibrium states for specific choices of parameter values. We propose to examine the effects of changing separately each price functions and unit transaction cost functions. We study the bifurcation problem (i.e., qualitative change in equilibrium states) as a parametrized variational inequality problem (VI). We conduct our analysis based on modeling the markets via a projected dynamical system (PDS), which is a type of constraint ordinary differential equations whose critical points are the market equilibrium states of the economic model. Numerical simulation for two examples is carried out to see if and when the behavior of these market steady states exhibits any qualitative change.

1 Introduction

The market equilibrium models considered in this work are spatial price equilibrium and market disequilibrium models [11, 14]. The extensive work on the spatial price equilibrium induced various improvements in both the formulation and the computation of this problem. In the formulation, it evolved from the formulation of linear complementarity [1], to the formulation via variational inequality theory as in [8]; then to PDS models as in [12]. In the computation, numerical methods and approximation algorithms are utilized to exploit the solutions at different settings and configurations [1, 12]. It also extended to a market disequilibrium case where the prices have been regulated [16]. The initial market disequilibrium problem appeared in [16]. Nagurney et al. [14] have introduced a new market equilibrium model with excess supply and demand, which extended from the spatial price equilibrium model presented in [11]. In this paper, we examine the possible structural changes of the behavior of the market equilibrium and disequilibrium problems under a parameter variation. Two frameworks are combined: the theory of PDS, and the theory of evolutionary variational inequality problems (EVI) (see for example [2, 5, 7] and

F. Etbaigha (✉) • M. Cojocaru
University of Guelph, Guelph, ON, Canada
e-mail: fetbaigh@uoguelph.ca; mcojocar@uoguelph.ca

© Springer International Publishing Switzerland 2016

J. Bélair et al. (eds.), *Mathematical and Computational Approaches in Advancing Modern Science and Engineering*, DOI 10.1007/978-3-319-30379-6_49

the references therein). The rest of the paper is organized as follows: in Sect. 2, we give a brief review of both market equilibrium and disequilibrium models; moreover, their variational inequalities forms and corresponding adjustment dynamics are also presented. Section 3 provides a study of bifurcation questions using the concept of EVI problem. Numerical simulations for two market equilibrium examples are presented in Sect. 4. The paper is concluded with a discussion and some suggestions for future work.

We assume the reader is familiar with the definitions of a convex cone, and those of the tangent and normal cone to a closed convex nonempty set (see for instance [3]).

2 Brief Review of Market Equilibrium Models

2.1 Market Equilibrium Model with Excess Supply and Demand (MESD)

The disequilibrium market model is presented in [14] and summarized below. It presented m supply markets and n demand markets which involved the production of a homogeneous commodity under perfect competition. The supply at market i is expressed as s_i and demand at market j as d_j. Further, the supply price at each supply market i is denoted by π_i, and ρ_j denotes the demand price at demand market j. All the supplies are grouped into a column vector $s \in R^m$, and the supply prices are grouped in a row vector $\pi \in R^m$. In the same context, the demands and demand prices are grouped respectively into a column vector $d \in R^n$ and a row vector $\rho \in R^n$. Moreover, the nonnegative commodity shipment attached with a supply and demand market pair (i,j) is denoted by Q_{ij}. Each trade between the pair (i,j) is associated with a unit transaction cost c_{ij}. The unit transaction costs which include the transportation cost and the tax are grouped into a row vector $c \in R^{mn}$, and the commodity shipments are grouped in a column vector $Q \in R^{mn}$. It is assumed that there are excess supply and excess demand at each market, denoted as u_i and v_j. All excess supplies and excess demands are grouped respectively into a column vector $u \in R^m$ and a row vector $v \in R^n$. Therefore, a feasible model requires the following constraints for all supplies, demands and shipments:

$$s_i = \sum_j Q_{ij} + u_i,\ i = 1, \dots m \text{ and } d_j = \sum_i Q_{ij} + v_j,\ j = 1, \dots .n. \tag{1}$$

Note that in [11] the equilibrium market is given as above with $u_i = 0$ and $v_j = 0$. Furthermore in [14], it is assumed that each supply price at the supply market is regulated by a fixed minimum supply price $\underline{\pi}_i$, called the price floor at supply market i. The fixed maximum demand price at demand market j is denoted by $\bar{\rho}_j$, called the price ceiling at demand market j. The supply price floors and the demand price ceilings are grouped into vectors $\underline{\pi} \in R^m$ and $\bar{\rho} \in R^n$, respectively. The vector, defined as $\tilde{\pi} \in R^{mn}$, contains m vectors of $\{\tilde{\pi}_i\}$, and each of these vectors consists of n components $\{\pi_i\}$. The vector $\tilde{\rho} \in R^{mn}$ consists of m vectors $\{\tilde{\rho}_j\}$ and each of these vectors has n components $\{\rho_1, \rho_2, \ldots, \rho_n\}$. In addition to the above restrictions, the supply, demand, the commodity shipment, excess supply and excess demand, which constitute the disequilibrium pattern must also satisfy the following conditions at all supply and demand markets:

$$\pi_i + c_{ij} \begin{cases} = \rho_j, & \text{if } Q_{ij} > 0, \\ \geq \rho_j, & \text{if } Q_{ij} = 0, \end{cases} \tag{2}$$

$$\pi_i \begin{cases} = \underline{\pi}_i, & \text{if } u_i > 0 \\ \geq \underline{\pi}_i, & \text{if } u_i = 0 \end{cases}, \quad \rho_j \begin{cases} = \bar{\rho}_j, & \text{if } v_j > 0, \\ \leq \bar{\rho}_j, & \text{if } v_j = 0. \end{cases} \tag{3}$$

The conditions in (2) are known as equilibrium conditions [11]. The conditions (3) are presented in detail in [14]. It is also considered that the supply price at any supply market i depends on the supplies at all the markets, that is $\pi = \pi(s)$. Similarly, the demand price at any demand market j depends on the demands at all the markets, that is $\rho = \rho(d)$. In the same context, the unit transaction cost associated with the pair (i, j) depends on the commodity shipments from i to j, that is $c = c(Q)$, where all π, ρ and c are smooth functions. The vectors $\hat{\pi} = \pi \in R^m$ and $\hat{\rho} = \rho \in R^n$ are also provided, where $\hat{\pi} = \hat{\pi}(Q, u)$ and $\hat{\rho} = \hat{\rho}(Q, v)$. The vector $\tilde{\hat{\pi}} \in R^{mn}$ consists of m vectors $\{\tilde{\hat{\pi}}_i\}$ and each these vectors contains n components $\{\hat{\pi}_i\}$ and the vector $\tilde{\hat{\rho}} \in R^{mn}$ consists of m vectors $\{\tilde{\hat{\rho}}_j\}$, where each vector has n components $\{\hat{\rho}_1, \hat{\rho}_2 \ldots, \hat{\rho}_n\}$.

2.2 *Variational Inequality Formulation and Adjustment Dynamics of MESD*

The pattern (Q^*, u^*, v^*) satisfies the conditions (2) and (3) governing the disequilibrium market problem if and only if it satisfies the VI problem

$$(\tilde{\hat{\pi}}(Q^*, u^*) + c(Q^*) - \tilde{\hat{\rho}}(Q^*, v^*)).(Q - Q^*) + (\hat{\pi}(Q^*, u^*) - \underline{\pi}).(u - u^*) +$$
$$(\bar{\rho} - \hat{\rho}(Q^*, v^*)).(v - v^*) \geq 0, \quad \forall (Q, u, v) \in K \; and \; K = R^{mn}_+ \times R^m_+ \times R^n_+. \tag{4}$$

Remark: note that in [11] the market equilibrium model without the excess supply and demand is formulated as a VI as follows:

$$\langle F(Q^*), Q - Q^* \rangle \geq 0 \;\; \forall \, Q \in K^1, \; K^1 = R^{mn}_+. \tag{5}$$

Thus, according to [12], it is known that the adjustment to spatial price equilibrium states in a VI problem can be obtained by studying the nonsmooth dynamical system:

$$\frac{dQ(t)}{dt} = P_{T_{KQ(t)}}(-F(Q(t))),\ Q(0) \in K^1, K^1 = R_+^{mn}, \tag{6}$$

where $F : K^1 \longrightarrow R^{mn}$ and $F_{ij}(Q) = \pi_i(s) + c_{ij}(Q) - \rho_j(d)$.

In the same manner, we can write the adjustment process of the MESD [7] as

$$\frac{d(Q(t), u(t), v(t))}{dt} = P_{T_{K(Q(t),u(t),v(t))}}(-F(Q(t), u(t), v(t))),\ \ K = R_+^{mn} \times R_+^m \times R_+^n. \tag{7}$$

For simplicity, we will write (Q, u, v) instead of $(Q(t), u(t), v(t))$. $F : K \longmapsto R^{mn} \times R^m \times R^n$ is defined by $F(Q, u, v) = (A(Q, u, v), G(Q, u), D(Q, v))$, and $A : K \longmapsto R^{mn}$, $G : R_+^{mn} \times R_+^m \longmapsto R^m$ and $D : R_+^{mn} \times R_+^n \longmapsto R^n$ are defined by: $A_{ij} = \tilde{\hat{\pi}}_i(Q, u) + c_{ij}(Q) - \tilde{\hat{\rho}}_j(Q, v)$, $G_i = \hat{\pi}_i(Q, u) - \underline{\pi}_i$, $D_j = \bar{\rho}_j - \hat{\rho}_j(Q, v)$. Let the vector $x = (Q, u, v) \in K$, and $F(x) = F(Q, u, v)$ then (7) can be written in the form,

$$\dot{x} = P_{T_{Kx(t)}}(-F(x(t))). \tag{8}$$

3 Bifurcations in MESD

3.1 Formulation of the Bifurcation Problem for Constrained Systems

Bifurcation theory for nonsmooth dynamical systems has not been studied as extensively as smooth dynamical systems (see for instance [15] and the references therein). However, the works in [9] and [10] give results for the existence of bifurcations in discontinuous Filippov systems and VI. The nonconventional bifurcations of fixed points and periodic solutions in Filippov systems have been addressed in [10]. In [9], it was shown that if V is the Hilbert space, $B : V \times R \longrightarrow V$ and K is closed convex subset of V then the VI of the form

$$\langle u - B(u, \lambda), v - u\rangle \geq 0 \ \ \forall v \in K \tag{9}$$

subject to $B(0, \lambda) = 0 \ \forall\ \lambda \in R$, $B(u, \lambda) = \lambda Bu + R(u)$, $R(u) = o(\|u\|)$ *as* $u \longrightarrow 0$ has a bifurcation point $(0, \lambda_0)$ for $\lambda_0 \in R$ provided there are a sequences (u_n, λ_n) of solutions of (9) such that $\|u_n\| \neq 0$ and as $n \longrightarrow \infty, \lambda_n \longrightarrow \lambda_0$ and $u_n \longrightarrow 0$, for all n. The above constrain can not be satisfied in our VI formulation, therefore (9) can not be applied here. Since PDS lies outside of the field of classical dynamical systems [13], one way to formulate and study the bifurcation problem is by using the concept of a parametrized VI problem, sometimes called an EVI.

Let us assume that an MESD is formulated as in Sect. 2, such that its solution(s) are given by equilibrium points of the dynamical system (7). Let $\lambda \in R$ and consider the system:

$$\frac{d(Q,u,v)}{dt} = P_{T_{K(Q,u,v)}}(-F(Q,u,v,\lambda)), \;\; K = R_+^{mn} \times R_+^m \times R_+^n, \tag{10}$$

which is equivalent to $\dot{x} = P_{T_{Kx(t)}}(-F(x(t),\lambda))$, where $x = (Q,u,v) \in K$ so that whenever $\lambda = 0$ we have the initial system (8).

We want to study the problem of finding points $x^* \in K$ so that [6]

$$P_{T_{K(x^*)}}(-F(x^*,\lambda)) = 0 \Leftrightarrow -F(x^*,\lambda) \in N_K(x^*), \tag{11}$$

for some interval of values of $\lambda \in [a,b]$. From (11), using the definition of a normal cone, we have to:

$$\text{find points } x^* \in K \text{ s.t. } \langle F(x^*,\lambda), y - x^* \rangle \geq 0, \forall y \in K. \tag{12}$$

In (12), F depends on x and λ and hence, we can reformulate it as a parametrized VI problem, or as an EVI problem (see for example [4, 5] for the current use of EVI models) where the bifurcation parameter λ replaces the "time" parameter t:

$$\langle F(x(\lambda),\lambda), y(\lambda) - x(\lambda) \rangle \geq 0, \forall y(\lambda) \in K(\lambda). \tag{13}$$

According to the EVI theory, all the functions will be in Hilbert space $L^2([a,b], R^{n+m+nm})$, then $F : [a,b] \times K \longrightarrow L^2([a,b], R^{n+m+nm})$ and the feasible set K can be written as

$$K = \{x \in L^2([a,b], R^{n+m+nm}) \,|\underline{0} \leq x(\lambda) \leq M, \; a.e \; in \; [a,b]\}, \; with \; x, y \in K. \tag{14}$$

Note that: $K(\lambda) = \{w \in R^{n+m+nm} |\; \underline{0} \leq w \leq M\}$, where M is a large.

Essentially, the bifurcation problem becomes an EVI problem where the bounds of the constraint set $K(\lambda)$ are fixed, and do not depend on the parameter λ. In order to solve such an EVI problem, one can use existing methods in the literature [5]. Cojocaru et al. in [5] showed that the linkage between the projected dynamical system and EVI time dependent problem where the critical points of the PDS are the same as the solutions of EVI problem. Using this approach, the PDS associated to λ dependent EVI can be formulated as

$$\frac{dx(\lambda,\tau)}{d\tau} = P_{T_{Kx(\lambda,\tau)}}(-F(x(\lambda,\tau),\lambda)), \tag{15}$$

where the time τ represents the time evolution to the equilibrium $x(\lambda,.)$. This formulation is employed to solve the problem in (13).

3.2 Bifurcations for MESD

The equilibrium problems are analyzed using different parameterization strategies. Initially, by introducing a parameter that represents the supply price floors, for which the system behavior did not change and the equilibrium is still unique. Secondly we impact the supply prices by introducing a parameter λ to represent the variation of the supplies on the supply price functions, the equilibrium is still unique and minor changes are observed only in the trajectories convergence for different values of λ. The same experiments are repeated for the demand prices and the same results are observed. These results suggest that the bifurcations are not likely to occur with such parametrization strategies. Parametrizing the unit transaction cost functions by introducing λ to represent the variation of the shipments will likely impact the market steady states behavior significantly such that $c = c(Q, \lambda)$. Another direction is to introduce λ into both the supply and demand prices functions such that $\pi = \pi(s, \lambda)$, and $\rho = \rho(d, \lambda)$. It is expected that this change will impact the market equilibria since changing both the supply and the demand prices at the same time will dramatically impact the equilibrium and disequilibrium conditions (2) and (3). The next section shows two examples of market equilibrium problems using this parametrization strategy.

It is known that whenever F is strictly monotone or beyond (strongly monotone, etc.) [11] as a function of x, the solution of the MESD problem is unique [13]. Thus, there is only one equilibrium point of (8). From the EVI theory [5, 7], whenever $F(x, \lambda)$ with $\lambda \in [a, b]$ remains strictly monotone, the parametrized MESD (13) still has a unique solution. It is therefore logical to seek values of λ which will lead to a non-strictmonotonic $F(x, \lambda)$. In these ranges of values we may find distinct behaviour of the equilibrium.

4 Examples

4.1 Example 1

We consider the example of a market equilibrium problem with two supply and two demand markets which has a unique equilibrium as shown in [11]. In order to detect if the change on the cost functions impacts the equilibrium, we modify the example and introduce λ in the cost function. At $\lambda = 0$ the equilibrium is still unique $(0.8, 2.96, 0, 0)$ since the vector field F is still strictly monotone. Additionally, the variables are adjusted to be λ dependent where $\lambda \in [a, b]$.

The supply price functions are defined by $\pi : R^2 \longrightarrow L^?([a, b], R^2)$,
$\pi_1(s(\lambda)) = 5s_1(\lambda) + s_2(\lambda) + 2, \quad \pi_2(s(\lambda)) = s_1(\lambda) + 2s_2(\lambda) + 3;$
the demand price functions are defined by $\rho : R^2 \longrightarrow L^2([a, b], R^2)$,
$\rho_1(d(\lambda)) = -2d_1(\lambda) - d_2(\lambda) + 28.75, \quad \rho_2(d(\lambda)) = -d_1(\lambda) - 4d_2(\lambda) + 41;$
and the unit transaction costs are defined by $c : [a, b] \times K \longrightarrow L^2([a, b], R^4)$,

$c_{11}(Q,\lambda) = Q_{11}(\lambda) + (0.5+\lambda)Q_{12}(\lambda) + 1$, $c_{12}(Q,\lambda) = (2+\lambda)Q_{12}(\lambda) + Q_{22}(\lambda) + 1.5$, $c_{21}(Q,\lambda) = (3-\lambda)Q_{21}(\lambda) + 2Q_{11}(\lambda) + 25$, $c_{22}(Q,\lambda) = (2-\lambda)Q_{22}(\lambda) + Q_{12}(\lambda) + 30$. The feasible set is $K = \{Q \in L^2([a,b],R^4) | \underline{0} \le Q(\lambda) \le \underline{5},\ a.e\ in\ [a,b]\}$.
We consider a vector field F defined as $F : [a,b] \times K \longrightarrow L^2([a,b],R^4)$ where
$F_{11}(Q(\lambda),\lambda) = \pi_1(s(\lambda)) + c_{11}(Q(\lambda),\lambda) - \rho_1(d(\lambda))$
$= 8Q_{11}(\lambda) + (6.5+\lambda)Q_{12}(\lambda) + 3Q_{21}(\lambda) + 2Q_{22}(\lambda) - 25.75.$
$F_{12}(Q(\lambda),\lambda) = \pi_1(s(\lambda)) + c_{12}(Q(\lambda),\lambda) - \rho_2(d(\lambda))$
$= 6Q_{11}(\lambda) + (11+\lambda)Q_{12}(\lambda) + 2Q_{21} + 6Q_{22}(\lambda) - 37.5.$
$F_{21}(Q(\lambda),\lambda) = \pi_2(s(\lambda)) + c_{21}(Q(\lambda),\lambda) - \rho_1(d(\lambda))$
$= 5Q_{11}(\lambda) + 2Q_{12}(\lambda) + (7-\lambda)Q_{21}(\lambda) + 3Q_{22}(\lambda) - 0.75.$
$F_{22}(Q(\lambda),\lambda) = \pi_2(s(\lambda)) + c_{22}(Q(\lambda),\lambda) - \rho_2(d(\lambda))$
$= 2Q_{11}(\lambda) + 6Q_{12}(\lambda) + 3Q_{21}(\lambda) + (8-\lambda)Q_{22}(\lambda) - 8.$
Our EVI problem is given by $\langle F(Q^*(\lambda),\lambda), Q(\lambda) - Q^*(\lambda)\rangle \ge 0, \forall Q(\lambda) \in K$, and the associated PDS can be written as in (15).

We examine the effect of changing λ in the cost functions considering $\lambda \in [0,8]$ to ensure that the costs are always nonnegative. We obtain the equilibrium at each value of λ by the method in [5] implemented in Matlab. For $0 \le \lambda < 8$, for any arbitrary initial condition, the system shows just one equilibrium corresponding to each value of λ. At $\lambda = 8$, the shipments converge to three boundary equilibria, which means that F is non-strictmonotonic. The numerical simulations of the system at $\lambda = 8$ are presented in the Figs. 1 and 2. Shipments at supply market 1 are

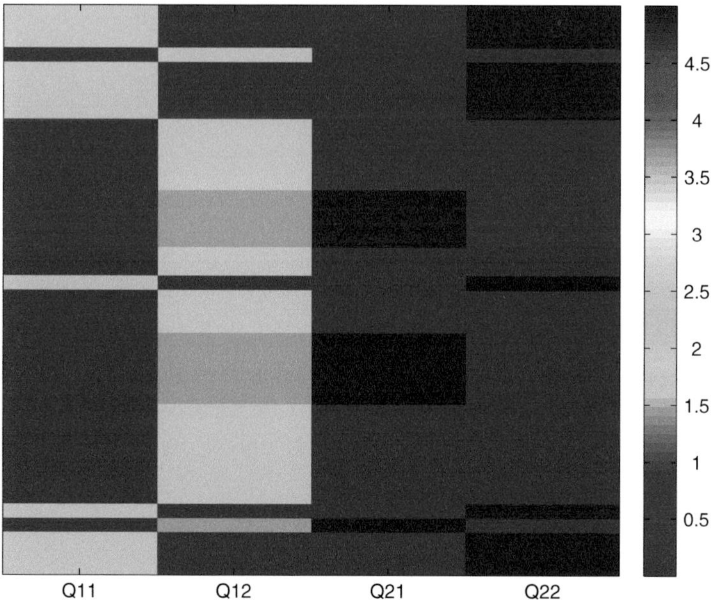

Fig. 1 Values of $(Q_{11}, Q_{12}, Q_{21}, Q_{22})$ at equilibrium are presented as a heatmap. The heatmap shows three equilibrium patterns, the equilibria are color-coded such that each repeated pattern is represented by the same color

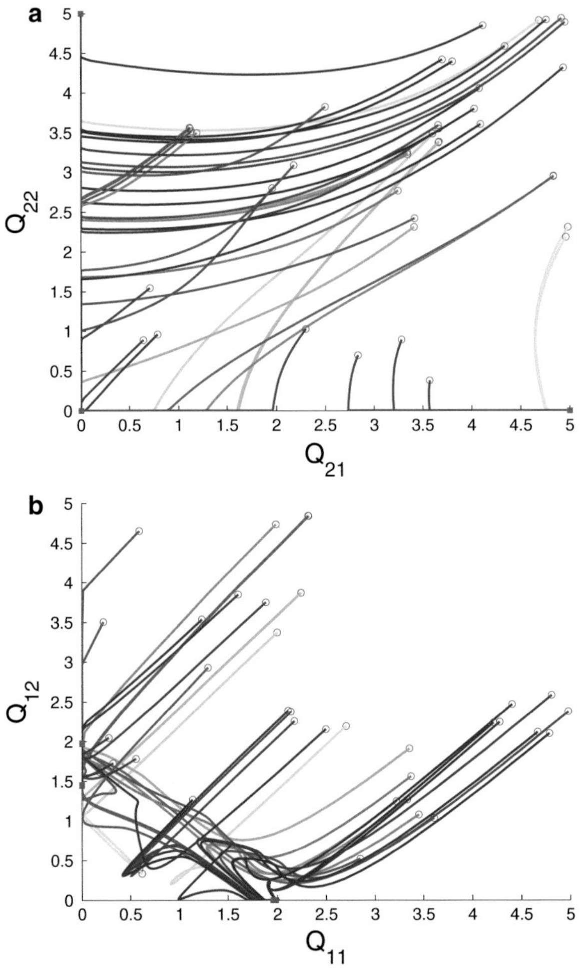

Fig. 2 Phase portraits for $\lambda = 8$, both figures show trajectories starting from 40 initial conditions (equilibria shown by *square markers*). There are three distinct equilibrium points

balanced at: $(Q_{11}, Q_{12}) \in \{(0, 1.44), (1.97, 0), (0, 1.97)\}$, and shipments at supply market 2 are balanced at: $(Q_{21}, Q_{22}) \in \{(5, 0), (0, 5), (0, 0)\}$ as shown in Fig. 2. As a result of the change in the number of equilibria, $\lambda = 8$ is a bifurcation value.

4.2 Example 2

Here we modify the above example to paramterize the supply and demand price functions. The example is extended to the case of excess supply and excess demand. Furthermore λ is introduced into both price functions. Without introducing λ the system shows one equilibrium $(1.6, 2.8, 0, 0, 0, 0, 0, 0)$ since F is still strictly monotone. The supply price functions are defined by
$\pi_1(s(\lambda), \lambda) = (5+\lambda)s_1(\lambda) + s_2(\lambda) + 12$, $\pi_2(s(\lambda), \lambda) = (1+\lambda)s_1(\lambda) + 2s_2(\lambda) + 30$.
The demand price functions are defined by
$\rho_1(d(\lambda), \lambda) = -2d_1(\lambda) - (1-\lambda)d_2(\lambda) + 45$,
$\rho_2(d(\lambda), \lambda) = -(1-\lambda)d_1(\lambda) - 4d_2(\lambda) + 55$.
The unit transaction cost functions are as in the Example 1 with $\lambda = 0$.
The feasible set is $K = \{x \in L^2([a,b], R^8) | \underline{0} \leq x(\lambda) \leq \underline{5},\ a.e\ in\ [a,b]\}$, where $x(\lambda) = (Q_{11}(\lambda), Q_{12}(\lambda), Q_{21}(\lambda), Q_{22}(\lambda), u_1(\lambda), u_2(\lambda), v_1(\lambda), v_2(\lambda))$. We consider the supply price floors at the markets are $\underline{\pi}_1 = 10$ and $\underline{\pi}_2 = 15$ and the demand price ceilings are $\bar{\rho}_1 = 45$ and $\bar{\rho}_2 = 55$. Then the vector field F is defined as
$F : [a,b] \times K \longrightarrow L^2([a,b], R^8)$; for simplicity we assume $Q(\lambda) = (x_1, x_2, x_3, x_4)$.
Then $F_{11}(Q(\lambda), u(\lambda), v(\lambda), \lambda) = (8+\lambda)x_1 + (6.5+\lambda)x_2 + (5+\lambda)u_1 + 3x_3 + (2-\lambda)x_4 + 2v_1 + u_2 + (1-\lambda)v_2 - 32$.
$F_{12}(Q(\lambda), u(\lambda), v(\lambda), \lambda) = (6+\lambda)x_1 + (11+\lambda)x_2 + (5+\lambda)u_1 + 6x_4 + 4v_2 + u_2 + (1-\lambda)v_1 + (2-\lambda)x_3 - 41.5$.
$F_{21}(Q(\lambda), u(\lambda), v(\lambda), \lambda) = (5+\lambda)x_1 + (1+\lambda)u_1 + (4+\lambda)x_2 + (3-\lambda)x_4 + 2v_1 + (1-\lambda)v_2 + 7x_3 + 2u_2$.
$F_{22}(Q(\lambda), u(\lambda), v(\lambda), \lambda) = 2x_1 + (5+\lambda)x_2 + (3-\lambda)x_3 + (1-\lambda)v_1 + 4v_2 + 6x_4 + (1+\lambda)u_1 + 2u_2 - 15$,
$(\pi_1 - \underline{\pi}_1, \pi_2 - \underline{\pi}_2) = ((5+\lambda)x_1 + (5+\lambda)x_2 + (5+\lambda)u_1 + x_3 + x_4 + u_2 + 2, (1+\lambda)x_1 + (1+\lambda)x_2 + (1+\lambda)u_1 + 2x_2 + 2x_4 + 2u_2 + 15)$
$(\bar{\rho}_1 - \rho_1, \bar{\rho}_2 - \rho_2) = (2x_1 + 2x_3 + 2v_1 + (1-\lambda)x_2 + (1-\lambda)x_4 + (1-\lambda)v_2, (1-\lambda)x_1 + (1-\lambda)x_3 + (1-\lambda)v_1 + 4x_2 + 4x_4 + v_2)$. Then the EVI problem can be written as $\langle F(Q^*(\lambda), u^*(\lambda), v^*(\lambda), \lambda), (Q(\lambda), u(\lambda), v(\lambda)) - (Q^*(\lambda), u^*(\lambda), v^*(\lambda))\rangle \geq 0$.

We set $\lambda \in [-7, 8]$ to ensure the prices are nonnegative. The experiments is carried out as discussed in Example 1. For each value of λ there exist only one equilibrium except for the value of $\lambda = -6$ at which two equilibria occur. Then at $\lambda = -6$, F is non-strictmonotonic, and $\lambda = -6$ is a bifurcation value. Figure 3 shows the two equilibria. Shipments at supply market 1 $(Q_{11}, Q_{12}) \in \{(0,5), (5,0)\}$ and shipments at supply market 2 $(Q_{21}, Q_{22}) \in \{(0, 0.83), (0, 2.4)\}$ are shown in

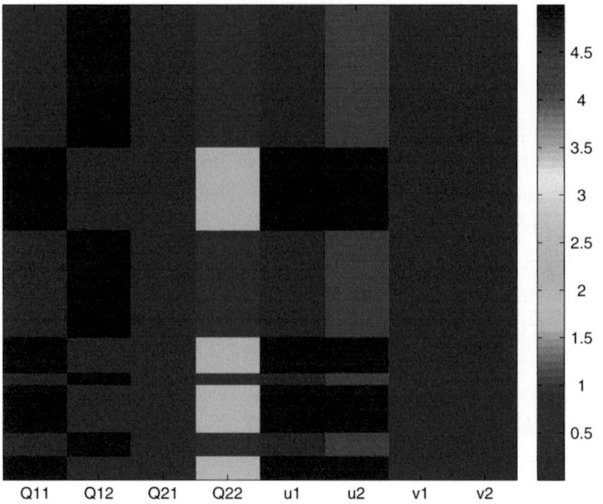

Fig. 3 The heatmap shows the equilibrium pattern $(Q_{11}, Q_{12}, Q_{21}, Q_{22}, u_1, u_2, v_1, v_2)$. Each value is represented by singular color, the number of equilibria in this case two which is indicated by switch in color

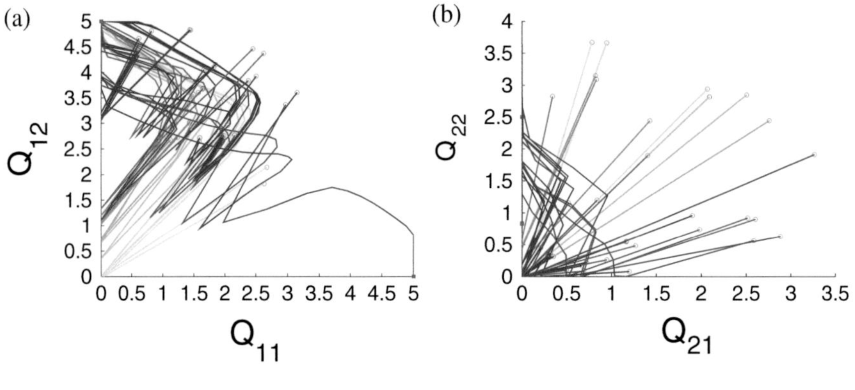

Fig. 4 Phase portraits for $\lambda = -6$ show the convergence of the shipments at each market. (**a**) (Q_{11}, Q_{12}) convergence. (**b**) (Q_{21}, Q_{22}) convergence

Fig. 4. The excess supply and demand at market 1 $(u_1, v_1) \in \{(0, 0), (4.9, 0)\}$ and the excess supply and demand at market 2 $(u_2, v_2) \in \{(4.18, 0), (5, 0)\}$ are presented in Fig. 5.

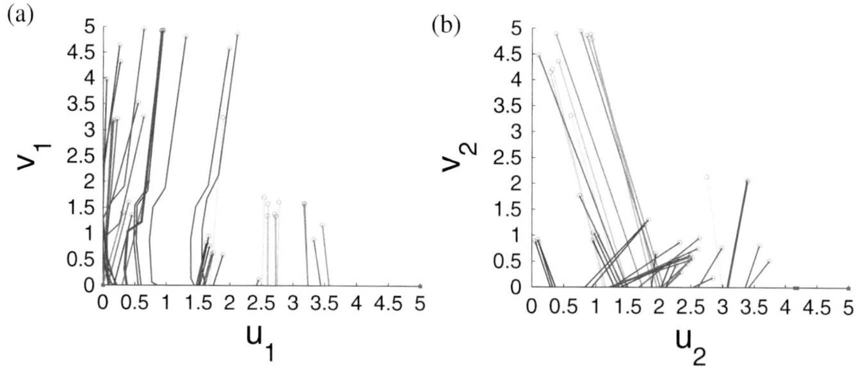

Fig. 5 Phase portraits for $\lambda = -6$ show the convergence of the excess supply and excess demand at each market. (**a**) (u_1, v_1) convergence. (**b**) (u_2, v_2) convergence

5 Conclusion

In this study we considered the bifurcation problem for the market equilibrium model as an EVI problem. Furthermore, we studied the impact of changing the supply price, demand price and the cost functions on the market equilibrium states. With both the cost functions and price functions, the effect of the variations of the parameter is seen on the number of equilibria occurring at specific values. The equilibrium states were obtained using trajectories of the associated projected dynamics. The empirical results on two examples showed that bifurcations occur in such systems. Further investigation is required to extend the applicability of the proposed method to other equilibrium models. Moreover, additional future work is to study the usage of nonlinear parameter dependencies to explore existence of bifurcations in the solution structure of market equilibrium problems.

References

1. Asmuth, R., Eaves, B.C., Peterson, E.L.: Computing economic equilibria on affine networks with lemke's algorithm. Math. Oper. Res. **4**(3), 209–214 (1979)
2. Barbagallo, A., Mauro, P.: Evolutionary variational formulation for oligopolistic market equilibrium problems with production excesses. J. Optim. Theory Appl. **155**(1), 288–314 (2012)
3. Cojocaru, M.-G.: Monotonicity and existence of periodic orbits for projected dynamical systems on hilbert spaces. Proc. Am. Math. Soc. **134**(3), 793–804 (2006)
4. Cojocaru, M.-G.: Double-layer dynamics theory and human migration after catastrophic events. In: Cojocaru, M.-G. (ed.) Nonlinear Analysis with Applications in Economics, Energy and Transportation, pp. 65–86. Bergamo University Press, Bergamo (2007)
5. Cojocaru, M.-G., Daniele, P., Nagurney, A.: Projected dynamical systems and evolutionary variational inequalities via hilbert spaces with applications1. J. Optim. Theory Appl. **127**(3), 549–563 (2005)

6. Cojocaru, M.-G., Jonker, L.: Existence of solutions to projected differential equations in hilbert spaces. Proc. Am. Math. Soc. **132**(1), 183–193 (2004)
7. Daniele, P.: Dynamic Networks and Evolutionary Variational Inequalities. Edward Elgar Publishing, Cheltenham/Northampton (2006)
8. Florian, M., Los, M.: A new look at static spatial price equilibrium models. Reg. Sci. Urban Econ. **12**(4), 579–597 (1982)
9. Le, V.K., Schmitt, K.: Global Bifurcation in Variational Inequalities: Applications to Obstacle and Unilateral Problems, vol. 123. Springer Science & Business Media, Berlin (1997)
10. Leine, R., Van Campen, D., Van de Vrande, B.: Bifurcations in nonlinear discontinuous systems. Nonlinear Dyn. **23**(2), 105–164 (2000)
11. Nagurney, A.: Network Economics: A Variational Inequality Approach, vol. 10. Springer Science & Business Media, Berlin (1998)
12. Nagurney, A., Takayama, T., Zhang, D.: Massively parallel computation of spatial price equilibrium problems as dynamical systems. J. Econ. Dyn. Control **19**(1), 3–37 (1995)
13. Nagurney, A., Zhang, D.: Projected Dynamical Systems and Variational Inequalities with Applications, vol. 2. Springer Science & Business Media, Berlin (2012)
14. Nagurney, A., Zhao, L.: Disequilibrium and variational inequalities. J. Comput. Appl. Math. **33**(2), 181–198 (1990)
15. Perko, L.: Differential Equations and Dynamical Systems, vol. 7. Springer Science & Business Media, Berlin (2013)
16. Thore, S.: Spatial disequilibrium. J. Reg. Sci. **26**(4), 661–675 (1986)

Pricing Options with Hybrid Stochastic Volatility Models

Glynis Jones and Roman Makarov

Abstract We introduce a hybrid stochastic volatility model where the asset price process follows the Heston model and interest rates are governed by a two-factor stochastic model. Two cases are considered. First, it is assumed that interest rates and asset prices are uncorrelated. The characteristic function method is used to derive semi-analytical pricing formulae for plain vanilla options. In the second case we introduce a correlation between the asset price process and the short rate process and use Monte Carlo simulations for pricing options. To reduce the stochastic error, we implement the control variate method where an estimator of the option value for the uncorrelated case is used as a control variate. The options are priced with a varying correlation coefficient. We observe that the control variate method allows us to speed up Monte Carlo computations by a factor with the magnitude of several hundreds. The efficiency of the method is higher for smaller values of the correlation coefficient. We then study the impact a correlation between the two processes has on option prices. It has been noticed that the call option price is an increasing function of the correlation coefficient.

1 Introduction

In 1973, Fischer Black, Myron Scholes and Robert Merton [2] introduced a technique for finding a closed-form solution for the price of a plain European option. The price, $V = V(t, S)$, of a European-style option is a function of time t and asset price S (at time t) that satisfies the partial differential equation

$$\frac{\partial V}{\partial t} + \frac{1}{2}\sigma^2 S^2 \frac{\partial^2 V}{\partial S^2} + rS\frac{\partial V}{\partial S} - rV = 0 \tag{1}$$

subject to the terminal condition $V(T, S) = f(S)$, where r is the risk-free interest rate, σ is the volatility of the stock, and $f(S)$ is the payoff at maturity T. Here we

G. Jones • R. Makarov (✉)
Department of Mathematics, Wilfrid Laurier University, 75 University Avenue West, Waterloo, ON, Canada
e-mail: j.glynis@gmail.com; rmakarov@wlu.ca

© Springer International Publishing Switzerland 2016

J. Bélair et al. (eds.), *Mathematical and Computational Approaches in Advancing Modern Science and Engineering*, DOI 10.1007/978-3-319-30379-6_50

assume no dividend on the stock. The solution to (1) is given by the discounted expectation of the terminal payoff:

$$V(t, S) = e^{-r(T-t)} E^{Q} \left[f(S_T) \mid S_t = S \right] , \tag{2}$$

where Q is the risk-neutral probability measure with the bank account $B_t = e^{rt}$ taken as a numeraire.

The Black–Scholes–Merton formula (2) assumes that the price of an underlying asset such as a stock, $\{S_t\}_{t \geq 0}$, follows geometric Brownian motion (under the risk-neutral measure Q):

$$dS_t = rS_t dt + \sigma S_t dW_t \, .$$

The main drawback of this model is that the volatility of the underlying risky asset is assumed to be constant. This assumption does not fit many factors observed on financial markets such as volatility smiles and skews. Option pricing theory has evolved since 1973. Other models that take into account many market phenomena have been developed. Examples of such models include models with stochastic volatility and stochastic interest rates. In this paper, we consider a hybrid model that combines a stochastic volatility model for asset prices with a stochastic short rate process (see [1, 8, 14] and references wherein). In particular, we develop a modification of the Heston model [9] with stochastic interest rates that are governed by a two-factor model (see [4]). We consider a two-factor modification of the Vasicek model [15] but other models such the Hull–White model [10] and the Cox–Ingersoll–Ross (CIR) model [5] can also be used.

To specify the hybrid model, we combine two systems of stochastic differential equations. The first one defines the price of a risky asset and its volatility. The second system defines the short rate process. As a result, we obtain a four-factor stochastic volatility model named here as the Heston–Vasicek model. This hybrid model fits in the class of affine diffusion processes (see [6]) for which a closed-form solution of the characteristic function exists. However, the derivation of the characteristic function for a hybrid multi-factor model with a full matrix of correlations is a challenging problem. In this paper we study a simpler approach. First, we assume that the stock price process and the interest rate process are uncorrelated. Fourier techniques are used to obtain no-arbitrage prices of European-style options. Second, we introduce a correlation between asset prices and interest rates and use the Monte Carlo method to simulate the coupled system of SDEs and to price options. To reduce the stochastic error, we use the control variate method (see, e.g., [7]), where an estimator of the option value for the uncorrelated case is used as a control variate.

The rest of this paper is organised as follows. In Sect. 2, we formulate the hybrid model and obtain characteristic functions for the uncorrelated case. A correlation between the asset price and interest rate processes is then introduced and options are priced using the Monte Carlo method. In Sect. 3, we discuss and analyse numerical results, and our conclusions are drawn in Sect. 4.

2 The Hybrid Model

In this section, we construct a four-factor hybrid stochastic volatility model where a stochastic interest rate is introduced into a stochastic volatility model. We define our model such that the log-price process, $\{X_t = \ln S_t\}_{t\geq 0}$, follows the Heston model and the short rate process, $\{r_t\}_{t\geq 0}$, is governed by the two-factor Vasicek model. Therefore, it is called the Heston–Vasicek model. The hybrid model consists of four coupled SDEs:

$$dX_t = (r_t - \frac{\nu_t}{2})dt + \sqrt{\nu_t}dW_t^s\,, \tag{3}$$

$$d\nu_t = \kappa(\theta - \nu_t)dt + \gamma\sqrt{\nu_t}dW_t^\sigma\,, \tag{4}$$

$$dr_t = (\theta + u_t - \bar{a}r_t)dt + \sigma_1 dW_t^r\,, \tag{5}$$

$$du_t = -\bar{b}u_t dt + \sigma_2 dW_t^u\,, \tag{6}$$

where W_t^s, W_t^σ, W_t^r, W_t^u are standard Brownian motions such that $dW_t^s dW_t^\sigma = \rho_{s\sigma}dt$ and $dW_t^r dW_t^u = \rho_{ru}dt$. The two-factor Vasicek model (5)–(6) can be written as a two-factor Gaussian model (the equivalence is proved in [4]):

$$dx_t = -ax_t dt + \sigma dW_t^x\,, \tag{7}$$

$$dy_t = -by_t dt + \eta dW_t^y\,, \tag{8}$$

$$r_t = x_t + y_t + \phi(t)\,, \tag{9}$$

where x_t and y_t are Gaussian processes, $\phi(t)$ is a deterministic function given by $\phi(t) = r_0 e^{-aT} + \frac{\theta}{a}(e^{at} - 1)$, and W^x and W^y are correlated standard Brownian motions with correlation coefficient ρ_{xy}. The parameters of (5)–(6) and (7)–(8) are related as follows:

$$\bar{a} = a, \quad \bar{b} = b,$$

$$\sigma_1 = \sqrt{\sigma^2 + \eta^2 + 2\rho_{xy}\sigma\eta}, \quad \sigma_2 = \eta(a - b),$$

$$\rho_{ru} = \frac{\sigma\rho_{xy} + \eta}{\sqrt{\sigma^2 + \eta^2 + 2\rho_{xy}\sigma\eta}}.$$

First, we assume that W^x and W^y are not correlated with W^s and W^σ. Thus, the correlation matrix for the four Brownian motions is given by

$$\begin{bmatrix} 1 & \rho_{s\sigma} & 0 & 0 \\ \rho_{s\sigma} & 1 & 0 & 0 \\ 0 & 0 & 1 & \rho_{xy} \\ 0 & 0 & \rho_{xy} & 1 \end{bmatrix}. \tag{10}$$

The option pricing formula for a standard European call takes the following form:

$$C(t,S) = E^{Q}_{t,r,S}\left[e^{-\int_t^T r_u du}(S_T - K)^+\right]. \tag{11}$$

Here, the expectation $E^{Q}_{t,r,S}[\ \cdot\ ]$ computed under the risk-neutral measure Q is conditional on $\{r_t = r,\ S_t = S\}$. The discounting factor cannot be pulled out since it is stochastic, so the mathematical expectation is computed as follows:

$$\begin{aligned} C(t,S) &= E^{Q}_{t,r,S}\left[e^{-\int_t^T r_u du}(S_T - K)^+\right] \\ &= E^{Q}_{t,r,S}\left[e^{-\int_t^T r_u du}(S_T - K)\mathbb{1}_{\{S_T>K\}}\right] \\ &= E^{Q}_{t,r,S}\left[e^{-\int_t^T r_u du}S_T\mathbb{1}_{\{S_T>K\}}\right] - E^{Q}_{t,r,S}\left[e^{-\int_t^T r_u du}K\mathbb{1}_{\{S_T>K\}}\right]. \end{aligned} \tag{12}$$

To calculate the conditional expectations in the r.h.s. of (12), we apply the change of numeraire technique. The first expectation can be simplified by using the asset price as a numeraire and the second expectation can be simplified by using the zero-coupon bond as a numeraire. So, we have

$$\begin{aligned} C(t,S) &= E^{Q_1}_{t,r,S}\left[\frac{B_T/B_t}{S_T/S_t}e^{-\int_t^T r_u du}S_T\mathbb{1}_{\{S_T>K\}}\right] \\ &\quad - E^{Q_2}_{t,r,S}\left[\frac{B_T/B_t}{Z(T,T)/Z(t,T)}e^{-\int_t^T r_u du}K\mathbb{1}_{\{S_T>K\}}\right] \\ &= SE^{Q_1}_{t,S}\left[\mathbb{1}_{\{S_T>K\}}\right] - KZ(t,T)E^{Q_2}_{t,S}\left[\mathbb{1}_{\{S_T>K\}}\right] \\ &= SQ_1(S_T > K \mid S_t = S) - KZ(t,T)Q_2(S_T > K \mid S_t = S) \\ &= SF_1 - KZ(t,T)F_2\,, \end{aligned} \tag{13}$$

where $B_t = e^{\int_0^t r_u du}$ is the bank account, $Z(t,T) = E^{Q}_{t,r}\left[e^{-\int_t^T r_u du}\right]$ is the time-t price of a unit zero-coupon bond with maturity T, Q_1 is the EMM (equivalent martingale measure) with the stock S_t used as a numeraire asset, and Q_2 is the T-forward EMM with the zero-coupon bond $Z(t,T)$ used as a numeraire.

As in [9], the characteristic functions need to be found for computing the probabilities $F_1 = Q_1(S_T > K \mid S_t = S)$ and $F_2 = Q_2(S_T > K \mid S_t = S)$. Since the interest rate in this model is no longer constant, the characteristic functions for the Heston model have to be modified to include the stochastic interest rate component. To determine the form of the new characteristic functions we start with the log-price process for the SDE (3) re-writing it as follows [11]:

$$dX_t = r_t\,dt + (-\frac{v_t}{2}dt + \sqrt{v_t}dW_t^s) = r_t\,dt + dX_t^H\,, \tag{14}$$

where $\{X_t^H\}_{t\geq 0}$ is the log-price process for the Heston model with interest $r = 0$. Integrating both sides gives us the following:

$$\int_t^T dX_s = \int_t^T r_s\,ds + \int_t^T dX_s^H \implies X_T = R_T + X_T^H\,,$$

where $R_T = \int_t^T r_u du$. Then, we have the characteristic functions $f_{HV_j}(\phi), j = 1, 2$ of the hybrid model as follows:

$$f_{HV_j}(\phi) = E_{t,r,X}^{Q_j}[e^{i\phi X_T}] = E_{t,r,X}^{Q_j}[e^{i\phi(R_T+X_T^H)}] = E_{t,r}^{Q_j}[e^{i\phi R_T}]E_{t,X}^{Q_j}[e^{i\phi X_T^H}]\,, \tag{15}$$

where we used the fact that R_T and X_T^H are independent. Therefore, the new characteristic functions admit the following form:

$$f_{HV_j}(\phi) = f_{V_j}(\phi)f_j(\phi),\ j = 1, 2\,,$$

where f_{V_j} is the characteristic function for the Vasicek model and f_j is the characteristic function for the Heston model (under Q_j).

The functions f_j, $j = 1, 2$ are already known [9, 13, 16] so we only need to determine the characteristic functions of the integrated short rate process, R_T, under EMMs Q_1 and Q_2, respectively,

$$f_{V_1}(\phi) = E_{t,r}^{Q_1}[e^{i\phi R_T}], \quad f_{V_2}(\phi) = E_{t,r}^{Q_2}[e^{i\phi R_T}]\,. \tag{16}$$

If the short rate r_t is a Gaussian process, then the integrated process R_t is Gaussian as well. Hence, we can use the identity $E[e^{i\phi Z}] = e^{i\phi\mu_z - \frac{\phi^2\sigma_z^2}{2}}$ for a normal random variable Z with mean μ_z and variance σ_z^2. The probability distribution of the short rate process does not change under the EMM Q_1. Therefore, Eqs. (7) and (8) can be used to find the mean and the variance of R_T under Q_1:

$$\mu_R = E[R_T] = -\frac{r_0}{a}e^{-aT} + \frac{r_0}{a} - \frac{\theta}{a^2} + \frac{\theta e^{-aT}}{a^2} + \frac{\theta T}{a},$$

$$\begin{aligned} V_R = Var(R_T) &= \frac{\sigma^2}{a}\left(T + \frac{2}{a}e^{-aT} - \frac{1}{2a}e^{-2aT} - \frac{3}{2a}\right) \\ &+ \frac{\eta^2}{b}\left(T + \frac{2}{b}e^{-bT} - \frac{1}{2b}e^{-2bT} - \frac{3}{2b}\right) \\ &+ \frac{2\rho_{xy}\sigma\eta}{ab}\left(T + \frac{e^{-aT}-1}{a} + \frac{e^{-bT}-1}{b} - \frac{e^{-(a+b)T}-1}{a+b}\right). \end{aligned}$$

Under the T-forward EMM Q_2 the short rate process, $r_t = x_t + y_t + \phi(t)$, is defined by the following SDEs [4]:

$$dx_t = \left[-ax_t - \tfrac{\sigma^2}{2}(1 - e^{-a(T-t)}) - \rho_{xy}\tfrac{\sigma\eta}{b}(1 - e^{-b(T-t)})\right] dt + \sigma dW_t^x \,, \tag{17}$$

$$dy_t = \left[-by_t - \tfrac{\eta^2}{2}(1 - e^{-b(T-t)}) - \rho_{xy}\tfrac{\sigma\eta}{a}(1 - e^{-a(T-t)})\right] dt + \eta dW_t^y \,. \tag{18}$$

Therefore, the mean and variance of R_T under the EMM Q_2 are, respectively,

$$\begin{aligned} E[R_T] &= x_{int} + y_{int} + \phi_{int} \,, \\ Var(R_T) &= \frac{\sigma^2}{a}\left(T + \frac{2}{a}e^{-aT} - \frac{1}{2a}e^{-2aT} - \frac{3}{2a}\right) \\ &+ \frac{\eta^2}{b}\left(T + \frac{2}{b}e^{-bT} - \frac{1}{2b}e^{-2bT} - \frac{3}{2b}\right) \\ &+ \frac{2\rho_{xy}\sigma\eta}{ab}\left(T + \frac{e^{-aT}-1}{a} + \frac{e^{-bT}-1}{b} - \frac{e^{-(a+b)T}-1}{a+b}\right), \end{aligned}$$

where

$$\begin{aligned} x_{int} &= E\left[\int_0^T x(s)ds\right] = -\frac{\sigma^2}{a}\frac{T^2}{2}(1 - e^{-aT}) - \frac{\rho_{s\sigma}\sigma\eta}{b}\frac{T^2}{2}(1 - e^{-bT}) \,, \\ y_{int} &= E\left[\int_0^T y(s)ds\right] = -\frac{\eta^2}{b}\frac{T^2}{2}(1 - e^{-bT}) - \frac{\rho_{s\sigma}\sigma\eta}{a}\frac{T^2}{2}(1 - e^{-aT}) \,, \\ \phi_{int} &= \int_0^T \phi(s)ds = -\frac{r_0}{a}e^{-aT} + \frac{r_0}{a} - \frac{\theta}{a^2} + \frac{\theta e^{-aT}}{a^2} + \frac{\theta T}{a} \,. \end{aligned}$$

Substituting these into Eq. (16) gives us the new characteristic function f_{V_2}. The price of the call option under the new model is as follows:

$$C(t, S) = SF_1 - KZ(t, T)F_2 \,, \text{ where} \tag{19}$$

$$F_j = \frac{1}{2} + \frac{1}{\pi}\int_0^\infty \text{Re}\left(f_{HV_j}(\phi)\frac{\exp(-i\phi \ln K)}{i\phi}\right) d\phi, \; j = 1, 2 \,, \tag{20}$$

and $Z(t, T)$ is the price of a zero-coupon bond. The integral in (20) can be evaluated numerically using a quadrature rule or the fast Fourier transform (FFT) method [13, 16]. Note that the price of the put option can be calculated using the put-call parity. Clearly, we are able to derived the pricing formula for European options in closed form (19)–(20) due to the assumption that the asset price process and the short rate process are uncorrelated.

We then move on to the correlated case. Let us assume that W^x is correlated with W^s, i.e. $dW_t^x dW_t^s = \hat{\rho} dt$ with $\hat{\rho} \in (-1, 1)$. Alternatively, we can introduce the correlation between W^y and W^s or between W^x and W^σ. We would like to see how $\hat{\rho}$ affects the option prices as it changes. The correlation matrix for the system of SDEs (3), (4), (7), (8) is as follows:

$$\begin{bmatrix} 1 & \rho_{s\sigma} & \hat{\rho} & \hat{\rho}\rho_{xy} \\ \rho_{s\sigma} & 1 & \hat{\rho}\rho_{s\sigma} & \hat{\rho}\rho_{xy}\rho_{s\sigma} \\ \hat{\rho} & \hat{\rho}\rho_{s\sigma} & 1 & \rho_{xy} \\ \hat{\rho}\rho_{xy} & \hat{\rho}\rho_{xy}\rho_{s\sigma} & \rho_{xy} & 1 \end{bmatrix}. \tag{21}$$

One can easily show that this matrix is positive definite for all $\rho_{s\sigma}, \rho_{xy}, \hat{\rho} \in (-1, 1)$. The Cholesky decomposition method can then be applied to generate correlated normal random variables. These correlated random variables are used in Monte Carlo simulations of the hybrid model.

As is seen from (21), the correlation coefficients $\mathrm{Corr}(W^s, W^y)$, $\mathrm{Corr}(W^\sigma, W^x)$, $\mathrm{Corr}(W^\sigma, W^y)$ cannot approach 1 or -1 as $|\hat{\rho}| \to 1$ since $\rho_{xy}, \rho_{s\sigma} \in (-1, 1)$. So, the selection of two processes between which the correlation is introduced affects the range for the correlation coefficient between another pair.

3 Computational Results and Analysis

We would like to calibrate the model using historical data and then use the parameters obtained for option pricing. First, the two-factor Gaussian model is calibrated using yield rates obtained from the US treasury website (see [3, 4]). The parameters obtained for the short rate model are as follows:

$$a = 0.2071, \; b = 2.6414, \; \sigma = 0.0100, \; \eta = 0.0117, \; \rho_{xy} = -0.8317, \; \theta = 0.0113.$$

Using these parameters, the Heston model has been calibrated using the least square method [13]. The historical prices of options on SPDR S&P 500 ETF (SPY) for June 10, 2010 were used with different strike prices and different maturities (option prices were obtained from the Bloomberg™ database). The initial asset value is $S_0 = 108.88$. The least squares method is used to minimize the difference between market prices and model prices of options. This method minimizes the sum of the squared errors (the market price minus the model price) made in every iteration. To achieve the best possible results, the same routine has been run multiple times with different starting points to get an optimal solution. The parameters that provide the best fit to market prices are

$$\nu_0 = 0.0648, \; \kappa = 0.0166, \; \theta_\nu = 0.3534, \; \gamma = 0.1223, \; \rho_{s\sigma} = -0.8380.$$

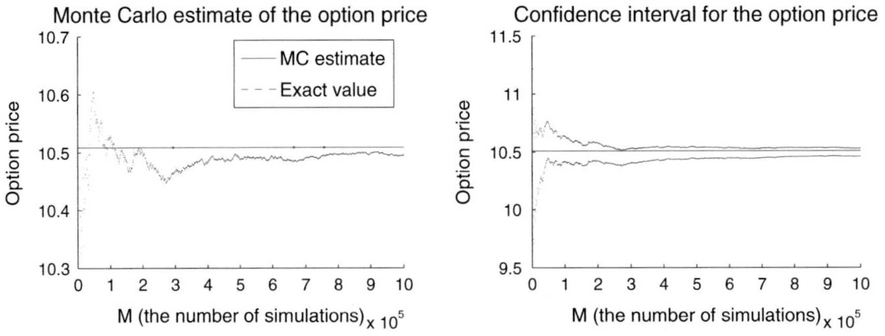

Fig. 1 Monte Carlo simulations for the at-the-money call option with strike price $K = 108.88$

Substituting these values into the option pricing formula (19), we can calculate no-arbitrage prices of call options for the uncorrelated case.

To verify the pricing formula (19), the exact option prices were compared with prices obtained using Monte Carlo simulations. The Monte Carlo prices are approaching the respective exact price in all our tests. For example, Fig. 1 demonstrates the convergence of Monte Carlo prices for the at-the-money call option (with $K = S_0$).

To construct Monte Carlo estimates for option prices, the Euler method (see, e.g., [12]) is used. Discrete-time approximations of sample paths, $(\hat{X}_i, \hat{v}_i, \hat{x}_i, \hat{y}_i) \approx (X_{i\Delta t}, v_{i\Delta t}, x_{i\Delta t}, y_{i\Delta t})$ with $i = 0, 1, \ldots, n-1$ and $\Delta t = \frac{T}{n}$, are generated using the following scheme:

$$
\begin{aligned}
\hat{X}_{i+1} &\leftarrow \hat{X}_i + \Delta t\,(\hat{r}_i - 0.5\max\{0, \hat{v}_i\}) + \sqrt{\max\{0, \hat{v}_i\}}\sqrt{\Delta t} z_s\,, \\
\hat{v}_{i+1} &\leftarrow \hat{v}_i + \Delta t \kappa(\theta_v - \hat{v}_i) + \gamma\sqrt{\max\{0, \hat{v}_i\}}\sqrt{\Delta t} z_\sigma\,, \\
\hat{x}_{i+1} &\leftarrow \hat{x}_i - a\hat{x}_i\Delta t + \sigma\sqrt{\Delta t} z_x\,, \\
\hat{y}_{i+1} &\leftarrow \hat{y}_i - b\hat{y}_i\Delta t + \eta\sqrt{\Delta t} z_y\,, \\
\hat{r}_{i+1} &\leftarrow \hat{x}_{i+1} + \hat{y}_{i+1} + r_0 e^{-a(i+1)\Delta t} - \frac{\theta}{a} e^{-a(i+1)\Delta t}\,.
\end{aligned}
$$

The initial conditions are $\hat{x}_0 = 0$, $\hat{y}_0 = 0$, $\hat{r}_0 = r_0$, and $\hat{X}_0 = \ln S_0$. For every time step, four correlated normal random variables, $(z_s, z_\sigma, z_x, z_y)$, are required; they are obtained using the Cholesky decomposition of the correlation matrix given by (10) or (21). A sample value of the discounted payoff, $e^{-\int_0^T r_u du}\left(e^{X_T} - K\right)^+$, is then computed. The trapezoidal rule is used to approximated the discounting factor.

Monte Carlo simulations usually have a large stochastic error associated with them. To reduce the variance of our Monte Carlo estimator of the option value we use the control variate method. The uncorrelated case (for which the option pricing formula is available in closed form) is used to construct the control variate. The same i.i.d. normal samples are used to calculate the estimates for both correlated and uncorrelated cases.

In Table 1 we examine how the option prices vary with $\hat{\rho}$. As $\hat{\rho}$ approaches 0, the option price gets closer to the analytical price calculated for $\hat{\rho} = 0$. We find that the option price increases as $\hat{\rho}$ increases for at-the-money ($K = 108.88$), in-the-money ($K \in \{100, 106\}$), and out-of-the-money ($K \in \{110, 115\}$) options. The reasoning behind this is that as interest rates go up, the return on stocks increases thereby increasing the value of a call option. We can say the call option price and the interest rates are positively correlated. Therefore, increasing the correlation coefficient between the interest rates and log-price processes increases the price of a call option as is seen from our simulations.

The ratio of the standard deviations is calculated (the standard deviation of the crude estimator over the standard deviation of the controlled estimator). This gives us the factor by which the control variate method improves the crude Monte Carlo method. This ratio ranges from 238 to 3846 depending on the values of $\hat{\rho}$ and K. As is seen from Table 1 and Fig. 2, the control variate method gives a significant improvement over the standard estimator when the correlation coefficient is close to zero.

4 Conclusion

In this paper we have successfully derived a semi-analytical formula for no-arbitrage pricing a standard European option under the hybrid four-factor Heston–Vasicek model, where the interest rate process and the log-asset price process are uncorrelated. We then studied the impact a correlation between the two processes had on option pricing using Monte Carlo simulations. It has been noticed that the price of a European call is an increasing function of the correlation coefficient. The use of a variance reduction method greatly improved our results for the correlated case as we can see from the ratios of standard deviations presented in Table 1. For future work, it will be interesting to study a hybrid model that combines the Heston model and the two-factor Hull–White model and compare its performance with the model studied in this paper.

Table 1 The impact of $\hat{\rho}$ on option prices

$\hat{\rho}$	$K = 100$	Std ratio	$K = 106$	Std ratio	$K = 108.88$	Std ratio	$K = 110$	Std ratio	$K = 115$	Std ratio
−0.9	13.0576	314.4654	9.3820	284.3563	7.8718	273.9726	7.3296	268.0965	5.2128	240.9639
−0.8	13.0597	389.1051	9.3843	357.1429	7.8742	340.1361	7.3321	332.2259	5.2153	298.5075
−0.7	13.0616	476.1905	9.3866	436.6812	7.8765	414.9378	7.3344	406.5041	5.2176	363.6364
−0.6	13.0636	477.0095	9.3889	537.6344	7.8789	510.2041	7.3368	500.0000	5.2200	448.4305
−0.5	13.0656	729.9270	9.3912	671.1409	7.8813	641.0256	7.3392	625.0000	5.2224	558.6592
−0.4	13.0676	934.5794	9.3935	862.0690	7.8837	826.4463	7.3416	813.0081	5.2248	719.4245
−0.3	13.0695	1282.0513	9.3958	1176.4706	7.8861	1136.3636	7.3440	1098.9011	5.2272	980.3922
−0.2	13.0715	1960.7843	9.3981	1785.7143	7.8885	1724.1379	7.3464	1666.6677	5.2296	1492.5373
−0.1	13.0735	3846.1538	9.4004	3571.4286	7.8909	3448.2759	7.3489	3448.2759	5.2320	2941.1765
0	13.0755	–	9.4027	–	7.8932	–	7.3513	–	5.2344	–
0.1	13.0775	3846.1538	9.4050	3571.4286	7.8956	3448.2759	7.3537	3448.2759	5.2368	2941.1765
0.2	13.0794	1923.0769	9.4073	1818.1818	7.8980	1724.1379	7.3560	1694.9153	5.2392	1492.5373
0.3	13.0814	1282.0513	9.4096	1176.4706	7.9004	1123.5955	7.3585	1098.9011	5.2416	980.3922
0.4	13.0834	934.5794	9.4119	862.0690	7.9028	819.6721	7.3608	806.4516	5.2440	735.2941
0.5	13.0854	724.6377	9.4142	666.6667	7.9051	636.9427	7.3632	625.0000	5.2464	555.5556
0.6	13.0873	584.7953	9.4165	534.7594	7.9075	510.2041	7.3656	500.0000	5.2487	444.4444
0.7	13.0893	476.1905	9.4187	434.7826	7.9099	414.9378	7.3680	404.8583	5.2511	361.0108
0.8	13.0913	389.1051	9.4210	355.8719	7.9123	340.1361	7.3704	332.2259	5.2535	296.7359
0.9	13.0933	312.5000	9.4233	286.5330	7.9146	272.4796	7.3728	266.6667	5.2559	238.6635

Calculations were done using the maturity time $T = 0.5$, the time step $\Delta t = 0.001$, and the number of sample paths $M = 1{,}000{,}000$

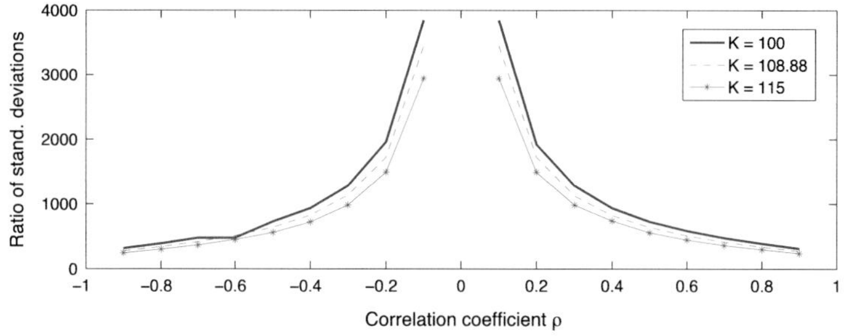

Fig. 2 The impact of the correlation coefficient $\hat{\rho}$ on the performance of the control variate methods

Acknowledgements R. Makarov wishes to acknowledge the generous support of the NSERC Discovery Grant program.

References

1. Ahlip, R., Rutkowski, M.: Pricing of foreign exchange options under the Heston stochastic volatility model and CIR interest rates. Quant. Financ. **13**(6), 955–966 (2013)
2. Black, F., Scholes, M.: The pricing of options and corporate liabilities. J. Pol. Econ. **81**(3), 637–654 (1973)
3. Blanchard, A.: The two-factor Hull-White model: pricing and calibration of interest rates derivatives. Accessed 28 Mar 2014
4. Brigo, D., Mercurio, F.: Interest Rate Models-Theory and Practice: With Smile, Inflation and Credit. Springer, Berlin/New York (2007)
5. Cox, J.C., Ingersoll, J.E., Ross, S.A.: A theory of the term structure of interest rates. Econometrica **53**(2), 385–407 (1985)
6. Duffie, D., Pan, J.: Transform analysis and asset pricing for affine jump-diffusions. Econometrica **68**(6), 1343–1376 (2000)
7. Glynn, P.W., Szechtman, R.: Some new perspectives on the method of control variates. In: Monte Carlo and Quasi-Monte Carlo methods, 2000 (Hong Kong), pp. 27–49. Springer, Berlin (2002)
8. Grzelak, L.A., Oosterlee, C.W.: On the Heston model with stochastic interest rates. SIAM J. Financ. Math. **2**(1), 255–286 (2011)
9. Heston, S.L.: A closed-form solution for options with stochastic volatility with applications to bond and currency options. Rev. Financ. Stud. **6**(2), 327–343 (1993)
10. Hull, J., White, A.: Pricing interest-rate-derivative securities. Rev. Financ. Stud. **3**(4), 573–592 (1990)
11. Kienitz, J., Kammeyer, H.: An implementation of the Hybrid-Heston-Hull-White model. Available at SSRN 1399389 (2009)
12. Kloeden, P.E., Platen, E.: Numerical Solution of Stochastic Differential Equations. Volume 23 of Applications of Mathematics (New York). Springer, Berlin (1992)
13. Rouah, F.D.: The Heston Model and Its Extensions in Matlab and C. Wiley, Hoboken (2013)
14. Van Haastrecht, A., Lord, R., Pelsser, A., Schrager, D.: Pricing long-maturity equity and FX derivatives with stochastic interest rates and stochastic volatility. Insur.: Math. Econ. **45**(3), 436–448 (2009)

15. Vasicek, O.: An equilibrium characterization of the term structure [reprint of J. Financ. Econ. **5**(2), 177–188 (1977)]. In: Financial Risk Measurement and Management. Volume 267 of International Library of Critical Writings in Economics, pp. 724–735. Edward Elgar, Cheltenham (2012)
16. Zhu, J.: Applications of Fourier Transform to Smile Modeling, 2nd edn. Springer Finance. Springer, Berlin (2010) Theory and implementation

Delay Stochastic Models in Finance

Anatoliy Swishchuk

Abstract The paper is devoted to an overview of delay stochastic models in finance and their applications to modeling and pricing of swaps. The volatility process is an important concept in financial modeling. This process can be stochastic or deterministic. In quantitative finance, we consider the volatility process to be stochastic as it allows to fit the observed market prices under consideration, as well as to model the risk linked with the future evolution of the volatility, which deterministic model cannot. Most stochastic dynamical systems (SDS) in finance describe the state of security (asset or volatility) as a value at time t,-instantaneous or current value at time t. We would like to take into account not only the current value at time t, but also values over some time interval $[t - \tau, t]$, where τ is a positive constant and is called the delay. In this way, we incorporate path-dependent history of the volatility under consideration. In this paper we will mainly focus on newly developed so-called delayed Heston model that significantly improve classical Heston model with respect to the market volatility surface fitting by 44 %. Review of some other delay stochastic models in finance will be given as well.

1 Introduction: Variance and Volatility Swaps

A security is a tradable financial asset of any kind, e.g., stock. A stock is a portion of ownership in a corporation. If $S(t)$ is a value of stock at time t, then $S(t - \tau)$ is a delayed stock price. Volatility is a measure for variation of price of a financial instrument (stock, etc.) over time. If $V(t)$ is the volatility at time t, then $V(t - \tau)$ is a delayed volatility.

One of the main problems in finance is pricing of volatility and variance (square of volatility) swaps. A forward contract or simply a forward is a contract between two parties to buy or sell an asset at a specified future time at a price agreed upon today. Volatility swaps are defined as forward contracts on future realized stock volatility. Variance swaps are similar contracts on variance, the square of the future volatility.

A. Swishchuk (✉)
University of Calgary, 2500 University Drive NW, Calgary, AB, T2N 1N4, Canada
e-mail: aswish@ucalgary.ca

© Springer International Publishing Switzerland 2016

J. Bélair et al. (eds.), *Mathematical and Computational Approaches in Advancing Modern Science and Engineering*, DOI 10.1007/978-3-319-30379-6_51

Volatility swaps allow investors to profit from the risks of an increase or decrease in future volatility of an index of securities or to hedge against these risks. If you think current volatility is low, for the right price you might want to take a position that profits if volatility increases.

Payoff of variance swap is defined as

$$N(V_R - K_{var}),$$

where V_R is the realized stock variance (quoted in annual terms) over the life of the contract, K_{var} is the strike price, N is the notional amount,

$$V_R := \frac{1}{T}\int_0^T V_s ds.$$

Price of variance swap is then:

$$P_{var} = E^Q\{e^{-rT}(V_R - K_{var})\},$$

where r is the risk-free discount rate corresponding to the expiration date T, and E^Q denotes the risk-neutral expectation.

Payoff of volatility swap is defined as

$$N(\sqrt{V_R} - K_{vol}),$$

where $\sqrt{V_R}$ is the realized stock volatility (quoted in annual terms) over the life of the contract, and

$$\sqrt{V_R} := \sqrt{\frac{1}{T}\int_0^T V_s ds}.$$

Price of volatility swap is then:

$$P_{vol}(x) = E^Q\{e^{-rT}(\sqrt{V_R} - K_{vol})\}.$$

If $dS(t)$ is a change of stock price at time t, then usually

$$\begin{aligned} dS(t) &= a(t, S(t))dt + V(t, S(t))dW(t) + b(t, S(t))dL(t) \\ &= \text{“drift”} + \text{“noise”} + \text{“random jumps”}, \end{aligned} \tag{1}$$

where $a(t, S(t))$ is a drift coefficient, $V(t, S(t))$ is a diffusion coefficient, $b(t, S(t))$ is a jump size coefficient, and $W(t)$ and $L(t)$ is a Brownian motion and a jump process (e.g., Lévy process, see [12]) (independent or correlated).

There are many papers (see, e.g., [1, 4, 5, 7, 8, 10, 11, 13]) in finance that consider $a \equiv a(t, S(t-\tau))$, $V \equiv V(t, S(t-\tau))$ or both parameters t and S depending on delay τ. In this paper, we will focus on volatility or variance models with delay in finance.

The volatility process $V(t, S(t))$ in (1) is an important concept in financial modeling. This process can be stochastic or deterministic. In quantitative finance, we consider the volatility process $V(t, S(t))$ to be stochastic as it allows to fit the observed market prices under consideration, as well as to model the risk linked with the future evolution of the volatility, which deterministic model cannot.

If $V(t, S(t))$ is the volatility at time t, then we consider some models for the change of $V(t)$ at time t, $dV(t)$:

$$\begin{aligned} dV(t, S(t)) &= c(t, V(t, S(t)))dt + d(t, V(t, S(t)))dW(t) + e(t, V(t, S(t)))dL(t) \\ &= \text{"drift"} + \text{"noise"} + \text{"random jumps"}. \end{aligned} \tag{2}$$

In delayed stochastic volatility models the coefficients c, d, e depend on delay τ or contain path-dependent history of volatility.

2 Some Delayed Stochastic Models in Finance: An Overview

Significance of path-dependency (a.k.a. delay) may be seen from the Fig. 1 below: the price of variance swap crucially depends on delay, where we used $S\&P60$ Canada Index as our real data (vertical line stands for delivery price, and two plane lines stand for jump intensity (left) and for delay (right)).

1. *Continuous-time GARCH model for Stochastic Volatility with Delay (see [9]).* We assume in this paper the following model for volatility $V(t, S(t))$:

$$\frac{dV(t, S(t))}{dt} = \gamma\sigma + \frac{\alpha}{\tau}\left[\int_{t-\tau}^{t} \sqrt{V(s, S(s))}dW(s)\right]^2 - (\alpha + \gamma)V(t, S(t)).$$

   Here, all the parameters $\alpha, \gamma, \tau, \sigma$ are positive constants and $0 < \alpha + \gamma < 1$.

   All the parameters of this model were inherited from its discrete-time analogue:

$$V_n = \gamma\sigma + \frac{\alpha}{l}\ln^2(S_{n-1}/S_{n-1-l}) + (1 - \alpha - \gamma)V_{n-1}, \quad l = \frac{\tau}{\Delta},$$

   which, in the special case $l = 1$, is a well-known GARCH(1,1) model for stochastic volatility without conditional mean of log-return (see [2, 3]). Here, S_n is a stock price at time n.

   We used this model to find the prices of variance and volatility swaps (see [14]).

2. *Multi-Factor Stochastic Volatility with Delay (see [15]).*

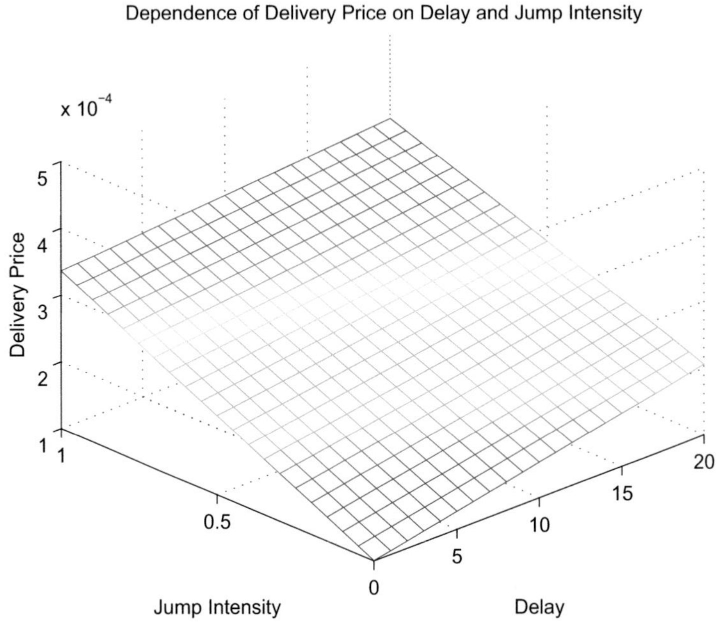

Fig. 1 Dependence of delivery price on delay and jump intensity (*S&P*60 Canada Index)

The main idea is to consider continuous-time GARCH model for stochastic volatility with coefficient σ above that itself is stochastic and depends on drift and noise $\sigma \equiv \sigma(t, chance)$.

In [15], CAMQ, we considered three two-factor and one three-factor stochastic volatility models with delay and we calculated the prices of variance swap for each model.

One of the examples is:

$$\begin{cases} \frac{dV(t)}{dt} = \gamma\sigma_t + \frac{\alpha}{\tau}\left[\int_{t-\tau}^{t}\sqrt{V(s)}dW(s)\right]^2 - (\alpha+\gamma)V(t). \\ d\sigma_t/\sigma_t = \xi dt + \beta dW_1(t), \end{cases}$$

where $S(t)$ is defined from the equation

$$dS(t) = \mu S(t)dt + \sqrt{V(t)}S_t dW(t),$$

$S_t := S(t-\tau)$ and Wiener processes $W(t)$ and $W_1(t)$ may be correlated.

3. *Stochastic Volatility with Delay and Jumps (see [16]).*

In this paper, the volatility satisfies the following equation:

$$\frac{dV(t,S(t))}{dt} = \gamma\sigma + \frac{\alpha}{\tau}\Big[\int_{t-\tau}^{t} \sqrt{V(s,S(s))}dW(s) + \int_{t-\tau}^{t} y_s dN(s)\Big]^2 - (\alpha+\gamma)V(t,S(t)),$$

where $W(t)$ is a Brownian motion, $N(t)$ is a Poisson process with intensity λ and y_t is the jump size at time t. We assume that $E[y_t] = A(t), E[y_s y_t] = C(s,t), s < t$ and $E[y_t^2] = B(t) = C(t,t)$, where $A(t), B(t), C(s,t)$ are all deterministic functions. We calculated the price of variance swap in this case.

4. *Lévy-based Stochastic Volatility with Delay (see [17]).* The stochastic volatility in this paper satisfies the following Lévy-driven equation:

$$\frac{dV(t,S(t))}{dt} = \gamma\sigma + \frac{\alpha}{\tau}\left[\int_{t-\tau}^{t} \sqrt{V(u,S(u)}dL(u)\right]^2 - (\alpha+\gamma)V(t,S(t))$$

where $L(t)$ is a Lévy process independent of $W(t)$. Here, $\sigma > 0$ is a mean-reverting level (or long-term equilibrium of $V(t,S_t)$), $\alpha,\gamma > 0$, and $\alpha+\gamma < 1$.

We also calculated the price of variance swap in this case.

3 Delayed Heston Model

Classical Heston model (see [6]) is one of the most popular stochastic volatility models in the industry as semi-closed formulas for vanilla option prices are available, few (five) parameters need to be calibrated, and it accounts for the mean-reverting feature of the volatility. In this section we will focus on newly developed so-called delayed Heston model that significantly improve classical Heston model with respect to the market volatility surface fitting by 44 %. In this model, we take into account not only current state of volatility at time t but also its past history over some interval $[t-\tau,t]$, where $\tau > 0$ is a constant and is called the delay. In this way, our model incorporates path-dependent history for volatility. We will show how to model and price variance and volatility swaps (forward contracts on variance and volatility) for the delayed Heston model and how to hedge volatility swaps using variance swaps (see [18]).

The main points of this section are motivation, advantage and goal:

- Motivation: to include past history (a.k.a. delay) of the variance (over some delayed time interval $[t-\tau,t]$);
- Advantage: Improvement of the Volatility Surface Fitting (44 % reduction of the calibration error) compare with Classical Heston model;
- Goal: to price and hedge volatility swaps.

We consider in a first approach adjusting the Heston drift by a deterministic function of time so that the expected value of the variance under our new delayed Heston model is equal to the one under 'delayed vol'. Our approach can therefore be seen as a variance 1st moment correction of the Heston model, in order to account for the delay. It is important to note that our model is a generalization of the classical Heston model (the latter corresponding to the zero delay case $\tau = 0$ of our model). We performed numerical tests to validate our approach. With recent market data (Sept. 30th 2011, underlying EURUSD), we performed the model calibration on the whole market vanilla option price surface (14 maturities from 1M to 10Y, 5 strikes ATM, 25 Delta Call/Put, 10 Delta Call/Put). The results show a significant (44 %) reduction of the average absolute calibration error compared to the Heston model (i.e. average of the absolute differences between market and model prices).

Further, we consider variance and volatility swaps hedging and pricing in our delayed Heston framework. These contracts are widely used in the financial industry and therefore it is relevant to know their price processes (how much they worth at each time t) and how we can hedge a position on them, i.e. theoretically cancel the risk inherent to holding one unit of them.

Using the fact that every continuous local martingale can be represented as a time-changed Brownian motion, as well as the Brockhaus and Long approximation (that allows to approximate the expected value of the square-root of an almost surely non negative random variable using a 2nd order Taylor expansion approach), we were able to derive closed formulas for variance and volatility swaps price processes. In addition, as variance swaps are relatively liquid instruments in the market (i.e. they can be easily bought and sold), we considered the question of hedging a position on a volatility swap using variance swaps in our framework.

We are able to derive a closed formula for the dynamic hedge ratio, i.e. the number of units of variance swaps to hold at each time in order to hedge a position on a volatility swap.

In the following subsections we give a list of main steps in pricing of variance and volatility swaps and hedging of volatility swaps for delayed Heston model.

3.1 Delayed Heston Model for Variance

As we mentioned before, our main motivation is to take into account past history of the varinace in its diffusion (over some delayed time interval $[t - \tau, t]$).

The Heston model is one of the most popular stochastic volatility models in the industry, as semi-closed formulas for vanilla option prices are available, few (five) parameters need to be calibrated, and it accounts for the mean-reverting feature of the volatility:

$$\begin{cases} dS_t = rS_t dt + \sqrt{V_t} S_t dW_t^Q \\ dV_t = [\gamma(\theta^2 - V_t)]dt + \delta\sqrt{V_t} dW_t^Q, \end{cases}$$

where S_t is a stock price, V_t is the stochastic variance (just square of volatility), W_t^Q is the Wiener process with respect to the risk-neutral probability Q and r is the interest rate.

We would like to take into account not only its current state (as it is the case in the Heston model) but also its past history over some interval $[t-\tau, t]$, where τ is a positive constant and is called the delay. Namely, at each time t, the immediate future volatility at time $t+\epsilon$ will not only depend on its value at time t but also on all its history over $[t-\tau, t]$. Namely, at each time t, the immediate future volatility at time $t+\epsilon$ will not only depend on its value at time t but also on all its history over $[t-\tau, t]$. We would like to mention that the non-Markovian continuous-time GARCH model (see [14])

$$\frac{dV_t}{dt} = \gamma(\theta^2 - V_t) + \alpha\left[\frac{1}{\tau}(\int_{t-\tau}^{t} \sqrt{V_s}dZ_s^Q - (\mu - r)\tau)^2 - V_t\right]$$

is not suitable in this case, as there is no closed-form solution available for this model. Therefore, we proposed the following Markovian delayed Heston model for variance (see [18]):

$$\begin{cases} dV_t = [\gamma(\theta^2 - V_t) + \epsilon_\tau(t)]dt + \delta\sqrt{V_t}dW_t^Q \\ \epsilon_\tau(t) := \alpha\left[\tau(\mu - r)^2 + \frac{1}{\tau}\int_{t-\tau}^{t} E^Q(V_s)ds - E^Q(V_t)\right]. \end{cases} \tag{3}$$

We can notice that the classical Heston model and 'delayed variance' in (3) are very similar in the sense that the expected values of the variances are the same – when we make the delay tends to 0 in 'delayed variance'. As mentioned before, the Heston framework is very convenient for practitioners, and therefore it is naturally tempting to adjust the Heston dynamics in order to incorporate – in some way – the delay introduced in 'delayed variance'. We note, that $\lim_{\tau\to 0}\sup_{t\in R_+}|\epsilon_\tau(t)| = 0$.

3.2 Variance & Volatility Swaps Pricing Techniques

We show in this section how to calculate the variance and volatility swaps.

Consider the realized variance: $V_R := \frac{1}{T}\int_0^T V_s ds$, where V_t is defined in (3), and suppose that

$$K_{var} = E^Q[V_R], \quad K_{vol} = E^Q[\sqrt{V_R}].$$

To calculate K_{var} we need only $E^Q[V_R]$, but for calculating K_{vol} we need more, due to Brockhaus and Long approximation:

$$E[\sqrt{Z}] \approx \sqrt{E[Z]} - \frac{Var[Z]}{8E[Z]^{3/2}}, \tag{4}$$

namely, $Var[V_R]$.

Using time-changed Brownian motion representation for continuous local martingales:
if $x_t := -(V_0 - \theta_\tau^2)e^{\gamma - \gamma_\tau t} + e^{\gamma t}(V_t - \theta_\tau^2)$, and
$dx_t = f(t, x_t)dW_t^Q, \quad x_t = \hat{W}_{T_t}, \quad T_t =< x >_t = \int_0^t f^2(s, x_s)ds$, where
$\theta_\tau^2 := \theta^2 + \frac{\alpha\tau(\mu - r)^2}{\gamma}, \quad \gamma_\tau := \alpha + \gamma + \frac{\alpha}{\gamma_\tau \tau}(1 - e^{\gamma_\tau \tau})$, then

$$V_t = \theta_\tau^2 + (V_0 - \theta_\tau^2)e^{-\gamma_\tau t} + e^{-\gamma t}\hat{W}_{T_t} = E^Q[V_t] + e^{-\gamma t}\hat{W}_{T_t}, \tag{5}$$

we get closed formula for Variance Swap and Volatility Swap fair strikes. Formula (5) for the variance is the time-change representation for the variance in the delayed Heston model and our main object in finding the variance and volatility swaps.

The parameter θ_τ^2 in (5) can be interpreted as the delayed-adjusted long-range variance. We note, that $\theta_\tau^2 \to \theta^2$ as $\tau \to 0$.

The parameter γ_τ in (5) can be interpreted as the delayed-adjusted mean-reverting speed. We note, that $\gamma_\tau \to \gamma$ as $\tau \to 0$.

3.3 *Volatility Swap Hedging*

Consider the Price Processes:

Variance Swap: $X_t(T) := E_t^Q[V_R]$, and Volatility Swap: $Y_t(T) := E_t^Q[\sqrt{V_R}]$, where

$V_R := \frac{1}{T}\int_0^T V_s ds$.

Portfolio containing 1 Volatility Swap and β_t Variance Swaps has the following form:

$$\Pi_t = e^{-r(T-t)}[Y_t(T) - K_{vol} + \beta_t(X_t(T) - K_{var})]$$

If $I_t := \int_0^t V_s ds$ is the accumulated variance at time t, then:

$$X_t(T) = E_t^Q[\frac{I_t}{T} + \frac{1}{T}\int_t^T V_s ds] := g(t, I_t, V_t)$$
$$Y_t(T) = E_t^Q[\sqrt{\frac{I_t}{T} + \frac{1}{T}\int_t^T V_s ds}] := h(t, I_t, V_t)$$

We compute the infinitesimal variations of these processes (using the fact that $X_t(T)$ and $Y_t(T)$ are martingales):

$$\begin{aligned} dX_t(T) &= \frac{\partial g}{\partial V_t}\delta\sqrt{V_t}dW_t^Q, \\ dY_t(T) &= \frac{\partial h}{\partial V_t}\delta\sqrt{V_t}dW_t^Q, \\ d\Pi_t &= r\Pi_t dt + e^{-r(T-t)}[\frac{\partial h}{\partial V_t} + \beta_t\frac{\partial g}{\partial V_t}]\delta\sqrt{V_t}dW_t^Q, \end{aligned} \tag{6}$$

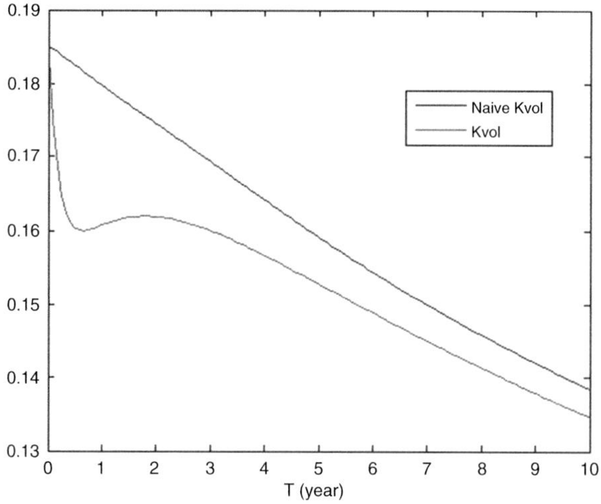

Fig. 2 Naive Volatility Swap vs. adjusted Volatility Swap strikes

hence, to keep our portfolio Π_t risk-neutral, we have to eliminate the coefficient at the random part in (6): $\Rightarrow$

$$\beta_t = -\frac{\frac{\partial h}{\partial V_t}}{\frac{\partial g}{\partial V_t}} = -\frac{\frac{\partial Y_t(T)}{\partial V_t}}{\frac{\partial X_t(T)}{\partial V_t}} \tag{7}$$

-hedge ratio.

We take the parameters that have been calibrated in above-mentioned section (vanilla options on September 30th 2011 for underlying EURUSD, maturities from 1M to 10Y, strikes ATM, 25D Put/Call, 10D Put/Call), namely, they are:

$(v_0, \gamma, \theta^2, \delta, c, \alpha, \tau) = (0.0343, 3.9037, 10^{-8}, 0.808, -0.5057, 71.35, 0.7821)$.

We plot below the naive Volatility Swap strike $\sqrt{K_{var}}$ and the adjusted Volatility Swap strike $\sqrt{K_{var}} - \frac{Var^Q(V_R)}{8K_{var}^{\frac{3}{2}}}$ along the maturity dimension, see Fig. 2, as well as the convexity adjustment $\frac{Var^Q(V_R)}{8K_{var}^{\frac{3}{2}}}$, see Fig. 3, respectively. Also, we plot initial hedge ratio $\beta_0(T)$ with respect to the formula (7) with $t = 0$, see Fig. 4.

4 Discussion and Some Open Problems

In this paper is we gave an overview of delay stochastic models in finance and their applications to modeling and pricing of swaps. Most stochastic dynamical systems in finance describe the state of security (asset or volatility) as a value at time t,-instantaneous or current value at time t. We took into account not only the

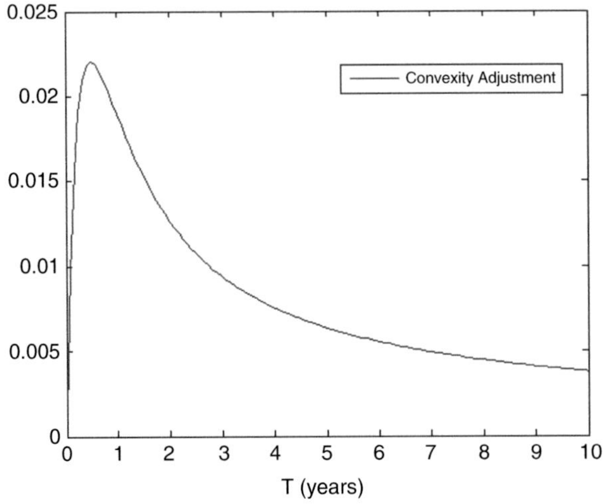

Fig. 3 Convexity adjustment (see (4))

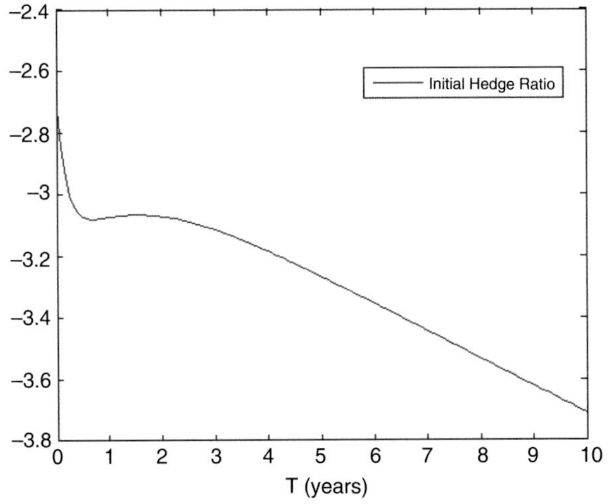

Fig. 4 Initial Hedge ratio $\beta_0(T)$

current value at time t, but also values over some time interval $[t-\tau, t]$, where τ is a positive constant and is called the delay. In this way, we incorporated path-dependent history of the security (asset or volatility) under consideration. In this paper we mainly focused on newly developed so-called delayed Heston model that significantly improve classical Heston model with respect to the market volatility surface fitting by 44 %. Review of some other delay stochastic models in finance has been given as well.

There are some very interesting open problems arising from the delay stochastic models in finance. One of them is modeling and pricing of covariance and correlation swaps in the case of two underlying assets with delayed Heston model for variances. Another problem is a comparison of volatility swap or other volatility derivatives pricing for delayed Heston model and Levy-based stochastic volatility model Levy-based stochastic volatility model. Also, there are some financial models that contain fractional Brownian motion as an uncertainty. It would be good to compare our delay stochastic models with fractional models. We shall leave all these problems for our future research papers.

References

1. Arriojas, M., Hu, Y., Mohammed, S-E.A., Pap, G.: A delayed Black and Scholes formula. Stoch. Anal. Appl. **25**, 471–492 (2007)
2. Bollerslev, T.: Generalized autoregressive conditional heteroskedasicity. J. Econom. **31**, 307–327 (1986)
3. Bollerslev, T., Chou, R., Kroner, K.: ARCH modeling in finance: a review of the theory and empirical evidence. J. Econom. **52**, 5–59 (1992)
4. Dokuchaev, N.: Modelling possibility of short-term forecasting of market parameters. Ann. Econ. Finance **16**(1), 43–161 (2015)
5. Dibeh, G., Harmanani, H.M.: A stochastic chartist fundamentalist model with time delays. Comput. Econ. **40**(2), 105–113 (2012)
6. Heston, S.: A closed-form solution for options with stochastic volatility with applications to bond and currency options. Rev. Financ. Stud. **6**(2), (1993), 327–343
7. Ivanov, A., Kazmerchuk, Yu., Swishchuk, A.: Theory, stochastic stability and applications of stochastic differential delay equations. A survey of recent results. Differ. Equ. Dyn. Syst. **11** (1–2), 55–115 (2003)
8. Kazmerchuk, Y.I., Swishchuk, A.V., Wu, J.H.: Black-Scholes Formula Revisited: Security Markets with Delayed Response, Bachelier Finance Society 2nd World Congress, Crete (2002)
9. Kazmerchuk, Y., Swishchuk, A., Wu, J.-H.: A continuous-time GARCH model for stochastic volatility with delay. Can. Appl. Math. Q. **13**(2), 123–148 (2005)
10. Kazmerchuk, Yu., Swishchuk, A., Wu, J.-H.: The pricing of options for security markets with delayed response. Math. Comput. Simul. **75**, 69–79 (2006)
11. Li, J.-C., Mei, D.-C.: The influences of delay time on the stability of a market model with stochastic volatility. Physica A: Stat. Mech. Appl. **392**(4), 763–772 (2013)
12. Sato, K.-I.: Lévy Processes and Infinitely Divisible Distributions. Cambridge University Press, Cambridge (1999)
13. Stoica, G.: A stochastic delay financial model. Proc. Am. Math. Soc. **133**(6) 1837–1841 (2004)
14. Swishchuk, A.: Modeling and pricing of variance swaps for stochastic volatilities with delay. WILMOTT Mag. (19), 63–73 (Sept. 2005)
15. Swishchuk, A.: Modeling and pricing of variance swaps for multi-factor stochastic volatilities with delay. Can. Appl. Math. Q. **14**(4), 439–467 (2006)
16. Swishchuk, A., Li, X.: Pricing of variance and volatility swaps for stochastic volatilities with delay and jumps. Int. J. Stoch. Anal. 27 (2011). http://dx.doi.org/10.1155/2011/435145
17. Swishchuk, A., Malenfant, K.: Variance swaps for local Levy based stochastic volatility with delay. Int. Rev. Appl. Financ. Issues Econ. **3**(2), 432–441 (2011)
18. Swishchuk, A., Vadori, N.: Smiling for the delayed volatility swaps. Wilmott Mag (November 2014)

Semi-parametric Time Series Modelling with Autocopulas

Antony Ware and Ilnaz Asadzadeh

Abstract In this paper we present an application of the use of autocopulas for modelling financial time series showing serial dependencies that are not necessarily linear. The approach presented here is semi-parametric in that it is characterized by a non-parametric autocopula and parametric marginals. One advantage of using autocopulas is that they provide a general representation of the auto-dependency of the time series, in particular making it possible to study the interdependence of values of the series at different extremes separately. The specific time series that is studied here comes from daily cash flows involving the product of daily natural gas price and daily temperature deviations from normal levels. Seasonality is captured by using a time dependent normal inverse Gaussian (NIG) distribution fitted to the raw values.

1 Introduction

In this study, autocopulas are used to characterise the joint distribution between successive observations of a scalar Markov chain. A copula joins a multivariate distribution to its marginals, and its existence is guaranteed by Sklar's theorem [8]. In particular, a Markov chain of first order with any given univariate margin can be constructed from a bivariate copula. A theoretical framework for the use of copulas for simulating time series was given by [3], who presented necessary and sufficient conditions for a copula-based time series to be a Markov process, but not necessarily a stationary one. They presented theorems specifying when time series generated using time varying marginal distributions and copulas are Markov processes. Joe [4] proposed a class of parametric stationary Markov models based on parametric copulas and parametric marginal distributions. Chen and Fan [2] studied the estimation of semiparametric stationary Markov models,

This work was supported by MITACS, Direct Energy and by an NSERC Discovery Grant.

A. Ware (✉) • I. Asadzadeh
University of Calgary, 2500 University Drive NW, Calgary, AB T2N 1N4, Canada
e-mail: aware@ucalgary.ca; iasadzad@ucalgary.ca

© Springer International Publishing Switzerland 2016

J. Bélair et al. (eds.), *Mathematical and Computational Approaches in Advancing Modern Science and Engineering*, DOI 10.1007/978-3-319-30379-6_52

using non-parametric marginal distributions with parametric copulas to generate stationary Markov processes.

The term 'autocopula' was first used to describe the unit lag self dependence structure of a univariate time series in [7], and we adopt the terminology here. We make use of the framework presented in [3] to produce Markov processes such that the marginal distribution changes over time. The main benefit of using autocopulas for univariate time series modelling is that the researcher is able to specify the unconditional (marginal) distribution of X_t separately from the time series dependence of X_t [6]. We apply a semiparametic method which is characterized by an empirical autocopula and a parametric time varying marginal distribution. This allows us to capture seasonal variations in a natural way. This is an important feature of our model, motivated by the fact that many financial time series exhibit seasonality, particularly those arising from energy and commodity markets.

The remainder of this paper is organized as follows. In Sect. 2, we introduce the data; Sect. 3 describes the model, including a review of copulas, and of the Normal Inverse Gaussian distribution. This section also includes details of the calibration and simulation procedures, and the final section presents some results.

2 The Data

The motivation for this project came from the desire to develop a parsimonious model that could help to capture so-called load-following (or swing) risk. This is one of the main sources of financial uncertainty for an energy retailer, and arises from the combination of retail customer consumption (volume) uncertainty and price uncertainty. Both volume (V) and price (P) are driven to a large extent by weather. In particular, average daily temperature is one of the main drivers of daily natural gas consumption in various North American markets: this in turn drives market prices through a supply and demand process.

Some of the load-following risk exposure can be hedged using gas forwards and temperature derivatives. A significant part of the risk exposure that cannot be easily hedged and is linked to the daily product between the weather deviation from normal (žF) and the daily price deviation from the expected value of the ex-ante forward price. To make this more specific, let $\overline{P}$ denote the last-traded forward monthly index price, and $\overline{V}$ the expected monthly average volume. Cash flows for the retailer depend on the product PV, and the uncertainty in this quantity can be written

$$(\overline{P} + \Delta P)(\overline{V} + \Delta V) - \overline{PV} = \overline{P}\Delta V + \overline{V}\Delta P + \Delta P \Delta V.$$

We can capture the first-order dependence between volume and weather deviations by writing $\Delta V = \beta \Delta W + \epsilon$, where ϵ represents higher-order dependence on ΔW as well as exogenous risk factors; modelling these other contributions was beyond the scope of this study. Forward instruments in weather and natural gas markets

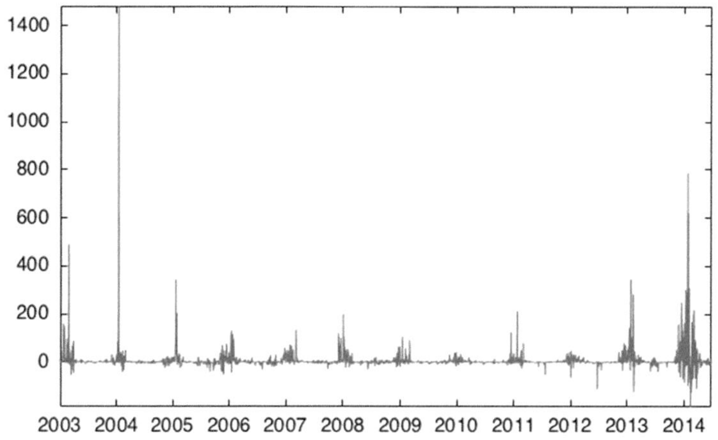

Fig. 1 Product of weather and gas price deviations ($\Delta P \Delta W$) in Algonquin over 2003–2014. Spikes correspond to combinations of high weather deviation from normal and high spot price deviation from next forward month

can then be used to hedge risks corresponding to the terms $\overline{P}\Delta V$ and $\overline{V}\Delta P$. It can be seen that the term $\Delta P \Delta W$ is an important component of unhedged risk in these cashflows. One approach to modelling this component would be to develop separate models for weather and natural gas prices (both daily and forward prices). However, because of the desire for parsimony, we instead seek a model that allows us to study the time-series $X_t = (\Delta P \Delta W)_t$, in order to estimate the range and probabilities of possible outcomes at the level of a complex portfolio of retail load obligations.

Here we focus on Algonquin Citygate gas prices and the Boston Logan station for weather data. The data cover the period 1 January 2003–31 June 2014 on a daily basis, and are shown in Fig. 1. The most dramatic feature of the graph is the presence of intermittent clusters of spikes, during which the gas prices rise from their approximate average daily value and at the same time temperature rises or falls drastically. These mostly occur during winter, although large deviations also occur at other times of the year. It is also clear that the marginal densities of these observations will not be well-represented by normal distributions.

3 The Model

Here we introduce the simulation model in more detail, providing a brief review of copulas, and the normal inverse Gaussian distribution, which we use for the marginal densities.

3.1 Copulas and Autocopulas

A copula[1] is a multivariate distribution function defined on a unit cube $[0, 1]^n$, with uniformly distributed marginals. In the following, we use copulas for the interdependence structure of time series and, for simplicity and the fact that we are interested in the first order lag interdependence, we focus on the bivariate case, although the approach can be used to capture dependence on higher order lags.

Let $F_{12}(x, y)$ be the joint distribution function of random variables X and Y whose marginal distribution functions, denoted as F_1 and F_2 respectively, are continuous. Sklar's theorem specifies that there exists a unique copula function $C(u, v) = F_{12}(F_1^{-1}(u), F_2^{-1}(v))$ that connects $F_{12}(x, y)$ to $F_1(x)$ and $F_2(y)$ via $F_{12}(x, y) = C(F_1(x), F_2(y))$. The information in the joint distribution $F_{12}(x, y)$ is decomposed into that in the marginal distributions and that in the copula function, where the copula captures the dependence structure between X and Y. Various families of parametric copulas are widely used (Gaussian, Clayton, Joe, Gumbel copulas, for example).

In a time series setting, we use a copula (or *autocopula*) to capture the dependence structure between successive observations. More generally, we have the following definition [7].

Definition 1 (Autocopula) Given a time series X_t and $\mathscr{L} = \{l_i \in \mathbb{Z}^+, i = 1, \ldots, d\}$ a set of lags, the autocopula $C_{X,\mathscr{L}}$ is defined as the copula of the $d + 1$ dimensional random vector $(X_t, X_{t-l_1}, \ldots, X_{t-l_d})$.

If a times series X_t is modelled with an autocopula model with unit lag, with autocopula function $C(u, v) = C_{X,1}(u, v)$, and (time-dependent) marginal CDF $F_t(x)$, then, for each t, the CDF of the conditional density of X_t given X_{t-1} can be expressed

$$F_{X_t|X_{t-1}}(x) = \frac{\partial C}{\partial u}\big(F_{t-1}(X_{t-1}), F_t(x)\big). \tag{1}$$

We will discuss issues related to calibration and simulation below.

Autocopula models include many familiar time series as special cases. For example, it is straightforward to show that an AR(1) process, $y_t = \alpha y_{t-1} + \beta + \sigma\epsilon(t)$, can be modelled using the autocopula framework using the marginal distribution $F_\infty(y) = \Phi\left(\frac{y-\beta/(1-\alpha)}{\sqrt{\sigma^2/(1-\alpha^2)}}\right)$ (where Φ denotes the standard normal CDF) and a Gaussian copula with mean $\mu = \beta/(1-\alpha)$ and covariance $\frac{\sigma^2}{1-\alpha^2}\begin{bmatrix} 1 & \alpha \\ \alpha & 1 \end{bmatrix}$.

Part of the motivation for the use of autocopulas in time series modelling is that, while correlation coefficients measure the general strength of dependence, they provide no information about how the strength of dependence may change across

[1] For more discussion on the theory of copulas and specific examples, see [5].

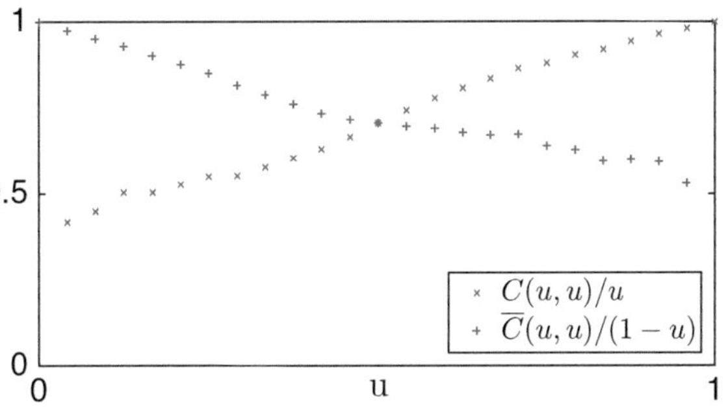

Fig. 2 Estimated values of the quantities $C(u,u)/(u)$ and $\overline{C}(u,u)/(1-u)$ showing lower and upper tail dependence in the observed values of $\Delta P\Delta W$

the distribution. For instance, in the dataset we consider here there is evidence of tail dependence, whereby correlation is higher near the tails of the distribution. We can quantify this using the following definition ([4], Section 2.1.10).

Definition 2 (Upper and Lower Tail Dependence) If a bivariate copula C is such that $\lim_{u\to 1}\overline{C}(u,u)/(1-u) = \lambda_U$ exists, where $\overline{C}(u,u) = 1 - C(1,u) - C(u,1) + C(u,u)$, then C has upper tail dependence if $\lambda_U \in (0,1]$ and no upper tail dependence if $\lambda_U = 0$. Similarly, if $\lim_{u\to 0} C(u,u)/(u) = \lambda_L$ exists, C has lower tail dependence if $\lambda_L \in (0,1]$ and no lower tail dependence if $\lambda_L = 0$

In Fig. 2 we show estimates of the quantities $C(u,u)/(u)$ and $\overline{C}(u,u)/(1-u)$, where here we use the order statistics of the time series $X_t = (\Delta P\Delta W)_t$ to generate a preliminary empirical proxy for the copula function C. It is clear from the figure that neither set of values tends towards zero in the limit $u \to 0$ or $u \to 1$, and we conclude that the data exhibit nonzero tail dependence.

3.2 *Time Varying Marginal Distribution*

As noted above, the marginal densities for our time series will not be normal. We found that the normal inverse Gaussian (NIG) distribution provided a more satisfactory fit. More information about this distribution and its applications can be found in [1]. Here we review its definition and properties.

3.2.1 Definition and Properties of the NIG Distribution

A non-negative random variable Y has an inverse Gaussian distribution with parameters $\alpha > 0$ and $\beta > 0$ if its density function is of the form

$$f_{\text{IG}}(y;\alpha,\beta) = \frac{\alpha}{\sqrt{2\pi\beta}} y^{-3/2} \exp\Big(-\frac{(\alpha-\beta y)^2}{2\beta y}\Big), \text{ for } y > 0.$$

A random variable X has an NIG distribution with parameters α, β, μ and δ if

$$X|Y = y \sim N(\mu + \beta y, y) \text{ and } Y \sim \mathrm{IG}(\delta\gamma, \gamma^2),$$

with $\gamma := \sqrt{\alpha^2 - \beta^2}$, $0 \leq |\beta| < \alpha$ and $\delta > 0$. We then write $X \sim \mathrm{NIG}(\alpha, \beta, \mu, \delta)$. Denoting by K_1 the modified Bessel function of the second kind, the density is given by

$$f_{\mathrm{NIG}}(x; \alpha, \beta, \mu, \delta) = \frac{\delta\alpha \exp\left(\delta\gamma + \beta(x - \mu)\right)}{\pi\sqrt{\delta^2 + (x - \mu)^2}} K_1\left(\alpha\sqrt{\delta^2 + (x - \mu)^2}\right).$$

There is a one-to-one map between the parameters of the NIG distribution and the mean, variance, skewness and kurtosis of the data. We first use moment matching to determine initial estimates for the parameters, which we use as starting values for a MLE search method. The corresponding fit to the data is shown in Fig. 3, where the best fitting normal density is also shown. It can be seen that the NIG fit is quite good. However, it is evident from Fig. 1 that the time series is strongly seasonal. We seek to capture this seasonality through the marginal densities by making one of the parameters of the NIG model time-dependent. This was achieved by assuming the parameter to be constant in each month, and maximizing the resulting joint likelihood across the entire data set. Of the four possible choices, δ gave the greatest improvement to the AIC and BIC, as shown in Table 1. The estimated values of δ are shown in Fig. 4, and the seasonal pattern that is evident in the original data is evident again here.

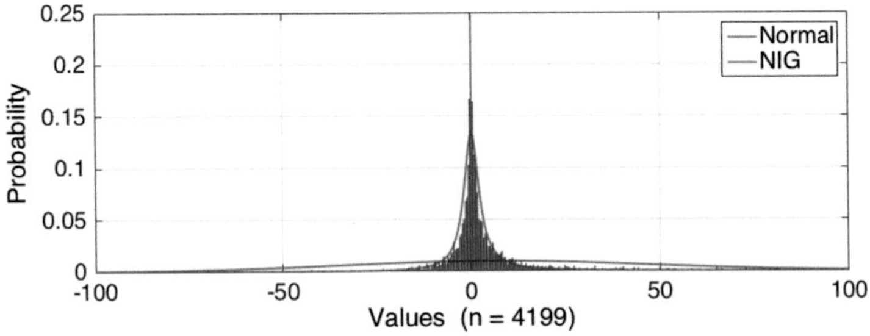

Fig. 3 Histogram of observed data ($\Delta P \Delta W$), with fitted normal distribution and NIG distribution. The estimated NIG parameters are: $\alpha = 0.0980$, $\beta = 0.0131$, $\mu = 0.0122$ and $\delta = 2.3799$

Table 1 Calibrated values of AIC, BIC, and Log likelihood

Model	AIC	BIC	Log likelihood
α be time varying	34,622.85	35,793.11	−17,133.42
β be time varying	36,280.04	37,450.30	−17,962.02
μ be time varying	35,850.14	37,020.40	−17,747.07
δ be time varying	33,495.75	34,666.02	−16,569.87

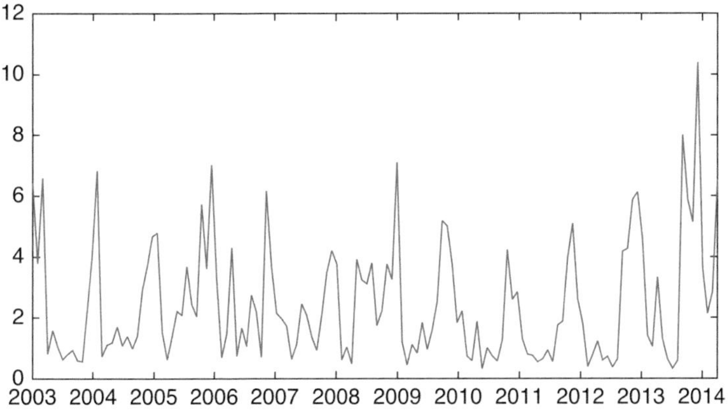

Fig. 4 Calibrated monthly values of δ from the combined NIG likelihood, $\alpha = 0.0293$, $\beta = 0.0205$, and $\mu = -0.0323$

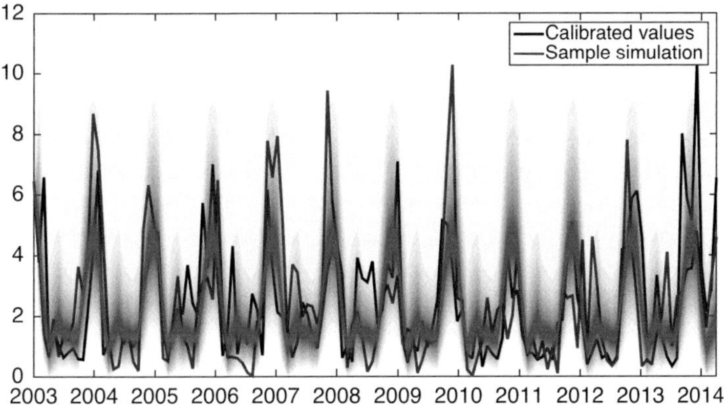

Fig. 5 Calibrated monthly values of δ, together with an example of a simulated path, as well as a colour contour plot of the quantiles from a large number of simulated paths. Seasonal mean and variance in (2) are linear combination of a constant and $\sin 2\pi nt$ and $\cos 2\pi nt$ for increasing n. The estimated parameters are as follows: $a = 0.31$, $b = (0.93, 0.30, -0.09, 0.22, -0.12, 0.032, -0.01, -0.03, -0.01)$, and $\sigma = (0.15, 0.09, 0.01)$

As can be seen in Fig. 4, the value of δ tends to be higher in winter and lower in summer. The time series of values appears to be mean reverting with seasonal mean and variance. We model the time series using a seasonal mean reverting process for $\nu_t = \sqrt{\delta_t}$:

$$\nu_{t+1} = a\nu_t + b(t) + \sigma(t)z_t, \tag{2}$$

where the z_t are independent standard normal samples. The mean and variance are estimated using periodic functions with periods from 1 year down to 3 months. Simulated and estimated values of δ_t are shown in Fig. 5. Twenty thousand paths were simulated using (2), and for each month the set of values was used to determine quantiles, which were then used to create the coloured patches shown in the figure. The darker patches correspond to quantiles nearer to the centre of the distribution, and the lighter patches to quantiles nearer the extremes.

Once we have values of δ_t, we can obtain the time varying cumulative distribution function and time varying density function. The NIG cumulative distribution function does not have a closed form solution, so we can compute the CDF using Gaussian quadrature to evaluate the following integral.

$$F(x_t; \alpha, \beta, \mu, \delta) = \int_{-\infty}^{x_t} f_{\mathrm{NIG}}(X_t; \alpha, \beta, \mu, \delta_t) dX_t \tag{3}$$

In next section we explain the procedure to calculate the empirical autocopulas and simulate cash flows.

3.3 Estimating the Empirical Autocopula

Having estimated the time-dependent NIG densities, we use these to produce a time series of values $V_t = F_t(X_t) \in [0, 1]$. If the marginal densities were exact, these would be uniformly distributed on [0, 1]. In practice, they will only be approximately uniform, and we generate an additional empirical marginal density and an empirical (auto)copula to capture the joint density of (V_t, V_{t-1}).

The empirical autocopula C is estimated by first estimating an empirical joint density for (V_t, V_{t-1}) in the form of a strictly increasing continuous function $\Phi(\cdot, \cdot)$ that is piecewise bilinear. The domain $[0, 1]^2$ is partitioned into rectangles containing approximately similar numbers of samples (V_{t-1}, V_t), and taking Φ to be the cumulative integral of the sum of indicator functions for these rectangles, scaled by the number of samples in each rectangle. Φ is then used to create strictly increasing piecewise linear marginal densities Φ_1 and Φ_2. The inverses of these densities are therefore also piecewise linear, and when composed with Φ they generate a piecewise bilinear copula function $C(u_1, u_2) = \Phi\big(\Phi_1^{-1}(u_1), \Phi_2^{-1}(u_2)\big)$.

This process is illustrated in Fig. 6. In Fig. 6a we plot the pairs of transformed values $\big(\Phi_1(V_{t-1}), \Phi_2(V_t)\big)$, together with the outlines of rectangles used to generate the piecewise bilinear function C. As mentioned, these rectangles contain roughly equal numbers of points; constructing the empirical autocopula in this way ensures that it is strictly increasing, and well-suited to enable the computations involved in time series simulation (see below) to be carried out efficiently. The resulting empirical autocopula C is shown in Fig. 6b. This function is binlinear on each of the rectangles shown in Fig. 6a, but is less regular than it looks. The corresponding joint density, $\frac{\partial^2 C}{\partial u_1 \partial u_2}(u_1, u_2)$, is shown in Fig. 6c. It can be seen that the density is higher near $(0, 0)$ and near $(1, 1)$, which is consistent with the tail dependency observed earlier.

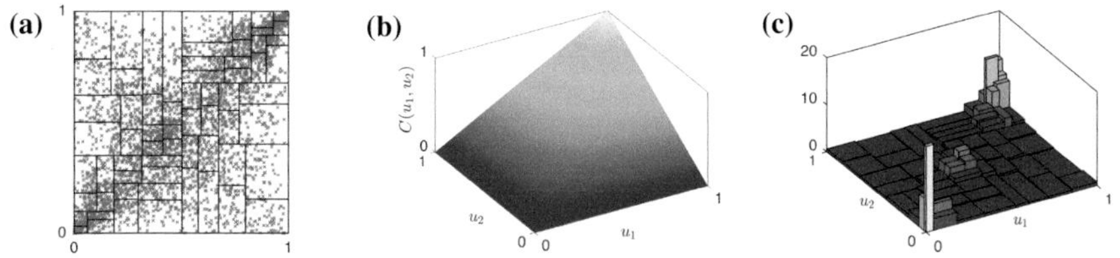

Fig. 6 Generation of the empirical autocopula. (**a**) Scatter plot of $\Phi_1(V_{t-1})$ against $\Phi_2(V_t)$. Each rectangle contains about 100 points (note that rectangles with around 25 points were used in the simulation). (**b**) Empirical autocopula $C(u_1, u_2)$ defined to be bilinear on each of the rectangles shown in (**a**). (**c**) The empirical density $\partial^2 C/\partial u_1 \partial u_2$, which is constant on each of the rectangles shown in (**a**)

3.4 Simulation of Time Series Using Autocopula

Armed with the time-dependent NIG densities $F_t(\cdot)$, the empirical marginal densities $F_{V,i}(\cdot)$ and the empirical autocopula $C(\cdot,\cdot)$, we can generate simulated values x_t as follows.

1. Given an initial value x_0, generate $v_0 = F_0(x_0)$.
2. For $t = 0, 1, \ldots,$ given v_t, generate v_{t+1}:
   a. Set $u_1 = \Phi_1(v_t)$.
   b. Given u_1, create the piecewise linear function $\underline{C}(u) := C(u_1, u)/u_1$.
   c. Set $u_2 = \underline{C}^{-1}(U)$, where U is an independent uniform random draw.
   d. Set $v_{t+1} = \Phi_2^{-1}(u_2)$.
3. For each $t > 0$, set $x_t = F_t^{-1}(v_t)$.

Here we have used the fact (already alluded to in (1)) that, if U_1 and U_2 are uniform random variables whose joint distribution is the copula $C(u_1, u_2)$, then, for $u_1 > 0$, the cumulative density function for U_2, conditional on $U_1 = u_1$, is

$$P[U_2 < u_2 | U_1 = u_1] = \frac{\partial C}{\partial u_2}(u_1, u_2) = \frac{C(u_1, u_2)}{u_1}.$$

The proof of this can be found in, for example, [3]. The fact that C is a piecewise bilinear function means that $\underline{C}$ will be piecewise linear. Moreover, the construction of the empirical copula as described in Sect. 3.3 ensures that it is an increasing function with a limited number of corners. Its inverse can then be constructed readily, and will also be an increasing piecewise linear function with a limited number of corners, and so can be evaluated with little computational effort. Indeed, in practice the computation of the final step in the above algorithm, the inversion of the time-dependent NIG densities, took more time than the copula-related computations.

4 Results

In Fig. 7 we show a 12-year sample time series for $\Delta P \Delta W$ computed as described in Sect. 3.4. In addition, we simulated around 700 independent time series and computed, for each month, the 99th percentile of values produced in that month across all simulations. What can be seen in the sample path is the same mixture of quiescent periods and periods with large deviations from zero. There is some evidence of 'clumps' of large deviations occurring in winter months, although this is less clear than in the original data (see Fig. 1). There is, nevertheless, an increased occurrence of large deviations in winter months, as can be seen from the plot of the 99th percentiles superimposed on the sample simulation, shown in Fig. 7.

Figure 8 reproduces the data from Fig. 2, with error bars corresponding to the 5th and 95th percentiles of the values obtained from the simulations, showing that the simulations have reproduced the observed tail dependence.

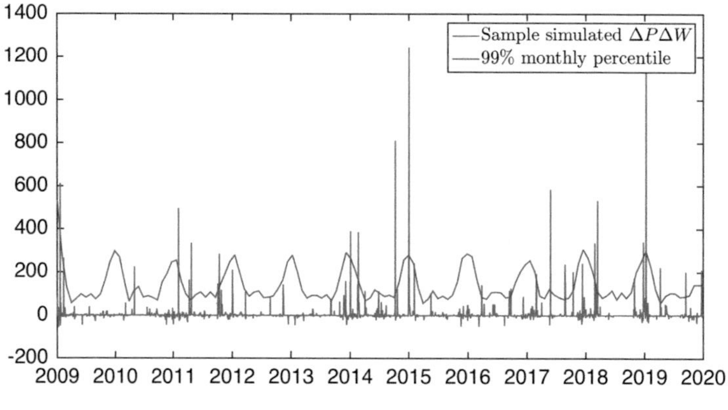

Fig. 7 One example of simulated daily values of $\Delta P \Delta W$, together with the 99th percentiles of collected monthly values from around 700 simulations

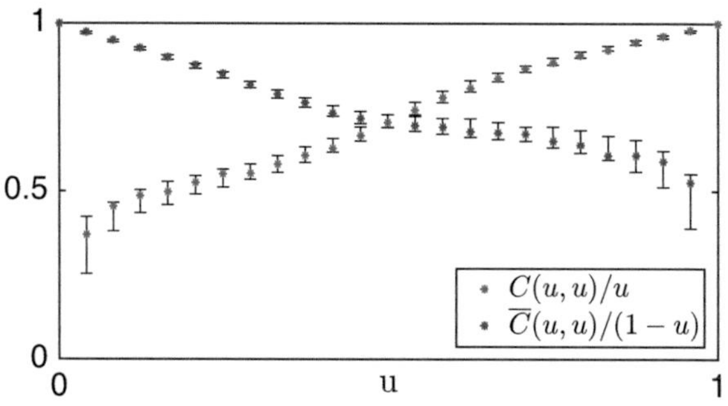

Fig. 8 Estimated values of the quantities $C(u,u)/(u)$ and $\overline{C}(u,u)/(1-u)$ for the original observations of $\Delta P \Delta W$. Also shown are error bars corresponding to the 5th and 95th percentiles of the values obtained from around 700 simulations

References

1. Barndorff-Nielsen, O.E., Mikosch, T., Resnick, S.I.: Lévy Processes: Theory and Applications. Springer Science & Business Media, New York (2001)
2. Chen, X., Fan, Y.: Estimation of copula-based semiparametric time series models. J. Econom. **130**(2), 307–335 (2006)
3. Darsow, W.F., Nguyen, B., Olsen, E.T., et al.: Copulas and Markov processes. Ill. J. Math. **36**(4), 600–642 (1992)
4. Joe, H.: Multivariate Models and Multivariate Dependence Concepts. CRC Press/Taylor & Francis Group, Boca Raton/London, New York (1997)
5. Nelsen, R.B.: An Introduction to Copulas, 2nd edn. Springer Science & Business Media, New York (2007)
6. Patton, A.J.: Copula–based models for financial time series. In: Handbook of Financial Time Series, pp. 767–785. Springer, Berlin/Heidelberg (2009)
7. Rakonczai, P., Márkus, L., Zempléni, A.: Autocopulas: investigating the interdependence structure of stationary time series. Methodol. Comput. Appl. Probab. **14**(1), 149–167 (2012)
8. Sklar, M.: Fonctions de répartition à n dimensions et leurs marges. Université Paris 8, (1959)

Optimal Robust Designs of Step-Stress Accelerated Life Testing Experiments for Proportional Hazards Models

Xaiojian Xu and Wan Yi Huang

Abstract Accelerated life testing (ALT) is broadly used to obtain reliability information of a product in a timely manner. The Cox's proportional hazards (PH) model is often utilized for reliability analysis. In this paper, we focus on designing ALT experiments for hazard rate prediction when a PH model is adopted. Due to the nature of prediction made from ALT experimental data, attained under the stress levels higher than the normal condition, extrapolation is encountered. In such case, the assumed model can not be tested. For possible imprecision in an assumed PH model, the method of construction for robust designs is explored. As an example, we investigate the robust designs for the situation where baseline hazard function in the fitted PH model is simple linear but the true baseline hazard function is actually quadratic. The optimal stress-changing times are determined by minimizing the asymptotic variance and asymptotic squared bias.

1 Introduction

Failure data are needed in order to quantify the life characteristics of a product. However, under normal design conditions such failure life data are very difficult to obtain for a product with high reliability within a reasonable time period. To overcome this problem, accelerate life testing (ALT) has been well-developed to shorten the lifetime of a product and quickly obtain failures.

In general, test units in accelerated life testing experiments are subjected to higher than normal design level of stresses. The use of such accelerating stresses for a particular material is established by engineering practice. In an ALT experiment, both censoring and time-dependent loading plan are often used. In this paper, we consider time-censoring and time-dependent step-stress loading plan. Time-censoring occurs when the experiment stops at a predefined censoring time. In time-dependent step-stress ALT, all test units are subjected to stress levels increasing by steps: first, all test units are subjected to a lower stress level for a

X. Xu (✉) • W.Y. Huang
Department of Mathematics and Statistics, Brock University, St. Catharines, ON L2S 3A1, Canada
e-mail: xxu@brocku.ca; wh13oq@brocku.ca

© Springer International Publishing Switzerland 2016

J. Bélair et al. (eds.), *Mathematical and Computational Approaches in Advancing Modern Science and Engineering*, DOI 10.1007/978-3-319-30379-6_53

specified length of time, then the stress level increases and the test continues for all unfailed test units, and so on.

To avoid invalid estimation, the range of stress levels should be restricted in a stress loading plan. The failure data obtained at accelerated conditions are extrapolated to estimate the characteristic of life distribution at normal design conditions.

In the literature of designing a step-stress ALT experiment, most work has been done for the cases when a simple step-stress plan is adopted, when the model is fully parametric, or when model assumed is exactly correct. To mention a few: Miller and Nelson [9] have first presented Q-optimal designs for simple step-stress tests with complete data from an exponential distribution. Their designs minimize the asymptotic variance (AVAR) of the maximum likelihood estimator (MLE) of the mean life at a normal design stress. Bai et al. [2] have extended the theory of Miller and Nelson [9] to censored data. They have constructed the optimal simple step-stress ALT with time-censoring. Bai and Kim [1] have presented an optimal simple step-stress ALT for a Weibull distribution under time-censoring. The optimal low stress level and stress change time are obtained by minimizing the AVAR of the MLE of a specified percentile at normal design stress. Fard and Li [5] have also presented a step-stress ALT for a Weibull distribution under time-censoring; however, their optimal stress-changing time is obtained by minimizing the AVAR of the MLE for reliability instead. Recent work on optimal designs for step-stress ALT has been reviewed by Hunt and Xu [6]. They have adopted a generalized Khamis-Higgins model for the effect of changing stress levels and also assumed that the lifetime of a test unit follows a Weibull distribution; however, their designs are also optimal when both the shape and scale parameters are considered being functions of the stress levels. The resulting optimal design chooses the stress-changing time in order to minimize AVAR of the MLE of reliability at the normal design stress level and at a pre-specified time.

Ma and Meeker [8] have extended the results of Bai and Kim [1] to provide a general method for multiple step-stress ALTs for a log location-scale family of distributions. They have presented an approach to calculate the large-sample approximate variance of the MLE, computed from step-stress ALT data, for a quantile of the failure time distribution at normal design conditions. Jiao [7] has first developed the simple step-stress ALT plan when a PH model is utilized for reliability prediction. Their optimal stress level is obtained by minimizing the variance of the MLE of hazard rate at the normal design stress level and over a pre-specified time period. Elsayed and Zhang [3] have presented an optimal simple step-stress ALT plan based on a PH model to obtain the most accurate reliability function estimates at normal design stress. They have also formulated a nonlinear programming problem to minimize AVAR of the hazard rate estimator over a prespecified time period at normal design stress.

The general theory for the asymptotic distribution of MLEs under model misspecification has been derived by White [11]. He has examined the consequences and detection of model misspecification when the maximum likelihood technique is used for estimation. The proposed quasi-MLE converges to a well-defined limit. In addition, the properties of their quasi-MLE and the information matrix have been exploited to yield several useful tests when model misspecification is suspected. Pascual [10] has presented the methodology for deriving the asymptotic distribution of MLEs of model parameters with constant stress ALT when the stress-life relationship is misspecified. When possible departures from an assumed ALT model are suspected, his ALT plans can provide protection against potential bias without much loss in efficiency. However, the methodology developed there can be used to derive robust designs only for constant stress ALTs.

In a complete general setting, robust designs for one-point extrapolation which is the case for ALT have been discussed in Wiens and Xu [12], for least squares estimation of a mean response. As Fang and Wiens [4] pointed out, "Extrapolation to regions outside of that in which observations are taken is, of course, an inherently risky procedure and is made even more so by an over-reliance on stringent model assumptions." The classical optimal designs minimize the variance alone. However, when the fitted models are incorrect, the estimation is biased. A robust design should be obtained in an optimal way so that even when the fitted model was not exactly correct, the designs can still be relatively efficient with a small bias.

The present paper extends previous work to construct optimal robust ALT designs by taking into account multiple step-stress, semi-parametric model assumption, and possible imprecision in the assumed model. Due to the nature of the prediction made from ALT experimental data, attained under the stress levels higher than the normal design condition, extrapolation is encountered. For possible imprecision in an assumed PH model, the method of construction for robust designs becomes significantly important. Therefore, we consider the situation where a PH model with a simple linear baseline hazard function is fitted; however, the true baseline hazard function is a quadratic function. Optimal robust designs will be obtained in order to protect against possible departure from the assumed model with a minimum loss of efficiency. We propose a two-stage design procedure, in Sect. 4, where the optimal stress-changing times are derived by minimizing the asymptotic squared bias (ABIAS^2) at the first stage and then minimizing the AVAR at the second stage.

2 Model Misspecification and Asymptotic Distribution of MLE Attained

The Cox's proportional hazards (PH) model is one of the most important means for predicting the lifetime of a product. This model provides a flexible method for identifying the effects of the covariates on failure rate. A PH model with a covariate (applied or transformed stress) independent of time is generally expressed as

$$\lambda(t; s) = \lambda_0(t) exp(\theta s), \tag{1}$$

where $\lambda_0(t)$ is the baseline hazard rate function at the time of t, s is a stress level used in an ALT test, and θ is a unknown parameter.

It is assumed that the true model is Model (1) with a quadratic baseline hazard function, denoted by M_T, i.e., Model (1) with $\lambda_0(t) = \kappa_0 + \kappa_1 t + \kappa_2 t^2$ and $\boldsymbol{\theta} = \alpha$. However, the fitting model is Model (1) with a linear baseline hazard function, denoted by M_F, i.e., Model (1) with $\lambda_0(t) = \gamma_0 + \gamma_1 t$ and $\boldsymbol{\theta} = \beta$.

The maximum likelihood method has been used for estimating the model parameters. Define $\boldsymbol{\zeta} = [\kappa_0, \kappa_1, \kappa_2, \alpha]^T$ and $\boldsymbol{\eta} = [\gamma_0, \gamma_1, \beta]^T$. Let $\ell(\boldsymbol{\zeta}; \xi)$ be the log-likelihood function under M_T, and $\ell(\boldsymbol{\eta}; \xi)$ be the log-likelihood function under M_F, both with the same design, ξ. The expected log-likelihood ratio under M_T over M_F is

$$I(\zeta : \eta) = E_{M_T}[\ell(\zeta; \xi) - \ell(\eta; \xi)]. \tag{2}$$

In this paper, we use a 3-step-stress design to demonstrate our design method; however, the method developed can be applied to any m-step-stress ALT ($m \geq 2$).

We let $f_F(t; s)$ and $G_F(t; s)$ be the probability density function and the cumulative distribution function at failure time t and stress level s under M_F, respectively. We also let c be the censoring time, and $F_i(t) = G_F(t; s_i)$, i =1, 2, 3. In addition, we define a such that the cumulative distribution function under stress level s_2, evaluated at a is the same as that under stress level s_1, evaluated at τ_1; and we define b such that the cumulative distribution function under stress level s_3, evaluated at b is the same as that under stress level s_2, evaluated at $a + \tau_2 - \tau_1$. For a 3-step-stress ALT design, the likelihood function from observed lifetimes t_i, $i = 1, \ldots, n$, is

$$L(\boldsymbol{\eta}; \xi) = \prod_{i=1}^{n_1} f_F(t_i; s_1) \prod_{i=1}^{n_2} f_F(x_i; s_2) \prod_{i=1}^{n_3} f_F(y_i; s_3) \prod_{i=1}^{n-\sum_{i=1}^{3} n_i} (1 - G_F(z; s_3)), \tag{3}$$

where $a = F_2^{-1}(F_1(\tau_1))$, $b = F_3^{-1}(F_2(a + \tau_2 - \tau_1))$, $x = a + t - \tau_1$, $y = b + t - \tau_2$, and $z = b + c - \tau_2$. Given Model (1) with $\lambda_0(t) = \gamma_0 + \gamma_1 t$ and $\boldsymbol{\theta} = \beta$, we have

$$f_F(t; s) = (\gamma_0 + \gamma_1 t)\exp(\beta s)\exp\left[-\left(\gamma_0 t + \frac{\gamma_1}{2}t^2\right)\exp(\beta s)\right],$$

$$G_F(t; s) = 1 - \exp\left[-\left(\gamma_0 t + \frac{\gamma_1}{2}t^2\right)\exp(\beta s)\right].$$

The negative log-likelihood function is

$$\begin{aligned}
-\ell(\boldsymbol{\eta}; \xi) = &-n_1\beta s_1 - \sum_{i=1}^{n_1} \ln(\gamma_0 + \gamma_1 t_i) - \left(\gamma_0 t_i + \frac{\gamma_1}{2}t_i^2\right)\exp(\beta s_1) \\
&-n_2\beta s_2 - \sum_{i=1}^{n_2} \left[\ln(\gamma_0 + \gamma_1 x_i) - \left(\gamma_0 x_i + \frac{\gamma_1}{2}x_i^2\right)\exp(\beta s_2)\right]
\end{aligned}$$

$$-n_3\beta s_3 - \sum_{i=1}^{n_3}\left[\ln\left(\gamma_0 + \gamma_1 y_i\right) - \left(\gamma_0 y_i + \frac{\gamma_1}{2} y_i^2\right)\exp\left(\beta s_3\right)\right]$$

$$+\left(n - \sum_{i=1}^{3} n_i\right)\left[\left(\gamma_0 z + \frac{\gamma_1}{2} z^2\right)\exp\left(\beta s_3\right)\right]. \tag{4}$$

Taking the derivative of (4) with respect to γ_0, γ_1 and β, we have

$$\frac{-\partial \ell\left(\boldsymbol{\eta};\xi\right)}{\partial \gamma_0} = -\sum_{i=1}^{n_1}\left[\frac{1}{\left(\gamma_0 + \gamma_1 t_i\right)} - t_i \exp\left(\beta s_1\right)\right]$$

$$-\sum_{i=1}^{n_2}\left[\frac{1}{\left(\gamma_0 + \gamma_1 x_i\right)} - x_i \exp\left(\beta s_2\right)\right]$$

$$-\sum_{i=1}^{n_3}\left[\frac{1}{\left(\gamma_0 + \gamma_1 y_i\right)} - y_i \exp\left(\beta s_3\right)\right]$$

$$+\left(n - \sum_{i=1}^{3} n_i\right)\left[z \exp\left(\beta s_3\right)\right], \tag{5}$$

$$\frac{-\partial \ell\left(\boldsymbol{\eta};\xi\right)}{\partial \gamma_1} = -\sum_{i=1}^{n_1}\left[\frac{t_i}{\left(\gamma_0 + \gamma_1 t_i\right)} - \frac{t_i^2}{2}\exp\left(\beta s_1\right)\right]$$

$$-\sum_{i=1}^{n_2}\left[\frac{x_i}{\left(\gamma_0 + \gamma_1 x_i\right)} - \frac{x_i^2}{2}\exp\left(\beta s_2\right)\right]$$

$$-\sum_{i=1}^{n_3}\left[\frac{y_i}{\left(\gamma_0 + \gamma_1 y_i\right)} - \frac{y_i^2}{2}\exp\left(\beta s_3\right)\right]$$

$$+\left(n - \sum_{i=1}^{3} n_i\right)\left[\frac{z^2}{2}\exp\left(\beta s_3\right)\right], \tag{6}$$

and

$$\frac{-\partial \ell\left(\boldsymbol{\eta};\xi\right)}{\partial \beta} = -n_1 s_1 + \sum_{i=1}^{n_1}\left[s_1\left(\gamma_0 t_i + \frac{\gamma_1}{2} t_i^2\right)\exp\left(\beta s_1\right)\right]$$

$$-n_2 s_2 + \sum_{i=1}^{n_2}\left[s_2\left(\gamma_0 x_i + \frac{\gamma_1}{2} x_i^2\right)\exp\left(\beta s_2\right)\right]$$

$$-n_3s_3 + \sum_{i=1}^{n_3}\left[s_3\left(\gamma_0 y_i + \frac{\gamma_1}{2}y_i^2\right)\exp\left(\beta s_3\right)\right]$$

$$+\left(n - \sum_{i=1}^{3} n_i\right)\left[s_3\left(\gamma_0 z + \frac{\gamma_1}{2}z^2\right)\exp\left(\beta s_3\right)\right]. \tag{7}$$

Let $\boldsymbol{\eta}^* = \left[\gamma_0^*, \gamma_1^*, \beta^*\right]^T$ be the value of $\boldsymbol{\eta}$ that minimizes (2), which is the solutions of $\gamma_0, \gamma_1, \beta$ by setting all equations in (5), (6), and (7) equal to 0. Suppose that the data are collected under design, ξ, with sample size n. The experimenter fits Model M_F to the data by using MLE method. We adopt quasi-MLE approach by taking n_i's to be independent although they are actually not. We let $\widehat{\boldsymbol{\eta}}$ denote the quasi-MLE of $\boldsymbol{\eta}$. By Theorem 3.2 of White [11], $\sqrt{n}\,(\widehat{\boldsymbol{\eta}} - \boldsymbol{\eta}^*)$ is asymptotically normal with mean $\mathbf{0}$ and covariance matrix $\mathbf{C}\left(\boldsymbol{\zeta};\boldsymbol{\eta}\right)$ which is defined as

$$\mathbf{C}\left(\boldsymbol{\zeta};\boldsymbol{\eta}\right) = \left[\mathbf{A}\left(\boldsymbol{\zeta};\boldsymbol{\eta}\right)\right]^{-1} \times \mathbf{B}\left(\boldsymbol{\zeta};\boldsymbol{\eta}\right) \times \left[\mathbf{A}\left(\boldsymbol{\zeta};\boldsymbol{\eta}\right)\right]^{-1}, \tag{8}$$

where

$$\mathbf{A}\left(\boldsymbol{\zeta};\boldsymbol{\eta}\right) = \left[-E_{M_T}\left(\frac{\partial^2 \ell\left(\boldsymbol{\eta},\xi\right)}{\partial\boldsymbol{\eta}\partial\boldsymbol{\eta}^T}\right)\right] \tag{9}$$

and

$$\mathbf{B}\left(\boldsymbol{\zeta};\boldsymbol{\eta}\right) = \left[E_{M_T}\left(\frac{\partial \ell\left(\boldsymbol{\eta},\xi\right)}{\partial\boldsymbol{\eta}} \cdot \frac{\partial \ell\left(\boldsymbol{\eta},\xi\right)}{\partial\boldsymbol{\eta}^T}\right)\right]. \tag{10}$$

3 Optimality Criteria

We consider to estimate the hazard rate over a given time period at the normal design stress level. We determine the optimal stress-changing times τ_1 and τ_2 in order to minimize the ABIAS^2 and AVAR of the MLE of a hazard function average over a specific period of time, $(0, T]$, under normal design stress level s_D. For a given design, the MLE estimator of the hazard rate at s_D can be obtained by:

$$\widehat{\lambda_{M_F}}\left(t;s_D\right) = \left(\widehat{\gamma_0} + \widehat{\gamma_1}t\right)\exp\left(\widehat{\beta}s_D\right). \tag{11}$$

The ABIAS^2 of (11) average over $(0, T]$ is

$$ABIAS^2\left[\int_0^T \left(\widehat{\lambda_{M_F}}\,(t; s_D)\,|M_T\right)dt\right]$$
$$= \left(\int_0^T \left(E_{M_T}\left[\widehat{\lambda_{M_F}}\,(t; s_D)\right] - \lambda_{M_T}\,(t; s_D)\right)dt\right)^2$$
$$\approx \left(\int_0^T \left(\begin{array}{c}\left(\gamma_0^* + \gamma_1^* t\right)\exp\left(\beta^* s_D\right) \\ -\left(\kappa_0 + \kappa_1 t + \kappa_2 t^2\right)\exp\left(\alpha s_D\right)\end{array}\right)dt\right)^2, \qquad (12)$$

and the AVAR of (11) average over $(0, T]$ is

$$\int_0^T AVAR\left[\widehat{\lambda_{M_F}}\,(t; s_D)\,|M_T\right]dt$$
$$= \frac{1}{n}\int_0^T \left[\frac{\partial\widehat{\lambda}}{\partial\widehat{\gamma_0}}\ \frac{\partial\widehat{\lambda}}{\partial\widehat{\gamma_1}}\ \frac{\partial\widehat{\lambda}}{\partial\widehat{\beta}}\right]\Bigg|_{\widehat{\eta}=\eta^*} C\left(\zeta : \eta = \eta^*\right)\left[\frac{\partial\widehat{\lambda}}{\partial\widehat{\gamma_0}}\ \frac{\partial\widehat{\lambda}}{\partial\widehat{\gamma_1}}\ \frac{\partial\widehat{\lambda}}{\partial\widehat{\beta}}\right]^T\Bigg|_{\widehat{\eta}=\eta^*} dt. \qquad (13)$$

4 Optimal Robust Designs

In ALT practice, a certain number of failures under each test stress level is often required. Such requirement is made to avoid that the stress changing times occur too soon to provide a reasonable step-stress ALT having the same number of stress levels as planned. Please see Elsayed and Zhang [3] for an example of simple step-stress ALT. We also take such practical requirement into our design consideration. These requirements on the minimum number of failures at each stress level are as follows:

(a) The expected number of failures at stress level s_1 has a minimum value MNF_1:

$$n\,\Pr\left[t \leq \tau_1\,|s_1\right] \geq MNF_1; \qquad (14)$$

(b) The expected number of failures at stress level s_2 has a minimum, called MNF_2:

$$(n - n_1)\,\Pr\left[a + t - \tau_1 \leq \tau_2\,|s_2\right] \geq MNF_2; \qquad (15)$$

(c) The expected number of failures at stress level s_3 has a minimum, named MNF_3:

$$(n - n_1 - n_2)\,\Pr\left[b + t - \tau_2 \leq c\,|s_3\right] \geq MNF_3, \qquad (16)$$

where n_i is the number of failures under s_i, $i = 1, 2, 3$.

Under the constraints of minimum number of failures at each stress level, we propose a two-stage optimal robust design procedure. At the first stage, we construct a benchmark design so that the estimation bias can be reduced to a minimal possible. At the second stage, we update the benchmark design to an optimal robust design in order to minimize the AVAR.

We illustrate our method of designing a 3-step-stress ALT experiment by revisiting an example discussed in Elsayed and Zhang [3] with initial values: $\kappa_0 = 0.0001$, $\kappa_1 = 0.5$, $\kappa_2 = 0.0015$, $\alpha = -3800$, and $T = 300$. We take s_2 be 197.5, the average of low ($s_1 = 145$) and high ($s_3 = 250$) stress levels. All other values of the parameters in the example remain the same.

We first determine the parameters needed for (12). The value of (γ_0^*, γ_1^*) can be determined by minimizing the absolute distance between quadratic baseline hazard function and linear baseline hazard function over the whole testing period, namely

$$(\gamma_0^*, \gamma_1^*) = \arg\min \int_0^T \left(\kappa_0 + \kappa_1 t + \kappa_2 t^2\right) - (\gamma_0 + \gamma_1 t)\, dt. \tag{17}$$

We have considered five scenarios of the constraints which are listed in Table 1.

At the first stage, the optimal τ_1 and τ_2 for our benchmark design can be determined in order to minimize (12) under each constraint. We denote ${}^B\xi_{C_k}$ as the optimal design obtained in benchmark under the given constraint C_k, where $k = 1, \cdots, 5$. At the second stage, we adopt the benchmark design parameters. We set the derivatives in (5), (6), and (7) to be zero, and solve for $\gamma_0, \gamma_1, \beta$. This is a constrained nonlinear problem. We solve it in an iterative way. First, we fix β at -3800, and search for $\widetilde{\gamma_0^*}$ and $\widetilde{\gamma_1^*}$ in positive ranges of γ_0 and γ_1 within their 95 % confidence intervals in order to maximize the log-likelihood function. Then, ${}^B\xi_{C_k}$ can be obtained for $\eta_B = [\widetilde{\gamma_0^*}, \widetilde{\gamma_1^*}, -3800]$ by minimizing (12). Second, with γ_0^* and γ_1^* obtained from the first step, we update our estimate of β to β^*. By iterating these two steps, we have $\boldsymbol{\eta}^* = \left[\gamma_0^*, \gamma_1^*, \beta^*\right]^T$. Then, the robust design can be obtained for η^*. We iterate these two steps, until either γ_0^*, γ_1^* or β^* remains unchanged or the difference between its values obtained from the current step and that from the previous step is sufficiently small within a pre-specified range. Finally, we can obtain the optimal τ_1 and τ_2 based on the most updated estimates for $\boldsymbol{\eta}^*$.

Table 1 Constraint cases for 3-step-stress ALT

Names of the constraints	Constraint parameters	Expected number of total failures
C_1	$MNF_1 = 40, MNF_2 = 20, MNF_3 = 10.$	70
C_2	$MNF_1 = 40, MNF_2 = 15, MNF_3 = 15.$	70
C_3	$MNF_1 = 30, MNF_2 = 30, MNF_3 = 10.$	70
C_4	$MNF_1 = 30, MNF_2 = 20, MNF_3 = 20.$	70
C_5	$MNF_1 = 40, MNF_2 = 10, MNF_3 = 10.$	60

Table 2 The optimal stress-changing times for both stages

First stage			Second stage			Efficiency
Design	τ_1	τ_2	Design	τ_1	τ_2	$\mathit{eff}_k\,(R, B)$
${}^B\xi_{C_1}$	160	201	${}^R\xi_{C_1}$	160	201	2.0758
${}^B\xi_{C_2}$	161	189	${}^R\xi_{C_2}$	91	119	2.9840
${}^B\xi_{C_3}$	148	198	${}^R\xi_{C_3}$	78	128	3.6900
${}^B\xi_{C_4}$	147	177	${}^R\xi_{C_4}$	77	107	12.4564
${}^B\xi_{C_5}$	177	203	${}^R\xi_{C_5}$	177	203	1.6514

We denote ${}^R\xi_{C_k}$ as the robust design obtained for three-step-stress ALT under the given constraint C_k, where $k = 1, \cdots, 5$. We define the efficiency of ${}^R\xi_{C_k}$ relative to ${}^B\xi_{C_k}$ in terms of AVAR as

$$\mathit{eff}_k\,(R, B) = \frac{AVAR\left({}^B\xi_{C_k}, \eta_B\right)}{AVAR\left({}^R\xi_{C_k}, \eta^*\right)}. \tag{18}$$

Our resulting robust designs, ${}^R\xi_{C_k}$, for both stages and their relative efficiencies are presented in Table 2.

5 Discussion and Future Research

The example above indicates that our two-stage design procedure has provided significant efficiency gains. The average efficiency of the optimal robust designs is as 4.57 times as that of their corresponding benchmarks. The minimum gain is 65 % for Scenario C_5. At a maximum, our resulting two-stage design for Scenario C_4 can be as 12.46 times efficient as its benchmark (Please see the value of $\mathit{eff}_k\,(R, B)$ in the second last row of Table 2). Compared C_1 to C_5, there only 10 required average failures less, but in result, the efficiency gain has increased from 65 % to 107 %.

Among the five scenarios considered in the example, the optimal robust designs obtained at the second stage for both C_1 and C_5 remains the same as their corresponding benchmarks attained at the first stage. We note that when the MNF_3 is much lower than MNF_1, the constraint has limited the design space for further minimizing the estimation variance at the second stage. For other scenarios, the optimal robust designs obtained after two stages are far different from their benchmarks. Both stress-changing times are much shorter for the optimal robust designs than its benchmarks. Consequently, the testing stress levels are increased more quickly and more failures are observed under higher stress levels, so that the variability in resulting estimation shall be reduced.

In general, we would recommend that for designing a multiple step-stress ALT, the experimenters should consider as less restriction on the total number of failures as possible, at the same time, keep the number of expected failures at each stress level as even as they can so that the estimation quality can be further improved.

In this paper, we focus on designing a multiple step-stress ALT experiment for hazard rate estimation; however, the proposed design procedure can be applied for estimation or prediction of any other quantity of interest, such as reliability of a product at a given time under the normal design conditions. Furthermore, the proposed method can also be extended with possible modification to an ALT experiment involved multiple stress factors. For this situation, a more general version of (1) can be considered as

$$\lambda(t;\mathbf{s}) = \lambda_0(t)\exp\left(\boldsymbol{\theta}^T\mathbf{s}\right), \tag{19}$$

where $\mathbf{s}$ is a column vector of multiple (applied or transformed) stresses used in ALT, and $\boldsymbol{\theta}$ is a column vector of unknown parameters. Optimal and robust designs for step-stress ALT, when (19) is adopted, will be investigated in our future research.

References

1. Bai, D.S., Kim, M.S.: Optimum simple step-stress accelerated life tests for Weibull distribution and type-I censoring. Nav. Res. Logist. **40**, 193–210 (1993)
2. Bai, D.S., Kim, M.S., Lee, S.H.: Optimum simple step-stress accelerated life tests with censoring. IEEE Trans. Reliab. **38**, 528–532 (1989)
3. Elsayed, E.A., Zhang, H.: Design of optimum simple step-stress accelerated life testing plans. In: Proceedings of the 2nd International IE Conference, Riyadh, Kingdom of Soudi Arabia (2005)
4. Fang, Z., Wiens, D.P.: Robust extrapolation designs and weights for biased regression models with heteroscedastic errors. Can. J. Stat. **27**, 751–770 (1999)
5. Fard, N., Li, C.: Optimal simple step stress accelerated life test design for reliability prediction. J. Stat. Plan. Inference **139**, 1799–1808 (2009)
6. Hunt, S., Xu, X.: Optimal design for accelerated life testing with simple step-stress plans. Int. J. Performability Eng. **8**, 575–579 (2012)
7. Jiao, L.: Optimal Allocations of Stress Levels and Test Units in Accelerated Life Tests. Rutgers University, New Brunswick (2001)
8. Ma, H., Meeker, W.Q.: Optimum step-stress accelerated life test plans for log-location-scale distributions. Nav. Res. Logist. **55**, 551–562 (2008)
9. Miller, R., Nelson, W.: Optimum simple step-stress plans for accelerated life testing. IEEE Trans. Reliab. **32**, 59–65 (1983)
10. Pascual, F.G.: Accelerated life test plans robust to misspecification of the stress-life relationship. Technometrics **48**, 11–25 (2006)
11. White, H.: Maximum likelihood estimation of misspecified models. Econometrica **50**, 1–25 (1982)
12. Wiens, D.P., Xu, X.: Robust designs for one-point extrapolation. J. Stat. Plan. Inference **138**, 1339–1357 (2008)

Detecting Coalition Frauds in Online-Advertising

Qinglei Zhang and Wenying Feng

Abstract Online advertising becomes to play a major role in the global advertising industry. Meanwhile, since publishers have strong incentives to maximize the number of views, clicks, and conversions on advertisements, the publisher fraud is a severer problem for advertisers and worth the endeavor to detect and prevent them. By reviewing the literature of the frauds in online advertising, the frauds can be categorized as non-coalition attacks and coalition attacks in general. In this paper, we attempt to mitigate the problem of coalition frauds by proposing a new hybrid detecting approach that identifies the coalition frauds from both economic and traffic perspectives. Moreover, we propose an algorithm to detect the coalition frauds efficiently with an inductive style and greedy strategy.

1 Introduction

With accordance to a recent report [21], online advertisements in 2014 account for nearly one-quarter of the global advertising spend. Figure 1 depicts the general process of the online advertising. The Internet publisher, the advertiser, and the ad intermediaries (i.e. ad exchangers) are the three key participants in the online advertising setting. The publishers make money through hosting websites with advertisements, and advertisers pay for having their ads displayed on publishers' websites. Ad-exchangers such as Google's DoubleClick [4] and Yahoo!'s Right-Media [18] are involved as brokers who pair the publishers' ad requests with the most profitable advertiser bid for the request. Since publishers earn revenue based on the number of views, clicks, and actions that are generated by the advertisement, publishers have strong incentives to maximize those numbers. While publishers can apply legitimate methods to attract more traffic to their websites, dishonest publishers attempt to generate invalid traffic to make more money. In particular, invalid (fraudulent) traffic is identified as impressions, clicks, or even actions that are not the result of genuine user interest [1]. The publisher fraud is a severer problem for advertiser and worth the endeavor to detect and prevent them.

Q. Zhang • W. Feng (✉)
Trent University, Peterborough, ON, K9J 7B8, Canada
e-mail: qzhang@trentu.ca; wfeng@trentu.ca

© Springer International Publishing Switzerland 2016

J. Bélair et al. (eds.), *Mathematical and Computational Approaches in Advancing Modern Science and Engineering*, DOI 10.1007/978-3-319-30379-6_54

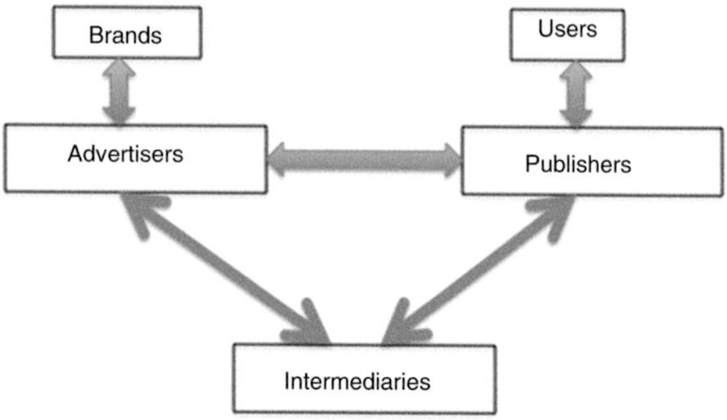

Fig. 1 The process of online advertising

One of the challenges for detecting the online advertising frauds is to determine characteristics that signal the potential of fraudulent traffics. Currently, neither practical systems nor academic research offer a standard mechanism for the fraud detection in online advertising. Moreover, although using data analysis techniques is an evident option, existing solutions are far from mature and there are many research gaps need to be tackled. Among various attacks in online advertising, coalition frauds are a type of attacks that become more and more prevalent nowadays. Since such attacks can be launched with lower cost while still difficult to be identified. In this paper, we attempt to mitigate the problem of coalition frauds by proposing a new detecting algorithm [7].

The paper is structured as follows. Section 2 reviews the related works of frauds in online advertising. In Sect. 3, we propose a hybrid detection approach for coalition frauds. We also discuss the effectiveness and efficiency of the proposed approach. We conclude and discuss the future work in Sect. 4.

2 Literature Review: Frauds in Online Advertising

With an extensive investigation in both academic and industrial context, we identify a collection of known attacks in online advertising. In particular, fraudulent publishers use two types of resources, machines, and web pages, to deploy an attack. The attacks can be performed either by the machines, or by the web pages, or by both. From the machine side, the attacks can be performed either by real human or by software. The fraudsters hire malicious users to constantly reload a page and click on ads. In order to make such attacks harder to trace and detect, some fraudsters abuse legitimate services to hide identifying information like cookies, IP address, etc., from HTTP requests. Fraudsters also resort to bot to simulate impressions, clicks, or conversions. Initially, those fraudulent bots sit on static machines and issue HTTP requests to certain URLs, which makes the identification of them relatively easy. Later on, more sophisticated techniques use anonymous proxies and malware-type bots to achieve the IP diversity of the traffic. On the side of web pages, the fraudulent publishers can use various of different mechanisms to commit frauds. For example, impression stuffing and keywords stuffing attacks inflate the number of impressions. Fraudulent publishers put excessive numbers of banners on their pages to get a large number of impressions for each page view. In particular, half a dozen or even more of different ad banners could present on a single page [5]. Sometimes, those banners are even stacked in front of each other, which make those back banners invisible. Moreover, to drive higher traffic to the page than it deserves, fraudsters invisibly use some high-value advertising keywords (e.g. in hidden HTML tags, or by text with the same color as the backgrounds). Furthermore, the publisher can also inflate the number of clicks with coercion attacks and force browser attacks. With coercion attacks, the content of ads are made invisible or modified to something more relevant to the user, and the users are instruct to perform clicks unknowingly on ads. The fraudulent publisher inserts additional HTML code that force the browser to click on the ads. With such client-side script, the users generate request implicitly and click on the ads stealthily when they loading the webpages.

2.1 The Classification of the Online Advertising Frauds

The classical way to detect publisher frauds is achieved by monitoring the performance of advertisements on sites. In other words, fraudulent traffic are identified as low quality traffic. However, such approach can be too aggressive that even legitimate traffic can be discarded. In addition, the malicious intents of fraudsters are not identified directly. Since the fraudsters can control the performance of advertisements loaded on their sites, they can easily fool the detection tools by mimicking validity metrics to avoid being detected.

To complement the classical traffic monitoring approach, more advanced techniques are proposed and applied to identify the pattern of fraudulent behaviors. More specifically, a classification of the fraudulent behaviours help to recognize a more generic pattern of frauds. A related work with a detailed discussion on the classification of online advertising frauds can be found in [14]. As aforementioned, fraudulent publishers use machines and sites to deploy an attack. Regardless of different attacking mechanisms (e.g. coercion, robots, etc.), the premise for data analysis approaches to detect publisher frauds is to identify the correlations between the attacking machines and the corresponding sites. There are two types of relationships between fraudulent machines and sites. In an one-to-one relationship, fraudsters only control their own machines and sites. On the other hand, in a many-to-many relationship, a group of fraudsters share their resources to perform the attacks. Consequently, the online advertising frauds can be categorized as two types: the former type of attacks is called the non-coalition attack, while the latter type is called the coalition attack.

With respect to the above classification, the key to identify fraudulent traffic is to identify the association between fraudulent websites and machines. More sophisticated attacks blur the strong correlation between the fraudulent sites and machines, which makes them difficult to be detected. In the case of non-coalition attacks, fraudsters only control their own machines and sites. An elementary attack with repeated views, clicks, or actions from one machine on one site can be easily detected and blocked. Various data analysis techniques have been proposed to detect such attacks, such as finding duplicates in data streams [12, 22], identifying abnormal source IPs that are shared by large number of users [19], and etc. To compromise those data analysis detection techniques, fraudsters need to develop sophisticated techniques (e.g., simulate fake visitors, or frequently change the identification of users) to avoid being detected. In other terms, to increase the difficulty of being detected, the cost of launching non-coalition attacks will also increase significantly. On the other hand, since the resources are shared by several fraudsters, coalition frauds are difficult to be identified while the cost of launching them is still relatively low.

2.2 Countermeasures for Frauds in Online Advertising

In the literature, various countermeasures have been explored to mitigate the frauds problem in online advertising. In [3], an alternative pricing model is proposed to remove the incentive of fraud. However, the approach modifies current pricing model, which requires a global effort and is not likely to happen in the near future.

Preventing and detection are two general measurements to mitigate fraudulent problems. In the context of online advertising, it is important to terminate the identified publishers from re-join the ad-networks. A typical preventing approach is to create a block list which specifies the signals that indicate fraudulent activities. By looking at different characteristics of HTTP requests (e.g. IP address, publisher ID, Customers ID), a potential fraudulent traffic might be filtered based on the maintained block list. One key challenge of the preventing approach is to determine characteristics that signal the potential of fraudulent traffics. The signal is considered as confidential information for the companies that run the ad network.

Detection techniques found in both practice and academic research can be generally categorized as positive and active detection. The active detection means that additional interactions are conducted with the web and/or users to verify the validity of the traffic. Y. Peng et al. [16] proposed an active detection technique, which combats malicious scripts clickbots by creating and validating an impression-click identifier. Another example of active detection is using bluff ads [6]. The strategy of bluff ads is to serve some ads that are purposely uninviting, and test the legitimacy of the individual click. However, the active detection is usually limited to address a specific type of fraud, which leads to an arms race between the fraudulent publishers and the detectors. On the other hand, passive detection mainly depends on applying data analysis techniques to the aggregate number of traffic logs.

Data mining techniques can be used for both signature-based detection (e.g. [9]), and anomaly-base detection (e.g. [19]). While signature-based detection works well for known malicious patterns, anomaly-based detection is more effective to identify

Table 1 Summary of the fraud detection techniques in the literature

	Signature-based	Anomaly-based	
Active	[6, 16]		Need interactions with the web and/or users
Positive	[9]	[8, 10, 12, 13, 15, 17, 19] and our new hybrid-approach	Use data analysis techniques to the traffic logs
	Work well for known malicious patterns	More effective to identify new fraudulent pattern	

fraudsters with updated types of attacks. Related research about passive fraud detection mainly focuses on developing different algorithms to recognize a specific behaviour pattern of attacks. For example, the pattern of duplicate clicks within a short period of time from the same visitor is recognized as one elementary attack. Algorithms about finding duplicates in data streams such as [12, 19] are applied to detect such an attack and its variation. Moreover, different features are proposed to distinguish the fraudulent behaviour patterns from normal behaviours. In [19], the number of users sharing the source IP is used as a metric to train the normal pattern of behaviors. Attacks are detected as an anomalous deviation from the expected publishers' IP size distribution. Recent efforts [9, 10, 17] have also been done by using classification algorithms based on multiple features to classify the valid and invalid traffic. For the coalition attacks, few published work [8, 13, 15, 20] have proposed algorithms to combat them. The techniques in [13, 15, 20] mainly based on the similarity among sites to identify the coalitions,while [8] aims to identify the coalitions from the economic incentive of attackers. We articulate the similarity-based and the gain-based metrics in detail later and adopt both of them in our detection technique. A summary of existing detection techniques in the literature is given in Table 1.

3 Detecting Coalition Frauds in Online Advertising

Unlike non-coalition attacks, the main challenge of detecting coalition attacks is to identify the correlations of multiple publishers instead of merely focusing on the pattern of a single publisher.

3.1 Identifying the Coalition Frauds

In order to identify the coalition groups, we can consider the problem from the economic perspective. In particular, there are two observations for the coalition groups: (1) the coalition groups inherently have higher ROI (return of investment) than normal, (2) the expect gain to each member increases while more fraudsters are involved.

The ratio of gain and cost can be used as an estimator of ROI. Therefore, the metric relates to the first observation can be expressed by

$$GPR(g) = \frac{W(g)}{R(g)} \geq \theta_{GPR}, \tag{1}$$

where $W(g)$ is the gain derived by the members of the coalition g, $R(g)$ denotes the amount of the resources of the coalition, and θ_{GPR} is a predefined threshold of the minimum GPR (Gain per Resource) for coalition groups.

A coalition has a high value of GPR, but not vice versa. In other words, a coalition group with some normal publishers or visitors may still have a high value of GPR. To capture the property of the second observation, we use the definition of GPR group. Formally, we define g is a GPR-GPOUP iff $\forall(g \mid g' \subset g : GPR(g') < GPR(g))$, where g' denotes any subgroup of g. Therefore, a coalition group can be defined in terms of GPR as follows:

Definition 1 (gain-based) A group of publishers and visitors, denoted as g, is called a coalition iff $GPR(g) \geq \theta_{GPR}$ and g is GPR group. Moreover, a coalition is a maximal coalition iff $\neg\, \exists(g'' \mid g \subset g'' : g''$ is a coalition$)$.

As another key character of coalition attacks, the traffic to an attacker's sites are from a relatively larger set of attacking machines that shared by many attackers. To capture the correlations between publishers, we can measure the traffic similarity among them. Several general similarity measurements can be applied in our context to capture the similarity metric.

Since the number of repeat visitors is an important indicator for detecting frauds in online advertising, we use a bag model to denote the visitors to a publisher. Unlike sets, elements can be duplicated in bags. Let B_{p_1} and B_{p_2} denote the bags of visitors to the publishers p_1 and p_2, respectively. The pairs of sites p_1 and p_2 have similar traffic iff

$$\text{Similarity}(p_1, p_1) = \frac{|B_{p_1} \sqcap B_{p2}|}{|B_{p_1} \sqcup B_{p_2}|} \geq \theta_{SIM}, \tag{2}$$

where $\sqcap$ and $\sqcup$ denote the intersection and union of bags. We use a pair (n, e) to represent an element in a bag, where n indicates how many times the e repeats in the bag. The definitions of $\sqcap$ and $\sqcup$ are given as follows:

$$\begin{aligned}
&\forall((n,e) \mid (n,e) \in A \sqcap B : \exists(m_1, m_2 \mid (m_1,e) \in A \wedge (m_2,e) \in B : n = min(m_1,m_2)))\\
&\forall((n,e) \mid (n,e) \in A \sqcup B : ((n,e) \in A \wedge \neg\, \exists(k \mid k \in N : (k,e) \in B))\\
&\qquad \vee((n,e) \in B \wedge \neg\, \exists(k \mid k \in N : (k,e) \in A))\\
&\qquad \vee(\exists(m_1, m_2 \mid (m_1,e) \in A \wedge (m_2,e) \in B : n = max(m_1,m_2)))
\end{aligned}$$

The lhs of the Equation (2) is a variant of Jaccard coefficient [2] capturing the traffic similarity of B_{p_1} and B_{p2}, and θ_{SIM} on the rhs denotes the minimum threshold of similarity for legitimate pairs of sites. Since two legitimate sites only have negligible similarity, it is unlikely that every pair of sites in a random group is similar. In other terms, the coalition attackers form a group that is pair-wise similar. Formally, the definition of coalition groups with regard to the traffic similarity is given as follows:

Definition 2 (similarity-based) A group of publishers g is a coalition group iff $\forall (p_1, p_2 \mid (p_1 \in g) \wedge (p_2 \in g) \wedge (p_1 \neq p_2) \ : \ \text{Similarity}(p_1, p_2) \geq \theta_{SIM})$

The aforementioned two approaches capture the key characters of coalition attacks from different perspectives. The gain-based approach calculates the economic gain of each group (see Definition 1), while the similarity based approach considers the traffic similarity within each group (see Definition 2). The reader can refer to [8] and [15] respectively for more details of these two approaches. However, a main challenge for both approaches is to define the proper threshold θ (see Equations 1 and 2). For example, some legitimate popular sites may have equal or even greater GPR than the coalition groups, and legitimate sites with similar traffic may be falsely identified as fraudsters. Therefore, to increase the confidence for detecting coalition attacks, it would be beneficial to consider the characters from both perspectives and combine both measurements. While the GPR is a measurement for a group of entities, the similarity is a measurement only for a pair of entities. In order to get much more confidence on the signal of coalitions, we further give the definition of coalitions as follows:

Definition 3 (hybrid) Let g denote a subset of all publishers, we say g is a coalition group iff

$$\forall (g' \mid g' < g \ : \ GPR(g') < GPR(g)) \wedge GPR(g) \geq \theta_{\text{GPR}}$$
$$\wedge \ \forall (p_1, p_2 \mid (p_1 \in g) \wedge (p_2 \in g) \wedge (p_1 \neq p_2) \ : \ \text{similarity}(p_1, p_2) \geq \theta_{\text{SIM}})$$

3.2 Detecting the Coalition Frauds

With accordance to the above definition of coalition frauds, we have three strategies to detect such coalitions. The first strategy is to detect the GPR groups that satisfy the GPR properties and their GPR value are greater than θ_{GPR}. Then we verify the pair-wise similarity property for each identified group. The second strategy is to calculate the similarity for all pairs and identify all pair-wised similar groups. For each identified group, we further verify the GPR properties of the group and its GPR value. The third strategy is to dynamically check the two properties while identifying the coalition groups. Such strategy will be the most efficient way, which is critical when dealing with big data. Common advanced analytics disciplines such as statistical analysis, data mining, and predictive analytics all can be used to deal with big data. Machine learning algorithms (e.g. neural network, decision tree) is especially useful for the analysis of big data, since it can explore the hidden characters with less reliance on human direction. On the other hand, the key neck-bottle of using machine learning algorithms in our context is the demand of first extracting related features for all potential groups, which is a computationally expensive process. However, the adaption of the third strategy is not straightforward. We should validate that the derived detection algorithm can correctly identify the frauds corresponding to the definition given in the previous section.

Given a group of publishers and visitors g, let $G_{click}(g)$ and $G_{similarity}(g)$ denote the click graph and the similarity graph associated with g, respectively. Before giving the detection algorithm, we first define two key properties used in the algorithm as follows:

- GPR-Core property: g is a GPR group $\Longrightarrow$
  $\forall(g' \mid G_{click}(g')$ is a connected subgraph of $G_{click}(g)$: g' is a GPR group)
- similar-clique property: g is a similar clique $\Longrightarrow$
  $\forall(g' \mid G_{similarity}(g')$ is a subgraph of $G_{similarity}(g)$: g' is a similar clique)

With accordance to the definition of coalition frauds given in Definition 3, we have the following proposition:

Proposition 1 *A group g is a coalition group iff* $GPR(g) \geq \theta_{GPR}$ *and g satisfies both the CPR-Core property and the similar-clique property.*

Due to the limit of space, we omit the proof of the proposition. Algorithm 1 depicts the pseudocode to detect coalition frauds with the hybrid approach. To combine the aforementioned two criteria dynamically, we apply an inductive style and greedy strategy in the algorithm, which can dramatically prune the search space and improve the efficiency of the detection technique. It can be proved that the GPR-Core and the similar-clique properties are anti-monotone. In other words, the GPR-Core and the similar-clique properties (used in Line 5 and 6, respectively) hold with the induction. Consequently, Proposition 1 indicates that the given algorithm can correctly detect the coalition frauds in our context.

Algorithm 1 Detecting coalition frauds

```
1: G ← calculate the base case value for gpr measurement
2: E ← calculate the pair similarity for the similarity measurement
3: for size in range(2, len(V)) do
4:     potential ← INDUCTIVE_KEY_GENERATOR(previous,V)
5:     current1 ← check the potential group with the GPR-Core property
6:     current2 ← check the potential group with the similarity-clique property
7:     current ← current1 ∩ current2
8:     coalition[size] ← append if it satisfies the GPR group threshold
9:     previous ← coalitions[size]
10: end for
11: max_coalitions ← MAXIUM(coaltions)
```

Another key challenge for both gain-based and similarity-based approaches is the complexity of the algorithm. Given the set of all publishers as P and visitors as V, the GPR approach require calculating the GPR for all potential subgroups of G, which complexity is exponential to $|P| + |V|$. On the other hand, the similarity approach requires to identify the pairwise similar group which is equivalent to discovering the maximal cliques in the sites' similarity graph. In general, finding maximal cliques in a graph also has exponential complexity. The problem of detecting coalitions is NP-hard in general. In practice, the problem of detection frauds in online advertising usually involves processing large sets of data. To further improve the efficiency of the detection algorithm, we adopt a MapReduce [11] computing model which can facilitate the parallel, distributed computing on multiple clusters. The lines of 1, 2, 4, 5, 6 and 11 in the pseudocode of Algorithm 1 are implemented with the MapReduce paradigm. However, the implementation of the algorithm is out the scope of this paper.

3.3 Preliminary Result

To evaluate the effectiveness of the proposed detecting techniques, we apply our detection technique to various traffic patterns in this section. As we mentioned earlier, the value of the threshold is critical when deciding whether or not a group is a coalition. To avoid the negative false alarms, both the parameters θ_{SIM} and θ_{GPR} should set to low values. However, the positive false alarms could increase when the values are too low. Existing techniques for detecting coalition approach (e.g. [8, 15]) usually determine the value of the threshold by making a trade-off between the positive and negative false alarms. Unlike those techniques with a singe metric, the hybrid approach adapted in this paper could minimize the rate of negative false alarms by setting the threshold to relatively low values. On the other hand, the rate of positive false alarms is reduced by combing the decisions from two different perspectives, Fig. 2 illustrates click graphs of three simplified traffic patterns. The nodes of p_1, p_2, and p_3 denote the publishers, while the nodes of v_1, v_2, v_3, and

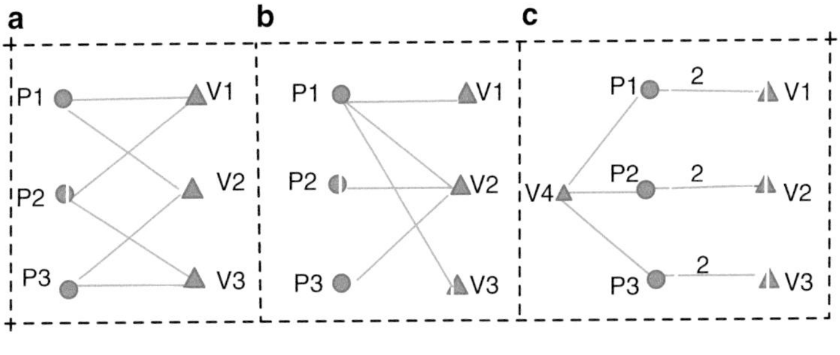

Fig. 2 Different traffic patterns with/without coalition Attacks

Table 2 The results of the three different detection techniques

	Similarity-based [15]	GPR-based [8]	Hybrid (our approach)
Case (a)	Fraud (✔)	Fraud (✔)	Fraud (✔)
Case (b)	Frauds (✖)	Normal (✔)	Normal (✔)
Case (c)	Normal (✔)	Frauds (✖)	Normal (✔)

v_4 denote the visitors. The case (a) is an example of coalition attacks that the three publishers share the resources. The case (b) and case (c) are two examples that the resources are not really shared by those publishers. However, the two latter cases are easy to be mis-identified since they show similar traffic metrics as coalition attacks. Table 2 shows the results of checking the three cases with different detection techniques. To identify the coalition attack of case (a), the parameter θ_{GPR} should be at least 1, and the parameter θ_{SIM} should be equal to or less than $\frac{1}{3}$. However, the case (b) is falsely identified as coalition frauds with $\theta_{SIM} = \frac{1}{3}$, while the case (c) is falsely identified with $\theta_{GPR} = 1$. But with the proposed hybrid approach, we can avoid such positive false alarms while remaining the precision for detecting coalition frauds.

4 Conclusions and Future Work

In this paper, we adapt a hybrid approach to identify the coalition frauds from two distinguish perspectives. To detect the coalition groups, we proposed an algorithm with inductive styles and greedy strategy, and discuss the correctness and efficiency of the algorithm. Finally, we evaluate the effectiveness of the proposed detection technique by applying it to several trial cases.

As further work, we can apply the proposed technique to different data sets to further evaluate the algorithm. Since there are no benchmark data sets for coalition frauds in online advertising, we can generate synthetic data sets by simulating the traffic of both normal and fraudulent traffics. We also can apply the algorithm to industrial data sets and validate the performance of the algorithm by using third party services that can label the traffic logs. Moreover, we can improve the precision of our hybrid detection technique by exploring the coalition frauds problems from more other perspectives and integrating them to our detection algorithm.

Acknowledgements The authors thank the referee for some helpful comments. The project was supported by the Mathematics of Information Technology and Complex Systems (MITACS) of Canada and the EQ Advertising Group Ltd.

References

1. Ad traffic quality resource center. http://www.google.com/intl/en_ALL/ads/adtrafficquality/index.html. Last accessed on Dec 2015
2. Charika, M.: Similarity estimation techniques from rounding algorithms. In: Proceedings of the 34th ACM STOC Symposium on Theory of Computing, Montreal, 9pp., 19–21 May 2002
3. Daswani, N., Mysen, C., Rao, V., Weis, S., Gharachorloo, K., Ghosemajumder, S., and the Google Ad Traffic Quality Team: Online Advertising Fraud. The paper is available at https://books.google.ca/books?id=U8bh6a9JvzOC
4. Doubleclick by google. https://www.doubleclickbygoogle.com/. Last accessed on Dec 2015
5. Edelman, B.G.: Securing online advertising: rustlers and sheriffs in the new wild west. In: Viega, J. (ed.) Beautiful Security, pp. 89–105. O'Reilly Media, Inc., Farnham (2009). Available at SSRN: http://ssrn.com/abstract=1267311 or http://dx.doi.org/10.2139/ssrn.1267311
6. Haddadi, H.: Fighting online click-fraud using bluff Ads. ACM SIGCOMM Comput. Commun. Rev. **40**(2), 21–25 (2010)
7. Jakobsson, M., Ramzan, Z. (ed.): Crimeware: Understanding New Attacks and Defenses (Symantec Press), pp. 325–354. Addison-Wesley Professional, Boston (2008)
8. Kim, C., Miao, H., Shim, K.: Catch: a detecting algorithm for coalition attacks of hit inflation in internet advertising. Inf. Syst. **36**(8), 1105–1123 (2011)
9. Kitts, B., Zhang, J.Y., Roux, A., Mills, R.: Click fraud detection with bot signatures. In: ISI 2013, Seattle, pp. 146–150 (2013)
10. Landergren, M.H.T.: Implementing best practices for fraud detection on an online advertising platform. Master's thesis, Chalmers University of Technology (2010)
11. MapReduce. http://hadoop.apache.org/docs/r1.2.1/mapred_tutorial.html. Last accessed on Dec 2015
12. Metwally, A., Agrawal, D., El Abbadi, A.: Duplicate detection in click streams. In: International World Wide Web Conference, WWW 2005, Chiba, pp. 12–21, 10–14 May. ACM 1-59593-046-9/05/0005 (2005)
13. Metwally, A., Agrawal, D., El Abbadi, A.: Using association rules for fraud detection in web advertising networks. In: Proceedings of the 31st VLDB Conference, Trondheim, pp. 169–180 (2005)
14. Metwally, A., Agrawal, D., El Abbadi, A., Zheng, Q.: On hit inflation techniques and detection in streams of web advertising networks. In: Proceedings of the 27th International Conference on Distributed Computing Systems, Toronto, p. 52, 25–27 June 2007. doi:10.1109/ICDCS.2007.124
15. Metwally, A., Agrawal, D., El Abbadi, A.: Detectives: detecting coalition hit inflation attacks in advertising networks streams. In: Proceedings of the 16th International Conference on World Wide Web, Banff, pp. 241–250, 8–12 May, 2007. doi:10.1145/1242572.1242606
16. Peng, Y., Zhang, L., Chang, J.M., Guan, Y.: An Effective Method for Combating Malicious Scripts Clickbots. Lecture Notes in Computer Science, vol. 5789, pp. 523–538. Springer, Berlin/Heidelberg (2009)
17. Perera, B.K.: A class imbalance learning approach to fraud detection in online advertising. Master's thesis, Masdar Institute of Science and Technology (2013)
18. Rightmedia from yahoo! https://advertising.yahoo.com/index.htm. Last accessed on Dec 2015
19. Soldo, F., Metwally, A.: Traffic anomaly detection based on the IP size distribution. In: Proceedings of the INFOCOM International Conference on Computer Communications, Joint Conference of the IEEE Computer and Communications Societies, Orlando, pp. 2005–2013, 25–30 Mar 2012
20. Springborn, K., Barford, P.: Impression fraud in online advertising via pay-per-view networks. In: SEC'13 Proceedings of the 22nd USENIX Conference on Security, Berkeley, pp. 211–226 (2013)
21. Zenithoptimedia. http://techcrunch.com/2014/04/07/. Last accessed on Dec 2015
22. Zhang, L., Guan, Y.: Detecting click fraud in pay-per-click streams of online advertising networks. In: Proceedings of the 28th International Conference on Distributed Computing Systems, Beijing, pp. 77–84, 17–20 June 2008. doi:10.1109/ICDCS.2008.98

Part V
New Challenges in Mathematical Modeling for Scientific and Engineering Applications

Circle Inversion Fractals

B. Boreland and H. Kunze

Abstract In 2000, Frame and Cogevina (Comput Graph 24(5):797–804, 2000) introduced a method for constructing fractals using circle inversion maps. The literature (Clancy and Frame, Fractals 3:689–699, 1995; Frame et al., A fractal representation of grammatical complexity in time series. Manuscript in Preparation, 1996; Frame and Cogevina, Comput Graph 24(5):797–804, 2000; Leys, Comput Graph 29(3):463–466, 2005) focuses on the graphical aspect of such fractals, without presenting a careful development of the underlying mathematical framework. In this paper, we present such a framework, making a strong connection to iterated function systems (IFS) theory. Our final result establishes that the set valued system of circle inversion maps induced by a collection of non-touching circles in the plane has a unique set attractor. We present a graphical example using the chaos game.

1 Introduction

Circle inversion was introduced in Apollonius of Pregas' book, Plane Loci, and has drawn interest in geometry. Mandelbrot introduced the early concepts of fractals in the 1970s and, in The Fractal Geometry of Nature, Mandelbrot discusses successive inversion with respect to a family of M circles. The literature applies the chaos game to circle inversion maps [2, 4] to explore the graphical aspect, while casually stating that there are contraction maps involved. As we will see in Sect. 2, a circle inversion map is only contractive on part of its domain, making clear the need for care. In this paper, we build the needed rigorous mathematical framework. In Sect. 2, we present some background concepts on circle inversion maps. In Sect. 3, we establish some contractivity results for a system of two circle inversion maps, which we extend in Sect. 4 to a general system of non-touching circles. In Sect. 4, we also establish the existence of a unique set attractor to the system of set-valued circle inversion maps. Approximations of this fractal set can be drawn using the chaos game; we provide an example in Sect. 5. We can use the already established content to now build a

B. Boreland (✉) • H. Kunze
Department of Mathematics and Statistics, University of Guelph, Guelph, ON, Canada
e-mail: bborelan@mail.uoguelph.ca; hkunze@uoguelph.ca

© Springer International Publishing Switzerland 2016

J. Bélair et al. (eds.), *Mathematical and Computational Approaches in Advancing Modern Science and Engineering*, DOI 10.1007/978-3-319-30379-6_55

framework to show why this random inversion algorithm arrives at a fractal, or in other words, converges to a set attractor.

2 Inversion in a Circle

Let $C \in \mathbb{R}^2$ be a solid circle with centre $\underset{\sim}{o}$ and boundary ∂C; then the following property is true.

Property 1 ∂C divides $\mathbb{R}^2$ into three pieces

- the interior of the circle, C_{int}
- the boundary of the circle, ∂C
- the exterior of the circle, C_{ext}

where $C = C_{int} \cup \partial C$

Definition 1 Given C in $\mathbb{R}^2$ with fixed centre $\underset{\sim}{o}$ and radius r we define the inversion map T on $\mathbb{R}^2 \setminus \{\underset{\sim}{o}\}$ by

$$||T(\underset{\sim}{x}) - \underset{\sim}{o}||\,||\underset{\sim}{x} - \underset{\sim}{o}|| = ||\Pi(\underset{\sim}{x}) - \underset{\sim}{o}||^2 = r^2, \forall \underset{\sim}{x} \in \mathbb{R}^2 \setminus \{\underset{\sim}{o}\} \tag{1}$$

where $||\cdot||$ is the Euclidean norm and $\Pi(\underset{\sim}{x})$ is the projection of $\underset{\sim}{x}$ onto ∂C along the radial ray from $\underset{\sim}{o}$ to $\underset{\sim}{x}$. Alternatively, we can write the inversion map as

$$d(T(\underset{\sim}{x}), \underset{\sim}{o})d(\underset{\sim}{x}, \underset{\sim}{o}) = d^2(\Pi(\underset{\sim}{x}), \underset{\sim}{o}) = r^2, \forall \underset{\sim}{x} \in \mathbb{R}^2 \setminus \{\underset{\sim}{o}\} \tag{2}$$

where $d(\underset{\sim}{x}, \underset{\sim}{y}) = ||\underset{\sim}{x} - \underset{\sim}{y}||$.

Lemma 1 *Given a circle C with inversion map T we have the following:*

i. $T : C_{int} \setminus \{\underset{\sim}{o}\} \longrightarrow C_{ext}$
ii. $T : C_{ext} \longrightarrow C_{int}$
iii. $T : \partial C \longrightarrow \partial C$

Proof Follows from the definition of T and Property 1.

When a point $\underset{\sim}{y} = T(\underset{\sim}{x})$ satisfies (2), $\underset{\sim}{y}$ is called the inverse of $\underset{\sim}{x}$ where $T(\underset{\sim}{y}) = \underset{\sim}{x}$ and $T(\underset{\sim}{b}) = \underset{\sim}{b}$, for $\underset{\sim}{b} \in \partial C$.

Definition 2 Given a collection of non touching circles C_i, $i = 1, \ldots, N$, with centres $\underset{\sim}{o}_i$ and radii r_i we define the inversion map T_i on $\mathbb{R}^2 \setminus \{\underset{\sim}{o}_i\}$ by

$$||T_i(\underset{\sim}{x}) - \underset{\sim}{o}_i||\,||\underset{\sim}{x} - \underset{\sim}{o}_i|| = ||\Pi_i(\underset{\sim}{x}) - \underset{\sim}{o}_i||^2 = r_i^2, \forall \underset{\sim}{x} \in \mathbb{R}^2, \underset{\sim}{x} \neq o_i \tag{3}$$

where $\Pi_i(\underset{\sim}{x})$ is the projection of $\underset{\sim}{x}$ onto ∂C_i along the radial ray from $\underset{\sim}{o}_i$ to $\underset{\sim}{x}$,

We can build (3) when $\underset{\sim}{x}$ and $T_i(\underset{\sim}{x})$ satisfy the expression

$$T_i(\underset{\sim}{x}) = \underset{\sim}{o}_i + \frac{||\Pi_i(\underset{\sim}{x}) - \underset{\sim}{o}_i||^2}{||\underset{\sim}{x} - \underset{\sim}{o}_i||^2}(\underset{\sim}{x} - \underset{\sim}{o}_i)$$

Any point $\underset{\sim}{x} \in \mathbb{R}^2$ can be represented in terms of the parametrization for the circle C_i:

$$\underset{\sim}{x} = ar_i\underset{\sim}{w}(t) + \underset{\sim}{o}_i \text{ for some } a \geq 0,\ t \in [0, 2\pi], \tag{4}$$

where $\underset{\sim}{w}(t) = (\cos(t), \sin(t))$. If $a < 1$ then $\underset{\sim}{x} \in C_{i,int}$, if $a = 1$ then $\underset{\sim}{x} \in \partial C_i$ and if $a > 1$ then $\underset{\sim}{x} \in C_{i,ext}$.

For $\underset{\sim}{x} \in \mathbb{R}^2 \setminus \{\underset{\sim}{o}_i\}$, we have $a > 0$, and we find that with some easy simplifications we obtain

$$T_i(\underset{\sim}{x}) = T_i(ar_i\underset{\sim}{w}(t) + \underset{\sim}{o}_i) = \underset{\sim}{o}_i + \frac{||r_i\underset{\sim}{w}(t)||^2}{||ar_i\underset{\sim}{w}(t)||^2}(ar_i\underset{\sim}{w}(t)) = \underset{\sim}{o}_i + \frac{1}{a}r_i\underset{\sim}{w}(t) \tag{5}$$

Thus, we see clearly why T_i is called a circle inversion map: under the action of T_i, the radial scaling factor $a_i > 0$ becomes a radial scaling factor of $\frac{1}{a_i}$.

3 Mathematical Framework for Two Non-touching Circles

We now develop a mathematical framework for a system of two non-touching circles. We begin by observing properties along the radial rays. We define some notation for use in upcoming results. Given a circle C_1 in $\mathbb{R}^2$ with centre $\underset{\sim}{o}_1$, let

$$\begin{aligned}
R_{\underset{\sim}{o}_1}(\underset{\sim}{o}_1, \underset{\sim}{b}_1) &= \text{semi infinite ray from } \underset{\sim}{o}_1 \text{ through } \underset{\sim}{b}_1 \in \partial C_1, \text{with endpoint } \underset{\sim}{o}_1 \\
R_{\underset{\sim}{o}_1,\infty}(\underset{\sim}{o}_1, \underset{\sim}{b}_1) &= \text{infinite line through } \underset{\sim}{o}_1 \text{ and } \underset{\sim}{b}_1 \\
R_{\underset{\sim}{o}_1,int}(\underset{\sim}{o}_1, \underset{\sim}{b}_1) &= R_{\underset{\sim}{o}_1}(\underset{\sim}{o}_1, \underset{\sim}{b}_1) \cap C_1 \\
R_{\underset{\sim}{o}_1,ext}(\underset{\sim}{o}_1, \underset{\sim}{b}_1) &= R_{\underset{\sim}{o}_1}(\underset{\sim}{o}_1, \underset{\sim}{b}_1) \setminus \{R_{\underset{\sim}{o}_1,int}(\underset{\sim}{o}_1, \underset{\sim}{b}_1), \underset{\sim}{b}_1\}
\end{aligned}$$

Lemma 2 *For $C_1 \in \mathbb{R}^2$ with centre $\underset{\sim}{o}_1$, $\underset{\sim}{b} \in \partial C_1$, $\underset{\sim}{b}_1 \in \partial C_1$, and T_1 the inversion map for C_1, we have the following:*

i. $T_1 : R_{\underset{\sim}{o}_1,int}(\underset{\sim}{o}_1, \underset{\sim}{b}_1) \setminus \{o_1\} \longrightarrow R_{\underset{\sim}{o}_1,ext}(\underset{\sim}{o}_1, \underset{\sim}{b}_1)$
ii. $T_1 : R_{\underset{\sim}{o}_1,ext}(\underset{\sim}{o}_1, \underset{\sim}{b}_1) \longrightarrow R_{\underset{\sim}{o}_1,int}(\underset{\sim}{o}_1, \underset{\sim}{b}_1)$
iii. $T_1(\underset{\sim}{b}) = \underset{\sim}{b}$
iv. $T_1^2 : R_{\underset{\sim}{o}_1,int}(\underset{\sim}{o}_1, \underset{\sim}{b}_1) \setminus \{o_1\} \longrightarrow R_{\underset{\sim}{o}_1,int}(\underset{\sim}{o}_1, \underset{\sim}{b}_1)$
v. $T_1^2 : R_{\underset{\sim}{o}_1,ext}(\underset{\sim}{o}_1, \underset{\sim}{b}_1) \longrightarrow R_{\underset{\sim}{o}_1,ext}(\underset{\sim}{o}_1, \underset{\sim}{b}_1)$
vi. $T_1^2(\underset{\sim}{a}) = \underset{\sim}{a}$

Proof (*i*), (*ii*) and (*iii*) follow from Lemma 1. (*iv*), (*v*) and (*vi*) follow by iterating Lemma 1 a second time.

Now, given two non-touching circles C_1 and C_2 in $\mathbb{R}^2$ with centres $\underset{\sim}{o}_1$ and $\underset{\sim}{o}_2$, respectively, define $(C_i)_{int}$, $(C_i)_{ext}$, ∂C_i, T_i and projection maps Π_i. Let

$$\begin{aligned}
R_{\underset{\sim}{o}_1\underset{\sim}{o}_2}(\underset{\sim}{x},\underset{\sim}{y}) &= \text{portion of the open line segment between } \underset{\sim}{o}_1 \text{ and } \underset{\sim}{o}_2 \text{ from } \underset{\sim}{x} \text{ to } \underset{\sim}{y} \\
R_{\underset{\sim}{o}_1\underset{\sim}{o}_2}(\underset{\sim}{o}_i,\underset{\sim}{b}_i) &= R_{\underset{\sim}{o}_1\underset{\sim}{o}_2}(\underset{\sim}{o}_1,\underset{\sim}{o}_2) \cap C_i \\
R_{\underset{\sim}{o}_1\underset{\sim}{o}_2}(\underset{\sim}{b}_1,\underset{\sim}{b}_2) &= R_{\underset{\sim}{o}_1\underset{\sim}{o}_2}(\underset{\sim}{o}_1,\underset{\sim}{o}_2) \backslash \{R_{\underset{\sim}{o}_1\underset{\sim}{o}_2}(\underset{\sim}{o}_i,\underset{\sim}{b}_i), i = 1, 2\}
\end{aligned}$$

Lemma 3 *For* $i, j \in 1, 2$ *with* $i \neq j$*, and the preceding set up, we have*

i. $T_i : R_{\underset{\sim}{o}_i\underset{\sim}{o}_j}(\underset{\sim}{o}_j,\underset{\sim}{b}_j) \longrightarrow R_{\underset{\sim}{o}_i\underset{\sim}{o}_j}(\underset{\sim}{o}_i,\underset{\sim}{b}_i)$
ii. $T_i : R_{\underset{\sim}{o}_i\underset{\sim}{o}_j}(\underset{\sim}{o}_i,\underset{\sim}{b}_i) \longrightarrow R_{\underset{\sim}{o}_i\underset{\sim}{o}_j}(\underset{\sim}{o}_j,\underset{\sim}{b}_i)$

Proof Both (*i*) and (*ii*) are easily proved using Lemmas 2(*ii*) and 2(*i*), respectively.

It is worth noting that in Lemma 3, the centres of the circles need to special consideration.

Theorem 1 *Let* C_1 *and* C_2 *be two circles in* $\mathbb{R}^2$ *with centres* $\underset{\sim}{o}_1$, $\underset{\sim}{o}_2$*, respectively. Then there exists a* $c \in [0, 1)$ *such that,*

$$d(T_1(\underset{\sim}{x}_1), T_1(\underset{\sim}{x}_2)) \leq c\, d(\underset{\sim}{x}_1, \underset{\sim}{x}_2),\ \forall \underset{\sim}{x}_1, \underset{\sim}{x}_2 \in R_{\underset{\sim}{o}_1\underset{\sim}{o}_2}(\underset{\sim}{o}_2,\underset{\sim}{b}_2)$$

Proof Introduce the x-axis so that $R_{\underset{\sim}{o}_1\underset{\sim}{o}_2}(\underset{\sim}{o}_2,\underset{\sim}{b}_2)$ lies along it. This means that $\underset{\sim}{x}_1$, $\underset{\sim}{b}_1$ and $\underset{\sim}{o}_1$ correspond to numbers along the x-axis; we think of T_1 as mapping numbers to numbers on the axis. Since $x_1 > o_1 \iff T_1(x_1) > o_1$, the circle inversion map along our x-axis is

$$T_1(x_1) = o_1 + \frac{(b_1 - o_1)^2}{(x_1 - o_1)}$$

Now,

$$T_1'(x_1) = -\frac{(b_1 - o_1)^2}{(x_1 - o_1)^2} \Longrightarrow T_1'(x_1) \leq \frac{(b_1 - o_1)^2}{(b_2 - o_1)^2}$$

By the Mean Value Theorem we get

$$\begin{aligned}
|T_1(x_1) - T_1(x_2)| &= |T_1'(\xi)||x_1 - x_2|, \xi \in \mathbb{R} \\
&\leq \frac{(b_1 - o_1)^2}{(b_2 - o_1)^2}|x_1 - x_2| \\
&= c\,|x_1 - x_2|, c < 1
\end{aligned}$$

Returning to the vector notation, we have proved there exists a $c \in [0, 1)$ such that,

$$d(T_1(\underset{\sim}{x_1}), T_1(\underset{\sim}{x_2})) \leq c\, d(\underset{\sim}{x_1}, \underset{\sim}{x_2}), \forall \underset{\sim}{x_1}, \underset{\sim}{x_2} \in R_{\underset{\sim}{o_1}\underset{\sim}{o_2}}(\underset{\sim}{o_2}, \underset{\sim}{b_2})$$

Theorem 1 leads to the following significant corollary, which follows from two applications of Theorem 1.

Corollary 1 *There exists a* $c \in [0, 1)$ *such that,*

$$d((T_1 \circ T_2)(\underset{\sim}{x_1}), (T_1 \circ T_2)(\underset{\sim}{x_2})) \leq c\, d(\underset{\sim}{x_1}, \underset{\sim}{x_2}), \forall \underset{\sim}{x_1}, \underset{\sim}{x_2} \in R_{\underset{\sim}{o_1}\underset{\sim}{o_2}}(\underset{\sim}{o_1}, \underset{\sim}{b_1})$$

In Corollary 1, $(R_{\underset{\sim}{o_1}\underset{\sim}{o_2}}(\underset{\sim}{o_1}, \underset{\sim}{b_1}), |\cdot|)$ is a complete metric space and $T_1 \circ T_2$ is a contraction map from $(R_{\underset{\sim}{o_1}\underset{\sim}{o_2}}(\underset{\sim}{o_1}, \underset{\sim}{b_1}), |\cdot|)$ to itself. Banach's Fixed Point Theorem applies: there exists a unique $\bar{\underset{\sim}{x}}$ on the radial ray in C_1 such that $(T_1 \circ T_2)(\bar{\underset{\sim}{x}}) = \bar{\underset{\sim}{x}}$. A similar statement can be made for $T_2 \circ T_1$ and the radial ray in C_2. Beginning at any point on the ray from $\underset{\sim}{o_1}$ to $\underset{\sim}{o_2}$, repeated alternating application of T_1 and T_2 approaches the two cycle consisting of these two fixed points.

4 Collections of Non-touching Circles

We present the following result in $\mathbb{R}^2$ with the understanding that this may be a two dimensional subspace of a higher dimensional setting. In this result we consider a collection of non-touching circles in the plane.

Lemma 4 *Given a circle* C_1 *with centre* $\underset{\sim}{o_1}$*, radii* r_1 *and inversion map* T_1,

$$d(T_1(\underset{\sim}{x_1}), T_1(\underset{\sim}{x_2})) = \frac{1}{a_1 a_2} d(\underset{\sim}{x_1}, \underset{\sim}{x_2}), \forall \underset{\sim}{x_1}, \underset{\sim}{x_2} \in \mathbb{R}^2 \setminus \{\underset{\sim}{o_1}\},$$

where, for $j = 1, 2$, $\underset{\sim}{x_j} = a_j r_1 \underset{\sim}{w_1}(t_j) + \underset{\sim}{o_1}$ *for some* $t_j \in [0, 2\pi]$, $a_j > 0$.

Proof We consider two cases.

Case 1: $\underset{\sim}{x_1}, \underset{\sim}{x_2}$ and $\underset{\sim}{o_1}$ lie on the same radial ray. Without loss of generality, we can place the three points on the x-axis at locations $\underset{\sim}{x_1} = (a_1 r_1, 0)$, $\underset{\sim}{x_2} = (a_2 r_1, 0)$ and $\underset{\sim}{o_1} = (0, 0)$. Then,

$$\begin{aligned} d(T_1(\underset{\sim}{x_1}), T_1(\underset{\sim}{x_2})) &= ||T_1(\underset{\sim}{x_1}) - T_1(\underset{\sim}{x_2})|| \\ &= \left\| \frac{||\Pi_1(\underset{\sim}{x_1})||^2}{||\underset{\sim}{x_1}||^2} \underset{\sim}{x_1} - \frac{||\Pi_1(\underset{\sim}{x_2})||^2}{||\underset{\sim}{x_2}||^2} \underset{\sim}{x_2} \right\| \\ &= \left\| \frac{r_1^2}{r_1^2 a_1^2}(a_1 r_1, 0) - \frac{r_1^2}{r_1^2 a_2^2}(a_2 r_1, 0) \right\| \end{aligned}$$

$$= \left| \left(\frac{1}{a_1} - \frac{1}{a_2} \right) r_1 \right| = \frac{1}{a_1 a_2} r_1 |a_2 - a_1|$$
$$= \frac{1}{a_1 a_2} d(\underset{\sim}{x}_1, \underset{\sim}{x}_2)$$

Case 2: $\underset{\sim}{x}_1, \underset{\sim}{x}_2$ and $\underset{\sim}{o}_1$ are not on the same radial ray. Once again the boundary ∂C_1 can be parametrized as in (4). There exists $t_1, t_2 \in [0, 2\pi]$, $t_2 > t_1$, without loss of generality, such that

$$\Pi_1(\underset{\sim}{x}_j) = r_1 \underset{\sim}{w}_1(t_j) + \underset{\sim}{o}_1 \text{ and } \underset{\sim}{x}_j = a_j r_1(\cos(t_j), \sin(t_j)) + \underset{\sim}{o}_1,\ j = 1, 2.$$

Note that this case includes the choice $t_2 = t_1 + \pi$, two points on the same diameter but on opposite sides of the centre. After straightforward calculations, we arrive at

$$d^2(\underset{\sim}{x}_1, \underset{\sim}{x}_2) = a_1^2 r_1^2 + a_2^2 r_1^2 - 2a_1 a_2 r_1^2 \cos(t_2 - t_1) \tag{6}$$

By (5),

$$T_1(\underset{\sim}{x}_j) = \underset{\sim}{o}_1 + \frac{1}{a_j} r_1 \underset{\sim}{w}(t_j)$$

Thus by using (5), (6) and some simplifying we have,

$$d^2(T_1(\underset{\sim}{x}_1), T_1(\underset{\sim}{x}_2)) = \frac{1}{a_1^2} r_1^2 + \frac{1}{a_2^2} r_1^2 - \frac{2}{a_1 a_2} r_1^2 \cos(t_2 - t_1)$$
$$= \frac{1}{a_1^2 a_2^2} d^2(\underset{\sim}{x}_1, \underset{\sim}{x}_2),$$

which upon taking the square root gives the desired result.

We will use Lemma 4 and the following definition when establishing the subsequent contractivity result.

Definition 3 Given a collection of non-touching circles C_i, $i = 1, \ldots, N$, with centres $\underset{\sim}{o}_i$ and radii r_i, for $\underset{\sim}{x} = a_i r_i \underset{\sim}{w}_i(t) + \underset{\sim}{o}_i \in \mathbb{R}^2$, $a_i > 0$, $t \in [0, 2\pi]$, we define

$$a_{i,min} = \min_{\underset{\sim}{x} \in C_j, j \neq i} \{a_i\} > 1$$

corresponding to the radial scaling of the closest point to $\underset{\sim}{o}_i$ in all other circles C_j, $j \neq i$.

Theorem 2 *T_i: $C_j \longrightarrow C_i$, $i \neq j$, is contractive. That is, there exists a $c \in [0,1)$ such that,*

$$d(T_i(\underset{\sim}{x}_1), T_i(\underset{\sim}{x}_2)) \leq c\, d(\underset{\sim}{x}_1, \underset{\sim}{x}_2),\ \forall \underset{\sim}{x}_1, \underset{\sim}{x}_2 \in C_j, i \neq j$$

Proof By Lemma 4 we have that

$$d(T_i(\underset{\sim}{x}_1), T_i(\underset{\sim}{x}_2)) = \frac{1}{a_1 a_2} d(\underset{\sim}{x}_1, \underset{\sim}{x}_2) \leq \frac{1}{a^2_{i,min}} d(\underset{\sim}{x}_1, \underset{\sim}{x}_2) = c\, d(\underset{\sim}{x}_1, \underset{\sim}{x}_2)$$

Since T_i: $C_{i,int} \rightarrow C_{i,ext}$ is expansive, we redefine the map as follows.

Definition 4 Given a collection of non-touching circles C_i, $i = 1, \ldots, N$, with centres $\underset{\sim}{o}_i$, define

$$X = \bigcup_{i=1}^{N} C_i$$

and, relative to each C_i, define

$$\overline{T}_i : X \rightarrow X$$

by

$$\overline{T}_i(\underset{\sim}{x}) = \begin{cases} T_i(\underset{\sim}{x}) \text{ if } \underset{\sim}{x} \in C_j, j \neq i \\ \underset{\sim}{x} \quad \text{if } \underset{\sim}{x} \in C_i \end{cases}$$

Theorem 3 *$\overline{T}_i : X \rightarrow X$ satisfies*

(*i*) $d(\overline{T}_i(\underset{\sim}{x}_1), \overline{T}_i(\underset{\sim}{x}_2)) \leq c\, d(\underset{\sim}{x}_1, \underset{\sim}{x}_2)$ *for some* $c \in [0,1)$, $\forall \underset{\sim}{x}_1 \in C_j, \underset{\sim}{x}_2 \in C_k, j, k \neq i$

(*ii*) $d(\overline{T}_i(\underset{\sim}{x}_1), \overline{T}_i(\underset{\sim}{x}_2)) = d(\underset{\sim}{x}_1, \underset{\sim}{x}_2)\ \forall \underset{\sim}{x}_1, \underset{\sim}{x}_2 \in C_i$

(*iii*) $d(\overline{T}_i(\underset{\sim}{x}_1), \overline{T}_i(\underset{\sim}{x}_2)) \leq c\, d(\underset{\sim}{x}_1, \underset{\sim}{x}_2)$ *for some* $c \in [0,1)$, $\forall \underset{\sim}{x}_1 \in C_i$, $\forall \underset{\sim}{x}_2 \in C_j, j \neq i$

Proof (i) follows from Theorem 2, (ii) follows immediately since $\overline{T}_i$ is the identity map in this case. For (iii), with $\underset{\sim}{x}_1 \in C_i$ and $\underset{\sim}{x}_2 \in C_j$, we have $a_2 > 1 > a_1 \geq 0$, and, for values $t_1, t_2 \in [0, 2\pi]$,

$$d^2(\underset{\sim}{x}_1, \underset{\sim}{x}_2) = r_i^2 \left(a_1^2 + a_2^2 - 2a_1 a_2 \cos(t_2 - t_1)\right), \tag{7}$$

using (6), and with some expanding and factoring we are left with

$$d^2(\overline{T}_i(\underset{\sim}{x}_1), \overline{T}_i(\underset{\sim}{x}_2)) = \frac{r_i^2}{a_2^2}\left(a_2^2 a_1^2 + 1 - 2a_1 a_2 \cos(t_2 - t_1)\right) \tag{8}$$

The bracketed expression in (8) is less than or equal to the bracketed expression in (7) provided that

$$a_2^2 a_1^2 + 1 \leq a_1^2 + a_2^2, \text{ where } a_2 > 1 > a_1 \geq 0. \tag{9}$$

Hence, letting $y = a_1^2$ and $z = a_2^2$, we consider the function

$$f(y, z) = y - yz + z.$$

If we show that $f(y, z) \geq 1$ for $y \in [0, 1)$ and $z > 1$, then (9) holds. We write

$$f(y, z) = y - yz + z = -y(z - 1) + (z - 1) + 1 = (z - 1)(1 - y) + 1$$

and we see that $f(y, z) > 1$ for $y < 1$ and $z > 1$, proving the result. This means that the statement in case (iii) of our theorem holds with $c = \frac{1}{a_2}$.

Now, using Theorem 3 we can prove the following contractivity result involving compositions of (non-touching) circle inversion maps.

Theorem 4 *For $i \neq j$, $\overline{T}_i \circ \overline{T}_j$: $X \longrightarrow X$ is contractive:*

$$d((\overline{T}_i \circ \overline{T}_j)(\underset{\sim}{x}_1), (\overline{T}_i \circ \overline{T}_j)(\underset{\sim}{x}_2)) \leq c\, d(\underset{\sim}{x}_1, \underset{\sim}{x}_2) \textit{ for some } c \in [0, 1), \forall \underset{\sim}{x}_1, \underset{\sim}{x}_2 \in X$$

Proof If $\underset{\sim}{x}_1, \underset{\sim}{x}_2 \in C_k, k \neq j$, then the result follows by applying Theorem 3(i) twice. If $\underset{\sim}{x}_1, \underset{\sim}{x}_2 \in C_j$, then the result follows by applying Theorem 3(ii) and 3(i). If $\underset{\sim}{x}_1 \in C_j, \underset{\sim}{x}_2 \in C_k, k \neq j$, then the result follows by applying Theorem 3(iii) and 3(i).

By Theorem 4 and Banach's Fixed Point Theorem, for each i we can say that there exists a unique $\bar{\underset{\sim}{x}}_j$ in C_i such that $(T_i \circ T_j)(\bar{\underset{\sim}{x}}_j) = \bar{\underset{\sim}{x}}_j$ for each choice of j. In fact, by Corollary 1 we know that each such fixed point lies on the appropriate radial ray, $R_{\underset{\sim}{O}_i\underset{\sim}{O}_j}(\underset{\sim}{O}_i, \underset{\sim}{b}_i)$.

We now formulate this multi-circle framework in the context of sets. For a set $A \subset \mathbb{R}^2$, define

$$\hat{\overline{T}}_i(A) = \{\overline{T}_i(\underset{\sim}{x}), \forall \underset{\sim}{x} \in A\}$$

Define the distance from a set A to a set B by

$$d(A,B) = \sup_{x \in A} d(x,B) = \sup_{x \in A} \inf_{y \in B} d(x,y)$$

The Hausdorff distance between the non-empty compact sets A and B is defined by

$$d_H(A,B) = \max\{d(A,B), d(B,A)\}$$

Let $\mathscr{H}(X)$ denote the set of all non-empty compact subsets of X. It is well known that $(\mathscr{H}(X), d_H)$ is complete.

Theorem 5 *For $i \neq j$,*

1. $\hat{\overline{T}}_i \circ \hat{\overline{T}}_j : \mathscr{H}(X) \longrightarrow \mathscr{H}(X)$
2. $\hat{\overline{T}}_i \circ \hat{\overline{T}}_j$ *is contractive on* $(\mathscr{H}(X), d_H)$

Proof The first claim follows because $\overline{T}_i \circ \overline{T}_j : \mathscr{H}(X) \to \mathscr{H}(X)$. The proof relies on the continuity of the map ([1], Lemma 2, page 80). By Theorem 4, we can denote by c_{ij} the contractivity factor with respect to d of $\overline{T}_i \circ \overline{T}_j$. For $A, B \in \mathscr{H}(X)$,

$$\begin{aligned}
& d_H((\hat{\overline{T}}_i \circ \hat{\overline{T}}_j)(A), (\hat{\overline{T}}_i \circ \hat{\overline{T}}_j)(B)) \\
&= \max\left\{ \sup_{\underset{\sim}{x}_1 \in (\hat{\overline{T}}_i \circ \hat{\overline{T}}_j)(A)} \inf_{\underset{\sim}{x}_2 \in (\hat{\overline{T}}_i \circ \hat{\overline{T}}_j)(B)} d(\underset{\sim}{x}_1, \underset{\sim}{x}_2), \sup_{\underset{\sim}{x}_1 \in (\hat{\overline{T}}_i \circ \hat{\overline{T}}_j)(B)} \inf_{\underset{\sim}{x}_2 \in (\hat{\overline{T}}_i \circ \hat{\overline{T}}_j)(A)} d(\underset{\sim}{x}_1, \underset{\sim}{x}_2) \right\} \\
&= \max\left\{ \sup_{\underset{\sim}{x}_1 \in A} \inf_{\underset{\sim}{x}_2 \in B} d((\overline{T}_i \circ \overline{T}_j)(\underset{\sim}{x}_1), (\overline{T}_i \circ \overline{T}_j)(\underset{\sim}{x}_2)), \right. \\
&\qquad \left. \sup_{\underset{\sim}{x}_1 \in B} \inf_{\underset{\sim}{x}_2 \in A} d((\overline{T}_i \circ \overline{T}_j)(\underset{\sim}{x}_1), (\overline{T}_i \circ \overline{T}_j)(\underset{\sim}{x}_2)) \right\} \\
&\leq c_{ij} \max\left\{ \sup_{\underset{\sim}{x}_1 \in A} \inf_{\underset{\sim}{x}_2 \in B} d(\underset{\sim}{x}_1, \underset{\sim}{x}_2), \sup_{\underset{\sim}{x}_1 \in B} \inf_{\underset{\sim}{x}_2 \in A} d(\underset{\sim}{x}_1, \underset{\sim}{x}_2) \right\} \\
&= c_{ij}\, d_H(A,B)
\end{aligned}$$

We reach our final theorem, which establishes the contractivity of the union of the maps in Theorem 5.

Theorem 6 *For* $A \in \mathscr{H}(X)$*, define* $\hat{\overline{T}}(A) = \bigcup_{\substack{i,j=1\\ i\neq j}}^{N} (\hat{\overline{T}}_i \circ \hat{\overline{T}}_j)(A)$*. Then*

1. $\hat{\overline{T}} : \mathscr{H}(X) \longrightarrow \mathscr{H}(X)$
2. $\hat{\overline{T}}$ *is contractive on* $(\mathscr{H}(X), d_H)$

Proof The first claim follows from Theorem 5 because the union is finite. For the second claim we have

$$\begin{aligned} d_H(\hat{\overline{T}}(A), \hat{\overline{T}}(B)) &= d_H\left(\bigcup_{\substack{i,j=1\\ i\neq j}}^{N} (\hat{\overline{T}}_i \circ \hat{\overline{T}}_j)(A), \bigcup_{\substack{i,j=1\\ i\neq j}}^{N} (\hat{\overline{T}}_i \circ \hat{\overline{T}}_j)(B) \right) \\ &\leq \max_{1\leq i,j\leq N} d_H((\hat{\overline{T}}_i \circ \hat{\overline{T}}_j)(A), (\hat{\overline{T}}_i \circ \hat{\overline{T}}_j)(B)), \text{ property of } d_H \\ &\leq \max_{1\leq i,j\leq N} c_{ij}\, d_H(A, B), \text{ by Theorem 5} \\ &= c\, d_H(A, B) \end{aligned}$$

Finally, by Theorem 6 and Banach's Fixed Point Theorem we can conclude that there exists a unique fixed point $\mathscr{A} \in \mathscr{H}(X)$, the set attractor of $\hat{\overline{T}}$, satisfying

$$\hat{\overline{T}}(\mathscr{A}) = \mathscr{A}$$

5 The Chaos Game

Following Sect. 3, we can use the Chaos Game to plot approximations of the set attractor $\mathscr{A}$ of $\hat{\overline{T}}$. Choose an initial point $\mathfrak{x}_0 \in X$. With equal probability select one of the composition maps $\overline{T}_i \circ \overline{T}_j$, $i \neq j$. Apply the map to $\mathfrak{x}_0$ to produce $\mathfrak{x}_1 \in X$ (in C_i, of course). Continue randomly selecting composition maps to produce the sequence $\{\mathfrak{x}_n\}_{n=0}^{\infty}$. If we start with a point on $\mathscr{A}$, all points are on $\mathscr{A}$. In practice, we stop after some large number of iterations, dependent upon the resolution of the plot we are producing.

An example of a fractal generated by a five-circle system is given in Fig. 1.

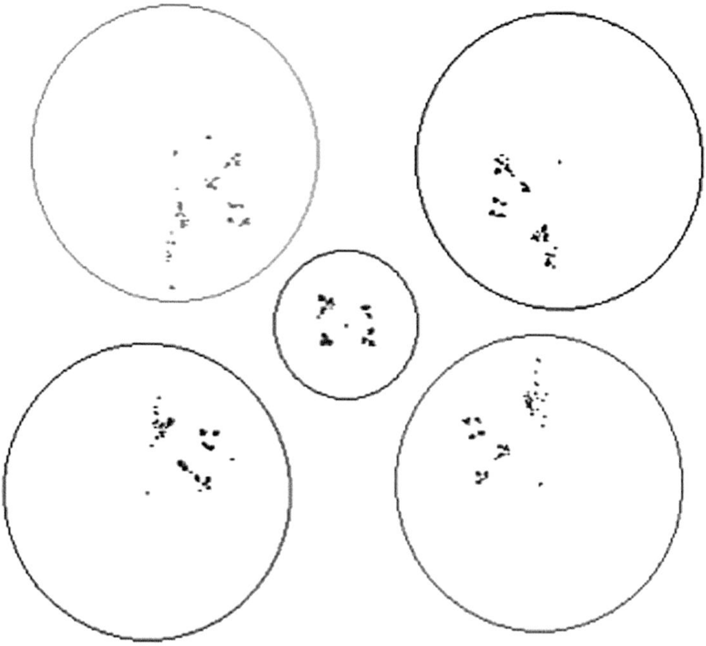

Fig. 1 Five circles used to generate a set attractor with 10,000 iterations of the chaos game

References

1. Barnsley, M.F.: Fractals Everywhere. Academic, London (1988)
2. Clancy, C., Frame, M.: Fractal geometry of restricted sets of circle inversion. Fractals **3**, 689–699 (1995)
3. Frame, M., Clancy, C., Peak, D.: A fractal representation of grammatical complexity in time series. Manuscript in Preparation (1996)
4. Frame, M., Cogevina, T.: An infinite circle inversion limit set fractal. Comput. Graph. **24**(5), 797–804 (2000)
5. Leys, J.: Sphere inversion fractals. Comput. Graph. **29**(3), 463–466 (2005)

Computation of Galois Groups in magma

Andreas-Stephan Elsenhans

Abstract We give an outline of a degree independent algorithm for the computation of Galois groups, as it is implemented in `magma` (van der Waerden BM, Algebra 1, Springer, Berlin 1960) (version 2.21). Further, we summarize several performance tests and list hard examples.

1 Introduction

Given a polynomial f with rational coefficients, one can ask for the Galois group of the splitting field (viewed as an extension of the rationals). This is a classical problem in algorithmic algebra. The first algorithm to solve it was described by van der Waerden in his famous book on algebra [12]. Later, more practical algorithms were given [2, Sec. 6.3].

A crucial point on the above question is the way one wants to present the result. One could give the Galois group as an abstract group. But this would be an incomplete answer, as this does not give the action on the splitting field. On the other hand, describing the action on the splitting field would require to construct it. And this would be impractically large in many cases.

A good compromise is to give complex or p-adic root approximations and the action of the Galois group on them. This is the starting point of the Stauduhar method [11]. The approach generalizes to any global field.

1.1 Previous Implementations

Several people implemented Galois group algorithms that used precomputed tables with all the combinatorial data necessary for the computation. Most notable is the implementation of K. Geißler that covered irreducible polynomials up to degree 23 [7].

A.-S. Elsenhans (✉)
Mathematisches Institut, Universität Paderborn, Warburger Str. 100, 33098 Paderborn, Germany
e-mail: elsenhan@math.uni-paderborn.de

© Springer International Publishing Switzerland 2016

J. Bélair et al. (eds.), *Mathematical and Computational Approaches in Advancing Modern Science and Engineering*, DOI 10.1007/978-3-319-30379-6_56

The first degree independent implementation that could also handle reducible polynomials was given by C. Fieker and J. Klüners [6]. However, this implementation ran out of memory or needed hours of CPU time with several polynomials of degree ≤ 30.

Recently, the author worked on bottlenecks of this package. The aim of this article is to give a summary of what is now possible.

2 Stauduhar's Step

Definition 1 Let $U \subset G \subset S_n$ be permutation groups. They act on $\mathbb{Z}[X_1, \ldots, X_n]$ by permutation of the variables.

A polynomial $I \in \mathbb{Z}[X_1, \ldots, X_n]$ is called a *U-invariant* if $\sigma I = I$ for all $\sigma \in U$. It is called a *relative invariant for* $U \subset G$ if its stabilizer in G is equal to U.

Lemma 1 *Relative invariants exist of all pairs of subgroups* $U \subset G \subset S_n$.

Proof Choose a monomial with trivial stabilizer in S_n. E.g., $m := X_1 X_2^2 \cdots X_{n-1}^{n-1}$. Then the orbit sum $\sum_{\sigma \in U} \sigma m$ is a relative invariant. □

Definition 2 Let $U \subset G$ be two groups. We denote by $G /\!/ U$ a set of coset representative of G/U.

With this preparation, we can describe the Stauduhar step [11].

Theorem 1 *Let $f \in \mathbb{Q}[T]$ be a degree n polynomial with distinct roots $r_1, \ldots, r_n$ in an arbitrary splitting field. Further, let $G \subset S_n$ be a permutation group that contains the Galois group of f, U be any subgroup of G and $I \in \mathbb{Z}[X_1, \ldots, X_n]$ a relative invariant for $U \subset G$. Finally, we assume that the values $(\sigma I)(r_1, \ldots, r_n)$ for $\sigma \in G /\!/ U$ are pairwise distinct.*

Then the Galois group of f is contained in U^σ if and only if $(\sigma I)(r_1, \ldots, r_n)$ is rational.

The above theorem leads to the following algorithm for the computation of Galois groups.

1. Choose a starting group (e.g., $G = S_n$).
2. Compute the conjugacy classes of the maximal subgroups of G.
3. For each subgroup class representative $U \subset G$ do the following:
   a. Compute a relative invariant I.
   b. Determine $(\sigma I)(r_1, \ldots, r_n)$ for $\sigma \in G /\!/ U$.
   c. Use the theorem above to find all the conjugates of U that contain the Galois group.
4. If no subgroup of G is found that contains the Galois group, return G as the Galois group.

5. Intersect all the subgroups found that contain the Galois group.
6. Redo all the steps above with the intersection as starting group.

Remark 1 There are various problems that have to be solved to make the above approach practical.

1. First, we need a good strategy to pick an initial group that is as small as possible. One approach for this is the use of subfields [7, Chap. 5.1]. In case of a reducible polynomial, one can compute the Galois group of each factor and take the direct product as a starting group [6].
2. In `magma`, the computation of maximal subgroups is done using the Cannon-Holt-algorithm [1].
3. How can we prove the rationality of $(\sigma I)(r_1, \ldots, r_n)$ when we only work with approximations of the roots? One way to do this is described in [7, Chap. 3.3]. This requires to work with p-adic precisions that are proportional to the index of the subgroup. This is practical only when the index is small.

   However, it is possible to work with a moderate p-adic precision and derive a heuristic result that has to be proven later.

3 The Use of the Frobenius

3.1 Using Cycle Types

Let a monic polynomial $f \in \mathbb{Z}[T]$ be given and a prime p that does not divide its discriminant. It is well known that the local Galois group of the p-adic splitting field of f is a subgroup of the Galois group of the global splitting field.

The local Galois group is generated by the Frobenius element. When we take the degrees of the irreducible factors of the reduction of f modulo p, we get the cycle type of the Frobenius element.

Doing this for several primes, we derive some information about the Galois group that can be used as follows:

1. If one gets sufficiently many different cycle types then one can prove that the Galois group is the full symmetric group S_n. More precisely, in the case that $f \in \mathbb{Z}[T]$ is irreducible, n is at least 8 and one has a cycle of prime length l with $\frac{n}{2} < l < n - 2$, the alternating group is contained in the Galois group. If, in addition, an odd permutation is detected then the Galois group is S_n.
2. Before we apply Stauduhar's step to a subgroup, we can check that it contains all the cycle types found.

3.2 Short Cosets

A more sophisticated way to use the local Galois group is provided by the so called short cosets [7, Chap. 5.2]. The idea behind them is that, when working with p-adic root approximations, the Frobenius Frob_p is known as an explicit permutation of the roots.

Thus, a necessary condition for U^σ to contain the Galois group is that it contains the Frobenius permutation. We call the remaining coset representatives

$$G/\!/_{\mathrm{Frob}_p} U := \{\sigma \in G/\!/U \mid \mathrm{Frob}_p \in U^\sigma\}$$

the *short cosets*. They can be computed without listing all the representatives $G/\!/U$ ([7, Algorithmus 5.12], [4]).

In many cases, the number of short cosets is very small. A naive explanation for this is the following:

$$\mathrm{Frob}_p \in U^\sigma \iff \mathrm{Frob}_p^{(\sigma^{-1})} \in U$$

Assuming the conjugates $\mathrm{Frob}_p^{(\sigma^{-1})}$ to be equidistributed in G, the probability to hit U is $\frac{1}{[G:U]}$. Thus, for large index subgroups this number is very small. Of course, this is just a very coarse heuristic.

4 The Invariants

To make the Stauduhar method run, we need a relative invariant for each pair of groups, the algorithm may run into. As there are 25,000 transitive permutation groups in degree 24, it is not practical to store them in a database. We have to compute them at run time. A more detailed description of the construction of the invariants if given in [3, 5–7]. The main tool for this is the usage of block systems.

Definition 3 A partition $B_1, \ldots, B_k$ of $\{1, \ldots, n\}$ is called a *system of blocks* for a transitive subgroup $G \subset \mathrm{S}_n$ if

$$\sigma B_i \in \{B_1, \ldots, B_k\}$$

for all $\sigma \in G$.

Theorem 2 *Let U be maximal in $G \subset \mathrm{S}_n$ and $\mathscr{B} = \{B_1, \ldots, B_k\}$ be a block system for U with $B_1 = \{1, \ldots, j\}$.*

1. *If $\mathscr{B}$ if not a block system for G then $\sum_{B\in\mathscr{B}}(\sum_{i\in B} X_i)^2$ is a relative invariant.*
2. *If $\mathscr{B}$ is a block system of G, denote by $\varphi_{\mathscr{B}}$ the action of G on the blocks.*

- *If* $\varphi_{\mathscr{B}}(U) \neq \varphi_{\mathscr{B}}(G)$ *then* $I_0(\sum_{i\in B_1} X_i, \ldots, \sum_{i\in B_k} X_i)$ *is a relative invariant. Here,* I_0 *is a relative invariant for* $\varphi_{\mathscr{B}}(U) \subset \varphi_{\mathscr{B}}(G)$.
- *If* $\mathrm{Stab}_U(B_1)|_{B_1} \neq \mathrm{Stab}_G(B_1)|_{B_1}$ *then* $\sum_{\sigma\in U/\!/\mathrm{Stab}_U(B_1)} I_0(X_1,..,X_j)^\sigma$ *is a relative invariant. Here,* I_0 *is a relative invariant for* $\mathrm{Stab}_U(B_1)|_{B_1} \subset \mathrm{Stab}_G(B_1)|_{B_1}$.

A proof of this is given in [7, Satz 6.14, 6.16].

If these constructions do not apply then we have to do a deeper inspection of the structure of the permutation groups. Here are a few examples:

Example 1

1. Consider the groups

$$H = \{(\sigma_1, \ldots, \sigma_{10}) \in \mathrm{A}_3^{10} \mid \sigma_1 \cdots \sigma_{10} = \mathrm{id}\} \rtimes \mathrm{S}_{10} = T_{30}^{5367}$$
$$\subset \mathrm{A}_3 \wr \mathrm{S}_{10} = T_{30}^{5407} = G\,.$$

A relative invariant is given by

$$I := \prod_{i=1}^{10} \left(X_{3i-2} + \zeta_3 X_{3i-1} + \zeta_3^2 X_{3i}\right)\,.$$

2. The above example has the extension $T_{30}^{5396} \subset T_{30}^{5421}$ by adding the generator $\sigma = (1\ 2)(4\ 5)\cdots(28\ 29)$ to both groups.
   A relative invariant is given by $I + I^\sigma$. In invariant theory, this construction of an invariant out of an invariant of a subgroup is known as the Reynolds operator.
3. Consider the groups

$$G = T_{30}^{4831} = (\mathbb{Z}/2\mathbb{Z})^{15} \rtimes \mathrm{GL}_4(\mathbb{F}_2) = \mathrm{S}_2 \wr \mathrm{GL}_4(\mathbb{F}_2) \subset \mathrm{S}_2 \wr \mathrm{A}_{15}$$

and $H = T_{30}^{3819} = N \rtimes \mathrm{GL}_4(\mathbb{F}_2) \subset G$. Here, N is the Hamming code in $\mathbb{F}_2^{15}$. Recall the combinatorial structure of $\mathrm{P}^3(\mathbb{F}_2)$ and the Hamming code:

- $\mathrm{P}^3(\mathbb{F}_2)$ has 15 points and 15 planes.
- The stabilizer of one plane is a subgroup $U_0 \subset \mathrm{GL}_4(\mathbb{F}_2)$ of index 15.
- The subgroup U_0 decomposes the points into two orbits of size 7 and 8.
- The Hamming code N is given as the subspace of $\mathbb{F}_2^{15}$. To construct it, we label the coordinates with the points of $\mathrm{P}^3(\mathbb{F}_2)$. The linear equations that define N are

$$\sum_{i\in \mathrm{P}^3(\mathbb{F}_2)\setminus E} C_i = 0$$

for all the planes $E \subset \mathrm{P}^3(\mathbb{F}_2)$.
- The automorphism group of the Hamming code is $\mathrm{GL}_4(\mathbb{F}_2)$.

- We have $\#G = 660{,}602{,}880$, $[G : H] = 16$.
- The important property of the subgroup U_0 is that it stabilizes one of the linear relations that define the Hamming code N.

The interested reader may consult [10, Sec. 3.3] for a more general introduction to the Hamming code.

To construct an invariant, we start with the preimage U of U_0 in G. In a first step, we construct an $H \cap U$-invariant that is not a U-invariant by putting

$$I_0 := \prod_{\sigma \in U /\!/ \mathrm{Stab}_U\{1,2\}} (X_2 - X_1)^\sigma .$$

This means, we turn the linear relation of N that is stabilized by U_0 into an invariant. Now, a direct application of the Reynolds operator gives us the invariant

$$I := \sum_{\sigma \in H /\!/ H \cap U} I_0^\sigma$$

as a sum of 15 products.

Experiment 1 To get an overview of the complexity of the invariants that may be used by our code, we use the database of transitive groups up to degree 30 [8] and apply our invariant constructions to all the maximal subgroups. We use the notation T_n^k for the k-th group of degree n in the database. The construction of all these invariants took about 1 h.

We use the cost function

$$(\text{Number of multiplications in one evaluation}) \cdot \deg(I_{G,U}) \cdot \min(45, [G : U])$$

to identify hard cases. They are listed in Table 1.

Example 2 When we compute the Galois group of a polynomial with Galois group T_{30}^{1153}, we run into the hardest case listed above.

To get the group of a test polynomial took about 10 s. About 1 s was used in the Stauduhar step that involves the hard invariant.

The computation was done on one core of an Intel i7-3770 processor running at 3.40 GHz.

Table 1 Pairs of groups with costly invariants

Groups	Index	$\deg(I)$	Evaluation costs
$P\Gamma L_2(\mathbb{F}_8) = T_9^{32} \subset A_9$	120	6	387 multiplications
$M_{24} = \mathrm{T}_{24}^{24,680} \subset A_{24}$	1,267,136,462,592,000	6	759 powers
$\mathrm{T}_{30}^{1153} \subset \mathrm{T}_{30}^{4863}$	20,160	8	5221 multiplications

Experiment 2 Using the database of polynomials [9], we can pick 1954 polynomials in degree 16 and 1117 polynomials in degree 20. One for each transitive group of degree 16 (resp. 20).

The total running time to get all the Galois groups for the degree 16 polynomials is 1094 s. The time for all the degree 20 polynomials is 907 s. The slowest examples are polynomials with large coefficients. They took 5 (resp. 13) s. This is explained by the larger p-adic precision that has to be used.

A slow example with small coefficients is $P_{455} := x^{16} - 11x^8 + 9$. The Galois group of this polynomial is T_{16}^{455}. It takes 2.2 s to compute it. This is explained by the fact that our method has to inspect 219 conjugacy classes of subgroups.

The computation was done on one core of an Intel i7-3770 processor running at 3.40 GHz.

5 Proving the Galois Group

The uses of heuristic p-adic precision and the short cosets result in an unproven Galois group. The first proof algorithm was described in [7, Chap. 5.3]. Our new proof algorithm is a slight variation.

The idea to confirm the results is that the Galois group predicts the factorizations of resolvent polynomials. We can compute these resolvents and check that the prediction is in fact correct. This proves that the Galois group is contained in the stabilizer of the factorization pattern. These set stabilizers can be computed in `magma`. This results in a subgroup of S_n that is proven to contain the Galois group. We repeat this with several resolvent polynomials until we run into a contradiction or confirm the heuristic result.

6 Conclusion

The Galois group package of `magma` (version 2.21) provides a degree independent implementation of the Stauduhar method. It can compute the Galois group of a rational degree 30 polynomial with moderate coefficient size in a few seconds on a modern standard PC. The initial result is heuristic. But, a proof algorithm is available as well.

References

1. Cannon, J., Holt, D.: Computing maximal subgroups of finite groups. J. Symb. Comput. **37**, 589–609 (2004)
2. Cohen, H.: A Course in Computational Algebraic Number Theory. Springer, Berlin/New York (1993)

3. Elsenhans, A.-S.: Invariants for the computation of intransitive and transitive Galois groups. J. Symb. Comput. **47**, 315–326 (2012)
4. Elsenhans, A.-S.: A note on short cosets. Exp. Math. **23**, 411–413 (2014)
5. Elsenhans, A.-S.: Improved methods for the construction of relative invariants for permutation groups (preprint, 2015)
6. Fieker, C., Klüners J.: Computation of Galois groups of rational polynomials. LMS J. Comput. Math. **17**, 141–158 (2014)
7. Geißler, K.: Berechnung von Galoisgruppen über Zahl- und Funktionenkörpern. Dissertation, Berlin (2003)
8. Hulpke, A.: Constructing transitive permutation groups. J. Symb. Comput. **39**(1), 1–30 (2005)
9. Klüners J., Malle G.: A database for field extensions of the rationals. LMS J. Comput. Math. **4**, 182–196 (2001)
10. van Lint, J.H.: Introduction to Coding Theory. Springer, Berlin/New York (1992)
11. Stauduhar, R.P.: The determination of Galois groups. Math. Comput. **27**, 981–996 (1973)
12. van der Waerden, B.L.: Algebra I. Springer, Berlin (1960)

Global Dynamics and Periodic Solutions in a Singular Differential Delay Equation

Anatoli F. Ivanov and Zari A. Dzalilov

Abstract Differential delay equation $\varepsilon\,[x'(t) + cx'(t-1)] + x(t) = f(x(t-1))$ is considered where $\varepsilon > 0$ and $c \in \mathbf{R}$ are parameters, and $f : \mathbf{R} \to \mathbf{R}$ is piecewise continuous. For small values of the parameter ε a connection is made to the continuous time difference equation $x(t) = f(x(t-1))$, which is further linked to the one-dimensional dynamical system $x \mapsto f(x)$. Two cases of the nonlinearity f are treated: when it is continuous and of the negative feedback with respect to a unique equilibrium, and when it is of the so-called Farrey-type with a single jump-discontinuity. Several properties are studied, such as continuous dependence of solutions on the singular parameter ε and the existence of periodic solutions. Open problems and conjectures are stated for the case of genuinely neutral equation, when $c \neq 0$.

1 Introduction

Differential delay equations (DDEs) of the form

$$\varepsilon\left[x'(t) + cx'(t-1)\right] + x(t) = f(x(t-1)), \quad t \geq 0 \tag{1}$$

appear as mathematical models of various real world phenomena. They were first mentioned, to the best of our knowledge, in paper [7] as exact reductions of nonlinear boundary value problems for one-dimensional wave equations modelling violin string oscillations. The very same idea of reduction was later used in paper [4] describing complex oscillations in electrical circuits with tunnel diodes. Note that in both papers the reduction is to the neutral type DDEs (1), when $c \neq 0$. This is a well known idea and approach that certain boundary value problems for hyperbolic partial differential equations can be exactly reduced to differential

A.F. Ivanov (✉)
Department of Mathematics, Pennsylvania State University, Lehman, PA 18627, USA
e-mail: aivanov@psu.edu

Z.A. Dzalilov
Federation University Australia, Mt Helen Campus, Ballarat, VIC 3350, Australia
e-mail: z.dzalilov@federation.edu.au

© Springer International Publishing Switzerland 2016

J. Bélair et al. (eds.), *Mathematical and Computational Approaches in Advancing Modern Science and Engineering*, DOI 10.1007/978-3-319-30379-6_57

difference equations of various types [5]. The retarded type DDEs (1), when $c = 0$, appear in numerous other applications, a partial list of which can be found e.g. in [1, 3, 6]. This case is more studied compared with the neutral case of $c \neq 0$. Very few results are available in the latter case. This paper is an attempt to initiate a comprehensive study of neutral differential delay equations (1) in one particular approach as a singular perturbation problem.

When $\varepsilon = 0$ Eq. (1) becomes a continuous time difference equation (DE) of the form

$$x(t) = f(x(t-1)), \quad t \geq 0. \tag{2}$$

The dynamics of the latter as $t \to \infty$ are largely determined by the interval map f, or the equivalent scalar difference equation

$$x_{n+1} = f(x_n), \quad n \in \mathbf{N}_0 := \mathbf{N} \cup \{0\}. \tag{3}$$

In this paper we attempt to relate several properties of solutions of DDEs (1) to the well-studied and understood properties of solutions of difference equations (2) and (3).

In Sect. 2 we list basic facts and recall necessary knowledge about differential delay equations (1), difference equations with continuous time (2), and interval maps associated with scalar difference equation (3). They all can be readily found in e.g. monographs [1, 2, 5, 6].

Section 3 contains our main results and is made of two subsections. Section 3.1 deals with the continuous dependence on the singular parameter $\varepsilon > 0$ between the solutions of DDE (1) and its limiting case $\varepsilon = 0$, the continuous time difference equation (2). Since the solutions to Eq. (2) are typically discontinuous at $t_k = k \in \mathbf{N}_0$ the closeness between the solutions of equations (1) and (2) can be proved everywhere on finite intervals $[0, T]$ except a small vicinity of the integer points t_k. In the case when the solutions to (2) are continuous the closeness is uniform on the entire interval $[0, T]$. The precise statements are given by Theorem 1 and its Corollary. Section 3.2 deals with the existence of periodic solutions to DDE (1) and their asymptotic shape as $\varepsilon \to 0+$. The case of continuous negative feedback and one-sided bounded f is considered in Sect. 3.2.1. The other case of this paper, when f is a Farey-type nonlinearity, is treated in Sect. 3.2.2. We show the existence of the so-called square-wave periodic solutions for small $\varepsilon > 0$ in both cases which correspond to cycles of the map f. The proof is similar in both cases, and it is outlined for the case of retarded DDEs (1), when $c = 0$.

Section 4 concerns some open questions and conjectures about the general neutral DDEs (1) when $c \neq 0$. It particular, it is a natural expectation that the periodicity results of Sects. 3.2.1 and 3.2.2 can be extended to this case. This conjecture is confirmed by preliminary numerical simulations. At present we do not have a rigorous proof of the periodicity in the neutral case.

This paper is a result of authors' presentation at the AMMCS-CAIMS Congress 2015 held at the WLU during 7–12 June 2015. It contains some recently obtained mathematical results which ideas and proofs are only outlined here, due to their length and complexity as well as to the space limitation of the proceedings. A full scale paper with all the mathematical details and proofs included is a forthcoming work.

2 Preliminaries

For arbitrary $\varphi(t) \in C^1([-1,0],\mathbf{R}) := X_1$ there exists unique solution $x_\varphi^\varepsilon(t)$ to Eq. (1) defined and continuous for all $t \geq -1$. The solution is differentiable everywhere for $t \geq -1$ except at the integer values $t_k = k \in \mathbf{N}_0$, where it is typically continuous only. The solution is formally obtained for $t > 0$ by *successive integration*.

Let $X_0 := C^0([-1,0),\mathbf{R})$ be a set of continuous initial functions ψ such that the limit $\lim_{t\to 0^-}\psi(t)$ exists. X_0 can be viewed as a restriction of the standard space of initial functions $C([-1,0],\mathbf{R})$ when its elements are considered on the reduced domain, the half open interval $[-1,0)$. Thus

$$X_0 := \{\psi \in C^0([-1,0),\mathbf{R}) \mid \lim_{s\to 0^-}\psi(s) = \psi(0) \text{ exists}\}.$$

For arbitrary $\psi \in X_0$ there exists unique solution $x_\psi(t)$ to Eq. (2) defined for all $t \geq -1$. The solution is formally obtained for $t \geq 0$ by *successive iterations*. The solution is typically discontinuous at the integer values $t_k = k \in \mathbf{N}_0$ with a jump discontinuity. In the case when the *contiguity condition*

$$\lim_{s\to 0^-}\psi(s) = f(\psi(-1)) \qquad \text{(cc)}$$

holds, and f is continuous, the solution $x_\psi(t)$ is also *continuous* for all $t \geq -1$.

We consider two distinct cases of the nonlinearity f in DDEs (1):

(H1) f is continuous and satisfies the negative feedback assumption

$$x\cdot f(x) < 0 \qquad \forall x \in \mathbf{R}\backslash\{0\}. \qquad (nf)$$

Also, f is one-sided bounded on $\mathbf{R}$: $f(x) \geq -M$ or $f(x) \leq M$ for all $x \in \mathbf{R}$ and some $M > 0$;

(H2) f is a Farey-type type nonlinearity defined by: $f(x) = mx + A$ if $x \leq 0$, and $f(x) = mx - B$ if $x > 0$, for some $0 < m < 1$ and $A, B > 0$.

The first case is motivated by a standard set of assumptions normally imposed for the retarded type DDEs (1) when $c = 0$; see e.g. [3] and further references therein. The second case is an exact reduction of a specific nonlinear boundary value problem for a scalar wave equation [5]. Though many aspects of the dynamics of equations (1) are studied in the case $c = 0$, virtually no comprehensive research is done and very little is known in the neutral case, when $c \neq 0$.

3 Main Results

3.1 Continuous Dependence on Singular Parameter ε

In this subsection we establish certain continuous dependence results on the singular parameter ε for the differential delay equation (1). We show that its solutions and solutions of the difference equation with continuous argument (2) are close on finite intervals in a properly chosen metric for all sufficiently small ε. Since the solutions to Eq. (1) are continuous for all $t \geq 0$, while the solutions to Eq. (2) are typically discontinuous at integer values $t_k = k \in \mathbf{N}_0$, we need a proper means of comparison of those solutions.

We shall use the uniform metric to measure the closeness between two functions $\alpha(t)$ and $\beta(t)$ defined on a set $S \subseteq \mathbf{R}$:

$$||\alpha - \beta||_S := \sup\{|\alpha(t) - \beta(t)|,\ t \in S\}.$$

For the differential delay equation (1) initial functions $\varphi_1, \varphi_2 \in C^1([-1,0], \mathbf{R}) = X_1$ are compared in the standard uniform metric:

$$||\varphi_1 - \varphi_2||_{[-1,0]} = \sup\{|\varphi_1(t) - \varphi_2(t)| + |\varphi_1'(t) - \varphi_2'(t)|,\ t \in [-1,0]\}.$$

For the difference equation (2) initial functions $\psi_1, \psi_2 \in C^0([-1,0), \mathbf{R}) = X_0$ are compared in the following uniform metric

$$||\psi_1 - \psi_2||_{[-1,0)} = \sup\{|\psi_1(t) - \psi_2(t)|,\ t \in [-1,0)\}.$$

Given $\varphi \in X_1$ and $\psi \in X_0$ we shall compare them on the initial interval $[-1,0]$ as follows:

$$||\varphi - \psi||_{[-1,0]} = \sup\{|\varphi(t) - \psi(t)|,\ t \in [-1,0],\ \text{where } \psi(0) := \lim_{t \to 0-} \psi(t)\}.$$

Given two initial functions, $\varphi \in X_1$ and $\psi \in X_0$, which are close on the initial interval $[-1,0]$, we would like to estimate the closeness between their respective solutions to Eqs. (1) and (2) on any finite interval $[0,T]$. Since $x_\varphi^\varepsilon(t)$ is continuous for all $t \geq 0$, while $x_\psi(t)$ is typically discontinuous at $t_k = k \in \mathbf{N}_0$, the closeness in the uniform metric on the interval $[0,T]$ cannot happen. However, the closeness takes place everywhere else, if the integer values $t_k = k$ are excluded on the interval $[0,T]$. For the purpose of such comparison we introduce the following notation

$$J_T^\sigma := [0,T] \setminus \bigcup_{k=0}^{[T]} U_\sigma(k),$$

where $U_\sigma(k) = (k - \sigma, k + \sigma)$ is the σ-neighborhood of the point $t_k = k$, and $[\cdot]$ is the integer value function.

Theorem 1 *Suppose $f \in C(\mathbf{R}, \mathbf{R})$ satisfies the Lipschitz condition, $|f(u) - f(v)| \leq K|u - v||$, and $\varphi \in X_1$ is fixed. For arbitrary $T > 0, \mu > 0$, and $\sigma > 0$ there exist $\delta > 0$ and $\varepsilon_0 > 0$ such that if $\psi \in X_0$ is such that $||\psi - \varphi||_{[-1,0]} < \delta$ then $||x_\psi - x^\varepsilon_\varphi||_{J^\sigma_T} < \mu$ for all $0 < \varepsilon < \varepsilon_0$.*

The proof of the theorem follows from the following three lemmas.

Lemma 1 *Suppose that $\psi_1, \psi_2 \in X_0$ are given, and let $x_{\psi_1}(t), x_{\psi_2}(t)$ be the corresponding solutions to difference equation (2). Then*

$$||x_{\psi_1} - x_{\psi_2}||_{[0,1]} \leq K||\psi_1 - \psi_2||_{[-1,0]}.$$

The proof follows from the Lipschitz continuity of function f.

Lemma 2 *Suppose that $\varphi_1, \varphi_2 \in X_1$ are given, and let $x^\varepsilon_{\varphi_1}(t), x^\varepsilon_{\varphi_2}(t)$ be the corresponding solutions to differential delay equation (1). Then there exists $\varepsilon_0 > 0$ such that for all $0 < \varepsilon < \varepsilon_0$ one has*

$$||x^\varepsilon_{\varphi_1} - x^\varepsilon_{\varphi_2}||_{[0,1]} \leq M_2||\varphi_1 - \varphi_2||_{[-1,0]}$$

for some constant $M_2 > 0$.

Proof The proof is similar to that of Lemma 2 of paper [3], page 185. Solutions to Eq. (1) also solve the following integral equation

$$x(t) = x(0)\exp\{-\frac{t}{\varepsilon}\} + \frac{1}{\varepsilon}\int_0^t f(x(s-1))\exp\{\frac{s-t}{\varepsilon}\}\,ds \qquad (4)$$
$$- c\int_0^t x'(s-1)\exp\{\frac{s-t}{\varepsilon}\}\,ds.$$

The reasoning and major steps of the proof of Lemma 2 should be repeated and applied now to the integral equation (4). We leave details to the reader. □

Lemma 3 *Suppose that $\varphi \in X_1$ is given, and let $x^\varepsilon_\varphi(t)$ and $x_\varphi(t)$ be the corresponding solutions to equations (1) and (2), respectively. Then for arbitrary $\mu > 0$ and $\sigma > 0$ there exists $\varepsilon_0 > 0$ such that for all $0 < \varepsilon < \varepsilon_0$ one has*

$$||x^\varepsilon_\varphi - x_\varphi||_{[\sigma,1]} \leq \mu.$$

Proof The proof follows ideas and repeats major steps of the proof of Lemma 3 of paper [3] when one estimates the difference $|x^\varepsilon_\varphi(t) - f(x(t-1))|$ for $t \in [0, 1]$. The integral equation (4) is used in the new calculations.

By integration by parts the last term of Eq. (4) one arrives at the following integral equation

$$x(t) = [x(0) + cx(-1)]\exp\{-\frac{t}{\varepsilon}\} \\ + \frac{1}{\varepsilon}\int_0^t [f(x(s-1)) + cx(s-1)]\exp\{\frac{s-t}{\varepsilon}\}\,ds - cx(t-1). \tag{5}$$

Using the identity

$$F(t) = \frac{1}{\varepsilon}\int_0^t F(t)\exp\{\frac{s-t}{\varepsilon}\}\,ds + \exp\{-\frac{t}{\varepsilon}\}F(t)$$

the difference $x^{\varepsilon}_{\varphi}(t) - x_{\varphi}(t) = x^{\varepsilon}_{\varphi}(t) - f(\varphi(t-1)), t \in [0,1]$, can be represented as

$$x^{\varepsilon}_{\varphi}(t) - f(\varphi(t-1)) = [\varphi(0) + c\varphi(-1) - f(\varphi(t-1)) - c\varphi(t-1)]\exp\{-\frac{t}{\varepsilon}\} \\ + \frac{1}{\varepsilon}\int_0^t [f(\varphi(s-1)) - f(\varphi(t-1))]\exp\{\frac{s-t}{\varepsilon}\}\,ds \\ + \frac{c}{\varepsilon}\int_0^t [\varphi(s-1) - \varphi(t-1)]\exp\{\frac{s-t}{\varepsilon}\}\,ds.$$

The rest of the estimates can be done very similar to those in Lemma 3 of [3]. We leave details to the reader. □

The proof of Theorem 1 now follows by induction and the triangle inequality

$$||x_{\psi} - x^{\varepsilon}_{\varphi}||_{J^{\sigma}_{T}} \leq ||x_{\psi} - x_{\varphi}||_{J^{\sigma}_{T}} + ||x_{\varphi} - x^{\varepsilon}_{\varphi}||_{J^{\sigma}_{T}}.$$

Corollary 1 *When the contiguity condition (cc) is satisfied Theorem 1 is valid in the uniform metric on the whole interval* $[0, T]$.

This essentially follows from the last chain of estimates in Lemma 3 for the difference $x^{\varepsilon}_{\varphi}(t) - f(\varphi(t-1)), t \in [0,1]$, where one can choose $\sigma = 0$.

3.2 *Periodicity*

This subsection deals with the existence of periodic solutions to differential delay equation (1). We consider the two cases of the nonlinearity f as described above by the hypotheses (**H1**) and (**H2**).

We first introduce a notion of periodic solutions to difference equation (2) associated with cycles of the corresponding map f.

Definition 1 (Square wave periodic solutions to DE (2)) Let $\{a_1, a_2, \ldots, a_N\}$ be an N-cycle of map f, $f(a_k) = a_{k+1}, k = 1, \ldots, N-1, f(a_N) = a_1$. Define the piece-wise constant function by

$$\mathrm{Sqwv}(t, a_1, \ldots, a_N) := a_k, \quad \text{for} \quad t \in [k-1, k), k = 1, 2, \ldots, N,$$

and extend it periodically to all $t \in \mathbf{R}$. The function $\mathrm{Sqwv}(t, a_1, \ldots, a_N), t \in \mathbf{R}$, is called the *square wave periodic solution* of difference equation (2) corresponding to the cycle $\{a_1, a_2, \ldots, a_N\}$.

We define next the convergence of a parameter dependent family of functions to the discontinuous function $\mathrm{Sqwv}(t, a_1, \ldots, a_N)$ on the interval $[0, N]$. Let a family $p_\varepsilon(t) \in C([0, N], \mathbf{R})$ with the parameter ε be given.

Definition 2 We say that $p_\varepsilon(t)$ converges to $\mathrm{Sqwv}(t, a_1, \ldots, a_N)$ on the interval $[0, N]$ as $\varepsilon \to 0^+$ if for arbitrary $\mu > 0$ and $\sigma > 0$ there exists $\varepsilon_0 > 0$ such that for all $0 < \varepsilon < \varepsilon_0$ one has $||\mathrm{Sqwv}(\cdot, a_1, \ldots, a_N) - p_\varepsilon(\cdot)||_{J_N^\sigma} < \mu$.

3.2.1 Negative Feedback Nonlinearity

Theorem 2 *Assume that **(H1)** is satisfied and $f'(0) < -1$. There exists $\varepsilon_0 > 0$ such that for all $0 < \varepsilon < \varepsilon_0$ DDE (1) has a periodic solution $x_p^\varepsilon(t)$ with the period $T = 2 + O(\varepsilon)$ where $O(\varepsilon) \to 0^+$ as $\varepsilon \to 0^+$. The periodic solution $x_p^\varepsilon(t)$ converges as $\varepsilon \to 0$ to one of the square wave periodic solutions* $\mathrm{Sqwv}(t, a_1, a_2)$ *of DE (2), where $\{a_1, a_2\}$ is a cycle of period two of the map f.*

Proof We outline main steps of the proof for the case $c = 0$, omitting the intermediate details due to the space limitation.

Consider the set of initial functions

$$X^* = \{\varphi \in C^0([-1, 0], \mathbf{R}) \mid \text{(i), (ii), (iii) are satisfied}\},$$

where the assumptions (i), (ii), and (iii) are specified as follows:

(i) φ is non-negative with $\varphi(-1) = 0, \varphi(t) > 0\ \forall t \in (-1, 0]$, and φ is bounded from above, $\varphi(t) \le M^*\ \forall t \in [0, 1]$ and some $M^* > 0$;

(ii) φ has an exponential growth in a right neighborhood of $t = -1$ given by:

$$\varphi(t) \ge \gamma_l(t) := K^*\left(1 - e^{-\frac{(t+1)\alpha}{\varepsilon}}\right) \quad \forall t \in [-1, -1 + \beta],$$

for some $K^* > 1, 0 < \alpha < 1$, and $\beta = O(\varepsilon)$;

(iii) φ is bounded away from zero, $\varphi(t) \ge \delta \quad \forall t \in [-1 + \beta, 0]$, for some sufficiently small and fixed $\delta > 0$.

Loosely speaking, the set X^* consists of non-negative bounded above initial functions such that all of them start at 0 when $t = -1$ (assumption (i)), they quickly grow with the exponential rate in a small right β-neighborhood of $t = -1$ beyond the value of $\delta > 0$ (assumption (ii)), and they remain bounded away from 0 by the value δ for the remainder $[-1+\beta, 0]$ of the initial interval $[-1, 0]$ (assumption (iii)).

Consider also a negatively symmetric to X^* set defined by

$$X_* := \{\psi \in C([-1,0], \mathbf{R}) | -\psi \in X^*\}.$$

When $c = 0$ solutions to Eq. (1) satisfy the following integral equation

$$x(t) = x(0)\exp\{-\frac{t}{\varepsilon}\} + \frac{1}{\varepsilon}\int_0^t f(x(s-1))\exp\{\frac{s-t}{\varepsilon}\}\, ds. \tag{6}$$

It is obtained from the integral equation (4) by setting $c = 0$.

By using (6) one can make precise forward estimations on the solution $x_\varphi^\varepsilon(t)$ for $\varphi \in X^*$. One can show that there exists $\varepsilon_0^1 > 0$ such that for every $\varphi \in X^*$ and all $0 < \varepsilon < \varepsilon_0^1$ there exists time $t_1 = t_1(\varphi) > 0$ such that the corresponding solution $x_\varphi^\epsilon(t)$ to DDE (1) has the properties:

(a) the segment $x_\varphi^\epsilon(t_1 + 1 + s), s \in [-1, 0]$, of the solution belongs to X_*;
(b) $0 < t_1 \leq P_1\varepsilon$ for some $P_1 > 0$.

Define now a mapping $\mathscr{F}_1$ on X^* with the values in X_* by the formula

$$\mathscr{F}_1(\varphi) := x_\varphi^\epsilon(t_1 + 1 + s), s \in [-1, 0].$$

Likewise, there exists $\varepsilon_0^2 > 0$ such that for every $\psi \in X_*$ and all $0 < \varepsilon < \varepsilon_0^2$ there exists time $t_2 = t_2(\psi) > 0$ such that the corresponding solution $x_\psi^\epsilon(t)$ to DDE (1) has the properties:

(c) the segment $x_\psi^\epsilon(t_2 + 1 + s), s \in [-1, 0]$, of the solution belongs to X^*;
(d) $0 < t_2 \leq P_2\varepsilon$ for some $P_2 > 0$.

Define a mapping $\mathscr{F}_2$ on X_* with the values in X^* by the formula

$$\mathscr{F}_2(\psi) := x_\psi^\epsilon(t_2 + 1 + s), s \in [-1, 0].$$

For arbitrary $\varphi \in X^*$ and all $0 < \varepsilon < \min\{\varepsilon_1, \varepsilon_2\} := \varepsilon_0$ define now the mapping $\mathscr{F}$ as the composition of the above mappings $\mathscr{F}_1$ and $\mathscr{F}_2$, $\mathscr{F} := \mathscr{F}_2 \circ \mathscr{F}_1$. $\mathscr{F}$ maps the convex set X^* into itself. It is a compact map, for when $c = 0$ the DDE (1) is a retarded type equation. Also the set $\mathscr{F}(X^*)$ is bounded due to the one-sided boundedness of the function f. Therefore, by the Schauder fixed point theorem there exists $\varphi_0 \in X^*$ such that $(\mathscr{F})(\varphi_0) = \varphi_0$. Clearly, that the corresponding solution $x_{\varphi_0}^\epsilon(t)$ is periodic. Its period is $T = 2 + t_1 + t_2 \leq 2 + (P_1 + P_2)\varepsilon$. The convergence $x_{\varphi_0}^\epsilon(t) \to \mathrm{Sqwv}(t, a_1, a_2)$, where $\{a_1, a_2\}$ is a 2-cycle of the map f, can be proved by using the continuous dependence on ε (Theorem 1) and the fact that its period $T = 2 + O(\varepsilon)$ is close to 2 for small $\varepsilon > 0$. Details are omitted due to their length. □

3.2.2 Farey-Type Nonlinearity

Theorem 3 *Assume that* **(H2)** *is satisfied, and let* $\{a_1, \ldots, a_N\}$ *be the unique globally attracting cycle of the map* f. *There exists* $\varepsilon_0 > 0$ *such that for all* $0 < \varepsilon < \varepsilon_0$ *DDE (1) has a periodic solution* $x_p^\varepsilon(t)$ *with the period* $T = N + O(\varepsilon)$, *where* $O(\varepsilon) \to 0$ *as* $\varepsilon \to 0$. *The periodic solution* $x_p^\varepsilon(t)$ *converges as* $\varepsilon \to 0$ *to the square wave periodic solution* $\mathrm{Sqwv}(t, a_1, \ldots, a_N)$ *of DE (2).*

Proof The idea of the proof is similar to that in the proof of Theorem 2. Associated with the unique cycle of the map f we construct a sequence of convex bounded sets $X_k^* \subset C([-1,0], \mathbf{R}), k = 1, \ldots, N$, and a sequence of compact maps $\mathscr{F}_k$ such that $\mathscr{F}_k : X_k^* \to X_{k+1}^* (N+1 := 1)$. The composition map $\mathscr{F} := \mathscr{F}_N \circ \cdots \circ \mathscr{F}_1$ then maps X_1^* into itself. By the Schauder fixed point theorem it has a fixed point $\varphi_0 \in X_1^*$ there. The corresponding solution $x_{\varphi_0}^\varepsilon(t)$ to DDE (1) is periodic. Its properties and the convergence to $\mathrm{Sqwv}(t, a_1, \ldots, a_N)$ as $\varepsilon \to 0$ follow from the construction of the sets X_k^* outlined below.

Let $\{a_1, \ldots, a_k, a_{k+1}, \ldots, a_N\}$ be the globally attracting cycle of the map f with $f(a_k) = a_{k+1}, k = 1, \ldots, N-1$ and $f(a_N) = a_1$. In order to be specific in the definition of X_k^* we assume that $a_{k+1} > a_k$. Define now X_k^* as follows:

$$X_k^* = \{\varphi \in C^0([-1,0], \mathbf{R}) \mid \text{(i), (ii) are satisfied}\},$$

where the assumptions (i) and (ii) are specified as follows:

(i) φ starts in a neighborhood of a_k with a fast exponential transition to a_{k+1} defined as: $\varphi(-1) = a_k + \delta$, and $\gamma_l(t) \le \varphi(t) \le \gamma_u(t) \quad \forall t \in [-1, -1+\beta]$, where

$$\gamma_l(t) = a_{k+1} + (a_k + \delta - a_{k+1})\, e^{-\frac{(t+1)\alpha}{\varepsilon}}, \qquad \gamma_u(t) = M + (a_k + \delta - M)\, e^{-\frac{(t+1)}{\varepsilon}}$$

with $\beta = O(\varepsilon), 0 < \alpha < 1$ and $\delta > 0$ is small and fixed;

(ii) φ stays in the δ-neighborhood of a_{k+1} for the rest of the time on the initial interval $[-1, 0]$:

$$a_{k+1} - \delta \le \varphi(t) \le a_{k+1} + \delta \quad \forall t \in [-1+\beta, 0].$$

In the opposite case of $a_{k+1} < a_k$ the inequalities in part (i) of the definition of X_k^* will have to be reversed, as well as the starting point for φ to be moved to $a_k - \delta$, $\varphi(-1) = a_k - \delta$.

By using the integral equation (6) one can show that there exists $\varepsilon_0^k > 0$ such that for all $0 < \varepsilon < \varepsilon_0^k$ and every $\varphi \in X_k^*$ there exists time $t_k > 0$ for its solution $x_\varphi^\varepsilon(t)$ such that the following holds:

(a) the segment $x_\varphi^\epsilon(t_k + 1 + s), s \in [-1, 0]$, of the solution belongs to X_{k+1}^*;
(b) $0 < t_k \le P_k \varepsilon$ for some $P_k > 0$.

Therefore, the map $\mathscr{F}_k : X_k^* \to X_{k+1}^*$ can be defined by

$$(\mathscr{F}_k\varphi)(t) := x_\varphi^\varepsilon(t_k + 1 + t), t \in [-1, 0].$$

The composite map $\mathscr{F} = \mathscr{F}_N \circ \cdots \circ \mathscr{F}_1$ has then a fixed point $\varphi_0 \in X_1^*$ which generates the periodic solution $x_{\varphi_0}^\varepsilon(t)$ to DDE (1) with the stated properties. □

4 Open Problems and Conjectures

The existence of periodic solutions in Theorem 2 is a well known fact due to Hadeler and Tomiuk [1, 2, 6]. The existence itself only requires the instability of the trivial solution $x \equiv 0$ and the one-sided boundedness of f. In the case of globally attracting cycle of period two for the map f, $\{a_1, a_2\}$, the square wave limiting shape of the periodic solutions was derived by Mallet-Paret and Nussbaum [2, 3]. We prove the asymptotic shape in the general case of the negative feedback assumption on f. The questions of uniqueness and asymptotic stability of the periodic solutions remain largely open problems. In fact, multiple periodic solutions and chaotic behaviors are shown to exist in the case of globally attracting two-cycle [3]. Therefore, to establish at least some sufficient conditions for the existence and stability of periodic solutions for the neutral DDE (1) appears to be quite challenging mathematical problem.

The theoretical results on periodic solutions of this paper can be proved for the retarded DDEs (1) at this time. The principal reasons being that the Schauder Fixed Point Theorem is used, with the essential requirements of the convexity and boundedness of the sets X^*, X_*, X_k^* and of the compactness of the shift operators $\mathscr{F}$ along the solutions of DDE (1) when $c = 0$. Though the estimates in the proofs of the Theorems 2 and 3 seem to remain valid also in the case $c \neq 0$ we are not aware of the existence of appropriate fixed point theorems that can be applied in the neutral case.

Conjecture 1 Theorems 2 and 3 remain valid for the case of neutral DDE (1), when $c \neq 0$.

Acknowledgements The first author would like to express his gratitude and appreciation for the support and hospitality extended to him during his visit and stay at the CIAO of the Federation University Australia, Ballarat, in December 2014–January 2015. This paper is a result of the collaborative research work initiated during the visit.

References

1. Erneux, T.: Applied Delay Differential Equations. Ser.: Surveys and Tutorials in the Applied Mathematical Sciences, vol. 3, 204 pp. Springer, New York (2009)
2. Hale, J.K., Lunel, S.M.V.: Introduction to Functional Differential Equations. Springer Applied Mathematical Sciences, vol. 99, 447 pp. Springer-Verlag, New York (1993)
3. Ivanov, A.F., Sharkovsky, A.N.: Oscillations in singularly perturbed delay equations. In: Jones, C.K.R.T., Kirchgraber, U., Walther, H.O. (eds.) Dynamics Reported (New Series), vol. 1, pp. 165–224. Springer, New York (1991)
4. Nagumo, J., Shimura, M.: Self-oscillation in a transmission line with a tunnel diode. Proc. IRE **49**(8), 1281–1291 (1961)
5. Sharkovsy, A.N., Maistrenko, Yu.L., Romanenko, E.Yu.: Difference Equations and Their Perturbations, vol. 250, 358 pp. Kluwer Academic Publishers, Dordrecht/Boston/London (1993)
6. Smith, H.: An Introduction to Delay Differential Equations with Applications to the Life Sciences. Ser.: Texts in Applied Mathematics, vol. 57, 172 pp. Springer, New York/Dordrecht/Heidelberg/London (2011)
7. Witt, A.A.: On the theory of the violin string. Zhurn. Tech. Fiz. **6**, 1459–1470 (1936, in Russian)

Localized Spot Patterns on the Sphere for Reaction-Diffusion Systems: Theory and Open Problems

Alastair Jamieson-Lane, Philippe H. Trinh, and Michael J. Ward

Abstract A new class of point-interaction problem characterizing the time evolution of spatially localized spots for reaction-diffusion (RD) systems on the surface of the sphere is introduced and studied. This problem consists of a differential algebraic system (DAE) of ODEs for the locations of a collection of spots on the sphere, and is derived from an asymptotic analysis in the large diffusivity ratio limit of certain singularly perturbed two-component RD systems. In Trinh and Ward (The dynamics of localized spot patterns for reaction-diffusion systems on the sphere. Nonlinearity Nonlinearity **29**(3), 766–806 (2016)), this DAE system was derived for the Brusselator and Schnakenberg RD systems, and herein we extend this previous analysis to the Gray-Scott RD model. Results and open problems pertaining to the determination of equilibria of this DAE system, and its relation to elliptic Fekete point sets, are highlighted. The potential of deriving similar DAE systems for more complicated modeling scenarios is discussed.

1 Introduction

Spatially localized patterns can occur for two-component reaction-diffusion (RD) systems in the singularly perturbed limit corresponding to a large diffusivity ratio between the two components in the system. In particular, on the surface of the unit sphere, the stability and dynamics of localized spot patterns, whereby the solution concentrates at discrete points on the sphere, have been analyzed recently for the singularly perturbed Brusselator and Schnakenberg RD systems (cf. [24, 27]). It was also shown that, in certain parameter regimes, these localized patterns can exhibit various instabilities, including either spot-shape instabilities, leading to spot self-replication events, or competition instabilities, leading to the annihilation of spots

A. Jamieson-Lane • M.J. Ward (✉)
Department of Mathematics, University of British Columbia, Vancouver, BC, Canada
e-mail: aja107@math.ubc.ca; ward@math.ubc.ca

P.H. Trinh
OCIAM, Mathematical Institute, University of Oxford, Oxford, UK
e-mail: trinh@maths.ox.ac.uk

© Springer International Publishing Switzerland 2016

J. Bélair et al. (eds.), *Mathematical and Computational Approaches in Advancing Modern Science and Engineering*, DOI 10.1007/978-3-319-30379-6_58

in the pattern. This analysis of [24] and [27] extends the previous studies of localized spot patterns on 2-D planar domains for related RD systems (cf. [6, 11, 29]).

By using formal asymptotic methods on the RD system in the singularly perturbed limit, it is possible to derive a differential algebraic ODE system characterizing the dynamics of spot patterns. This new class of dynamically interacting particle system has some common features with the well-known ODE system characterizing the dynamics of Eulerian point vortices in fluid mechanics. In this latter context, there has been an intense study of the dynamics and equilibria of point vortices on the sphere over the past three decades (cf. [2, 3, 10, 18, 19, 23]). Similar to the original asymptotic derivation of the limiting point vortex problem in [3] starting with the Euler equations of fluid mechanics, the main result for spot dynamics in [27] for the Brusselator and Schnakenberg model, and herein for the Gray-Scott RD model, provides a reduced dynamical system for the time evolution of the centres of the localized spots on the sphere.

Motivated by specific questions related to the development of biological patterns on both stationary and time-evolving surfaces (cf. [5, 12, 17, 20, 21]), there have been many numerical studies of RD patterns on the sphere and other compact manifolds (cf. [1, 13, 14, 28] see also the references therein). Prior analytical studies of RD pattern formation on the surface of the sphere, have focused on using weakly nonlinear and equivariant bifurcation theory to derive normal form equations characterizing the development of small amplitude spatial patterns that bifurcate from a spatially uniform steady-state (cf. [4, 7, 16]). However, as a result of the typical high degree of degeneracy of the eigenspace associated with spherical harmonics of large mode number, these amplitude equations typically consist of a rather large coupled set of nonlinear ODEs. The latter are known to have an intricate subcritical bifurcation structure (cf. [4, 7, 16]). As a result, the preferred spatial pattern that emerges from an interaction of these weakly nonlinear modes is difficult to predict theoretically. This intrinsic difficulty is accentuated for RD systems where there is a large diffusivity ratio, which effectively yields a large aspect ratio system where center manifold analysis is of more limited use [25]. For such large aspect ratio RD systems, there is typically a rather wide band of unstable modes [24, 27], and so the prediction of pattern development based on the conventional paradigm of using both a Turing and a weakly nonlinear analysis is not generally possible.

However, it is in this singular limit of a large diffusivity ratio that localized spot patterns robustly appear from a transient process starting from small random perturbations of a spatially uniform state [24]. A discussion of results and open problems relating to the study of such "far-from equilibrium patterns" is the topic of this short article. In particular, in certain cases the equilibria of this DAE system for spot dynamics have a close relationship to the classical problem in approximation theory of determining a set of elliptic Fekete points, which are the globally minimizer of the discrete logarithmic energy for N points on the sphere.

The outline of this brief article is as follows. In Sect. 2, we briefly present the DAE system for the dynamics of spots for the Brusselator models as derived in [27]. In Sect. 3 we discuss some results and open questions related to determining equilibria for these DAE systems. New results for the equilibria of patterns with either 9 or 10 spots are presented. In Sect. 4 we give a new result for the DAE dynamics for spot patterns for the well-known Gray-Scott model. Finally, in Sect. 5 we list a few open problems related to deriving similar DAE dynamics for spot interactions for more complicated models.

2 Dynamics of Spots on the Sphere

Under the assumption of a large diffusivity ratio and a small "fuel" supply, the Brusselator model of [22] posed on the surface of the sphere, can be scaled into the following system for $u = u(\mathbf{x}, t)$ and the inhibitor $v = v(\mathbf{x}, t)$ (cf. [24, 27]):

$$\frac{\partial u}{\partial t} = \varepsilon^2 \Delta_S u + \varepsilon^2 \mathrm{E} - u + f u^2 v \,, \qquad \tau \frac{\partial v}{\partial t} = \Delta_S v + \varepsilon^{-2} \left(u - u^2 v \right) , \tag{1}$$

for some $\mathcal{O}(1)$ constants $\mathrm{E} > 0$, $\tau > 0$, and $0 < f < 1$. We refer to E as the "fuel" parameter. Here Δ_S is the Laplace-Beltrami operator on the sphere.

Spatial patterns for which u concentrates as $\varepsilon \to 0$ at a discrete set of points $\mathbf{x}_1, \ldots, \mathbf{x}_N$ on the sphere are called spot patterns. For $\varepsilon \to 0$, we have $u = \mathcal{O}(1)$ in the core of the spot, where $|\mathbf{x} - \mathbf{x}_j| = \mathcal{O}(\varepsilon)$, and $u \sim \varepsilon^2 \mathrm{E}$ away from the spot centers where $|\mathbf{x} - \mathbf{x}_j| = \mathcal{O}(1)$. Then for $\varepsilon \to 0$, the effect of the localized spots on the inhibitor field v in (1) is to introduce a sum of Dirac-delta "forces" where the strength of the "force" induced by the spot at $\mathbf{x}_j$ is proportional to S_j (see [24] for details). As such, v can be represented as a superposition of the well-known source neutral Green's function for the sphere. In this way, in [24] a quasi-equilibrium spot pattern with frozen locations $\mathbf{x}_1, \ldots, \mathbf{x}_N$ was constructed using the method of matched asymptotic expansions by formulating a nonlinear algebraic system for the spot locations $\mathbf{x}_1, \ldots, \mathbf{x}_N$ and the spot strengths $S_1, \ldots, S_N$. The linear stability of this quasi-equilibrium spot pattern to $\mathcal{O}(1)$ time-scale perturbations was analyzed in [24].

Provided that the quasi-equilibrium spot pattern is linearly stable, the slow dynamics of the spot pattern on the long time scale $\sigma = \varepsilon^2 t$ was derived in [27]. The collective coordinates characterizing this slow dynamics are the spot locations $\mathbf{x}_1, \ldots, \mathbf{x}_N$ and their corresponding spot source strengths $S_1, \ldots, S_N$, that both evolve slowly on the long time-scale $\sigma = \varepsilon^2 t$. The slow dynamics derived in [27] is a differential algebraic system of ODEs as given by the following result:

Principal Result 1 (Slow spot dynamics (cf. [27])) *Let $\varepsilon \to 0$. Provided that there are no $\mathcal{O}(1)$ time-scale instabilities of the quasi-equilibrium spot pattern, the time-dependent spot locations, $\mathbf{x}_j$ for $j = 1, \ldots, N$, on the surface of the sphere vary* *on the slow time-scale* $\sigma = \varepsilon^2 t$, *and satisfy the dynamics*

$$\frac{d\mathbf{x}_j}{d\sigma} = \frac{2}{\mathscr{A}_j(S_j)} \left(\mathrm{I} - \mathscr{Q}_j\right) \sum_{\substack{i=1\\ i\neq j}}^{N} \frac{S_i \mathbf{x}_i}{|\mathbf{x}_i - \mathbf{x}_j|^2}, \qquad \mathscr{Q}_j \equiv \mathbf{x}_j \mathbf{x}_j^T, \qquad j = 1, \ldots, N, \tag{2a}$$

coupled to the constraints for $S_1, \ldots, S_N$ *in terms of* $\mathbf{x}_1, \ldots, \mathbf{x}_N$ *given by the roots of a nonlinear algebraic system involving the Green's matrix* $\mathscr{G}$

$$\mathscr{N}(\mathbf{S}) \equiv \left[\mathrm{I} - \nu(\mathrm{I} - \mathscr{E}_0)\mathscr{G}\right]\mathbf{S} + \nu(\mathrm{I} - \mathscr{E}_0)\boldsymbol{\chi}(\mathbf{S}) - \frac{2\mathrm{E}}{N}\mathbf{e} = \mathbf{0}. \tag{2b}$$

Here I *is* $N \times N$ *identity matrix,* $(\mathscr{E}_0)_{ij} = \frac{1}{N}$, $(\mathbf{S})_i = S_i$, $(\boldsymbol{\chi}(\mathbf{S}))_i = \chi(S_i)$, $(\mathscr{G})_{ij} = \log|\mathbf{x}_i - \mathbf{x}_j|$ *for* $i \neq j$ *and* $(\mathscr{G})_{ii} = 0$, $(\mathbf{e})_i = 1$, *and* $\nu = -1/\log\varepsilon \ll 1$.

From (2) the spot locations are coupled to the spot strengths by (2b), yielding a DAE system for the slow spot evolution. Since the spot strengths can be calculated in terms of the locations from (2b), the DAE system has index 1. One readily established feature of the DAE system (2) is that it is invariant under an orthogonal transformation, corresponding to a rotation of the spots on the sphere.

In this DAE system (2), there are two functions $\chi(S_j)$ and $\mathscr{A}_j(S_j) < 0$, which depend only on S_j and the Brusselator parameter f, that must be determined numerically in terms of the local profile of the spot $\mathbf{x}_j$ (cf. [27]). These are shown in Fig. 1 for a few values of f. In the limit $\varepsilon \to 0$, and for $\mathrm{E} = \mathscr{O}(1)$, the only possible $\mathscr{O}(1)$ time-scale instability of the quasi-equilibrium spot pattern is a linear instability of the local spot profile to a peanut-shape if $S_j > \Sigma_2(f)$. This linear instability is found in [24] to lead to a nonlinear spot self-replication event. The threshold values $\Sigma_2(f)$ for spot self-replication for a few values of f are given in the caption of Fig. 1.

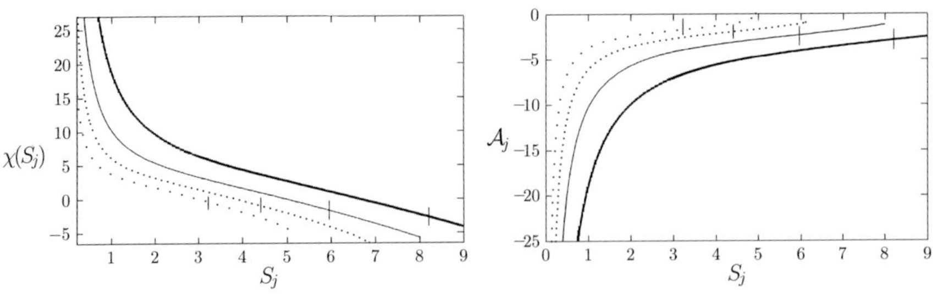

Fig. 1 *Left panel*: the function $\chi(S_j; f)$ in (2) for $f = 0.4$ (*heavy solid*), $f = 0.5$ (*solid*), $f = 0.6$ (*dotted*), and $f = 0.7$ (*widely spaced dots*). The spot self-replication threshold $\Sigma_2(f)$ for S_j is shown by the thin vertical lines in this figure. If $S_j > \Sigma_2(f)$ the local spot profile is linearly unstable to a peanut-splitting instability (cf. [24]). The threshold values are $\Sigma_2(0.4) \approx 8.21$, $\Sigma_2(0.5) \approx 5.96$, $\Sigma_2(0.6) \approx 4.41$, and $\Sigma_2(0.7) \approx 3.23$. *Right panel*: the function $\mathscr{A}_j(S_j) < 0$ in (2) with the same labels as in the left panel

3 Equilibria of the DAE Dynamics and Open Questions

In this section we discuss some previous results obtained in [27] as well as some new results for the equilibria of (2) that have large basins of attraction to initial conditions. We consider patterns for small values of N which $S_j = \mathcal{O}(1)$ as $\nu \to 0$. A few open problems are mentioned.

To determine the possible equilibrium spot configurations of (2) with large basins of attraction for $N \geq 3$ when $\mathrm{E} = \mathcal{O}(1), f$, and ν are given, we performed numerical simulations of (2) for both pre-specified and for randomly generated uniformly distributed initial conditions for the spot locations on the surface of the sphere. To generate N initial points that are uniformly distributed with respect to the surface area we let h_ϕ and h_θ be uniformly distributed random variables in $(0, 1)$ and define spherical coordinates $\phi = 2\pi h_\phi$ and $\theta = \cos^{-1}(2h_\theta - 1)$. Newton's method was used to solve (2b) for the initial set of N points. If the Newton iterates failed to converge, indicating that no quasi-equilibrium exists for the initial configuration of spots, a new randomly generated initial configuration was generated. The DAE dynamics (2a) was then implemented by using an adaptive time-step ODE solver coupled to a Newton iteration scheme to compute the spot strengths.

In the simulations below we took $f = 0.5$ and $\varepsilon = 0.02$. We remark that whenever $\mathbf{e} = (1, \ldots, 1)^T$ is an eigenvector of the Green's matrix $\mathcal{G}$, then the constraint (2b) admits a solution where $\mathbf{S} = S_c\mathbf{e}$. For such an equal spot–source strength pattern, the equilibrium spatial configuration of spots for (2) is independent of E, f, and ν.

In our discussion below, we refer to a ring pattern as a collection of N equally-spaced spots lying on an equator of the sphere. We refer to an $(N - 2) + 2$ pattern as a spot pattern consisting of two antipodal spots with the remaining $N - 2$ spots equally-spaced on the equatorial mid-plane between the two polar spots.

The simulations of [27] of 50 randomly generated initial spot for $N = 3, \ldots, 8$ yielded the following results for equilibria of (2) with large basin of attractions:

- $N = 3$: three equally-spaced spots that lie on a plane through the center of the sphere. (Common spot strength pattern.)
- $N = 4$: four spots centered at the vertices of a regular tetrahedron. (Common spot strength pattern.)
- $N = 5, 6, 7$: an $(N - 2) + 2$ pattern consisting of a pair of antipodal spots, with the remaining $N - 2$ spots equally-spaced on the equatorial mid-plane between the two polar spots. (Two different spot strengths for polar and mid-plane spots.)
- $N = 8$: a "twisted cuboidal" shape, consisting of two parallel rings of four equally-spaced spots, with the rings symmetrically placed above and below an equator. The spots are phase shifted by 45° between each ring. The perpendicular distance between the two planes is ≈ 1.12924 as compared to a minimum distance of ≈ 1.1672 between neighboring spots on the same ring, so that the pattern does not form a true cube. (Common spot strength pattern.)

We remark that for the case $N = 2$ it was shown in Lemma 5 of [27] that any two initial spots on the sphere will become antipodal in the long-time limit $\sigma \to \infty$. This was done by deriving a simple ODE for the angle $\gamma(\sigma)$ between the spot centers $\mathbf{x}_1$ and $\mathbf{x}_2$ at time σ, as measured from the center of the sphere, i.e. $\mathbf{x}_2^T\mathbf{x}_1 = \cos\gamma$, and establishing from this ODE that $\gamma \to \pi$ as $\sigma \to \infty$ for any $\gamma(0)$.

In [27] the linear stability of ring configuration of spots was studied numerically. The following conjecture was formulated in [27] based on numerical experiments.

- A ring pattern of $N = 3$ is orbitally stable, but is unstable if $N \geq 4$.
- For $N = 4, 5, 6, 7$, an $(N-2) + 2$ pattern is orbitally stable, but such a pattern is unstable if $N \geq 8$.

More specifically for $N = 3$, the numerical computations of [27] suggest that a ring solution is *orbitally stable* to small random perturbations in the spot locations in the sense that as time increases the perturbed spot locations become colinear on a nearby (tilted) ring. For $N \geq 4$, a similar small, but otherwise arbitrary, perturbation of the spot locations on the ring leads to a breakup of the ring pattern. Similar an $(N-2) + 2$ pattern for $N = 8$ breaks up and forms a twisted cuboidal shape.

Open Problem: *Establish analytically these results for the equilibria of spot patterns for* $N = 3, \ldots, 8$ *using group theory methods for ODEs. Analyze the linear stability of ring patterns by using an approach similar to that done in [2] for the corresponding problem of the linear stability of Eulerian point vortices on a ring.*

For larger values of N it becomes increasingly difficult to visualize the symmetries of the final equilibrium pattern that emerges under the DAE system (2) from initial data. More specifically, it becomes increasingly challenging to find a rotation matrix to put the pattern in a standard reference configuration. We now discuss two new results for $N = 9$ and $N = 10$ not obtained in [27]. For even larger values of N point-matching algorithms from computer science may be useful for classifying the symmetries of the final pattern.

For $N = 9$ the equilibrium state of (2) with a large basin of attraction for initial conditions is the pattern shown in the lower right subfigure of Fig. 2. Our simulations with 50 random initial configurations have shown that the limiting pattern consists of 3 parallel planes of 3 spots each. The spots on the equatorial plane and the other two planes are phase-shifted 60° (see the caption of Fig. 2 for details.)

For $N = 10$ the equilibrium state of (2) with a large basin of attraction for initial conditions is the pattern shown in the lower right subfigure of Fig. 3. The equilibrium pattern consists of two polar spots together with two parallel planes with four equally-spaced spots on each plane. The relative phase-shift of the spots on the two planes is 45° (see the caption of Fig. 3 for details).

A classical problem of point configurations on the sphere is the problem of finding the global minimizer of the discrete logarithmic energy $V \equiv -\sum\sum_{i\neq j} \log|\mathbf{x}_i - \mathbf{x}_j|$ on the sphere where $|\mathbf{x}_j| = 1$, which also has applications to minimizing the mean first passage time for Brownian motion on the sphere (cf. [8]). Such optimizing configurations are called elliptic Fekete point sets. By comparing our results for equilibrium spot configurations with the optimal energies V of elliptic

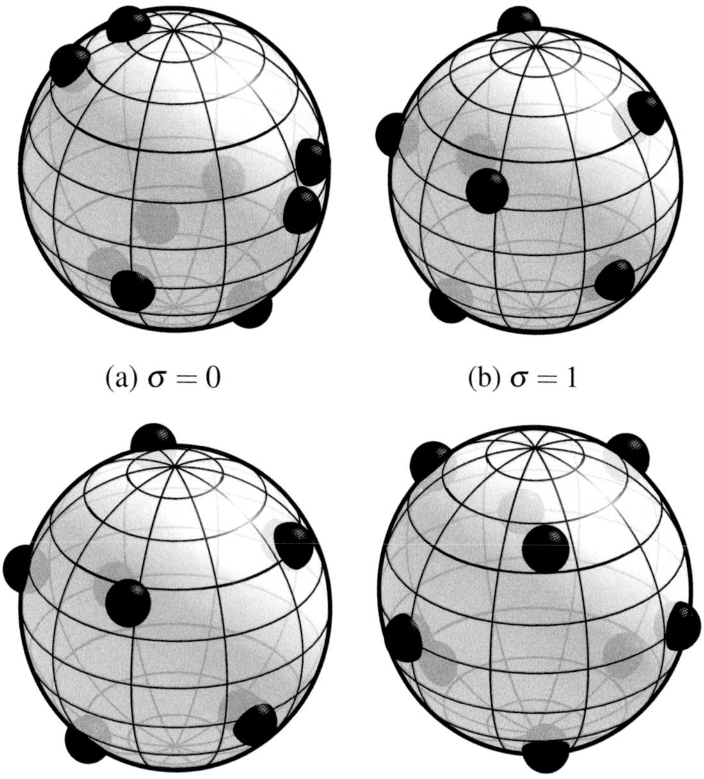

(a) $\sigma = 0$ (b) $\sigma = 1$ (c) $\sigma = 3$ (d) $\sigma = \infty$

Fig. 2 The evolution of a 9-spot pattern at different time σ for the Brusselator (2) when $f = 0.5$, $\mathrm{E} = 18$, and $\epsilon = 0.02$. (**a**) The initial state $\sigma = 0$. (**b**) $\sigma = 1$. (**c**) $\sigma = 3$. (**d**) The computed steady state after a suitable rotation. The steady-state consists of 3 planes of 3 spots each. The spots on the equatorial plane and the other two planes are phase-shifted 60°. The distance d between the equatorial plane and each of the other two planes is $d \approx 0.7014$. The common value of the spot strengths on the equatorial plane differs from that of the other 6 spots

Fekete point sets, as given in Table 1 of [26], we conclude that our equilibrium spot configurations for $N = 3, \ldots, 10$ having a large basin of attraction are indeed elliptic Fekete point sets. As a remark, if we were to set $S_j = 1$ in (2) and ignore the constraint (2b), then it is readily seen upon introducing Lagrange multipliers that local and global minima of the discrete energy V are stable equilibria of the simplified DAE dynamics.

Open Problem: *Explore computationally for $N \geq 10$ whether there is a relationship between elliptic Fekete point sets and equilibria of the full DAE dynamics (2) that have large basins of attraction of initial conditions.*

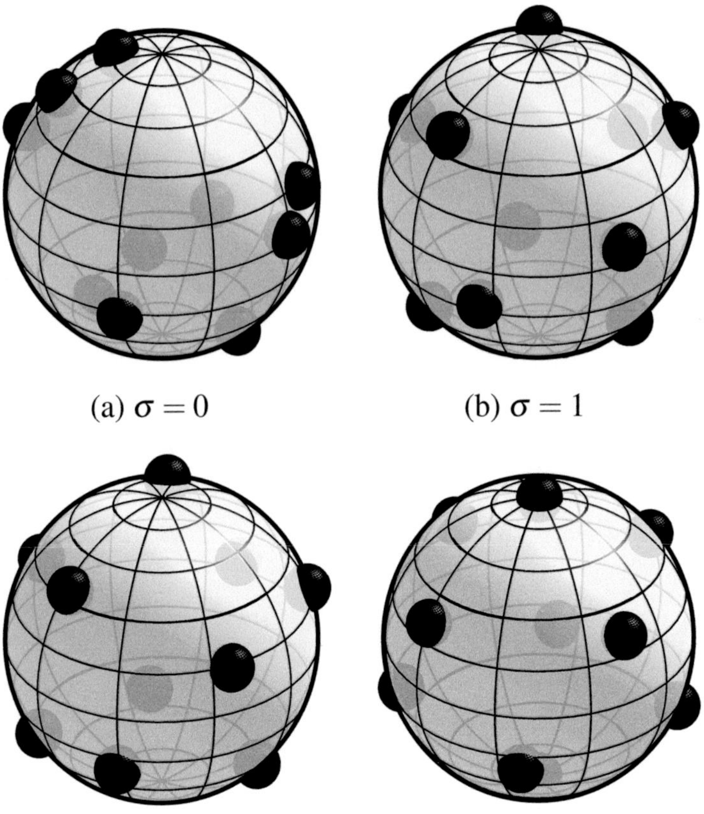

(a) $\sigma = 0$ (b) $\sigma = 1$ (c) $\sigma = 3$ (d) $\sigma = \infty$

Fig. 3 The evolution of a 10-spot pattern at different times σ for the Brusselator (2) when $f = 0.5$, $E = 22$, and $\epsilon = 0.02$. (**a**) The initial state $\sigma = 0$. (**b**) $\sigma = 1$. (**c**) $\sigma = 3$. (**d**) The computed steady state after a suitable rotation. The steady-state consists of two polar spots together with two parallel planes of 4 equidistantly spaced spots. The relative phase-shift of the spots on the two planes is $45°$. The distance d between the equator and either of the two planes is $d \approx 0.4234$

4 Dynamics of Spots on the Sphere: The Gray-Scott Model

In this section we give a new result for the slow dynamics of spots on the unit sphere for the well-known Gray-Scott RD model (cf. [6]).

$$v_t = \varepsilon^2 \Delta_S v - v + Buv^2 \,, \quad \tau u_t = D \Delta_s u + (1 - u) - uv^2 \,. \tag{3}$$

Although the analysis of spot dynamics for this problem follows the methodology done in [27] for the Brusselator model, this new analysis requires the reduced-wave

Green's function $G(\mathbf{x};\boldsymbol{\xi})$ on the sphere satisfying

$$\Delta_S G - \frac{1}{D} G = -\delta(\mathbf{x} - \boldsymbol{\xi}), \tag{4a}$$

$$G(\mathbf{x};\boldsymbol{\xi}) \sim -\frac{1}{2\pi}\log|\mathbf{x} - \boldsymbol{\xi}| + R + o(1), \quad \text{as} \quad \mathbf{x} \to \boldsymbol{\xi}, \tag{4b}$$

for some R independent of $\boldsymbol{\xi}$. In terms of this Green's function we can derive the following result for the slow dynamics of a collection of spots for the GS model.

Principal Result 2 (Slow spot dynamics for the GS model) *Let $\varepsilon \to 0$, and assume that $B = \mathcal{O}(\varepsilon/\nu)$, where $\nu = -1/\log\varepsilon$. Then, provided that there are no $\mathcal{O}(1)$ time-scale instabilities of the quasi-equilibrium spot pattern, the time-dependent spot locations, $\mathbf{x}_j$ for $j = 1,\ldots,N$, on the surface of the sphere vary on the slow time-scale $\sigma = \varepsilon^2 t$, and satisfy the dynamics for $j = 1,\ldots,N$,*

$$\frac{d\mathbf{x}_j}{d\sigma} = -2\pi\varepsilon^2\gamma_j(S_j)\left(\mathrm{I} - \mathcal{Q}_j\right)\sum_{\substack{i=1\\ i\neq j}}^{N} S_i \nabla_{\mathbf{x}} G(\mathbf{x}_j;\mathbf{x}_i), \qquad \mathcal{Q}_j \equiv \mathbf{x}_j\mathbf{x}_j^T, \tag{5a}$$

coupled to the nonlinear constraints for $S_1,\ldots,S_N$ in terms of $\mathbf{x}_1,\ldots,\mathbf{x}_N$ given by

$$\mathcal{B} = S_j(1 + 2\pi\nu R) + \nu\chi(S_j) + 2\pi\nu\sum_{\substack{i=1\\ i\neq j}}^{N} S_i G(\mathbf{x}_j;\mathbf{x}_i), \quad j = 1,\ldots,k. \tag{5b}$$

where $\nu = -1/\log\varepsilon$ and $\mathcal{B} \equiv \nu B\varepsilon^{-1}\sqrt{D} = \mathcal{O}(1)$.

In this system $\chi(S_j)$ and $\gamma(S_j)$ depend on the core problem near the spot and are specific to the GS model. Since the reduced-wave Green's function and its regular part R is not available in simple explicit form, and can only be written in terms of the Legendre function (see [24]), it is more challenging to investigate the dynamics and equilibria of spot patterns on the sphere for the GS model. This topic requires further investigation.

5 DAE Spot Dynamics in Complex Models: Open Problems

There are several possible extensions of the methodology for deriving and analyzing localized spot patterns for other scenarios.

We remark that explicit DAE dynamics for spot patterns is only possible when the source-neutral Green's function , satisfying $\Delta_s G = |\Omega|^{-1} - \delta(x - x_0)$ is explicitly available. Such a Green's function is well-known for the sphere, and this fact is key to deriving (2). However, recently in [9], this Green's function has been provided analytically for a particular class of surfaces of revolution. For this class, DAE dynamics on a manifold of varying curvature, and the possibility of pinning of localized spot, can be analyzed.

The second possible extension of the modeling framework is to allow for spatial heterogeneity in the fuel supply, so that $\mathrm{E} = \mathrm{E}(\mathbf{x})$. The corresponding outer solution v will involve a sum of Dirac distributions, one near each spot, together with a term v_p of the form $v_p = \int_\Omega \left(\mathrm{E}(\xi) - \bar{\mathrm{E}}\right) G(x;\xi)\, d\xi$, where $\bar{\mathrm{E}}$ denotes the spatial average of E over the sphere. In this way, it can be shown that the corresponding DAE dynamics will yield a nonlocal system for the evolution of the spots.

Finally, a more intricate method to introduce spatial heterogeneity in (1) is to consider spot patterns on the sphere for a model that couples passive bulk-diffusion in the interior of the sphere to the Brusselator PDE on the surface. In this context, the fuel supply E represents an exchange, or flux, between the surface-bound concentrations and their bulk counterparts. This coupling should lead to rich behavior in the DAE dynamics for spots on the sphere. This paradigm of studying coupled surface-bulk models is becoming more prominent in scientific computation and in applications (cf. [14, 15]).

Acknowledgements PHT thanks Lincoln College, Oxford and the Zilkha Trust for generous funding. MJW gratefully acknowledges grant support from NSERC.

References

1. Barreira, R., Elliott, C.A., Madzvamuse, A.: The surface finite element method for pattern formation on evolving biological surfaces. J. Math. Biol. **63**, 1095–1109 (2011)
2. Boatto, S., Cabral, H.E.: Nonlinear stability of a latitudinal ring of point vortices on a non-rotating sphere. SIAM J. Appl. Math. **64**(1), 216–230 (2003)
3. Bogomolov, V.A.: Dynamics of vorticity at a sphere. Fluid Dyn. (USSR) **6**, 863–870 (1977)
4. Callahan, T.K.: Turing patterns with $O(3)$ symmetry. Physica D **188**(1), 65–91 (2004)
5. Chaplain, M.A.J, Ganesh, M., Graham, I.G.: Spatio-temporal pattern formation on spherical surfaces: numerical simulation and application to solid tumour growth. J. Math. Biol. **42**(5), 387–423 (2001)
6. Chen, W., Ward, M.J.: The stability and dynamics of localized spot patterns in the two-dimensional Gray-Scott model. SIAM J. Appl. Dyn. Syst. **10**(2), 582–666 (2011)
7. Chossat, P., Lauterbach, R., Melbourne, I.: Steady-sate bifurcation with $O(3)$ symmetry. Arch. Rat. Mech. Anal. **113**, 313–376 (1990)
8. Coombs, D., Straube, R., Ward, M.J.: Diffusion on a sphere with localized traps: mean first passage time, eigenvalue asymptotics, and Fekete points. SIAM J. Appl. Math. **70**(1), 302–332 (2009)
9. Dritschel, D.G., Boatto, S.: The motion of point vortices on closed surfaces. Proc. R. Soc. A **471**, 20140890 (2015)
10. Kidambi, R., Newton, P.K.: Motion of three vortices on the sphere. Physica D **116**(1–2), 143–175 (1998)
11. Kolokolnikov, T., Ward, M.J., Wei, J.: Spot self-replication and dynamics for the Schnakenberg model in a two-dimensional domain. J. Nonlinear Sci. **19**(1), 1–56 (2009)
12. Kondo, S., Asai, R.: A reaction–diffusion wave on the skin of the marine angelfish Pomacanthus. Nature **376**, 765–768 (1995)

13. Landsberg, C., Voigt, V.: A multigrid finite element method for reaction-diffusion systems on surfaces. Comput. Vis. Sci. **13**, 177–185 (2010)
14. Macdonald, C.B., Merriman, B., Ruuth, S.J.: Simple computation of reaction-diffusion processes on point clouds. Proc. Natl. Acad. Sci. U.S.A. **110**(23), 9209–9214 (2013)
15. Madzvamuse, A., Chung, A.H.W., Venkataraman, V.: Stability analysis and simulations of coupled bulk-surface reaction diffusion systems. Proc. R. Soc. A **471**, 20140546 (2015)
16. Matthews, P.C.: Pattern formation on a sphere. Phys. Rev. E **67**(3), 036206 (2003)
17. Nagata, W., Harrison, L.G., Wehner, S.: Reaction-diffusion models of growing plant tips: Bifurcations on hemispheres. Bull. Math. Biol. **6**(4), 571–607 (2003)
18. Newton, P.K.: The N-Vortex Problem: Analytical Techniques. Springer, New York (2001)
19. Newton, P.K., Sakajo, T.: Point vortex equilibria and optimal packings of circles on a sphere. Proc. R. Soc. A **467**, 1468–1490 (2011)
20. Painter, K.J.: Modelling of pigment patterns in fish. In: Maini, P.K., Othmer, H.G. (eds.) Mathematical Models for Biological Pattern Formation. IMA Volumes in Mathematics and Its Applications, vol. 121, pp. 58–82. Springer-Verlag, New York (2000)
21. Plaza, R.G., Sánchez-Garduño, F., Padilla, P., Barrio, R.A., Maini, P.K.: The effect of growth and curvature on pattern formation. J. Dyn. Diff. Equ. **16**(4), 1093–1121 (2004)
22. Prigogine, I., Lefever, R.: Symmetry breaking instabilities in dissipative systems II. J. Chem. Phys. **48**(4), 1695–1700 (1968)
23. Roberts, G.: Stability of relative equilibria in the planar n-vortex problem. SIAM J. Appl. Dyn. Syst. **12**(2), 1114–1134 (2013)
24. Rozada, I., Ruuth, S., Ward, M.J.: The stability of localized spot patterns for the Brusselator on the sphere. SIAM J. Appl. Dyn. Syst. **13**(1), 564–627 (2014),
25. Sewalt, L., Doelman, A., Meijer, H., Rottschafer, V., Zagarias, A.: Tracking pattern evolution through extended center manifold reduction and singular perturbations. Physica D **298**, 48–67 (2015)
26. Stortelder, W., de Swart, J., Pintér, J.: Finding elliptic Fekete point sets: two numerical solution approaches. J. Comput. Appl. Math. **130**(1–2), 205–216 (1998)
27. Trinh, P., Ward, M.J.: The dynamics of localized spot patterns for reaction-diffusion systems on the sphere. Nonlinearity **29**(3), 766–806 (2016)
28. Varea, C., Aragón, J.L., Barrio, R.A.: Turing patterns on a sphere. Phys. Rev. E. **60**(4), 4588–4592 (1999)
29. Wei, J., Winter, M.: Stationary multiple spots for reaction-diffusion systems. J. Math. Biol. **57**(1), 53–89 (2008)

Continuous Dependence on Modeling in Banach Space Using a Logarithmic Approximation

Matthew Fury, Beth Campbell Hetrick, and Walter Huddell

Abstract We prove regularization for certain ill-posed problems in Banach space by considering small changes in the model. The problems are of the form $du/dt = Au,\ 0 \leq t < T,\ u(0) = \chi$ where $-A$ is assumed to generate a uniformly bounded holomorphic semigroup $\{e^{-zA} : \Re z \geq 0\}$. Continuous dependence on modeling is established by considering an approximate problem where the operator A is replaced by an operator $f_\beta(A)$, $\beta > 0$ which approximates A in some sense as the parameter β tends to zero. The particular logarithmic approximation we apply originates from a modified quasi-reversibility method recently introduced by Boussetila and Rebbani.

1 Introduction

Ill-posed problems have been the focus of considerable attention over the last few decades. In this paper we focus on the abstract Cauchy problem

$$\begin{aligned} \frac{du}{dt} &= Au\ , \quad 0 \leq t < T\ , \\ u(0) &= \chi \end{aligned} \tag{1}$$

where $-A$ is the infinitesimal generator of a uniformly bounded holomorphic semigroup $\{S(z) = e^{-zA} : \Re z \geq 0\}$ in a Banach space $(X, \|\cdot\|)$, $\chi \in X$, and u is a function $u : [0, T] \to X$. This problem is generally ill-posed in that a solution may not exist, or if it does, it may not depend continuously on the data. To estimate

M. Fury (✉)
Penn State Abington, 1600 Woodland Road, Abington, PA 19001, USA
e-mail: maf44@psu.edu

B. Campbell Hetrick
Gettysburg College, 300 North Washington Street, Gettysburg, PA 17325, USA
e-mail: bcampbel@gettysburg.edu

W. Huddell
Eastern University, 1300 Eagle Road, St. Davids, PA 19087, USA
e-mail: whuddell@eastern.edu

© Springer International Publishing Switzerland 2016

J. Bélair et al. (eds.), *Mathematical and Computational Approaches in Advancing Modern Science and Engineering*, DOI 10.1007/978-3-319-30379-6_59

the solution, we introduce an approximate well-posed problem

$$\frac{dv}{dt} = f_\beta(A)v\,, \quad 0 \le t < T\,, \tag{2}$$
$$v(0) = \chi$$

where $\beta > 0$ and $f_\beta(A)$ approximates A. We prove continuous dependence on modeling; i.e., that if $f_\beta(A)$ is a suitable approximation of A, then a solution of the ill-posed problem (if it exists) is appropriately close to the solution of the well-posed model problem. This approach, introduced by Lattes and Lions in [11], has been used with several different approximations f_β. Examples include $f_\beta(A) = A - \beta A^2$, used in [2, 3, 5, 11, 12, 14], and $f_\beta(A) = A(I + \beta A)^{-1}$, as used in [3, 5, 10, 13]. However, both of these cases induce an error of order $e^{C/\beta}$, which is problematic in extending results to the nonlinear case. In this paper we use a different approximation introduced by Boussetila and Rebbani [4]:

$$f_\beta(A) = -\frac{1}{pT}\ln(\beta + e^{-pTA})\,, \quad \beta > 0\,, \quad p \ge 1\,. \tag{3}$$

Boussetila and Rebbani use this approximation in Hilbert space (see also [15] and [8]); Huang [9] extends the results to Banach space. In both cases, the authors use the quasi-reversibility method: they use the solution to a well-posed model problem to approximate the initial value for the original problem, then solve the original problem backward. They show that the final value of the solution obtained in this manner depends continuously on the initial data. Our work differs in that we prove directly continuous dependence on the model. Specifically, we show that

$$\|u(t) - v_\beta(t)\| \le C(\sqrt{\beta})^{1-\frac{t}{T}} M^{\frac{t}{T}}\,, \quad 0 \le t < T \tag{4}$$

where $u(t)$ is an assumed solution of (1), $v_\beta(t)$ is the solution of (2) using the logarithmic function f_β in (3), and C and M are constants independent of β.

The problems above share the same initial data, but we note that in practice the observed data may differ. Thus we are interested in the approximate problem (2) with χ replaced by χ_δ where $\delta > 0$ yields a change in the initial data satisfying $\|\chi - \chi_\delta\| \le \delta$. In this context, we directly obtain regularization results by applying the estimate (4).

In Sect. 2, we introduce the functional calculus used for $f_\beta(A)$ and prove several auxiliary results. In Sect. 3, we use these results to prove continuous dependence on modeling (Theorem 1). In Sect. 4, we prove regularization for the problem (1). Finally, Sect. 5 demonstrates the theory of this paper applied to certain partial differential equations. Below, $B(X)$ denotes the space of bounded linear operators on X, and $\rho(A)$ the resolvent set of A. Also, a classical solution $u(t)$ of (1) is a function $u : [0, T] \to X$ such that $u(t) \in Dom(A)$ for $0 < t < T$, $u \in C[0, T] \cap C^1(0, T)$, and u satisfies (1) in X.

2 Approximation

In order to make sense of the operator $f_\beta(A)$, we employ a functional calculus by deLaubenfels [6]. By the fact that $-A$ generates $\{S(z) = e^{-zA} : \Re z \geq 0\}$, define the function

$$G(s, A) = \frac{1}{\pi} \int_{-\infty}^{\infty} \frac{1 - \cos(sr)}{r^2} e^{irA} dr , \quad s \geq 0 .$$

It may be shown that $G(s, A)$ is a continuous function in s mapping into $B(X)$. Also while $\|G(s, A)\| \leq s \left(\sup_{\Re z \geq 0} \|e^{-zA}\|\right)$, by switching within equivalent norms, we may assume without loss of generality that $\|G(s, A)\| \leq s$ for all $s \geq 0$ (see [9]). From this emerges deLaubenfels's functional calculus

$$f(A) = \left[\lim_{t\to\infty} f(t)\right] I + \int_0^{\infty} f''(s) G(s, A) ds \tag{5}$$

for $f \in AC_r^1[0, \infty) := \{h \circ g \ : \ h \in AC^1[0, 1]\}$ where $g(t) = (1 + t)^{-1}$ and $AC^1[0, 1] := \{f \ : \ f'$ exists and is absolutely continuous on $[0, 1]\}$. In [9], Huang provides a concise layout of applying (5) in order to find a formula for $f_\beta(A)$ defined by (3). Inserting $f(s) = -\frac{1}{pT} \ln(\beta + e^{-pTs})$ in (5) yields

$$f_\beta(A) = -\frac{1}{pT} \ln \beta - \int_0^{\infty} \frac{\beta p T e^{-pTs}}{(\beta + e^{-pTs})^2} G(s, A) ds .$$

Also, Huang shows that $f_\beta(A)$ is a bounded operator on X satisfying $\|f_\beta(A)\| \leq -\frac{3}{pT} \ln \beta$ for $0 < \beta < (\sqrt{5} - 1)/2$.

Lemma 1 *Let $-A$ be the infinitesimal generator of a uniformly bounded holomorphic semigroup $\{S(z) = e^{-zA} : \Re z \geq 0\}$ on X and let $0 < \beta < (\sqrt{5} - 1)/2$. Then*

$$\| - Ax + f_\beta(A)x\| \leq \frac{8\beta}{pT} \|S^{-1}(2pT)x\|$$

for all $x \in Dom(S^{-1}(2pT)) = Ran(S(2pT))$.

Proof Note that in the statement of the lemma we have used implicitly the fact that $S(t)$ is a bounded, *injective* operator for each $t \geq 0$ (cf. [7, Lemma 3.1]). Now, define for $s \geq 0$,

$$h(s) = (-s - \frac{1}{pT} \ln(\beta + e^{-pTs}))e^{-2pTs} = -\frac{1}{pT} \ln(\beta e^{pTs} + 1)e^{-2pTs} .$$

Using the fact that $\ln(x+1) \le x$ for $x \ge 0$, we have

$$|h''(s)| = \left| \frac{4\beta^2 pT e^{2pTs} + 3\beta pT e^{pTs}}{(\beta e^{pTs}+1)^2} - 4pT \ln(\beta e^{pTs}+1) \right| e^{-2pTs}$$
$$\le \left(\frac{4\beta^2 pT e^{2pTs} + 3\beta pT e^{pTs}}{(\beta e^{pTs}+1)^2} + 4\beta pT e^{pTs} \right) e^{-2pTs} \le 8\beta pT e^{-pTs} \,.$$

Then for $x \in Dom(S^{-1}(2pT)) = Ran(S(2pT)) \subseteq Dom(A)$,

$$\| -Ax + f_\beta(A)x\| = \|(-A + f_\beta(A))e^{-2pTA} S^{-1}(2pT)x\|$$
$$= \| \int_0^\infty h''(s) G(s,A) S^{-1}(2pT)x \, ds \|$$
$$\le \int_0^\infty 8\beta pTs \, e^{-pTs} \|S^{-1}(2pT)x\| \, ds = \frac{8\beta}{pT} \|S^{-1}(2pT)x\| \,.$$

□

In light of Lemma 1, for $x \in Dom(A)$, define the operator $g_\beta(A)$ in X by

$$g_\beta(A)x = -Ax + f_\beta(A)x \,.$$

Proposition 1 *Let $-A$ be the infinitesimal generator of a uniformly bounded holomorphic semigroup $\{S(z) = e^{-zA} : \Re z \ge 0\}$ on X and let $0 < \beta < (\sqrt{5}-1)/2$. Then $g_\beta(A)$ generates a C_0 semigroup $\{e^{tg_\beta(A)}\}_{t\ge 0}$ on X satisfying*

$$\|e^{tg_\beta(A)}\| \le 2\left(\frac{t}{pT} \ln\left(\frac{pT}{\beta t}\right) + 1 \right) \quad \textit{for} \quad 0 < t \le T \,.$$

Proof For $s \ge 0$ and $0 < t \le T$, define the function $h(t,s) = (\beta e^{pTs}+1)^{-t/pT}$. Also, let $s_\beta = \frac{1}{pT} ln(\frac{pT}{\beta t})$. Then for $0 < t \le T$ and $x \in X$,

$$\|e^{tg_\beta(A)}x\| = \| \int_0^\infty \frac{\partial^2}{\partial s^2} h(t,s) G(s,A)x \, ds\|$$
$$= \| \int_0^\infty \frac{tpT\beta e^{pTs}(\frac{t}{pT}\beta e^{pTs} - 1)}{(\beta e^{pTs}+1)^{\frac{t}{pT}+2}} G(s,A)x \, ds\|$$
$$\le \int_0^{s_\beta} \frac{tpT\beta s \, e^{pTs}(-\frac{t}{pT}\beta e^{pTs} + 1)}{(\beta e^{pTs}+1)^{\frac{t}{pT}+2}} \|x\| \, ds + \int_{s_\beta}^\infty \frac{tpT\beta s \, e^{pTs}(\frac{t}{pT}\beta e^{pTs} - 1)}{(\beta e^{pTs}+1)^{\frac{t}{pT}+2}} \|x\| \, ds$$

$$= \left(\frac{2\beta e^{pTs_\beta}(ts_\beta + 1) + 2}{(\beta e^{pTs_\beta} + 1)^{\frac{t}{pT}+1}} - \frac{1}{(\beta+1)^{\frac{t}{pT}}} \right) \|x\|$$

$$< 2\|x\| \left(\frac{\ln(\frac{pT}{\beta t})}{(\frac{pT}{t} + 1)^{\frac{t}{pT}+1}} + \frac{1}{(\frac{pT}{t}+1)^{\frac{t}{pT}}} \right) \le 2\|x\| \left(\frac{t}{pT} \ln\left(\frac{pT}{\beta t}\right) + 1 \right) .$$

Finally, for any $y \in X$, it may be shown by a dominated convergence argument that $\lim_{t\to 0^+} \int_0^\infty \frac{\partial^2}{\partial s^2}\{(h(t,s)-1)e^{-pTs}\}G(s,A)y\, ds = 0$. In other words, $\|e^{tg_\beta(A)}e^{-pTA}y - e^{-pTA}y\| \to 0$ as $t \to 0^+$ for all $y \in X$. The result

$$\|e^{tg_\beta(A)}x - x\| \to 0 \quad \text{as} \quad t \to 0^+ \quad \text{for all} \quad x \in X \tag{6}$$

then follows from the density of $Ran(e^{-pTA})$ in X together with the fact that $\frac{t}{pT}\ln(\frac{pT}{\beta t}) \to 0$ as $t \to 0^+$. □

Lemma 2 *For sufficiently small $\beta > 0$,*

$$\|e^{tg_\beta(A)}\| \le \frac{2}{\sqrt{\beta}}\left(\sqrt{\frac{t}{pT}} + 1\right) \quad \textit{for} \quad 0 \le t \le T .$$

Proof By Proposition 1 and the fact that $x\ln(\frac{1}{x}) \le \sqrt{x}$ for $x > 0$, we have

$$\|e^{tg_\beta(A)}\| \le 2\left(\frac{1}{\beta}\sqrt{\frac{\beta t}{pT}} + 1\right) = \frac{2}{\sqrt{\beta}}\left(\sqrt{\frac{t}{pT}} + \sqrt{\beta}\right) \le \frac{2}{\sqrt{\beta}}\left(\sqrt{\frac{t}{pT}} + 1\right)$$

for $0 < t \le T$. This estimate also holds for sufficiently small β when $t = 0$ by (6).

3 Continuous Dependence on Modeling

The results proved in Sect. 2 will be used to establish continuous dependence on modeling. For this, we assume $u(t)$ is a solution of (1) and let $v_\beta(t)$ be the unique solution of (2). We extend these solutions into the complex plane and apply the Three Lines Theorem. Following [7], for $\varepsilon > 0$ define a family of bounded operators

$$C_\varepsilon = \frac{1}{2\pi i}\int_{\Gamma_\phi} e^{-\varepsilon w^2}(w - A)^{-1} dw$$

where Γ_ϕ is a complex contour contained within $\rho(A)$, running from $\infty e^{i\phi}$ to $\infty e^{-i\phi}$ with $0 < \phi < \frac{\pi}{2}$. It may be shown that $\{C_\varepsilon\}_{\varepsilon>0}$ is a strongly continuous holomorphic semigroup on X generated by $-A^2$.

Lemma 3 *Let $\varepsilon > 0$ and let $u(t)$ be a classical solution of* (1). *Then*

$$C_\varepsilon e^{tf_\beta(A)}\chi = e^{tg_\beta(A)}C_\varepsilon u(t) \quad \textit{for all} \quad t \in [0,T] .$$

Proof The result follows from uniqueness of solutions to well-posed problems since each item is a classical solution of (2) with initial data χ replaced by $C_\varepsilon \chi$. □

We also require the following based on work of Agmon and Nirenberg [1].

Lemma 4 ([1, p. 148]) *Let $\phi(z)$ be a complex function which is bounded and continuous on $S = \{z = \xi + i\eta \mid \xi \in [0,T], \eta \in \mathbb{R}\}$. For $\alpha = t + ir \in S$, define*

$$\Phi(\alpha) = -\frac{1}{\pi}\int\int_S \phi(z)\left(\frac{1}{z-\alpha} + \frac{1}{\bar{z}+1+\alpha}\right) d\xi d\eta .$$

Then $\Phi(\alpha)$ is absolutely convergent, $\bar{\partial}\Phi(\alpha) = \phi(\alpha)$ where $\bar{\partial}$ denotes the Cauchy-Riemann operator, and there exist constants $K > 0$ and $L > T$ such that

$$\int_{-\infty}^{\infty}\left|\frac{1}{z-\alpha} + \frac{1}{\bar{z}+1+\alpha}\right| d\eta \le K\left(1 + \ln\frac{L}{|\xi - t|}\right) \quad \textit{if} \quad \xi \neq t .$$

Theorem 1 *Let $-A$ be the infinitesimal generator of a uniformly bounded holomorphic semigroup $\{S(z) = e^{-zA} : \Re z \ge 0\}$ on X and let $0 < \beta < (\sqrt{5}-1)/2$. Let $u(t)$ and $v_\beta(t)$ respectively be classical solutions of* (1) *and* (2) *and assume that $u(t) \in Dom(S^{-1}(2pT)) = Ran(S(2pT))$ and $\|S^{-1}(2pT)u(t)\| \le M'$ for all $t \in [0,T]$. Then there exist constants C and M, each independent of β, such that*

$$\|u(t) - v_\beta(t)\| \le C(\sqrt{\beta})^{1-\frac{t}{T}} M^{\frac{t}{T}} \quad \textit{for} \quad 0 \le t < T .$$

Proof Define S to be the complex strip $S = \{t + ir \mid t \in [0,T],\ r \in \mathbb{R}\}$. Since e^{irA} is a bounded operator on X for every $r \in \mathbb{R}$, we may define for $\alpha = t + ir \in S$,

$$\phi_\varepsilon(\alpha) = e^{irA}C_\varepsilon(u(t) - v_\beta(t)) = e^{irA}C_\varepsilon(u(t) - e^{tf_\beta(A)}\chi) .$$

Also, following work of Agmon and Nirenberg [1], for $\alpha = t + ir \in S$ define

$$\Phi_\varepsilon(\alpha) = -\frac{1}{\pi}\int\int_S \bar{\partial}\phi_\varepsilon(z)\left(\frac{1}{z-\alpha} + \frac{1}{\bar{z}+1+\alpha}\right) d\xi d\eta ,$$

where $z = \xi + i\eta \in S$ and $\bar{\partial}$ denotes the Cauchy-Riemann operator $\bar{\partial} = \frac{1}{2}\left(\frac{\partial}{\partial t} + i\frac{\partial}{\partial r}\right)$. In view of Lemma 4, we will show that $\bar{\partial}\phi_\varepsilon(\alpha)$ is bounded and continuous on S.

Note, using the fact that $Dom(AC_\varepsilon) = X$ (cf. [7, Propoisition 2.10]), we have

$$\frac{\partial}{\partial t}\phi_\varepsilon(\alpha) = \mathrm{e}^{\mathrm{i}rA}C_\varepsilon\left(\frac{d}{dt}u(t) - \frac{d}{dt}\mathrm{e}^{tf_\beta(A)}\chi\right) = \mathrm{e}^{\mathrm{i}rA}C_\varepsilon(Au(t) - f_\beta(A)\mathrm{e}^{tf_\beta(A)}\chi)\,,$$

$$\frac{\partial}{\partial r}\phi_\varepsilon(\alpha) = \frac{\partial}{\partial r}\mathrm{e}^{\mathrm{i}rA}C_\varepsilon(u(t) - \mathrm{e}^{tf_\beta(A)}\chi) = \mathrm{e}^{\mathrm{i}rA}(\mathrm{i}A)C_\varepsilon(u(t) - \mathrm{e}^{tf_\beta(A)}\chi)\,.$$

Therefore,

$$\begin{aligned}2\,\bar{\partial}\phi_\varepsilon(\alpha) &= \mathrm{e}^{\mathrm{i}rA}C_\varepsilon(Au(t) - f_\beta(A)\mathrm{e}^{tf_\beta(A)}\chi) - \mathrm{e}^{\mathrm{i}rA}AC_\varepsilon(u(t) - \mathrm{e}^{tf_\beta(A)}\chi)\\ &= \mathrm{e}^{\mathrm{i}rA}(A - f_\beta(A))C_\varepsilon\mathrm{e}^{tf_\beta(A)}\chi\,.\end{aligned}$$

Recall, $\{C_\varepsilon\}_{\varepsilon>0}$ is a holomorphic semigroup satisfying $C_\varepsilon x \to x$ as $\varepsilon \to 0$ for every $x \in X$. Hence for small ε, we may set $C = \sup_{0<\varepsilon<1}\|C_\varepsilon\|$. Also, set $J = \sup_{\Re z\geq 0}\|\mathrm{e}^{-zA}\|$. Then by Lemma 3, Lemma 2, Lemma 1, and stabilizing condition,

$$\begin{aligned}&\|\mathrm{e}^{\mathrm{i}rA}(A - f_\beta(A))C_\varepsilon\mathrm{e}^{tf_\beta(A)}\chi\| = \|\mathrm{e}^{\mathrm{i}rA}\mathrm{e}^{tg_\beta(A)}(A - f_\beta(A))C_\varepsilon u(t)\|\\ &\leq J\frac{2}{\sqrt{\beta}}\left(\sqrt{\frac{t}{pT}} + 1\right)\frac{8\beta}{pT}\|S^{-1}(2pT)C_\varepsilon u(t)\| \leq J\frac{2}{\sqrt{\beta}}\left(\sqrt{\frac{t}{pT}} + 1\right)\frac{8\beta}{pT}CM'\end{aligned}$$

since C_ε commutes with A. Hence we have shown

$$\|\bar{\partial}\phi_\varepsilon(\alpha)\| \leq C'\sqrt{\beta} \tag{7}$$

where C' is a constant independent of β, ε, and α. Thus, $\bar{\partial}\phi_\varepsilon(\alpha)$ is bounded on S. It is also readily shown that $\bar{\partial}\phi_\varepsilon(\alpha)$ is continuous on S. Hence, by Lemma 4, $\Phi_\varepsilon(\alpha)$ is absolutely convergent, $\bar{\partial}\Phi_\varepsilon(\alpha) = \bar{\partial}\phi_\varepsilon(\alpha)$ and there exist constants $K > 0$ and $L > T$ such that

$$\int_{-\infty}^{\infty}\left|\frac{1}{z-\alpha} + \frac{1}{\bar{z}+1+\alpha}\right|d\eta \leq K\left(1 + \ln\frac{L}{|\xi - t|}\right) \quad \text{if} \quad \xi \neq t\,.$$

Next, define $w_\varepsilon : S \to \mathbb{C}$ by

$$w_\varepsilon(\alpha) = x^*\phi_\varepsilon(\alpha) - x^*\Phi_\varepsilon(\alpha)\,,$$

where x^* is in the dual space of X. We will show that the Three Lines Theorem may be applied to w_ε. First for $\alpha = t + \mathrm{i}r \in S$, using standard properties of semigroups and our stabilizing assumptions, we have

$$\|\phi_\varepsilon(\alpha)\| \le J\|C_\varepsilon u(t) - C_\varepsilon e^{tf_\beta(A)}\chi\| = J\|C_\varepsilon u(t) - e^{tg_\beta(A)}C_\varepsilon u(t)\|$$

$$= J\|(I - e^{tg_\beta(A)})C_\varepsilon u(t)\| = J\| - \int_0^t \frac{d}{ds}(e^{sg_\beta(A)}C_\varepsilon u(t))ds\|$$

$$= J\| - \int_0^t e^{sg_\beta(A)}g_\beta(A)C_\varepsilon u(t)ds\| \le J\int_0^t \frac{2}{\sqrt{\beta}}\left(\sqrt{\frac{s}{pT}} + 1\right)\frac{8\beta}{pT}CM'ds \quad (8)$$

showing that $\phi_\varepsilon(\alpha)$ is bounded on S. Next, for $z = \xi + i\eta \in S$, by Lemma 4 and (7),

$$\|\Phi_\varepsilon(\alpha)\| \le \frac{1}{\pi}\int_{-\infty}^{\infty}\int_0^T C'\sqrt{\beta}\left|\frac{1}{z-\alpha} + \frac{1}{\bar{z}+1+\alpha}\right| d\xi d\eta$$

$$\le \frac{1}{\pi}\int_0^T C'\sqrt{\beta}\left[K\left(1 + \ln\frac{L}{|\xi - t|}\right)\right]d\xi = \widetilde{K}\sqrt{\beta}\int_0^T \left(1 + \ln\frac{L}{|\xi - t|}\right)d\xi \quad (9)$$

where $\widetilde{K}$ is a constant. It follows that w_ε is bounded on S. Furthermore, w_ε is continuous on S as well, and by Lemma 4, for α in the interior of S, $\bar{\partial}w_\varepsilon(\alpha) = x^*\bar{\partial}\phi_\varepsilon(\alpha) - x^*\bar{\partial}\Phi_\varepsilon(\alpha) = 0$ showing that w_ε is analytic on the interior of S.

Therefore, by the Three Lines Theorem,

$$|w_\varepsilon(t)| \le M(0)^{1-\frac{t}{T}}M(T)^{\frac{t}{T}}$$

for $0 \le t \le T$, where $M(t) = \sup_{r\in\mathbb{R}}|w_\varepsilon(t + ir)|$. Looking at the sides of the strip S, we have from (9),

$$M(0) = \sup_{r\in\mathbb{R}}|w_\varepsilon(ir)| \le \|x^*\|\|\phi_\varepsilon(ir)\| + \|x^*\|\|\Phi_\varepsilon(ir)\|$$

$$= \|x^*\|\|e^{irA}(u(0) - v_\beta(0))\| + \|x^*\|\|\Phi_\varepsilon(ir)\| = \|x^*\|\|\Phi_\varepsilon(ir)\|$$

$$\le \widetilde{K}\sqrt{\beta}\|x^*\|\int_0^T \left(1 + \ln\frac{L}{\xi}\right)d\xi\ .$$

Next by (8), we have $\|\phi_\varepsilon(T + ir)\| \le J'$ where J' is a constant independent of β, ε, and r since $\beta < 1$. Hence, together with (9) we have that

$$M(T) = \sup_{r\in\mathbb{R}}|w_\varepsilon(T + ir)| \le \|x^*\|\|\phi_\varepsilon(T + ir)\| + \|x^*\|\|\Phi_\varepsilon(T + ir)\|$$

$$\le \|x^*\|\left[J' + \widetilde{K}\int_0^T \left(1 + \ln\frac{L}{|\xi - t|}\right)d\xi\right]$$

where again we have used the fact that $\beta < 1$. Therefore

$$|w_\varepsilon(t)| \le \|x^*\| \left[\widetilde{K}\sqrt{\beta}\int_0^T \left(1+\ln\frac{L}{\xi}\right)d\xi\right]^{1-\frac{t}{T}} \left[J' + \widetilde{K}\int_0^T \left(1+\ln\frac{L}{|\xi - t|}\right)d\xi\right]^{\frac{t}{T}}.$$

Taking the supremum over all $x^* \in X$ with $\|x^*\| \le 1$, we have

$$\|\phi_\varepsilon(t) - \Phi_\varepsilon(t)\| \le \left[\widetilde{K}\sqrt{\beta}M_1\right]^{1-\frac{t}{T}} \left[J' + \widetilde{K}M_2\right]^{\frac{t}{T}} = C(\sqrt{\beta})^{1-\frac{t}{T}}M^{\frac{t}{T}}$$

where C and M are constants independent of both β and ε. Again using (9),

$$\begin{aligned}\|C_\varepsilon(u(t) - v_\beta(t))\| &= \|\phi_\varepsilon(t)\| \le \|\phi_\varepsilon(t) - \Phi_\varepsilon(t)\| + \|\Phi_\varepsilon(t)\| \\ &\le C(\sqrt{\beta})^{1-\frac{t}{T}}M^{\frac{t}{T}} + \widetilde{K}\sqrt{\beta}\int_0^T \left(1+\ln\frac{L}{|\xi - t|}\right)d\xi \\ &\le C(\sqrt{\beta})^{1-\frac{t}{T}}M^{\frac{t}{T}}\end{aligned}$$

for a possibly different value of C still independent of β. Here the bound on the right is independent of ε, so we let $\varepsilon \to 0$ to obtain

$$\|u(t) - v_\beta(t)\| \le C(\sqrt{\beta})^{1-\frac{t}{T}}M^{\frac{t}{T}}.$$

4 Regularization for Problem (1)

We prove regularization for problem (1) using our estimate from Theorem 1.

Definition 1 [10, Definition 3.1] A family $\{R_\beta(t) \mid \beta > 0,\ t \in [0, T]\} \subseteq B(X)$ is called a family of regularizing operators for the problem (1) if for each solution $u(t)$ of (1) with initial data χ and for any $\delta > 0$, there exists $\beta(\delta) > 0$ such that

1. $\beta(\delta) \to 0$ as $\delta \to 0$,
2. $\|u(t) - R_\beta(t)\chi_\delta\| \to 0$ as $\delta \to 0$ for each $t \in [0, T]$ whenever $\|\chi - \chi_\delta\| \le \delta$.

Theorem 2 *Let $-A$ be the infinitesimal generator of a uniformly bounded holomorphic semigroup $\{S(z) = e^{-zA} : \Re z \ge 0\}$ on X and let $f_\beta(A)$ be defined by* (3). *Then $\{R_\beta(t) := e^{tf_\beta(A)} \mid 0 < \beta < (\sqrt{5} - 1)/2,\ t \in [0, T]\}$ is a family of regularizing operators for problem* (1).

Proof Let $u(t)$ be a classical solution of (1) satisfying $\|S^{-1}(2pT)u(t)\| \le M'$ for all $t \in [0, T]$ and let $\|\chi - \chi_\delta\| \le \delta$. Note that since $f_\beta(A) \in B(X)$, $e^{tf_\beta(A)}$ satisfies

$$\|e^{tf_\beta(A)}\| \le e^{t\|f_\beta(A)\|} \le e^{-\frac{3t}{pT}\ln\beta} = \beta^{-\frac{3t}{pT}}$$

for all $t \in [0, T]$, for $0 < \beta < (\sqrt{5} - 1)/2$.

First let $t \in [0, T)$ and choose $\beta = \delta^{\frac{p}{6}}$. Then $\beta \to 0$ as $\delta \to 0$ and from Theorem 1,

$$\begin{aligned}\|u(t) - R_\beta(t)\chi_\delta\| &\leq \|u(t) - e^{tf_\beta(A)}\chi\| + \|e^{tf_\beta(A)}\chi - e^{tf_\beta(A)}\chi_\delta\| \\ &= \|u(t) - v_\beta(t)\| + \|e^{tf_\beta(A)}(\chi - \chi_\delta)\| \\ &\leq C(\sqrt{\beta})^{1-\frac{t}{T}} M^{\frac{t}{T}} + \beta^{-\frac{3t}{pT}}\delta \\ &= C(\delta^{\frac{p}{12}})^{1-\frac{t}{T}} M^{\frac{t}{T}} + \delta^{1-\frac{t}{2T}} \to 0 \quad \text{as} \quad \delta \to 0\,.\end{aligned}$$

In the case $t = T$, from (8) and (9) the estimate $\|u(T) - v_\beta(T)\| \leq N\sqrt{\beta}$ can easily be obtained where N is a constant independent of β. Then as above, still with $\beta = \delta^{\frac{p}{6}}$,

$$\|u(T) - R_\beta(T)\chi_\delta\| \leq N\sqrt{\beta} + \beta^{-\frac{3}{p}}\delta = N\delta^{\frac{p}{12}} + \sqrt{\delta} \to 0 \quad \text{as} \quad \delta \to 0\,.$$

5 Example

For an example, we use that of deLaubenfels in [6]. Let T be the unit circle in the complex plane and for $1 \leq p \leq \infty$, let $H^p(T)$ be the set of all functions in $L^p(T)$ that can be extended to a holomorphic function on the unit disc. Define $\mathscr{A} = H^\infty(T) \cap C(T)$ and let $A = -\mathrm{i}\frac{d}{d\theta}$. Then $\mathrm{i}A = \frac{d}{d\theta}$ is the generator of the familiar translation group on $\mathscr{A}$

$$(e^{\mathrm{i}rA}\psi)(e^{\mathrm{i}\theta}) = \psi(e^{\mathrm{i}(\theta+r)})\,, \quad r \in \mathbb{R}$$

for $\psi \in H^1(T)$. This group extends to a uniformly bounded holomorphic semigroup $\{e^{-zA} : \Re z \geq 0\}$ as shown in [6].

In this setting, (1) becomes the ill-posed partial differential equation

$$\begin{aligned}u_t &= -\mathrm{i}u_\theta\,, \quad 0 \leq t < T\,, \\ u(e^{\mathrm{i}\theta}, 0) &= \phi(e^{\mathrm{i}\theta})\,.\end{aligned}$$

By Theorem 2, $R_\beta(t) = e^{tf_\beta(A)} = (\beta + e^{-pTA})^{-\frac{t}{pT}}$. If $\phi \in \mathscr{A}$, deLaubenfels in [6] shows $G(s, A)$ in this case to be given by

$$G(s, A)\phi = \sum_{k=0}^{[s]} (s-k)\hat{\phi}(k) g_k$$

where $g_k(e^{i\theta}) = e^{ik\theta}$ and $\hat{\phi}(k) = \frac{1}{2\pi}\int_{-\pi}^{\pi}\phi(e^{i\theta})g_k(e^{-i\theta})d\theta$. Defining the function $h(s) = (\beta+e^{-pTs})^{-\frac{t}{pT}}$ and employing the functional calculus (5), we find for $\phi \in \mathscr{A}$

$$R_\beta(t)\phi = h(A)\phi = \beta^{-\frac{t}{pT}} + \int_0^\infty h''(s)G(s,A)\phi ds$$

$$= \quad \beta^{-\frac{t}{pT}} + \int_0^\infty \frac{-tpTe^{-pTs}}{(\beta+e^{-pTs})^{\frac{t}{pT}+1}}\left[1 - \frac{e^{-pTs}(t+pT)}{pT(\beta+e^{-pTs})}\right]\sum_{k=0}^{[s]}(s-k)\hat{\phi}(k)g_k ds\,.$$

Acknowledgements The authors would like to thank Rhonda J. Hughes for her guidance and also the 2015 AMMCS-CAIMS Congress.

References

1. Agmon, S., Nirenberg, L.: Properties of solutions of ordinary differential equations in Banach space. Commun. Pure Appl. Math. **16**, 121–151 (1963)
2. Ames, K.A.: On the comparison of solutions of related properly and improperly posed Cauchy problems for first order operator equations. SIAM J. Math. Anal. **13**, 594–606 (1982)
3. Ames, K.A., Hughes, R.J.: Structural stability for ill-posed problems in Banach space. Semigroup Forum **70**, 127–145 (2005)
4. Boussetila, N., Rebbani, F.: A modified quasi-reversibility method for a class of ill-posed Cauchy problems. Georgian Math. J. **14**, 627–642 (2007)
5. Campbell Hetrick, B.M., Hughes, R.J.: Continuous dependence results for inhomogeneous ill-posed problems in Banach space. J. Math. Anal. Appl. **331**, 342–357 (2007)
6. deLaubenfels, R.: Functional calculus for generators of uniformly bounded holomorphic semigroups. Semigroup Forum **38**, 91–103 (1989)
7. deLaubenfels, R.: Entire solutions of the abstract Cauchy problem. Semigroup Forum **42**, 83–105 (1991)
8. Fury, M.: Modified quasi-reversibility method for nonautonomous semilinear problems, Ninth Mississippi state conference on differential equations and computational simulations. Electron. J. Differ. Equ. Conf. **20**, 99–121 (2013)
9. Huang, Y.: Modified quasi-reversibility method for final value problems in Banach spaces. J. Math. Anal. Appl. **340**, 757–769 (2008)
10. Huang, Y., Zheng, Q.: Regularization for a class of ill-posed Cauchy problems. Proc. Am. Math. Soc. **133**, 3005–3012 (2005)
11. Lattes, R., Lions J.L.: The Method of Quasi-Reversibility, Applications to Partial Differential Equations. American Elsevier, New York (1969)
12. Miller, K.: Stabilized quasi-reversibility and other nearly-best-possible methods for non-well-posed problems. In: Symposium on Non-Well-Posed Problems and Logarithmic Convexity. Lecture Notes in Mathematics, vol. 316, pp. 161–176. Springer, Berlin (1973)
13. Showalter, R.E.: The final value problem for evolution equations. J. Math. Anal. Appl. **47**, 563–572 (1974)
14. Trong, D.D., Tuan, N.H.: Regularization and error estimates for non homogeneous backward heat problems. Electron. J. Differ. Equ. **4**, 1–10 (2006)
15. Trong, D.D., Tuan, N.H.: Stabilized quasi-reversibility method for a class of nonlinear ill-posed problems. Electron. J. Differ. Equ. **84**, 1–12 (2008)

Solving Differential-Algebraic Equations by Selecting Universal Dummy Derivatives

Ross McKenzie and John D. Pryce

Abstract A common way of making a high index DAE amenable to numerical solution is that of index reduction. A classical way of reducing a DAE's index is the dummy derivative method of Mattsson and Söderlind, however for many problems this method only provides a local index 1 DAE. Using the Signature Matrix based structural analysis of Pryce to inform the dummy derivative method we present a way to make this reduction global, where instead of picking new dummy derivatives at run time and thus changing the overall structure of the problem you instead have to update a list of parameters.

1 Introduction

We consider DAEs of the form:

$$f_i(t,\ \text{the } x_j \text{ and derivatives of them}), \qquad i = 1, \ldots, n \tag{1}$$

where $x_j(t), j = 1, \ldots, n$ are state variables and functions of some independent variable t, usually considered to be time. Usually to solve such a problem differentials of some of the equations are added to the system, these equations are called the *hidden constraints*. A DAE has an associated *index*, mainly we are concerned with the differential index, which is the minimum number of differentiations required to obtain an ODE. Generally speaking the higher the index the more difficult the numerical solution—there exist good solvers for index 1 systems and as such we would like to reduce any DAE to an equivalent (same solution set) index 1 formulation. We are interested in reducing the index of our DAE via the Dummy Derivative (henceforth DD method) method introduced in [4] by using the structural analysis of the Signature Matrix method introduced in [2] (henceforth SA) to

R. McKenzie (✉) • J.D. Pryce
School of Mathematics, Cardiff University, Cardiff, CF24 4AG, Wales, UK
e-mail: mckenzierl@cardiff.ac.uk; prycejd1@cardiff.ac.uk

© Springer International Publishing Switzerland 2016

J. Bélair et al. (eds.), *Mathematical and Computational Approaches in Advancing Modern Science and Engineering*, DOI 10.1007/978-3-319-30379-6_60

identify the hidden constraints and otherwise inform the DD method more than is possible than in the classical approach of DDs via the Pantelides algorithm [1]. The DD method introduces a choice of derivatives of variables to be considered algebraic so that our enlarged system containing the hidden constraints is made square. This process is inherently local because it relies on the non-singularity of potentially dynamic matrices. We show in Sect. 2 how to pick a global index 1 system for a simple example, in Sect. 3 we demonstrate how this approach extends to any DAE on which Pryce's SA works and in Sect. 4 we present numerical results.

2 Avoiding Dummy Pivoting for a Simple Example

Example 1 Consider the infamous simple pendulum:

$$\left.\begin{aligned} f_1(t) &= x''(t) + \lambda(t)x(t) &&= 0 \\ f_2(t) &= y''(t) + \lambda(t)y(t) - g &&= 0 \\ f_3(t) &= x^2(t) + y^2(t) - L^2 &&= 0 \end{aligned}\right\} \tag{2}$$

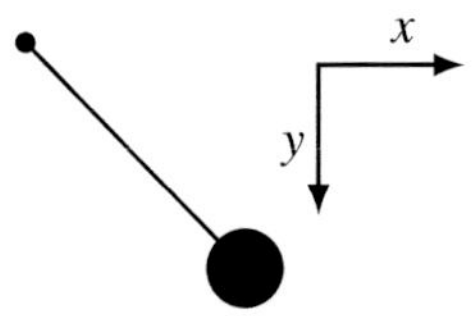

Pendulum bob, mass=1

where g and L are gravitational acceleration and length of the rod respectively. The problem's signature matrix as specified by the Signature Matrix Method [2] is:

$$\Sigma = \begin{array}{c} \\ f_1 \\ f_2 \\ f_3 \\ d_j \end{array}\!\!\begin{array}{c} \begin{array}{ccc} x & y & \lambda \end{array} \\ \left(\begin{array}{ccc} 2^{\bullet} & -\infty & 0^{\circ} \\ -\infty & 2^{\circ} & 0^{\bullet} \\ 0^{\circ} & 0^{\bullet} & -\infty \end{array}\right) \\ \begin{array}{ccc} 2 & 2 & 0 \end{array} \end{array}\begin{array}{c} c_i \\ 0 \\ 0 \\ 2 \\ \\ \end{array}$$

Note: there are two highest value transversals (henceforth HVTs) for the simple pendulum, marked by • and ∘ in the signature matrix above. The offsets **c** and **d** tell us how many times to differentiate each equation and the associated derivative order for the variables that will be found. We now proceed to carry out the DD algorithm in [4] but note whilst the matrices we use are the same the numbering is changed so as to line up with the SA notation. The DD algorithm proceeds in stages, using matrices $G^{[\kappa]}$ and $H^{[\kappa]}$ for $\kappa = 0, 1, \ldots$ where $G^{[0]}$ is the $n \times n$ system Jacobian $\mathbf{J} = \partial f_i^{(c_i)} / \partial x_j^{(d_j)}$. Deleting appropriate rows of $G^{[\kappa]}$ gives $H^{[\kappa]}$. Deleting appropriate columns of $H^{[\kappa]}$ to form a nonsingular matrix gives $G^{[\kappa+1]}$. This example illustrates the process; for space reasons we refer to [4] for details. We have an initial Jacobian (with highest order equations and variables) and a secondary non-square Jacobian:

$$G^{[0]} = \begin{array}{c} \\ f_1 \\ f_2 \\ f_3'' \\ d_j \end{array} \begin{array}{c} \begin{array}{ccc} x'' & y'' & \lambda \end{array} \\ \begin{pmatrix} 1 & 0 & x \\ 0 & 1 & y \\ 2x & 2y & 0 \end{pmatrix} \\ \begin{array}{ccc} 2 & 2 & 0 \end{array} \end{array} \begin{array}{c} c_i \\ 0 \\ 0 \\ 2 \\ \\ \end{array} \qquad \text{and} \qquad H^{[0]} = \begin{array}{c} \\ f_3'' \end{array} \begin{array}{c} \begin{array}{ccc} x'' & y'' & \lambda \end{array} \\ \begin{pmatrix} 2x & 2y & 0 \end{pmatrix} \end{array}.$$

We have to select 1 column from $H^{[0]}$ to get a square non-singular matrix $G^{[1]}$. We cannot choose column 3 because it is structurally 0, as is always the case with columns corresponding to non-differentiated variables. We thus either choose column 1 or column 2. Consider instead that we want to eliminate the need to choose between any variables and instead pick both, since any choice may become invalid (due to a subsequent G matrix becoming locally ill conditioned, i.e. as x or $y \to 0$) and it will be expensive to pivot between potential systems as this happens for a general DAE. If we want to eliminate the choice of candidate dummy derivatives at this stage we will need to add equations to the DAE so that we can choose a square submatrix of $H^{[0]}$ that contains the columns corresponding to all candidate dummy derivatives, i.e. x and y. This is achieved by adding an equation of form:

$$Z_1 := \alpha x' + \beta y' - z_1 = 0 \tag{3}$$

to the original DAE, where α and β are some parameters that will be chosen at run time so the row using new equations (newly introduced equations of the form $Z_i = 0$) in $G^{[1]}$ is roughly orthogonal to the row using old equations (equations part of the original DAE including hidden constraints) in $G^{[1]}$ and z_1 is some new variable to solve for. We would like the $H^{[0]}$ for our new DAE to be of the form:

$$H^{[0]} = \begin{array}{c} \\ f_3'' \\ Z_1' \end{array} \begin{array}{c} \begin{array}{cccc} x'' & y'' & \lambda & z_1' \end{array} \\ \begin{pmatrix} 2x & 2y & 0 & 0 \\ \alpha & \beta & 0 & -1 \end{pmatrix} \end{array}$$

so that it is now possible to pick x'' and y'' as DDs for all time and update α and β along the solution to keep the resulting $G^{[1]}$ matrix well-conditioned. Note: checking the condition number of the corresponding G matrix and using a Gram-Schmidt or QR type procedure for 'new' rows if the matrix becomes ill conditioned would be a reasonable way of updating α and β dynamically. Consider however the new DAEs signature matrix and canonical (element-wise minimum) offsets:

$$\Sigma = \begin{array}{c} \\ f_1 \\ f_2 \\ f_3 \\ Z_1 \\ d_j \end{array}\begin{array}{c} \begin{array}{cccc} x & y & \lambda & z_1 \end{array} \\ \left(\begin{array}{ccc|c} 2^\bullet & -\infty & 0^\circ & -\infty \\ -\infty & 2^\circ & 0^\bullet & -\infty \\ 0^\circ & 0^\bullet & -\infty & -\infty \\ \hline 1 & 1 & -\infty & 0 \end{array}\right) \\ \begin{array}{cccc} 2 & 2 & 0 & 0 \end{array} \end{array}\begin{array}{c} c_i \\ 0 \\ 0 \\ 2 \\ 0 \\ \\ \end{array}$$

where the block structure is highlighted, see [3]. Unfortunately we see that entries in positions $(4,1)$ and $(4,2)$ are both structurally zero (meaning $d_j - c_i \neq \sigma_{i,j}$), so won't appear in $H^{[0]}$, see [5]. There is a key inequality that comes from the SA for finding the offsets $\mathbf{c}$ and $\mathbf{d}$:

$$d_j - c_i \geq \sigma_{i,j} \qquad \text{with equality on a HVT.} \tag{4}$$

Noting this one can see that changing $d_4 = c_4$ (because entry $(4,4)$ must be on a HVT) to 1 will not affect the other offsets. Due to our choice of Z_1 we see that $d_4 = c_4 = 1$ makes the entries in positions $(4,1)$ and $(4,2)$ structurally non zero. If we write down $H^{[1]}$ for this new system (i.e. the DAE using equations f_1,f_2,f_3 and Z_1) we get:

$$H^{[1]} = \begin{array}{c} \\ f_3' \end{array}\begin{array}{c} \begin{array}{cc} x' & y' \end{array} \\ \left(\begin{array}{cc} 2x & 2y \end{array}\right) \end{array};$$

again we add an equation to the system so that we can always choose the variables that are DD candidates at this stage:

$$Z_2 := \gamma x + \delta y - z_2 = 0. \tag{5}$$

So that our new DAE's signature matrix with canonical offsets is:

$$\Sigma = \begin{array}{c} \\ f_1 \\ f_2 \\ f_3 \\ Z_1 \\ Z_2 \\ d_j \end{array}\begin{array}{c} \begin{array}{ccccc} x & y & \lambda & z_1 & z_2 \end{array} \\ \left(\begin{array}{ccc|c|c} 2^\bullet & -\infty & 0^\circ & -\infty & -\infty \\ -\infty & 2^\circ & 0^\bullet & -\infty & -\infty \\ 0^\circ & 0^\bullet & -\infty & -\infty & -\infty \\ \hline 1 & 1 & -\infty & 0 & -\infty \\ 0 & 0 & -\infty & -\infty & 0 \end{array}\right) \\ \begin{array}{ccccc} 2 & 2 & 0 & 0 & 0 \end{array} \end{array}\begin{array}{c} c_i \\ 0 \\ 0 \\ 2 \\ 0 \\ 0 \\ \\ \end{array}$$

Now to have all necessary entries of $H^{[0]}$ and $H^{[1]}$ structurally non zero we need to set $c_4 = d_4 = 1$ and $c_5 = d_5 = 2$. Note that adding these equations to the system does not change the original solution of the DAE, because the equations we add form a new block dependent on the original DAE's (block 1 above) solution. If we now consider the DD stages of our new system we get:

$$H^{[0]} = \begin{array}{c} \\ f_3'' \\ Z_1' \\ Z_2'' \\ d_j \end{array} \begin{array}{c} \begin{array}{ccccc} x'' & y'' & \lambda & z_1' & z_2'' \end{array} \\ \left(\begin{array}{ccc|c|c} 2x & 2y & 0 & 0 & 0 \\ \alpha & \beta & 0 & -1 & 0 \\ \gamma & \delta & 0 & 0 & -1 \end{array}\right) \\ \begin{array}{ccccc} 2 & 2 & 0 & 1 & 2 \end{array} \end{array} \begin{array}{c} c_i \\ 2 \\ 1 \\ 2 \\ \end{array}$$

We see that a valid choice is x'', y'' and z_2''—what we're doing is adding DDs for all variables in an old block that doesn't contain an equation introduced at this stage in the original system and 'borrowing' DDs from the new blocks that do use an equation introduced at this stage in the original system. We're forcing ourselves in to a scheme that is unobtainable by performing a block decomposition of the new DAE and doing DDs on each block. When proceeding to find $H^{[1]}$ we see that the equation introduced for stage 1 is removed and we select x' and y' as DDs. Now note, that since the second equation we introduced is almost the antiderivative of the first we can condense both in to only one additional equation and finish with a DAE that has the following signature matrix and valid offsets and now only selects x'', x', y'' and y' as DDs. The reason this is possible to do is perhaps not clear at first glance (since whilst such a reduction is clearly structurally valid it may fail numerically). Note that due to Griewank's Lemma [2] our new DAE's DD Jacobians $G^{[1]}$ and $G^{[2]}$ are equal, so if a set of parameters works for stage 1 it will also work for stage 2.

$$\Sigma = \begin{array}{c} \\ f_1 \\ f_2 \\ f_3 \\ Z_2 \\ d_j \end{array} \begin{array}{c} \begin{array}{cccc} x & y & \lambda & z_2 \end{array} \\ \left(\begin{array}{ccc|c} 2^{\bullet} & -\infty & 0^{\circ} & -\infty \\ -\infty & 2^{\circ} & 0^{\bullet} & -\infty \\ 0^{\circ} & 0^{\bullet} & -\infty & -\infty \\ 0 & 0 & -\infty & 0 \end{array}\right) \\ \begin{array}{cccc} 2 & 2 & 0 & 2 \end{array} \end{array} \begin{array}{c} c_i \\ 0 \\ 0 \\ 2 \\ 2 \\ \end{array}$$

3 Finding 'Universal' Dummy Derivatives in General

Before giving a general algorithm for adding such equations to the system we need the following definition:

Definition 1 Let the m_κ and n_κ be the number of rows and columns in the DD matrix $H^{[\kappa-1]}$ respectively.

We have the following algorithm to find a static selection of states (Note: we only suggest to carry out the algorithm below in practice if a static selection cannot already be found.):

Algorithm 1 The 'Universal' Dummy Derivative Algorithm

```
1: When finding the matrix G^[κ]:
2: if m_κ = n_κ
3:     Proceed as normal
4: else
5:     Create a vector say, candidatelist, of variables occurring in H^[κ−1], ...
6:     with structurally non zero columns, to order d_j − κ
7:     S = size(candidatelist) − m_κ
8:     for j = 1, ..., S (using a new vector, parameterlist, in each equation)
9:         Add an equation to the DAE of the form:
10:        Σ_{i=1}^{size(candidatelist)} candidatelist(i) × parameterlist(i) − v_j = 0
11:        Set each new equation's offsets equal to κ
12:        Add DDs to the system for the differentials of entries in candidatelist
13:        if κ > 1
14:            for j = 1, ..., S
15:                Add DDs to the system for v_j^(2), ..., v_j^(max c_i)
16:        Form the now non-square G^[κ] comprising of all rows for variables in candidatelist
17:        Proceed with DD algorithm, i.e. form H^[κ] and proceed to the next stage
18: Tidy up final system:
19: Check if any equation introduced is the antiderivative (barring new variables and parameters)
    of one introduced at a later κ stage.
20: Remove all such equations and any corresponding new variables and DDs from the system.
```

Note: The final check at the end is not needed to keep a static selection of dummy derivatives, but can reduce the size of the resulting index 1 problem, as seen in Sect. 2. Note also line 15 where we introduce extra dummy derivatives for the new variables—we need the new equations at some stage κ but we do not need them at previous κ stages and will instead solve their block based DD system at those stages, which by construction is always square. Let us consider the following example:

Example 2 Consider a problem with signature and initial H matrix:

$$\Sigma = \begin{array}{c} \\ f_1 \\ f_2 \\ f_3 \\ f_4 \\ f_5 \\ d_j \end{array} \begin{array}{c} \begin{array}{ccccc} x_1 & x_2 & x_3 & x_4 & x_5 \end{array} \\ \left(\begin{array}{ccccc} 2 & & & & 1 \\ & 2 & & 2 & \\ 1 & & 0 & & \\ & & 0 & 0 & \\ & 1 & & & 1 \end{array}\right) \\ \begin{array}{ccccc} 3 & 2 & 2 & 2 & 2 \end{array} \end{array} \begin{array}{c} c_i \\ 1 \\ 0 \\ 2 \\ 2 \\ 1 \\ \\ \end{array} \quad \text{and} \quad H^{[0]} = \begin{array}{c} \\ f_1' \\ f_3'' \\ f_4'' \\ f_5' \end{array} \begin{array}{c} \begin{array}{ccccc} x_1''' & x_2'' & x_3'' & x_4'' & x_5'' \end{array} \\ \left(\begin{array}{ccccc} \bullet & 0 & 0 & 0 & \bullet \\ \bullet & 0 & \bullet & 0 & 0 \\ 0 & 0 & \bullet & \bullet & 0 \\ 0 & \bullet & 0 & 0 & \bullet \end{array}\right) \end{array}$$

In $H^{[0]}$ $\bullet$ is being used to denote a structurally non-zero entry. We have

$$\text{candidatelist} = (x_1'', x_2', x_3', x_4', x_5') \qquad (6)$$

and $S = 1$. This makes sense: there are only two possible choices of $G^{[1]}$. All variables that appear in a $G^{[1]}$ are also in candidatelist, either we have

$$G^{[1]} = \begin{array}{c} \\ f_1 \\ f_3' \\ f_4' \\ f_5 \end{array} \begin{array}{c} \begin{array}{cccc} x_1'' & x_2' & x_3' & x_4' \end{array} \\ \begin{pmatrix} \bullet & 0 & 0 & 0 \\ \bullet & 0 & \bullet & 0 \\ 0 & 0 & \bullet & \bullet \\ 0 & \bullet & 0 & 0 \end{pmatrix} \end{array} \qquad \text{or} \qquad G^{[1]} = \begin{array}{c} \\ f_1 \\ f_3' \\ f_4' \\ f_5 \end{array} \begin{array}{c} \begin{array}{cccc} x_1'' & x_3' & x_4' & x_5' \end{array} \\ \begin{pmatrix} \bullet & 0 & 0 & \bullet \\ \bullet & \bullet & 0 & 0 \\ 0 & \bullet & \bullet & 0 \\ 0 & 0 & 0 & \bullet \end{pmatrix} \end{array}.$$

We add an equation of the form:

$$Z_1 := \alpha x_1'' + \beta x_2' + \gamma x_3' + \delta x_4' + \epsilon x_5' - v_1 = 0 \tag{7}$$

to the DAE and choose x_1''', x_2'', x_3'', x_4'' and x_5'' to be 'Universal' DDs, setting $d_6 = c_6 = 1$.

We form the following non square Jacobian:

$$G^{[1]} = \begin{array}{c} \\ f_1 \\ f_3' \\ f_4' \\ f_5 \end{array} \begin{array}{c} \begin{array}{ccccc} x_1'' & x_2' & x_3' & x_4' & x_5' \end{array} \\ \begin{pmatrix} \bullet & 0 & 0 & 0 & \bullet \\ \bullet & 0 & \bullet & 0 & 0 \\ 0 & 0 & \bullet & \bullet & 0 \\ 0 & \bullet & 0 & 0 & \bullet \end{pmatrix} \end{array} \qquad \text{and get} \qquad H^{[1]} = \begin{array}{c} \\ f_3' \\ f_4' \end{array} \begin{array}{c} \begin{array}{ccccc} x_1'' & x_2' & x_3' & x_4' & x_5' \end{array} \\ \begin{pmatrix} \bullet & 0 & \bullet & 0 & 0 \\ 0 & 0 & \bullet & \bullet & 0 \end{pmatrix} \end{array}$$

again, this system is not square, we have

$$\text{candidatelist} = (x_1', x_3, x_4) \tag{8}$$

and $S = 1$ again. We add an equation of the form:

$$Z_2 := \zeta x_1' + \eta x_3 + \theta x_4 - vv_1 = 0 \tag{9}$$

to the DAE and choose x_1'', x_3', x_4' and vv_1'' to be 'Universal' DDs, setting $d_7 = c_7 = 2$.

Let us just confirm our choice of DDs is valid (given the potential need to update our parameters to keep any structurally non singular G matrix numerically non singular). Our enlarged system has signature matrix and initial H matrix:

$$\Sigma = \begin{array}{c} \\ f_1 \\ f_2 \\ f_3 \\ f_4 \\ f_5 \\ Z_1 \\ Z_2 \\ d_j \end{array} \begin{array}{c} \begin{array}{ccccccc} x_1 & x_2 & x_3 & x_4 & x_5 & v_1 & vv_1 \end{array} \\ \left(\begin{array}{ccccc|c|c} 2 & & & & 1 & & \\ & 2 & & 2 & & & \\ 1 & & 0 & & & & \\ & & 0 & 0 & & & \\ & 1 & & & 1 & & \\ \hline 2 & 1 & 1 & 1 & 1 & 0 & \\ \hline 1 & & 0 & 0 & & & 0 \end{array}\right) \\ \begin{array}{ccccccc} 3 & 2 & 2 & 2 & 2 & 1 & 2 \end{array} \end{array} \begin{array}{c} c_i \\ 1 \\ 0 \\ 2 \\ 2 \\ 1 \\ 1 \\ 2 \\ \\ \end{array} \quad \text{and} \quad H^{[0]} = \begin{array}{c} \\ f_1' \\ f_3'' \\ f_4'' \\ f_5' \\ Z_1' \\ Z_2'' \end{array} \begin{array}{c} \begin{array}{ccccccc} x_1''' & x_2'' & x_3'' & x_4'' & x_5'' & v_1' & vv_1'' \end{array} \\ \left(\begin{array}{ccccc|c|c} \bullet & 0 & 0 & 0 & \bullet & 0 & 0 \\ \bullet & 0 & \bullet & 0 & 0 & 0 & 0 \\ 0 & 0 & \bullet & \bullet & 0 & 0 & 0 \\ 0 & \bullet & 0 & 0 & \bullet & 0 & 0 \\ \hline \alpha & \beta & \gamma & \delta & \epsilon & -1 & 0 \\ \hline \zeta & 0 & \eta & \theta & 0 & 0 & -1 \end{array}\right) \end{array}.$$

Where as expected a valid $G^{[1]}$ is:

$$G^{[1]} = \begin{array}{c} \\ f_1 \\ f_3' \\ f_4' \\ f_5 \\ Z_1 \\ Z_2' \end{array} \begin{array}{c} \begin{array}{cccccc} x_1'' & x_2' & x_3' & x_4' & x_5' & vv_1' \end{array} \\ \left(\begin{array}{ccccc|c} \bullet & 0 & 0 & 0 & \bullet & 0 \\ \bullet & 0 & \bullet & 0 & 0 & 0 \\ 0 & 0 & \bullet & \bullet & 0 & 0 \\ 0 & \bullet & 0 & 0 & \bullet & 0 \\ \alpha & \beta & \gamma & \delta & \epsilon & 0 \\ \hline \zeta & 0 & \eta & \theta & 0 & -1 \end{array}\right) \end{array} \quad \text{yielding} \quad H^{[1]} = \begin{array}{c} \\ f_3' \\ f_4' \\ Z_2' \end{array} \begin{array}{c} \begin{array}{cccccc} x_1'' & x_2' & x_3' & x_4' & x_5' & vv_1' \end{array} \\ \left(\begin{array}{ccccc|c} \bullet & 0 & \bullet & 0 & 0 & 0 \\ 0 & 0 & \bullet & \bullet & 0 & 0 \\ \hline \zeta & 0 & \eta & \theta & 0 & -1 \end{array}\right) \end{array}$$

again, as expected a valid choice for $G^{[2]}$ is:

$$G^{[2]} = \begin{array}{c} \\ f_3 \\ f_4 \\ Z_2 \end{array} \begin{array}{c} \begin{array}{ccc} x_1' & x_3 & x_4 \end{array} \\ \left(\begin{array}{ccc} \bullet & \bullet & 0 \\ 0 & \bullet & \bullet \\ \zeta & \eta & \theta \end{array}\right) \end{array}$$

where the algorithm terminates.

One may think a way to tidy up the above system would be to increase $c_6 = d_6 = 2$ and remove the final equation and variable. Structurally this will still work, but numerically we may fall in to trouble as $G^{[2]}$ may be singular.

We now go about proving the above algorithm provides a valid DD scheme with same solution as the original problem. First we need the following definition:

Definition 2 (Transversals) A transversal for an $m \times n$ matrix, with $m \leq n$ is a set of m tuples (i, j) so that $i \in \{1, \ldots, m\}$ and $j \in \{1, \ldots, n\}$ where no i or j is repeated.

Theorem 1 *If a variable's derivative does not appear in any potential DD scheme the corresponding entries in the H matrix are all* 0.

This result is easily derived from the following:

Theorem 2 *Given any rectangular matrix with more columns than rows, and of structural full row rank every non empty column contains an element of some transversal.*

Proof Assume for the sake of contradiction there exists a non empty column that does not contain any elements that belong to a transversal. Without loss of generality rearrange so that a non 0 entry is in the bottom right corner, say position (m_κ, n_κ). Since we have a valid DD scheme at the stage (i.e. the system is of full row rank) we know there exists some transversal, T say. This transversal contains some element (m_κ, J), for some $J \in \{1, \ldots, n_\kappa - 1\}$. We can therefore form a new transversal of the form $\{T\ (m_\kappa, J)\} \cup (m_\kappa, n_\kappa)$.

Theorem 1 is why we consider all non empty columns in our candidatelist. We now seek to justify our removal of equations in the last part of the algorithm above.

Theorem 3 *If a scheme produced by the above algorithm yields equations that can be removed the solution is the same as the scheme without removing those equations.*

Proof Structurally it is clear that such removal still leaves us with the same potential choice of DDs, since if such a removal is possible we will have (at least) structurally identical equations at some stage κ, (at least one) of which is being treated as its own block system until its new variable is needed to make a square G matrix. Numerically if such a reduction is possible then the size of the candidatelist vector must remain unchanged. Since we only remove equations in our DD stages we have two possibilities, either the difference between the size of candidatelist and m_κ has stayed the same, or it has increased. If it has stayed the same then the proposed removal of equations gives us the same structural matrix, and by Griewank's Lemma the same numerical matrix. If it has reduced we will be adding new equations that are not the antiderivatives of equations previously introduced. In which case we can still make our corresponding G matrix non-singular by varying only the new parameters introduced at this stage, since we know the rows are linearly independent.

Note: The above proof yields a necessary condition for introducing equations that can be removed that could shorten our algorithm. If at some stage $c_i - \kappa > 1$ for each i considered at that stage then at the next stage we must introduce equations which are the 'antiderivatives' of the ones introduced at this stage. We could shorten Algorithm 1 by introducing such a condition. Finally we have the following:

Theorem 4 *The above algorithm provides an always static selection of DDs that is always valid (provided one chooses suitable parameters throughout integration) and has same solution as the original DAE.*

Proof Theorems 1 and 3 give us that we have a valid DD scheme, hence all that is left to prove is that such equation additions do not change the solution set. As illustrated in above example the inclusion of additional equations in the described manner is equivalent to adding more dependent blocks to the system, which will not change the original block's solution.

Note: 'same solution' in the above Theorem may be confusing on a first read since the reformulated DAE is of a larger size. We mean that the value of any $x_i(t)$ in the original DAE is also the value of that same $x_i(t)$ in the reformulated DAE.

4 Preliminary Numerical Results for Universal Dummy Derivatives

We solve the simple pendulum index 1 universal DD reformulation given below (where x_d denotes the DD corresponding to selecting x' as a DD in the algorithm above, x_{dd} denotes x'' and similarly for y):

$$\left.\begin{aligned}
f_1(t) &= x_{dd}(t) + \lambda(t)x(t) &&= 0\\
f_2(t) &= y_{dd}(t) + \lambda(t)y(t) - g &&= 0\\
f_3(t) &= x^2(t) + y^2(t) - L^2 &&= 0\\
Z_2(t) &= \gamma x(t) + \delta y(t) - z_2(t) &&= 0\\
f_3'(t) &= 2x(t)x_d(t) + 2y(t)y_d(t) &&= 0\\
Z_2'(t) &= \gamma x_d(t) + \delta y_d(t) - z_2'(t) &&= 0\\
f_3''(t) &= 2x(t)x_{dd}(t) + 2x_d^2(t) + 2y(t)y_{dd}(t) + 2y_d^2(t) &&= 0\\
Z_2''(t) &= \gamma x_{dd}(t) + \delta y_{dd}(t) - z_2''(t) &&= 0
\end{aligned}\right\} \tag{10}$$

in MATLAB using ode45 (using variable step size with initial conditions $x = 6$, $y = -8$, $x' = y' = 0$ and parameters $L = 10$ and $G = 9.81$) by reformulating the problem as an ODE in z_2 and z_2' and switching parameters whenever the angle between (x, y) and (γ, δ) becomes small. We compare the solution with one produced by DAETS [6], an accurate order 30 Taylor series solution, see Fig. 1.

We briefly present some information on switching the system (changing γ and β), a full discussion is left to a future work. If we choose our switching condition the angle between (γ, δ) and (x, y) being less that $\pi/4$, i.e. trying to keep the G matrices well conditioned then the change in energy from $t = 0$ to $t = 100$ grows to around 10^{-5}. If we instead switch after every time to step to a new (γ, δ) orthogonal to (x, y) this change is greatly reduced to around 10^{-12}. One could also attempt to solve the new 'Universal' DD system by coming up with a continuous choice of parameters so that the G matrices are globally non-singular, however in general this will likely be difficult to achieve so we do not present results for doing this with the simple pendulum.

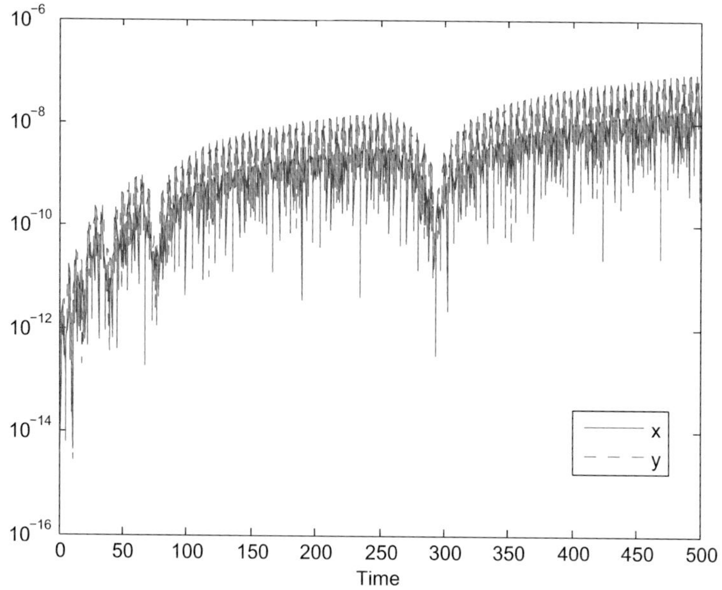

Fig. 1 Difference between the solution of the 'Universal' Dummy Derivative index 1 formulation of the simple pendulum solved in MATLAB by reformulating to an ODE and using ode45 and the solution to the original index 3 formulation solved via an order 30 Taylor Series method using DAETS

5 Conclusion

We have given and demonstrated by examples a way of finding dummy derivatives such that one does not have to change the structure of the reduced index 1 problem when solving high index DAEs. Instead we have changed the problem from one of structural pivoting of variables to numerical switching of parameters, which we believe to be much easier for a general DAE code to handle, since one can exploit underlying structures of the index 1 formulation throughout a simulation. In a future work we aim to more formally develop a good switching condition for an arbitrary DAE or discover a way of choosing continuous parameters for an arbitrary DAE.

References

1. Pantelides, C.C.: The consistent initialization of differential-algebraic systems. SIAM J. Sci. Stat. Comput. **9**, 213–231 (1988)
2. Pryce, J.D.: A simple structural analysis method for DAEs. BIT Numer. Math. **41**(2), 364–394 (2001)
3. Pryce, J.D., Nedialkov, N.S., Tan, G.: DAESA—a matlab tool for structural analysis of differential-algebraic equations: theory. ACM TOMS **41**(2), 1–20 (2015)

4. Mattsson, S.E., Söderlind, G.: Index reduction in differential-algebraic equations using dummy derivatives. SIAM J. Sci. Comput. **14**(3), 677–692 (1993)
5. McKenzie, R., Pryce, J.D.: Structural analysis and dummy derivatives: some relations. In: Cojocaru, M.G. (ed.) Interdisciplinary Topics in Applied Mathematics, Modeling and Computational Science, pp. 293–299. Springer, Cham (2015)
6. Nedialkov, N.S., Pryce, J.D.: DAETS User Guide. 2008–2009

On a Topological Obstruction in the Reach Control Problem

Melkior Ornik and Mireille E. Broucke

Abstract This paper explores aspects of the Reach Control Problem (RCP) to drive the states of an affine control system to a facet of a simplex without first exiting from other facets. In analogy with the problem of nonlinear feedback stabilization, we investigate a topological obstruction that arises in solving the RCP by continuous state feedback. The problem is fully solved in this paper for the case of two and three dimensions.

1 Problem Statement

This paper studies a topological obstruction that arises in solving the Reach Control Problem (RCP) using continuous state feedback. We consider a simplex $\mathcal{S} := \text{co}\{v_0, \ldots, v_n\}$ with vertices $\{v_0, \ldots, v_n\}$ and facets $\{\mathcal{F}_0, \ldots, \mathcal{F}_n\}$. Each facet is indexed according to the vertex it does not contain. Facet $\mathcal{F}_0$ is called the *exit facet*. Let $\mathbf{h_j}$ be the normal vector to facet $\mathcal{F}_j$ pointing outside $\mathcal{S}$. Define $I = \{1, \ldots, n\}$. Let $I(x) \subseteq \{0, 1, \ldots, n\}$ be the minimal set of indices such that $x \in \text{co}\{v_i, \mid i \in I(x)\}$. That is, x is in the interior of $\text{co}\{v_i \mid i \in I(x)\}$.

We consider the affine control system on $\mathcal{S}$

$$\dot{x} = Ax + Bu + a\,, \tag{1}$$

where $x \in \mathbb{R}^n$ and $u \in \mathbb{R}^m$ where $1 \leq m < n$. Let $\mathcal{B} := \text{Im}(B)$ and $\mathcal{O} := \{x \in \mathbb{R}^n \mid Ax + a \in \mathcal{B}\}$. Let $\phi_u(t, x_0)$ denote the trajectory of (1) starting at x_0 under a feedback $u(x)$. The problem is to find a state feedback $u(x)$ such that all trajectories $\phi_u(\cdot, x_0)$ starting in $\mathcal{S}$ exit $\mathcal{S}$ through $\mathcal{F}_0$ in finite time without first leaving $\mathcal{S}$. The RCP has been extensively studied [4–6]. The purpose of this paper is to announce a topological obstruction in the RCP, paralleling the analogous problem arising in the problem of continuous state feedback stabilization [3], and to give preliminary results on the problem for low dimensional systems.

M. Ornik (✉) • M.E. Broucke
Department of Electrical & Computer Engineering, University of Toronto, Toronto, ON M5S 3G4, Canada
e-mail: melkior.ornik@scg.utoronto.ca; broucke@control.utoronto.ca

© Springer International Publishing Switzerland 2016

J. Bélair et al. (eds.), *Mathematical and Computational Approaches in Advancing Modern Science and Engineering*, DOI 10.1007/978-3-319-30379-6_61

A parallel study of the same problem was made in [10] (some preliminary work appeared in [8]). The contributions of this paper significantly differ from [10]. This paper primarily uses retraction theory to study the case of $\dim(\mathcal{O}_\mathcal{S}) = n - 1$. This leads to a simple solution in low dimensions of n. Studies of low-dimensional systems are prevalent [1]. On the other hand, [10] uses homotopy theory to study the case of $\dim(\mathcal{B}) = 2$. While the conclusions of [10] could also potentially lead to a solution in low dimensions of n, this situation is not explored in [10]. Furthermore, the elegant cone condition for the topological obstruction developed in this paper, $\mathcal{B} \cap \mathrm{cone}(\mathcal{O}_\mathcal{S}) = \mathbf{0}$, does not make an appearance at all in [10]; its place is taken by a more involved result based on null-homotopic maps on a circle.

A supplement to this paper is found in [11] where we present supporting results and a study of the case of an obstruction using affine feedback. The results of this paper for the case of $n = 2, 3$, Theorems 3 and 4, could be obtained from [11]. However, while it is tempting to forgo more advanced topological methods in favour of the brute force linear algebra arguments in [11], we insist on the importance of the topological approach in order to have a hope of generalizing the results to higher dimensions.

For each $x \in \mathcal{S}$, we define the cone

$$\mathcal{C}(x) = \{\mathbf{y} \in \mathbb{R}^n \mid \mathbf{h_j} \cdot \mathbf{y} \leq 0, j \in I \backslash I(x)\} . \tag{2}$$

In other words, $\mathcal{C}(x)$ is the set of all vectors $\mathbf{y}$ which, when attached at x, point into $\mathcal{S}$ or through the exit facet $\mathcal{F}_0$. (We note that if $x \in \mathrm{Int}(\mathcal{S})$, $\mathcal{C}(x) = \mathbb{R}^n$.) In order for the trajectory $\phi_u(t, x_0)$ to not leave $\mathcal{S}$ through any facet except the exit facet, we require [6]:

$$\frac{d\phi_u}{dt} = Ax + Bu(x) + a \in \mathcal{C}(x) , \quad x \in \mathcal{S} . \tag{3}$$

In addition to this necessary condition, if $u(x)$ solves the RCP then there are no closed-loop equilibria in $\mathcal{S}$. The equilibria of an affine system can only lie in the affine space $\mathcal{O}$, and for all $x \in \mathcal{S} \cap \mathcal{O}$ and $u \in \mathbb{R}^m$, $Ax + Bu + a \in \mathcal{B}$. Defining the closed-loop vector field $f(x) = Ax + Bu(x) + a$, the previous statements suggest that a necessary condition to solve the RCP by continuous state feedback is: there exists a non-vanishing continuous map $f(x)$ on the set $\mathcal{S} \cap \mathcal{O}$ such that $f(x) \in \mathcal{B} \cap \mathcal{C}(x)$. Motivated by Brockett's work [3], we will say that if such a function does not exist, the system contains a *topological obstruction*.

Define $\mathcal{O}_\mathcal{S} = \mathcal{S} \cap \mathcal{O}$. We define $\mathrm{cone}(\mathcal{O}_\mathcal{S}) = \cap_{x \in \mathcal{O}_\mathcal{S}} \mathcal{C}(x)$. For the remainder of the paper we assume that $\mathcal{O}_\mathcal{S} \neq \emptyset$. We will also assume $v_0 \notin \mathcal{O}_\mathcal{S}$, as well as $1 \leq \dim(\mathcal{O}_\mathcal{S}) \leq n - 1$. The cases of $\dim(\mathcal{O}_\mathcal{S}) = 0, n$ and $v_0 \in \mathcal{O}_\mathcal{S}$ are trivial to analyze. For the sake of completeness, this analysis was formally done in Lemma 9, Lemma 10 and Corollary 11 of [11].

We study the following problem.

Problem 1 Let $\mathcal{S}$, $\mathcal{B}$, $\mathcal{O}$, and $\mathcal{O}_{\mathcal{S}}$ be as above. Does there exist a continuous map $f : \mathcal{O}_{\mathcal{S}} \to \mathcal{B}\backslash\{\mathbf{0}\}$ such that for every $x \in \mathcal{O}_{\mathcal{S}}, f(x) \in \mathcal{C}(x)$?

2 Main Results

This section presents the main results on solving Problem 1. We show that if $\dim(\mathcal{O}_{\mathcal{S}}) = n - 1$, then it is possible to characterize the solution of Problem 1 in terms of a smaller polytope $\mathcal{O}'_{\mathcal{S}}$, and $\mathcal{O}'_{\mathcal{S}}$ will be amenable to a complete analysis of the problem in low dimensions. The main result is presented in Theorem 1. The consequences of Theorem 1 to low dimensional systems are presented in Theorem 3.

Let us assume $\mathcal{O}_{\mathcal{S}}$ is $(n-1)$-dimensional. According to [7, 9], this means $\mathcal{S}$ is cut by $\mathcal{O}$ into two parts: one part containing v_0 and $p \geq 0$ other vertices, and the other containing the other $n - p \geq 1$ vertices of $\mathcal{S}$. W.l.o.g. we assume $\{v_0, v_1, \ldots, v_p\}$ are on one side of $\mathcal{O}_{\mathcal{S}}$ and $\{v_{p+1}, \ldots, v_n\}$ are on the other side, where we assume vertices of $\mathcal{S}$ on $\mathcal{O}_{\mathcal{S}}$ are in the set $\{v_{p+1}, \ldots, v_n\}$. The vertices of $\mathcal{O}_{\mathcal{S}}$ lie on those edges of $\mathcal{S}$ connecting v_i's which are on different sides of $\mathcal{O}_{\mathcal{S}}$. Thus, we employ the notation o_{ij} to denote a vertex of $\mathcal{O}_{\mathcal{S}}$ with $I(o_{ij}) = \{i, j\}$. If there are no vertices of $\mathcal{S}$ on $\mathcal{O}_{\mathcal{S}}$, then $\mathcal{O}_{\mathcal{S}}$ has $(p + 1)(n - p)$ vertices [7], but if $\mathcal{O}_{\mathcal{S}}$ contains r vertices of $\mathcal{S}$, then $\mathcal{O}_{\mathcal{S}}$ has $(p + 1)(n - p) - pr$ vertices. At this point we introduce a mild abuse of notation with the convention that if $v_j \in \mathcal{O}_{\mathcal{S}}$, then $o_{ij} = v_j$ for all $i = 0, \ldots, p$.

Let us introduce the following notation. Let

$$\{i_1, i_2, \ldots, i_k | j_1, j_2, \ldots, j_l\} = \mathrm{co}\{o_{i_\alpha j_\beta} : 1 \leq \alpha \leq k, 1 \leq \beta \leq l\}.$$

We observe that since $I(o_{ij}) = \{i, j\}$, if $x \in \{i_1, i_2, \ldots, i_k | j_1, j_2, \ldots, j_l\}$ then $I(x) \subseteq \{i_1, i_2, \ldots, i_k\} \cup \{j_1, j_2, \ldots, j_l\}$. Also observe that $\mathcal{O}_{\mathcal{S}} = \{0, \ldots, p | p + 1, \ldots, n\}$.

Lemma 1 *Let $\mathcal{O}_{\mathcal{S}} = \{0, \ldots, p | p+1, \ldots, n\}$, $\mathcal{A} = \{i_1, \ldots, i_k | j_1, \ldots, j_l\} \subseteq \mathcal{O}_{\mathcal{S}}$, and $\mathcal{A}' = \{i'_1, \ldots, i'_{k'} | j'_1, \ldots, j'_{l'}\} \subseteq \mathcal{O}_{\mathcal{S}}$. Let $L = \{i_1, \ldots, i_k\} \cap \{i'_1, \ldots, i'_{k'}\}$. Analogously, let $R = \{j_1, \ldots, j_l\} \cap \{j'_1, \ldots, j'_{l'}\}$. Then, $\mathcal{A} \cap \mathcal{A}' = \{L|R\}$.*

Proof By definition every vertex of $\{L|R\}$ is a vertex of $\mathcal{A}$ and of $\mathcal{A}'$, so $\{L|R\} \subseteq \mathcal{A} \cap \mathcal{A}'$. Conversely, suppose $x \in \mathcal{A} \cap \mathcal{A}'$. Since $x \in \mathcal{A}$, $I(x) \subseteq \{i_1, \ldots, i_k, j_1, \ldots, j_l\}$ and since $x \in \mathcal{A}'$, $I(x) \subseteq \{i'_1, \ldots, i'_{k'}, j'_1, \ldots, j'_{l'}\}$. Hence, $I(x) \subseteq L \cup R$, where we use the fact that $\{i_1, \ldots, i_k\} \cap \{j'_1, \ldots, j'_{l'}\} = \emptyset$ and $\{i'_1, \ldots, i'_{k'}\} \cap \{j_1, \ldots, j_l\} = \emptyset$. It follows $x \in \{L|R\}$. □

Before getting to the crux of the problem, let us introduce the notions of a homeomorphism and a retraction. Let $\mathcal{X}$ and $\tilde{\mathcal{X}}$ be topological spaces. $\mathcal{X}$ and $\tilde{\mathcal{X}}$ are *homeomorphic* if there exists a continuous bijection $h : \mathcal{X} \to \tilde{\mathcal{X}}$ which has a continuous inverse. Furthermore, if $\mathcal{A}$ is a subspace of $\mathcal{X}$, a continuous map $r : \mathcal{X} \to \mathcal{A}$ is a *retraction* if $r|_{\mathcal{A}} \equiv id$.

Example 1 Let us consider the case of $n = 3$, with $\mathcal{O}_S$ a quadrilateral with vertices $o_{02} \in \mathrm{co}\{v_0, v_2\}$, $o_{03} \in \mathrm{co}\{v_0, v_3\}$, $o_{12} \in \mathrm{co}\{v_1, v_2\}$, and $o_{13} \in \mathrm{co}\{v_1, v_3\}$. By definition, $\mathcal{C}(o_{02}) \subseteq \mathcal{C}(o_{12})$ and $\mathcal{C}(o_{03}) \subseteq \mathcal{C}(o_{13})$. In fact, one can easily show that the cone of any point on the edge $\overline{o_{12}o_{02}}$ will be larger than $\mathcal{C}(o_{02})$, and analogously for the edge $\overline{o_{13}o_{03}}$. Thus, we reach the idea that if a continuous function satisfying Problem 1 exists on the convex set $\mathrm{co}\{o_{02}, o_{03}\}$ containing the most restrictive cones, then that function can easily be extended to the entire $\mathcal{O}_S$.

Theorem 1 will serve to prove the above claim. The procedure outlined in the proof of Theorem 1, adapted to this example, is as follows. If a function f satisfying Problem 1 can be defined on the edge $\overline{o_{02}o_{03}}$, we can also define it on the edge $\overline{o_{12}o_{02}}$ by $f(x) = f(o_{02})$ and on the edge $\overline{o_{13}o_{03}}$ by $f(x) = f(o_{03})$. We note that such f is non-zero and satisfies the cone condition $f(x) \in \mathcal{C}(x)$ because $\mathcal{C}(o_{02}) \subseteq \mathcal{C}(o_{12})$ and $\mathcal{C}(o_{03}) \subseteq \mathcal{C}(o_{13})$. So far f has been defined on three edges of $\mathcal{O}_S$. Then f can be defined on the remainder of $\mathcal{O}_S$, which consists of its interior as well as the relative interior of the edge $\overline{o_{12}o_{13}}$, by using a retraction r — a continuous map from $\mathcal{O}_S$ to the three edges of $\mathcal{O}_S$ on which f is already defined, such that r is identity on those three edges. More formally, it can be shown, as in Theorem 1, that the function $f(x) = f(r(x))$ exists and solves Problem 1.

Theorem 1 (Dimension Reduction) *Let* $\dim \mathcal{O}_S = n-1$, $v_0 \notin \mathcal{O}_S$, *and* $p > 0$. *Define* $V'_{\mathcal{O}_S} = \{o \in V_{\mathcal{O}_S} \mid (\exists j \in \{1, \dots, n\}) o \in \mathrm{co}\{v_0, v_j\}\}$, *and let* $\mathcal{O}'_S = \mathrm{co}(V'_{\mathcal{O}_S})$. *Then the answer to Problem 1 is affirmative if and only if it is affirmative for* $\mathcal{O}'_S$.

Proof Since $V'_{\mathcal{O}_S} \subseteq V_{\mathcal{O}_S}$ it follows that $\mathcal{O}'_S \subseteq \mathcal{O}_S$. Thus, if there exists $f : \mathcal{O}_S \to \mathcal{B}\backslash\{\mathbf{0}\}$ solving Problem 1 then $f|_{\mathcal{O}'_S} : \mathcal{O}'_S \to \mathcal{B}\backslash\{\mathbf{0}\}$ also solves Problem 1. Conversely, suppose there exists $f' : \mathcal{O}'_S \to \mathcal{B}\backslash\{\mathbf{0}\}$ solving Problem 1. From our notational convention, $\mathcal{O}_S = \{0, 1, \dots, p \mid p+1, \dots, n\}$, $V'_{\mathcal{O}_S} = \{o_{0(p+1)}, \dots, o_{0n}\}$, and $\mathcal{O}'_S = \{0 \mid p+1, \dots, n\}$.

We now proceed with the main topological argument. Informally, we build a skeleton of $\mathcal{O}_S$, starting with $\mathcal{O}'_S$, and adding in each step additional edges and faces of $\mathcal{O}_S$ until in the last step all of $\mathcal{O}_S$ is added. We then use topological methods to show that Problem 1 for the set obtained in each step can be reduced to the same problem applied to the set from the previous step, thus going back from $\mathcal{O}_S$ to $\mathcal{O}'_S$.

We build a skeleton of $\mathcal{O}_S$ as follows. Let $\mathcal{O}^1_S = \mathcal{O}'_S$, and for all $2 \leq k \leq n$, let

$$\mathcal{O}^k_S = \mathcal{O}^{k-1}_S \cup \bigcup_{\substack{0<i_1<\dots<i_\alpha\leq p,\\ p<j_1<\dots<j_\beta\leq n,\\ \alpha+\beta=k,\alpha,\beta\geq 1}} \{0, i_1, \dots, i_\alpha | j_1, \dots, j_\beta\}.$$

Observe that each $\mathcal{H} = \{0, i_1, \dots, i_\alpha \mid j_1, \dots, j_\beta\}$ is a closed, convex polytope of some dimension d, so it is homeomorphic to the closed ball $\mathbb{B}^d$, and its boundary is homeomorphic to the $(d-1)$-dimensional sphere $\mathbb{S}^{d-1}$ [2]. We claim that

$$\partial\{0, i_1, \dots, i_\alpha | j_1, \dots, j_\beta\} \setminus \mathcal{O}^{k-1}_S = \mathrm{Int}(\{i_1, \dots, i_\alpha | j_1, \dots, j_\beta\}). \tag{4}$$

There are three points to the proof of the claim.

(i) $\{i_1,\ldots,i_\alpha|j_1,\ldots,j_\beta\} \in \partial\{0,i_1,\ldots,i_\alpha|j_1,\ldots,j_\beta\}$. To show that, we note that if $x \in \{i_1,\ldots,i_\alpha|j_1,\ldots,j_\beta\}$, then

$$I(x) \subseteq \{i_1,\ldots,i_\alpha,j_1,\ldots,j_\beta\} \subseteq \{0,i_1,\ldots,i_\alpha,j_1,\ldots,j_\beta\},$$

so $x \in \{0,i_1,\ldots,i_\alpha|j_1,\ldots,j_\beta\}$. Moreover, $\partial\{0,i_1,\ldots,i_\alpha|j_1,\ldots,j_\beta\}$ consists of points with $|I(x)| \le \alpha + \beta$.

(ii) $\mathrm{Int}(\{i_1,\ldots,i_\alpha|j_1,\ldots,j_\beta\}) \cap \mathcal{O}_\mathcal{S}^{k-1} = \emptyset$. Assume $x \in \mathrm{Int}(\{i_1,\ldots,i_\alpha|j_1,\ldots,j_\beta\})$. Then $0 \notin I(x)$ and $I(x) = \alpha + \beta = k$. However, if $x \in \mathcal{O}_\mathcal{S}^{k-1}$, then either $0 \in I(x)$ or $|I(x)| \le k-1$.

(iii) $\partial\{0,i_1,\ldots,i_\alpha|j_1,\ldots,j_\beta\} \setminus \mathrm{Int}(\{i_1,\ldots,i_\alpha|j_1,\ldots,j_\beta\}) \subseteq \mathcal{O}_\mathcal{S}^{k-1}$. This follows because if $x \in \partial\{0,i_1,\ldots,i_\alpha|j_1,\ldots,j_\beta\} \setminus \mathrm{Int}(\{i_1,\ldots,i_\alpha|j_1,\ldots,j_\beta\})$, then either $0 \in I(x)$ and $|I(x)| \le k$ so $x \in \mathcal{O}_\mathcal{S}^{k-1}$; or $0 \notin I(x)$ and $|I(x)| \le k-1$, so again $x \in \mathcal{O}_\mathcal{S}^{k-1}$.

What we have shown so far is that $\{0,i_1,\ldots,i_\alpha|j_1,\ldots,j_\beta\}$ is homeomorphic to a closed ball, and $\partial\{0,i_1,\ldots,i_\alpha|j_1,\ldots,j_\beta\} \cap \mathcal{O}_\mathcal{S}^{k-1}$ is homeomorphic to its boundary sphere $\mathbb{S}^{d-1}$ with an open connected set $\mathrm{Int}(\{i_1,\ldots,i_\alpha|j_1,\ldots,j_\beta\})$ of dimension $d-1$ cut out of it. Now we show there exists a retraction from $\mathbb{B}^d$ to the punctured sphere $\mathbb{S}^{d-1} \setminus \mathcal{P}$, where $\mathcal{P}$ is homeomorphic to an open ball of dimension $d-1$. Since $\mathcal{P}$ has dimension $d-1$, $\mathbb{S}^{d-1}$ is split into two connected parts: $\mathcal{P}$ and $\mathbb{S}^{d-1}\setminus\mathcal{P}$. The retraction argument is standard in topology. We provide the proof since it is integral to our results.

First, let us note that $\mathbb{B}^d$ is homeomorphic to the upper half-ball $\mathbb{B}^{+d} = \{x \in \mathbb{B}^d : x_1 \ge 0\}$. The precise homeomorphism is not difficult to find, but one can simply imagine taking the ball and flattening its lower half. Now, our sphere $\mathbb{S}^{d-1}$ was mapped by this to the boundary of $\mathbb{B}^{+d}$. Furthermore, without loss of generality,[1] we can assume that the closed part $\mathbb{S}^{d-1}\setminus\mathcal{P}$ makes up the bottom of the half-ball: $\{x \in \mathbb{B}^d : x_1 = 0\}$, while the open part $\{x \in \mathbb{B}^d : x_1 > 0\}$ corresponds to $\mathcal{P}$.

Let us define the function $r'_\mathcal{H} : \mathbb{B}^{+d} \to \{x \in \mathbb{B}^d : x_1 = 0\}$ by $r'_\mathcal{H}(x_1,x_2,\ldots,x_n) = (0,x_1,x_2,\ldots,x_n)$. Clearly, this is a valid retraction and thus, we have obtained a retraction from $\mathbb{B}^{+d}$ to $\{x \in \mathbb{B}^d : x_1 = 0\}$. Now, using the fact that $\mathbb{B}^{+d}$ is homeomorphic to $\mathbb{B}^d$, while the same homeomorphism takes $\{x \in \mathbb{B}^d : x_1 = 0\}$ to $\mathbb{S}^{d-1}\setminus\mathcal{P}$, we know there thus exists a retraction $r''_\mathcal{H} : \mathbb{B}^d \to \mathbb{S}^{d-1}\setminus\mathcal{P}$. Finally, reminding ourselves that there exists a homeomorphism between $\mathcal{H}$ and $\mathbb{B}^d$ which takes $\mathbb{S}^{d-1}\setminus\mathcal{P}$ to $\partial\mathcal{O}_\mathcal{S} \cap \mathcal{O}_\mathcal{S}^{k-1}$, by "pushing" $r''_\mathcal{H}$ through that homeomorphism, we obtain a retraction $r_\mathcal{H} : \{0,i_1,\ldots,i_\alpha|j_1,\ldots,j_\beta\} \to \partial\{0,i_1,\ldots,i_\alpha|j_1,\ldots,j_\beta\} \cap \mathcal{O}_\mathcal{S}^{k-1}$.

Now we glue these retractions to each other. In order to do that, we need to know that for $\mathcal{H}$'s with constant $\alpha + \beta = k$, all the different retractions $r_\mathcal{H} : \{0,i_1,\ldots,i_\alpha|j_1,\ldots,j_\beta\} \to \partial\{0,i_1,\ldots,i_\alpha|j_1,\ldots,j_\beta\} \cap \mathcal{O}_\mathcal{S}^{k-1}$ agree on the

[1]Really, this is done through another homeomorphism: this time, imagine, before flattening the ball, choosing the part that needs to be flattened to be $\mathbb{S}^{d-1}\setminus\mathcal{P}$.

intersections of their domains. That is, if $\mathcal{H} \neq \mathcal{H}'$, then $r_{\mathcal{H}}|_{\mathcal{H}\cap\mathcal{H}'} \equiv r_{\mathcal{H}'}|_{\mathcal{H}\cap\mathcal{H}'}$. (The claim is obvious if $\mathcal{H} = \mathcal{H}'$.) Let $\mathcal{H} = \{0, i_1, \dots, i_\alpha | j_1, \dots, j_\beta\}$. Analogously, let $\mathcal{H}' = \{0, i'_1, \dots, i'_{\alpha'} | j'_1, \dots, j'_{\beta'}\}$. We noted in Lemma 1 that $\mathcal{H} \cap \mathcal{H}' = \{0, \{i_1, \dots, i_\alpha\} \cap \{i'_1, \dots, i'_{\alpha'}\} | \{j_1, \dots, j_\beta\} \cap \{j'_1, \dots, j'_{\beta'}\}\}$. Since we assumed that $\mathcal{H} \neq \mathcal{H}'$, there needs to be an element in $\{i_1, \dots, i_\alpha, j_1, \dots, j_\beta\}$ which is not an element of the set $\{i'_1, \dots, i'_{\alpha'}, j'_1, \dots, j'_{\beta'}\}$ and vice versa (note that both of those sets have k elements, so one cannot be a subset of the other).

Thus, $\mathcal{H} \cap \mathcal{H}'$ will not contain more than $k-1$ non-zero vertices of $\mathcal{S}$ in its notation, and hence it will be in both $\mathcal{O}_{\mathcal{S}}^{k-1}$ (by the definition of $\mathcal{O}_{\mathcal{S}}^{k-1}$), and in $\partial\mathcal{H}$ (as none of its elements can be in the interior of $\mathcal{H}$: the expansion as a convex sum of every element in the interior needs to contain every vertex mentioned in the notation of $\mathcal{H}$). Analogously, $\mathcal{H} \cap \mathcal{H}' \in \partial\mathcal{H}' \cap \mathcal{O}_{\mathcal{S}}^{k-1}$, which is the image of the retraction $r_{\mathcal{H}'}$.

Hence, we know that $r_{\mathcal{H}'}|_{\mathcal{H}\cap\mathcal{H}'}$ is an identity map, and so is $r_{\mathcal{H}}|_{\mathcal{H}\cap\mathcal{H}'}$. Thus, these two retractions can indeed be glued together. By iterating this procedure for all $\mathcal{H}$, we obtain a glued retraction $r^k : \mathcal{O}_{\mathcal{S}}^k \to \mathcal{O}_{\mathcal{S}}^{k-1}$ which takes each k-dimensional edge in $\mathcal{O}_{\mathcal{S}}^k$ to its boundary. Let us note what this retraction does. For every point $x \in \mathcal{O}_{\mathcal{S}}^k$, if x is also in $\mathcal{O}_{\mathcal{S}}^{k-1}$, it will not do anything. Thus, $\mathcal{C}(r^k(x)) = \mathcal{C}(x)$.

If $x \notin \mathcal{O}_{\mathcal{S}}^{k-1}$, then either x is in the interior of some $\mathcal{H} = \{0, i_1, \dots, i_\alpha | j_1, \dots, j_\beta\}$ that is being added to $\mathcal{O}_{\mathcal{S}}^k$, or it is in the interior of $\{i_1, \dots, i_\alpha | j_1, \dots, j_\beta\}$. Now, if x is in the interior of $\{0, i_1, \dots, i_\alpha | j_1, \dots, j_\beta\}$, r^k maps it to a point in the boundary of $\mathcal{H}$. In that case, it is easy to verify that $\mathcal{C}(r^k(x)) \subseteq \mathcal{C}(x)$. This has formally been done in Lemma 6 of [11].

If x is in the interior of $\{i_1, \dots, i_\alpha | j_1, \dots, j_\beta\}$, then $I(x) = \{i_1, \dots, i_\alpha, j_1, \dots, j_\beta\}$. On the other hand, $\mathcal{H} = \{0, i_1, \dots, i_\alpha | j_1, \dots, j_\beta\}$, so for any point $y \in \mathcal{H}$, $I(y) \subseteq \{i_1, \dots, i_\alpha, j_1, \dots, j_\beta\}$. Thus, $\mathcal{C}(y) \subseteq \mathcal{C}(x)$. Thus, specifically $\mathcal{C}(r^k(x)) \subseteq \mathcal{C}(x)$.

In all three cases, we deduce that $\mathcal{C}(r^k(x)) \subseteq \mathcal{C}(x)$. By composing $r = r^2 \circ r^3 \circ \dots \circ r^n$, we obtain a retraction $r : \mathcal{O}_{\mathcal{S}} = \mathcal{O}_{\mathcal{S}}^n \to \mathcal{O}_{\mathcal{S}}^1 = \mathcal{O}'_{\mathcal{S}}$. Define $f(x) = f'(r(x))$. We obtained a nowhere vanishing function f on $\mathcal{O}_{\mathcal{S}}$ such that $f(x) = f'(r^2(r^3(\dots(r^n(x))\dots)))$. Thus, $f(x)$ is contained in $\mathcal{C}(r^2(r^3(\dots(r^n(x))\dots))) \subseteq \mathcal{C}(r^2(r^3(\dots(r^{n-1}(x))\dots))) \subseteq \dots \subseteq \mathcal{C}(x)$. This function satisfies the conditions of Problem 1. □

We now proceed to resolving Problem 1 for the case of $n = 2$ and $n = 3$. We have previously assumed that $\dim \mathcal{O}_{\mathcal{S}} \neq 0$ and $\dim \mathcal{O}_{\mathcal{S}} \neq n$, as these cases are simple to analyze. The case of $n = 2$ is thus reduced to $\dim \mathcal{O}_{\mathcal{S}} = 1$. As we have also required $1 \leq \dim \mathcal{B} < n$, we conclude that $\dim \mathcal{B} = 1$. However, the case of $\dim \mathcal{B} = 1$ is resolved by Theorem 1 in [12].

This resolves the case of $n = 2$, as well as $\dim \mathcal{B} = 1$. The only cases that remain are when $n = 3$, $\dim \mathcal{B} = 2$, and $\dim \mathcal{O}_{\mathcal{S}}$ is either 1 or 2. We will see that, when $\dim \mathcal{O}_{\mathcal{S}} = 1$, an argument based on linear algebra applies. On the other hand, a purely topological argument applies when $\dim \mathcal{O}_{\mathcal{S}} = 2$.

First we examine why a sufficiently high dimension for $\mathcal{B}$ resolves Problem 1.

Lemma 2 *Suppose $\mathcal{O}_{\mathcal{S}} = \text{co}\{o_1, \ldots, o_{\kappa+1}\}$ where the o_i's are the vertices of $\mathcal{O}_{\mathcal{S}}$. If there exists a linearly independent set $\{\mathbf{b_i} \in \mathcal{B} \cap \mathcal{C}(o_i) \mid i = 1, \ldots, \kappa + 1\}$, then the answer to Problem 1 is affirmative.*

Proof Let $f : \mathcal{O}_{\mathcal{S}} \to \mathcal{B}$ be defined by $f(\sum_{i=1}^{\kappa+1} \alpha_i o_i) = \alpha_i \mathbf{b_i}$, where $\sum \alpha_i = 1$ and $\alpha_i \geq 0$. Necessarily $f(x) \neq \mathbf{0}$ for $x \in \mathcal{O}_{\mathcal{S}}$ for otherwise the $\mathbf{b_i}$'s would be linearly dependent. Also, by a standard convexity argument $f(x) \in \mathcal{C}(x)$, $x \in \mathcal{O}_{\mathcal{S}}$. □

The following is the key result in the case of $\dim \mathcal{O}_{\mathcal{S}} = 1$.

Lemma 3 *Let $n = 3$, $\dim \mathcal{B} = 2$, and let o_1 and o_2 be vertices of $\mathcal{O}_{\mathcal{S}}$. Then there exist linearly independent vectors $\{\mathbf{b_1}, \mathbf{b_2} \mid \mathbf{b_i} \in \mathcal{B} \cap \mathcal{C}(o_i)\}$. Moreover, if $\mathcal{O}_{\mathcal{S}} = \text{co}\{o_1, o_2\}$, the answer to Problem 1 is affirmative.*

Proof First we assume $o_1 \in \text{Int}(\mathcal{F}_i)$ for some $i \in \{0, 1, 2, 3\}$. By the definition of $\mathcal{C}(o_1)$, it is a closed half space or $\mathbb{R}^3$, so there exist linearly independent vectors $\mathbf{b_{11}}, \mathbf{b_{12}} \in \mathcal{B} \cap \mathcal{C}(o_1)$. We claim $\mathcal{B} \cap \mathcal{C}(o_2) \neq \mathbf{0}$. If $o_2 \in \text{Int}(\mathcal{F}_i)$ for some $i \in \{0, 1, 2, 3\}$ then the argument above proves the claim. Instead, assume w.l.o.g. that $o_2 \in \mathcal{F}_1 \cap \mathcal{F}_2$. Then $\mathcal{C}(o_2) = \{\mathbf{y} \in \mathbb{R}^3 | \mathbf{h_1} \cdot \mathbf{y} \leq 0, \mathbf{h_2} \cdot \mathbf{y} \leq 0\}$. Let $\mathcal{B} = \text{Ker}(M^T)$ for some $M \in \mathbb{R}^{3\times1}$. Finding $\mathbf{0} \neq \mathbf{y} \in \mathcal{B} \cap \mathcal{C}(o_2)$ is equivalent to solving

$$\begin{bmatrix} \mathbf{h_1}^T \\ \mathbf{h_2}^T \\ M^T \end{bmatrix} \mathbf{y} = \begin{bmatrix} s_1 \\ s_2 \\ 0 \end{bmatrix} \tag{5}$$

for some $s_1, s_2 \in \mathbb{R}_0^-$ and $\mathbf{y} \neq \mathbf{0}$. Because $\{\mathbf{h_1}, \mathbf{h_2}\}$ are linearly independent, $\text{rank}(H) \geq 2$, where H is the matrix appearing on the left hand side of equation 5. If $\text{rank}(H) = 3$, then let

$$[\mathbf{y_1}\ \ \mathbf{y_2}] = H^{-1} \begin{bmatrix} -1 & 0 \\ 0 & -1 \\ 0 & 0 \end{bmatrix}.$$

Since $(-1, 0, 0)$ and $(0, -1, 0)$ are linearly independent, $\mathbf{y_1}$ and $\mathbf{y_2}$ are linearly independent as well.

Next, assume $\text{rank}(H) = 2$. In other words, $M = c_1 \mathbf{h_1} + c_2 \mathbf{h_2}$ for some $c_1, c_2 \in \mathbb{R}$. Then, by taking $s_1 = s_2 = 0$, equation (5) reduces to $[\mathbf{h_1}\ \mathbf{h_2}]^T y = 0$. By the rank-nullity theorem, there exists $\mathbf{y} \neq \mathbf{0}$ satisfying this equation. Moreover, if w.l.o.g. $v_0 = 0$, then $\mathbf{y} \in \mathcal{F}_1 \cap \mathcal{F}_2 = \text{co}\{v_0, v_3\}$, and we can take $\mathbf{y} = v_3$. We have shown there exist linearly independent $\mathbf{b_{11}}, \mathbf{b_{12}} \in \mathcal{C}(o_1)$ and there exists $0 \neq \mathbf{b_2} \in \mathcal{C}(o_2)$. We claim at least one of the pairs $\{\mathbf{b_{11}}, \mathbf{b_2}\}$ and $\{\mathbf{b_{12}}, \mathbf{b_2}\}$ is linearly independent. For otherwise there exist $c_1, c_2 \in \mathbb{R}$ such that $\mathbf{b_2} = c_1 \mathbf{b_{11}} = c_2 \mathbf{b_{12}}$, implying $\mathbf{b_{11}}$ and $\mathbf{b_{12}}$ are linearly independent, a contradiction. We conclude there exists a linearly independent set $\{\mathbf{b_1}, \mathbf{b_2} \mid \mathbf{b_i} \in \mathcal{C}(o_i)\}$.

Next we assume neither o_1 nor o_2 lies in the interior of a facet. W.l.o.g. suppose $o_1 \in \mathcal{F}_1 \cap \mathcal{F}_2$ and $o_2 \in \mathcal{F}_1 \cap \mathcal{F}_3$. If either $\mathcal{C}(o_1)$ or $\mathcal{C}(o_2)$ contains two linearly

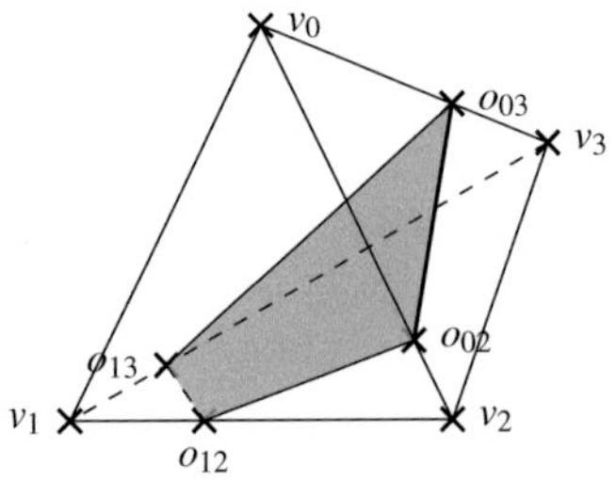

Fig. 1 The set $\mathcal{O}_\mathcal{S}$ for Example 1. The edge from o_{02} to o_{03} forms $\mathcal{O}'_\mathcal{S}$

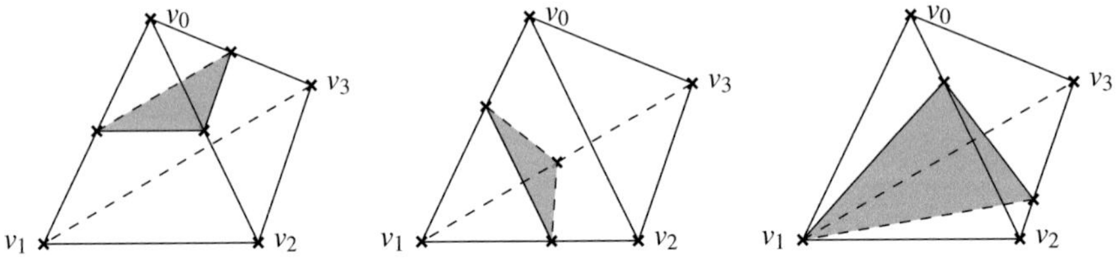

Fig. 2 Three of the four possible configurations of set $\mathcal{O}_\mathcal{S}$ for $n = 3$ and $\dim \mathcal{O}_\mathcal{S} = 2$, with the fourth one given in Fig. 1. The leftmost configuration is addressed by Theorem 2, while the other two can be reduced using Theorem 1

independent vectors, then by the previous argument, we are done. Otherwise, by the previous argument again $v_3 \in \mathcal{C}(o_1)$ and $v_2 \in \mathcal{C}(o_2)$. Since $\{v_2, v_3\}$ are linearly independent, we are done. Finally, if $\mathcal{O}_\mathcal{S} = \text{co}\{o_1, o_2\}$, then by Lemma 2 the answer to Problem 1 is affirmative. □

The remaining case to study is when $n = 3$, $\dim \mathcal{O}_\mathcal{S} = 2$, and $\dim \mathcal{B} = 2$. Assuming $v_0 \notin \mathcal{O}_\mathcal{S}$ (which is a trivial case discussed in Lemma 10 of [11]), there are four topologically distinct cases for $\mathcal{O}_\mathcal{S}$, depending on the way $\mathcal{O}$ cuts $\mathcal{S}$. These are given in Figs. 1 and 2. In Fig. 1, $\mathcal{O}_\mathcal{S}$ is a quadrangle. In that case, $p = 1$; that is, there are two vertices of $\mathcal{S}$ on each side of $\mathcal{O}$. Then we can apply Theorem 1 to reduce $\mathcal{O}_\mathcal{S}$ to $\mathcal{O}'_\mathcal{S}$, and according to the construction in the proof, $\mathcal{O}'_\mathcal{S}$ has dimension 1, and we can apply Lemma 3. Similarly, in the cases given in middle and the rightmost configuration of Fig. 2, we can apply Theorem 1 to reduce $\mathcal{O}_\mathcal{S}$ to $\mathcal{O}'_\mathcal{S}$ with $\dim \mathcal{O}'_\mathcal{S}$ being either 0 or 1, respectively. Finally, in the situation given by the leftmost configuration of Fig. 2, we draw upon a proof method already utilized in [4], which is based on Sperner's lemma. Here we employ a variant found in [13].

Lemma 4 *Let $\mathcal{P} = \text{co}\{w_1, \ldots, w_{n+1}\}$ be an n-dimensional simplex. Furthermore, let $\{\mathcal{Q}_1, \ldots, \mathcal{Q}_{n+1}\}$ be a collection of sets covering $\mathcal{P}$ such that*

(P1) Vertex $w_i \in \mathcal{Q}_i$ and $w_i \notin \mathcal{Q}_j$ for $j \neq i$.
(P2) If w.l.o.g. $x \in \text{co}\{w_1, \ldots, w_l\}$ for some $1 \leq l \leq n+1$, then $x \in \mathcal{Q}_1 \cup \cdots \cup \mathcal{Q}_l$.

Then $\bigcap_{1=1}^{n+1} \overline{\mathcal{Q}}_i \neq \emptyset$.

Theorem 2 *Let $n = 3$ and suppose $\mathcal{O}_\mathcal{S} = \text{co}\{o_1, o_2, o_3\}$ with $v_0 \notin \mathcal{O}_\mathcal{S}$ and $o_i \in (v_0, v_i]$, $i = 1, 2, 3$. The answer to Problem 1 is affirmative if and only if*

$$\mathcal{B} \cap \text{cone}(\mathcal{O}_\mathcal{S}) \neq \mathbf{0}.$$

Proof Sufficiency is clear: if $\mathbf{0} \neq \mathbf{b} \in \mathcal{B} \cap \text{cone}(\mathcal{O}_\mathcal{S})$, a constant function $f(x) = \mathbf{b}$ satisfies Problem 1. For necessity, suppose there exists $f : \mathcal{O}_\mathcal{S} \to \mathcal{B} \setminus \{\mathbf{0}\}$ such that $f(x) \in \mathcal{C}(x)$, $x \in \mathcal{O}_\mathcal{S}$. By way of contradiction suppose $\mathcal{B} \cap \text{cone}(\mathcal{O}_\mathcal{S}) = \mathbf{0}$. Since $o_i \in (v_0, v_i]$, $i = 1, 2, 3$, we have

$$\text{cone}(\mathcal{O}_\mathcal{S}) = \{\mathbf{y} \in \mathbb{R}^n \mid \mathbf{h_j} \cdot \mathbf{y} \leq 0\,, j = 1, 2, 3\}\,.$$

Define the sets

$$\mathcal{Q}_i := \{x \in \mathcal{O}_\mathcal{S} \mid \mathbf{h_i} \cdot f(x) > 0\}\,, \quad i = 1, 2, 3\,. \tag{6}$$

Now we verify the conditions of Lemma 4.

Firstly, we claim that $\{\mathcal{Q}_i\}$ cover $\mathcal{O}_\mathcal{S}$. For suppose not. Then there exists $x \in \mathcal{O}_\mathcal{S}$ such that $h_j \cdot f(x) \leq 0$, $j = 1, 2, 3$. Hence $f(x) \in \mathcal{B} \cap \text{cone}(\mathcal{O}_\mathcal{S})$, so $f(x) = \mathbf{0}$, a contradiction to f being non-vanishing on $\mathcal{O}_\mathcal{S}$. Secondly, we verify property (P1). We claim that $o_i \in \mathcal{Q}_i$ for $i = 1, 2, 3$. For suppose not. Then $h_i \cdot f(x) \leq 0$. Additionally, because $f(o_i) \in \mathcal{C}(o_i)$, $h_j \cdot f(x) \leq 0$, $j \in \{1, 2, 3\} \setminus \{i\}$. We conclude $f(o_i) \in \mathcal{B} \cap \text{cone}(\mathcal{O}_\mathcal{S})$, so $f(o_i) = \mathbf{0}$, a contradiction. Next we claim $o_i \notin \mathcal{Q}_j$, $j \neq i$. This is immediate since $f(o_i) \in \mathcal{C}(o_i)$ implies $h_j \cdot f(o_i) \leq 0$, $j \neq i$. Thirdly, we verify property (P2). Suppose w.l.o.g. (by reordering the indices $\{1, 2, 3\}$) $x \in \text{co}\{o_1, \ldots, o_r\}$ for some $1 \leq r \leq 3$. We claim $x \in \mathcal{Q}_1 \cup \cdots \cup \mathcal{Q}_r$. For suppose not. Then $h_j \cdot f(x) \leq 0$, $j = 1, \ldots, r$. Also, it is easily verified that $\mathcal{C}(x) = \{y \in \mathbb{R}^n \mid h_j \cdot y \leq 0\,, j = r + 1, \ldots, 3\}$. Thus, $h_j \cdot f(x) \leq 0$, $j = r + 1, \ldots, 3$. Hence, $f(x) \in \mathcal{B} \cap \text{cone}(\mathcal{O}_\mathcal{S})$, so $f(x) = \mathbf{0}$, a contradiction to f being non-vanishing on $\mathcal{O}_\mathcal{S}$.

We have verified (P1)-(P2) of Lemma 4. Applying the lemma, there exists $\overline{x} \in \bigcap_{i=1}^{3} \overline{\mathcal{Q}_i}$; that is, $h_j \cdot f(\overline{x}) \geq 0$, $j = 1, 2, 3$. We conclude that $-f(\overline{x}) \in \mathcal{B} \cap \text{cone}(\mathcal{O}_\mathcal{S})$, so $f(\overline{x}) = \mathbf{0}$, a contradiction. □

The following result finally resolves Problem 1 in all cases of interest.

Theorem 3 *Let $\mathcal{S}$, $\mathcal{B}$ and $\mathcal{O}_\mathcal{S}$ be as above, and let $n \in \{2, 3\}$. If $n = 3$, $\dim \mathcal{B} = 2$ and $\mathcal{O}_\mathcal{S}$ does not satisfy the conditions of Theorem 2, then the answer to Problem 1 is affirmative. Otherwise, the answer to Problem 1 is affirmative if and only if $\mathcal{B} \cap \text{cone}(\mathcal{O}_\mathcal{S}) \neq \mathbf{0}$.*

Proof The discussion prior to Lemma 2, as well as Lemma 3 and Theorem 2, covered all the cases except for the one where $\dim \mathcal{B} = \dim \mathcal{O}_\mathcal{S} = 2$ and $\mathcal{O}_\mathcal{S}$ does not satisfy the conditions of Theorem 2. However, in that case, as described prior to Lemma 4, Theorem 1 reduces $\mathcal{O}_\mathcal{S}$ to $\mathcal{O}'_\mathcal{S}$ of dimension 0 or 1. Applying Lemma 3, we now obtain that the answer to Problem 1 is affirmative. □

An interesting modification of Problem 1 requires that f be not only continuous, but also affine. This corresponds to a classic problem of designing an affine state feedback, applied to RCP. We note that in most of the configurations considered above, the same claims that work for Problem 1 can also be used in the affine case. The only significantly different case is $\dim \mathcal{O}_S = \dim \mathcal{B} = 2$, i.e., the situation covered by Theorem 1. In this case, the full analysis of the problem of affine obstruction can be done using linear algebra. Such an analysis is not difficult, but is computationally long. It is presented in full in Section 3.2.2 of [11]. With the results contained in [11] (in fact, Theorem 16 in [11] is essentially the same as our Theorem 3), we reach the following theorem:

Theorem 4 *Let* $n \in \{2, 3\}$. *The problem of affine obstruction is solvable if and only if Problem 1 is solvable.*

It is not known if such a result holds in general. Hence, we end this paper with the following conjecture.

Conjecture 1 *Let* $n \geq 4$. *The problem of affine obstruction is solvable if and only if Problem 1 is solvable.*

3 Conclusion

This paper introduces a topological obstruction to solving the RCP via continuous state feedback. The results show an interplay between linear algebra-based arguments regarding the number of control inputs and purely topological arguments regarding a cone condition on $\mathcal{B}$. We show that for $n = 2$ and $n = 3$ these two properties together fully characterize when a topological obstruction arises.

References

1. Bemporad, A., Morari, M., Dua, V., Pistikopulos, E.N.: The explicit linear quadratic regulator for constrained systems. Automatica **38**(1), 3–20 (2002)
2. Bredon, G.E.: Topology and Geometry. Springer, New York (1997)
3. Brockett, R.W.: Asymptotic stability and feedback stabilization. In: Brockett, R.W., Millman, R.S., Sussmann, H.J. (eds.) Differential Geometric Control Theory, pp. 181–191. Birkhauser, Boston (1983)
4. Broucke, M.E.: Reach control on simplices by continuous state feedback. SIAM J. Control Optim. **48**(5), 3482–3500 (2010)
5. Broucke, M.E., Ganness, M.: Reach control on simplices by piecewise affine feedback. SIAM J. Control and Optim. **52**(5), 3261–3286 (2014)
6. Habets, L.C.G.J.M., van Schuppen, J.H.:A control problem for affine dynamical systems on a full-dimensional polytope. Automatica **40**(1), 21–35 (2004)
7. Kettler, P.C.: Vertex and Path Attributes of the Generalized Hypercube with Simplex Cuts. University of Oslo, Oslo (2008)

8. Mehtaa, K.: A topological obstruction in a control problem. Master's thesis, University of Toronto (2012)
9. Min, C.: Simplicial isosurfacing in arbitrary dimension and codimension. J. Comput. Phys. **190**(1), 295–310 (2003)
10. Ornik, M., Broucke, M.E.: A topological obstruction to reach control by continuous state feedback. In: This work has now been published in 54th IEEE Conference on Decision and Control (CDC), 2015, pp. 2258-2263, doi: 10.1109/CDC.2015.7402543. For full citation, see here: http://ieeexplore.ieee.org/xpls/abs_all.jsp?arnumber=7402543&tag=1
11. Ornik, M., Broucke, M.E.: Some results on an affine obstruction to reach control. arXiv:1507.02957 [math.OC] (2015)
12. Semsar Kazerooni, E., Broucke, M.E.: Reach controllability of single input affine systems. IEEE Trans. Autom. Control **59**(3), 738–744 (2014)
13. Talman, D.: Intersection theorems on the unit simplex and the simplotope. In: Imperfections and Behavior in Economic Organizations, pp. 257–278. Springer, New York (1994)

Continuous Approaches to the Unconstrained Binary Quadratic Problems

Oksana Pichugina and Sergey Yakovlev

Abstract Two continuous approaches to discrete problems over sets inscribed into a sphere are presented. They use an analytic description of the sets and a convex extension of objective functions from the sets onto Euclidean space. The first approach is the Branch and Bound Polyhedral-Spherical Method (B&BPSM) for "two-layer sets", which utilizes the sets representation as an intersection of a polyhedron with a sphere and the global minimizers on a sphere. The concept of a functional representation of a discrete set as an intersection of surfaces is introduced and used, for constructing penalty functions, in the second approach – the Penalty Method based on Functional Representations (FRPM). The methods are applied to the Unconstrained Binary Quadratic Problem. For the purpose, the representation of the binary set as a "touching set" of smooth surfaces is derived.

1 Introduction

The Unconstrained Binary Quadratic Problem (UBQP):

$$z = f(x) \rightarrow \min, \tag{1}$$

$$f(x) = x^T A x + c^T x, \tag{2}$$

$$x \in B_n = \{0^n, 1^n\} \tag{3}$$

is a famous NP-hard problem known by numerous applications and a variety of exact and heuristic methods (the solution methods alongside with different applications on graphs, facility and resources allocation problems, clustering and partitioning problems, variations of assignment problems, etc. see in details in [4]). The methods are roughly divisible into combinatorial and continuous. The division

O. Pichugina (✉)
Department of Mathematics & Statistics, Brock University, St. Catharines, ON, Canada
e-mail: op13ci@brocku.ca

S. Yakovlev
Department of IT & Protection of Information, National University of Internal Affairs, Kharkov, Ukraine
e-mail: svsyak@mail.ru

© Springer International Publishing Switzerland 2016

J. Bélair et al. (eds.), *Mathematical and Computational Approaches in Advancing Modern Science and Engineering*, DOI 10.1007/978-3-319-30379-6_62

is not very precise since the continuous techniques are often used for solution of subproblems in Branch & Bound, Branch & Cut and other combinatorial algorithms. In [3] connection between continuous and combinatorial approaches to discrete problems are discussed, and an overview of the continuous ones is given. The last approaches are typically classified into continuous relaxation and reformulation based. Among the relaxation techniques there are polyhedral and semi-definite relaxations. For instance, the semi-definite programming is applicable to the UBQP after reformulation of (1), (2) and (3) as a linear problem in $R^{n \times n}$ over the cut polytope. Since the analytic description of the polytope is unknown, its semi-definite relaxation is used [2] and efforts of researchers are focused on strengthening this basic relaxation [5].

The UBQP equivalent reformulations depend on ways how the discrete constraint (3) is rewritable as continuous one, e.g.:

- Way 1 [3]: $x_i + y_i = 1,\ x_i \cdot y_i = 0,\ x_i, y_i \geq 0,\ i \in J_n = \{1, \ldots, n\}$, transform the UBQP into a constrained quadratic problem in R^{2n};
- Way 2 [3, 4, 6]: replacement of (3) by (4) allows to consider the unconstrained problem (5) instead of the UBQP.

The next two approaches utilizing (4) reduce the UBQP to concave or convex problems and are based on the following property: if $f(x)$ satisfies (7), then $\exists \Lambda_{min}, \Lambda_{max}, \Lambda_{min} \leq \Lambda_{max}$ such that $\Phi^{1,2}(x, \lambda)$ are concave $\forall \lambda \geq \Lambda_{max}$ and convex $\forall \lambda \leq \Lambda_{min}$.

$$x_i(1 - x_i) = 0,\ i \in J_n. \tag{4}$$

$$\Phi^l(x, \lambda) = f(x) + \lambda \sum_i (x_i(1 - x_i))^l \to \min. \tag{5}$$

$$\mathbf{0} \leq x \leq \mathbf{1}, \tag{6}$$

$$f(x) \in C^2(R^n),\ \partial^2 f(x)/(\partial x_i \partial x_j) < \infty,\ i, j. \tag{7}$$

For example, concave methods [3] are applied to (1), (2) and (3) equivalent reformulation (2), (5), (6) ($l = 1$) where λ is chosen to provide concavity of (5). The smoothing methods [6] utilize (5), (6) ($l = 2$) in a penalty function; use the boxing constraints (6) in forming barriers and providing convexity of the function for a range of λ.

We continue study of the continuous approaches to binary optimization based on nonlinear reformulations at the same space. For the purpose, we introduce the concepts of an extension of a function from sets onto larger sets (see Sect. 2) and a functional representation of discrete sets (see Sect. 2.3). The presented in Sects. 2.2 and 2.3 methods operate with sets inscribed into a sphere and convex extensions from the sets. Ways of constructing the extensions, in particular for (2), are described in Sect. 2.1. The Branch and Bound Polyhedral-Spherical Method (B&BPSM) for “two-layer sets”, combining branch and bound techniques with polyhedral and spherical relaxations, is described and specified for the UBQP in Sect. 2.2. The Penalty Method based on Functional Representations (FRPM), presented in Sect. 2.3, is applicable to any sets representable as intersection of surfaces. In particular, for the binary set (3) a new representation as a touching set of a sphere and an order four surface is given and recommended to use in the FRPM.

2 Main Part

In the section we consider the following generalization of the UBQP – the problem (1) over a finite discrete set inscribed into a hypersphere:

$$x \in E \subset R^n, \tag{8}$$

$$|E| = N < \infty, \tag{9}$$

$$E \subset S_r(a), \tag{10}$$

$$S_r(a) = \{x \in R^n : (x-a)^2 = r^2\}. \tag{11}$$

There is a number of combinatorial sets inscribed into a sphere, such as the binary set B_n [6, 12], the permutation and polypermutation sets, some classes of the partial permutations and combinations [8]. So, the class (1), (8), (9), and (10) includes all unconstrained problems over these sets, in particular, the UBQP.

The set (9), (10) is vertex located, i.e. E coincides with the vertex set of the corresponding polyhedron:

$$E = vert\, P \text{ where } P = conv\, E \text{ is the polyhedron.} \tag{12}$$

Definition 1 ([11]) Function $F(x)$ is an *extension* of $f(x)$ from M onto $M' \supset M$ if it is defined over M' and coincides with $f(x)$ over M: $F(x) = f(x),\ x \in M$.

Depending on a type of the function $F(x)$ the $f(x)$-extension can be continuous, differentiable, smooth, convex or concave, quadratic, etc.

Theorem 1 ([11]) *If $f(x)$ is defined on a vertex located set E, then there exists $F(x)$, which is a convex differentiable extension (CDE) of $f(x)$ from E onto R^n.*

By (12), Theorem 1 is applicable to the sets (9)–(10). Therefore, we can assume that, instead of the original function (1), its CDE from E onto R^n is considered.

A CDE can be constructed in different way, e.g., by using the next result:

Theorem 2 ([10]) *If $f(x)$ satisfies* (7) *then*

$$\exists K :\ \forall k > K\ F(x) = f(x) + k\left((x-a)^2 - r^2\right) \tag{13}$$

is a CDE of $f(x)$ from $S_r(a)$ onto R^n.

Remark 1 In addition, in [10] the lower estimated bound of K is given.

This property of reducing the original problem (1), (8), (9), and (10) to optimization of a convex function, $F(x) \underset{E}{\rightarrow} \min$, seems useful but the problem is still discrete. The convexity can be really helpful if continuous optimization is applicable to it.

The methods presented in Sects. 2.2 and 2.3 are especially effective for the functions and the combinatorial sets with the known solutions of the following subproblems:

1. an analytic description of the polyhedron P: $P = \{x : Ax \leq b\}$;
2. a linear problem over E: $x^{LP} = \underset{x \in E}{argmin}\, c^T x$;
3. the global solution of (1) on a sphere:

$$z^S = \min_{x \in S_r(a)} f(x),\ x^S = \underset{x \in S_r(a)}{argmin} f(x); \tag{14}$$

4. the minimal circumscribed sphere around E, S^{min}.

Alongside with (10), these four cases can simplify the solution of the next problems considered typically complicated:

- *minimization* of $f(x)$ *over a polyhedron*:

$$z^P = \min_{x \in P} f(x),\ x^P = \underset{x \in P}{argmin} f(x). \tag{15}$$

  Knowledge of the exact solution of the subproblem 2 allows to utilize the modification of conditional gradient methods such as the Frank-Wolfe algorithm, where linear direction-finding subproblems are solved explicitly [12];
- *search of a projection* of an arbitrary point $x \in R^n$ onto a discrete set E, $y = Pr_E x$. This problem for sets satisfying (10) is reduced to the linear subproblem 2.

Note that for the UBQP the optimization approaches are applicable effectively. Indeed, the explicit solution of the subproblems: #1 is the unit hypercube (6); #2 – is $x_i^{LP} = 0$, if $c_i \geq 0$, otherwise 0 $(i \in J_n)$; #3 – is presented in [1]; #4 – is:

$$S^{min} = S_{\frac{\sqrt{n}}{2}}(\mathbf{0.5})\,. \tag{16}$$

2.1 Convex Extensions of Quadratic Functions

We present two ways of constructing convex extensions of (2) from B_n onto R^n:

Way 1: Notice that $\Phi^1(x, \lambda)$ in (5) is rewritable as

$$\Phi^1(x, \lambda) = f(x) - \lambda\left((x - \mathbf{0.5})^2 - n/4\right). \tag{17}$$

By (16), (17) is an extension of any $f(x)$ from S^{min} and $\forall E \subseteq S^{min}$ onto R^n.

Using Theorem 2 and taking $\lambda = -k$ in (13), we obtain that (17) is the CDE of (7) $\forall \lambda < \Lambda_{min} = -K$, where K can be estimated according to Remark 1.

In particular, (17) is a CDE of (2) from B_n onto R^n if $\lambda < \Lambda_{min}$, where the bound Λ_{min} can be found exactly: Λ_{min} is the minimal eigenvalue of the matrix A.

Way 2: The convex extensions can be formed specifically for a discrete set. For instance, a convex extension of (2) from B_n onto R^n can be constructed as follows:

- the function (2) is represented as a sum of a convex linear form $f_1(x)$ and a quadratic form $f_2(x)$:

$$f_1(x) = \sum_{i=1}^{n} c_i x_i,\ f_2(x) = \sum_{i=1}^{n} \sum_{i \le j}^{n} a_{ij} x_i x_j = \sum_{i=1}^{n} \sum_{i \le j}^{n} |a_{ij}| \cdot (\pm x_i x_j);$$

- the convex extensions are formed for each nonconvex term $-|a_{ii}| \cdot x_i^2$ and $\pm|a_{ij}| \cdot x_i x_j$ using the property (4): $\forall i \in J_n,\ x_i^2 \underset{B_n}{=} x_i$. For the both types of nonconvex terms the convex extensions are formed as follows:

  (a) if $i = j$ and $a_{ii} < 0$, then the CDE of the component $f_{ii} = -x_i^2$ is $F_{ii} = -x_i$;
  (b) if $i \neq j$ and $a_{ij} \neq 0$, then the CDE F_{ij} of the component $f_{ij} = \pm x_i x_j$ is:
  $f_{ij} = \pm x_i x_j = \frac{1}{2}\left(\left(x_i \pm x_j\right)^2 - x_i^2 - x_j^2\right) \underset{B_n}{=} \frac{1}{2}\left(\left(x_i \pm x_j\right)^2 - x_i - x_j\right) = F_{ij}$;

- all these convex extensions are incorporated to the final CDE $F(x)$ of (2):

$$F(x) = f(x) + D\sum_{i=1}^{n} x_i^2 - D\sum_{i=1}^{n} x_i, \text{ where } D = \sum_{a_{ii}<0} |a_{ii}| + \sum_{a_{ij}\neq 0, i<j} |a_{ij}|.$$

2.2 *The Polyhedral Spherical Method*

The discrete problem (1), (8)–(10) is equivalent to the $f(x)$-optimization problem (1) over intersection of a sphere (11) and a polyhedron (12), $f(x) \underset{P\cap S_r(a)}{\rightarrow} \min$. It yields two types of continuous relaxations: **Relaxation 1** – optimization over a polyhedron, which is convex problem (15); **Relaxation 2** – optimization over a sphere, which is nonconvex problem (14).

The Polyhedral-Spherical Method (PSM) is the cutting-plane method where cuttings of infeasible parts of a hypersphere are conducted by polyhedron facets.

In [9, 12] the approximating version of this method was presented and applied to quadratic optimization over the permutation set and the binary set, respectively. This version essentially uses: (a) the convex extensions of a quadratic function; (b) simplicity of search of all stationary points of a convex quadratic function on a sphere; (c) easiness of solving (15) over the hypercube (6). The search of

$x^{**} = \underset{E}{argmin} f(x)$ is restricted to facets cutting the stationary points. Then the problem is reduced to search of these points and solving the subproblems of type (14), (15) on these facets, etc. As a result, a series of optimization problems of different dimensions with convex objective functions over sets inscribed into spheres is solved.

In [7] the method is generalized for any sets inscribed into a sphere and is named the Polyhedral-Spherical Method.

In [12] the exact version for (1), (2), and (3) is also presented, which utilizes the decomposition of B_n into two subsets located on parallel facets of (6).

In this paper, we modify the exact method and generalize it, in the Branch&Bound PSM (B&BPSM), to any "two-layer" combinatorial set, i.e. it allows the decomposition: $\forall a, b : ax = b$ is a P-facet $dim\, conv(\{x \in E : ax \neq b\}) < dim\, P$.

The B&BPSM Algorithm for two-layer sets

Step 1. The unconstrained convex problem (1) is solved, $z^* = \underset{R^n}{min} f(x) = f(x^*)$, and x^* is projected onto E, $y^* = Pr_E x^*$. The initial upper z^u and lower z^l estimated bounds are: $z^l = z^*$, $z^u = z(y^*) = z^*_y$.

Step 2. Depending on location of x^* we choose: **Scheme 1** – solving the Relaxation 2 problems at each step – if $x^* \in P$ ($x^P = x^*$); **Scheme 2** – solution of Relaxation 1–2 problems at each iteration – if $x^* \notin P$.

Step 3. Initialization $H = \{\emptyset\}$, $B^H = B^{\{\emptyset\}}$ is the root of the search tree. The candidate solution set $\mathbf{B} = B^{\{\emptyset\}}$.

Step 4. Take the branch $B^H \in \mathbf{B}$ with the least lower bound and $|H| < n$. If the choice is not unique, then among $B^H \in \{B^h, B^{h'}\}$ the branch B^h is examined first, $B^{h'}$ – next.

If for B^H the relaxation problems were solved, go to Step 4.2, otherwise to Step 4.1:

Step 4.1. For B^H solve the relaxation and projection problems on $R^{n-|H|}$-facet depending on the scheme: (a) for Scheme 1 – $x^{S,H}$, $y^{S,H}$, the bounds are $z^{l,H} = z_x^{S,H}$, $z^u = min(z^u, z_y^{S,H})$; (b) for Scheme 2 – $x^{S,H}$, $x^{P,H}$, $y^{S,H}$, $y^{P,H}$, the bounds are $z^{l,H} = max(z_x^{S,H}, z_x^{P,H})$, $z^u = min(z^u, z_y^{S,H}, z_y^{P,H})$;

Step 4.2. Let:

- $\Pi(x^{S,H}) = \{x : ax = b\}$ be a cutting plane for the point $x^{S,H}$ nearest to x^P such that $\{x : ax \leq b\}$ is a right cut of $x^{S,H}$, $E^H = E \cap \Pi(x^{S,H})$;
- $\Pi'(x^{S,H}) = \{x : a'x = b'\}$ be the second layer, $E^{H'} = E \backslash E^H$, facet.

The sets $E^H, E^{H'}$ form two branches from the node B^H – $B^{\{H,|H+1|\}}$ and $B^{\{H,|H+1|'\}}$ - with the same lower bound $z^{u,B^{\{H,|H+1|\}}} = z^{u,B^{\{H,|H+1|'\}}} = z^{u,B^H}$. They are added to $\mathbf{B}$ instead of B^H, $\mathbf{B} = \mathbf{B} \backslash B^H \cup \{B^{\{H,|H+1|\}}, B^{\{H,|H+1|'\}}\}$.

Step 4 is repeated until the candidate solutions with lower bound less than z^u remain.

The B&BPSM for the UBQP. Step 4.2 becomes: $\Pi(x^{S,H}) = \{x : x_i = b\}$. The rule for cutting plane choice on Step 4.2 becomes: $\Pi(x^{S,H}) = \{x : x_i = b\}$, $\Pi'(x^{S,H}) = \{x : x_i = 1 - b\}$ where $b \in \{0, 1\}$, $i \in J_n$ is chosen depending on which one of the restrictions $x_i \geq 0$ or $x_i \leq 1$ is mostly violated at $x^{S,H}$. Respectively, new branches $B^{\{H,|H+1|\}}$, $B^{\{H,|H+1|'\}}$ correspond to fixing the coordinate x_i.

Example 1 Solve the following problem: $f(x) = 33.1 \cdot x_1^2 + 23.7 \cdot x_2^2 + 21.6 \cdot x_3^2 - 11.4 \cdot x_1x_2 + 38.8 \cdot x_1x_3 - 15.6 \cdot x_2x_3 - 52 \cdot x_1 - 4 \cdot x_2 - 39 \cdot x_3 \to \min$, $x \in B_3$.

Solution outline:

1. $n = 3$, $f(x)$ is convex, $x^* = -\frac{1}{2}A^{-1}c = (0.51, 0.40, 0.59)$, $y^* = (1, 0, 1)$. The initial bounds are $z^l = z^* = -25.54$, $z^u = z_y^* = 2.5$;
2. x^* is an interior point of the unit hypercube ($x^P = x^*$, $y^P = y^*$). The results of the relaxation Scheme 1: $x^S = (0.06, 0.51, 1.25)$, $y^S = (0, 1, 1)$. The bounds are $z^l = z^S = -21.30$, $z^u = min(z^u, z_y^S) = min(2.50, -13.30) = -13.30$ (the record occurs at $x^u = (0, 1, 1)$).
3. $H = \{\emptyset\}$, the right cut for $x^{S,H} = x^S$ is uniquely defined by $x_3 \leq 1$. Respectively, x_3 is fixed and exploring of the search tree starts with the branch $B^{\{H,|H+1|\}} = B^1 : x_3 = 1$, the next one is $B^{\{H,|H+1|'\}} = B^{1'} : x_3 = 0$.
4. The reduction is conducted ($H = \{1\}$) and, at first, the Relaxation 2 problem on the plane $x_3 = 1$ is solved yielding $x^{S1} = (-0.09, 0.11, 1)$, $z^{l1} = z_x^{S1} = -17.7$. $y^{S1} = (0, 0, 1)$, $z_y^{S1} = -17.4 > z^u$; new record $z^u = -17.4$ occurs at $x^u = (0, 0, 1)$.

   The right cut for x^{S1} is $x_1 \geq 0$ yields two branches with $H = \{H, |H+1|\} = \{1, 2\}$, $B^{12} : x_1 = 0, x_3 = 1$ and $B^{12'} : x_1 = 1, x_3 = 1$, with fixed x_1 and the lower bound $z^{l12} = z^{l12'} = z^{l1} = -17.7$.
5. The same is conducted for $B^{1'}$: $x^{S1'} = (1.04, 0.04, 0.00)$, $z_x^{S1'} = -18.92$; $y^{S1'} = (1, 0, 0)$, $z_y^{S1'} = -18.9$. New record is $z^u = -18.9$ at $x^u = (1, 0, 0)$ and the bound is $z^{l1'} = -18.92$. The right $x^{S1'}$-cut is $x_1 \leq 1$ yields the branches $B^{1'2} : x_1 = 1, x_3 = 0$ and $B^{1'2'} : x_1 = 0, x_3 = 0$, with the lower bound $z^{l1'2} = z^{l1'2'} = z^{l1'} = -18.9$.
6. The least lower bound -18.92 corresponds to $B^{1'2}$ and $B^{1'2'}$, which are examined consecutively. First, for $H = \{1', 2\}$ Relaxation 2 is conducted. Since, $|H| = n - 1$, $B^{1'2}$ corresponds to 1-sphere, where, in addition to $y^{S1'} = (1, 0, 0)$, it is enough to analyse another endpoint $t^{S1'} = (1, 1, 0)$ of the hypercube edge, $z_t^{S1'} = -10.6$. $z^{l1'2} = min(z_y^{S1'}, z_t^{S1'}) = min(-18.9, -10.6) = -18.9$. For $H = \{1', 2'\}$ and $B^{1'2'}$ the result of Relaxation 2 is $x^{S1'2'} = (0, 0, 0)$. Since $z^{l1'2'} = z^{S1'2'} = 0 > z^u$, $B^{1'2'}$ is discarded.
7. No bud nodes with less lower bound than $z^u = -18.9$ left. The examination of the search tree is terminated. The solution is $x^{**} = (1, 0, 0)$, $z^{**} = -18.9$.

Illustration of the solution is shown in Figs. 1, 2, and 3.

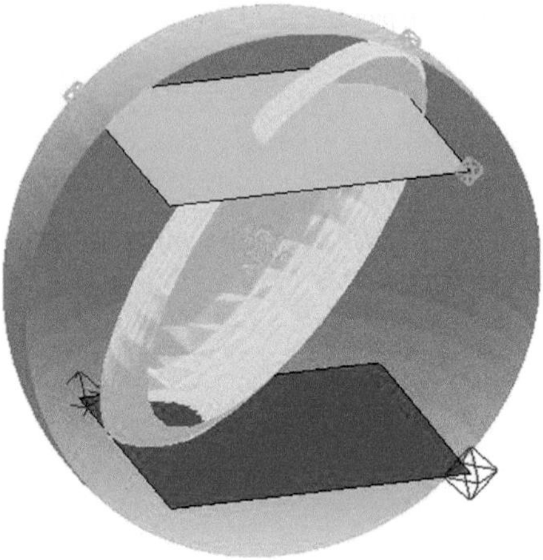

Fig. 1 The touching the level surface at x^S. In Figs. 2 and 3 the reduced problems solving at Step 4 are demonstrated, namely, the ellipsoids yielding x^{S1}, $x^{S1'}$ alongside with the projections y^{S1}, $y^{S1'}$, their cuts and the branches of the next level

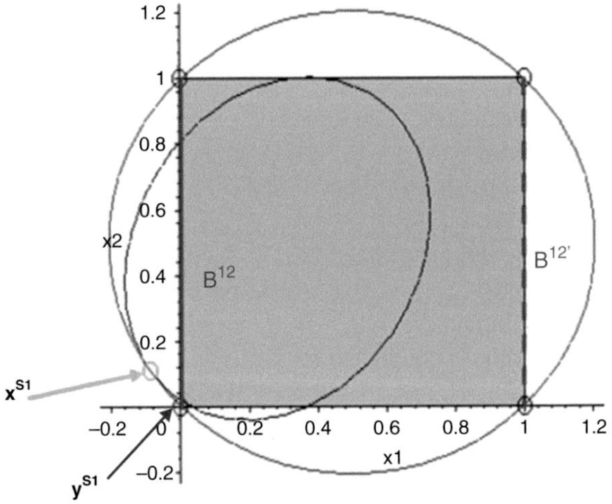

Fig. 2 B^1 exploration

2.3 The Penalty Method Based on Functional Representations

Definition 2 The following system of functions:

$$\mathbf{f}(x) = \{f_i(x)\}_{i \in J_m} \tag{18}$$

is called a *functional representation* of a set $M \subset R^n$ if

$$x^0 \in M \Leftrightarrow f_i(x^0) = 0 \ \forall f_i(x) \in \mathbf{f}(x). \tag{19}$$

Similar to the extensions of functions, depending on a type of functions in (18), the sets' representations can be continuous, differentiable, convex, smooth, etc.

Fig. 3 $B^{1'}$ exploration

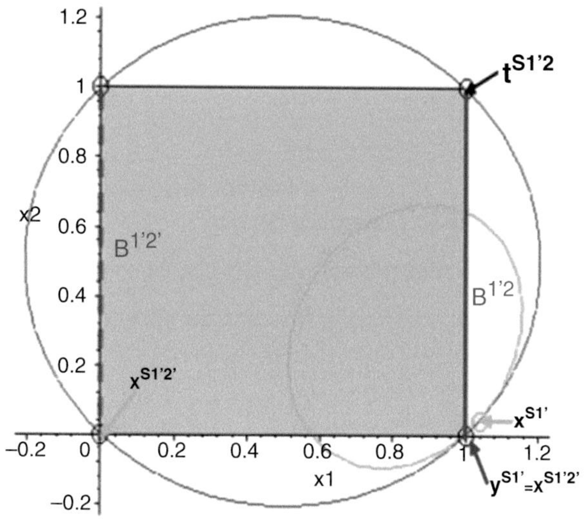

Definition 3 The representation (18) is *irreducible* if it is impossible to extract anything from (18) without violating (19): $\forall j \in J_m\ M \neq \{x : f_j(x) = 0\}_{f_i(x)\in \mathbf{f}(x), i\neq j}$.

The underlying idea of the Penalty Method based on Functional Representations (FRPM) is constructing and applying irreducible differential representations of the initial discrete set E in penalty optimization.

Suppose, for E its differentiable representation is known. Then, by (19), the discrete problem (1), (8) is transformed into the constrained continuous problem (1) subject to $\mathbf{f}(x) = \mathbf{0}$. After placing these constraints in a penalty function, say quadratic, the problem becomes unconstrained one of minimizing the differentiable function:

$$\Phi(x,\lambda) = f(x) + \lambda \sum_{i=1}^{m} \left(f^i(x)\right)^2 \underset{\lambda>0}{\rightarrow} min. \tag{20}$$

The problem (20) is solved for an increasing sequence

$$\lambda \in \Lambda = \{\lambda_k : \lambda_k < \lambda_{k+1}, \lambda_0 = 0\}_{k\in J_K^0 = J_K \cup \{0\}}. \tag{21}$$

Its solutions gradually approach a local minimizer x^* of (1): $x^k \underset{k\to\infty}{\rightarrow} x^*$. Notice that $\Phi(x,\lambda)$ in (20) is an extension of (1) onto R^n and a sum of a convex objective function (the first term) and a nonconvex penalty function (the second term). Hence $\exists k^0 : \Phi(x,\lambda_k)$ is convex for $k \leq k^0$ and the solution of (20) is unique, otherwise is not and the rest of the search is held in the vicinity $N(x^*)$ of x^*.

For the problem (1), (8), (9), and (10), by using Theorem 2, the region of $\Phi(x,\lambda)$-convexity can be increased and, respectively, the vicinity $N(x^*)$ can be narrowed by switching from (20) to the following extension of (1) from E onto R^n:

$$\Phi(x, \lambda, \mu) = f(x) + \mu((x-a)^2 - r^2) + \lambda \sum_{i=1}^{m} \left(f^i(x)\right)^2 \ (\lambda, \mu > 0). \qquad (22)$$

The problem (22) can be solved for two increasing sequences, (21) and $\{\mu_j : \mu_j < \mu_{j+1}\}_j$, in the way described below:

The FRPM Algorithm for (1), (8) based on (22).

Step 0. Set $j = 0$, $\mu_0 = 0$, $\varepsilon > 0$, the upper and lower bounds $z^l = -\infty$, $z^u = \infty$;
Step 1. Solve (22) for (21), $\mu = \mu_j$, and obtain $\{x^{jk}\}_k$, $x^{j*} = x^{jK}$. $z^l = \max(z^l, z^{j*})$, $z^u = \min_k(z^u, f(Pr_E(x^{jk}))$;
Step 2. For (22), by Remark 1, find the lower bound μ_j^0: $\forall \mu > \mu_j^0$ $\Phi(x, \lambda_K, \mu)$ is convex. Solve (22) for $\lambda = \lambda_K$, $\mu \in M_j = \{\mu_{j'} > 0 : \mu_{j'} < \mu_{j'+1}, \mu_{J'} > \mu_j^0\}_{j' \in J_{j'}}$, the solutions are $\{x^{jK}\}_j$, $x^{*K} = x^{J'K}$. $z^l = \max(z^l, z^{*K})$, $z^u = \min_j(z^u, f(Pr_E(x^{jK}))$;
Step 3. Set $f(x) = \Phi(x, \lambda_K, \mu_{J'})$;
Step 4. Repeat Steps 1–3. The process terminates if $z^u - z^l < \varepsilon$.

Assuming that $n > 3$, we classify functional representations of discrete sets depending on a number of functions in (18) as follows: (a) *intersecting* functional representation if $m = n$ implying that E is formed as an intersection of n surfaces; (b) *touching* functional representation if $m = 2$, hence, E is the set of touching points of two surfaces; (c) otherwise *mixed* functional representation.

Deriving functional representations of combinatorial sets is an interesting on its own problem, because it provides better understanding the sets topological structure. Among this class of problems finding the touching representations is a problem of particular interest since it investigates extremal properties of the sets. Indeed, let assume that a touching representation of E is known and the functions in (18) are enumerated such that the surface $S^2 = \{f_2(x) = 0\}$ is inscribed into $S^1 = \{f_1(x) = 0\}$. Then E is the set of local minimizers of $f_2(x)$ on the surface S^1.

Touching sets representations, especially differentiable ones, are useful in optimization, e.g., they allows usage in the FRPM, instead of (22), the penalty function:

$$\Phi(x, \boldsymbol{\lambda}, \boldsymbol{\mu}) = f(x) + \sum_{i=1}^{2} \mu_i f^i(x) + \sum_{i=1}^{2} \lambda_i \left(f^i(x)\right)^2, \qquad (23)$$

and managing a few penalty parameters $\mu_1, \mu_2, \lambda_1, \lambda_2$ for obtaining the solutions closer to exact one.

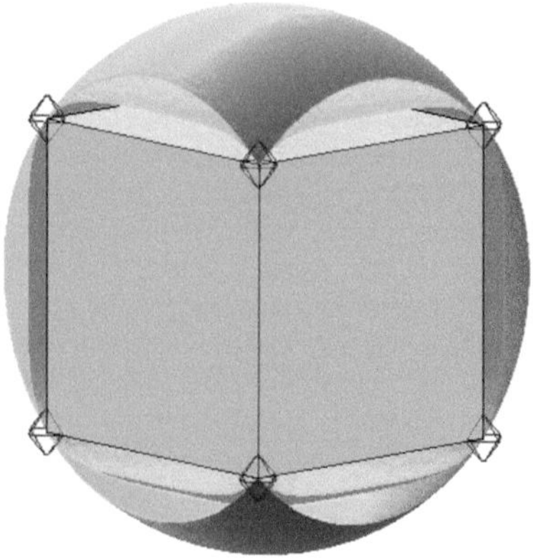

Fig. 4 Illustration of the touching representations (24). The Fig. 4 shows the unit hypercube inscribed into the order four surface S^2. In turn, S^2 is inscribed into the sphere $S^1 = S^{min}$. The binary set B_3 is the "touching set" of S^1, S^2

The Functional Representations of B_n . We show that the FRPM based on any of (22), (23) can be used for (1), (3):

- Equation (22) is applicable since the irreducible intersecting representation $f_i(x) = x_i - x_i^2, i \in J_n$, is derived directly from (4) and B_n is inscribed into the sphere (16).
- Equation (23) can be used for the following irreducible touching differentiable representation of B_n, which we derived (the geometric interpretation is shown in Fig. 4):

$$f_1(x) = \sum_{i=1}^{n} (x_i - 0.5)^2 = n/4, f_2(x) = \sum_{i=1}^{n} (x_i - 0.5)^4 = n/16. \tag{24}$$

We recommend for the UBQP utilizing (24) in the penalty function (23) of (1), (2), and (3).

3 Conclusion

Two approaches, the B&BPSM and FRPM, to discrete optimization over sets inscribed into a sphere are presented. For the UBQP they use analytic descriptions of B_n as an intersection of a hypercube with a sphere and as a touching set; convex extensions of quadratic functions. The approaches can be applied to other sets with known analytic description of a convex hull, a circumscribed surface, a functional representation, and to any function with known the global minimizer on the surface.

References

1. Dahl, J.: Convex optimization in signal processing and communications. Ph.D. dissertation. Department of Communication Technology, Aalborg University (2003)
2. Helmberg, C., Rendl, F.: Solving quadratic (0,1)-problems by semidefinite programs and cutting planes. Math. Program. **82**(3), 291–315 (1998)
3. Hillier, F.S., Appa, L., Pitsoulis, L., Williams, H.P., Pardalos, P.M., Prokopyev, O.A., Busygin, S.: Continuous approaches for solving discrete optimization problems. In: Handbook on Modelling for Discrete Optimization, pp. 1–39. Springer, New York (2006)
4. Kochenberger, G., Hao, J.-K., Glover, F., Lewis, M., Lu, Z., Wang, H., Wang, Y.: The unconstrained binary quadratic programming problem: a survey. J. Comb. Optimiz. **1**, 58–81 (2014)
5. Krislock, N., Malick, J., Roupin, F.: Improved semidefinite bounding procedure for solving Max-Cut problems to optimality. Math. Program. **143**(1–2), 61–86 (2014)
6. Murray, W., Ng, K.-M.: An algorithm for nonlinear optimization problems with binary variables. Comput. Optimiz. Appl. **47**(2), 257–288 (2008)
7. Pichugina, O., Yakovlev, S.: Polyhedral-spherical approach to solving some classes of combinatorial optimization problems. In: Proceedings of the 6th International School-Seminar on Decision Theory, Uzhgorod, pp. 152–153 (2012)
8. Stoyan, Y.G., Yemets, O.: Theory and Methods of Euclidean Combinatorial Optimization. ISSE, Kiev (1993)
9. Stoyan, Y.G., Yakovlev, S.V., Parshin, O.V.: Quadratic optimization on combinatorial sets in R^n. Cybern. Syst. Anal. **27**(4), 562–567 (1991)
10. Stoyan, Y., Yakovlev, S., Emets, O., Valuiskaya, O.: Construction of convex continuations for functions defined on a hypersphere. Cybern. Syst. Anal. **34**(2), 176–184 (1998)
11. Yakovlev, S.: The theory of convex continuations of functions at the vertices of convex polygons. Comput. Math. Math. Phys. **34**(7), 959–965 (1994)
12. Yakovlev, S., Grebennik, I.: Certain classes of optimization problems on set distributions and their properties. Izvestiya Vysshikh Uchebnikh Zavedenii Mathematika **11**, 74–86 (1991)

Fixed Point Techniques in Analog Systems

Diogo Poças and Jeffery Zucker

Abstract Analog computation is concerned with continuous rather than discrete spaces. Most of the physical processes arising in nature are modeled by differential equations, either ordinary (example: spring/mass/damper system) or partial (example: heat diffusion). In analog computability, the existence of an effective way to obtain solutions (either exact or approximate) of these systems is essential.

We develop a framework in which the solutions can be seen as fixed points of certain operators on continuous data streams, using the framework of Fréchet spaces. We apply a fixed point construction to retrieve these solutions and present sufficient conditions on the operators and initial inputs to ensure existence and uniqueness of these corresponding fixed points.

1 Introduction

Analog computation, as conceived by Kelvin [10], Bush [1], and Hartree [4], is a form of experimental computation with physical systems called analog devices or analog computers. Historically, data are represented by measurable physical quantities, including lengths, shaft rotation, voltage, current, resistance, etc., and the analog devices that process these representations are made from mechanical or electromechanical or electronic components [5, 7, 9].

The main objects of our study are *analog networks* or *analog systems*, [6, 11–13], whose main components are described as follows:

$$\textit{Analog network} = \textit{data} + \textit{time} + \textit{channels} + \textit{modules}.$$

We will model *data* as elements from a topological vector space $\mathscr{A}$ that is actually a Fréchet space. We will use the nonnegative real numbers as a continuous model of *time* $\mathbb{T} = [0, \infty)$. Each *channel* carries a continuous stream, represented as a function $u : \mathbb{T} \to \mathscr{A}$ (this space is denoted by $C(\mathbb{T}, \mathscr{A})$). Each *module* M is

D. Poças (✉) • J. Zucker
Department of Mathematics and Statistics, McMaster University, Hamilton, ON, Canada
e-mail: pocasd@math.mcmaster.ca; zucker@cas.mcmaster.ca

© Springer International Publishing Switzerland 2016

J. Bélair et al. (eds.), *Mathematical and Computational Approaches in Advancing Modern Science and Engineering*, DOI 10.1007/978-3-319-30379-6_63

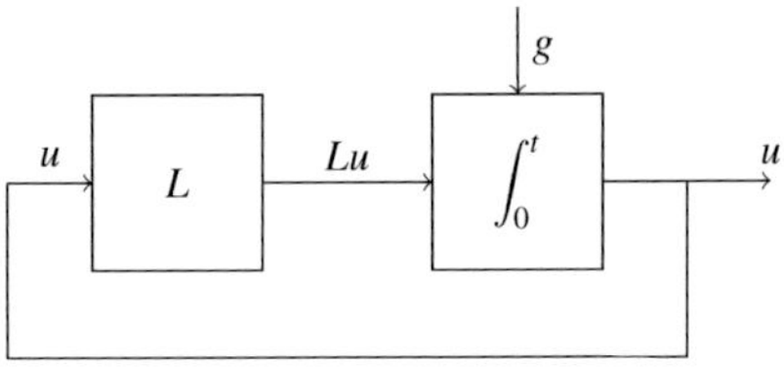

Fig. 1 Analog network (with two modules and three channels) for the time evolution problem

specified by a *stream function*

$$F_M : C(\mathbb{T}, \mathscr{A})^k \times \mathscr{A}^\ell \to C(\mathbb{T}, \mathscr{A}).$$

In this way, we can think of analog networks as directed graphs where modules are nodes and channels are edges (see Fig. 1). We will use analog systems to study the time evolution problem (also called the Cauchy problem).

Definition 1 (Time evolution problem, [3]) For a given initial condition $g \in \mathscr{A}$ and an operator $L : \mathscr{A} \to \mathscr{A}$, the *time evolution problem* is given by the system

$$\begin{cases} \frac{du}{dt} = Lu, \ t \in \mathbb{T}; \\ u(0) = g. \end{cases} \tag{1}$$

We look for a solution $u \in C(\mathbb{T}, \mathscr{A})$.

To construct an analog system that represents the time evolution problem, we can simply integrate the differential equation (1) to obtain

$$u(t) = g + \int_0^t Lu(s)ds =: F_g(u), \tag{2}$$

where we use the right hand side to define an operator $F_g : C(\mathbb{T}, \mathscr{A}) \to C(\mathbb{T}, \mathscr{A})$, which can be computed using an analog network with two modules. Introducing a feedback to implement the equality, we obtain the analog system of Fig. 1.

We can informally define a specification of the analog network as a tuple of streams describing the data on all channels which satisfy the equations given by the modules. We can then observe the equivalence between the notions of (a) solutions to the time evolution problem of Definition 1; (b) specifications of the analog system of Fig. 1; and (c) fixed points of the operator F_g of Equation (2). Henceforth we will focus on the last notion. Our goal is to provide sufficient conditions on F_g that ensure existence and uniqueness of fixed points, as well as the existence of a constructive method to obtain fixed points when they exist.

In Sect. 2 we introduce Fréchet spaces, which form the framework for our problem at hand. In Sect. 3 we assume analyticity of g to prove local existence and convergence of fixed points (Theorem 2). In Sect. 4 we first extend our results to global existence and convergence (Theorem 3) and then extend our constructive method for different choices of initial input (Theorem 5). Finally, we turn to uniqueness of fixed points and prove it for certain operators (Theorem 6).

2 Fréchet Spaces

For the remainder of this paper, we use the following notation:

Notation 1 *$\mathscr{A}$ is the space of infinitely differentiable functions, $\mathscr{A} = C^\infty(\mathbb{R})$.*

Notation 2 *The operator $L : \mathscr{A} \to \mathscr{A}$ is given by $Lu = \alpha\partial_x u$, for some $\alpha \in \mathbb{R}$.*

Thus, the operator F_g becomes

$$F_g(u)(t,x) = g(x) + \alpha \int_0^t \partial_x u(s,x)ds. \tag{3}$$

As we know, $\mathscr{A} = C^\infty(\mathbb{R})$ is not complete under the supremum norm; however, for each $x_0 \in \mathbb{R}$, $X \in \mathbb{R}^+$ and $k \in \mathbb{N}$, we can define a pseudonorm by

$$\|f\|_{X,x_0,k} = \sup_{|x-x_0|\leq X} \left| \frac{\partial^k f}{\partial x^k}(x) \right|.$$

It turns out that the notion of interest is that of Fréchet space, which we now briefly review (a detailed exposition can be found in [8, Ch. V]).

Definition 2 (Fréchet space [8]) A Fréchet space is a topological vector space X whose topology is induced by a countable family of pseudonorms $\{\|\cdot\|_\alpha\}_{\alpha\in A}$. Moreover,

- the family $\{\|\cdot\|_\alpha\}_{\alpha\in A}$ separates points, that is,

$$u = 0 \text{ if and only if } \|u\|_\alpha = 0 \text{ for all } \alpha;$$

- X is complete with respect to $\{\|\cdot\|_\alpha\}_{\alpha\in A}$, that is, for every sequence (x_n) which is Cauchy with respect to each pseudonorm $\|\cdot\|_\alpha$, there exists $x \in X$ such that (x_n) converges to x with respect to each pseudonorm $\|\cdot\|_\alpha$.

Example 1 The space $\mathscr{A} = C^\infty(\mathbb{R})$ of infinitely differentiable functions in $\mathbb{R}$ is a Fréchet space with the countable family of pseudonorms given by

$$\|f\|_{N,k} = \sup_{-N\leq x\leq N} \left| \frac{\partial^k f}{\partial x^k}(x) \right|, \tag{4}$$

for $N, k \in \mathbb{N}$.

Example 2 The space $C(\mathbb{T}, \mathscr{A})$ of continuous streams is also a Fréchet space, with the countable family of pseudonorms given by

$$\|u\|_{M,N,k} = \sup_{0 \le t \le M} \sup_{-N \le x \le N} \left| \frac{\partial^k u}{\partial x^k}(t,x) \right|, \tag{5}$$

for $M, N, k \in \mathbb{N}$.

We can see that the family of pseudonorms in Example 2 is closely related to the family in Example 1. In fact, this illustrates a useful property of Fréchet spaces; in general, the space of continuous streams over a Fréchet space is itself a Fréchet space. In other words, Fréchet spaces work well with the operation of taking continuous streams.

Proposition 1 (New Fréchet spaces from old) *If $\mathscr{A}$ is a Fréchet space with a countable family of pseudonorms $\{\|\cdot\|_\alpha\}_{\alpha \in A}$, then so is $C(\mathbb{T}, \mathscr{A})$ with the countable family of pseudonorms $\{\|\cdot\|_{M,\alpha}\}_{M \in \mathbb{N}, \alpha \in A}$, where*

$$\|u\|_{M,\alpha} = \sup_{0 \le t \le M} \|u(t)\|_\alpha.$$

Even though Fréchet spaces, as they stand, are not necessarily normed spaces, we can define a metric from the pseudonorms, under which these spaces are complete.

Proposition 2 *Given a Fréchet space, we can define a complete metric from the family of pseudonorms which induces the same topology.*

For a proof, see [8, Ch. V], in particular Theorem V.5.

The usefulness of complete metric spaces is evident due to the following.

Theorem 1 (Banach fixed point theorem) *Given a complete metric space (X, d), suppose that $T : X \to X$ is a contracting operator in the sense that there exists $0 \le \lambda < 1$ with*

$$d(T(x), T(y)) \le \lambda d(x, y) \textit{ for all } x, y \in X.$$

Then T has a unique fixed point x^*. Moreover, for all $x_0 \in X$ the sequence of iterations $x_n := T^n(x_0)$ converges to x^*.

3 Local Convergence Theorem

Consider the space $C(\mathbb{T}, \mathscr{A})$. Take any (arbitrary but fixed) $x_0 \in \mathbb{R}$, $X \in \mathbb{R}^+$, $T \in \mathbb{T}$. Then, for any $k \in \mathbb{N}$, we have a pseudonorm

$$\|u\|_{T,X,x_0,k} = \sup_{\substack{0 \le t \le T \\ |x - x_0| \le X}} \left| \frac{\partial^k u}{\partial x^k}(t,x) \right|. \tag{6}$$

Fig. 2 Compact rectangles

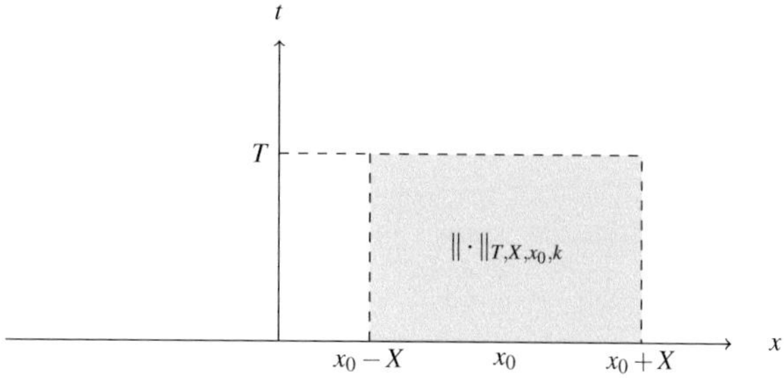

Observe that we are taking suprema on compact rectangles of the form $[0,T] \times [x_0 - X, x_0 + X]$ (see Fig. 2). The reason for taking suprema on compact rectangles will be made clear shortly with Theorem 2 (local convergence theorem). We also observe that, for each compact rectangle $\mathbb{X}' = [0,T]\times[x_0-X, x_0+X]$, we can define the space of *compact continuous streams* $C([0,T], C^\infty(x_0-X, x_0+X))$. Clearly, any function in $C(\mathbb{T}, \mathscr{A})$ can be mapped to a function in $C([0,T], C^\infty(x_0 - X, x_0 + X))$ via the restriction $u \mapsto u\restriction_{\mathbb{X}'}$. Moreover, $C([0,T], C^\infty(x_0 - X, x_0 + X))$ can be seen to be a Fréchet space with the family of pseudonorms $\|\cdot\|_{T,X,x_0,k}$ given by (6). Note that x_0, X and T are *fixed* and the indexing is on $k \in \mathbb{N}$.

Finally, observe that the operator $F_g : C(\mathbb{T}, \mathscr{A}) \to C(\mathbb{T}, \mathscr{A})$ has a restriction $F_g\restriction_{\mathbb{X}'}$ to the space $C([0,T], C^\infty(x_0 - X, x_0 + X))$.

Our next step is to prove contraction inequalities, which play an important role in fixed point techniques.

Lemma 1 (Contraction inequalities) *Consider the Fréchet space $C(\mathbb{T}, \mathscr{A})$ with pseudonorms $\|\cdot\|_{T,X,x_0,k}$ given by* (6). *Let $g \in C^\infty(\mathbb{R})$ and $F_g : C(\mathbb{T}, \mathscr{A}) \to C(\mathbb{T}, \mathscr{A})$ be given by* (3). *Then, for any $u, v \in C(\mathbb{T}, \mathscr{A})$, any pseudonorm $\|\cdot\|_{T,X,x_0,k}$ and any $m \in \mathbb{N}$, we have the following bound:*

$$\|F_g^m(u) - F_g^m(v)\|_{T,X,x_0,k} \quad \leq \quad \frac{(|\alpha|T)^m}{m!}\|u - v\|_{T,X,x_0,k+m}. \tag{7}$$

Proof By induction on m. □

Let us see how we can use these bounds in a proof.

Theorem 2 (Local Fréchet space convergence theorem) *Consider the Fréchet space $C(\mathbb{T}, \mathscr{A})$ with pseudonorms $\|\cdot\|_{T,X,x_0,k}$ given by* (6). *Take an initial input $u_0 \in C(\mathbb{T}, \mathscr{A})$ and initial condition $g \in C^\infty(\mathbb{R})$. Assume also that $u_0 = 0$ and g is analytic at x_0 with some radius of convergence*[1] *R. Let $F_g : C(\mathbb{T}, \mathscr{A}) \to C(\mathbb{T}, \mathscr{A})$ be given by* (3). *Then, for any $T, X \in \mathbb{R}^+$ such that $|\alpha|T + X < R$, the sequence (u_m)*

[1]Or equivalently, that g has a holomorphic extension on a disk of the complex plane with center x_0 and radius R; see Remark 1.

given by $u_m = F_g^m(u_0)$ *converges in the rectangle* $\mathbb{X}' = [0,T] \times [x_0 - X, x_0 + X]$ *to a fixed point of* $F_g \restriction_{\mathbb{X}'}$.

Proof To ease the exposition we introduce the pseudonorms on g given by

$$\|g\|_{X,x_0,k} = \sup_{|x-x_0|\le X} \left| \frac{\partial^k g}{\partial x^k}(x) \right|, \quad \text{for} \quad x_0 \in \mathbb{R},\ X \in \mathbb{R}^+,\ k \in \mathbb{N}. \tag{8}$$

Since g is analytic at x_0 with radius of convergence R, there is a sequence of real coefficients (a_j) such that, for all $x \in (x_0 - R, x_0 + R)$,

$$g(x) = \sum_{j=0}^{\infty} a_j (x - x_0)^j.$$

It also follows that $\limsup_{n\to\infty} \sqrt[n]{|a_n|} \le \frac{1}{R}$ (see Footnote 1). Moreover, we have the following bound, for any $X < R$:

$$\|g\|_{X,x_0,k} = \left| \sup_{|x-x_0|<X} \sum_{j=0}^{\infty} \frac{(j+k)!}{j!} a_{j+k}(x-x_0)^j \right| \le \sum_{j=0}^{\infty} \frac{(j+k)!}{j!} |a_{j+k}| X^j. \tag{9}$$

Let $T, X \in \mathbb{R}^+$ such that $|\alpha|T + X < R$. We show that (u_m) is a Cauchy sequence with respect to the pseudonorm $\|\cdot\|_{T,X,x_0,k}$. First observe that

$$\begin{aligned}
\sum_{m=0}^{\infty} \|u_{m+1} - u_m\|_{T,X,x_0,k} &= \sum_{m=0}^{\infty} \|F_g^m(g) - F_g^m(0)\|_{T,X,x_0,k} \\
&\overset{1}{\le} \sum_{m=0}^{\infty} |\alpha|^m \frac{T^m}{m!} \|g\|_{X,x_0,k+m} \\
&\overset{2}{\le} \sum_{m=0}^{\infty} \sum_{j=0}^{\infty} |\alpha|^m T^m X^j |a_{k+m+j}| \frac{(k+m+j)!}{m!j!} \\
&\overset{3}{=} \sum_{s=0}^{\infty} \sum_{m=0}^{s} |\alpha|^m T^m X^{s-m} |a_{k+s}| \frac{(k+s)!}{m!(s-m)!} \\
&\overset{4}{=} \sum_{s=0}^{\infty} (|\alpha|T + X)^s |a_{k+s}| \frac{(k+s)!}{s!},
\end{aligned}$$

where (1) is justified by the Contraction Inequalities (Lemma 1), (2) by equation (9), (3) by rearranging the sum and adding over diagonals $s = m + j$ and (4) by taking the binomial expansion of $(|\alpha|T + X)^s$.

By the root test, the above series is convergent, since

$$\limsup_{s\to\infty} \sqrt[s]{(|\alpha|T+X)^s|a_{k+s}|\frac{(k+s)!}{s!}} = (|\alpha|T+X)\cdot\limsup_{n\to\infty}\sqrt[n]{|a_n|}\cdot 1 < \frac{R}{R} = 1.$$

Since the series is convergent, it follows that, for $i<j$,

$$\|u_j - u_i\|_{T,X,x_0,k} \leq \sum_{m=i}^{j-1}\|u_{m+1}-u_m\|_{T,X,x_0,k} \leq \sum_{m=i}^{\infty}\|u_{m+1}-u_m\|_{T,X,x_0,k} \underset{i\to\infty}{\longrightarrow} 0.$$

Hence (u_m) is a Cauchy sequence with respect to the pseudonorm $\|\cdot\|_{T,X,x_0,k}$. Since this holds for all $k\in\mathbb{N}$ and $C([0,T],C^\infty(x_0-X,x_0+X))$ is complete, it follows that (u_m) has a limit in $\mathbb{X}'$. Now, using continuity of $F_g\restriction_{\mathbb{X}'}$, we conclude that this limit must be a fixed point of $F_g\restriction_{\mathbb{X}'}$. □

Remark 1 The reader should distinguish between the following two concepts:

- the existence of a holomorphic function, defined in a disk of the complex plane $\mathbb{C}$, which coincides with g at the real axis $\{y=0\}$;
- the convergence of the construction $u_m = F_g^m(0)$ to a fixed point u, defined in a rectangle of $\mathbb{T}\times\mathbb{R}$, which coincides with g at initial time $\{t=0\}$.

As seen from Theorem 2, the existence of a holomorphic extension implies convergence to a fixed point. Both these functions (the holomorphic extension and the fixed point) can be depicted by planar diagrams, and both can be seen as extensions of g (see Fig. 3). However, these functions, and the domains which they inhabit, are substantially different.

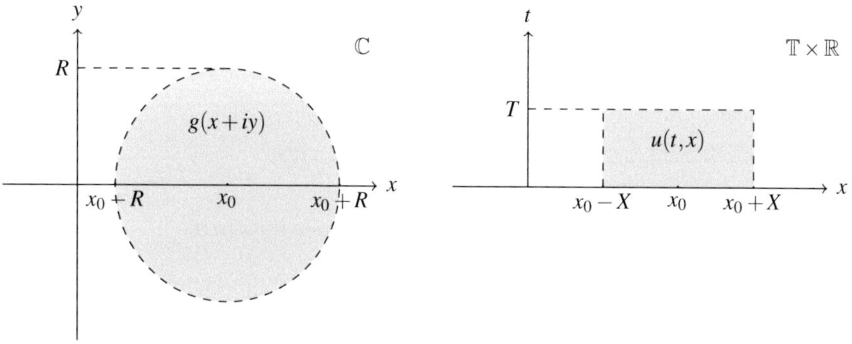

Fig. 3 On the *left*: a function $g(x+iy)$ of type $\mathbb{C}\to\mathbb{C}$, defined in a disk, that coincides with g at $\{y=0, x_0-R<x<x_0+R\}$. On the *right*: a fixed point $u(t,x)$ of type $\mathbb{T}\times\mathbb{R}\to\mathbb{R}$, defined in a rectangle, that coincides with g at $\{t=0, x_0-X<x<x_0+X\}$. The rectangle and disk dimensions follow the relation $|\alpha|T+X<R$

4 Global Convergence Theorems

As an immediate corollary of Theorem 2, we have:

Theorem 3 (First global Fréchet space convergence theorem) *Consider the Fréchet space $C(\mathbb{T}, \mathscr{A})$ with pseudonorms $\|\cdot\|_{T,X,x_0,k}$ given by (6). Take an initial input $u_0 \in C(\mathbb{T}, \mathscr{A})$ and initial condition $g \in C^\infty(\mathbb{R})$. Assume also that $u_0 = 0$ and g is entire (i.e. has a holomorphic extension to the complex plane). Let $F_g : C(\mathbb{T}, \mathscr{A}) \to C(\mathbb{T}, \mathscr{A})$ be given by (3). Then the sequence (u_m) given by $u_m = F_g^m(u_0)$ converges to a fixed point of F.*

Proof Since g is entire, it is analytic at any x_0 with any radius of convergence R. Thus, by Theorem 2, the sequence u_m converges to a fixed point on any compact rectangle $[0, T] \times [x_0 - X, x_0 + X]$. Therefore, we have convergence of u_m for any pseudonorm $\|\cdot\|_{T,X,x_0,k}$, so that we have convergence in $C(\mathbb{T}, \mathscr{A})$. □

The next step is to generalize Theorem 3 to a larger class of initial functions u_0 (other that $u_0 = 0$). We do that proof in two steps: first assume $g = 0$ to establish sufficient conditions on u_0; then consider the more general case $g \in C^\infty(\mathbb{R})$.

Definition 3 We say that a function $u \in C(\mathbb{T}, \mathscr{A})$ is *uniformly entire* if $u(t, x) = \sum_{j=0}^{\infty} a_j(t)x^j$ for some sequence of functions $(a_j) \in C(\mathbb{R})$ such that

$$\lim \sqrt[j]{\sup_{0 \le t \le T} |a_j(t)|} = 0 \text{ for all } T \in \mathbb{T}.$$

The motivation for this terminology is that, for such a function u, the section $x \mapsto u(t, x)$ is entire for all t, and the convergence $\sqrt[j]{|a_j(t)|} \to 0$ is uniform in t.

Theorem 4 *Consider the Fréchet space $C(\mathbb{T}, \mathscr{A})$ with the family of pseudonorms $\|\cdot\|_{T,X,x_0,k}$ given by (6). Let also $u_0 \in C(\mathbb{T}, \mathscr{A})$ be an initial input, and $g \in C^\infty(\mathbb{R})$ be an initial condition. We assume in addition that u_0 is uniformly entire and $g = 0$. Let $F_0 : C(\mathbb{T}, \mathscr{A}) \to C(\mathbb{T}, \mathscr{A})$ be given by*

$$F_0(u)(t, x) = \alpha \int_0^t \partial_x u(s, x) ds. \tag{10}$$

Then the sequence (u_m) given by $u_m = F_0^m(u_0)$ converges to 0.

Proof To ease the exposition we introduce the pseudonorms on a_j given by

$$\|a_j\|_T = \sup_{0 \le t \le T} |a_j(t)|, \text{ for } T \in \mathbb{T}. \tag{11}$$

We show that $\sum_m \|u_m\|_{T,X,0,k}$ is a convergent series for any pseudonorm $\|\cdot\|_{T,X,x_0,k}$ with $x_0 = 0$. We have that (see proof of Theorem 2)

$$\begin{aligned}
\sum_{m=0}^{\infty} \|u_m\|_{T,X,0,k} &= \sum_{m=0}^{\infty} \|F_0^m(u_0) - F_0^m(0)\|_{T,X,0,k} \\
&\leq \sum_{m=0}^{\infty} \frac{|\alpha|^m T^m}{m!} \|u_0\|_{T,X,0,k+m} \\
&= \sum_{m=0}^{\infty} \frac{|\alpha|^m T^m}{m!} \sup_{\substack{0\leq t\leq T \\ |x|\leq X}} \left| \sum_{j=0}^{\infty} \frac{(j+k+m)!}{j!} a_{j+k+m}(t) x^j \right| \\
&\leq \sum_{m=0}^{\infty} \sum_{j=0}^{\infty} |\alpha|^m T^m X^j \frac{(j+k+m)!}{m! j!} \|a_{j+k+m}\|_T \\
&= \sum_{s=0}^{\infty} \sum_{m=0}^{s} |\alpha|^m T^m X^{s-m} \frac{(k+s)!}{m!(m-s)!} \|a_{k+s}\|_T \\
&= \sum_{s=0}^{\infty} (|\alpha| T + X)^s \frac{(k+s)!}{s!} \|a_{k+s}\|_T .
\end{aligned}$$

By the root test, the above series is convergent, since

$$\sqrt[s]{(|\alpha|T + X)^s \frac{(k+s)!}{s!}} \underset{s\to\infty}{\longrightarrow} |\alpha|T + X$$

and $\sqrt[s]{\|a_{k+s}\|_T} \underset{s\to\infty}{\longrightarrow} 0$ by assumption. Therefore $\sum \|u_m\|_{T,X,0,k}$ is convergent, so that $\|u_m\|_{T,X,0,k} \underset{m\to\infty}{\longrightarrow} 0$ and thus u_m converges to 0, as we wanted to prove. □

We now combine Theorems 3 and 4 to prove our most general result.

Theorem 5 (Second global Fréchet space convergence theorem) *Consider the Fréchet space $C(\mathbb{T}, \mathscr{A})$ with pseudonorms $\|\cdot\|_{T,X,x_0,k}$ given by* (6). *Take an initial input $u_0 \in C(\mathbb{T}, \mathscr{A})$ and initial condition $g \in C^\infty(\mathbb{R})$. Assume also that u is uniformly entire and that g is entire. Let $F_g : C(\mathbb{T}, \mathscr{A}) \to C(\mathbb{T}, \mathscr{A})$ be given by* (3). *Then the sequence (u_m) given by $u_m = F^m(u_0)$ converges to a fixed point of F.*

Proof Let $F_g, F_0 : C(\mathbb{T}, \mathscr{A}) \to C(\mathbb{T}, \mathscr{A})$ be given by (3), (10). We observe that, for any $u, v \in C^{0,\infty}(\mathbb{X})$ we have

$$F_g(u+v) = g + \alpha \int_0^t (u+v)_x ds = g + \alpha \int_0^t u_x ds + \alpha \int_0^t v_x ds = F_g(u) + F_0(v).$$

We can then infer that $u_1 = F_g(u_0) = F_g(0 + u_0) = F_g(0) + F_0(u_0)$. Also, $u_2 = F_g(u_1) = F_g(F_g(0) + F_0(u_0)) = F_g^2(0) + F_0^2(u_0)$, and, in general,

$$u_m = F_g^m(0) + F_0^m(u_0).$$

By Theorem 3, $(F_g^m(0))$ converges to a fixed point of F_g. By Theorem 4, $(F_0^m(u_0))$ converges to 0. Therefore, (u_m) and $(F_g^m(0))$ have the same limit. In particular, (u_m) converges to a fixed point of F_g. □

A nice consequence of the proof is that it allows us to also establish uniqueness in a certain class of functions.

Theorem 6 (Uniqueness of uniformly entire fixed points) *Consider the Fréchet space $C(\mathbb{T}, \mathscr{A})$ with pseudonorms $\|\cdot\|_{T,X,x_0,k}$ given by* (6). *Take an initial condition $g \in C^\infty(\mathbb{R})$ and assume also that g is entire. Let $F_g : C(\mathbb{T}, \mathscr{A}) \to C(\mathbb{T}, \mathscr{A})$ be given by* (3). *Then there is at most one uniformly entire fixed point of F_g.*

Proof Let u be any uniformly entire fixed point of F_g. By the proof of Theorem 5, we know that $u = F_g^m(u) = F_g^m(0) + F_0^m(u)$. Since $(F_0^m(u_0))$ converges to 0, we get that $(F_g^m(0))$ converges to u. Thus any uniformly entire fixed point of F_g must coincide with the limit of $(F_g^m(0))$. □

5 Conclusion and Further Research

In this paper we have seen how to study solutions to differential equations as outputs of analog networks, and how to obtain them using fixed point techniques. The example we have considered ($L = \alpha\partial_x$) is a well-known problem whose solution can be obtained analytically by taking a Fourier transform. However, the method presented in this paper provides a different perspective which is suitable for analog computability and the study of analog systems as in [6, 11–13], where Fréchet spaces clearly provide a natural framework.

We intend to investigate this approach (fixed points in Fréchet spaces) in more general settings. In fact, as a next step one can look at a more general operator $L : \mathscr{A} \to \mathscr{A}$ using higher-order derivatives, for example with bounds of the form

$$\|Lu\|_{T,X,x_0,k} \leq C\|\partial_x^\ell u\|_{T,X,x_0,k} \leq C\|u\|_{T,X,x_0,k+\ell}. \tag{12}$$

We observe that analyticity of the initial condition g is not enough to ensure existence of solutions. A counterexample is given by the heat equation $u_t = u_{xx}$ with initial condition $g(x) = \frac{1}{1-x}$. Even though g is analytic near zero, the solution fails to be analytic at a neighborhood of the origin (see [2]). Thus, the general case may require different tools such as explicit bounds on the pseudonorms of g.

We also plan to investigate properties of the fixed points, such as *computability*, *continuity* and *stability*, as functions of the parameters and initial conditions.

References

1. Bush, V.: The differential analyzer. A new machine for solving differential equations. J. Frankl. Inst. **212**(4), 447–488 (1931)
2. Egorov, Y.V., Shubin, M.A.: Foundations of the Classical Theory of Partial Differential Equations. Springer-Verlag, Berlin (1998)
3. Hadamard, J.: Lectures on Cauchy's Problem in Linear Partial Differential Equations. Dover, New York (1952)
4. Hartree, D.R.: Calculating Instruments and Machines. Cambridge University Press, Cambridge (1950)
5. Holst, P.A.: Svein Rosseland and the Oslo Analyzer. IEEE Ann. Hist. Comput. **18**(4), 16–26 (1996)
6. James, N.D., Zucker, J.I.: A class of contracting stream operators. Comput. J. **56**, 15–33 (2013)
7. Johansson, M.: Early analog computers in Sweden – with examples from Chalmers University of Technology and the Swedish aerospace industry. IEEE Ann. Hist. Comput. **18**(4), 27–33 (1996)
8. Reed, M., Simon, B.: Methods of Modern Mathematical Physics: Functional Analysis. Academic Press Inc., San Diego (1980)
9. Small, J.S.: General-purpose electronic analog computing: 1945–1965. IEEE Ann. Hist. Comput. **15**(2), 8–18 (1993)
10. Thompson, W., Tait, P.G.: Treatise on Natural Philosophy, vol. 1, (Part I), 2nd edn. Cambridge University Press, Cambridge (1880)
11. Tucker, J.V., Zucker, J.I.: Computability of analog networks. Theor. Comput. Sci. **371**, 115–146 (2007)
12. Tucker, J.V., Zucker, J.I.: Continuity of operators on continuous and discrete time streams. Theor. Comput. Sci. **412**, 3378–3403 (2011)
13. Tucker, J.V., Zucker, J.I.: Computability of operators on continuous and discrete time streams. Computability **3**, 9–44 (2014)

A New Look at Dummy Derivatives for Differential-Algebraic Equations

John D. Pryce and Ross McKenzie

Abstract We show the dummy derivatives index reduction method for DAEs, introduced in 1993 by Mattsson & Söderlind, is a particular case of the Pryce Σ-method solution scheme. We give a pictorial display of the underlying block triangular form.

This approach gives a simple general method to cast the reduced system in semi-explicit index 1 form, combining order reduction and index reduction in one process.

It also shows each DD scheme for a given DAE is uniquely described by an integer "DDspec" vector $\boldsymbol{\delta}$.

The method is illustrated by an example.

We give various reasons why, contrary to common belief, converting further from semi-explicit index 1 form to an explicit ODE, can be a good idea for numerical solution.

1 Introduction

We give a brief summary, wherein terms in *slanted* font are defined more fully later.

The dummy derivatives (DDs) method for a differential-algebraic equation (DAE), described as a way to reduce it to a (locally) equivalent DAE of index 1, was introduced by Mattsson and Söderlind [2] (here, MS denotes this paper or its authors).

MS use the Pantelides method [4] for structural analysis (SA) of the DAE; this can be replaced by the Pryce *signature matrix method* (Σ-method) [5] which can do the same analysis more simply, and is more powerful in that Pantelides only applies to DAEs of first order. Hence the DAEs to which DDs, as described by MS, can be applied are precisely the *SA-friendly* DAEs: those for which SA (Pantelides or Pryce) *succeeds*, by giving a nonsingular *System Jacobian* at some point.

The batches of derivatives formed by DDs are exactly those in stages of the Σ-method's *standard solution scheme*, but in reverse order, as proved in [3]. This paper follows up this connection. Adapting notation in [5], we give a concise description

J.D. Pryce (✉) • R. McKenzie
School of Mathematics, Cardiff University, Cardiff CF24 4AG, Wales, UK
e-mail: prycejd1@cardiff.ac.uk; mckenzier1@cardiff.ac.uk

© Springer International Publishing Switzerland 2016

J. Bélair et al. (eds.), *Mathematical and Computational Approaches in Advancing Modern Science and Engineering*, DOI 10.1007/978-3-319-30379-6_64

of the block-triangular form (BTF) of dependencies between batches, plus a pictorial view. Each DD scheme for a given DAE is uniquely described by an integer *DD-specification vector* $\boldsymbol{\delta} = (\delta_1, \ldots, \delta_n)$. The DAE resulting from DDs is generally called "index 1" but has the somewhat stronger property of being an *implicit ODE*.

MS ignores the conceptually trivial but practically important issue of reduction to order 1, needed when giving a DAE to a solver such as DASSL [1]. Our approach unifies index and order reduction into one process.

2 Basic Theory

DAEs and index As in MS (but with a different notation), we consider DAEs of the general form

$$f_i(\,t,\ \text{the } x_j \text{ and derivatives of them}\,) = 0, \quad i = 1, \ldots, n, \tag{1}$$

where the $x_j(t),\ j = 1, \ldots, n$ are state variables that are functions of an independent variable t, usually regarded as time.

The classic differentiation index ν_d is the largest number of times some equation f_i of the system *must* be differentiated w.r.t. t to produce an enlarged system of equations, solvable as an algebraic system to give an equivalent ordinary differential equation system (ODE) for the state variables $x_j(t)$.

Example 1 To contrast the above with the DDs viewpoint below, consider the DAE

$$0 = A := x - u(t), \quad 0 = B := \dot{x} - y, \qquad \text{(dot denotes } \mathrm{d}/\mathrm{d}t\text{)}, \tag{2}$$

where u is a given forcing function. It is an edge-case (in computer science terms) in having a unique solution—zero degrees of freedom—but the general ν_d definition still applies. One must differentiate B once to give an equation with $\dot{y}$ in it; now $\ddot{x}$ appears, so differentiate A twice to produce $\ddot{x}$ which can then be eliminated. Thus $\nu_d = 2$, since combining $\ddot{x} - \dot{y} = 0$ and $\ddot{x} - \ddot{u}(t) = 0$ with the original set one obtains an ODE

$$\dot{x} = y, \quad \dot{y} = \ddot{u}(t). \tag{3}$$

Then (3) has the same solution as (2) through any *consistent* point of (2). But (3) has DOF $= 2$ so its general solution $x = u(t) + \alpha + \beta t,\ y = \dot{u}(t) + \beta$ (with α, β arbitrary scalars) includes many beside the unique solution $x = u(t),\ y = \dot{u}(t)$ of (2).

(DOF means the phrase "degrees of freedom", DOF is the corresponding number.)

Solving the ODE numerically as a proxy for the DAE results in progressive drift from the DAE's solution manifold, which in the above example, as a subset of the ODE's (t, x, y) state space, is the set $\mathcal{M} = \{\,(t, x, y) \in \mathbb{R}^3 \mid x = u(t), y = \dot{u}(t)\,\}$.

By contrast the DDs method gives a system genuinely equivalent to the original, but with no extra DOF that cause numerical drift. It is not concerned with *producing an ODE* in all the x_j but, like the Σ-method, seeks merely the least number of differentiations that allows us to *find* all the x_j as functions of t. In Example 1 the answer is clearly to differentiate A once and B not at all, and rearrange to get

$$x = u(t), \quad y = \dot{u}(t), \tag{4}$$

which unlike (3) has no degrees of freedom, i.e. shows the unique solution. Strictly by DD rules one obtains (4) via a DAE with three equations and variables, namely

$$0 = A = x - u(t), \quad 0 = B = x' - y, \quad 0 = \dot{A} = x' - \dot{u}(t),$$

where the "genuine" derivative $\dot{x}$ has been replaced by the dummy derivative x'.

Σ-method summary The Σ-method finds the $n \times n$ *signature matrix* $\Sigma = (\sigma_{ij})$, where σ_{ij} is the order of the highest derivative of x_j in f_i, if x_j occurs in f_i, and $-\infty$ if not. A *transversal* is a set T of n matrix positions (i,j), with just one in each row and each column, and we seek a *highest-value transversal* (HVT), such that the number $\mathrm{Val}(T) = \sum_{(i,j)\in T} \sigma_{ij}$ is maximised. We find corresponding dual variables, the *offsets*, integer n-vectors $\mathbf{c} = (c_1, \dots, c_n)$ and $\mathbf{d} = (d_1, \dots, d_n)$ satisfying

$$d_j - c_i \geq \sigma_{ij}, \qquad \text{with equality on a HVT.} \tag{5}$$

By default we choose the *canonical* offsets, the unique elementwise smallest ones having $\min_i c_i = 0$. Then c_i says how many times equation $f_i = 0$ is to be differentiated, and (5) fulfils the requirement stated in Sect. 2.2 of MS, that "the differentiated problem is structurally nonsingular with respect to its highest-order derivatives"—i.e., $x_j^{(d_j)}$ occurs in $f_i^{(c_i)}$ for each (i,j) in some transversal.

The Σ-method's *standard solution scheme* (SSS) says: let $k_\mathrm{d} = -\max_j d_j$ and $k_\mathrm{c} = -\max_i c_i$, then

$$\begin{array}{l} \text{for } k = k_\mathrm{d}, k_\mathrm{d}+1, \dots \text{ do} \\ \quad \text{Solve the } m_k \text{ equations:} \quad f_i^{(k+c_i)} = 0 \quad \text{for those } i \text{ such that } k + c_i \geq 0 \\ \quad \text{for the } n_k \text{ unknowns:} \quad x_j^{(k+d_j)} \quad \text{for those } j \text{ such that } k + d_j \geq 0, \\ \quad \text{using already known values from previous (with smaller } k\text{) stages.} \end{array} \tag{6}$$

The sizes m_k, n_k satisfy $m_k \leq n_k$, i.e., each set of equations is underdetermined or square. Clearly, m_k, n_k increase with k, and $= n$ for $k \geq 0$. The iteration starts at $k = k_\mathrm{d}$ because $m_k = 0$ for $k < k_\mathrm{c}$, $n_k = 0$ for $k < k_\mathrm{d}$, and $k_\mathrm{d} \leq k_\mathrm{c}$.

When solving a DAE by Taylor series expansion, k goes to some positive expansion order. In the DDs context we stop at $k = 0$. Here, a *consistent point* of the DAE shall mean a solution of (6) up to stage $k = 0$, i.e. values of $x_j^{(q)}$ for $q = 0 : d_j, j = 1 : n$, that satisfy $f_i^{(p)} = 0$ for $p = 0 : c_i,\ i = 1 : n$. A fuller definition is in [5].

Henceforth we assume SA *succeeds*—or the DAE is *SA-friendly*—meaning the $n \times n$ System Jacobian

$$\mathbf{J} = \left(\partial f_i^{(c_i)} / \partial x_j^{(d_j)}\right)_{i,j=1:n}. \tag{7}$$

is nonsingular at some consistent point. Then, the value $\mathrm{Val}(T)$ of a HVT, which also equals $\sum_j d_j - \sum_i c_i$, is the *number of degrees of freedom* DOF of the DAE; also

$$\text{the } m_k \times n_k \text{ Jacobian } \mathbf{J}_k = \left(\partial f_i^{(k+c_i)} / \partial x_j^{(k+d_j)}\right) \text{ of (6) is of full row rank,} \tag{8}$$

and $\mathbf{J}_k$ is the sub-matrix (rows where $k + c_i \geq 0$, columns where $k + d_j \geq 0$) of $\mathbf{J}$.

An *implicit ODE* is an SA-friendly DAE with $\mathbf{c} = 0$, so one can solve to get an ODE without differentiating any equations. For instance the DAE $\dot{x} + \dot{y} = 1$, $x - y = 0$ is index 1 but not an implicit ODE, while $\dot{x} + y = 1$, $x - y = 0$ is an implicit ODE.

Example 2 Figure 1 shows the DAE of the simple pendulum in Cartesian coordinates, with signature matrix Σ annotated with the canonical offsets, and its HVTs marked $^\bullet$, $^\circ$. The Jacobian $\mathbf{J}$ is annotated in the style of MS, also showing the offsets.

For conciseness we use notation introduced in [5]. Write x_{jl} to mean the lth derivative $x_j^{(l)}$ of variable $x_j = x_j(t)$, similarly f_{il} to mean the lth derivative $f_i^{(l)}$ of function f_i. Thus for an *index-pair* (i, l) or (j, l) unless said otherwise it is assumed that i, j, l are integers with $i, j \in 1:n$ and $l \geq 0$.

For a set S of index-pairs let $\mathbf{x}_S$ be the vector comprising all the x_{jl} for $(j, l) \in S$, listed in some arbitrary but fixed order. It may mean a vector of symbolic variables, or real numbers, or functions of t, depending on context. Similarly, let $\mathbf{f}_S$ be the vector comprising all the f_{il} for $(i, l) \in S$, in some arbitrary but fixed order.

For a given DAE with offsets c_i, d_j and for any integer k define index-pair sets

$$I_k = \{\, (i, l) \mid l = k + c_i \geq 0 \,\}, \qquad J_k = \{\, (j, l) \mid l = k + d_j \geq 0 \,\}. \tag{9}$$

Write $|X|$ for the number of elements in a finite set X. Then the sizes m_k, n_k in (6) are given by $|I_k| = m_k$ and $|J_k| = n_k$.

Further define $I_{\leq k}$ to be the union of I_l for all $l \leq k$, and similarly $I_{<k}$, $J_{\leq k}$, $J_{<k}$. Then the SSS, equation (6), taken up to $k = 0$, can be written

$$\begin{array}{c} \textsc{Pend} \\ 0 = A = \ddot{x} + x\lambda \\ 0 = B = \ddot{y} + y\lambda - G \\ 0 = C = x^2 + y^2 - L^2 \end{array}, \quad \Sigma = \begin{array}{c|ccc|c} & x & y & \lambda & c_i \\ \hline A & 2^\bullet & & 0^\circ & 0 \\ B & & 2^\circ & 0^\bullet & 0 \\ C & 0^\circ & 0^\bullet & & 2 \\ \hline d_j & 2 & 2 & 0 & \end{array}, \quad \mathbf{J} = \begin{array}{c|ccc|} & \ddot{x} & \ddot{y} & \lambda \\ \hline A & 1 & & x \\ B & & 1 & x \\ \ddot{C} & 2x & 2y & \end{array}.$$

Fig. 1 Pendulum DAE with signature matrix and Jacobian

k	-2	-1	0	1	$\cdots$
I_k			(1,0)	(1,1)	
			(2,0)	(2,1)	$\cdots$
	(3,0)	(3,1)	(3,2)	(3,3)	
$\mathbf{f}_{I_k}$	(C)	$(\dot{C})$	$(A,B,\ddot{C})$	$(\dot{A},\dot{B},\dddot{C})$	$\cdots$

k	-2	-1	0	1	$\cdots$
J_k	(1,0)	(1,1)	(1,2)	(1,3)	
	(2,0)	(2,1)	(2,2)	(2,3)	$\cdots$
			(3,0)	(3,1)	
$\mathbf{x}_{J_k}$	(x,y)	$(\dot{x},\dot{y})$	$(\ddot{x},\ddot{y},\lambda)$	$(\dddot{x},\dddot{y},\dot{\lambda})$	$\cdots$

Fig. 2 I_k, J_k, $\mathbf{f}_{I_k}$ and $\mathbf{x}_{J_k}$ for the pendulum, using offsets $\mathbf{c} = (0, 0, 2)$ and $\mathbf{d} = (2, 2, 0)$. For example, J_{-1} is the set $\{(1, 1), (2, 1)\}$ so that $\mathbf{x}_{J_{-1}} = (x_{11}, x_{21})$, which is the same as $(\dot{x}, \dot{y})$

$$\begin{aligned}&\text{for } k = k_{\mathrm{d}}, k_{\mathrm{d}}{+}1, \dots, 0 \text{ do}\\ &\qquad \text{solve } \mathbf{f}_{I_k}(t, \mathbf{x}_{J_{\le k}}) = 0, \quad \text{or equivalently } \mathbf{f}_{I_k}(t, \mathbf{x}_{J_{<k}};\ \mathbf{x}_{J_k}) = 0, \quad \text{for } \mathbf{x}_{J_k}.\end{aligned} \tag{10}$$

The second form separates the "already known" t and $\mathbf{x}_{J_{<k}}$ from the "to be found" $\mathbf{x}_{J_k}$.

For example, Fig. 2 tabulates I_k, J_k, $\mathbf{f}_{I_k}$ and $\mathbf{x}_{J_k}$ for the pendulum, taking x_1, x_2, x_3 to mean x, y, λ and f_1, f_2, f_3 to mean A, B, C, in the order given.

3 DDs as a Particular Case of the SSS

In the "vanilla" SSS, all components in $\mathbf{x}_{J_k}$ are on an equal footing. For numerical solution, in those stages $k < 0$ where an underdetermined system is to be solved because $m_k < n_k$, typically a trial $\mathbf{x}_{J_k}$ vector is projected on the manifold defined by $\mathbf{f}_{I_k} = 0$, using a Gauss-Newton method or similar, to obtain the accepted solution.

Example 3 For the pendulum, stage $k = -2$ consists in projecting a trial value of $\mathbf{x}_{J_{-2}} = (x_{10}, x_{20}) = (x, y)$ on the zero set of $\mathbf{f}_{I_{-2}} = (f_{30}) = (h)$, i.e. on the circle $x^2 + y^2 = L^2$ in the x, y plane. Stage $k = -1$ consists in projecting a trial value of $\mathbf{x}_{J_{-1}} = (x_{11}, x_{21}) = (\dot{x}, \dot{y})$ on the zero set of $\mathbf{f}_{I_{-1}} = (f_{31}) = (\dot{h})$, i.e. on the line $x\dot{x} + y\dot{y} = 0$ in the $\dot{x}, \dot{y}$ plane defined by the previously found values of x and y.

In DDs, the components in $\mathbf{x}_{J_k}$ are *not* on an equal footing. The method as presented in MS starts with the high-order derivatives $x_j^{(d_j)}$ and works down, at each stage selecting some derivatives to be dummy, which essentially means they are *found*, as functions of others. For the pendulum, cf. [2, Sect. 4, Example 3], when $y \neq 0$ one can choose $\ddot{y}, \dot{y}$ to be dummy derivatives y'', y', giving a reduced DAE where everything is a function of x and $\dot{x}$; when $x \neq 0$ one can do the reverse.

From the Σ-method angle it is natural to go the opposite way, starting with the low-order derivatives. DDs then becomes a particular way to perform the SSS, as we now describe. It is proved in [6] that for a given DAE, the set of DD schemes constructible by this "forward" method is identical with the set constructible by the classical "reverse" method.

For brevity let all the unknowns that are found by the SSS in stages $k = k_d : 0$ be called *items*. That is, an item is any $x_j^{(q)}$ with $0 \le q \le d_j, j \in 1:n$. Equivalently, the items are precisely the components of $\mathbf{x}_{J_{\le 0}}$.

The forward DDs method We assume henceforth that the functions f_i are sufficiently smooth for all needed uses of the Implicit Function Theorem (IFT). For each stage k the $m_k \times n_k$ system (10) is of full row rank by (8). Hence we can find subsets $F_k \subseteq J_k$ and $S_k = J_k \setminus F_k$ such that, denoting $n_k - m_k$ by DOF_k

$$|F_k| = m_k, \quad |S_k| = \text{DOF}_k,$$

and the $m_k \times m_k$ sub-matrix of $\mathbf{J}_k$ defined by

$$\mathbf{G}_k = (\text{columns of } \mathbf{J}_k \text{ corresponding to } F_k), \tag{11}$$

is *nonsingular* at the current consistent point. ($\mathbf{G}_k$ is the same as $G^{[\kappa]}$ in MS, where $\kappa = 1 - k$.) Then by the IFT the m_k items forming the vector $\mathbf{x}_{F_k}$ can locally be found as functions of the remaining DOF_k items, which form $\mathbf{x}_{S_k}$.

DOF_k is the number of DOF *introduced at solution stage k*. E.g, for the pendulum, $n_k - m_k = 2 - 1 = 1$ for $k = -2$ and $k = -1$. That is, one DOF is introduced at each of these stages: an arbitrary position on the circle, and an arbitrary velocity.

The subvector $\mathbf{x}_{F_k}$ of $\mathbf{x}_{J_k}$ comprises the items *found* at this stage as functions of the items in $\mathbf{x}_{S_k}$. We call the latter *state items*. We call the sequence of F_k, which defines all the found items, a *solving scheme*.

The *state vector* is the vector of all state items, that is

$$\mathbf{x}_S, \quad \text{where } S = \bigcup_{k_d \le k \le -1} S_k. \tag{12}$$

Note we stop at $k = -1$ because S_0 is empty. The components of $\mathbf{x}_S$ are the state variables of the implicit ODE, to which this solving scheme reduces the DAE. All the remaining items are "found" and form the *found vector*

$$\mathbf{x}_F, \quad \text{where } F = \bigcup_{k_d \le k \le 0} F_k. \tag{13}$$

Since each vector $\mathbf{x}_{J_k}$ is the concatenation of $\mathbf{x}_{F_k}$ and $\mathbf{x}_{S_k}$, the SSS in the form (10), combined with the IFT used in a solving scheme, becomes

$$\begin{aligned} &\text{for } k = k_d, k_d{+}1, \dots, 0 \text{ do} \\ &\quad \text{solve } \mathbf{f}_{I_k}(t, \mathbf{x}_{S_{<k}}, \mathbf{x}_{F_{<k}};\ \mathbf{x}_{S_k}, \mathbf{x}_{F_k}) = 0 \text{ for } \mathbf{x}_{F_k} \text{ as a function of } \mathbf{x}_{S_{\le k}} \text{ and } \mathbf{x}_{F_{<k}}. \end{aligned} \tag{14}$$

This set of equations has a block-triangular form (BTF) such that, by repeated use of the IFT, and repeated substitution, one has

Theorem 1 *Each found item can be expressed as a function of the state vector* $\mathbf{x}_S$.

Proof We show this for the case $k_d = -2$, from which the general proof is clear.

For $k=-2$, $\mathbf{x}_{S_{<k}}$ and $x_{F_{<k}}$ are empty vectors, so we find $\mathbf{x}_{F_{-2}}$ as a function of $\mathbf{x}_{S_{-2}}$. For $k = -1$, we find $\mathbf{x}_{F_{-1}}$ as a function of $\mathbf{x}_{S_{-1}}$, $\mathbf{x}_{S_{-2}}$ and $\mathbf{x}_{F_{-2}}$, which by substituting $\mathbf{x}_{F_{-2}}$ becomes a function of $\mathbf{x}_{S_{\le -1}}$. For $k = 0$, we find $\mathbf{x}_{F_0}$ as a function of $\mathbf{x}_{S_0}$, $\mathbf{x}_{S_{-1}}$, $\mathbf{x}_{S_{-2}}$, $\mathbf{x}_{F_{-1}}$ and $\mathbf{x}_{F_{-2}}$, which by substituting $\mathbf{x}_{F_{-1}}$ and $\mathbf{x}_{F_{-2}}$ becomes a function of $\mathbf{x}_{S_{\le 0}}$.

Since $\mathbf{x}_{S_{\le 0}}$ equals $\mathbf{x}_S$, of which $\mathbf{x}_{S_{\le -1}}$ and $\mathbf{x}_{S_{\le -2}}$ are subvectors, the result follows.

Numerically, this involves one root-finding at each k-stage, taking given input values of $\mathbf{x}_S$ and using the nonsingular matrices $\mathbf{G}_{k_d}, \mathbf{G}_{k_d+1}, \dots, \mathbf{G}_0 = \mathbf{J}$ in turn.

Not all solving schemes are useful in practice. As so far described, for the pendulum one can choose state items y at stage -2 and $\dot{x}$ at stage -1, which does not lead to an implicit ODE. The extra rule needed to make a useful scheme is

Definition 1 A *dummy derivative scheme* (DD scheme) is a solving scheme for which, if an item is a derivative of a found item, then it is also a found item. Equivalently, the projections $\hat{F}_k = \{j \mid (j,l) \in F_k\}$, of F_k on the j component, increase with k.

For instance, in the preceding paragraph, choosing y and $\dot{x}$ as state items violates this constraint, since x must be a found item while its derivative $\dot{x}$ is not.

Some derivative of each x_j is a found item, since stage $k = 0$ is an $n \times n$ system that solves for each leading derivative $x_j^{(d_j)}$. Hence there is a unique least integer $\delta_j \in 0 : d_j$ such that $x_j^{(\delta_j)}$ is a found item. The vector $\boldsymbol{\delta} = (\delta_1, \dots, \delta_n)$ is the *DD-specification vector*, or DDspec, of a DD scheme. Then

- An $x_j^{(l)}$ with $l < \delta_j$ is a state item.
- An item $x_j^{(l)}$ with $l > \delta_j$ is necessarily a found item and the derivative of a found item. It is a *dummy derivative*—this accords with the MS definition.
- The remaining n items $x_j^{(\delta_j)}$ are *loop-closers* if $\delta_j > 0$, as they create the relations in the reduced DAE that make an ODE, and *algebraic items* if $\delta_j = 0$ (see below).

The total number of state items equals DOF, and there are δ_j state items for each j. Hence any DDspec vector satisfies $0 \le \boldsymbol{\delta} \le \mathbf{d}$ (elementwise), and $\sum_j \delta_j = \text{DOF}$. Not every $\boldsymbol{\delta}$ with these properties defines a DD scheme. E.g., for the pendulum, $\boldsymbol{\delta} = (2, 0, 0)$, $(0, 2, 0)$ and $(1, 1, 0)$ obey these constraints but only the first two "work".

Figure 3 shows the process pictorially. On the left is the general scheme for the case $k_d = 2$. Each set of equations $\mathbf{f}_{I_k} = 0$ has its output (what is solved for) above it, and downward lines from it to the previously computed data it uses as input. On the right is the more detailed data for the particular case of Example 4 below. These are created by MATLAB functions, in the second case using structural data produced by DAESA's function `daeSA` from the MATLAB code for the DAE in Example 4.

The diagram uses natural notation for derivatives, e.g. x_2' (MATLAB graphics is poor at dots) instead of the x_{21} notation used in Example 4.

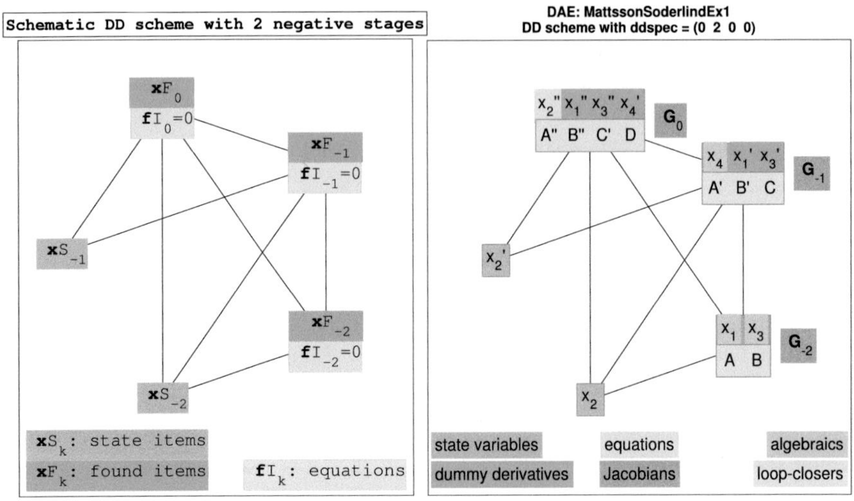

Fig. 3 DDs process picture. *Left*: general scheme, $k_d = 2$. *Right*: specific picture for Example 4

Scheme (14), whether done conceptually or numerically, splits the x_j into "differential variables" having $\delta_j > 0$, and "algebraic variables" having $\delta_j = 0$. E.g., if there are variables u, v, w whose δ's are $2, 1, 0$, then u, v are differential and w is algebraic. Two loop-closer equations create a 2nd-order ODE $\ddot{u} = U(u, \dot{u}, v)$, $\dot{v} = V(u, \dot{u}, v)$. After one solves the ODE to get $u, \dot{u}, v$ as functions of t, the relation $w{=}W(u, \dot{u}, v)$ gives w. As another example, the DAE in (2) has empty differential part.

Depending on DAE structure, one might find higher derivatives along the way, e.g. if doing DDs for the pendulum with state variables $x, \dot{x}$, one cannot find the algebraic variables y and λ without also finding $\dot{y}$ and $\ddot{y}$ at the same time.

4 Preparing for Numerical Solution

Converting to First Order Implicit ODE How to reduce the result of DDs to *first order* seems not well explained in the literature. Online Modelica tutorials commonly order-reduce *before* SA and DDs: clumsy and giving a larger final system than necessary. In fact this task fits almost trivially into DDs, in the following steps.

In Step 1, write down the SSS equations (6) up to stage 0—this is independent of any DD scheme. We use x_{jl} notation to show the items count as unrelated variables. There are DOF more variables than equations, $N{=}(n{+}\sum_j d_j)$ versus $M{=}(n{+}\sum_i c_i)$.

In Step 2, make the system "square" by adding DOF new equations

$$\dot{x}_{jl} = x_{j,(l+1)}, \tag{15}$$

one for each state variable x_{jl} in $\mathbf{x}_S$, to say what its derivative "really is". If $x_{j,(l+1)}$ is a state item, this implements order-reduction; otherwise it is a loop-closer.

In Step 3, rearrange the variables with the state vector $\mathbf{x}_S$ followed by the found vector $\mathbf{x}_F$, see (12) and (13), to get the result of DDs in the form

$$\dot{\mathbf{x}}_S = \mathbf{E}(t, \mathbf{x}_S, \mathbf{x}_F), \qquad 0 = \mathbf{F}(t, \mathbf{x}_S, \mathbf{x}_F). \tag{16}$$

Function $\mathbf{E}$ is very simple: its components are all the $x_{j,(l+1)}$'s of (15); it does not really depend on t. Function $\mathbf{F}$ is actually $\mathbf{f}_{I_{\le 0}}$. Write this as the concatenation of $\mathbf{f}_{I_k}$ listed $k = k_d, k_d+1, \dots, 0$, and $\mathbf{x}_F$ as the concatenation of $\mathbf{x}_{F_k}$ in the same order.

Then $\partial\mathbf{F}/\partial\mathbf{x}_F$ is block lower triangular with the nonsingular matrices $\mathbf{G}_k$ on its block diagonal, so it is nonsingular. Hence the two equations (16) form a *semi-explicit index-1* DAE, see [1, p. 34]. From this easily follows:

Theorem 2 *System (16) is SA-friendly with offsets c_i all 0, so it is an implicit ODE.*

Example 4 MS use the following linear, constant coefficient DAE [2, Example 1]:

$$\begin{array}{l} \text{MSEXAMPLE1} \\ 0 = A = x_1 + x_2 \qquad\qquad + u_1(t) \\ 0 = B = x_1 + x_2 + x_3 \qquad + u_2(t) \\ 0 = C = x_1 \qquad\quad + \dot{x}_3 + x_4 + u_3(t) \\ 0 = D = 2\ddot{x}_1 + \ddot{x}_2 + \ddot{x}_3 + \dot{x}_4 + u_4(t) \end{array}, \quad \Sigma = \begin{array}{c} \\ A \\ B \\ C \\ D \\ d_j \end{array}\!\!\begin{array}{c} \begin{array}{cccc} x_1 & x_2 & x_3 & x_4 \end{array} \\ \left[\begin{array}{cccc} 0^\bullet & 0^\circ & & \\ 0 & 0^\bullet & 0^\circ & \\ 0 & & 1^\bullet & 0^\circ \\ 2^\circ & 2 & 2 & 1^\bullet \end{array}\right] \\ \begin{array}{cccc} 2 & 2 & 2 & 1 \end{array} \end{array}\!\!\begin{array}{c} c_i \\ 2 \\ 2 \\ 1 \\ 0 \\ \\ \end{array} \quad \mathbf{J} = \begin{array}{c} \\ \ddot{A} \\ \ddot{B} \\ \dot{C} \\ D \end{array}\!\!\begin{array}{c} \begin{array}{cccc} \ddot{x}_1 & \ddot{x}_2 & \ddot{x}_3 & \dot{x}_4 \end{array} \\ \left[\begin{array}{cccc} 1 & 1 & 0 & 0 \\ 1 & 1 & 1 & 0 \\ 0 & 0 & 1 & 1 \\ 2 & 1 & 1 & 1 \end{array}\right] \end{array}.$$

The $u_i(t)$ are given forcing functions. The two HVTs of Σ are shown $^\bullet$ and $^\circ$. We have $\mathbf{c} = (2, 2, 1, 0)$ and $\mathbf{d} = (2, 2, 2, 1)$, so $\mathbf{J} = \partial(\ddot{A}, \ddot{B}, \dot{C}, D)/\partial(\ddot{x}_1, \ddot{x}_2, \ddot{x}_3, \dot{x}_4)$. It is nonsingular, so SA succeeds and there are 2 DOF.

Step 1, using the x_{il} notation, writes down $M = 9$ equations in $N = 11$ variables, grouped here by stages $k = -2, -1, 0$:

$$\left.\begin{array}{l} 0 = A = x_{10} + x_{20} \qquad\qquad + u_1(t) \\ 0 = B = x_{10} + x_{20} + x_{30} \qquad + u_2(t) \end{array}\right\} \tag{17}$$

$$\left.\begin{array}{l} 0 = \dot{A} = x_{11} + x_{21} \qquad\qquad + \dot{u}_1(t) \\ 0 = \dot{B} = x_{11} + x_{21} + x_{31} \qquad + \dot{u}_2(t) \\ 0 = C = x_{10} \qquad\quad + x_{31} + x_{40} + u_3(t) \end{array}\right\} \tag{18}$$

$$\left.\begin{array}{l} 0 = \ddot{A} = x_{12} + x_{22} \qquad\qquad + \ddot{u}_1(t) \\ 0 = \ddot{B} = x_{12} + x_{22} + x_{32} \qquad + \ddot{u}_2(t) \\ 0 = \dot{C} = x_{11} \qquad\quad + x_{32} + x_{41} + \dot{u}_3(t) \\ 0 = D = 2x_{12} + x_{22} + x_{32} + x_{41} + u_4(t) \end{array}\right\} \tag{19}$$

There are two possible DDspecs, $\boldsymbol{\delta} = (2, 0, 0, 0)$ and $(0, 2, 0, 0)$. Following MS, we choose the latter, so the state variables are x_{20} and x_{21}. Step 2 now defines two equations; the upper one does order reduction and the lower is a loop-closer:

$$\dot{\mathbf{x}}_S = \begin{bmatrix} \dot{x}_{20} \\ \dot{x}_{21} \end{bmatrix} = \begin{bmatrix} x_{21} \\ x_{22} \end{bmatrix} = \mathbf{E}(t, \mathbf{x}_S, \mathbf{x}_F). \tag{20}$$

Step 3 splits the variables into $\mathbf{x}_S = (x_{20};\ x_{21})^T$ and $\mathbf{x}_F = (x_{10}, x_{30};\ x_{11}, x_{31}, x_{40};\ x_{12}, x_{22}, x_{32}, x_{41})^T$, where the semicolons group into stages $k = -2, -1$ for $\mathbf{x}_S$ and $-2, -1, 0$ for $\mathbf{x}_F$. Then (17), (18), and (19) jointly define the equations

$$0 = \mathbf{F}(t, \mathbf{x}_S, \mathbf{x}_F), \tag{21}$$

and (20), (21) give the desired semi-explicit index 1 form.

The overall Jacobian is below, where a blank means an all-zero block of the appropriate size, and $\mathbf{G}_0 = \mathbf{J}$ with sub-matrices $\mathbf{G}_{-1} = \left[\begin{smallmatrix} 1&0&0\\ 1&1&0\\ 0&1&1 \end{smallmatrix}\right]$, $\mathbf{G}_{-2} = \left[\begin{smallmatrix} 1&0\\ 1&1 \end{smallmatrix}\right]$.

$$\frac{\partial(\mathbf{E},\mathbf{F})}{\partial(\mathbf{x}_S,\mathbf{x}_F)} = \begin{array}{r} \\ \\ \mathbf{E}=\begin{bmatrix} x_{21}\\ x_{22}\end{bmatrix} \\ \mathbf{f}_{I_{-2}}=\begin{bmatrix} A\\ B\end{bmatrix} \\ \mathbf{f}_{I_{-1}}=\begin{bmatrix} \dot A\\ \dot B\\ C\end{bmatrix} \\ \mathbf{f}_{I_0}=\begin{bmatrix} \ddot A\\ \ddot B\\ \dot C\\ D\end{bmatrix} \end{array} \begin{array}{cccc} \mathbf{x}_S & \mathbf{x}_{F_{-2}} & \mathbf{x}_{F_{-1}} & \mathbf{x}_{F_0} \\ [x_{20}, x_{21}] & [x_{10}, x_{30}] & [x_{11}, x_{31}, x_{40}] & [x_{12}, x_{22}, x_{32}, x_{41}] \\ \begin{bmatrix} 0&1\\ 0&0 \end{bmatrix} & & & \begin{bmatrix} 0&0&0&0\\ 0&1&0&0 \end{bmatrix} \\ \begin{bmatrix} 1&0\\ 1&0 \end{bmatrix} & \mathbf{G}_{-2} & & \\ \begin{bmatrix} 0&1\\ 0&1\\ 0&0 \end{bmatrix} & \begin{bmatrix} 0&0\\ 0&0\\ 1&0 \end{bmatrix} & \mathbf{G}_{-1} & \\ & & \begin{bmatrix} 0&0&0\\ 0&0&0\\ 1&0&0\\ 0&0&0 \end{bmatrix} & \mathbf{G}_0 \end{array}. \tag{22}$$

This formulation was confirmed by giving it to the MATLAB implicit solver `ode15i`, for certain $u_i(t)$ and initial values, and comparing with the analytic solution.

Converting to explicit ODE Conceptually it is obvious from the nonsingularity of $\partial\mathbf{F}/\partial\mathbf{x}_F$ in the DAE (16) that near a consistent point we can convert it to an explicit ODE $\dot{\mathbf{y}} = \mathbf{f}(t, \mathbf{y})$—solve the second equation for $\mathbf{x}_F$ and substitute in the first. It may seem strange to do this numerically, in view of the impressive track record of codes such as DASSL [1], for solving DAEs of the form $\mathbf{G}(t, \mathbf{z}, \dot{\mathbf{z}}) = \mathbf{0}$.

But there are reasons for doing so. Some applications generate models with n in the thousands and only a few DOF, and for which stiffness is not a problem. Then it makes sense to convert to a small explicit ODE and solve by, say, a Runge–Kutta code, using far less working memory than would a DAE code. The ESI-CyDesign Modelica system does numerical solution this way, having found it more efficient for their typical models. The Numerical Algorithms Group (NAG) Ltd have recently put a *reverse communication* RK code into their library, for just such uses.

Also, ways to compute the Jacobians $\mathbf{G}_k$ for an SA-friendly DAE are well known, while the full Jacobian of the DASSL-style formulation—in the example, the below-diagonal blocks of (21)—is messier to find; but see the discussion in Sect. 5.

Further (Andreas Rauh, Rostock, personal communication 2015), some calculations of robust design optimisation in control theory are far easier if the governing DAE (which is SA-friendly) has been converted to an explicit ODE.

5 Discussion

We have shown the dummy derivatives method is a particular case of the Σ-method's standard solution scheme, and displayed the underlying BTF pictorially. This gives a simple method to cast the DD'ed system as a semi-explicit index 1 DAE, hence an implicit ODE, combining order reduction and index reduction in one process. We pointed out that there may be compelling reasons for the further step of converting to an explicit ODE before numerical solution.

Step 3 of the method in Sect. 4 is optional—being an implicit ODE is independent of how the equations and variables are arranged. Thus different DD schemes for a DAE are distinguished purely by their sets of added equations (15). This set is deducible immediately from the DDspec vector $\boldsymbol{\delta}$. All this, except the vector $\boldsymbol{\delta}$, is implicit in Mattsson and Söderlind's paper, but the simplicity of the algorithm—hence of programming it—does not seem to be known in the literature.

A Jacobian such as (21) is used for Newton iterations inside a DASSL-style solver. In this context it seems one can ignore its "messy to compute" below-diagonal blocks, i.e. set them to 0, assuming a sufficiently good initial guess. Namely, as $\mathbf{G}_{-2}$ is correct it will make the $\mathbf{x}_{F_{-2}}$ component converge quadratically. Once this has happened, as $\mathbf{G}_{-1}$ is correct it will now make the $\mathbf{x}_{F_{-1}}$ component converge quadratically; and so on. In the context of solving Example 4 by `ode151` we have checked that this (surely well known) fact is true, but we have done no experiments to validate it on harder, nonlinear, problems.

References

1. Brenan, K.E., Campbell, S.L., Petzold, L.R.: Numerical Solution of Initial-Value Problems in Differential-Algebraic Equations, 2nd edn. SIAM, Philadelphia (1996)
2. Mattsson, S.E., Söderlind, G.: Index reduction in differential-algebraic equations using dummy derivatives. SIAM J. Sci. Comput. **14**(3), 677–692 (1993)
3. McKenzie, R.: Structural analysis based dummy derivative selection for differential-algebraic equations. Technical report, Cardiff University (2015). Submitted to BIT Numerical Analysis
4. Pantelides, C.C.: The consistent initialization of differential-algebraic systems. SIAM. J. Sci. Stat. Comput. **9**, 213–231 (1988)
5. Pryce, J.D.: A simple structural analysis method for DAEs. BIT Numer. Math. **41**(2), 364–394 (2001)
6. Pryce, J.D.: A simple approach to Dummy Derivatives for DAEs. Technical report, Cardiff University, July 2015. In preparation

New Master-Slave Synchronization Criteria of Chaotic Lur'e Systems with Time-Varying-Delay Feedback Control

Kaibo Shi, Xinzhi Liu, Hong Zhu, and Shouming Zhong

Abstract This study focuses on the issue of designing a time-delay output feedback controller for master-slave synchronization of chaotic Lur'e systems (CLSs). The time delay is assumed to be a time-varying continuous function which is bounded below and above by positive constants. By constructing an appropriate Lyapunov-Krasovskii functional (LKF), a novel delay-dependent synchronization condition is obtained. Besides, by employing a new free-matrix-based inequality (FMBI), the desired controller gain matrix can be achieved by solving a set of linear matrix inequalities (LMIs). Finally, one numerical example of Chua's circle is given to illustrate the effectiveness and advantages of the proposed results.

1 Introduction

During the past few decades, the chaotic synchronization problem has attracted increasing attention due to its extensive applications in many fields including secure communication, physical, chemical and ecological systems, human heartbeat regulation, and so on [1–4]. As is well-known that many nonlinear systems can be modeled precisely in the form of Lur'e systems, such as Chua's circuit, network systems and hyper chaotic attractors, which include a feedback connection of

K. Shi (✉) • H. Zhu
School of Automation Engineering, University of Electronic Science and Technology of China, Chengdu, Sichuan 611731, P.R. China
e-mail: skbs111@163.com; zhuhong@uestc.edu.cn

X. Liu
Department of Applied Mathematics, University of Waterloo, Waterloo, ON, N2L 3G1, Canada
e-mail: xzliu@uwaterloo.ca

S. Zhong
School of Mathematical Sciences, University of Electronic Science and Technology of China, Chengdu, 611731, P.R. China
e-mail: zhongsm@uestc.edu.cn

© Springer International Publishing Switzerland 2016

J. Bélair et al. (eds.), *Mathematical and Computational Approaches in Advancing Modern Science and Engineering*, DOI 10.1007/978-3-319-30379-6_65

a linear system and a nonlinear element satisfying the sector condition [5–7]. Therefore, master-slave synchronization for CLSs has been a focused research topic.

Recently, the synchronization problem of chaotic systems with time delay has been intensively investigated due to the unavoidable signal propagation delay frequently encountered in remote master-slave synchronization scheme [8–10]. Especially, many delay-independent and delay-dependent synchronization criteria have been derived by constructing an appropriate LKF in [11]. Compared with the results in [11], less conservative synchronization criteria were obtained and some fairly simple algebraic conditions are derived for easier verification in [12]. Based on Lyapunov method and LMIs approach, the authors in [13] further generalized and improved the proposed results in [11, 12]. However, in order to obtain sufficient conditions for master-slave synchronization, the authors in [11, 13] employed model transformation, which leads to some conservatism for inducing additional terms. Based on the free weighting matrix approach, a delay-dependent synchronization condition is obtained in [14]. By using a delay-partition approach, several delay-dependent synchronization criteria are derived in the form of LMIs in [15, 16]. By constructing an appropriate LKF including the information of time-varying delay range, new delay-range-dependent synchronization criteria for Lur'e systems are established in [17]. Besides, it should be noted that only constant delay is considered in [11–16, 18]. In practice, as everyone knows that the range of time-varying delay non-zero lower bound is often encountered, and such systems are referred to as interval delayed systems.

Motivated by the issues discussed above, the delay-dependent master-slave synchronization problem of CLSs with time-varying-delay feedback control is investigated in this paper. By taking full advantage of the information of time-varying-delay range and nonlinear term of the error system, a less conservative delay-dependent synchronization criterion is obtained. In addition, based on an appropriate LKF combined with a new FMBI, an explicit expression of the desired control law can be obtained in terms of LMIs. Finally, one numerical example of Chua's circle is presented to demonstrate the effectiveness and advantages of the design methods.

Notation Notations used in this paper are fairly standard: $\mathbb{R}^n$ denotes the n-dimensional Euclidean space, $\mathbb{R}^{n\times m}$ the set of all $n \times m$ dimensional matrices; $\mathbf{I}$ the identity matrix of appropriate dimensions, $\mathbf{A}^T$ the matrix transposition of the matrix $\mathbf{A}$. By $\mathbf{X} > 0$ (respectively $\mathbf{X} \geq 0$), for $\mathbf{X} \in \mathbf{R}^{n\times n}$, we mean that the matrix $\mathbf{X}$ is real symmetric positive definite (respectively, positive semi-definite); $diag\{r_1, r_2, \cdots, r_n\}$ denotes block diagonal matrix with diagonal elements $r_i, i = 1, \cdots, n$, the symbol $*$ represents the elements below the main diagonal of a symmetric matrix, $Sym\{\mathbf{M}\}$ is defined as $Sym\{\mathbf{M}\} = \frac{1}{2}(\mathbf{M} + \mathbf{M}^T)$.

2 Preliminaries

First, consider the following master-slave synchronization scheme of two identical chaotic Lur'e systems with time-varying delay feedback control:

$$\mathscr{M}: \begin{cases} \dot{\mathbf{x}}(t) = \mathbf{A}\mathbf{x}(t) + \mathbf{B}\varphi(\mathbf{D}\mathbf{x}(t)), \\ \mathbf{p}(t) = \mathbf{H}\mathbf{x}(t), \end{cases} \tag{1}$$

$$\mathscr{S}: \begin{cases} \dot{\mathbf{y}}(t) = \mathbf{A}\mathbf{y}(t) + \mathbf{B}\varphi(\mathbf{D}\mathbf{y}(t)) + \mathbf{u}(t), \\ \mathbf{q}(t) = \mathbf{H}\mathbf{y}(t), \end{cases} \tag{2}$$

$$\mathscr{C}: \mathbf{u}(t) = \mathbf{K}(\mathbf{p}(t-d(t)) - \mathbf{q}(t-d(t))), \tag{3}$$

which consists of master system $\mathscr{M}$, slave system $\mathscr{S}$ and controller $\mathscr{C}$. $\mathscr{M}$ and $\mathscr{S}$ with $\mathbf{u}(t) = 0$ are identical chaotic time-delay Lur'e systems with state vectors $\mathbf{x}(t)$, $\mathbf{y}(t) \in \mathbb{R}^n$, outputs of subsystems $\mathbf{p}(t)$, $\mathbf{q}(t) \in \mathbb{R}^l$, respectively, $\mathbf{u}(t) \in \mathbb{R}^n$ is the slave system control input, and $A \in \mathbb{R}^{n\times n}$, $B \in \mathbb{R}^{n\times n_d}$, $D \in \mathbb{R}^{n_d\times n}$, $H \in \mathbb{R}^{l\times n}$ are known real matrices, $K \in \mathbb{R}^{n\times l}$ is the time-varying delay controller gain matrix to be designed. It is assumed that $\varphi(\cdot)$ is the nonlinear function in the feedback path.

Assumption A Time-varying delay $d(t)$ is differential function and satisfies the following condition:

$$0 < d_L \le d(t) \le d_U, \quad \dot{d}(t) \le \mu < 1. \tag{4}$$

where d_L, d_U and μ are three real constants.

Assumption B The nonlinear function $\varphi_s(\alpha)$ satisfies the following sector condition: $(s = 1, \cdots, n_d)$

$$\varphi_s(\alpha) \in K_{[k_s^-, k_s^+]} = \{\varphi_s(\alpha) \mid \varphi_s(0) = 0, k_s^-\alpha^2 \le \alpha\varphi_s(\alpha) \le k_s^+\alpha^2, \alpha \ne 0\} \tag{5}$$

Next, given the synchronization schemes (1), (2) and (3), the synchronization error is defined as $\mathbf{r}(t) = \mathbf{x}(t) - \mathbf{y}(t)$, and we can get the following synchronization error system:

$$\dot{\mathbf{r}}(t) = \mathbf{A}\mathbf{r}(t) + \mathbf{B}\boldsymbol{\zeta}(\mathbf{D}\mathbf{r}(t), \mathbf{y}(t)) - \mathbf{K}\mathbf{H}\mathbf{r}(t-d(t)), \tag{6}$$

where $\boldsymbol{\zeta}(\mathbf{D}\mathbf{r}(t), \mathbf{y}(t)) = \boldsymbol{\varphi}(\mathbf{D}\mathbf{r}(t) + \mathbf{D}\mathbf{y}(t)) - \boldsymbol{\varphi}(\mathbf{D}\mathbf{y}(t))$. Let $\mathbf{D} = [\mathbf{d}_1, \cdots, \mathbf{d}_{n_d}]^T$ with $\mathbf{d}_s \in \mathbb{R}^n$ $(s = 1, 2, \cdots, n_d)$. Under Assumption B, it is easy to obtain that $\boldsymbol{\zeta}_s(\mathbf{d}_s^T\mathbf{r}, \mathbf{y})$ satisfies the following condition

$$k_s^- \le \frac{\boldsymbol{\zeta}_s(\mathbf{d}_s^T\mathbf{r}, \mathbf{y})}{\mathbf{d}_s^T\mathbf{r}} = \frac{\boldsymbol{\varphi}_s(\mathbf{d}_s^T(\mathbf{r}+\mathbf{y})) - \boldsymbol{\varphi}_s(\mathbf{d}_s^T\mathbf{y})}{\mathbf{d}_s^T\mathbf{r}} \le k_s^+, \forall \mathbf{r}, \mathbf{y} \in \mathbb{R}^n, \mathbf{d}_s^T\mathbf{r} \ne 0. \tag{7}$$

Lemma 1 ([19]) *For differentiable signal* $\mathbf{r}(t)$ *in* $[a,b] \rightarrow \mathbb{R}^n$, *for symmetric matrices* $\mathbf{R} \in \mathbb{R}^{n\times n}$, *and* $\mathbf{Z}_1 \in \mathbb{R}^{3n\times 3n}$, $\mathbf{Z}_3 \in \mathbb{R}^{3n\times 3n}$, *and any matrices* $\mathbf{Z}_2 \in \mathbb{R}^{3n\times 3n}$, *and* $\mathbf{N}_1 = \begin{bmatrix} \mathbf{N}_{11} \\ \mathbf{N}_{12} \\ \mathbf{N}_{13} \end{bmatrix}$, $\mathbf{N}_2 = \begin{bmatrix} \mathbf{N}_{21} \\ \mathbf{N}_{22} \\ \mathbf{N}_{23} \end{bmatrix} \in \mathbb{R}^{3n\times n}$ *satisfying* $\bar{\mathbf{R}} = \begin{bmatrix} \mathbf{Z}_1 & \mathbf{Z}_2 & \mathbf{N}_1 \\ * & \mathbf{Z}_3 & \mathbf{N}_2 \\ * & * & \mathbf{R} \end{bmatrix} > 0$, *the following inequality holds:*

$$-\int_a^b \dot{\mathbf{r}}^T(s)\mathbf{R}\dot{\mathbf{r}}(s)ds \leq \boldsymbol{\varpi}^T[\Omega_0 + Sym\{\mathbf{N}_1\Sigma_1 + \mathbf{N}_2\Sigma_2\}]\varpi,$$

where $\Omega_0 = (b-a)(\mathbf{Z}_1 + \frac{1}{3}\mathbf{Z}_3)$, $\Sigma_1 = [\mathbf{I}, -\mathbf{I}, \mathbf{0}]$, $\Sigma_2 = [-\mathbf{I}, -\mathbf{I}, 2\mathbf{I}]$,
$\varpi = [\mathbf{r}^T(a), \mathbf{r}^T(b), \frac{1}{b-a}\int_a^b \mathbf{r}^T(s)ds]^T$.

Remark 1 It should be noted that the set of slack variables in the above inequality can provide great freedom in deriving less conservative stability conditions, which it is possible to obtain a much sharper bound. It is easy to prove that some well-known integral inequalities are special cases of this one in [19]. For example, if we let $\mathbf{N}_1 = [\mathbf{Y}^T, 0]^T$, $\mathbf{N}_2 = 0$, $\mathbf{Z}_1 = diag\{\mathbf{X}, 0\}$, $\mathbf{Z}_2 = 0$ and $\mathbf{Z}_3 = 0$, this integral inequality can reduce to the integral inequality in [20]. And if we let $\mathbf{N}_1 = \frac{1}{b-a}[-\mathbf{R}, \mathbf{R}, 0]^T$, $\mathbf{N}_2 = \frac{3}{b-a}[\mathbf{R}, \mathbf{R}, -2\mathbf{R}]^T$, $\mathbf{Z}_1 = \mathbf{N}_1^T\mathbf{R}^{-1}\mathbf{N}_1$, $\mathbf{Z}_2 = \mathbf{N}_1^T\mathbf{R}^{-1}\mathbf{N}_2$, $\mathbf{Z}_3 = \mathbf{N}_2^T\mathbf{R}^{-1}\mathbf{N}_2$, this integral inequality also becomes the celebrated Wirtinger integral inequality [21].

3 Main Results

Theorem 1 *Under Assumptions A and B, for given scalars* μ, d_L, d_U, *the error system (6) is globally asymptotically stable if there exist positive matrices* $\mathbf{P}$, $\mathbf{R}_i$ $(i = 1, \cdots, 5)$, *any positive definite diagonal matrices* $\mathbf{G} = diag\{g_1, \cdots, g_{n_d}\}$, $\mathbf{L} = diag\{l_1, \cdots, l_{n_d}\}$, $\mathbf{W} = diag\{w_1, \cdots, h_{n_d}\}$, *symmetrical matrices* $\mathbf{X}_1$, $\mathbf{Y}_1$, $\mathbf{Z}_1$, $\mathbf{X}_3$, $\mathbf{Y}_3$ *and* $\mathbf{Z}_3$, *and any matrices* $\mathbf{X}_2$, $\mathbf{Y}_2$, $\mathbf{Z}_2$, $\mathbf{N}_i$ $(i = 1, 2, \cdots, 6)$, $\mathbf{N}$ *and* $\mathbf{U}$ *with the appropriate dimensions, such that the following symmetric linear matrix inequalities hold:*

$$\begin{bmatrix} \mathbf{X}_1 & \mathbf{X}_2 & \mathbf{N}_1 \\ * & \mathbf{X}_3 & \mathbf{N}_2 \\ * & * & \mathbf{R}_4 \end{bmatrix} > 0, \begin{bmatrix} \mathbf{Y}_1 & \mathbf{Y}_2 & \mathbf{N}_3 \\ * & \mathbf{Y}_3 & \mathbf{N}_4 \\ * & * & \mathbf{R}_5 \end{bmatrix} > 0 \quad and \quad \begin{bmatrix} \mathbf{Z}_1 & \mathbf{Z}_2 & \mathbf{N}_5 \\ * & \mathbf{Z}_3 & \mathbf{N}_6 \\ * & * & \mathbf{R}_5 \end{bmatrix} > 0, \tag{8}$$

$$\Psi < 0, \tag{9}$$

where

$$\begin{aligned}\Psi &= \mathbf{e}_1(\mathbf{R}_1+\mathbf{R}_3-\mathbf{D}\mathbf{K}^-\mathbf{W}\mathbf{K}^+\mathbf{D}^T)\mathbf{e}_1^T + e_2(d_L\mathbf{R}_4+d_{UL}R_5)e_2^T - \mathbf{e}_3(\mathbf{R}_1-\mathbf{R}_2)\mathbf{e}_3^T \\ &- \mathbf{e}_4\mathbf{R}_2\mathbf{e}_4^T - (1-\mu)\mathbf{e}_5\mathbf{R}_3\mathbf{e}_5^T - 2\mathbf{e}_6\mathbf{W}\mathbf{e}_6^T + Sym\{\mathbf{e}_1\mathbf{P}\mathbf{e}_2 + \mathbf{e}_1\mathbf{D}(\mathbf{K}^- + \mathbf{K}^+)\mathbf{e}_6^T \\ &+ e_2\mathbf{D}^T(\mathbf{G}-\mathbf{L})e_6^T + e_1\mathbf{D}^T[\mathbf{K}^+\mathbf{L}-\mathbf{K}^-\mathbf{G}]\mathbf{D}e_2^T + [\mathbf{e}_1x+\mathbf{e}_2y]\mathbf{N}[-\mathbf{e}_2^T+\mathbf{A}\mathbf{e}_1^T+\mathbf{B}\mathbf{e}_6^T \\ &- \mathbf{K}\mathbf{H}\mathbf{e}_4^T]\} + [\mathbf{e}_1,\mathbf{e}_3,\mathbf{e}_7]\left[d_L\left(\mathbf{X}_1+\frac{1}{3}\mathbf{X}_3\right) + Sym\{\mathbf{M}_1\Pi_1+\mathbf{M}_2\Pi_2\}\right][\mathbf{e}_1,\mathbf{e}_3,\mathbf{e}_7]^T \\ &+ [\mathbf{e}_3,\mathbf{e}_4,\mathbf{e}_8]\left[d_{UL}\left(\mathbf{Y}_1+\frac{1}{3}\mathbf{Y}_3\right) + Sym\{\mathbf{M}_3\Pi_1+\mathbf{M}_4\Pi_2\}\right][\mathbf{e}_3,\mathbf{e}_4,\mathbf{e}_8]^T \\ &+ [\mathbf{e}_4,\mathbf{e}_5,\mathbf{e}_9]\left[d_{UL}\left(\mathbf{Z}_1+\frac{1}{3}\mathbf{Z}_3\right) + Sym\{\mathbf{M}_5\Pi_1+\mathbf{M}_6\Pi_2\}\right][\mathbf{e}_4,\mathbf{e}_5,\mathbf{e}_9]^T,\end{aligned}$$

Moreover, the gain matrix of state estimator is given by $\mathbf{K}=\mathbf{N}^{-1}\mathbf{U}$.

Proof Consider the following a newly augmented LKF for the error system (6):

$$\mathbf{V}(\mathbf{r}_t) = \mathbf{V}_1(\mathbf{r}_t) + \mathbf{V}_2(\mathbf{r}_t) + \mathbf{V}_3(\mathbf{r}_t), \tag{10}$$

where

$$\begin{aligned}\mathbf{V}_1(\mathbf{r}_t) &= \mathbf{r}^T(t)\mathbf{P}\mathbf{r}(t) + \int_{t-h_L}^{t}\mathbf{r}^T(s)\mathbf{R}_1\mathbf{r}(s)ds \\ &+ \int_{t-d_U}^{t-d_L}\mathbf{r}^T(s)\mathbf{R}_2\mathbf{r}(s)\mathbf{d}s + \int_{t-d(t)}^{t}\mathbf{r}^T(s)\mathbf{R}_3\mathbf{r}(s)ds,\end{aligned} \tag{11}$$

$$\mathbf{V}_2(\mathbf{r}_t) = \int_{-d_L}^{0}\int_{t+\theta}^{t}\dot{\mathbf{r}}^T(s)\mathbf{R}_4\dot{\mathbf{r}}(s)dsd\theta + \int_{-d_U}^{-d_L}\int_{t+\theta}^{t}\dot{\mathbf{r}}^T(s)\mathbf{R}_5\dot{\mathbf{r}}(s)dsd\theta, \tag{12}$$

$$\mathbf{V}_3(\mathbf{r}_t) = 2\sum_{s=1}^{n_d} g_s\int_0^{d_s^T\mathbf{r}(t)}[\zeta_s(\theta)-k_s^-\theta]d\theta + 2\sum_{s=1}^{n_d} l_s\int_0^{d_s^T\mathbf{r}(t)}[k_s^+\theta-\zeta_s(\theta)]d\theta, \tag{13}$$

Taking the derivative of $\mathbf{V}(\mathbf{x}_t)$ along the trajectory of the error system (6) yields:

$$\begin{aligned}\dot{\mathbf{V}}_1(r_t) &= 2\mathbf{r}^T(t)\mathbf{P}\dot{\mathbf{r}}(t) + \mathbf{r}^T(t)(\mathbf{R}_1+\mathbf{R}_3)\mathbf{r}(t) - \mathbf{r}^T(t-d_L)(\mathbf{R}_1-\mathbf{R}_2)\mathbf{r}(t-d_L) \\ &- \mathbf{r}^T(t-d_U)\mathbf{R}_2\mathbf{r}(t-d_U) - (1-\dot{d}(t))\mathbf{r}^T(t-d(t))\mathbf{R}_3\mathbf{r}(t-d(t)) \\ &\le 2\mathbf{r}^T(t)\mathbf{P}\dot{\mathbf{r}}(t) + \mathbf{r}^T(t)(\mathbf{R}_1+\mathbf{R}_3)\mathbf{r}(t) - \mathbf{r}^T(t-d_L)(\mathbf{R}_1-\mathbf{R}_2)\mathbf{r}(t-d_L) \\ &- \mathbf{r}^T(t-d_U)\mathbf{R}_2\mathbf{r}(t-d_U) - (1-\mu)\mathbf{r}^T(t-d(t))\mathbf{R}_3\mathbf{r}(t-d(t)),\end{aligned} \tag{14}$$

$$\begin{aligned}\dot{\mathbf{V}}_2(\mathbf{r}_t) =&\dot{\mathbf{r}}^T(t)(d_L\mathbf{R}_4 + d_{UL}R_5)\dot{\mathbf{r}}(t)\\ &-\int_{t-d_L}^{t}\dot{\mathbf{r}}^T(s)\mathbf{R}_4\dot{\mathbf{r}}(s)ds - \int_{t-d_U}^{t-d_L}\dot{\mathbf{r}}^T(s)\mathbf{R}_5\dot{\mathbf{r}}(s)ds\\ =&\dot{\mathbf{r}}^T(t)(d_L\mathbf{R}_4 + d_{UL}\mathbf{R}_5)\dot{\mathbf{r}}(t) - \int_{t-d_L}^{t}\dot{\mathbf{r}}^T(s)\mathbf{R}_4\dot{\mathbf{r}}(s)ds\\ &-\int_{t-d(t)}^{t-d_L}\dot{\mathbf{r}}^T(s)\mathbf{R}_5\dot{\mathbf{r}}(s)ds - \int_{t-d_U}^{t-d(t)}\dot{\mathbf{r}}^T(s)\mathbf{R}_5\dot{\mathbf{r}}(s)ds, \end{aligned}\tag{15}$$

By using Lemma 1, we can have

$$\begin{aligned}&-\int_{t-d_L}^{t}\dot{\mathbf{r}}^T(s)\mathbf{R}_4\dot{\mathbf{r}}(s)ds\\ &\le \varpi_1^T(t)\left[d_L\left(\mathbf{X}_1 + \frac{1}{3}\mathbf{X}_3\right) + Sym\{\mathbf{M}_1\Pi_1 + \mathbf{M}_2\Pi_2\}\right]\varpi_1(t),\end{aligned}\tag{16}$$

$$\begin{aligned}&-\int_{t-d(t)}^{t-d_L}\dot{\mathbf{r}}^T(s)\mathbf{R}_5\dot{\mathbf{r}}(s)ds\\ &\le \varpi_2^T(t)\left[d_{UL}\left(\mathbf{Y}_1 + \frac{1}{3}\mathbf{Y}_3\right) + Sym\{\mathbf{M}_3\Pi_1 + \mathbf{M}_4\Pi_2\}\right]\varpi_2(t),\end{aligned}\tag{17}$$

$$\begin{aligned}&-\int_{t-d_U}^{t-d(t)}\dot{\mathbf{r}}^T(s)\mathbf{R}_5\dot{\mathbf{r}}(s)ds\\ &\le \varpi_3^T(t)\left[d_{UL}\left(\mathbf{Z}_1 + \frac{1}{3}\mathbf{Z}_3\right) + Sym\{\mathbf{M}_5\Pi_1 + \mathbf{M}_6\Pi_2\}\right]\varpi_3(t),\end{aligned}\tag{18}$$

where $d_{UL} = d_U - d_L$, $\varpi_1^T(t) = [\mathbf{r}^T(t), \mathbf{r}^T(t-d_L), \frac{1}{d_L}\int_{t-d_L}^{t}\mathbf{r}^T(s)ds]$, $\varpi_2^T(t) = [\mathbf{r}^T(t-d_L), \mathbf{r}^T(t-d(t)), \frac{1}{d(t)-d_L}\int_{t-d_L}^{t}\mathbf{r}^T(s)ds]$,$\Pi_1 = [\mathbf{I}, -\mathbf{I}, \mathbf{0}]$ $\varpi_3^T(t) = [\mathbf{r}^T(t-d(t)), \mathbf{r}^T(t-d_U), \frac{1}{d_U-d(t)}\int_{t-d_L}^{t}\mathbf{r}^T(s)ds]$, $\Pi_2 = [-\mathbf{I}, -\mathbf{I}, 2\mathbf{I}]$.

$$\begin{aligned}\dot{\mathbf{V}}_3(\mathbf{r}_t) =&2\sum_{s=1}^{n_d}(g_s - l_s)d_s^T\dot{r(t)}\zeta_s(d_s^T r(t)) + 2\sum_{s=1}^{n_d}d_s^T\dot{r(t)}(l_s k_s^+ - g_s k_s^-)\eth_s^T r(t)\\ =&2\dot{\mathbf{r}}^T(t)\mathbf{D}^T(\mathbf{G}-\mathbf{L})\zeta(\mathbf{D}\mathbf{r}(t)) + 2\mathbf{r}^T(t)\mathbf{D}^T[\mathbf{K}^+\mathbf{L} - \mathbf{K}^-\mathbf{G}]\mathbf{D}\dot{\mathbf{r}}(t),\end{aligned}\tag{19}$$

Next, for any scalars x and y, and arbitrary matrix $\mathbf{N}$ with appropriate dimensions, we can obtain

$$0 = 2[\dot{\mathbf{r}}^T(t)x + \mathbf{r}^T(t)y]\mathbf{N}[-\dot{\mathbf{r}}(t) + \mathbf{A}\mathbf{r}(t) + \mathbf{B}\zeta(\mathbf{D}\mathbf{r}(t), \mathbf{y}(t)) - \mathbf{K}\mathbf{H}\mathbf{r}(t - d(t))], \tag{20}$$

From Eq. (7), for any positive diagonal matrix $\mathbf{W} = diag\{w_1, \cdots, w_{n_d}\}$, it yields that

$$\begin{aligned} &-2\sum_{s=1}^{n_d}[\zeta_s(\mathbf{d}_s^T\mathbf{r}(t), \mathbf{y}(t)) - k_s^- \mathbf{d}_s^T\mathbf{r}(t)]w_s[\zeta_s(\mathbf{d}_s^T\mathbf{r}(t), \mathbf{y}(t)) - k_s^+ \mathbf{d}_s^T\mathbf{r}(t)] \\ =& -2\mathbf{r}^T(t)\mathbf{D}\mathbf{K}^-\mathbf{W}\mathbf{K}^+\mathbf{D}^+\mathbf{r}(t) + 2r^T(t)\mathbf{D}(\mathbf{K}^- + \mathbf{K}^+)\mathbf{W}\zeta(\mathbf{D}^T\mathbf{r}(t), \mathbf{y}(t)) \\ &- 2\zeta^T(\mathbf{D}^T\mathbf{r}(t), \mathbf{y}(t))\mathbf{W}\zeta(\mathbf{D}^T\mathbf{r}(t), \mathbf{y}(t)), \end{aligned} \tag{21}$$

Combining Eqs. (14), (15), (16), (17), (18), (19), (20), and (21) yields

$$\mathbf{V}(\mathbf{r}_t) \leq \xi^T(t)\Psi\xi(t), \tag{22}$$

where $\xi^T(t) = [\mathbf{r}^T(t), \dot{\mathbf{r}}^T(t), \mathbf{r}^T(t - d_L), \mathbf{r}^T(t - d(t)), \mathbf{r}^T(t - d_U), \zeta^T(\mathbf{D}^T\mathbf{r}(t), \mathbf{y}(t)), \varpi_1^T(t), \varpi_2^T(t), \varpi_3^T(t)]$,

From (9), we can have $\mathbf{V}(t, \mathbf{r}_t) \leq \varepsilon \parallel r(t) \parallel^2$ holds for any sufficiently small $\varepsilon > 0$.

Therefore, this implies the error system (6) is globally asymptotically stable. The proof is completed. □

4 Numerical Example

In this section, a numerical simulation example is given to show the effectiveness and correctness of the main results derived above.

Consider Chua's circuit example and the system equation in [14, 16–18] is given by

$$\begin{cases} \dot{x}_1(t) = \alpha(x_2(t) - h(x_1(t))), \\ \dot{x}_2(t) = x_1(t) - x_2(t) + x_3(t), \\ \dot{x}_3(t) = -\beta x_2(t), \end{cases} \tag{23}$$

with nonlinear characteristic

$$h(x_1(t)) = m_1x_1(t) + 0.5(m_0 - m_1)[|x_1(t) + c| - |x_1(t) - c|],$$

and parameters $m_0 = -\frac{1}{7}$, $m_1 = \frac{2}{7}$, $\alpha = 9$, $\beta = 14.28$, and $c = 1$. The system can be written in chaotic Lur'e system framework (1) with

$$\mathbf{A} = \begin{bmatrix} -\alpha m_1 - 1 & \alpha & 0 \\ 1 & -2 & 1 \\ 0 & -\beta & -1 \end{bmatrix}, \mathbf{B} = \begin{bmatrix} -\alpha(m_0 - m_1) \\ 0 \\ 0 \end{bmatrix}, \mathbf{D} = \mathbf{H} = \begin{bmatrix} 1 \\ 0 \\ 0 \end{bmatrix}.$$

Under the above conditions and $\mu = 0$, the maximum upper bounds on the allowable delays of h obtained from the above references [14, 16–18] and Theorem 1 are listed in Table 1. Thus, it is also to see that our result is more effective than the recently reported ones.

In addition, set $\varphi(\alpha) = 0.5(|\alpha+1|-|\alpha-1|)$, $h = 0.191$, $x(0) = [-0.2, -0.3, 0.2]$ and $y(0) = [0.5, 0.1, -0.6]$, the simulation results are shown in Figs. 1, 2, 3, 4, 5, and 6 for the above gain matrix. The trajectories of the master-slave systems with

Table 1 Maximum allowed delays h and the best gain matrices $\mathbf{K}$ for $\mu = 0$

Method	[18]	[14]	[17]	[16]	Theorem 1
h	0.141	0.180	0.183	0.185	0.188
K	$\begin{bmatrix} 6.0229 \\ 1.3367 \\ -2.1264 \end{bmatrix}$	$\begin{bmatrix} 3.9125 \\ 0.9545 \\ -3.8273 \end{bmatrix}$	$\begin{bmatrix} 4.1455 \\ 0.9250 \\ -4.2596 \end{bmatrix}$	$\begin{bmatrix} 4.0779 \\ 0.9087 \\ -4.3430 \end{bmatrix}$	$\begin{bmatrix} 4.6512 \\ 0.5992 \\ -4.5177 \end{bmatrix}$

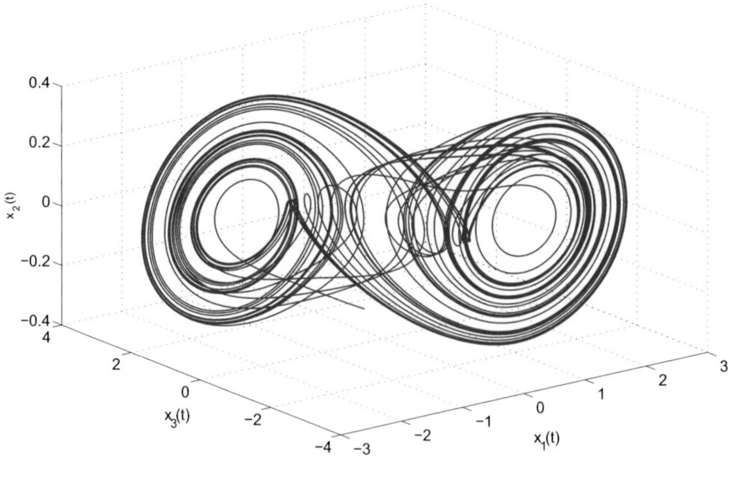

Fig. 1 State trajectories of master system $\mathbf{x}(t)$

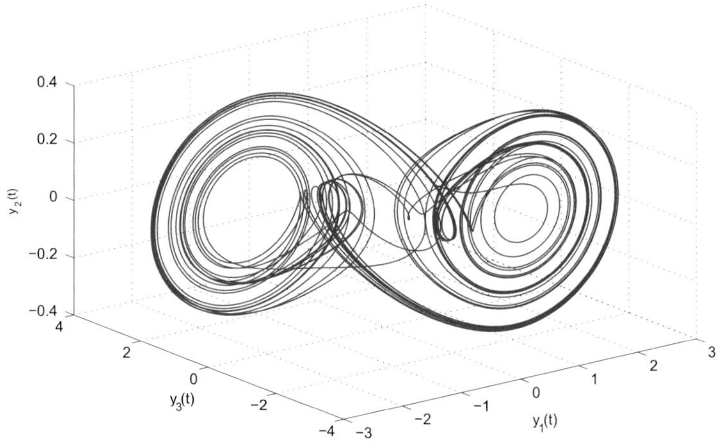

Fig. 2 State trajectories of slave system $\mathbf{y}(t)$ without $\mathbf{u}(t)$

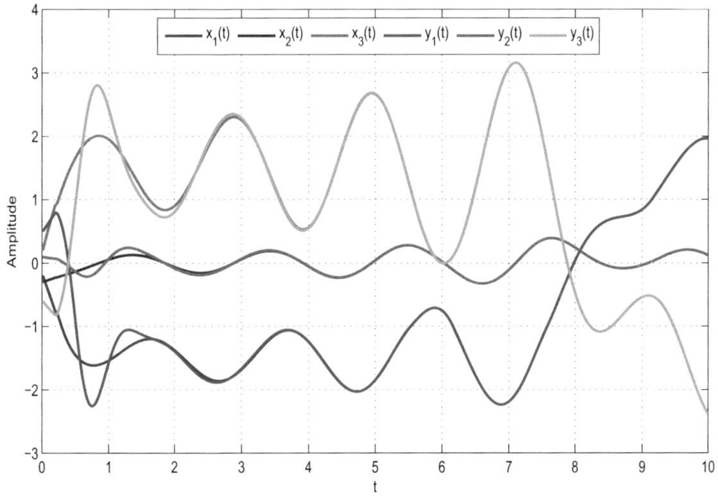

Fig. 3 State trajectories of master-slave systems $\mathbf{x}(t)$ and $\mathbf{y}(t)$ with $\mathbf{u}(t)$

$\mathbf{u}(t) = 0$ are shown in Figs. 1 and 2. Under the above gain matrix $\mathbf{K}$, the responses of the state $\mathbf{x}(t)$ and $\mathbf{y}(t)$, the error signal $\mathbf{r}(t)$, outputs of subsystems $\mathbf{p}(t)$ and $\mathbf{q}(t)$ are represented in Figs. 3, 4, 5, and 6, respectively. Therefore, it is clear to show that the synchronization error is tending asymptotically to zero.

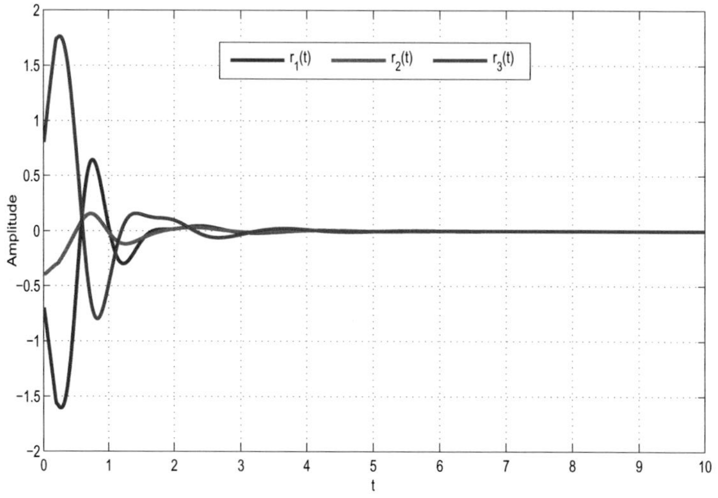

Fig. 4 State trajectories of synchronization error system $\mathbf{r}(t)$

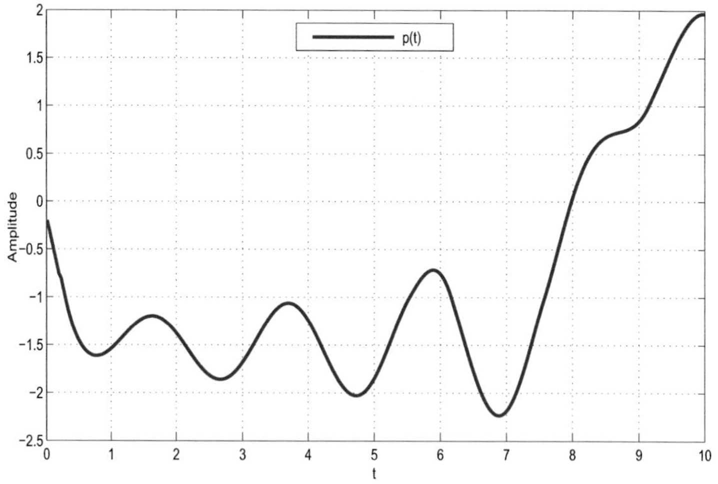

Fig. 5 State trajectories of output vector $\mathbf{p}(t)$

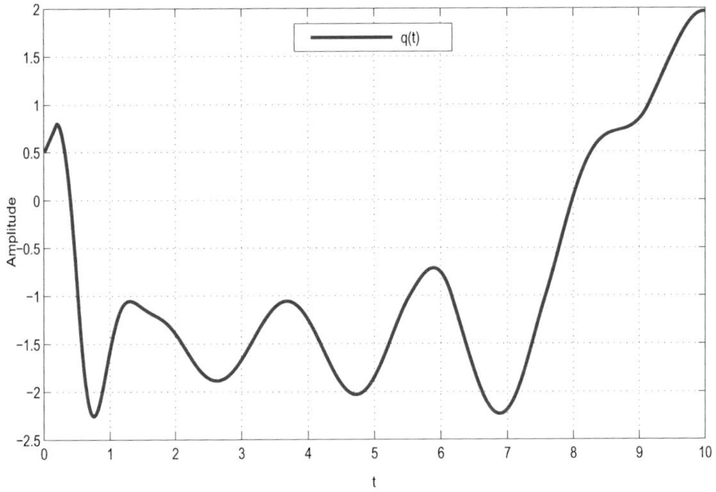

Fig. 6 State trajectories of output vector $\mathbf{q}(t)$

5 Conclusions

This paper consider the delay-dependent master-slave synchronization problem of CLSs with time-varying-delay feedback control. First, by choosing an augmented LKF, a novel delay-dependent synchronization criterion is derived. Second, by employing a new FMBI, which has been proved to be lesser conservative than the existing integral inequalities and showed to have a great potential efficient in the actual operation, the desired controller gain matrix can be achieved by solving a set of LMIs. Finally, one numerical example of Chua's circle is given to illustrate the effectiveness and advantages of the proposed results. The foregoing methods have the potential to be useful for the further study of CLSs. Meanwhile, it is expected that these approaches can be further used for other time-delay systems.

Acknowledgements This work was supported by National Basic Research Program of China (2010CB732501), National Natural Science Foundation of China (61273015), The National Defense Pre-Research Foundation of China (Grant No. 9140A27040213DZ02001), The Program for New Century Excellent Talents in University (NCET-10-0097).

References

1. Liao, T.L., Tsai, S.H.: Adaptive synchronization of chaotic systems and its application to secure communications. Chaos, Solitons Fractals **11**, 1387–1396 (2000)
2. Lu, J.N., Wu, X.Q., Lü, J.H.: Synchronization of a unified chaotic system and the application in secure communication. Phys. Lett. A **305**, 365–370 (2002)

3. Grzybowski, J.M.V., Rafikov, M., Balthazar, J.M.: Synchronization of the unified chaotic system and application in secure communication. Commun. Nonlinear Sci. Numer. Simul. **14**, 2793–2806 (2009)
4. Voss, H.: Anticipating chaotic synchronization. Phys. Rev. E **61**, 5115–5119 (2000)
5. Mkaouar, H., Boubaker, O.: Chaos synchronization for master slave piecewise linear systems: application to Chua's circuit. Commun. Nonlinear Sci. Numer. Simul. **17**, 1292–1302 (2012)
6. Gámez-Guzmán, L., Cruz-Hernández, C., López-Gutiérrez, R.M., García-Guerrero, E.E.: Synchronization of Chua's circuits with multi-scroll attractors: application to communication. Commun. Nonlinear Sci. Numer. Simul. **14**, 2765–2775 (2009)
7. Lü, J.H., Murali, K., Sinha, S., Leung, H., Aziz-Alaoui, M.A.: Generating multi-scroll chaotic attractors by thresholding. Phys. Lett. A **372**, 3234–3239 (2008)
8. Stepp, N.: Anticipation in feedback-delayed manual tracking of a chaotic oscillator. Exp. Brain Res. **198**, 521–525 (2009)
9. Pyragas, K., Pyragas, T.: Extending anticipation horizon of chaos synchronization schemes with time-delay coupling. Philos. Trans. R. Soc. A **368**, 305–317 (2010)
10. Milton, J.G.: The delayed and noisy nervous system: implications for neural control. J. Neural Eng. **8**, 065005 (2011)
11. Yalcin, M.E., Suykens, J.A.K., Vandewalle, J.: Master-slave synchronization of Lur'e systems with time-delay. Int. J. Bifurc. Chaos **11**(6), 1707–1722 (2001)
12. Liao, X.X., Chen, G.R.: Chaos synchronization of general Lur'e systems via time-delay feedback control. Int. J. Bifurc. Chaos **13**(1), 207–213 (2003)
13. Cao, J.D., Li, H.X., Daniel, W.C.H.: Synchronization criteria of Lur'e systems with time-delay feedback contro. Chaos, Solitons Fractals **23**, 1285–1298 (2005)
14. He, Y., Wen, G.L., Wang, Q.G.: Delay-dependent synchronization criterion for Lur'e systems with delay feedback control. Int. J. Bifurc. Chaos **16**(10), 3087–3091 (2006)
15. Ding, K., Han, Q.L.: Master-slave synchronization criteria for horizontal platform systems using time delay feedback control. J. Sound Vib. **330**(11), 2419–2436 (2011)
16. Ge, C., Hua, C.C., Guan, X.P.: Master-slave synchronization criteria of Lur'e systems with time-delay feedback control. Appl. Math. Comput. **244**, 895–902 (2014)
17. Li, T., Yu, J.J., Wang, Z.: Delay-range-dependent synchronization criterion for Lur'e systems with delay feedback control. Commun. Nonlinear Sci. Numer. Simul. **14**(5), 1796–1803 (2009)
18. Han, Q.L.: New delay-dependent synchronization criteria for Lur'e systems using time delay feedback control. Phys. Lett. A **360**(4), 563–569 (2007)
19. Zeng, H.B., He, Y., Wu, M., She, J.H.: Stability of time-delay systems via Wirtinger-based double integral inequality. IEEE Trans. Autom. Control. **60**(10), 2768–2772 (2015)
20. Zhang, X.M., Wu, M., She, J.H., He, Y.: Delay-dependent stabilization of linear systems with time-varying state and input delays. Automatica **41**, 1405–1412 (2005)
21. Seuret, A., Gouaisbaut, F.: Wirtinger-based integral inequality: application to time-delay systems. Automatica **49**, 2860–2866 (2013)

Robust Synchronization of Distributed-Delay Systems via Hybrid Control

Peter Stechlinski and Xinzhi Liu

Abstract Drive and response systems which exhibit time-delays and uncertainties are synchronized using hybrid control. Classes of dwell-time satisfying switching rules are identified under which synchronization can be achieved in a robust manner. The theoretical results are established using multiple Lyapunov functions and Halanay-like inequalities.

1 Introduction

Synchronization problems arise in secure communications [1], multi-agent networks [2], chaotic systems [3], neural networks [4], and many other important applications. Many real-world phenomena exhibit time-delays (e.g., see [5, 6]) and uncertainties in parameter measurements and data input [7], which can have significant impacts on control strategies for synchronization. The main focus of the present article is to investigate the synchronization of uncertain systems with distributed delays and nonlinear perturbations. The approach here is to use a combination of switching and impulsive control. This type of hybrid control, which can have a number of advantages over continuous control (see [8–10] for details), leads to a switched system model for the error between the drive and response systems.

Switched systems, which are a type of hybrid system, are governed by a mixture of continuous/discrete dynamics and logic-based switching, and have applications in a wide-range of problems in applied mathematics, engineering, and computer science [10]. In the switched systems literature, finding classes of switching rules which guarantee a performance goal (e.g., stability, synchronization, etc.) is a major area of research (e.g., see [8, 10–14]), where concepts such as multiple Lyapunov functions and dwell-time satisfying switching rules are used. This line of research has been extended to switched systems with time-delays composed of stable and unstable modes by way of Halanay-like inequalities (e.g., see [15–20]). Robust control (e.g., see [21, 22]) is important to take into account since relatively small

P. Stechlinski (✉) • X. Liu
University of Waterloo, Waterloo, ON, N2L 3G1, Canada
e-mail: pstechli@uwaterloo.ca; xzliu@uwaterloo.ca

© Springer International Publishing Switzerland 2016

J. Bélair et al. (eds.), *Mathematical and Computational Approaches in Advancing Modern Science and Engineering*, DOI 10.1007/978-3-319-30379-6_66

uncertainties can lead to instability (or, desynchronization in the present context), and has been studied in the time-delay switched systems literature (e.g., see [7, 23–25]).

Motivated by the above discussion, the main objective of this article is to extend the literature by studying the robust hybrid synchronization of systems with distributed delays. Contributions include establishing a switched and impulsive Halanay-like inequality using multiple Lyapunov functions to prove synchronization that is robust to uncertainties, and developing verifiable synchronization conditions in the face of time-delays. Classes of dwell-time, average dwell-time, and periodic switching rules are identified under which synchronization can be achieved via hybrid control. Methods are developed in the present article to deal with switching desynchronization (where impulsive control is crucial), nonlinear perturbations, as well as impulsive disturbances. The results are generalizable to stability studies of switched systems composed of a mixture of stable and unstable modes. The theoretical findings are augmented with numerical simulations.

2 Problem Formulation

Let $\mathbb{R}_+$ denote the set of non-negative real numbers, let $\mathbb{N}$ denote the set of positive integers, and let $\mathbb{R}^n$ denote the Euclidean space of n-dimensions (equipped with the Euclidean norm $\|\cdot\|$). For a positive constant τ, denote $PC = PC([-\tau, 0], \mathbb{R}^n)$ to be the space of piecewise continuous functions mapping $[-\tau, 0]$ to $\mathbb{R}^n$, equipped with the norm $\|\psi\|_\tau := \sup_{-\tau\leq s\leq 0} \|\psi(s)\|$. Let $\lambda_{\max}(\cdot)$ and $\lambda_{\min}(\cdot)$ denote the maximum and minimum eigenvalues of a symmetric matrix, respectively.

In the spirit of Guan et al. [14], consider the following drive system:

$$\dot{x}(t) = (A + \tilde{A}(t))x(t) + (C + \tilde{C}(t)) \int_{-\tau}^{0} x(t+s)ds + F(t, x_t), \tag{1}$$

where $\tau > 0$ is an upper bound on the distribution of time-delays; $x_t \in PC$ is defined by $x_t(s) := x(t+s)$ for $-\tau \leq s \leq 0$; $F : \mathbb{R} \times PC \to \mathbb{R}^n$ satisfies $F(t, 0) \equiv 0$ for all $t \geq t_0$; $A, C \in \mathbb{R}^{n\times n}$ are known; $\tilde{A}, \tilde{C}$ are uncertainties of the form $\tilde{A}(t) := \Gamma_A \Xi_A(t) \Upsilon_A$, $\tilde{C}(t) := \Gamma_C \Xi_C(t) \Upsilon_C$, where Γ_A, Γ_C, Υ_A, Υ_C are known constant matrices of appropriate dimensions and Ξ_A, Ξ_C are time-dependent uncertainty matrices.

The response system is given by

$$\dot{y}(t) = (A + \tilde{A}(t))y(t) + (C + \tilde{C}(t)) \int_{-\tau}^{0} y(t+s)ds + F(t, y_t) + u, \tag{2}$$

where $u := u_1 + u_2$ is a controller designed as follows:

$$u_1(t,x,y) := \sum_{k=1}^{\infty} \left[(H_{i_k} + \tilde{H}_{i_k}(t))e(t) + (G_{i_k} + \tilde{G}_{i_k}(t)) \int_{-\tau}^{0} e(t+s)ds \right] \mathbf{1}_{\mathscr{I}_k}(t),$$

$$u_2(t,x,y) := \sum_{k=1}^{\infty} (E_{i_k} + \tilde{E}_{i_k}(t))e(t)\delta(t - t_k^-),$$

where $e := y - x$ is the error between the drive and response systems; H_i, G_i, E_i, $i = 1, \ldots, m$ (for some positive integer m) are known constant control matrices, $\tilde{H}_i$, $\tilde{G}_i$, $\tilde{E}_i$, $i = 1, \ldots, m$, are time-dependent uncertainties of the form $\tilde{H}_i(t) := \Gamma_{H,i}\Xi_{H,i}(t)\Upsilon_{H,i}$, $\tilde{G}_i(t) := \Gamma_{G,i}\Xi_{G,i}(t)\Upsilon_{G,i}$, $\tilde{E}_i(t) := \Gamma_{E,i}\Xi_{E,i}(t)\Upsilon_{E,i}$, where $\Gamma_{H,i}$, $\Gamma_{G,i}$, $\Gamma_{E,i}$, $\Upsilon_{H,i}$, $\Upsilon_{G,i}$, $\Upsilon_{E,i}$ are known constant matrices of appropriate dimensions and $\Xi_{H,i}$, $\Xi_{G,i}$, $\Xi_{E,i}$ are time-dependent uncertainty matrices; $i_k \in \mathscr{P} := \{1, \ldots, m\}$ for each $k \in \mathbb{N}$, where m is a positive integer; $\mathscr{I}_k := [t_{k-1}, t_k)$ is called the switching or hybrid interval, $\mathbf{1}_{\mathscr{I}_k}(\cdot)$ is the indicator function, and $\delta(\cdot)$ is the generalized Dirac delta function. Under this construction, u_1 and u_2 are switching and impulsive controllers containing some uncertainties, respectively (see, e.g., [14] for a similar hybrid control construction without time-delays or uncertainties present).

The error system can be written as the following hybrid system:

$$\begin{aligned} \dot{e}(t) &= A_\sigma(t)e(t) + C_\sigma(t) \int_{-\tau}^{0} e(t+s)ds + F(t, y_t) - F(t, x_t), && t \neq t_k, \\ e(t) &= (I + E_\sigma(t^-))e(t^-), && t = t_k, \\ e_{t_0} &= \phi_0, && k = 1, 2, \ldots, \end{aligned} \tag{3}$$

where $t_0 \in \mathbb{R}$ is the initial time; $\phi_0 := y_{t_0} - x_{t_0} \in PC$ is the initial error; $A_i(t) := A + \tilde{A}(t) + H_i + \tilde{H}_i(t)$, $C_i(t) := C + \tilde{C}(t) + G_i + \tilde{G}_i(t)$, and $E_i(t) := E_i + \tilde{E}_i(t)$ for each $i = 1, \ldots, m$. The switching rule $\sigma : [t_0, +\infty) \to \mathscr{P}$ is a piecewise constant function that maps each switching interval $[t_{k-1}, t_k)$ to an integer in $\mathscr{P}$, i.e., it selects the active mode of (3). Denote the set of all such switching rules by $\mathscr{S}$.

Definition 1 The uncertainties are said to be admissible if they are continuous matrix functions satisfying $\|\Xi_A(t)\| < 1$, $\|\Xi_C(t)\| < 1$, $\|\Xi_{H,i}(t)\| < 1$, $\|\Xi_{G,i}(t)\| < 1$, and $\|\Xi_{E,i}(t)\| < 1$ for all $t \geq t_0$, $i = 1, \ldots, m$. The drive system (1) and the response system (2) are said to be *robustly synchronized* (RS) if $\lim_{t\to\infty} \|e(t)\| = 0$ for all $\phi_0 \in PC$ and admissible uncertainties. The systems are said to be *robustly exponentially synchronized* (RES) if there exist $\gamma > 0$, $C > 0$ such that, for $t \geq t_0$, $\|e(t)\| \leq C\|\phi_0\|_\tau \exp[-\gamma(t - t_0)]$ for all $\phi_0 \in PC$ and admissible uncertainties.

Remark 1 If $\sigma \in \mathscr{S}_{\text{inf-dwell}} := \{\sigma \in \mathscr{S} : \exists\zeta > 0 \text{ s.t. } \inf_{k\in\mathbb{N}}\{t_k - t_{k-1}\} \geq \zeta\}$, F_i is composite-PC in the sense of Ballinger and Liu [26], continuous in its first variable, and locally Lipschitz in its second variable for each $i = 1, \ldots, m$, then there exists a unique solution of (3) by [26] and using the method of steps over the hybrid moments [27].

Given the hybrid control matrices $\{(H_i, \tilde{H}_i, G_i, \tilde{G}_i, E_i, \tilde{E}_i) : i \in \mathscr{P}\}$, the goal of this article is to design the hybrid time mode sequences (i.e., the set $\{(i_k, t_k) : k \in \mathbb{N}\}$) such that the drive and response systems achieve robust synchronization.

2.1 Main Results

Consider the following notions of activation time and total number of switches.

Definition 2 Let $t^1, t^2 \in \mathbb{R}$ such that $t^2 \geq t^1$ and let $A \subset \mathscr{P}$. Denote the total activation time of the modes in A by $T_A(t^1, t^2) := \int_{t^1}^{t^2} \mathbf{1}_A(\sigma(t))dt$. Denote the total number of switches to a mode in A by $\Phi_A(t^1, t^2) := |\{t_k : \sigma(t_k) \in A, t^1 \leq t_k < t^2\}|$.

Zhu established the following Halanay-like lemma.

Lemma 1 ([18]) *Let $a, b > 0$. Assume that $u : [t_0 - \tau, \infty) \to \mathbb{R}_+$ satisfies $\dot{u}(t) \leq b\|u_t\|_\tau - au(t)$ for $t \geq t_0$. If $b - a \geq 0$ then $u(t) \leq \|u_{t_0}\|_\tau \exp[(b-a)(t-t_0)]$ for $t \geq t_0$, while if $b - a < 0$ then there exists a positive constant η satisfying $\eta + b\exp(\eta\tau) - a < 0$ such that $u(t) \leq \|u_{t_0}\|_\tau \exp[-\eta(t-t_0)]$ for $t \geq t_0$.*

For use in the main theorem, the following Halanay-like result is proved.

Proposition 1 *Let $a_i, b_i, d_i, h_i \geq 0$ for $i = 1, \ldots, m$. Assume that $u : [t_0 - \tau, +\infty) \to \mathbb{R}_+$ satisfies*

$$\begin{cases} \dot{u}(t) \leq b_\sigma \|u_t\|_\tau - a_\sigma u(t), & t \neq t_k, \quad t \geq t_0, \\ u(t) \leq d_\sigma u(t^-) + h_\sigma \|u_t\|_\tau, & t = t_k, \quad k = 1, 2, \ldots, \end{cases} \tag{4}$$

for some $\sigma \in \mathscr{S}$ which satisfies $t_k - t_{k-1} \geq \tau$ for all $k \in \mathbb{N}$. Then, for $t \geq t_0$,

$$u(t) \leq \|u_{t_0}\|_\tau \left(\prod_{j=1}^{\Phi_{\mathscr{P}}(t_0,t)} \delta_{i_j} \right) \exp\left\{ \sum_{i \in \mathscr{P}_u} \lambda_i T_{\{i\}}(t_0, t) - \sum_{i \in \mathscr{P}_s} \eta_i \tilde{T}_{\{i\}}(t_0, t) \right\}, \tag{5}$$

where $\tilde{T}_{\{i\}}(t_0, t) := T_{\{i\}}(t_0, t) - \Phi_{\{i\}}(t_0, t)\tau$, $\lambda_i := b_i \max_{i \in \mathscr{P}}\{1/\delta_i, 1\} - a_i$, $\delta_i := d_i + h_i e^{\xi\tau}$, $\xi := \max_{i \in \mathscr{P}_s}\{\xi_i\}$, $\xi_i > 0$ is chosen for $i \in \mathscr{P}_s$ so that $\xi_i + b_i e^{\xi_i \tau} - a_i < 0$, $\mathscr{P}_u := \{i \in \mathscr{P} : \lambda_i \geq 0\}$, $\mathscr{P}_s := \{i \in \mathscr{P} : \lambda_i < 0\}$, and $\eta_i > 0$ is chosen for $i \in \mathscr{P}_s$ so that $\eta_i + b_i \max_{i \in \mathscr{P}}\{1/\delta_i, 1\}\exp(\eta_i\tau) - a_i < 0$.

Proof Lemma 1 implies that, for $t \in [t_0, t_1)$,

$$u(t) \leq \begin{cases} \|u_{t_0}\|_\tau \exp[\lambda_{i_1}(t - t_0)], & \text{if } i_1 \in \mathscr{P}_u, \\ \|u_{t_0}\|_\tau \exp[-\eta_{i_1}(t - t_0)], & \text{if } i_1 \in \mathscr{P}_s, \end{cases}$$

where $\|u_{t_0}\|_\tau = \sup_{-\tau\le s\le 0} u(t_0+s)$. Let

$$w(t) := \|u_{t_0}\|_\tau \left(\prod_{j=1}^{\Phi_{\mathscr{P}}(t_0,t)} \delta_{i_j}\right) \exp\left[\Psi(t_0,t)\right]$$

where

$$\Psi(t_0,t) := \sum_{i\in\mathscr{P}_u} \lambda_i T_{\{i\}}(t_0,t) - \sum_{i\in\mathscr{P}_s} \eta_i \tilde{T}_{\{i\}}(t_0,t).$$

Suppose the result holds for $t\in[t_{k-1},t_k)$, that is, $u(t)\le w(t)$. We claim that $u(t)\le w(t)$ for $t\in[t_k,t_{k+1})$. If not, then there exists a time $t^*\in[t_k,t_{k+1})$ such that $u(t^*)=w(t^*)$, $u(t)\le w(t)$ for all $t\in[t_k,t^*)$ and for any $\varepsilon>0$ there exists a time $t_\varepsilon\in(t^*,t^*+\varepsilon)$ such that $u(t_\varepsilon)>w(t_\varepsilon)$. Suppose that $i_{k+1}\in\mathscr{P}_s$, then

$$\begin{aligned}\dot u(t^*) &\le \beta_{i_{k+1}} \sup_{-\tau\le s\le 0} u(t^*+s) - \alpha_{i_{k+1}} u(t^*),\\ &\le \beta_{i_{k+1}} \max_{i\in\mathscr{P}}\{\delta_i,1\}\|u_{t_0}\|_\tau \left(\prod_{j=1}^{k-1}\delta_{i_j}\right)\exp[\Psi(t_0,t_k)]\exp[-\eta_{i_{k+1}}(t^*-\tau-t_k)]\\ &\quad - \alpha_{i_{k+1}}\delta_{i_k}\|u_{t_0}\|_\tau \left(\prod_{j=1}^{k-1}\delta_{i_j}\right)\exp[\Psi(t_0,t_k)]\exp[-\eta_{i_{k+1}}(t^*-t_k)],\end{aligned}$$

and so

$$\dot u(t^*) \le \left[\beta_{i_{k+1}} \max_{i\in\mathscr{P}}\left\{\frac{1}{\delta_i},1\right\}\exp(\eta_{i_{k+1}}\tau) - \alpha_{i_{k+1}}\right]\|u_{t_0}\|_\tau \left(\prod_{j=1}^{k}\delta_{i_j}\right)\exp(\Psi(t_0,t^*)).$$

It follows that $\dot u(t^*) \le -\eta_{i_{k+1}} w(t^*) = \dot w(t^*)$, a contradiction to the definition of t_ε. On the other hand, if $i_{k+1}\in\mathscr{P}_u$, then

$$\begin{aligned}\dot u(t^*) &\le \beta_{i_{k+1}} \sup_{-\tau\le s\le 0} u(t^*+s) - \alpha_{i_{k+1}} v(t^*),\\ &\le \beta_{i_{k+1}} \max_{i\in\mathscr{P}}\{\delta_i,1\}\|u_{t_0}\|_\tau \left(\prod_{j=1}^{k-1}\delta_{i_j}\right)\exp[\Psi(t_0,t_k)]\exp[\lambda_{i_{k+1}}(t^*-t_k)]\\ &\quad - \alpha_{i_{k+1}}\delta_k\|u_{t_0}\|_\tau \left(\prod_{j=1}^{k-1}\delta_{i_j}\right)\exp[\Psi(t_0,t_k)]\exp[\lambda_{i_{k+1}}(t^*-t_k)],\end{aligned}$$

and therefore,

$$\dot{u}(t^*) \leq \left[\beta_{i_{k+1}} \max_{i \in \mathscr{P}} \left\{\frac{1}{\delta_i}, 1\right\} - \alpha_{i_{k+1}}\right] \|u_{t_0}\|_\tau \left(\prod_{j=1}^{k} \delta_{i_j}\right) \exp[\Psi(t_0, t^*)].$$

Thus, $\dot{u}(t^*) \leq \lambda_{i_{k+1}} w(t^*) = \dot{w}(t^*)$, a contradiction. The result holds by induction.

The nonlinear perturbations are assumed to satisfy the following condition.

Assumption 1 Assume there exist $\vartheta_1, \vartheta_2 \geq 0$ such that, for all $(t, \psi, \phi) \in \mathbb{R}_+ \times PC \times PC$, $\|F(t, \psi) - F(t, \phi)\| \leq \vartheta_1 \|\psi(0) - \phi(0)\| + \vartheta_2 \int_{-\tau}^{0} \|\psi(s) - \phi(s)\| ds$.

The following lemmata are required for the main result.

Lemma 2 (Matrix Cauchy Inequality) *For any symmetric positive definite matrix* $W \in \mathbb{R}^{n\times n}$ *and* $x, y \in \mathbb{R}^n$, $2x^T y \leq x^T W x + y^T W^{-1} y$.

Lemma 3 ([22]) *For any positive scalar* ε, *matrices* U, V, W *with* $W^T W \leq I$, $UWV + V^T W^T U^T \leq \varepsilon U U^T + \varepsilon^{-1} V^T V$.

Lemma 4 ([28]) *For any positive definite symmetric matrix* $W \in \mathbb{R}^{n\times n}$, *nonnegative scalar* v, *and* $w : [0, v] \to \mathbb{R}^n$, $\left(\int_0^v w(s)ds\right)^T W \left(\int_0^v w(s)ds\right) \leq v \int_0^v w^T(s) W w(s) ds$.

Lemma 5 ([23]) *Let* P *be a symmetric positive definite matrix and let* M, U, V *and* $W(t)$ *be real matrices of appropriate dimensions, with* $W(t)$ *being a matrix function. Then, for any* $\varepsilon > 0$ *such that* $P^{-1} - \varepsilon U U^T$ *is positive definite and* $W(t)^T W(t) \leq I$, $(M + UW(t)V)^T P (M + UW(t)V) \leq M(P^{-1} - \varepsilon U U^T)^{-1} M + \varepsilon^{-1} V^T V$ *for all* $t \in \mathbb{R}$.

We are now in a position to present the main synchronization result.

Theorem 1 *Suppose that Assumption 1 holds. For* $i = 1, \ldots, m$, *let* $P_i \in \mathbb{R}^{n\times n}$ *be positive definite symmetric matrices and let* $\varepsilon_{A,i}, \varepsilon_{C,i}, \varepsilon_{G,i}, \varepsilon_{H,i}$ *be positive constants such that* $P_i^{-1} - \varepsilon_{E,i} \Gamma_{E,i}^T \Gamma_{E,i}^T$ *is positive definite. Let*

$$\begin{aligned}
\beta_i :=& \tau^2 \lambda_{max}(P_i^{-1}(C + G_i)^T P_i (C + G_i)) + 2\tau \|P_i\| \vartheta_2 \lambda_{max}(P_i^{-1}) \\
&+ \tau^2 [\varepsilon_{G,i} \lambda_{max}(P_i^{-1} \Gamma_{G,i}^T \Gamma_{G,i}) + \varepsilon_{C,i} \lambda_{max}(P_i^{-1} \Gamma_C^T \Gamma_C)], \\
\alpha_i :=& -1 - 2\|P_i\| \vartheta_1 \lambda_{max}(P_i^{-1}) - \lambda_{max}(P_i^{-1}[(A + H_i)^T P_i + P_i(A + H_i) + \varepsilon_{A,i}^{-1} P_i \Upsilon_A \Upsilon_A^T P_i \\
&+ \varepsilon_{A,i} \Gamma_A^T \Gamma_A + \varepsilon_{H,i}^{-1} P_i \Upsilon_{H,i} \Upsilon_{H,i}^T P_i + \varepsilon_{H,i} \Gamma_{H,i}^T \Gamma_{H,i} + \varepsilon_{C,i}^{-1} P_i \Upsilon_C^T \Upsilon_C P_i + \varepsilon_{G,i}^{-1} P_i \Upsilon_{G,i}^T \Upsilon_{G,i} P_i]), \\
\mu_i :=& \frac{\lambda_{max}(P_i)}{\lambda_{min}(P_i)}, \quad \mu := \max_{i \in \mathscr{P}} \mu_i, \quad \lambda_i := \beta_i \max_{i \in \mathscr{P}} \left\{\frac{1}{\delta_i}, 1\right\} - \alpha_i, \\
\delta_i :=& \mu \lambda_{max}(P_i^{-1}[(I + E_i)^T (P_i^{-1} - \varepsilon_{E,i} \Gamma_{E,i}^T \Gamma_{E,i})^{-1} (I + E_i) + \varepsilon_{E,i}^{-1} \Upsilon_{E,i} \Upsilon_{E,i}^T]),
\end{aligned}$$

where $\eta_i > 0$, $i \in \mathscr{P}_s$, *satisfy* $\eta_i + \beta_i \max_{i \in \mathscr{P}}\{1/\delta_i, 1\} \exp(\eta_i \tau) - \alpha_i < 0$. *Let* $\mathscr{P}_u := \{i \in \mathscr{P} : \lambda_i \geq 0\}$, $\mathscr{P}_s := \{i \in \mathscr{P} : \lambda_i < 0\}$, $\lambda^+ := \max_{i \in \mathscr{P}_u} \lambda_i$,

$\lambda^- := \min_{i\in\mathscr{P}_s}\eta_i$, $T^+(t_0,t) := T_{\mathscr{P}_u}(t_0,t)$, $T^-(t_0,t) := T_{\mathscr{P}_s}(t_0,t) - \tau\Phi_{\mathscr{P}_s}(t_0,t)$, $\delta := \max_{i\in\mathscr{P}}\delta_i$. *Then, the following results hold:*

(i) If there exist $\chi > 0$, $\nu > 0$ *such that* $\tau \le t_k - t_{k-1} \le \chi$ *for all* $k \in \mathbb{N}$, $\delta < 1$, $T^+(t_0,t) \le \nu T^-(t_0,t)$, *and* $\ln(\delta)/\chi + \nu\lambda^+ - \lambda^- < 0$, *then RES is achieved.*

(ii) If there exist $\zeta > 0$, $\nu > 0$ *such that* $\tau \le \zeta \le t_k - t_{k-1}$ *for all* $k \in \mathbb{N}$, $\delta > 1$, $T^+(t_0,t) \le \nu T^-(t_0,t)$, *and* $\ln(\delta)/\zeta + \nu\lambda^+ - \lambda^- < 0$, *then RES is achieved.*

(iii) If there exist $\gamma > 1$, $\chi_k > 0$ *for* $k \in \mathbb{N}$ *such that* $\tau \le t_k - t_{k-1} \le \chi_k$, $\mathscr{P}_s \equiv \emptyset$, *and* $\ln(\gamma\delta_{i_k}) + \lambda_{i_k}\chi_k \le 0$ *for all* $k \in \mathbb{N}$, *then RS is achieved.*

(iv) If there exist $\gamma > 1$, $\zeta_k > 0$ *for* $k \in \mathbb{N}$ *such that* $\tau \le \zeta_k \le t_k - t_{k-1}$, $\mathscr{P}_u \equiv \emptyset$, *and* $\ln(\gamma\delta_{i_k}) - \eta_{i_k}(\zeta_k - \tau) \le 0$ *for* $k \in \mathbb{N}$, *then RS is achieved.*

(v) If there exist $t_{avg} > 0$, $N_0 > 0$ *such that* $k - 1 \le N_0 + (t - t_0)/t_{avg}$ *for any* $t \in [t_{k-1}, t_k)$, $\delta > 1$, *and* $\ln(\delta) - \eta t_{avg} < 0$, *then RS is achieved.*

(vi) If there exist $\rho_i > 0$ *for* $i \in \mathscr{P}$ *such that, for* $t \in [t_{k-1}, t_k)$, $\sigma(t) = k$ *and* $\sigma(t+\rho) = \sigma(t)$, *where* $\rho_i := t_i - t_{i-1}$, $\rho := \rho_1 + \rho_2 + \ldots + \rho_m$, *then RS is achieved if*

$$\sum_{i=1}^{m}\ln\delta_i + \sum_{i\in\mathscr{P}_u}\lambda_i\rho_i - \sum_{i\in\mathscr{P}_s}\eta_i(\rho_i - \tau) < 0. \tag{6}$$

Proof Let $V_i(e) := e^T P_i e$ for $i = 1, \ldots, m$, and let $\tilde{e}(t) := \int_{-\tau}^{0} e(t+s)ds$. The time-derivative of V_i along the ith mode of (3) for $t \ne t_k$ is given by

$$\frac{dV_i}{dt}(e(t)) = e^T(t)[(A + \tilde{A}(t) + H_i + \tilde{H}_i(t))^T P_i + P_i(A + \tilde{A}(t) + H_i + \tilde{H}_i(t))]e(t)$$
$$+ 2e^T(t)P_i(C + \tilde{C}(t) + G_i + \tilde{G}_i(t))\tilde{e}(t) + 2e^T(t)P_i[F_i(t,y_t) - F_i(t,x_t)].$$

By Lemma 2, $2e^T(t)P_i(C+G_i)\tilde{e}(t) \le e^T(t)P_ie(t) + \tilde{e}^T(t)(C+G_i)^TP_i(C+G_i)\tilde{e}(t)$. Furthermore, Lemma 3 gives that

$$e^T(t)(\tilde{A}^T(t)P_i + P_i\tilde{A}(t))e(t) \le \varepsilon_{A,i}^{-1}e^T(t)P_i\Upsilon_A\Upsilon_A^TP_ie(t) + \varepsilon_{A,i}e^T(t)\Gamma_A^T\Gamma_Ae(t),$$

$$e^T(t)(\tilde{H}_i^T(t)P_i + P_i\tilde{H}_i(t))e(t) \le \varepsilon_{H,i}^{-1}e^T(t)P_i\Upsilon_{H,i}\Upsilon_{H,i}^TP_ie(t) + \varepsilon_{H,i}e^T(t)\Gamma_{H,i}^T\Gamma_{H,i}e(t),$$

$$2e^T(t)P_i(\tilde{C}(t) + \tilde{G}_i(t))\tilde{e}(t) \le \varepsilon_{C,i}^{-1}e^T(t)P_i\Upsilon_C^T\Upsilon_CP_ie(t) + \varepsilon_{C,i}\tilde{e}^T(t)\Gamma_C^T\Gamma_C\tilde{e}(t)$$
$$+ \varepsilon_{G,i}^{-1}e^T(t)P_i\Upsilon_{G,i}^T\Upsilon_{G,i}P_ie(t) + \varepsilon_{G,i}\tilde{e}^T(t)\Gamma_{G,i}^T\Gamma_{G,i}\tilde{e}(t).$$

Lemma 4 implies that

$$\tilde{e}^T(t)(C+G_i)^TP_i(C+G_i)\tilde{e}(t) \le \tau\int_{-\tau}^{0} e^T(t+s)(C+G_i)^TP_i(C+G_i)e(t+s)ds,$$

and can be applied similarly to the terms $\tilde{e}^T(t)\Gamma_{C,i}^T\Gamma_{C,i}\tilde{e}(t)$ and $\tilde{e}^T(t)\Gamma_{G,i}^T\Gamma_{G,i}\tilde{e}(t)$.

By Assumption 1,

$$2e^T(t)P_i[F(t,y_t)-F(t,x_t)] \leq 2\|e(t)\|\|P_i\|\left(\vartheta_1\|e(t)\|+\vartheta_2\int_{-\tau}^{0}\|e(t+s)\|ds\right).$$

At the switching and impulsive moment $t = t_k$,

$$\begin{aligned}
V_{i_{k+1}}(e(t_k)) &\leq \mu e^T(t_k^-)(I+E_{i_k}+\tilde{E}_{i_k}(t_k^-))^T P_{i_k}(I+E_{i_k}+\tilde{E}_{i_k}(t_k^-))e(t_k^-)\\
&\leq \mu e^T(t_k^-)[(I+E_{i_k})^T(P_i^{-1}-\varepsilon_{E,i_k}\Gamma_{E,i_k}^T\Gamma_{E,i_k})^{-1}(I+E_{i_k})]e(t_k^-)\\
&\quad + \mu e^T(t_k^-)[\varepsilon_{E,i_k}^{-1}\Upsilon_{E,i_k}\Upsilon_{E,i_k}^T]e(t_k^-),
\end{aligned}$$

by Lemma 5. Then, using $\int_{-\tau}^{0} e^T(t+s)P_ie(t+s)ds \leq \tau \sup_{-\tau\leq s\leq 0} e^T(t+s)P_ie(t+s)$ and the well-known fact that for any positive definite matrix $P \in \mathbb{R}^{n\times n}$, symmetric matrix $Q \in \mathbb{R}^{n\times n}$, and $x \in \mathbb{R}^n$, $\lambda_{\min}(P^{-1}Q)x^TPx \leq x^TQx \leq \lambda_{\max}(P^{-1}Q)x^TPx$, it follows that $v(t) := V_\sigma(e(t))$ satisfies the conditions of Proposition 1 with $b_i := \beta_i$, $a_i := \alpha_i$, $h_i := 0$, $d_i := \delta_i$.

To prove case (i), note that $\phi_{\mathscr{P}}(t_0,t) = k-1$ for $t \in [t_{k-1},t_k)$, and $t_k - t_{k-1} \leq \chi$ implies that $t - t_0 \leq k\chi$ for $t \in [t_{k-1},t_k)$. Equation (5) implies that

$$\begin{aligned}
v(t) &\leq \|v_{t_0}\|_\tau \exp\left[(k-1)\ln\delta + \lambda^+T^+(t_0,t) - \lambda^-T^-(t_0,t)\right] \qquad (7)\\
&\leq \|v_{t_0}\|_\tau(1/\delta)\exp\left[(t-t_0)\ln(\delta)/\chi + \lambda^+T^+(t_0,t) - \lambda^-T^-(t_0,t)\right]\\
&\leq \|v_{t_0}\|_\tau(1/\delta)\exp\left[(t-t_0)\ln(\delta)/\chi + (\nu\lambda^+ - \lambda^-)T^-(t_0,t)\right]\\
&\leq \|v_{t_0}\|_\tau(1/\delta)\exp\left[(\ln(\delta)/\chi + \nu\lambda^+ - \lambda^-)(t-t_0)/M\right]
\end{aligned}$$

for some $M > 0$ since $\sup_{t\geq t_0}\{(t-t_0)/T^-(t_0,t)\}$ is finite. Robust exponential synchronization of the drive and response systems follows. Since $\tau \leq \zeta \leq t_k - t_{k-1}$ implies that $t - t_0 \geq k\zeta$ for $t \in [t_{k-1},t_k)$, case (ii) is proved by similar arguments used to show case (i).

To prove case (iii), RS is implied from Eq. (5) since, for $t \in [t_{k-1},t_k)$,

$$\begin{aligned}
v(t) &\leq \|v_{t_0}\|_\tau \exp\left[\sum_{j=1}^{k-1}\ln\delta_{i_j} + \sum_{i\in\mathscr{P}_u}\lambda_iT_{\{i\}}(t_0,t)\right]\\
&\leq \|v_{t_0}\|_\tau \exp\left[\ln\delta_{i_1} + \lambda_{i_1}\chi + \ln\delta_{i_2} + \lambda_{i_2}\chi + \ldots + \ln\delta_{i_{k-1}} + \lambda_{i_{k-1}}\chi + \lambda_{i_k}\chi\right].
\end{aligned}$$

That is, $v(t) \leq c\gamma^{-k}$ where $c := \|v_{t_0}\|_\tau/\min_{i\in\mathscr{P}}\delta_i$. Case (iv) is proved similarly.

Beginning from Eq. (7), case (v) implies that, for $t \in [t_{k-1},t_k)$,

$$\begin{aligned}
v(t) &\leq \|v_{t_0}\|_\tau(\delta)^{N_0}\exp[(t-t_0)\ln(\delta)/t_{\text{avg}} + (\nu\lambda^+ - \lambda^-)T^-(t_0,t)]\\
&\leq \|v_{t_0}\|_\tau(\delta)^{N_0}\exp[(\ln(\delta)/t_{\text{avg}} + \nu\lambda^+ - \lambda^-)(t-t_0)/M],
\end{aligned}$$

implying RS. For case (vi), Eq. (5) implies that, for $j \in \mathbb{N}$,

$$\begin{aligned}
& v(t_0 + j\omega) \\
& \leq \|v_{t_0}\|_\tau \exp\left[\sum_{i=1}^{jm} \ln \delta_i + \sum_{i \in \mathscr{P}_u} \lambda_i T_{\{i\}}(t_0, t_0 + j\omega) - \sum_{i \in \mathscr{P}_s} \eta_i \tilde{T}_{\{i\}}(t_0, t_0 + j\omega)\right] \\
& = \|v_{t_0}\|_\tau \exp\left[j \sum_{i=1}^{m} \ln \delta_i + j \sum_{i \in \mathscr{P}_u} \lambda_i T_{\{i\}}(t_0, t_0 + \omega) - j \sum_{i \in \mathscr{P}_s} \eta_i \tilde{T}_{\{i\}}(t_0, t_0 + \omega)\right] \\
& \leq \|v_{t_0}\|_\tau \exp\left[j\left(\sum_{i=1}^{m} \ln \delta_i + \sum_{i \in \mathscr{P}_u} \lambda_i \rho_i - \sum_{i \in \mathscr{P}_s} \eta_i(\rho_i - \tau)\right)\right].
\end{aligned}$$

Since $v(t)$ is also bounded on any compact interval, the result follows.

3 Example

Consider system (3) with $n = 2$, $\mathscr{P} = \{1, 2, 3, 4\}$, $t_0 = 0$, distributed delay $\tau = 0.1$,

$$A = \begin{pmatrix} 0 & -1 \\ -1 & 0 \end{pmatrix}, \quad C = \begin{pmatrix} 2 & 0.5 \\ 0 & 1.4 \end{pmatrix},$$

$$H_1 = \begin{pmatrix} 3.1 & 0 \\ 0 & 3.1 \end{pmatrix}, \quad H_2 = \begin{pmatrix} 1.8 & 0.1 \\ -0.1 & 1.6 \end{pmatrix}, \quad H_3 = \begin{pmatrix} -7.1 & -1.3 \\ 1.3 & -22.3 \end{pmatrix}, \quad H_4 = \begin{pmatrix} -15 & 2 \\ -2 & -8.8 \end{pmatrix},$$

$$G_1 = \begin{pmatrix} 1.05 & 1.1 \\ 1.11 & 1.04 \end{pmatrix}, \quad G_2 = \begin{pmatrix} 0.4 & 0.4 \\ 0.3 & 0.1 \end{pmatrix}, \quad G_3 = \begin{pmatrix} 0.1 & 0.1 \\ 0.1 & 0.1 \end{pmatrix}, \quad G_4 = \begin{pmatrix} 0.12 & 0.13 \\ 0.3 & 0.1 \end{pmatrix},$$

$$E_1 = \begin{pmatrix} 0.1 & 0.1 \\ 0.2 & 0.1 \end{pmatrix}, \quad E_2 = \begin{pmatrix} -0.4 & 0 \\ 0.01 & -0.4 \end{pmatrix}, \quad E_3 = \begin{pmatrix} -0.2 & 0 \\ 0 & -0.2 \end{pmatrix}, \quad E_4 = \begin{pmatrix} -0.3 & -0.1 \\ -0.1 & -0.3 \end{pmatrix}.$$

For $\psi \in PC$, let $F(\psi) = (0, 0.1 \ln(\cosh(\psi_1(0))))$. Let $\Upsilon_{A,i} = \Upsilon_{C,i} = \Upsilon_{H,i} = \Upsilon_{G,i} = \Upsilon_{E,i} = I$, $\Gamma_{A,i} = \Gamma_{C,i} = \Gamma_{H,i} = \Gamma_{G,i} = \Gamma_{E,i} = 0.01I$, $\Xi_{A,i}(t) = \Xi_{C,i}(t) = \Xi_{H,i}(t) = \Xi_{G,i}(t) = \Xi_{E,i}(t) = \cos(5t)I$. Suppose the initial conditions are given by $x_{t_0}(s) = (-14, -20)$ and $y_0(s) = (22, 33)$ for $-\tau \leq s \leq 0$, and suppose that the switching rule is periodic with $\rho_1 = 0.4$, $\rho_2 = 0.4$, $\rho_3 = 0.2$, and $\rho_4 = 1$ (i.e., $\omega = 2$).

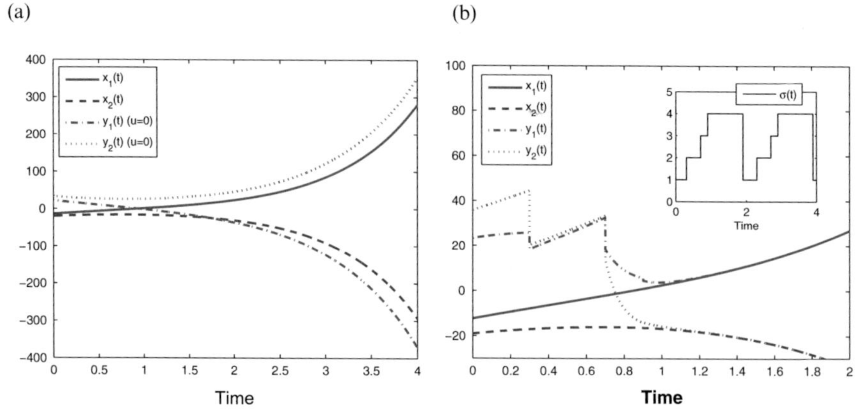

Fig. 1 Simulations of drive system (1) and response system (2). (**a**) Trajectories without hybrid control. (**b**) Trajectories with hybrid control

Choose

$$P_1 = \begin{pmatrix} 1 & 0 \\ 0 & 1 \end{pmatrix}, \; P_2 = \begin{pmatrix} 1 & 0 \\ 0 & 1 \end{pmatrix}, \; P_3 = \begin{pmatrix} 0.354 & -0.020 \\ -0.020 & 0.204 \end{pmatrix}, \; P_4 = \begin{pmatrix} 0.253 & -0.037 \\ -0.037 & 0.374 \end{pmatrix},$$

and $\varepsilon_{A,i} = \varepsilon_{H,i} = \varepsilon_{C,i} = \varepsilon_{H,i} = \varepsilon_{E,i} = 10$ for $i \in \mathscr{P}$, $\vartheta_1 = 0.1$, $\vartheta_2 = 0$. Then, $\beta_1 = 0.021$, $\beta_2 = 0.006$, $\beta_3 = 0.060$, $\beta_4 = 0.059$, $\alpha_1 = -5.602$, $\alpha_2 = -6.812$, $\alpha_3 = 12.866$, $\alpha_4 = 11.867$, $\mu = 1.768$, $\delta_1 = 1.070$, $\delta_2 = 0.825$, $\delta_3 = 2.087$, $\delta_4 = 1.028$. Hence $\lambda_1 = 5.623$, $\lambda_2 = 6.818$, $\lambda_3 = -12.806$, $\lambda_4 = -11.808$, so that $\mathscr{P}_s = \{3, 4\}$ and $\mathscr{P}_u = \{1, 2\}$. Choose $\eta_1 = 11$, $\eta_2 = 12$, to get

$$\sum_{i=1}^{4} \ln \delta_i + \sum_{i \in \mathscr{P}_u} \lambda_i \rho_i - \sum_{i \in \mathscr{P}_s} \eta_i (\rho_i - \tau) = -5.485,$$

which implies robust synchronization of the drive and response systems by Theorem 1. See Fig. 1 for an illustration.

4 Conclusions

The sets $\mathscr{P}_s$ and $\mathscr{P}_u$ in Theorem 1 represent the synchronizing and desynchronizing modes, respectively, while the conditions $\delta < 1$ and $\delta > 1$ represent impulsive control and possible impulsive perturbations, respectively. Each theorem condition establishes threshold conditions involving the mode activation times, mode synchronization/desynchronization effects, and impulsive effects, for various classes of switching rules (i.e., dwell-time (i)–(iv), average dwell-time (v), and periodic switching (vi)). The results found are robust to model uncertainties, nonlinear perturbations, and impulsive perturbations. One possible future direction is to consider the robust hybrid synchronization from an optimal control point of view.

Acknowledgements This research was financially supported by the Natural Sciences and Engineering Research Council of Canada (NSERC).

References

1. Khadra, A., Liu, X., Shen, X.: Application of impulsive synchronization to communication security. IEEE Trans. Circuits Syst. I: Fundam. Theory Appl. **50**(3), 341 (2003)
2. Olfati-Saber, R., Murray, R.M.: Consensus problems in networks of agents with switching topology and time-delays. IEEE Trans. Autom. Control **49**(9), 1520 (2004)
3. Li, C., Liao, X., Yang, X., Huang, T.: Impulsive stabilization and synchronization of a class of chaotic delay systems. J. Nonlinear Sci. **15**(4), 043 (2005)
4. Li, P., Cao, J., Wang, Z.: Robust impulsive synchronization of coupled delayed neural networks with uncertainties. Phys. A: Stat. Mech. Appl. **373**, 261 (2007)
5. Lakshmikantham, V., Rao, M.R.M.: Theory of Integro-Differential Equations. Gordon and Breach Science Publishers, Amsterdam (1995)
6. Burton, T.: Volterra Integral and Differential Equations. Elsevier B.V., Amsterdam (2005)
7. Liu, J., Liu, X., Xie, W.C.: Delay-dependent robust control for uncertain switched systems with time-delay. Nonlinear Anal. Hybrid Syst. **2**(1), 81 (2008)
8. Liberzon, D.: Switching in Systems and Control. Birkhauser, Boston (2003)
9. Davrazos, G., Koussoulas, N.T.: A review of stability results for switched and hybrid systems. In: Proceedings of 9th Mediterranean Conference on Control and Automation, Dubrovnik (2001)
10. Evans, R.J., Savkin, A.V.: Hybrid Dynamical Systems. Birkhauser, Boston (2002)
11. Shorten, R., Wirth, F., Mason, O., Wulff, K., King, C.: Stability criteria for switched and hybrid systems. SIAM Rev. **49**(4), 545 (2007)
12. Hespanha, J.P., Morse, A.S.: Stability of switched systems with average dwell-time. In: Proceedings of the 38th IEEE Conference on Decision and Control, Phoenix, vol. 3, p. 2655 (1999)
13. Guan, Z.H., Hill, D., Shen, X.: On hybrid impulsive and switching systems and application to nonlinear control. IEEE Trans. Autom. Control **50**(7), 1058 (2005)
14. Guan, Z.H., Hill, D., Yao, J.: A hybrid impulsive and switching control strategy for synchronization of nonlinear systems and application to Chua's chaotic circuit. Int. J. Bifurc. Chaos **16**(1), 229 (2006)
15. Alwan, M.S., Liu, X.: On stability of linear and weakly nonlinear switched systems with time delay. Math. Comput. Model. **48**(7–8), 1150(2008)
16. Zhang, Y., Liu, X., Shen, X.: Stability of switched systems with time delay. Nonlinear Anal. Hybrid Syst. **1**(1), 44 (2007)
17. Yang, C., Zhu, W.: Stability analysis of impulsive switched systems with time delays. Math. Comput. Model. **50**(7–8), 1188 (2009)
18. Zhu, W.: Stability analysis of switched impulsive systems with time delays. Nonlinear Anal. Hybrid Syst. **4**(3), 608 (2010)
19. Niamsup, P.: Stability of time-varying switched systems with time-varying delay. Nonlinear Anal. Hybrid Syst. **3**(4), 631 (2009)
20. Wang, X., Li, X.: A generalized Halanay inequality with impulse and delay. Adv. Dev. Technol. Int. **1**(2), 9 (2012)

21. Wang, Y., Xie, L., de Souza, C.E.: Robust control of a class of uncertain nonlinear systems. Syst. Control Lett. **19**, 139 (1992)
22. de Souza, C.E., Li, X.: Delay-dependent robust H-infinity control of uncertain linear state-delayed systems. Automatica **35**, 1313 (1999)
23. Chen, W.H., Zheng, W.X.: Robust stability and H-infinity control of uncertain impulsive systems with time-delay. Automatica **45**, 109 (2009)
24. Xiang, Z., Chen, Q.: Robust reliable control for uncertain switched nonlinear systems with time delay under asynchronous switching. Appl. Math. Comput. **216**(3), 800 (2010)
25. Liu, X., Zhong, S., Ding, X.: Robust exponential stability of impulsive switched systems with switching delays: a Razumikhin approach. Commun. Nonlinear Sci. Numer. Simul. **17**, 1805 (2012)
26. Ballinger, G., Liu, X.: Existence and uniqueness results for impulsive delay differential equations. Dyn. Contin. Discret. Impuls. Syst. **5**(3), 579 (1999)
27. Liu, J., Liu, X., Xie, W.C.: Input-to-state stability of impulsive and switching hybrid systems with time-delay. Automatica **47**, 899 (2011)
28. Hien, L.V., Phat, V.N.: Exponential stabilization for a class of hybrid systems with mixed delays in state and control. Nonlinear Anal. Hybrid Syst. **3**(3), 259 (2009)

Regularization and Numerical Integration of DAEs Based on the Signature Method

Andreas Steinbrecher

Abstract Modeling and simulation of dynamical systems often leads to differential-algebraic equations (DAEs). Since direct numerical integration of DAEs in general leads to instabilities and possibly non-convergence of numerical methods, a regularization is required. We present three approaches for the regularization of DAEs that are based on the Signature method. Furthermore, we present a software package suited for the proposed regularizations and illustrate its efficiency on two examples.

1 Introduction

Modeling and simulation of dynamical systems is an important issue in many industrial applications and often leads to systems of *differential-algebraic equations (DAEs)*. In this paper, we consider nonlinear DAEs of the form

$$F(t, z, \dot{z}) = 0, \tag{1}$$

with sufficiently smooth functions $F : \mathbb{I} \times \mathbb{R}^n \times \mathbb{R}^n \to \mathbb{R}^m$ and $z : \mathbb{I} \to \mathbb{R}^n$, with a compact time interval $\mathbb{I} \subset \mathbb{R}$.

In general, DAEs are not suited for direct numerical integration due to the so called *hidden constraints*, which restrict the solution but are not explicitly stated as equations. Hidden constraints can be determined by a certain number of differentiation of (parts of) the DAE. This number of differentiations gives a classification of difficulties in the treatment of DAEs, see [2, 4, 5, 12].

To guarantee a stable and robust numerical integration, a regularization of DAEs is required. The basic idea of an approach to obtain a regularization for DAEs is to consider the original system together with a sufficient number of its derivatives. Then a system with the same solution set as the original system can be derived that now contains all (hidden) constraints explicitly as algebraic equations. Then it can

A. Steinbrecher (✉)
Institute of Mathematics MA 4-5, TU Berlin, Straße des 17. Juni 136, 10623 Berlin, Germany
e-mail: anst@math.tu-berlin.de

© Springer International Publishing Switzerland 2016

J. Bélair et al. (eds.), *Mathematical and Computational Approaches in Advancing Modern Science and Engineering*, DOI 10.1007/978-3-319-30379-6_67

be guaranteed that all (hidden) constraints are satisfied within the numerical solution and instabilities or drift from the solution manifold are avoided.

In most modeling and simulation tools, the current state-of-the-art to regularize DAEs is to use some kind of structural analysis based on the sparsity pattern of the system, to obtain necessary information for a regularization. The advantage of a structural analysis in comparison to classical algebraic regularizations is that fast algorithms based on graph theory can be applied. Usually, the *Pantelides algorithm* [9] in combination with the *Dummy Derivative Method* [6] is used. In this paper, we present regularization approaches based on the *Signature method (Σ-method)* [10].

In Sect. 3, we discuss regularization techniques that are based on the information provided by the Σ-method which is reviewed in Sect. 2. Two of the regularization techniques yield overdetermined systems of DAEs. Then, in Sect. 4 we shortly discuss the software package `QUALIDAES` for the numerical integration of (overdetermined) quasi-linear DAEs. In Sect. 5 we illustrate the applicability of `QUALIDAES` in combination with the three proposed regularized formulations.

2 Signature Method: Structural Analysis of DAEs

The Σ-method, see [7, 8, 10] and also [11], can be applied to regular nonlinear DAEs of arbitrary high order p of the form $F(t, z, \dot{z}, \ldots, z^{(p)}) = 0$. We denote by F_i the ith component of the vector-valued function F and by z_j the jth component of the vector z. Then, the Σ-method consists of the following steps:

1. Building the *signature matrix* $\Sigma = [\sigma_{ij}]_{i,j=1,\ldots,n}$ with

$$\sigma_{ij} = \begin{cases} \text{highest order of derivative of } z_j \text{ in } F_i, \\ -\infty \text{ if } z_j \text{ does not occur in } F_i. \end{cases}$$

2. Finding a *highest value transversal (HVT)* of Σ, i.e., a *transversal* $T = \{(1, j_1), (2, j_2), \ldots, (n, j_n)\}$, where $(j_1, \ldots, j_n)$ is a permutation of $(1, \ldots, n)$, with maximal *value* $\mathrm{Val}(T) := \sum_{(i,j)\in T} \sigma_{ij}$.
3. Computing the *offset vectors* $c = [c_i]_{i=1,\ldots,n}$ and $d = [d_j]_{j=1,\ldots,n}$ with $c_i, d_j \in \mathbb{N}_0$ such that

$$d_j - c_i \geq \sigma_{ij} \;\text{ for all }\; i, j = 1, \ldots, n, \;\text{ and }\; d_j - c_i = \sigma_{ij} \;\text{ for all }\; (i, j) \in T. \quad (2)$$

4. Forming the *Σ-Jacobian* $\mathfrak{J} = [\mathfrak{J}_{ij}]_{i,j=1,\ldots,n}$ with

$$\mathfrak{J}_{ij} := \begin{cases} \dfrac{\partial F_i}{\partial z_j^{(\sigma_{ij})}} & \text{if } d_j - c_i = \sigma_{ij}, \\ 0 & \text{otherwise.} \end{cases}$$

5. Building the *reduced derivative array*

$$\mathscr{F}(t,\mathscr{Z})=\begin{bmatrix}\mathscr{F}_1(t,\mathscr{Z})\\ \vdots\\ \mathscr{F}_n(t,\mathscr{Z})\end{bmatrix}=0 \text{ with } \mathscr{F}_i(t,\mathscr{Z})=\begin{bmatrix}F_i(t,\mathscr{Z})\\ \frac{\mathrm{d}}{\mathrm{d}t}F_i(t,\mathscr{Z})\\ \vdots\\ \frac{\mathrm{d}^{c_i}}{\mathrm{d}t^{c_i}}F_i(t,\mathscr{Z})\end{bmatrix} \text{ for } i=1,\ldots n, \tag{3}$$

with $\mathscr{Z}^T=\left[z_1\ \dot{z}_1\ \ldots\ z_1^{(d_1)}\ \ldots\ z_n\ \dot{z}_n\ \ldots\ z_n^{(d_n)}\right]$.

6. Success check: if $\mathscr{F}(t,\mathscr{Z})=0$, considered locally as an algebraic system, has a solution $(t^*,\mathscr{Z}^*)\in\mathbb{I}\times\mathbb{R}^{n+\sum_{i=1}^n d_i}$, and $\mathfrak{J}$ is nonsingular at $(t^*,\mathscr{Z}^*)$, then $(t^*,\mathscr{Z}^*)$ is a consistent point and the method succeeds.

If the Σ-method succeeds, the DAE (1) is called *structurally regular* with the *structural index* ν_S defined as $\nu_S:=\max_i c_i$ if all $d_j>0$ or $\nu_S:=\max_i c_i+1$ if some $d_j=0$. Note that the success check of the Σ-method is performed locally at a fixed point $(t^*,\mathscr{Z}^*)$, such that the result may hold only locally in a neighborhood. The HVT as well as the offset vectors c and d are not uniquely defined. However, there exists a unique element-wise smallest solution, the so-called *canonical offsets*. These are uniquely determined and independent of the chosen HVT, see [10].

The reduced derivative array (3) contains all necessary equations to extract all hidden constraints by use of analytical manipulations. This forms the basis for all in the following proposed regularization approaches.

Example 1 Let us consider the simple pendulum (Fig. 1) of mass $m>0$, length $\ell>0$ under gravity $\mathfrak{g}$, see also [10, 11]. The state of the pendulum is described by the position x and y, the velocities v and w of the mass point, and the Lagrange multiplier λ. The system equations are given by

$$0=F(z,\dot{z})=\begin{bmatrix}F_1(z,\dot{z})\\ F_2(z,\dot{z})\\ F_3(z,\dot{z})\\ F_4(z,\dot{z})\\ F_5(z,\dot{z})\end{bmatrix}=\begin{bmatrix}\dot{x}-v\\ \dot{y}-w\\ m\dot{v}+2x\lambda\\ m\dot{w}+2y\lambda+m\mathfrak{g}\\ x^2+y^2-\ell^2\end{bmatrix} \tag{4, 5}$$

where $z=\left[x\ y\ v\ w\ \lambda\right]^T$. The signature matrix Σ for (4) and the canonical offset vectors c and d are given by

$$\Sigma=\begin{bmatrix}\boxed{1}&-\infty&\boxed{0}&-\infty&-\infty\\ -\infty&\boxed{1}&-\infty&\boxed{0}&-\infty\\ 0&-\infty&\boxed{1}&-\infty&\boxed{0}\\ -\infty&0&-\infty&\boxed{1}&\boxed{0}\\ \boxed{0}&\boxed{0}&-\infty&-\infty&-\infty\end{bmatrix},\quad c=\begin{bmatrix}1\\1\\0\\0\\2\end{bmatrix},\quad d=\begin{bmatrix}2\\2\\1\\1\\0\end{bmatrix}, \tag{6}$$

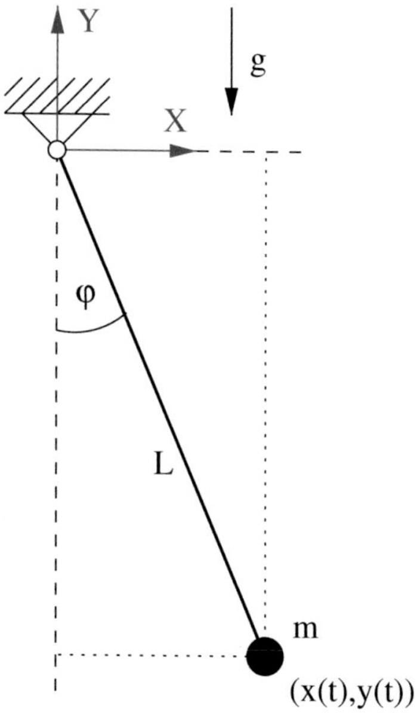

Fig. 1 Simple pendulum

where the two possible HVTs are marked by light and dark gray boxes. The corresponding Σ-Jacobian $\mathfrak{J}$ and the reduced derivative array (3) takes the form

$$\mathfrak{J} = \begin{bmatrix} 1 & 0 & -1 & 0 & 0 \\ 0 & 1 & 0 & -1 & 0 \\ 0 & 0 & m & 0 & 2x \\ 0 & 0 & 0 & m & 2y \\ 2x & 2y & 0 & 0 & 0 \end{bmatrix}, \qquad 0 = \mathscr{F}(t, \mathscr{X}) = \begin{bmatrix} \dot{x} - v \\ \ddot{x} - \dot{v} \\ \dot{y} - w \\ \ddot{y} - \dot{w} \\ m\dot{v} + 2x\lambda \\ m\dot{w} + 2y\lambda + m\mathfrak{g} \\ x^2 + y^2 - \ell^2 \\ 2x\dot{x} + 2y\dot{y} \\ 2x\ddot{x} + 2\dot{x}^2 + 2y\ddot{y} + 2\dot{y}^2 \end{bmatrix} \tag{7}$$

with $\mathscr{X}^T = \begin{bmatrix} x \; \dot{x} \; \ddot{x} \; y \; \dot{y} \; \ddot{y} \; v \; \dot{v} \; w \; \dot{w} \; \lambda \end{bmatrix}$. Since $\det(\mathfrak{J}) = -4m(x^2 + y^2) = -4m\ell^2 \neq 0$, the Σ-Jacobian $\mathfrak{J}$ is nonsingular at every consistent point, and the Σ-method succeeds with $\nu_S = \max_i c_i + 1 = 3$. ◁

3 Regularization Approaches for DAEs

In the following, we discuss three different regularization approaches for structurally regular DAEs (1) with $m = n$. These regularization approaches mainly use the reduced derivative array (3) provided by the Σ-method and differ in its degree of analytical preprocessing and possible degree of automation of the approach.

3.1 Regularization via State Selection

The reduced derivative array (3) consists of $M = \sum_i c_i + n$ equations in n unknowns $z_1, \ldots, z_n$. Thus, to obtain the same number of equations and unknowns, we have to introduce $\sum_i c_i$ new variables.

For that we choose an HVT T^* from the set of all existing HVTs. Then for each $j = 1, \ldots, n$ we have a unique i with $(i,j) \in T^*$, and a regularization can be obtained from the reduced derivative array (3) by replacing the derivatives $z_j^{(\sigma_{ij}+1)}, \ldots, z_j^{(d_i)}$ with the algebraic variables $\omega_j^{\sigma_{ij}+1}, \ldots, \omega_j^{d_i}$ if $d_i > \sigma_{ij}$. We obtain the *structurally extended (StE) formulation*

$$\mathscr{S}(t, z, \dot{z}, \omega) = 0 \tag{8}$$

with $\omega^T = \left[\omega_1^T \cdots \omega_n^T\right]$ and $\omega_j^T := \left[\omega_j^{\sigma_{ij}+1} \cdots \omega_j^{d_i}\right] \in \mathbb{R}^{c_i}$ if $d_j > \sigma_{ij}$ or $\omega_j := [\cdot] \in \mathbb{R}^0$ if $d_j = \sigma_{ij}$ as new algebraic variables. Note that $d_j = \sigma_{ij} + c_i$ by (2).

The obtained structurally extended formulation (8) is of increased size, not unique, since it depends on the chosen HVT T^*, and in general it is only valid locally in a neighborhood of a consistent point. Furthermore, it may lead to an inappropriate regularization, if an inappropriate HVT T^* is used (see Example 2). To choose a suitable HVT, i.e., one that is valid in a preferably large neighborhood of a consistent point, we define the (local) *weighting coefficient* for each HVT T as $\kappa_T := \prod_{(i,j)\in T} |\mathfrak{J}_{ij}|$ and choose one with largest value κ_T, i.e., we choose T^* as an HVT of Σ with $\kappa_{T^*} = \max_T(\kappa_T)$. In [11] it is shown that the structurally extended system (8) is (locally) a regular system of structural index $\nu_S \leq 1$ if T^* is chosen as described above. This formulation has locally the same set of solutions for the original unknowns z as the original DAE (1). Due to the local validity, the determination of a new regularized formulation during the (numerical) integration may be necessary. In this case, the integration has to be interrupted and to be restarted with the newly regularized model equations. Unfortunately, this leads to an increase in simulation time and influences the obtained precision negatively.

This approach completely can be automated using automatic or symbolic differentiation tools. Therefore, an analytical preprocessing is not necessary.

In the proposed method, the selection of variables for which derivatives are replaced is directly prescribed by the HVT T^* and the offset vectors. Once the Σ-method is done, this selection of variables is easy to achieve and requires no further numerical computations. In contrast, in the Dummy Derivative Method, see [6]. As a result, the two approaches might result in different regularized systems.

Example 2 The system (4) is regular and of structural index $\nu_S = 3$. The reduced derivative array (7) consists of $M = \sum c_i + n = 9$ equations in $n = 5$ unknowns.

For the HVT $T^* = T_1$ marked by the light gray boxes in the signature matrix (6) (assuming that $\kappa_{T_1} = 4x^2 \geq \kappa_{T_2} = 4y^2$), we have to introduce new algebraic variables as in the following table:

For j	with i (s.t. $(i,j) \in T^*$)	replace derivatives $z_j^{(\sigma_{ij}+1)}, \ldots, z_j^{(d_j)}$	with new algebraic variables $\omega_j^{\sigma_{ij}+1}, \ldots, \omega_j^{d_j}$
1	5	$z_1^{(1)}, z_1^{(2)}$	ω_1^1, ω_1^2
2	2	$z_2^{(2)}$	ω_2^2
3	1	$z_3^{(1)}$	ω_3^1
4	4	$\emptyset$	$\emptyset$
5	3	$\emptyset$	$\emptyset$

Then we construct the structurally extended formulation as

$$\mathscr{S}_1(t,z,\dot{z},\omega) = \begin{bmatrix} \omega_1^1 - v \\ \omega_1^2 - \omega_3^1 \\ \dot{y} - w \\ \omega_2^2 - \dot{w} \\ m\omega_3^1 + 2x\lambda \\ m\dot{w} + 2y\lambda + mg \\ x^2 + y^2 - \ell^2 \\ 2x\omega_1^1 + 2y\dot{y} \\ 2x\omega_1^2 + 2(\omega_1^1)^2 + 2y\omega_2^2 + 2\dot{y}^2 \end{bmatrix} = 0 \text{ with } z = \begin{bmatrix} x \\ y \\ v \\ w \\ \lambda \end{bmatrix}, \ \omega = \begin{bmatrix} \omega_1^1 \\ \omega_1^2 \\ \omega_2^2 \\ \omega_3^1 \end{bmatrix}. \tag{9}$$

For the HVT $T^* = T_2$ marked by the dark gray boxes in the signature matrix (6) (assuming that $\kappa_{T_2} = 4y^2 \geq \kappa_{T_1} = 4x^2$), we have to introduce new algebraic variables as in the following table:

For j	with i (s.t. $(i,j) \in T^*$)	replace derivatives $z_j^{(\sigma_{ij}+1)}, \ldots, z_j^{(d_j)}$	with new algebraic variables $\omega_j^{\sigma_{ij}+1}, \ldots, \omega_j^{d_j}$
1	1	$z_1^{(2)}$	ω_1^2
2	5	$z_2^{(1)}, z_2^{(2)}$	ω_2^1, ω_2^2
3	3	$\emptyset$	$\emptyset$
4	2	$z_4^{(1)}$	ω_4^1
5	4	$\emptyset$	$\emptyset$

Then we construct the structurally extended formulation as

$$\mathcal{S}_2(t,z,\dot{z},\omega) = \begin{bmatrix} \dot{x} - v \\ \omega_1^2 - \dot{v} \\ \omega_2^1 - w \\ \omega_2^2 - \omega_4^1 \\ m\dot{v} + 2x\lambda \\ m\omega_4^1 + 2y\lambda + m\mathfrak{g} \\ x^2 + y^2 - \ell^2 \\ 2x\dot{x} + 2y\omega_2^1 \\ 2x\omega_1^2 + 2\dot{x}^2 + 2y\omega_2^2 + 2(\omega_2^1)^2 \end{bmatrix} = 0 \text{ with } z = \begin{bmatrix} x \\ y \\ v \\ w \\ \lambda \end{bmatrix}, \ \omega = \begin{bmatrix} \omega_1^2 \\ \omega_2^1 \\ \omega_2^2 \\ \omega_4^1 \end{bmatrix}. \tag{10}$$

Applying the Σ-method to (9) and (10), we obtain a structurally regular system with structural index $\nu_S = 1$ as long as $x \neq 0$ while (10) forms a structurally regular system with structural index $\nu_S = 1$ as long as $y \neq 0$. ◁

3.2 *Regularization via Algebraic Derivative Arrays*

As we have seen in the previous section, the structurally extended regularization is only valid locally. Therefore, it may be necessary to switch to another structurally extended formulation within a numerical integration (also known as *dynamic state selection*). This influences the time integration negatively. As a remedy in this section we propose an extended regularization.

As mentioned above, the reduced derivative array (3) contains all necessary equations to obtain all hidden constraints. Now we replace all derivatives of all variables by algebraic variables, i.e., we replace $z_i^{(j)}$ by ω_i^j for $i = 1, \ldots, n$ and $j = 1, \ldots, d_i$, in the reduced derivative array to obtain the *algebraic derivative array*

$$0 = \mathcal{F}(t,z,\omega) = \mathcal{F}(t,z_1,\omega_1,\ldots,z_n,\omega_n) \text{ with } \omega_i = \begin{cases} [\omega_i^1,\ldots,\omega_i^{d_i}]^T & \text{if } d_i > 1, \\ [\cdot] & \text{if } d_i = 0. \end{cases}$$

From this it is possible to extract the set of all hidden constraints in an algebraic way which, in particular, can be done automatically within the numerical integration.

In combination with the original DAE (1) we obtain the *algebraic derivative array (ADA) formulation*

$$0 = \mathcal{A}(t,z,\dot{z},\omega) = \begin{bmatrix} F(t,z,\dot{z}) \\ \mathcal{F}(t,z,\omega) \end{bmatrix} \tag{11}$$

which forms an overdetermined DAE for z and ω and has the same set of solutions for the original unknowns z as the original DAE (1). The obtained algebraic derivative array formulation is of more increased size as the structurally extended formulation but globally valid such that the regularized formulation has to be determined only once before the numerical integration. This approach also can be automated using automatic or symbolic differentiation tools. Therefore, an analytical preprocessing is not necessary.

Example 3 Let us use the model equations of the simple pendulum to illustrate the regularization based on the algebraic derivative array formulation. The reduced derivative array is given in (7). Replacing all occurring derivatives with algebraic variables, i.e., $\dot{x} \to \omega_1^1$, $\ddot{x} \to \omega_1^2$, $\dot{y} \to \omega_2^1$, $\ddot{y} \to \omega_2^2$, $\dot{v} \to \omega_3^1$, and $\dot{w} \to \omega_4^1$, leads to the algebraic derivative array

$$0 = \mathscr{F}(t, z, \omega) = \begin{bmatrix} \omega_1^1 - v \\ \omega_1^2 - \omega_3^1 \\ \omega_2^1 - w \\ \omega_2^2 - \omega_4^1 \\ m\omega_3^1 + 2x\lambda \\ m\omega_4^1 + 2y\lambda + mg \\ x^2 + y^2 - L^2 \\ 2(x\omega_1^1 + y\omega_2^1) \\ 2(x\omega_1^2 + (\omega_1^1)^2 + y\omega_2^2 + (\omega_2^1)^2) \end{bmatrix}. \tag{12}$$

Together with the original DAE (4), we get the algebraic derivative array formulation (11) with $F(t, z, \dot{z})$ given in (4) and $\mathscr{F}(t, z, \omega)$ given in (12). This corresponds to an overdetermined set of DAEs containing 14 equations for 11 unknowns. ◁

3.3 *Regularization via Overdetermined Formulation*

In the previous sections, we did obtain regularizations that increased in its size more then necessary. Therefore, in the numerical integration more computational time is needed than necessary.

As mentioned in Sect. 2 the reduced derivative array (3) allows the determination of the *set of (hidden) constraints*

$$0 = H(t, z)$$

in an analytical preprocessing. In combination with the original DAE (1), we obtain the set

$$0 = \mathscr{O}(t, z, \dot{z}) = \begin{bmatrix} F(t, z, \dot{z}) \\ H(t, z) \end{bmatrix},$$

which forms the *regularized overdetermined (OVD) formulation* for z and which has the same set of solutions as the original DAE (1). The obtained overdetermined formulation is of increased number of equations for the same unknown variables as in the original DAE. For more details see [12].

Example 4 For the simple pendulum, see Example 1, we get the set of (hidden) constraints as

$$0 = H(t, z) = \begin{bmatrix} x^2 + y^2 - \ell^2 \\ 2xv + 2yw \\ v^2 + w^2 - \frac{2}{m}(x^2 + y^2)\lambda - yg \end{bmatrix} \tag{13}$$

and consequently the regularization via overdetermined formulations is determined by (1), (13) as a set containing eight equations for five unknowns. ◁

4 The Software Package QUALIDAES

The software package QUALIDAES (QUAsi LInear DAE Solver) is suited for the numerical integration of (overdetermined) quasi-linear DAEs of the form

$$\begin{bmatrix} E(z,t) \\ \hline 0 \end{bmatrix} \dot{z} = \begin{bmatrix} k(z,t) \\ \hline G(z,t) \end{bmatrix} \quad \begin{array}{l} \text{differential part,} \\ \text{algebraic constraints.} \end{array}$$

QUALIDAES is implemented in FORTRAN. The discretization scheme is the 3-stage Runge-Kutta RADAU IIa of order 5 and uses adapted decomposition methods w.r.t. the overdeterminedness to solve the (hidden) constraints (numerically) precisely, while the differential part is solved in an 'approximative sense'.

Further features of QUALIDAES are variable step size control, continuous output, and check and correction of the initial values with respect to its consistency. The specification of the system DAEs can be done in FORTRAN source code or alternatively in MODELICA using a simple MODELICA-parser. For MODELICA a Matlab interface is provided, see [1]. For more information to QUALIDAES we refer to [13].

5 Numerical Results

In the following, we illustrate the efficiency of QUALIDAES with the proposed regularization techniques on two examples. The first one is the simple pendulum, and the second one is a mass-spring-chain with path constraints.

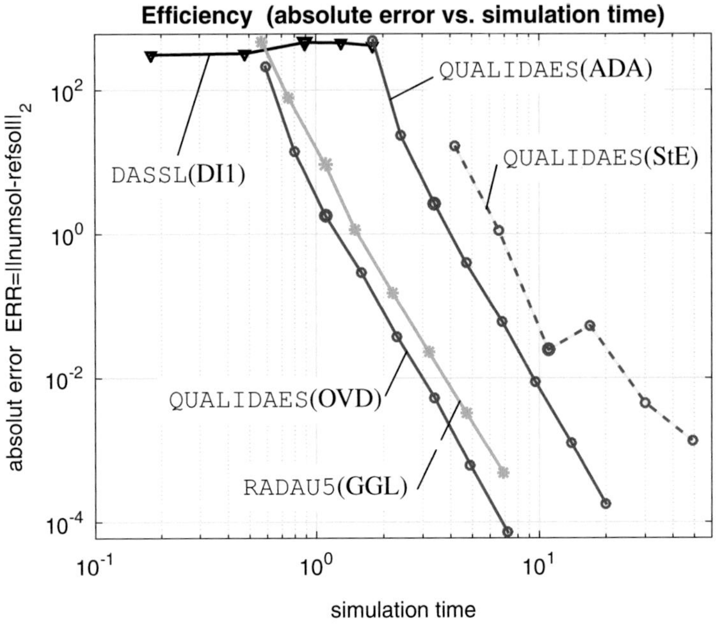

Fig. 2 Efficiency (simulation time vs. absolute error for different prescribed tolerances)

The numerical integrations are done on an AMD Phenom(tm) II X6 1090T, 3210 MHz, 16 GB RAM, openSuSE 13.1 (Linux 3.11.10), GNU Fortran compiler gcc version 4.8.1, no compiler options.

Example 5 The model equations of the simple pendulum, see Example 1, are stated in (4) and have structural index $\nu_S = 3$ and differentiation index $\nu_d = 3$. The mass and length are given by $m = 1$ and $\ell = 1$ while we use the gravitational acceleration $\mathfrak{g} = 13.7503716373294544$. In this case the exact solution has a period of 2 s which allows the comparison of the accuracy every period.

The simulation is done for the time domain $\mathbb{I} = [0\,\text{s}, 2000\,\text{s}]$ with initial values $z(0) = \begin{bmatrix} 1 & 0 & 0 & 0 & 0 \end{bmatrix}^T$. To show the efficiency (see Fig. 2) of the numerical integration with QUALIDAES in comparison we use the widely used solvers RADAU5, with the Gear-Gupta-Leimkuler (GGL) formulation [3] and DASSL with the d-index-1 (DI1) formulation consisting the first four equations of (4) and the last equation of (13).

The numerical solution QUALIDAES(OVD) offers the best efficiency, i.e., less computational effort and small absolute error of the numerical solution. This numerical solution is slightly more efficient as the numerical result obtained with RADAU5(GGL) followed by the numerical solutions of QUALIDAES(ADA) and QUALIDAES(StE). The large time consumption for the integration of the StE-form lies in its local validity as discussed in Example 2. Within one period of 2 s it has to be switched between (9) and (10) four times. Furthermore, due to the increased size of the ADA-form and the StE-form both are less efficient than the OVD-

Fig. 3 Mass-spring-chain

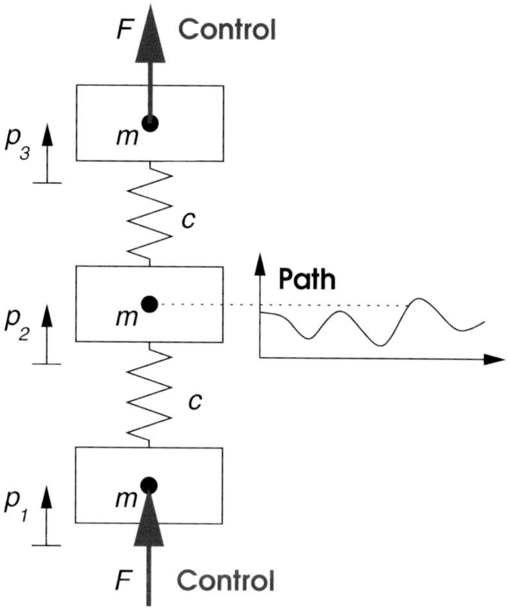

form. The numerical integration using DASSL is not successful at all due to the stability properties of BDF methods, the occurring drift off effect by use of the DI1-form, where the constraints on position level and on velocity level are lost, and the large time domain $\mathbb{I}$. Nevertheless, the maximally obtained precision is excellent for QUALIDAES(OVD) and QUALIDAES(ADA). ◁

Example 6 In the following we are interested in a path following problem of a mass-spring-chain as in Fig. 3. On the first and the last body the force F is applied while the center body has to follow a prescribed path (14g). We have p_i and v_i as the position and velocity of body $i = 1, 2, 3$ and F as unknown variables. The model equations are given by

$$\dot{p}_1 = v_1, \tag{14a}$$

$$\dot{p}_2 = v_2, \tag{14b}$$

$$\dot{p}_3 = v_3, \tag{14c}$$

$$m\dot{v}_1 = -c(p_1 - p_2) + F, \tag{14d}$$

$$m\dot{v}_2 = c(p_1 - p_2) - c(p_2 - p_3), \tag{14e}$$

$$m\dot{v}_3 = c(p_2 - p_3) + F, \tag{14f}$$

$$0 = p_2 - \sin(t) \tag{14g}$$

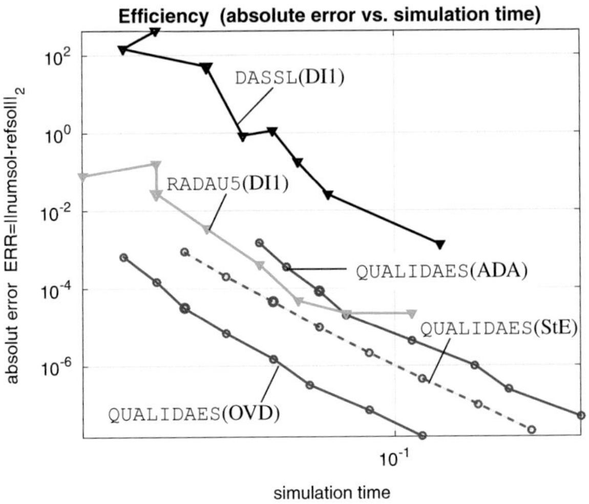

Fig. 4 Efficiency (simulation time vs. absolute error for different prescribed tolerances)

with structural index $\nu_S = 5$ and differentiation index $\nu_d = 5$. Therefore, we have hidden constraints up to level 4, i.e., up to the 4th derivative of (several) model equations are necessary to determine all hidden constraints. We use $m = 1$ and $c = 1/6$ with the exact solution $p_2(t) = \sin(t)$, $v_2(t) = \cos(t)$, $p_1(t) = p_3(t) = -2\sin(t)$, $v_1(t) = v_3(t) = -2\cos(t)$, $F(t) = 3\sin(t)/2$. The simulation is done for the time do-main $\mathbb{I} = [0\,\text{s}, 500\,\text{s}]$.

In the same style as in the last example, the efficiency of the numerical solutions is illustrated in Fig. 4.

Again the numerical solution QUALIDAES(OVD) offers the best efficiency followed by the numerical solution QUALIDAES(StE). In contrast to the previous example, the efficiency using the StE-form is better since the StE-form is valid for the whole time domain and has not to be switched. Furthermore, the numerical solution QUALIDAES(ADA) offers a good efficiency. The efficiency of RADAU5(DI1) is similar, but the obtained precision is not as good as for the previous results. The reason is that we used the DI1-form, where the hidden constraints of level 4 are included instead of (14g). Therefore, the (hidden) constraints up to level 3 are lost, and the solution drifts away from these lost constraints. Nevertheless, the maximally obtained precision is excellent for QUALIDAES with all regularizations proposed in Sect. 3, i.e., for the StE-form, the ADA-form, and the OVD-form. ◁

6 Summary

The aim of this article was to discuss several approaches for the regularization of differential-algebraic equations that benefit numerical integration. For this purpose, we have proposed three different regularization approaches which end in the structurally extended formulation, the algebraic derivative array formulation, and the regularized overdetermined formulation. All these regularization approaches are based on the signature method, which was reviewed in Sect. 2, and two of them require numerical integration methods suited for overdetermined differential-algebraic equations. We also briefly introduced the software package `QUALIDAES` and illustrated its efficiency for two examples.

Acknowledgements This work has been supported by the European Research Council through Advanced Grant MODSIMCONMP.

References

1. Altmeyer, R., Steinbrecher, A.: Regularization and Numerical Simulation of Dynamical Systems Modeled with Modelica. Preprint 29-2013, Institut für Mathematik, Berlin (2013)
2. Brenan, K.E., Campbell, S.L., Petzold, L.R.: Numerical Solution of Initial-Value Problems in Differential Algebraic Equations. Classics in Applied Mathematics, vol. 14. SIAM, Philadelphia (1996)
3. Gear, C.W., Leimkuhler, B., Gupta, G.K.: Automatic integration of Euler-Lagrange equations with constraints. J. Comput. Appl. Math. **12–13**, 77–90 (1985)
4. Hairer, E., Wanner, G.: Solving Ordinary Differential Equations II – Stiff and Differential-Algebraic Problems, 2nd edn. Springer, Berlin (1996)
5. Kunkel, P., Mehrmann, V.: Differential-Algebraic Equations. Analysis and Numerical Solution. EMS Publishing House, Zürich, 2006.
6. Mattsson, S.E., Söderlind, G.: Index reduction in differential-algebraic equations using dummy variables. SIAM J. Sci. Comput. **14**, 677–692 (1993)
7. Nedialkov, N., Pryce, J., Tan, G.: DAESA – a Matlab tool for structural analysis of DAEs: software. Technical report CAS-12-01-NN, Department of Computing and Software, McMaster University, Hamilton (2012)
8. Nedialkov, N.S., Pryce, J.D.: Solving differential-algebraic equations by Taylor series (i): computing Taylor coefficients. BIT Numer. Math. **45**(3), 561–591 (2005)
9. Pantelides, C.C.: The consistent initialization of differential-algebraic systems. SIAM J. Sci. Stat. Comput. **9**(2), 213–231 (1988)
10. Pryce, J.D.: A simple structural analysis method for DAEs. BIT **41**(2), 364–394 (2001)
11. Scholz, L., Steinbrecher, A.: Regularization of DAEs based on the signature method. BIT Numer. Math. 1–22 (2015). doi:10.1007/s10543-015-0565-x
12. Steinbrecher, A.: Numerical solution of quasi-linear differential-algebraic equations and industrial simulation of multibody systems. Ph.D. thesis, Technische Universität Berlin (2006)
13. Steinbrecher, A.: QUALIDAES: a software package for the numerical integration of quasi-linear differential-algebraic equations. Technical report, Institut für Mathematik, Technische Universität Berlin, Berlin (in preparation)

Symbolic-Numeric Methods for Improving Structural Analysis of Differential-Algebraic Equation Systems

Guangning Tan, Nedialko S. Nedialkov, and John D. Pryce

Abstract Systems of differential-algebraic equations (DAEs) are generated routinely by simulation and modeling environments, such as MapleSim and those based on the Modelica language. Before a simulation starts and a numerical method is applied, some kind of structural analysis is performed to determine which equations to be differentiated, and how many times. Both Pantelides's algorithm and Pryce's Σ-method are equivalent in the sense that, if one method succeeds in finding the correct index and producing a nonsingular Jacobian for a numerical solution procedure, then the other does also. Such a success occurs on many problems of interest, but these structural analysis methods can fail on simple, solvable DAEs and give incorrect structural information including the index. This article investigates Σ-method's failures and presents two symbolic-numeric conversion methods for fixing them. Both methods convert a DAE on which the Σ-method fails to a DAE on which this SA may succeed.

1 Introduction

We consider differential-algebraic equations (DAEs) of the general form

$$f_i(\, t, \text{ the } x_j \text{ and derivatives of them}\,) = 0, \quad i = 1, \ldots, n, \tag{1}$$

G. Tan (✉)
School of Computational Science and Engineering, McMaster University, 1280 Main Street West, Hamilton, ON, L8S 4L8, Canada
e-mail: tang4@mcmaster.ca

N.S. Nedialkov
Department of Computing and Software, McMaster University, 1280 Main Street West, Hamilton, ON, L8S 4L8, Canada
e-mail: nedialk@mcmaster.ca

J.D. Pryce
School of Mathematics, Cardiff University, Senghennydd Road, Cardiff CF24 4AG, Wales, UK
e-mail: prycejd1@cardiff.ac.uk

© Springer International Publishing Switzerland 2016

J. Bélair et al. (eds.), *Mathematical and Computational Approaches in Advancing Modern Science and Engineering*, DOI 10.1007/978-3-319-30379-6_68

where the $x_j(t)$, $j = 1, \ldots, n$ are state variables that are functions of an independent variable t, usually regarded as time.

Pryce's structural analysis (SA), the Σ-method [10], determines for (1) its structural index, number of degrees of freedom (DOF), variables and derivatives that need initial values, and the constraints of the DAE. These SA results can help decide how to apply an index reduction algorithm [3], perform a regularization process [12], or design a solution scheme for a Taylor series method [5–7]. The Σ-method is equivalent to Pantelides's algorithm [9]: they both produce the same structural index [10, Theorem 5.8], which is an upper bound for the differentiation index, and often both indices are the same [10].

The Σ-method succeeds on many problems of practical interest, producing a nonsingular System Jacobian. However, this SA can fail—hence Pantelides's algorithm can fail as well—on some simple, solvable DAEs, producing an identically singular System Jacobian.

We investigate such SA's failures and present two symbolic-numeric conversion methods for fixing them. After each conversion, provided some conditions are satisfied, the value of the signature matrix is guaranteed to decrease. We conjecture that such a decrease should result in a better problem formulation of a DAE, so that the SA may produce a nonsingular System Jacobian and hence succeed.

Section 2 summarizes the Σ-method. Section 3 describes SA's failures. Section 4 presents our two conversion methods, each of which is illustrated with an example therein. Section 5 presents concluding remarks.

This article is a succinct version of the technical report [13]; the reader is referred to it for more detailed explanations and examples.

2 Summary of the Σ-Method

This SA method [10] constructs for a DAE (1) an $n \times n$ *signature matrix* Σ, whose (i,j) entry σ_{ij} is either an integer ≥ 0, the order of the highest derivative to which variable x_j occurs in equation f_i, or $-\infty$ if x_j does not occur in f_i.

A *highest-value transversal* (HVT) of Σ is a set T of n positions (i,j) with one position in each row and each column, such that the sum of the corresponding entries is maximized. This sum is the *value of* Σ, written $\mathrm{Val}(\Sigma)$. If $\mathrm{Val}(\Sigma)$ is finite, then the DAE is *structurally well posed* (SWP); otherwise it is *structurally ill posed.*

We assume henceforth the SWP case. Using the HVT, we find n *equation* and n *variable offsets* $c_1, \ldots, c_n$ and $d_1, \ldots, d_n$, respectively, which are integers satisfying

$$c_i \geq 0 \quad \text{for all } i; \qquad d_j - c_i \geq \sigma_{ij} \quad \text{for all } i,j \text{ with equality on a HVT}\,. \tag{2}$$

These offsets are *valid* but not unique. There exists an elementwise smallest solution of (2) termed the *canonical offsets* [10].

We associate with given valid offsets an $n \times n$ *System Jacobian* $\mathbf{J}$, defined as

$$\mathbf{J}_{ij} = \begin{cases} \partial f_i / \partial x_j^{(\sigma_{ij})} & \text{if } d_j - c_i = \sigma_{ij}, \text{ and} \\ 0 & \text{otherwise}. \end{cases} \tag{3}$$

If $\mathbf{J}$ is nonsingular at a *consistent point* as defined in [5], then we say the Σ-method *succeeds* and there exists (locally) a unique solution through this point [10].

In the success case, the canonical offsets determine the *structural index* and the *number of DOF*, defined as

$$\nu_S = \max_i c_i + \begin{cases} 1 & \text{if } \min_j d_j = 0 \\ 0 & \text{otherwise} \end{cases}$$

$$\text{and} \quad \text{DOF} = \text{Val}(\Sigma) = \sum_{(i,j)\in T} \sigma_{ij} = \sum_j d_j - \sum_i c_i \,.$$

(DOF means "degrees of freedom", while DOF is the corresponding number.)

Example 1 We illustrate[1] the above concepts with the simple pendulum DAE of differentiation index 3.

$$\begin{aligned} 0 = f_1 &= x'' + x\lambda \\ 0 = f_2 &= y'' + y\lambda - g \\ 0 = f_3 &= x^2 + y^2 - L^2 \end{aligned} \qquad \Sigma = \begin{array}{c} \\ f_1 \\ f_2 \\ f_3 \\ d_j \end{array} \begin{array}{ccc|c} x & y & \lambda & c_i \\ 2^\bullet & & 0^\circ & 0 \\ & 2^\circ & 0^\bullet & 0 \\ 0^\circ & 0^\bullet & & 2 \\ 2 & 2 & 0 & \end{array} \qquad \mathbf{J} = \begin{array}{c} \\ f_1 \\ f_2 \\ f_3 \end{array} \begin{array}{ccc} x & y & \lambda \\ 1 & & x \\ & 1 & y \\ 2x & 2y & \end{array}$$

The state variables are x, y, and λ; G is gravity and $L > 0$ is the length of the pendulum. There are two HVTs of Σ, marked with $\bullet$ and $\circ$, respectively. A blank in Σ denotes $-\infty$, and a blank in $\mathbf{J}$ denotes 0.

Since $\det(\mathbf{J}) = -2(x^2 + y^2) = -2L^2 \neq 0$, the System Jacobian is nonsingular, and the SA succeeds. The structural index is $\nu_S = \min_i c_i + 1 = 2 + 1 = 3$ (because $\min_j d_j = d_3 = 0$), which equals the differentiation index. The number of DOF is $\text{DOF} = \text{Val}(\Sigma) = \sum_j d_j - \sum_i c_i = 4 - 2 = 2$.

[1] When we present a DAE example, we show its signature matrix Σ, canonical offsets c_i and d_j, and the associated System Jacobian $\mathbf{J}$.

3 Structural Analysis's Failure

We say that the Σ-method *fails*, if a DAE (1) has a finite $\mathrm{Val}(\Sigma)$ and an identically singular $\mathbf{J}$. In the failure case, the SA reports $\mathrm{Val}(\Sigma)$ as an "apparent" DOF but not a meaningful one.

In this article, we focus on the case where such an identically singular $\mathbf{J}$ is structurally nonsingular. That is, there exists a HVT T of Σ such that $\mathbf{J}_{ij}$ is generically nonzero for all $(i,j) \in T$.

Example 2 We illustrate a failure case with the following DAE[2] in [1, p. 23].

$$0 = f_1 = x' + ty' - g_1(t) \qquad 0 = f_2 = x + ty - g_2(t)$$

$$\Sigma = \begin{array}{c|cc|c} & x & y & c_i \\ \hline f_1 & 1^\bullet & 1 & 0 \\ f_2 & 0 & 0^\bullet & 1 \\ \hline d_j & 1 & 1 & \end{array} \qquad \mathbf{J} = \begin{array}{c|cc} & x & y \\ \hline f_1 & 1 & t \\ f_2 & 1 & t \end{array}$$

The SA fails since $\det(\mathbf{J}) = 0$. Now $\mathbf{J}$ is identically singular but structurally nonsingular: on a HVT marked by $\bullet$, each of $\mathbf{J}_{11} = 1$ and $\mathbf{J}_{22} = t$ is generically nonzero.

One simple fix is to replace f_1 by $\overline{f}_1 = -f_1 + f_2'$, which results in the system below; cf. [3, Example 5].

$$0 = \overline{f}_1 = y + g_1(t) - g_2'(t) \qquad 0 = f_2 = x + ty - g_2(t)$$

$$\overline{\Sigma} = \begin{array}{c|cc|c} & x & y & c_i \\ \hline \overline{f}_1 & & 0^\bullet & 0 \\ f_2 & 0^\bullet & 0 & 0 \\ \hline d_j & 0 & 0 & \end{array} \qquad \overline{\mathbf{J}} = \begin{array}{c|cc} & x & y \\ \hline \overline{f}_1 & & 1 \\ f_2 & 1 & t \end{array}$$

Now that $\det(\overline{\mathbf{J}}) = -1 \neq 0$, the SA succeeds. Notice $\mathrm{Val}(\overline{\Sigma}) = 0 < 1 = \mathrm{Val}(\Sigma)$.

Another simple fix is to introduce a new variable $z = x + ty$ and eliminate x in f_1 and f_2.

$$0 = \overline{f}_1 = -y + z' - g_1(t) \qquad 0 = \overline{f}_2 = z - g_2(t)$$

$$\overline{\Sigma} = \begin{array}{c|cc|c} & y & z & c_i \\ \hline \overline{f}_1 & 0^\bullet & 1 & 0 \\ \overline{f}_2 & & 0^\bullet & 1 \\ \hline d_j & 0 & 1 & \end{array} \qquad \overline{\mathbf{J}} = \begin{array}{c|cc} & y & z \\ \hline \overline{f}_1 & -1 & 1 \\ \overline{f}_2 & & 1 \end{array}$$

For this resulting DAE, $\det(\overline{\mathbf{J}}) = -1 \neq 0$, and the SA succeeds. After solving for y and z, we can obtain $x = z - ty$. This fix also gives $\mathrm{Val}(\overline{\Sigma}) = 0 < 1 = \mathrm{Val}(\Sigma)$.

The above two manipulations illustrate the two conversion methods presented in Sect. 4, respectively.

[2] In the original formulation, the driving functions are f_1, f_2. Here they are renamed g_1, g_2.

4 Conversion Methods

We present two conversion methods for systematically fixing SA's failures. The first method is based on replacing an existing equation by a linear combination of some equations and derivatives of them. We call this method the linear combination (LC) method and describe it in Sect. 4.1. The second method is based on substituting newly introduced variables for some expressions and enlarging the system. We call this method the expression substitution (ES) method and describe it in Sect. 4.2.

We only present the main features of these methods; the reader is referred to [13] for more details and especially the proofs of Theorems 1 and 2 below.

Given a DAE (1), we assume it has a finite $\mathrm{Val}(\Sigma)$ and a System Jacobian $\mathbf{J}$ that is identically singular but structurally nonsingular. We also assume the equations in (1) are sufficiently differentiable, so that our methods fit into the Σ-method theory.

After a conversion, we obtain a system with signature matrix $\overline{\Sigma}$ and System Jacobian $\overline{\mathbf{J}}$. If $\mathrm{Val}(\overline{\Sigma})$ is finite and $\overline{\mathbf{J}}$ is still identically singular, then we can perform another conversion using either of the methods, provided the corresponding conditions are satisfied. Suppose a sequence of conversions produces a solvable DAE with $\mathrm{Val}(\overline{\Sigma}) \geq 0$ and a generically nonsingular $\overline{\mathbf{J}}$. Given the fact that each conversion reduces some $\mathrm{Val}(\Sigma)$ by at least one, the total number of conversions does not exceed the value of the original signature matrix.

If the resulting system is structurally ill posed after a conversion, that is, $\mathrm{Val}(\overline{\Sigma}) = -\infty$, then we say the original DAE is ill posed [13].

4.1 Linear Combination Method

Let u be a nonzero n-vector function in the cokernel coker($\mathbf{J}$) of $\mathbf{J}$, that is, $\mathbf{J}^T u$ is identically zero. We consider $\mathbf{J}$ and u as functions of t, the x_j's and derivatives of them.

Let $\sigma\left(x_j, u\right)$ denote the order of the highest derivative to which x_j occurs in the expressions defining the function u, or $-\infty$ if x_j does not occur in u at all.

Denote

$$I = \{\, i \mid u_i \not\equiv 0 \,\}, \quad \underline{c} = \min_{i \in I} c_i, \quad \text{and} \quad L = \left\{\, i \in I \mid c_i = \underline{c} \,\right\}. \tag{4}$$

Here $u_i \not\equiv 0$ means that u_i is generically nonzero. The LC method is based on the following theorem.

Theorem 1 *If*

$$\sigma\left(x_j, u\right) < d_j - \underline{c} \qquad \textit{for all } j = 1, \ldots, n\,, \tag{5}$$

and we replace an f_l with $l \in L$ by

$$\overline{f}_l = \sum_{i \in I} u_i f_i^{(c_i - \underline{c})}, \tag{6}$$

then Val($\overline{\Sigma}$) < Val(Σ), where $\overline{\Sigma}$ is the signature matrix of the resulting DAE.

We call (5) the condition for applying the LC method.

Example 3 We illustrate this method with the following problem:

$$\begin{aligned} 0 = f_1 &= -x_1' + x_3 \qquad & 0 = f_3 &= x_1 x_2 + g_1(t) \\ 0 = f_2 &= -x_2' + x_4 \qquad & 0 = f_4 &= x_1 x_4 + x_2 x_3 + x_1 + x_2 + g_2(t)\,, \end{aligned}$$

where g_1 and g_2 are given driving functions.

$$\Sigma = \begin{array}{c} \\ f_1 \\ f_2 \\ f_3 \\ f_4 \\ d_j \end{array} \begin{array}{cccc|c} x_1 & x_2 & x_3 & x_4 & c_i \\ 1^\bullet & & 0 & & 0 \\ & 1 & & 0^\bullet & 0 \\ 0 & 0^\bullet & & & 1 \\ \boxed{0} & \boxed{0} & 0^\bullet & 0 & 0 \\ 1 & 1 & 0 & 0 & \end{array} \qquad \mathbf{J} = \begin{array}{c} \\ f_1 \\ f_2 \\ f_3 \\ f_4 \end{array} \begin{array}{cccc} x_1 & x_2 & x_3 & x_4 \\ -1 & & 1 & \\ & -1 & & 1 \\ x_2 & x_1 & & \\ & & x_2 & x_1 \end{array}$$

Here Val(Σ) = 1. A shaded entry $\boxed{\sigma_{ij}}$ in Σ marks a position (i,j) where $d_j - c_i > \sigma_{ij} \geq 0$ and hence $\mathbf{J}_{ij} = 0$ by the formula (3) for $\mathbf{J}$. The SA fails here since $\mathbf{J}$ is identically singular (but structurally nonsingular).

We choose $u = (x_2, x_1, 1, -1)^T \in$ coker($\mathbf{J}$). Then (4) becomes $I = \{1, 2, 3, 4\}$, $\underline{c} = 0$, and $L = \{1, 2, 4\}$. The condition (5) holds since

$$\begin{aligned} \sigma(x_1, u) &= 0 < 1 - 0 = d_1 - \underline{c}, \quad & \sigma(x_2, u) &= 0 < 1 - 0 = d_2 - \underline{c}\,, \\ \sigma(x_3, u) &= -\infty < 0 - 0 = d_3 - \underline{c}, \quad & \sigma(x_4, u) &= -\infty < 0 - 0 = d_4 - \underline{c}\,. \end{aligned}$$

Selecting $l = 4 \in L$ for example, we replace f_4 by

$$\overline{f}_4 = \sum_{i \in I} u_i f_i^{(c_i - \underline{c})} = x_2 f_1 + x_1 f_2 + f_3' - f_4 = -x_1 - x_2 + g_1'(t) - g_2(t)\,.$$

The resulting DAE is $0 = (f_1, f_2, f_3, \overline{f}_4)$ with the SA results as follows.

$$\overline{\Sigma} = \begin{array}{c} \\ f_1 \\ f_2 \\ f_3 \\ \overline{f}_4 \\ d_j \end{array} \begin{array}{cccc|c} x_1 & x_2 & x_3 & x_4 & c_i \\ 1 & & 0^\bullet & & 0 \\ & 1 & & 0^\bullet & 0 \\ 0 & 0^\bullet & & & 1 \\ 0^\bullet & 0 & & & 1 \\ 1 & 1 & 0 & 0 & \end{array} \qquad \overline{\mathbf{J}} = \begin{array}{c} \\ f_1 \\ f_2 \\ f_3 \\ \overline{f}_4 \end{array} \begin{array}{cccc} x_1 & x_2 & x_3 & x_4 \\ -1 & & 1 & \\ & -1 & & 1 \\ x_2 & x_1 & & \\ -1 & -1 & & \end{array}$$

Now $\text{Val}(\overline{\Sigma}) = 0 < 1 = \text{Val}(\Sigma)$. The SA succeeds at all points where $\det(\overline{\mathbf{J}}) = x_2 - x_1 \neq 0$.

From (4) and (6), we can recover the replaced equation f_l by

$$f_l = \left(\overline{f}_l - \textstyle\sum_{i\in I\setminus\{l\}} u_i f_i^{(c_i-\underline{c})}\right) / u_l \,.$$

Provided $u_l \neq 0$ for all t in the interval of interest, it is not difficult to show that the original DAE and the resulting one have the same solution (if there exists one); see [13, § 5.3].

4.2 Expression Substitution Method

Let v be a nonzero n-vector function in the kernel of $\mathbf{J}$; write $v \in \ker(\mathbf{J})$. Denote

$$\begin{aligned} J &= \{\, j \mid v_j \not\equiv 0 \,\}, \quad s = |J| \,, \\ M &= \{\, i \mid d_j - c_i = \sigma_{ij} \text{ for some } j \in J \,\}, \quad \text{and} \quad \overline{c} = \max_{i\in M} c_i \,. \end{aligned} \tag{7}$$

We choose an $l \in J$, and introduce $s-1$ new variables

$$y_j = x_j^{(d_j-\overline{c})} - \frac{v_j}{v_l} x_l^{(d_l-\overline{c})} \qquad \text{for all } j \in J \setminus \{l\} \,. \tag{8}$$

In each f_i with $i \in M$, we

$$\begin{aligned} &\text{substitute} \quad \left(y_j + \frac{v_j}{v_l} x_l^{(d_l-\overline{c})}\right)^{(\overline{c}-c_i)} \\ &\text{for every} \quad x_j^{(\sigma_{ij})} \quad \text{with } \sigma_{ij} = d_j - c_i \text{ and } j \in J \setminus \{l\} \,. \end{aligned} \tag{9}$$

Denote by $\overline{f}_i$ the equations that result from these substitutions, and write $\overline{f}_i = f_i$ for $i \notin M$. Using (8), we append to these $\overline{f}_i$'s the equations

$$0 = g_j = -y_j + x_j^{(d_j-\overline{c})} - \frac{v_j}{v_l} x_l^{(d_l-\overline{c})} \quad \text{for all } j \in J \setminus \{l\} \tag{10}$$

that prescribe the substitutions. Hence the resulting enlarged system consists of

$$\begin{aligned} &\text{equations} \quad 0 = \left(\overline{f}_1, \ldots, \overline{f}_n\right) \quad \text{and} \quad 0 = g_j \quad \text{for all } j \in J \setminus \{l\} \\ &\text{in variables} \quad x_1, \ldots, x_n \qquad\quad \text{and} \quad y_j \qquad\ \text{for all } j \in J \setminus \{l\} \,. \end{aligned}$$

Key to applying the ES method is the following theorem.

Theorem 2 *Let J, s, M, and $\bar{c}$ be as defined in (7). Assume*

$$\sigma\left(x_j, v\right) \begin{cases} < d_j - \bar{c} & \text{if } j \in J \\ \le d_j - \bar{c} & \text{otherwise}, \end{cases} \qquad \text{and} \quad d_j - \bar{c} \ge 0 \quad \text{for all } j \in J\,. \tag{11}$$

For any $l \in J$, if we

- *introduce $(s-1)$ new variables as defined in (8),*
- *perform substitutions in f_i for all $i \in M$ by (9), and*
- *append the equations g_j in (10),*

then $Val(\overline{\Sigma}) < Val(\Sigma)$, where $\overline{\Sigma}$ is the signature matrix of the resulting DAE.

We call (11) the conditions for applying the ES method.

Example 4 We illustrate the ES method with the artificially constructed DAE below.

$$\begin{aligned} 0 &= f_1 = x_1 + e^{-x_1' - x_2 x_2''} + h_1(t) \\ 0 &= f_2 = x_1 + x_2 x_2' + x_2^2 + h_2(t) \end{aligned} \qquad \Sigma = \begin{array}{c} \\ f_1 \\ f_2 \\ d_j \end{array} \begin{array}{cc|c} x_1 & x_2 & c_i \\ 1^\bullet & 2 & 0 \\ 0 & 1^\bullet & 1 \\ 1 & 2 & \end{array} \qquad \mathbf{J} = \begin{array}{c} \\ f_1 \\ f_2 \end{array} \begin{bmatrix} x_1 & x_2 \\ -\alpha & -\alpha x_2 \\ 1 & x_2 \end{bmatrix}$$

Here h_1 and h_2 are given driving functions, and $\alpha = e^{-x_1' - x_2 x_2''}$. Obviously $\det(\mathbf{J}) = 0$ and the SA fails.

Suppose we choose $v = (x_2, -1)^T \in \ker(\mathbf{J})$. Then (7) becomes

$$J = \{1, 2\}, \quad s = |J| = 2, \quad M = \{1, 2\}, \quad \text{and} \quad \bar{c} = \max_{i \in M} c_i = 1\,.$$

We can apply the ES method as the conditions (11) hold:

$$\begin{aligned} \sigma\left(x_1, v\right) &= -\infty \le -1 = 1 - 1 - 1 = d_1 - \bar{c} - 1, \quad d_1 - \bar{c} = 1 - 1 = 0 \ge 0\,, \\ \sigma\left(x_2, v\right) &= \ \ 0 \ \ \le \ 0 = 2 - 1 - 1 = d_2 - \bar{c} - 1, \quad d_2 - \bar{c} = 2 - 1 = 1 \ge 0\,. \end{aligned}$$

For example, we choose $l = 2 \in J$. Now $J \setminus \{l\} = \{1\}$. Using (8) and (10), we introduce a new variable

$$y_1 = x_1^{(d_1 - \bar{c})} - \frac{v_1}{v_2} x_2^{(d_2 - \bar{c})} = x_1^{(1-1)} - \frac{x_2}{-1} x_2^{(2-1)} = x_1 + x_2 x_2'\,,$$

and append the equation $0 = g_1 = -y_1 + x_1 + x_2 x_2'$. Then we substitute $(y_1 - x_2 x_2')'$ for x_1' in f_1 to obtain $\bar{f}_1$, and substitute $y_1 - x_2 x_2'$ for x_1 in f_2 to obtain $\bar{f}_2$. The resulting DAE and its SA results are shown below.

$$\begin{aligned} 0 &= \bar{f}_1 = x_1 + e^{-y_1' + x_2'^2} + h_1(t) \\ 0 &= \bar{f}_2 = y_1 + x_2^2 + h_2(t) \\ 0 &= g_1 = -y_1 + x_1 + x_2 x_2' \end{aligned} \qquad \overline{\Sigma} = \begin{array}{c} \\ \bar{f}_1 \\ \bar{f}_2 \\ g_1 \\ d_j \end{array} \begin{array}{ccc|c} x_1 & x_2 & y_1 & c_i \\ 0 & 1 & 1^\bullet & 0 \\ & 0^\bullet & 0 & 1 \\ 0^\bullet & 1 & \boxed{0} & 0 \\ 0 & 1 & 1 & \end{array} \qquad \overline{\mathbf{J}} = \begin{array}{c} \\ \bar{f}_1 \\ \bar{f}_2 \\ g_1 \end{array} \begin{bmatrix} x_1 & x_2 & y_1 \\ 1 & 2x_2'\beta & -\beta \\ & 2x_2 & 1 \\ 1 & x_2 & \end{bmatrix}$$

Here $\beta = e^{-y_1' + x_2'^2}$. Now $\mathrm{Val}(\overline{\Sigma}) = 1 < 2 = \mathrm{Val}(\Sigma)$. The SA succeeds at all points where $\det(\overline{\mathbf{J}}) = 2\beta(x_2 + x_2') - x_2 \neq 0$.

From the steps of applying the ES method, we can "undo" the expression substitutions to recover the original DAE. Similar to the LC method, the ES method also guarantees that, provided $v_l \neq 0$ for all t in some interval of interest, the original DAE and the resulting one have equivalent solutions (if any) in a natural sense; cf. [13, § 6.3].

Our experience suggests that it is effective to attempt the LC method first, and if its condition (5) is violated, then we try the ES method. Also, it is desirable to choose an $l \in L$ [resp. $l \in J$] in the LC [resp. ES] method, such that an u_l [resp. v_l] *never* becomes zero. For example, it can be a nonzero constant, $x_1^2 + 1$, or $2 + \cos x_2$. Such a choice of l guarantees that the resulting DAE is "equivalent" to the original one—that is, they *always* have the same solution, if there exists one.

We show below that the LC method does not succeed on the DAE in Example 4 because the condition (5) is not satisfied.

Choose $u = \left(1, \alpha\right)^T \in \mathrm{coker}(\mathbf{J})$, where $\alpha = e^{-x_1' - x_2 x_2''}$. Then (4) becomes

$$I = \left\{\, i \mid u_i \neq 0 \,\right\} = \left\{1, 2\right\}, \quad \underline{c} = \min_{i \in I} c_i = 0, \text{ and } L = \left\{\, i \in I \mid c_i = \underline{c} \right\} = \left\{1\right\}.$$

Since x_1 and x_2 occur of orders 1 and 2 in u, respectively, we have

$$\sigma\left(x_1, u\right) = 1 \nless 1 - 0 = d_1 - \underline{c} \quad \text{and} \quad \sigma\left(x_2, u\right) = 2 \nless 2 - 0 = d_2 - \underline{c}\,.$$

Hence (5) is not satisfied. Choosing $l = 1 \in I$ for example and replacing f_1 by

$$\overline{f}_1 = u_1 f_1 + u_2 f_2' = x_1 + h_1(t) + \alpha\left(1 + x_1' + x_2 x_2'' + (x_2')^2 + 2x_2 x_2' + h_2'(t)\right)$$

results in the DAE $0 = \left(\overline{f}_1, f_2\right)$.

$$\overline{\Sigma} = \begin{array}{c|cc|c} & x_1 & x_2 & c_i \\ \hline \overline{f}_1 & 1^\bullet & 2 & 0 \\ f_2 & 0 & 1^\bullet & 1 \\ \hline d_j & 1 & 2 & \end{array} \qquad \overline{\mathbf{J}} = \begin{array}{c} \\ \overline{f}_1 \\ f_2 \end{array}\!\!\begin{array}{c} \begin{array}{cc} x_1 & x_2 \end{array} \\ \begin{bmatrix} -\gamma\alpha & -\gamma\alpha x_2 \\ 1 & x_2 \end{bmatrix} \end{array}$$

Here $\gamma = x_1' + x_2 x_2'' + (x_2')^2 + 2x_2 x_2' + h_2'(t)$. The SA fails still, since $\overline{\mathbf{J}}$ is identically singular. Now $\mathrm{Val}(\overline{\Sigma}) = \mathrm{Val}(\Sigma) = 2$.

5 Conclusions

We proposed two symbolic-numeric conversion methods for improving the Σ-method. They convert a DAE with a finite $\mathrm{Val}(\Sigma)$ and an identically singular (but structurally nonsingular) System Jacobian to a DAE that may have a generically nonsingular System Jacobian. A conversion guarantees that both DAEs have (at least locally) the same solution. The conditions for applying these methods can be checked automatically, and the main result of a conversion is $\mathrm{Val}(\overline{\Sigma}) < \mathrm{Val}(\Sigma)$, where $\overline{\Sigma}$ is the signature matrix of the resulting DAE.

An implementation of these methods requires

(1) computing a symbolic form of a System Jacobian **J**,
(2) finding a vector in coker(**J**) [respectively ker(**J**)],
(3) checking the LC condition (5) [respectively ES conditions (11)], and
(4) generating the equations for the resulting DAE.

These symbolic computations may seem expensive. However, for DAEs whose **J** can be permuted into a block-triangular form [8, 11], we can locate the diagonal blocks that are singular and then apply our methods to these blocks only, instead of working on the whole DAE. This approach is presently under development.

We combine MATLAB's Symbolic Math Toolbox [14] with our structural analysis software DAESA [8, 11], and have built a prototype code that applies our conversion methods automatically. We aim to incorporate them in a future version of DAESA.

With our prototype code, we have applied our methods on numerous DAEs. They are either arbitrarily constructed to be "SA-failure cases" for our investigations, or borrowed from the existing literature: Campbell and Griepentrog's robot arm [2], the transistor amplifier and the ring modulator [4], and the linear constant coefficient DAE in [12]. Our conversion methods succeed in fixing all these solvable DAEs; see [13] and especially Appendix B thereof. We believe that our assumptions and conditions are reasonable for practical problems, and that these methods can help make the Σ-method more reliable.

Finally, we conjecture that reducing $\mathrm{Val}(\Sigma)$ tends to give a better formulation of a DAE from the SA's perspective and then a nonsingular System Jacobian.

Acknowledgements The authors acknowledge with thanks the financial support for this research: GT is supported in part by the Ontario Research Fund, Canada, NSN is supported in part by the Natural Sciences and Engineering Research Council of Canada, and JDP is supported in part by the Leverhulme Trust, the UK.

References

1. Brenan, K., Campbell, S., Petzold, L.: Numerical Solution of Initial-Value Problems in Differential-Algebraic Equations, 2nd edn. SIAM, Philadelphia (1996)
2. Campbell, S.L., Griepentrog, E.: Solvability of general differential-algebraic equations. SIAM J. Sci. Comput. **16**(2), 257–270 (1995)
3. Mattsson, S.E., Söderlind, G.: Index reduction in differential-algebraic equations using dummy derivatives. SIAM J. Sci. Comput. **14**(3), 677–692 (1993)
4. Mazzia, F., Iavernaro, F.: Test set for initial value problem solvers. Technical report 40, Department of Mathematics, University of Bari (2003). http://pitagora.dm.uniba.it/~testset/
5. Nedialkov, N.S., Pryce, J.D.: Solving differential-algebraic equations by Taylor series (I): computing Taylor coefficients. BIT Numer. Math. **45**(3), 561–591 (2005)

6. Nedialkov, N.S., Pryce, J.D.: Solving differential-algebraic equations by Taylor series (II): computing the system Jacobian. BIT Numer. Math. **47**(1), 121–135 (2007)
7. Nedialkov, N.S., Pryce, J.D.: Solving differential-algebraic equations by Taylor series (III): the DAETS code. JNAIAM J. Numer. Anal. Ind. Appl. Math. **3**, 61–80 (2008)
8. Nedialkov, N.S., Pryce, J.D., Tan, G.: Algorithm 948: DAESA—a Matlab tool for structural analysis of differential-algebraic equations: software. ACM Trans. Math. Softw. **41**(2), 12:1–12:14 (2015). doi:10.1145/2700586
9. Pantelides, C.: The consistent initialization of differential-algebraic systems. SIAM. J. Sci. Stat. Comput. **9**, 213–231 (1988)
10. Pryce, J.D.: A simple structural analysis method for DAEs. BIT Numer. Math. **41**(2), 364–394 (2001)
11. Pryce, J.D., Nedialkov, N.S., Tan, G.: DAESA—a Matlab tool for structural analysis of differential-algebraic equations: theory. ACM Trans. Math. Softw. **41**(2), 9:1–9:20 (2015). doi:10.1145/2689664
12. Scholz, L., Steinbrecher, A.: A combined structural-algebraic approach for the regularization of coupled systems of DAEs. Technical report 30, Reihe des Instituts für Mathematik Technische Universität Berlin, Berlin (2013)
13. Tan, G., Nedialkov, N., Pryce, J.: Symbolic-numeric methods for improving structural analysis of differential-algebraic equation systems. Technical report, Department of Computing and Software, McMaster University, 1280 Main St. W., Hamilton, ON, L8S4L8, Canada (2015). 84pp. Download link:http://www.cas.mcmaster.ca/cas/0reports/CAS-15-07-NN.pdf
14. The MathWorks, Inc.: Matlab symbolic math toolbox (2015). https://www.mathworks.com/help/symbolic/index.html

Pinning Stabilization of Cellular Neural Networks with Time-Delay Via Delayed Impulses

Kexue Zhang, Xinzhi Liu, and Wei-Chau Xie

Abstract This paper studies the pinning stabilization problem of time-delay cellular neural networks. A new pinning delayed-impulsive controller is proposed to stabilize the networks. Based on a Razumikhin-type stability result, a global exponential stability of time-delay cellular neural networks with delayed impulses is constructed. It is shown that the global exponential stabilization of delayed cellular neural networks can be effectively realized by controlling a small portion of units in the networks via delayed impulses. Numerical simulations are provided to illustrate the theoretical result.

1 Introduction

Cellular neural networks (CNNs), introduced by Chua and Yang in [1, 2], have attracted the attentions of several researchers in recent years. This is mainly due to their broad applications in many areas, including image processing and pattern recognition (see, e.g., [2, 3]), data fusion [4], odor classification [5], and solving partial differential equations [6].

In real-world applications, it is inevitable for the existence of time delay in the processing and transmission of signals among units of CNNs. Hence, it is practical to investigate CNNs with time-delay (DCNNs) (see, e.g., [7–12]). Stability of DCNNs, as a prerequisite for their applications, has been studied extensively in the past decades, and various control methods have been introduced to stabilize DCNNs, such as intermittent control [9], sliding mode control [10], impulsive control [11], and sampled-data control [12]. Among these control algorithms, the impulsive control method has been proved to be an effective approach to stabilize DCNNs. The control mechanism of this method is to control the unit states of

K. Zhang (✉) • X. Liu
Department of Applied Mathematics, University of Waterloo, Waterloo, ON, N2L 3G1, Canada
e-mail: k57zhang@uwaterloo.ca; xzliu@uwaterloo.ca

W.-C. Xie
Department of Civil and Environmental Engineering, University of Waterloo, Waterloo, ON, N2L 3G1, Canada
e-mail: xie@uwaterloo.ca

© Springer International Publishing Switzerland 2016
J. Bélair et al. (eds.), *Mathematical and Computational Approaches in Advancing Modern Science and Engineering*, DOI 10.1007/978-3-319-30379-6_69

the CNN with small impulses, which are small samples of the state variables of the CNN, at a sequence of discrete moments. Since the time delay in unavoidable in sampling and transmission of the impulsive information is dynamical systems, many control problems of dynamical systems have been investigated via delayed impulses in recent years, such as stabilization of stochastic functional systems [13] and synchronization of dynamical networks [14].

The regular impulsive control method to stabilize a CNN is to control each unit of the network to tame the unit dynamics to approach a steady state (i.e., equilibrium point). However, a neural network is normally composed of a large number of units, and it is expensive and infeasible to control all of them. Motivated by this practical consideration, the idea of controlling a small portion of units, named pinning control, was introduced in [15, 16], and many pinning impulsive control algorithms have been reported for many control problems of dynamical networks (see, e.g., [17–23]) It is worth noting that no time delay is considered in these pinning impulsive controllers proposed in the above literatures. However, it is natural and essential to consider the delay effects when processing the impulse information in the controller.

Due to the cost effective advantage of impulsive control method and pinning control strategy and the wide existence of time delay, it is practical to investigate the pinning impulsive control approach that takes into account of delays. However, to our best knowledge, no such result has been reported for stabilization of DCNNs. Therefore, in this paper, we propose a novel pinning delayed-impulsive controller for the DCNNs. The remainder of this paper is organized as follows. In Sect. 2, we formulate the problem and introduce the pinning delayed-impulsive control algorithm. In Sect. 3, a global exponential stability result of the impulsive DCNNs is obtained. Then, in Sect. 4, a numerical example is considered to illustrate the theoretical result. Finally, conclusions are stated in Sect. 5.

2 Preliminaries

Let $\mathbb{N}$ denote the set of positive integers, $\mathbb{R}$ the set of real numbers, $\mathbb{R}^+$ the set of nonnegative real numbers, and $\mathbb{R}^n$ the n-dimensional real space equipped with the Euclidean norm. For $a, b \in \mathbb{R}$ with $a < b$ and $S \subseteq \mathbb{R}^n$, we define

$$\mathscr{PC}([a,b],S) = \Big\{\psi : [a,b] \to S \Big| \psi(t) = \psi(t^+),\ \text{for any } t \in [a,b);\ \psi(t^-) \text{ exists in } S,\ \text{for any } t \in (a,b];\ \psi(t^-) = \psi(t) \text{ for all but at most a finite number of points } t \in (a,b]\Big\},$$

where $\psi(t^+)$ and $\psi(t^-)$ denote the right and left limit of function ψ at t, respectively. For a given constant $\tau > 0$, the linear space $\mathscr{PC}([-\tau,0],\mathbb{R}^n)$ is equipped with the norm defined by $||\psi||_\tau = \sup_{s\in[-\tau,0]} ||\psi(s)||$, for $\psi \in \mathscr{PC}([-\tau,0],\mathbb{R}^n)$.

Consider the following CNN with delays:

$$\dot{x} = -Cx(t) + Af(x(t)) + Bf(x(t - \tau_1)), \tag{1}$$

where $x = (x_1, x_2, \ldots, x_n)^T \in \mathbb{R}^n$, and x_i is the state of the ith unit; n denotes the number of units in CNN (1); $f(x(t)) = (f_1(x_1(t)), f_2(x_2(t)), \ldots, f_n(x_n(t)))^T$ and $f_j(x_j(t))$ denotes the output of the jth unit at time t; $A = [a_{ij}]_{n\times n}$ and $B = [b_{ij}]_{n\times n}$; constants a_{ij} and b_{ij} represent the strengths of connectivity between units i and j at time t and $t - \tau_1$, respectively; τ_1 corresponds to the transmission delay when processing information from the jth unit; $C = \text{Diag}\{c_1, c_2, \ldots, c_n\}$ and constant c_i denotes the rate at which the ith unit will reset its potential when disconnected with other units of the network.

Throughout this paper, we make the following assumptions:

(A) $f(0) = 0$ and there exists a constant L such that $||f(u) - f(v)|| \leq L||u - v||$ for all $u, v \in \mathbb{R}^n$.

The Lipschtiz condition on the nonlinear activation function f_i has been widely considered due to its significance in the application of CNNs (see, e.g., [2, 24, 25]). Based on the assumption, system (1) admits the trivial solution.

The objective of this paper is to design the following delay-dependent pinning impulsive controller to exponentially stabilize CNN (1):

$$U_i(t, x_i) = \begin{cases} \sum_{k=1}^{\infty}[\gamma_1 x_i(t) + \gamma_2 x_i(t - \tau_2)]\delta(t - t_k^-), \ i \in \mathfrak{D}_k^l, \\ 0, \ i \notin \mathfrak{D}_k^l, \end{cases} \tag{2}$$

for $i = 1, 2, \ldots, n$, where γ_1 and γ_2 are impulsive control gains to be determined; $\tau_2 > 0$ denotes the time delay in controller U_i; the impulsive instant sequence $\{t_k\}$ satisfies $\{t_k\} \subseteq \mathbb{R}$, $0 \leq t_0 < t_1 < \ldots < t_k < \ldots$, and $\lim_{k\to\infty} t_k = \infty$; $\delta(\cdot)$ is the Dirac Delta function, and $x_i(t_k^-)$ denotes the left limit of x_i at time t_k. Let l denote the number of units to be pinned at each impulsive instant, and the index set $\mathfrak{D}_k^l$ is defined as follows: reorder states $x_1(t_k^-), x_2(t_k^-), \ldots, x_n(t_k^-)$ such that $|x_{p1}(t_k^-)| \geq |x_{p2}(t_k^-)| \geq \ldots |x_{pl}(t_k^-)| \geq |x_{pl+1}(t_k^-)| \geq \ldots \geq |x_{pn}(t_k^-)|$, then define $\mathfrak{D}_k^l = \{p1, p2, \ldots, pl\}$. The pinning impulsive control mechanism can be explained as follows: at each impulsive instant, we only control l units that have larger deviations from the trivial state than the remaining $n - l$ units.

The definition of $\mathfrak{D}_k^l$ is borrowed from [26]. Since no delay has been considered in the pinning impulsive controller in [26], the pinning algorithm introduced in [26] can be treated as a special case of our pinning delayed-impulsive control strategy (i.e., $\gamma_2 = 0$).

The impulsive controlled CNN (1) can be written in the form of an impulsive system:

$$\begin{cases} \dot{x}(t) = -Cx(t) + Af(x(t)) + Bf(x(t - \tau_1)), \ t \in [t_{k-1}, t_k), \\ \Delta x_i(t_k) = \gamma_1 x_i(t_k^-) + \gamma_2 x_i(t_k - \tau_2), \ \ i \in \mathfrak{D}_k^l, \ k \in \mathbb{N}, \\ x_{t_0} = \phi, \end{cases} \tag{3}$$

where $\phi = (\phi_1, \phi_2, \ldots, \phi_n)^T$ and $\phi_i \in \mathscr{PC}([a, b], \mathbb{R})$ is the initial function, $\Delta x_i(t_k) = x_i(t_k^+) - x_i(t_k^-)$, and x_{t_0} is defined by $x_{t_0}(s) = x(t_0 + s)$ for $s \in [-\tau, 0]$ and $\tau = \max\{\tau_1, \tau_2\}$. Then the pinning impulsive stabilization problem of CNN (1) is transformed into the stability problem of impulsive system (3).

Definition 1 The trivial solution of system (3) is said to be globally exponentially stable (GES), if there exist constants $\alpha > 0$ and $M \geq 1$ such that, for any initial data $x_{t_0} = \phi$, the following inequality holds,

$$||x(t, t_0, \phi)|| \leq M||\phi||_\tau e^{-\alpha(t-t_0)}, \text{ for all } t \geq t_0.$$

3 Stabilization

In this section, we will use a Razumikhin-type stability criterion to construct verifiable conditions for the GES of impulsive CNN (3). First, a class of Lyapunov functions for impulsive systems is introduced.

Definition 2 Function $V : \mathbb{R}^+ \times \mathbb{R}^n \to \mathbb{R}^+$ is said to belong to the class ν_0 if the following is true:

(1) V is continuous in each of the sets $[t_{k-1}, t_k) \times \mathbb{R}^n$, and for each $x \in \mathbb{R}^n$, $t \in [t_{k-1}, t_k)$, and $k \in \mathbb{N}$, $\lim_{(t,y)\to(t_k^-,x)} V(t, y) = V(t_k^-, x)$ exists;
(2) $V(t, x)$ is locally Lipschiz in all $x \in \mathbb{R}^n$, and for all $t \geq t_0$, $V(t, 0) \equiv 0$.

To establish a stability result for system (3), the following lemma is required.

Lemma 1 *Assume that there exist $V \in \nu_0$, and constants $p, q, c, w_1, w_2, \rho_1 > 0$, and $\rho_2 \geq 0$, such that*

(i) $w_1||x||^p \leq V(t, x) \leq w_2||x||^p$, *for* $t \in \mathbb{R}^+$ *and* $x \in \mathbb{R}^n$;
(ii) for $t \in [t_{k-1}, t_k)$, $\psi \in \mathscr{PC}([-\tau, 0], \mathbb{R}^n)$, *and* $k \in \mathbb{N}$,

$$V'(t, \psi(0)) \leq cV(t, \psi(0)),$$

whenever $V(t + s, \psi(s)) < qV(t, \psi(0))$ *for all* $s \in [-\tau, 0]$;
(iii) for $k \in \mathbb{N}$ *and* $\psi \in \mathscr{PC}([-\tau, 0], \mathbb{R}^n)$,

$$V(t_k, \psi(0) + I_k(t_k, \psi)) \leq \rho_1 V(t_k^-, \psi(0)) + \rho_2 \sup_{s\in[-\tau,0]} \{V(t_k^- + s, \psi(s))\};$$

(iv) $q > \frac{1}{\rho_1+\rho_2} > e^{cd}$, *where* $d = \sup_{k\in\mathbb{N}}\{t_{k+1} - t_k\}$.

Then the trivial solution of system (3) is GES.

Lemma 1 is a direct consequence of Theorem 3.1 in [27]. Next, verifiable conditions will be constructed for the GES of system (3) by utilizing a quadratic Lyapunov function.

Theorem 1 *If the following inequality is satisfied*

$$\rho e^{cd} < 1, \tag{4}$$

where $\rho = 1 - \frac{l}{n} + \left(\sqrt{\frac{l}{n}}|1+\gamma_1| + |\gamma_2|\right)^2$, $c = -2\min_i\{c_i\} + 2L(||A|| + \frac{||B||}{\sqrt{\rho}}) > 0$, *and* $d = \sup_{k\in\mathbb{N}}\{t_{k+1} - t_k\}$, *then the trivial solution of system (3) is GES.*

Proof Consider the Lyapunov function $V(x) = x^T x$. Note that condition (i) of Lemma 1 is satisfied with $w_1 = w_2 = 1$ and $p = 2$. Taking the time-derivative along solutions of (3) yields

$$\begin{aligned}
\dot{V}(x) &= 2x^T(t)\dot{x}(t)\\
&= 2x^T(t)\Big[-Cx(t) + Af(x(t)) + Bf(x(t-\tau_1))\Big]\\
&\le \big(-2c_{\min} + 2||A||L\big)V(x(t)) + 2||B||L||x(t)||||x(t-\tau_1)||\\
&\le \big(-2c_{\min} + 2||A||L + ||B||L\varepsilon^{-1}\big)V(x(t)) + \varepsilon||B||LV(x(t-\tau_1)),
\end{aligned} \tag{5}$$

where $c_{\min} = \min_i\{c_i\}$ and constant $\varepsilon > 0$. It can be seen from (4) that there exists a constant $q > 0$ such that

$$q > \frac{1}{\rho} > e^{\bar{c}d}, \tag{6}$$

where $\bar{c} = -2\min_i\{c_i\} + 2L(||A|| + \sqrt{q}||B||)$. If $V(x(t+s)) < qV(x(t))$ for all $s \in [-\tau, 0]$, then we can obtain from (5) that

$$\dot{V}(x) \le \big[-2c_{\min} + 2||A||L + ||B||L(\varepsilon^{-1} + q\varepsilon)\big]V(x(t)). \tag{7}$$

Define h as a function of ε: $h(\varepsilon) = -2c_{\min} + 2||A||L + ||B||L(\varepsilon^{-1} + q\varepsilon)$. Then, for $\varepsilon > 0$, function h attains its minimum value $\bar{c}$ at $\varepsilon = \frac{1}{\sqrt{q}}$ (that is, $h'(\varepsilon) = 0$ at $\varepsilon = \frac{1}{\sqrt{q}}$). Therefore, we can get from (7) that $\dot{V}(x) \le \bar{c}V(x)$.

Given a constant $\xi > 0$, letting $\bar{\gamma}_1 = 1 - \frac{l}{n}[1 - (1+\xi)(1+\gamma_1)^2]$, then

$$\begin{aligned}
(1-\bar{\gamma}_1)\sum_{i\notin\mathfrak{D}_k^l} x_i^2(t_k^-) &\le (1-\bar{\gamma}_1)(n-l)\min_{i\in\mathfrak{D}_k^l}\{x_i^2(t_k^-)\}\\
&= l[\bar{\gamma}_1 - (1+\xi)(1+\gamma_1)^2]\min_{i\in\mathfrak{D}_k^l}\{x_i^2(t_k^-)\}\\
&\le [\bar{\gamma}_1 - (1+\xi)(1+\gamma_1)^2]\sum_{i\in\mathfrak{D}_k^l} x_i^2(t_k^-),
\end{aligned}$$

i.e.,

$$(1+\xi)(1+\gamma_1)^2\sum_{i\in\mathfrak{D}_k^l}x_i^2(t_k^-)+\sum_{i\notin\mathfrak{D}_k^l}x_i^2(t_k^-)\le\bar{\gamma}_1\sum_{i=1}^{n}x_i^2(t_k^-).$$

Then, for $t=t_k$, we have

$$\begin{aligned}
V(x(t_k)) &= \sum_{i\in\mathfrak{D}_k^l}x_i^2(t_k)+\sum_{i\notin\mathfrak{D}_k^l}x_i^2(t_k)\\
&= \sum_{i\in\mathfrak{D}_k^l}[(1+\gamma_1)x_i(t_k^-)+\gamma_2x_i(t_k-\tau_2)]^2+\sum_{i\notin\mathfrak{D}_k^l}x_i^2(t_k)\\
&\le \sum_{i\in\mathfrak{D}_k^l}[(1+\xi)(1+\gamma_1)^2x_i^2(t_k^-)+(1+\xi^{-1})\gamma_2^2x_i^2(t_k-\tau_2)]\\
&\quad+\sum_{i\notin\mathfrak{D}_k^l}x_i^2(t_k)\\
&\le (1+\xi)(1+\gamma_1)^2\sum_{i\in\mathfrak{D}_k^l}x_i^2(t_k^-)+\sum_{i\notin\mathfrak{D}_k^l}x_i^2(t_k)\\
&\quad+(1+\xi^{-1})\gamma_2^2\sum_{i=1}^{n}x_i^2(t_k-\tau_2)\\
&\le \bar{\gamma}_1V(x(t_k^-))+(1+\xi^{-1})\gamma_2^2V(x(t_k-\tau_2))\\
&\le \rho_1V(x(t_k^-))+\rho_2\sup_{s\in[-\tau,0]}\{V(x(t_k^-+s))\},
\end{aligned}$$

where $\rho_1=\bar{\gamma}_1$, $\rho_2=(1+\xi^{-1})\gamma_2^2$, and constant $\xi>0$ to be determined to minimize the value of $\rho_1+\rho_2$.

Let $\bar{h}(\xi)=\frac{l}{n}(1+\gamma_1)^2\xi+\gamma_2^2\xi^{-1}$, then, for $\xi=|\frac{\gamma_2}{1+\gamma_1}|\sqrt{\frac{n}{l}}$, we have $\bar{h}'(\xi)=0$, i.e., $\bar{h}(\xi)$ attains its minimum for $\xi>0$. Hence,

$$\rho=\min_{\xi>0}\{\rho_1+\rho_2\}=1-\frac{l}{n}+\left(\sqrt{\frac{l}{n}}|1+\gamma_1|+|\gamma_2|\right)^2.$$

Based on the above discussion, we can conclude that all the conditions of Lemma 1 are satisfied. Thus, the trivial solution of system (3) is GES. □

Remark 1 Parameter ρ is related to impulsive control gains γ_1, γ_2 and the ratio l/n. It can be seen from (4) that, the fewer units are controlled at impulsive instants, the more frequently the impulsive controllers need to be added to the network.

4 Numerical Simulations

In this section, we will consider an example to demonstrate our theoretical result. In order to observe the pinning control process clearly, we will investigate a CNN with only two units in the following example.

Example 1 Consider CNN (1) with $n = 2$, $c_1 = c_2 = -1$, $\tau_1 = 1$,

$$A = \begin{bmatrix} 2 & -0.1 \\ -5 & 3 \end{bmatrix}, \quad B = \begin{bmatrix} -1.5 & -0.1 \\ -0.2 & -2.5 \end{bmatrix},$$

and $f(x) = (f_1(x_1), f_2(x_2))^T$ with $f_1(\cdot) = f_2(\cdot) = \tanh(\cdot)$. The chaotic attractor of CNN (1) is shown in Fig. 1.

We consider two types of impulsive controllers:

(1) $l = 1$, i.e., impulsive control one unit at each impulsive instant. Let $t_k - t_{k-1} = 0.03$, $\tau_2 = 1$, $\gamma_1 = -0.868$, and $\gamma_2 = 0.2$, then (4) is satisfied. Thus, Theorem 1 implies that the trivial solution of (3) is GES. See Fig. 2 for numerical simulations.
(2) $l = 2$, i.e., impulsive control two units at each impulsive instant. Let $t_k - t_{k-1} = 0.08$, and $\tau_2, \gamma_1, \gamma_2$ are the same as those in the first scenario, then (4) is satisfied and Theorem 1 implies that the trivial solution of (3) is GES. Numerical results are shown in Fig. 3.

The initial data in Figs. 2 and 3 is chosen the same as that in Fig. 1, and the red dot denotes the state x at initial time $t = 0$. The vertical (or horizontal) lines in Fig. 2a represent the state jump of x_2 (or x_1) while the other state is unchanged. Since both units are controlled in Fig. 3a, no vertical and horizontal lines can be observed. It can be seen from Fig. 2 that different unit may be controlled at different impulsive instants. This is consistent with our pinning algorithm of controlling the unit which has the largest state deviation with the equilibrium. However, it is more practical to control one specific unit at all impulsive instants. Next, we apply the pinning impulsive controller to the first and second unit at all impulsive instants respectively, and numerical results are shown in Fig. 4a, b. The impulsive control

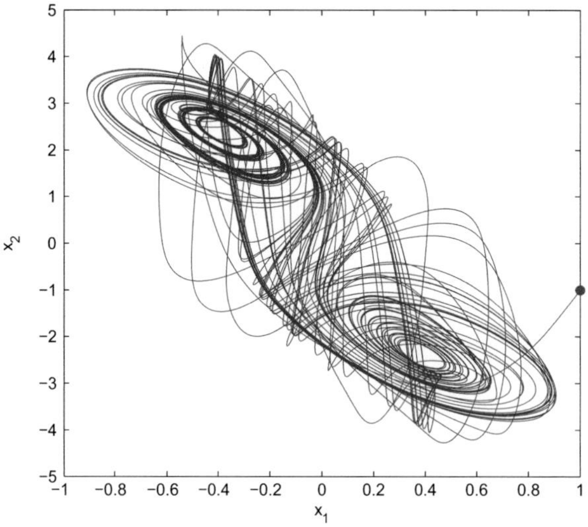

Fig. 1 Chaotic behavior of CNN (1) with the parameters given in Example 1. The initial data for this simulation is $\phi(s) = [1, -1]^T$ for $s \in [-\tau_1, 0]$, and the *red dot* denotes the state x at the initial time $t = 0$

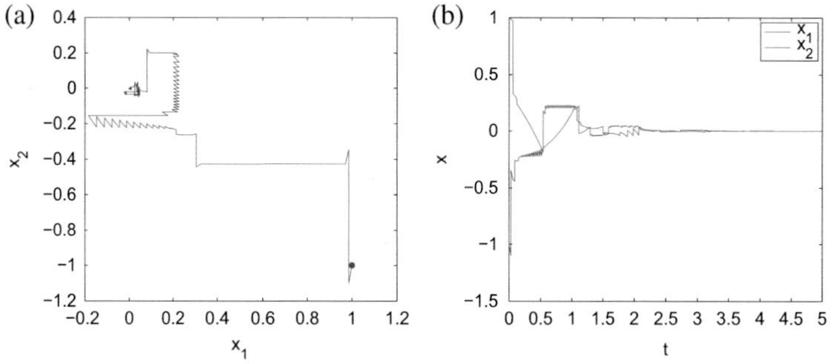

Fig. 2 Impulsive control one unit of CDN (1) at each impulsive instant. (**a**) Phase portrait. (**b**) State trajectories

gains γ_1, γ_2, and the impulsive sequence $\{t_k\}$ are chosen the same as those in Fig. 2. Figure 4 implies that stabilization cannot be realized via this type of pinning strategy with the given parameters, and more strict conditions may be required to guarantee the stability which will be investigated in our future research.

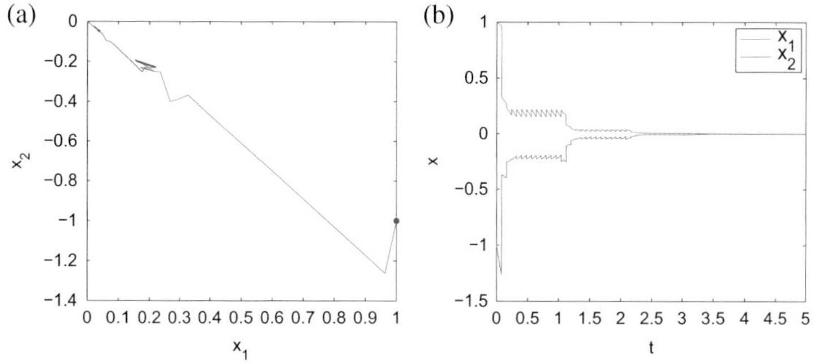

Fig. 3 Impulsive control both units of CDN (1) at each impulsive instant. (**a**) Phase portrait. (**b**) State trajectories

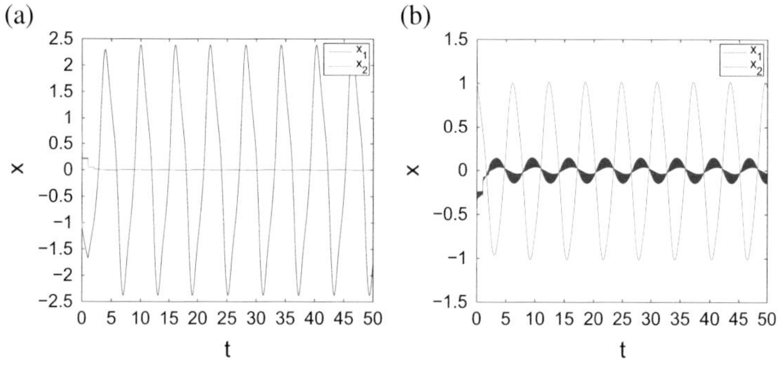

Fig. 4 Impulsive control one specific unit of CDN (1) through all the impulsive instants. (**a**) Control the first unit. (**b**) Control the second unit

5 Conclusion

We have studied the stabilization problem of cellular neural networks with time-delay. An impulsive controller that takes into account both the pinning control algorithm and time delays has been proposed. Sufficient conditions for the global exponential stability of cellular neural networks with time-delay and delayed impulses have been derived by using a Razumikhin-type stability result. Our result has shown that the delayed cellular neural networks can be exponentially stabilized by pinning controlling a small portion of units at each impulsive instant. Numerical simulations have been provided to demonstrate the effectiveness of our theoretical result.

References

1. Chua, L.O., Yang, L.: Cellular neural networks: theory. IEEE Trans. Circuits Syst. **35**(10), 1257–1272 (1988)
2. Chua, L.O., Yang, L.: Cellular neural networks: applications. IEEE Trans. Circuits Syst. **35**(10), 1273–1290 (1988)
3. Jamrozik, W.: Cellular neural networks for welding arc thermograms segmentation. Infrared Phys. Technol. **66**, 18–28 (2014)
4. Chatziagorakis, P., Sirakoulis, G.Ch., Lygouras, J.N.: Design automation of cellular neural networks for data fusion applications. Microprocess. Microsyst. **36**(1), 33–44 (2012)
5. Ayhan, T., Yalçin, M.E.: An application of small-world cellular neural networks on odor classification. Int. J. Bifurc. Chaos **22**(01), 1250013 (12pp.) (2012)
6. Hadada, K., Piroozmanda, A.: Application of cellular neural network (CNN) method to the nuclear reactor dynamics equations. Ann. Nucl. Energy **34**(5), 406–416 (2007)
7. Civalleri, P.P., Gilli, M., Pandolfi, L.: On stability of cellular neural networks with delay. IEEE Trans. Circuits Syst. I: Fundam. Theory Appl. **40**(3), 157–165 (1993)
8. Mohamad, S., Gopalsamy, K.: Exponential stability of continuous-time and discrete-time cellular neural networks with delays. Appl. Math. Comput. **135**(1), 17–38 (2003)
9. Yu, J., Hu, C., Jiang, H., Teng, Z.: Exponential lag synchronization for delayed fuzzy cellular neural networks via periodically intermittent control. Math. Comput. Simul. **82**(5), 895–908 (2012)
10. Gan, Q., Xu, R., Yang, P.: Synchronization of non-identical chaotic delayed fuzzy cellular neural networks based on sliding mode control. Commun. Nonlinear Sci. Numer. Simul. **17**(1), 433–443 (2012)
11. Wang, Q., Liu, X.: Impulsive stabilization of cellular neural networks with time delay via Lyapunov functionals. J. Nonlinear Sci. Appl. **1**(2), 72–86 (2008)
12. Kalpana, M., Balasubramaniam, P.: Stochastic asymptotical synchronization of chaotic Markovian jumping fuzzy cellular neural networks with mixed delays and the Wiener process based on sampled-data control. Chin. Phys. B **22**(7), 078401 (10pp.) (2013)
13. Cheng, P., Deng, F., Yao, F.: Exponential stability analysis of impulsive functional differential equations with delayed impulses. Commun. Nonlinear Sci. Numer. Simul. **19**(6), 2104–2114 (2014)
14. Zhang, W., Tang, Y., Miao, Q., Fang, J.: Synchronization of stochastic dynamical networks under impulsive control with time delays. IEEE Trans. Neural Netw. Learn. Syst. **25**(10), 1758–1768 (2014)
15. Wang, X.F., Chen, G.: Pinning control of scale-free dynamical networks. Phys. A: Stat. Mech. Appl. **310**(3–4), 521–531 (2002)
16. Li, X., Wang, X.F., Chen, G.: Pinning a complex dynamical netwrok to its equilibrium. IEEE Trans. Circuits Syst. I: Regul. Pap. **51**(10), 2074–2087 (2004)
17. Lu, J., Kurths, J., Cao, J., Mahdavi, N., Huang, C.: Synchronization control for nonlinear stochastic dynamical networks: pinning impulsive strategy. IEEE Trans. Neural Netw. Learn. Syst. **23**(2), 285–292 (2012)
18. Mahdavi, N., Menhaj, M.B., Kurths, J.: Pinning impulsive synchronization of complex dynamical networks. Int. J. Bifurc. Chaos **22**(10), 1250239 (14pp.) (2012)
19. Hu, A., Xu, Z.: Pinning a complex dynamical network via impulsive control. Phys. Lett. Sect. A: Gen. Atom. Solid State Phys. **374**(2), 186–190 (2009)
20. Zhou, J., Wu, Q., Xiang, L.: Pinning complex delayed dynamical networks by a single impulsive controller. IEEE Trans. Circuits Syst. I: Regul. Pap. **58**(12), 2882–2893 (2011)
21. Lu, J., Wang, Z., Cao, J.: Pinning impulsive stabilization of nonlinear dynamical networks with time-varying delay. Int. J. Bifurc. Chaos **22**(7), 1250176 (12pp.) (2012)
22. Yang, X., Cao, J., Yang, Z.: Synchronization of coupled reaction-diffusion neural networks with time-varying delays via pinning-impulsive controller. SIAM J. Control Optim. **51**(5), 3486–3510 (2013)

23. Zhang, K., Liu, X., Xie, W.C.: Impulsive control and synchronization of spatiotemporal chaos in the Gray Scott model. In: Interdisciplinary Topics in Applied Mathematics, Modeling and Computational Science. Springer Proceedings in Mathematics & Statistics, pp. 549–555. Springer, Cham (2015)
24. Trivedi, A.R., Datta, S., Mukhopadhyay, S.: Application of Silicon-Germanium source Tunnel-FET to enable ultralow power cellular neural network-based associative memory. IEEE Trans. Electron Devices **61**(11), 3707–3715 (2014)
25. Yildiz, N., Cesur, E., Kayaer, K., Tavsanoglu, V., Alpay, M.: Architecture of a fully pipelined real-time cellular neural network emulator. IEEE Trans. Circuits Syst. I: Regul. Pap. **62**(1), 130–138 (2015)
26. Lu, J., Kurths, J., Cao, J., Mahadavi, N., Huang, C.: Sychronization control for nonlinear stochastic dynamical networks: pinning impulsive strategy. IEEE Trans. Netw. Learn. Syst. **23**(2), 285–292 (2012)
27. Cheng, P., Deng, F., Yao, F.: Exponential stability analysis of impulsive stochastic functional differential systems with delayed impulses. Commun. Nonlinear Sci. Numer. Simul. **19**(6), 2104–2114 (2014)

Convergence Analysis of the Spectral Expansion of Stable Related Semigroups

Yixuan Zhao and Pierre Patie

Abstract The purpose of this note is to carry out a convergence analysis of the spectral representation, derived recently in Patie and Savov (Spectral expansions of non-self-adjoint generalized Laguerre semigroups, submitted, 2015), of some non-self-adjoint Markovian semigroups related to spectrally negative α-stable Lévy processes conditioned to stay positive. More specifically, we start by performing an error analysis for the spectral type series expansions. Moreover, these semigroups are closely related to a class of invariant semigroups, whose speed of convergence to equilibrium has been studied in Patie and Savov (Spectral expansions of non-self-adjoint generalized Laguerre semigroups, submitted, 2015). Our second aim is to carry out a numerical analysis on the convergence rate which is illustrated with two examples.

1 Introduction

The class of Lévy stable and related processes have attracted much attention over the past decades. In particular, this class of processes have been shown to be an appropriate statistical description for a large number of phenomena. Such applications include, e.g., the study of turbulence in physics [4], seismic series and earthquakes in geography [6], risk process in financial mathematics [10], and primary sequences of protein-like copolymers in biomedical science [3]. Therefore, being able to explicitly represent the transition kernel and the solution to the Cauchy problem associated to stable related semigroups has become a critical issue in many areas. In our context, we consider a spectrally negative α-stable process $Z = (Z_t)_{t\geq 0}$, with $\alpha \in (1,2)$. It means that Z is a process with stationary and independent increments, having no positive jumps, and its law is characterized by

Y. Zhao (✉)
Cornell University ORIE Department, 288 Rhodes Hall, Cornell University, Ithaca, NY, USA
e-mail: yz645@cornell.edu

P. Patie
Cornell University ORIE Department, 206 Rhodes Hall, Cornell University, Ithaca, NY, USA
e-mail: pp396@orie.cornell.edu

© Springer International Publishing Switzerland 2016

J. Bélair et al. (eds.), *Mathematical and Computational Approaches in Advancing Modern Science and Engineering*, DOI 10.1007/978-3-319-30379-6_70

its characteristic exponent which takes the form, for $t > 0, \Re(z) > 0$,

$$\ln \mathbb{E}[e^{zZ_t}] = z^\alpha t = \int_{-\infty}^0 (e^{zy} - 1 - zy)\frac{|y|^{-\alpha-1}}{\Gamma(-\alpha)}dy$$

where throughout Γ stands for the Gamma function. Next, let $K^0 = (K_t^0)_{t\geq 0}$ be the semigroup of the process Z killed upon entering the negative half-line. That is, for $f \in B_b(\mathbb{R}^+)$, the set of bounded Borelian functions on $\mathbb{R}^+ = [0, \infty)$, and for any $t, x > 0$, K_t^0 is defined by

$$K_t^0 f(x) = \mathbb{E}_x[f(Z_t)\mathbb{I}_{\{t<T_0\}}],$$

where $T_0 = \inf\{t > 0;\ Z_t < 0\}$. By [2, Section 3.2], $p_{\alpha-1}(x) = x^{\alpha-1}, x > 0$, is an excessive function for K^0, i.e. $p_{\alpha-1} \geq 0$ on $\mathbb{R}^+$ and for all $t > 0$, $K_t^0 p_{\alpha-1}(x) \leq p_{\alpha-1}(x)$. Hence one may construct a semigroup $K^\uparrow = (K_t^\uparrow)_{t\geq 0}$ through Doob's h-transform as follows,

$$K_t^\uparrow f(x) = p_{\alpha-1}^{-1}(x) K_t^0 (p_{\alpha-1} f)(x),\ x > 0,$$

which turns out to be the Feller semigroup of the α-stable process conditioned to stay positive, and will be referred to as α-SP+ throughout . It has been shown in [2] that the infinitesimal generator of $K^\uparrow$, denoted by $L^\uparrow$, can be expressed in terms of an Erdélyi-Kober type operator in the form, for a smooth function f,

$$L^\uparrow f(x) = x^{-\alpha} D_{0+,1,-1}^\alpha f(x) = x^{1-\alpha}\frac{d^2}{dx^2}\int_0^x \frac{y^{\alpha-1}f(y)}{(x-y)^{\alpha-1}}\frac{dy}{\Gamma(2-\alpha)}.$$

It should be noted that $K^\uparrow$ is also a $\frac{1}{\alpha}$-self-similar semigroup as in the definition of [5], i.e. $K_t^\uparrow f(cx) = K_{c^{-1/\alpha}t}^\uparrow \mathrm{d}_c f(x)$ for any $c > 0$, where $\mathrm{d}_c f(x) = f(cx)$. Hence it is easily seen that the semigroup $K = (K_t)_{t\geq 0}$ defined by

$$K_t f(x) = K_t^\uparrow f \circ p_\alpha(x^{\frac{1}{\alpha}})$$

is a 1-self-similar semigroup, as $p_\alpha(x) = x^\alpha$ is a homeomorphism of $\mathbb{R}^+$ to $\mathbb{R}^+$. For sake of presentation, it will be more convenient to work with the semigroup K. According to a celebrated result by Lamperti [5], there exists a bijection between the set of convolution semigroups on $\mathbb{R}$ and the one of positive self-similar Feller semigroups. In particular, it has been shown, see [2, 8], that K is associated via the Lamperti mapping to the semigroup of a spectrally negative Lévy process $\xi = (\xi_t)_{t\geq 0}$, with Laplace exponent ψ given, for $z \in \mathbb{C}_+ = \{z \in \mathbb{C};\ \Re(z) > 0\}$, by

$$\log \mathbb{E}[e^{z\xi_1}] = \psi(z) = \frac{\Gamma(\alpha z + \alpha)}{\Gamma(\alpha z)}. \tag{1}$$

Another semigroup of interests is the so-called associated *generalized Laguerre (gL) semigroup* denoted by $P = (P_t)_{t\geq 0}$ and defined as follows

$$P_t f(x) = K_{e^t-1} f \circ \mathrm{d}_{e^{-t}}(x), \quad x > 0. \tag{2}$$

Moreover, P admits a stationary measure with absolutely continuous density $\nu(x)dx$, and by [9, Theorem 1.6], P can be extended uniquely to a strongly continuous contraction semigroup, also denoted by P, on the weighted Hilbert space $L^2(\nu)$ endowed with the norm $||f||_\nu^2 = \int_0^\infty f^2(x)\nu(x)dx$. From Bertoin and Yor [1], the integer moments of ν are given by $\int_0^\infty x^n \nu(x)dx = \frac{\prod_{k=1}^n \psi(k)}{n!} = \frac{\Gamma(\alpha n+\alpha)}{\Gamma(\alpha)n!}$, which yields that its Mellin transform is

$$\mathcal{M}_\nu(s) = \int_0^\infty x^{s-1}\nu(x)dx = \frac{\Gamma(\alpha s)}{\Gamma(\alpha)\Gamma(s)}, \quad s \in \mathbb{C}_+. \tag{3}$$

Hence standard Mellin inversion technique gives

$$\nu(x) = \sum_{n=0}^{\infty} \frac{(-1)^{n+1}\sin(\frac{n\pi}{\alpha})\Gamma(\frac{n}{\alpha}+1)}{\Gamma(\alpha)\pi n!} x^{\frac{n}{\alpha}}, \tag{4}$$

which, by Stirling approximation, defines a function analytical on $\mathbb{C}_+$ and by [9, Theorem 2.1], is a positive probability density on $\mathbb{R}^+$. It can be easily shown that the semigroup P is non-self-adjoint in $L^2(\nu)$ and thus the spectral theorem can not be applied. However, Patie and Savov [9] have recently developed an original approach to provide a spectral representation of the class of gL semigroups, which enables us to write $P_t f$ and $K_t f$ as infinite series expansions under some mild conditions on f and t. In particular, we shall observe that when f is a power function, the series expansion is reduced to finitely many terms and thus can be evaluated precisely. Hence the first aim of this paper is to perform a numerical convergence analysis for the series expansion. Moreover, in [9], an interesting study on the speed of convergence of a gL semigroup to its equilibrium is presented. In particular, depending on the subspace of $L^2(\nu)$ one considers, one can observe either the hypocoercivity phenomena, that is bounds of the form Ce^{-t} or a non-classical bound of the form $C(e^{2t}-1)^{-1/2}$ for some $C > 0$. Therefore, the second objective of this paper is to carry out a numerical analysis on the speed of convergence for $P_t f$ to νf in $L^2(\nu)$ as $t \to \infty$, where $\nu f = \int_0^\infty f(x)\nu(x)dx$. Such numerical plots can serve as illustrations for the theoretical results in [9].

In this paper, we also study another class of semigroups called the *Gauss-Laguerre semigroup*, denoted by $G = (G_t)_{t\geq 0}$ and defined as follows. For any $\alpha \in (1,2)$ and $\beta \in [1-\frac{1}{\alpha-1}, \infty)$, we introduce the Gauss-Laguerre operator defined for a smooth function f and $x > 0$, by

$$\mathbf{L}_{\alpha,\beta} f(x) = (\mathrm{d}_{\alpha,\beta} - x)f'(x) + \frac{\sin((\alpha-1)\pi)}{\pi} x \int_0^1 f''(xy) g_{\alpha,\beta}(y)dy, \tag{5}$$

where $\mathrm{d}_{\alpha,\beta} = \frac{\Gamma(\alpha\beta+\alpha-\beta)}{\Gamma(\alpha\beta+1-\beta)}$ and

$$g_{\alpha,\beta}(y) = \frac{\Gamma(\alpha-1)}{\beta+\frac{1}{\alpha-1}+1} y^{\beta+\frac{1}{\alpha-1}+1} {}_2F_1(\alpha\beta+\alpha-\beta,\alpha;\alpha\beta+\alpha-\beta+1;y^{\frac{1}{\alpha-1}}), \tag{6}$$

with ${}_2F_1$ the Gauss hypergeometric function. Then it is shown in [7, Theorem 1.1] that $\mathbf{L}_{\alpha,\beta}$ is the generator of a non-self-adjoint contraction Markov semigroup G in $L^2(e_{\alpha,\beta})$, where

$$e_{\alpha,\beta}(x) = \frac{x^{\beta+\frac{1}{\alpha-1}-1} e^{-x^{\frac{1}{\alpha-1}}}}{\Gamma((\alpha-1)\beta+1)}, \quad x > 0, \tag{7}$$

is its unique invariant measure. The motivation for studying this class of semigroups is that while they share many similar properties with P, the expressions of their spectral expansions are much simpler and easier to evaluate. In this paper, we will also study the speed of convergence to equilibrium for G.

2 The α-SP+ Semigroup and Its Associated gL Semigroup

2.1 Spectral Expansions and Convergence to Equilibrium

We start by introducing a few notations. For $x \in \mathbb{R}^+$, let

$$\mathscr{P}_n(x) = \sum_{k=0}^{n} (-1)^k \frac{\binom{n}{k}\Gamma(\alpha)k!}{\Gamma(\alpha k+\alpha)} x^k, \tag{8}$$

$$\mathscr{W}_n(x) = \frac{(x^n \nu(x))^{(n)}}{n!} = \frac{1}{n!}\sum_{k=0}^{\infty} \frac{(-1)^{k+1}\sin(\frac{k\pi}{\alpha})\Gamma(n+\frac{k}{\alpha}+1)}{\Gamma(\alpha)\pi k!} x^{\frac{k}{\alpha}}, \tag{9}$$

where, for a smooth function f, $f^{(n)}(x) = \frac{d^n f(x)}{dx^n}$, and we let $\mathscr{V}_n(x) = \frac{\mathscr{W}_n(x)}{\nu(x)}$, which, by [9], is in $L^2(\nu)$. Finally, we let ρ_α to be the largest solution to the equation $(1-\rho)^{\frac{1}{\alpha-1}} \cos\left(\frac{\arcsin(\rho)}{\alpha-1}\right) = \frac{1}{2}$, and set

$$T_{\pi_\alpha} = -\log\sin\frac{(\alpha-1)\pi}{2} \text{ and } T_{\pi_\alpha,\rho_\alpha} = \max\left(T_{\pi_\alpha}, 1+\frac{1}{\rho_\alpha}\right).$$

Then we have the following theorem.

Theorem 1

(a) *For any* $f \in L^2(\nu)$, $x > 0$ *and* $t > T_{\pi_\alpha,\rho_\alpha}$, *we have in* $L^2(\nu)$,

$$P_t f(x) = \sum_{n=0}^{\infty} e^{-nt} \langle f, \mathcal{V}_n \rangle_\nu \mathcal{P}_n(x), \tag{10}$$

$$K_t f(x) = \sum_{n=0}^{\infty} (t+1)^{-n} \left\langle f \circ \mathrm{d}_{(t+1)}, \mathcal{V}_n \right\rangle_\nu \mathcal{P}_n(x). \tag{11}$$

Particularly, recalling that $p_m(x) = x^m$, *we have for any* $t > 0$,

$$K_t p_m(x) = x^m + \sum_{n=1}^{m} \frac{\Gamma(\alpha m + \alpha)}{\Gamma(\alpha(m-n)+\alpha) n!} x^{m-n} t^n. \tag{12}$$

(b) *There exists* $C_\nu > 0$ *and an integer* $k \geq 0$ *such that for any* $f \in L^2(\nu)$ *and* $t > T_{\pi_\alpha,\rho_\alpha}$,

$$\|P_t f - \nu f\|_\nu \leq C_\nu 2^{\frac{k}{2}} \sqrt{\left(\frac{1}{e^{2(t-T_{\pi_\alpha,\rho_\alpha})} - 1}\right)^{(k)}} \|f - \nu f\|_\nu. \tag{13}$$

Moreover, when $f = p_m$, *we have, for all* $t > 0$,

$$\|P_t p_m - \nu p_m\|_\nu \leq \frac{\Gamma(\alpha m + \alpha)}{\Gamma(\alpha)(m!)^2} e^{-t} \|p_m - \mathbf{e} p_m\|_{\mathbf{e}}, \tag{14}$$

where $\mathbf{e}(x) = e^{-x}, x > 0$.

Remark 1 Note that (12) indeed coincides with the expression of the entire moments given by Bertoin and Yor [1, Proposition 1], where the form of ψ as in their theorem is given by (1).

Proof We first observe that $\lim_{u\to\infty} u^{-\alpha}\psi(u) = \lim_{u\to\infty} u^{-\alpha}\frac{\Gamma(\alpha u+\alpha)}{\Gamma(\alpha u)} = C'_\alpha$ for some $C'_\alpha > 0$, therefore the spectral expansion (10) follows from [9, Theorem 1.9(1)]. Next, by applying a simple deterministic time change on (2), we get that for any $f \in L^2(\nu)$ and $t > 0$,

$$K_t f(x) = P_{\log(t+1)} f \circ \mathrm{d}_{(t+1)}(x),$$

which further yields (11) from (10). Moreover, by [9, Proposition 9.3], the Mellin transform of $\mathcal{W}_n$ is, for $s \in \mathbb{C}_+$,

$$\mathcal{M}_{\mathcal{W}_n}(s) = \frac{(-1)^n \Gamma(s)}{\Gamma(n+1)\Gamma(s-n)} \mathcal{M}_\nu(s) = \frac{(-1)^n \Gamma(\alpha s)}{\Gamma(\alpha)\Gamma(n+1)\Gamma(s-n)}. \tag{15}$$

Hence taking $f = p_m$, we have

$$\langle p_m \circ \mathrm{d}_{t+1}, \mathscr{V}_n \rangle_\nu = (t+1)^m \mathscr{M}_{\mathscr{W}_n}(m+1) = \frac{(-1)^n (t+1)^m \Gamma(\alpha m + \alpha)}{\Gamma(\alpha)\Gamma(n+1)\Gamma(m+1-n)},$$

which, since Γ has a pole at any negative integers, vanishes when $n \geq m+1$. Then (11) becomes a finite series and easy algebra yields the representation (12). For the speed of convergence, (13) comes from [9, Theorem 1.9(3c)]. Moreover, by [9, Theorem 2.9], P intertwins with $Q = (Q_t)_{t \geq 0}$, the classical Laguerre semigroup, which is self-adjoint on $L^2(\mathbf{e})$ with invariant measure $\mathbf{e}(dx)$. More specifically, there exists a multiplicative kernel Λ such that $P_t \Lambda f = \Lambda Q_t f$ for any $f \in L^2(\mathbf{e})$, and we have for all $m \in \mathbb{N}$, $\Lambda p_m(x) = \frac{m!}{\prod_{k=1}^m \frac{\psi(k)}{k}} p_m(x) = \frac{\Gamma(\alpha)(m!)^2}{\Gamma(\alpha m+\alpha)} p_m(x)$. Hence we see that $\Lambda^{-1} p_m(x) = \frac{\Gamma(\alpha m+\alpha)}{\Gamma(\alpha)(m!)^2} p_m(x)$ and by [9, Theorem 1.9(3a)], we have

$$\|P_t p_m - \nu p_m\|_\nu \leq \|\Lambda^{-1} p_m - \mathbf{e}\Lambda^{-1} p_m\|_{\mathbf{e}} = \frac{\Gamma(\alpha m + \alpha)}{\Gamma(\alpha)(m!)^2} e^{-t} \|p_m - \mathbf{e} p_m\|_{\mathbf{e}}$$

which completes the proof. □

2.2 Numerical Illustrations

In this section, we first provide the plots of $e_m(N)$, the relative truncation error of (12) up to N terms which is formally defined by

$$e_m(N) = \left| \frac{\sum_{n=N+1}^m (t+1)^{-n} \langle f \circ \mathrm{d}_{(t+1)}, \mathscr{W}_n \rangle \mathscr{P}_n(x)}{\sum_{n=0}^m (t+1)^{-n} \langle f \circ \mathrm{d}_{(t+1)}, \mathscr{W}_n \rangle \mathscr{P}_n(x)} \right|,$$

for N running from 1 up to $m - 1 > 0$. Figures 1, 2, and 3 depict the plots for $(m = 20, x = 1), (m = 40, x = 1), (m = 40, x = 5)$, respectively. Each plot consists of three experiments $\alpha = 1.2, 1.5, 1.8.$, and in each figure, the left panel plots $\log e_m(N)$ with respect to N while the right panel plots $e_m(N)$. It can be seen from the plots that $e_m(N)$ exhibits a super-exponential decay, and the relative error becomes negligible within only a few terms. Comparing Fig. 1 with Fig. 2, we observe that keeping the value of x constant, the convergence is slower for larger m (for example, 6 terms are sufficient for a fairly good convergence when $m = 20$, but not when $m = 40$), which is indeed expected since there are more non-zero terms in the series summation (12) when m gets large. Comparing Fig. 2 with Fig. 3, we also observe that the rate of convergence among different values of α's depends on the value of x. When $x = 1$, $\alpha = 1.2$ converges the fastest and $\alpha = 1.8$ converges the slowest. But when $x = 5$, $\alpha = 1.5$ converges the fastest but $\alpha = 1.2$ becomes the slowest. Hence the relation of convergence rate

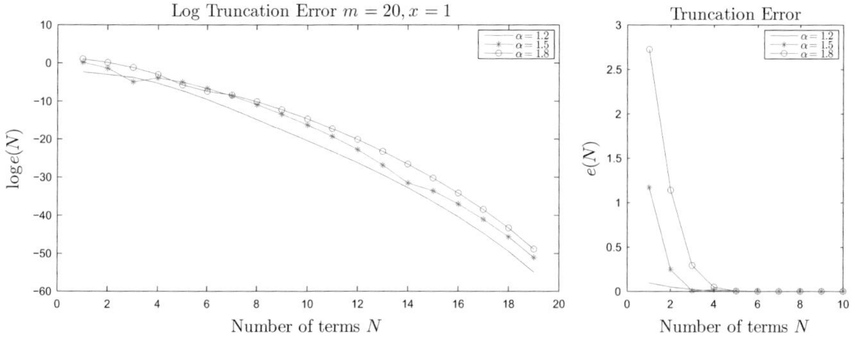

Fig. 1 Truncation error plots for $m = 20, x = 1, \alpha = 1.2, 1.5, 1.8$

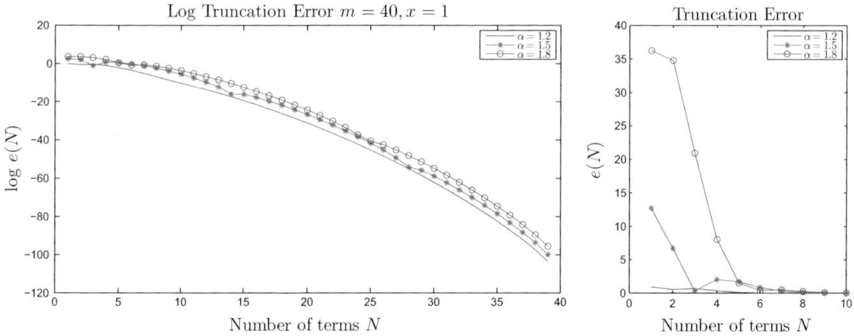

Fig. 2 Truncation error plots for $m = 40, x = 1, \alpha = 1.2, 1.5, 1.8$

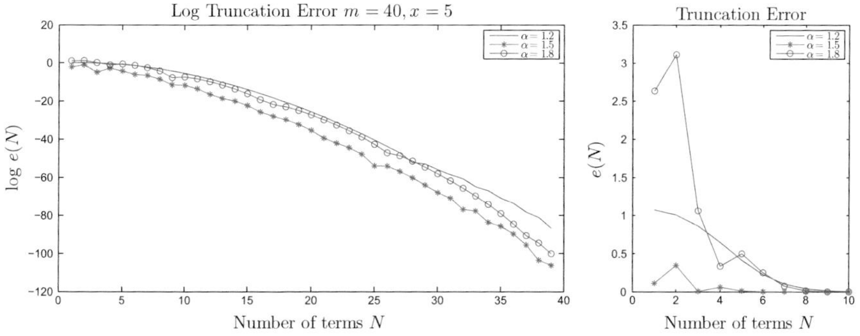

Fig. 3 Truncation error plots for $m = 40, x = 5, \alpha = 1.2, 1.5, 1.8$

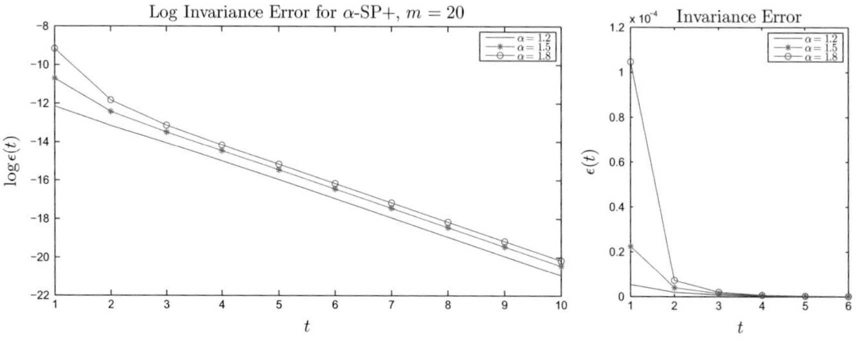

Fig. 4 Invariance error plots for $m = 20, \alpha = 1.2, 1.5, 1.8$

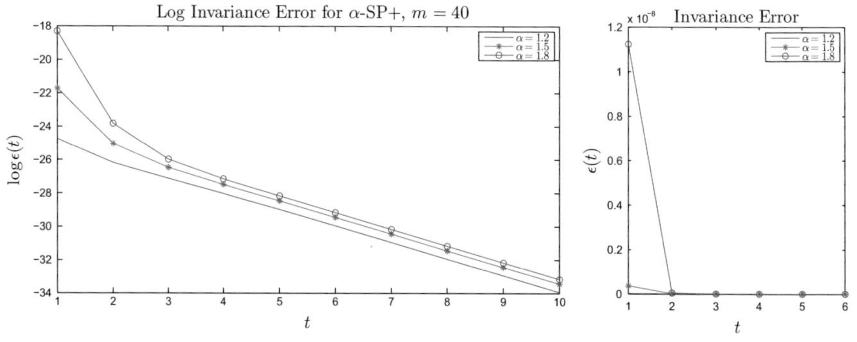

Fig. 5 Invariance error plots for $m = 40, \alpha = 1.2, 1.5, 1.8$

with respect to α is inconclusive. Next, we provide some numerical plots on the speed of convergence to equilibrium for the semigroup P. Figures 4 and 5 depict the plots for $\epsilon(t) = \frac{\|P_t p_m - \nu p_m\|_\nu}{\|p_m - \mathbf{e} p_m\|_\mathbf{e}}$ with $m = 20$ and $m = 40$, respectively. Each plot consists of three experiments $\alpha = 1.2, 1.5, 1.8$, and each experiment runs for $t = 1$ to 10. In each figure, the left panel plots $\log \epsilon(t)$ while the right panel plots $\epsilon(t)$. From the figures, it is easy to observe exponential decrease for each case. Under the same conditions, by comparing $\alpha = 1.2, 1.5, 1.8$, we can see that the convergence becomes faster as α increases. Moreover, by comparing $m = 20$ with $m = 40$, we also observe that the convergence rate increases as m grows, which can be attributed to the fact that $m+1$ is indeed the number of non-zero terms in the series summation. In other words, the convergence is faster using more terms for evaluation.

3 Convergence to Equilibrium of the Gauss-Laguerre Semigroup

The spectral expansion of G can be found in [7] and in this section, we only state the speed of convergence to equilibrium as in the following theorem.

Theorem 2 *Denoting* $T_\alpha = -\log(2^{\alpha-1}-1)$, *there exists a constant* $C_\alpha > 0$ *such that for all* $f \in L^2(e_{\alpha,\beta})$ *and* $t > T_\alpha$,

$$\|G_t f - e_{\alpha,\beta} f\|_{e_{\alpha,\beta}} \le C_\alpha \sqrt{\frac{1}{e^{2(t-T_\alpha)}-1}}\|f - e_{\alpha,\beta} f\|_{e_{\alpha,\beta}}. \tag{16}$$

In particular, when $f = p_m$, *we have for all* $t > 0$,

$$\|G_t p_m - e_{\alpha,\beta} p_m\|_{e_{\alpha,\beta}} \le \frac{\Gamma((\alpha-1)(m+\beta)+1)}{m!\Gamma((\alpha-1)\beta+1)} e^{-t}\|p_m - \mathbf{e}p_m\|_{\mathbf{e}}. \tag{17}$$

Proof By [9, Example 3.3], G corresponds to a spectrally negative Lévy process $\xi^{\alpha,\beta}$ with Laplace exponent $\psi^{\alpha,\beta}(u) = u\frac{\Gamma((\alpha-1)(u+\beta)+1)}{\Gamma((\alpha-1)(u+\beta-1)+1)}$, hence (16) follows from [9, Theorem 1.9 (3b)]. Moreover, as in the previous case, G also intertwins with the classical Laguerre semigroup Q via an intertwining kernel denoted by $\Lambda_{\alpha,\beta}$, i.e. we have $G_t\Lambda_{\alpha,\beta} f = \Lambda_{\alpha,\beta} Q_t f$ for all $f \in L^2(\mathbf{e})$. Furthermore, by [7, Proposition 3.2] and its proof, we have for all $m \in \mathbb{N}$,

$$\Lambda_{\alpha,\beta} p_m(x) = \frac{m!\Gamma((\alpha-1)\beta+1)}{\Gamma((\alpha-1)(m+\beta)+1)} p_m(x).$$

Hence $\Lambda^{-1}_{\alpha,\beta} p_m(x) = \frac{\Gamma((\alpha-1)(m+\beta)+1)}{m!\Gamma((\alpha-1)\beta+1)} p_m(x)$ and by [9, Theorem 1.9(3a)], we have

$$\|G_t p_m - e_{\alpha,\beta} p_m\|_{e_{\alpha,\beta}} \le e^{-t}\|\Lambda^{-1}_{\alpha,\beta} p_m - \mathbf{e}\Lambda^{-1}_{\alpha,\beta} p_m\|_{\mathbf{e}} = \frac{\Gamma((\alpha-1)(m+\beta)+1)}{m!\Gamma((\alpha-1)\beta+1)} e^{-t}\|p_m - \mathbf{e}p_m\|_{\mathbf{e}}.$$

This completes the proof of this proposition. □

Now we proceed to give the numerical plots for the convergence of $G_t p_m$ to $e_{\alpha,\beta} p_m$. Figures 6 and 7 depict the plots $\epsilon(t) = \frac{\|G_t p_m - e_{\alpha,\beta} p_m\|_{e_{\alpha,\beta}}}{\|p_m - \mathbf{e}p_m\|_{\mathbf{e}}}$ with $m = 20$ and $m = 40$ respectively. Each plot consists of three experiments $(\alpha, \beta) = (1.2, -4), (1.5, -1), (1.8, 0)$ and each experiment runs for $t = 1$ to 10. In each figure, the left panel plots $\log \epsilon(t)$ while the right panel plots $\epsilon(t)$.

It is easy to observe that in all cases, we still have exponential decay, which verifies our bound (17). Moreover, we still observe faster convergence when m gets larger, which can be explained exactly as the previous α-SP+ example since for larger value of m, there are more terms used in the series evaluation.

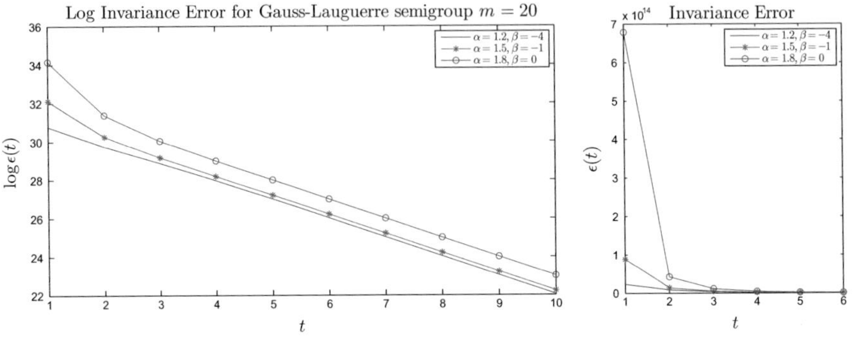

Fig. 6 Invariance error plots for $m = 20$, $(\alpha, \beta) = (1.2, -4), (1.5, -1), (1.8, 0)$

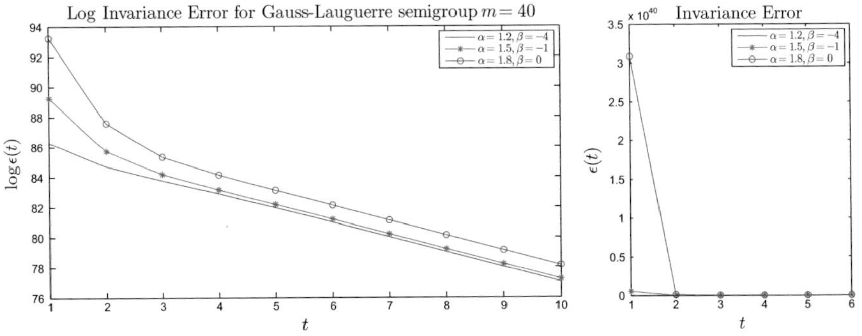

Fig. 7 Invariance error plots for $m = 40$, $(\alpha, \beta) = (1.2, -4), (1.5, -1), (1.8, 0)$

Acknowledgements The authors would like to thank an anonymous referee for insightful suggestions which improved an earlier version of the paper.

References

1. Bertoin, J., Yor, M.: On the entire moments of self-similar Markov processes and exponential functionals of Lévy processes. In: Annales de la faculté des sciences de Toulouse 6e série, tome 11, pp. 33–45 (2002)
2. Caballero, M.E., Chaumont, L.: Conditioned stable Lévy processes and the Lamperti representation. J. Appl. Probab. **43**(4), 967–983 (2006)
3. Govorun, E.N., Ivanov, V.A., Khokhlov, A.R., Khalatur, P.G., Borovinsky, A.L., Grosberg, A.Y.: Primary sequences of proteinlike copolymers: Lévy–flight-type long-range correlations. Phys. Rev. E **64**, 040903–040903-4 (2001)
4. Kolmogorov, A.N.: Dissipation of energy in isotropic turbulence. Dokl. Akad. Nauk SSSR **32**, 19–21 (1941)
5. Lamperti, J.: Semi-stable Markov processes. I. In: z. Wahrscheinlichkeitstheorie verw. Geb. 22, pp. 205–225. Springer (1972)

6. Posadas, A., Morales, J., Vidal, F., Sotolongo-Costa, O., Antoranz, J.C.: Continuous time random walks and South Spain seismic series. J. Seismol. **6**, 61–67 (2002)
7. Patie, P., Savov, M.: Cauchy problem of the non-self-adjoint Gauss-Laguerre semigroups and uniform bounds of generalized Laguerre polynomials (2013, Submitted)
8. Patie, P., Simon, T.: Intertwining certain fractional operators. Potent. Anal. **36**, 569–587 (2012)
9. Patie, P., Savov, M.: Spectral expansions of non-self-adjoint generalized Laguerre semigroups (2015, Submitted)
10. Yang, H., Zhang, L.: Spectrally negative Lévy processes with applications in risk theory. Adv. Appl. Prob. **33**, 281–291 (2001)

Erratum to: Examining the Role of Social Feedbacks and Misperception in a Model of Fish-Borne Pollution Illness

Michael Yodzis

Erratum to:
© Springer International Publishing Switzerland 2016
J. Bélair et al. (eds.), *Mathematical and Computational Approaches in Advancing Modern Science and Engineering*, DOI 10.1007/978-3-319-30379-6_32

Erratum DOI 10.1007/978-3-319-30379-6_71

The authorship of the present article has been revised so that Michael Yodzis is now listed as the only author, to make clear that Dr. Chris T. Bauch and Dr. Madhur Anand were not contributors to this publication. The present contribution is an extended and revised version of the publication 'Coupling fishery dynamics, human health and social learning in a model of fish-borne pollution exposure' by M. Yodzis, C.T. Bauch and M. Anand, Sustainability Science 11, pp. 179-192 (2016).

The results are extended to include a more complete explanation of the model's dependence on its initial conditions, and the sensitivity analysis is enlarged to study the influence of the ecological parameters on the model's qualitative outcomes. Any mistakes or omissions are the sole responsibility of Michael Yodzis.

The updated online version of the original chapter can be found at DOI 10.1007/978-3-319-30379-6_32

© Springer International Publishing Switzerland 2016
J. Bélair et al. (eds.), *Mathematical and Computational Approaches in Advancing Modern Science and Engineering*, DOI 10.1007/978-3-319-30379-6_71

Index

© Springer International Publishing Switzerland 2016

J. Bélair et al. (eds.), *Mathematical and Computational Approaches in Advancing Modern Science and Engineering*, DOI 10.1007/978-3-319-30379-6